TEACHER'S GUIDE

Volume B

# IMPACT MATHEMATICS

## Algebra and More

**Course 1**


**Developed by Education Development Center, Inc.**

**Senior Project Director:** Cynthia J. Orrell

**Senior Curriculum Developers:** Michelle Manes, Sydney Foster, Daniel Lynn Watt, Ricky Carter, Joan Lukas, Kristen Herbert

**Curriculum Developers:** Haim Eshach, Phil Lewis, Melanie Palma, Peter Braunfeld, Amy Gluckman, Paula Pace

**Special Contributors:** Faye Nisonoff Ruopp, Elizabeth D. Bjork



McGraw Hill Glencoe

New York, New York   Columbus, Ohio   Chicago, Illinois   Peoria, Illinois   Woodland Hills, California

Glencoe

The McGraw·Hill Companies

The algebra content for *Impact Mathematics* was adapted from the series *Access to Algebra*, by Neville Grace, Jayne Johnston, Barry Kissane, Ian Lowe, and Sue Willis. Permission to adapt this material was obtained from the publisher, Curriculum Corporation of Level 5, 2 Lonsdale Street, Melbourne, Australia.



Send all inquiries to:
Glencoe/McGraw-Hill
8787 Orion Place
Columbus, OH 43240

ISBN 0-07-860911-9

4 5 6 7 8 9 10 058/055 12 11 10 09 08 07 06

## Impact Mathematics Project Reviewers

Education Development Center appreciates all the feedback from the curriculum specialists and teachers who participated in review and testing.

**Special thanks to:**

Peter Braunfeld
Professor of Mathematics Emeritus
University of Illinois

Sherry L. Meier
Assistant Professor of Mathematics
Illinois State University

Judith Roitman
Professor of Mathematics
University of Kansas

---

Marcie Abramson
Thurston Middle School
Boston, Massachusetts

Denise Airola
Fayetteville Public Schools
Fayetteville, Arizona

Chadley Anderson
Syracuse Junior High School
Syracuse, Utah

Jeanne A. Arnold
Mead Junior High
Elk Grove Village, Illinois

Joanne J. Astin
Lincoln Middle School
Forrest City, Arkansas

Jack Beard
Urbana Junior High
Urbana, Ohio

Chad Cluver
Maroa-Forsyth Junior High
Maroa, Illinois

Robert C. Bieringer
Patchogue-Medford School Dist.
Center Moriches, New York

Susan Coppleman
Nathaniel H. Wixon Middle School
South Dennis, Massachusetts

Sandi Curtiss
Gateway Middle School
Everett, Washington

Alan Dallman
Amherst Middle School
Amherst, Massachusetts

Sharon DeCarlo
Sudbury Public Schools
Sudbury, Massachusetts

David P. DeLeon
Preston Area School
Lakewood, Pennsylvania

Jacob J. Dick
Cedar Grove School
Cedar Grove, Wisconsin

Sharon Ann Dudek
Holabird Middle School
Baltimore, Maryland

Cheryl Elisara
Centennial Middle School
Spokane, Washington

Patricia Elsroth
Wayne Highlands Middle School
Honesdale, Pennsylvania

Dianne Fink
Bell Junior High
San Diego, California

Terry Fleenore
E.B. Stanley Middle School
Abingdon, Virginia

Kathleen Forgac
Waring School
Beverly, Massachusetts

Steven J. Fox
Bendle Middle School
Burton, Michigan

Kenneth L. Goodwin Jr.
Middletown Middle School
Middletown, Delaware

Fred E. Gross
Sudbury Public Schools
Sudbury, Massachusetts

Penny Hauben
Murray Avenue School
Huntingdon, Pennsylvania

Jean Hawkins
James River Day School
Lynchburg, Virginia

Robert Kalac
Butler Junior High
Frombell, Pennsylvania

Robin S. Kalder
Somers High School
Somers, New York

Darrin Kamps
Lucille Umbarge Elementary
Burlington, Washington

Sandra Keller
Middletown Middle School
Middletown, Delaware

Pat King
Holmes Junior High
Davis, California

Kim Lazarus
San Diego Jewish Academy
La Jolla, California

Ophria Levant
Webber Academy
Calgary, Alberta
Canada

Mary Lundquist
Farmington High School
Farmington, Connecticut

Ellen McDonald-Knight
San Diego Unified School District
San Diego, California

Ann Miller
Castle Rock Middle School
Castle Rock, Colorado

Julie Mootz
Ecker Hill Middle School
Park City, Utah

Jeanne Nelson
New Lisbon Junior High
New Lisbon, Wisconsin

DeAnne Oakley-Wimbush
Pulaski Middle School
Chester, Pennsylvania

Tom Patterson
Ponderosa Jr. High School
Klamath Falls, Oregon

Maria Peterson
Chenery Middle School
Belmont, Massachusetts

Lonnie Pilar
Tri-County Middle School
Howard City, Michigan

Karen Pizarek
Northern Hills Middle School
Grand Rapids, Michigan

Debbie Ryan
Overbrook Cluster
Philadelphia, Pennsylvania

Sue Saunders
Abell Jr. High School
Midland, Texas

Ivy Schram
Massachusetts Department of Youth Services
Massachusetts

Robert Segall
Windham Public Schools
Willimantic, Connecticut

Kassandra Segars
Hubert Middle School
Savannah, Georgia

Laurie Shappee
Larson Middle School
Troy, Michigan

Sandra Silver
Windham Public Schools
Willimantic, Connecticut

Karen Smith
East Middle School
Braintree, Massachusetts

Kim Spillane
Oxford Central School
Oxford, New Jersey

Carol Struchtemeyer
Lexington R-5 Schools
Lexington, Missouri

Kathy L. Terwelp
Summit Public Schools
Summit, New Jersey

Laura Sosnoski Tracey
Somerville, Massachusetts

Marcia Uhls
Truesdale Middle School
Wichita, Kansas

Vendula Vogel
Westridge School for Girls
Pasadena, California

Judith A. Webber
Grand Blanc Middle School
Grand Blanc, Michigan

Sandy Weishaar
Woodland Junior High
Fayetteville, Arkansas

Tamara L. Weiss
Forest Hills Middle School
Forest Hills, Michigan

Kerrin Wertz
Haverford Middle School
Havertown, Pennsylvania

Anthony Williams
Jackie Robinson Middle School
Brooklyn, New York

Deborah Winkler
The Baker School
Brookline, Massachusetts

Lucy Zizka
Best Middle School
Ferndale, Michigan

# CONTENTS

# Program Philosophy

In developing *Impact Mathematics: Algebra and More,* we, the authors at Education Development Center, Inc., have relied on our collective experiences as teachers, parents, and former students. Our main goal is to offer a curriculum that respects the background and knowledge of middle school teachers, recognizes the competence and energy of middle school students, and addresses the need for intellectually challenging and inclusive mathematics materials. With *Impact Mathematics,* we have combined the best of what is known as "reform" curricula with the best of "traditional" curricula, incorporating more active involvement on the part of students in making sense of important mathematical ideas.

With middle grades teachers and students in mind, we have created a comprehensive curriculum for Grades 6 through 8 that completes a full year of algebra by the end of Grade 8. While the number and operations, geometry, and data and probability strands were created especially for this program, the algebra strand is based on the highly successful Australian program *Access to Algebra,* developed by Curriculum Corporation.

The rewarding and interesting introduction to algebra offered by this program can help develop and maintain students' ongoing interest in all areas of mathematics. The materials created for *Impact Mathematics* follow the *Access to Algebra* material in style: use of narrative and realistic contexts, personalization in the form of cartoons in which middle grades students explain how they approach problems, and opportunities for students to choose or create their own problems.

## Conceptual Understanding and Basic Skills

Discussions regarding mathematics learning in both professional circles and the popular media might lead you to believe that teaching for conceptual understanding and teaching basic skills are mutually exclusive. But, in fact, the opposite is true. Conceptual understanding and basic skills are not opposing interests; they go hand in hand and support each other.

*Impact Mathematics* makes the big ideas as well as the important skills of mathematics accessible to middle school students. It presents mathematical ideas intact, not broken down into bite-sized bits that lack the big idea. *Impact Mathematics* helps students both build new mathematical ideas and see how these new ideas relate to ideas they have already developed. In this way, *Impact Mathematics* takes a conceptual approach.

At the same time, *Impact Mathematics* recognizes that for students to be able to use the new ideas and procedures effectively, they need practice. Practice need not be the enemy of learning; the enemy of learning is mindless drill. Instead, practice can encourage students to stay interested in the mathematical concepts. *Impact Mathematics* provides plenty of opportunity for practice, but with variety and contrast to keep students' attention focused.

## Algebraic Focus in a Comprehensive Program

*Impact Mathematics* is a comprehensive program including number and operations, proportional reasoning, geometry, probability, and data, with a focus on the development of algebraic thinking. The program takes a developmental approach to algebra. Student understanding of the algebra strand—interwoven with and related to the other mathematical strands—evolves over a three-year period, allowing the ideas and skills to develop and become familiar over time.

Most students develop strong algebraic ideas in the early years of elementary school, but they don't acquire ways of expressing and manipulating them in algebraic terms until later, when algebra is formally taught. For example, young children know how to share \$36 among three people by first distributing the ten-dollar bills and then distributing the ones. Later, if children learn a standard method for dividing $3\overline{)36}$, they may see again that the process is like dividing $3\overline{)30}$, then dividing $3\overline{)6}$, and finally adding the results. If this process is written out as $\frac{36}{3} = \frac{30}{3} + \frac{6}{3}$, that concise statement contains an important idea about adding fractions and an even more general algebraic idea. Students who understand *why* $\frac{36}{3} = \frac{30}{3} + \frac{6}{3}$ know that the sum of $\frac{30}{3}$ and $\frac{6}{3}$ must be $\frac{36}{3}$, and not $\frac{36}{6}$. The idea, expressed more generally, is $\frac{a}{3} + \frac{b}{3} = \frac{a+b}{3}$, and even more generally, is $\frac{a}{c} + \frac{b}{c} = \frac{a+b}{c}$, and so leads to the distributive law of division over addition.

Our approach in *Impact Mathematics* is to start with algebra as a notation for "generic" arithmetic, a description of processes that students understand. Later, algebra also becomes a handy language for "unlocking secrets" (equation solving) and building mathematical models. By the end of Course 3, students will have learned both to express functions using variables and to graph these functions. They will have also learned how to use variables to set up and solve equations, as well as how to factor some familiar polynomials, and to understand the origin and use of the quadratic formula.

## Use of Manipulatives and Calculators

Manipulatives and calculators can be powerful tools for teaching and learning mathematics. There is, however, much discussion and controversy about the appropriateness of their use. As the authors of *Impact Mathematics,* we believe that when manipulatives and calculators are used, they must be used to support the content learning. More specifically, we consider the important mathematical ideas first and then determine whether manipulatives or calculators can be used in learning those ideas more completely.

We believe it's critical that students develop good number sense and calculation skills before they work extensively with calculators. For example, we incorporate graphing calculators in Course 3 to explore families of functions, but only *after* students have a firm idea of how to graph "parent" functions by hand. Graphing technology can then be used to allow students to graph more complex functions, analyze their behavior, and compare representations. Similar to our philosophy of integrating skills with understanding, we believe that students need experiences with pencil and paper along with graphing technology.

## Organization by Content

*Impact Mathematics* often uses applications to help develop a particular mathematical concept or place it in context. However, *Impact Mathematics* remains organized by mathematical content, not by contexts. This organization helps both teacher and student keep the mathematical ideas at the fore, easily recognizable and never buried or lost in the settings. While the mathematical focus shifts with each chapter, the *Impact Mathematics* approach offers opportunities to connect topics to one another so that earlier learning is not abandoned as new ideas are introduced.

## Developing Concepts in Varied Contexts

The contexts used for developing concepts and practicing skills include real-world applications, as well as mathematical settings such as number puzzles, and the world of the imagination such as a factory that uniformly resizes objects using stretching machines. Sometimes, *Impact Mathematics* provides exercises that are *not* set in contexts or integrated into word problems precisely so that students can focus on the mathematical ideas, undistracted by surrounding material.

## A Final Note

The unique power of mathematics stems from the world of the imagination in which one envisions triangles with perfectly straight sides, or two-dimensional objects embedded in perfectly smooth planes. In the real world, all objects are three dimensional (even a line drawn on paper has thickness, or it wouldn't be visible!), all lines are irregular, and all surfaces are pitted. Likewise, all measurements are only approximations, and no physical object can have an irrational length. Our minds reason well precisely because we can ignore irregularities and focus instead on the essential features. We can reason about quantities that no physical ruler can measure but that we can "measure" with our mental rulers. In sum, we reason well because we can abstract reality.

We, the authors of *Impact Mathematics,* recognize that all people, from early childhood on, do reason abstractly, and that what grows over time is both their ability to recognize the abstractions, and the formality with which they are able to express abstractions. We also recognize that mathematics, while not simply common sense, is rooted in common sense. Mathematics is a human product that has developed as an extension and a codification of ways of thinking that are natural to us all. Students must not think of mathematics as a departure from natural, logical thinking. To that aim *Impact Mathematics* is written to help students use and sharpen their own logical thinking, learn to be comfortable with the abstractions that give mathematics its power, develop their ideas and mathematical imagination, and acquire the skills that support all that good thinking and the ability to express it clearly to others.

We hope you will enjoy teaching and learning with these materials.

# Scope and Sequence

## Number and Operations

| Topics | Course 1 | | | | | | | | | | Course 2 | | | | | | | | | | Course 3 | | | | | | | | | |
|---|---|---|---|---|---|---|---|---|---|---|---|---|---|---|---|---|---|---|---|---|---|---|---|---|---|---|---|---|---|---|
| | 1 | 2 | 3 | 4 | 5 | 6 | 7 | 8 | 9 | 10 | 1 | 2 | 3 | 4 | 5 | 6 | 7 | 8 | 9 | 10 | 1 | 2 | 3 | 4 | 5 | 6 | 7 | 8 | 9 | 10 |
| **Numbers and Number Sense** | Develop | | | | | | | | | | Develop | | | | | | | | | | Review & Extend | | | | | | | | | |
| Whole Numbers | C | F | C | | | | C | C | | | C | C | C | C | | | C | C | | | | | C | | | | C | C | | |
| Signed Numbers | C | F | | | | | | | C | | | | | F | C | | | | | | C | | C | | | | | | | |
| Exponents and Roots | | | | | | | F | | | | | F | F | F | C | | | | F | | | F | F | | | | F | | | |
| **Rationals and Irrationals** | Develop | | | | | | | | | | Review & Extend | | | | | | | | | | Review & Extend | | | | | | | | | |
| Fractions and Decimal Concepts | | F | F | C | | | C | | | | | | | | | | | C | | | | | | | | C | | | | |
| Percents | | | | F | | | | | | | | | C | | C | | C | F | | | | | C | | | C | | | | |
| **Ratios and Rates** | Expose | | | | | | | | | | Develop | | | | | | | | | | Review & Extend | | | | | | | | | |
| Meaning and Representations | | C | | C | | | | | | C | | C | | | F | F | C | F | | | F | | | | C | | | | | |
| Proportions | | | | C | | | | | C | C | | C | | | F | F | C | F | | | | | | | C | | | | | |
| **Algorithms and Operations** | Review & Extend | | | | | | | | | | Review & Extend | | | | | | | | | | Review & Extend | | | | | | | | | |
| Fractions | C | F | F | F | | | | C | | | C | | | | | | | C | | | | | | | | C | | | | |
| Decimals | | F | F | F | | | | C | | | C | | | | | | | | | | | | | | | | | | | |
| Signed Numbers | | | | | | | | | | | | | | F | | | | | | | | | | | | | | | | |

## Algebra

| Topics | Course 1 | | | | | | | | | | Course 2 | | | | | | | | | | Course 3 | | | | | | | | | |
|---|---|---|---|---|---|---|---|---|---|---|---|---|---|---|---|---|---|---|---|---|---|---|---|---|---|---|---|---|---|---|
| | 1 | 2 | 3 | 4 | 5 | 6 | 7 | 8 | 9 | 10 | 1 | 2 | 3 | 4 | 5 | 6 | 7 | 8 | 9 | 10 | 1 | 2 | 3 | 4 | 5 | 6 | 7 | 8 | 9 | 10 |
| **Algebraic Representations** | Develop | | | | | | | | | | Develop | | | | | | | | | | Develop | | | | | | | | | |
| Coordinate Graphs | | C | | | F | F | | C | F | | C | | C | F | F | F | C | C | F | C | F | F | F | F | F | | | F | | F |
| Tables and Graphs | C | C | | | F | F | | F | F | C | C | | C | F | F | F | | | F | F | F | F | F | F | | C | C | F | | F |
| **Algebraic Reasoning** | Develop | | | | | | | | | | Develop | | | | | | | | | | Develop | | | | | | | | | |
| Patterns and Numeric Forms | F | F | | C | C | C | | F | F | | F | F | F | F | F | F | C | C | F | F | F | F | F | C | | F | F | F | | F |
| Properties and Rules | F | C | C | | | | F | F | C | | F | | F | F | C | F | | | | | C | C | F | C | | F | F | C | | |
| **Functions and Relations** | Expose | | | | | | | | | | Develop | | | | | | | | | | Develop | | | | | | | | | |
| Linear Expressions/Equations | | | | | C | C | | | F | | F | | | C | F | F | C | F | F | | F | | | F | | | | F | | F |
| Quadratic Expressions/Equations | | | | | | | | | C | | C | C | C | | C | | | | F | | | F | F | F | | | F | F | | C |
| Exponential Expressions/Equations | | | | | C | C | | C | C | | | | F | C | | | | | F | | | | F | F | | F | | F | | C |
| Rational Expressions/Equations | | | | | | | | | | | | | | | | | | C | C | | | | | F | | F | | C | | C |

**F** = **This topic is a Focus of Instruction in this chapter.**

**C** = **This topic is Connected to the content of the chapter and is either reviewed in this chapter or informally introduced.**

**Expose:** Ideas are introduced at an informal concrete level and will be fully developed later in the program.

**Develop:** Ideas are formalized and fully developed.

**Review & Extend:** Ideas are reviewed and used to extend understanding of related ideas.

## Geometry

| Topics | Course 1 | | | | | | | | | | Course 2 | | | | | | | | | | Course 3 | | | | | | | | | |
|---|---|---|---|---|---|---|---|---|---|---|---|---|---|---|---|---|---|---|---|---|---|---|---|---|---|---|---|---|---|---|
| | 1 | 2 | 3 | 4 | 5 | 6 | 7 | 8 | 9 | 10 | 1 | 2 | 3 | 4 | 5 | 6 | 7 | 8 | 9 | 10 | 1 | 2 | 3 | 4 | 5 | 6 | 7 | 8 | 9 | 10 |
| **Two-Dimensional Shapes** | **Develop** | | | | | | | | | | **Review & Extend** | | | | | | | | | | **Review & Extend** | | | | | | | | | |
| Polygons | F | C | C | | | | | F | | | | C | | | | | C | C | | | | | | | C | | | | | |
| Quadrilaterals | F | | | | | | | F | | | | C | | | | | C | C | | | | | | | C | | | | | |
| Triangles | F | | | | | | | F | | | | C | | C | | | F | F | | | | | | | C | | | | | |
| Angles | C | | | | | | | F | | | | | | | | | C | C | | | | | | | C | | | | | |
| **Geometric Relationships** | **Expose** | | | | | | | | | | **Develop** | | | | | | | | | | **Review & Extend** | | | | | | | | | |
| Congruence | | | | | | | | | | | | | | | | | F | C | | | | | | | C | | | | | |
| Similarity | | | | | | | C | | | C | | C | | C | C | | F | F | | | C | | | | C | | | | | |
| **Three-Dimensional Figures** | **Expose** | | | | | | | | | | **Develop** | | | | | | | | | | **Review & Extend** | | | | | | | | | |
| Spatial Visualization | | | | | | | | | | | | F | | | | | C | | | | | | | | | | | | | |
| 3-D Solids | | | | | | | | | | | | F | | | | | C | | | | | | | | | | | | | |
| **Measurement** | **Develop** | | | | | | | | | | **Develop** | | | | | | | | | | **Review & Extend** | | | | | | | | | |
| Perimeter and Area | | C | C | | | | | F | | | | F | | C | | | F | C | | | | | | | C | | | | | |
| Surface Area and Volume | | | | | | | | C | | | | F | | | | | F | C | | | | | | | C | | | | | |
| **Coordinate Geometry** | **Develop** | | | | | | | | | | **Develop** | | | | | | | | | | **Develop** | | | | | | | | | |
| Coordinate Representations | | | | | F | F | | C | C | | C | | F | F | F | F | C | C | F | C | F | F | F | F | C | | | F | | F |
| Transformations | | | | | | | | | | | | C | | C | | | F | | | | | | | | F | | | C | | |

## Data and Probability

| Topics | Course 1 | | | | | | | | | | Course 2 | | | | | | | | | | Course 3 | | | | | | | | | |
|---|---|---|---|---|---|---|---|---|---|---|---|---|---|---|---|---|---|---|---|---|---|---|---|---|---|---|---|---|---|---|
| | 1 | 2 | 3 | 4 | 5 | 6 | 7 | 8 | 9 | 10 | 1 | 2 | 3 | 4 | 5 | 6 | 7 | 8 | 9 | 10 | 1 | 2 | 3 | 4 | 5 | 6 | 7 | 8 | 9 | 10 |
| **Data Analysis** | **Develop** | | | | | | | | | | **Develop** | | | | | | | | | | **Review & Extend** | | | | | | | | | |
| Graphs and Displays | C | C | | C | F | F | | | | F | | | | | | | | | C | F | C | | | | | | | | | |
| Modeling and Analysis | C | C | | | F | F | | | | F | | C | | | C | | | | | F | C | | | | | | | | | F |
| Statistical Measures | | | | | | F | | | | C | | | | | | | | | | F | | | | | | | | | C | |
| Surveys and Sampling | | | | | | | | | | F | | | | | | | | C | | F | | | | | | | | | C | |
| **Probability** | **Develop** | | | | | | | | | | **Develop** | | | | | | | | | | **Develop** | | | | | | | | | |
| Basic Concepts and Rules | | | | | | | | | | F | | | | | | | | | | F | | | | | | | | | F | |
| Counting Methods | | | | | | | | | | | | | | | | | | | | F | | | | | | | | | F | |
| Experiments and Simulations | | | | | | | | | | F | | | | | | | | | | F | | | | | | | | | F | |

# Expectations

## Entrance Expectations for Course 1

***What students should know as they begin Course 1***

### Algebra

- Are familiar with some relationships in tabular form such as input/output boxes
- Have some limited experience with variables

### Geometry

- Know the names of common geometric figures
- Identify figures with line symmetry
- Measure lengths and are familiar with both customary and metric measures of length
- Find the perimeter of figures with straight-line sides
- Find the areas of rectangles

### Number and Operation

- Are proficient with whole number arithmetic
- Are proficient with decimal addition and subtraction
- Have multiplied decimals but are not proficient
- Know the algorithm for finding equivalent fractions but may not understand why it works
- Are proficient with writing decimals as fractions
- Know decimal equivalents for $\frac{1}{4}$, $\frac{1}{2}$, and $\frac{3}{4}$ and for fractions with denominators that are powers of 10
- Add and subtract fractions with the same denominator
- Have been exposed to addition and subtraction of fractions with unlike denominators but may not be proficient
- Have been exposed to multiplication of fractions but are not proficient
- Have seen percents but know only simple things about them

### Data and Probability

- Interpret and create bar graphs and pictographs

# Exit Expectations for Course 1

*What students should know as they finish Course 1*

## Algebra

- Understand the concept of a variable
- Solve simple one- and two-step equations with the variable on one side only

## Geometry

- Understand area and perimeter and have committed important formulas to memory
- Give reasonable estimates for angle measures and measure angles with a protractor
- Plot points in the first quadrant

## Number and Operation

- Are proficient with fraction and decimal operations
- Move efficiently among fraction, decimal, and percent representations

## Data and Probability

- Conduct simple experiments to determine experimental probabilities
- Calculate theoretical probabilities in simple situations with a small number of equally likely outcomes
- Calculate measures of central tendency
- Interpret bar graphs, line graphs, circle graphs, line plots, and stem plots

# The Instructional Cycle

*Impact Mathematics* is designed to actively engage students in their own learning. To facilitate the learning and teaching process, *Impact Mathematics* is designed around a three-step instructional cycle.

## Introduce

Each multiday lesson begins with a class discussion, activity, or problem designed to introduce the mathematics and help set a context for learning. To help guide the introduction, **Explore** activities and **Think & Discuss** questions are provided in the student materials.

## Develop

Each lesson in *Impact Mathematics* is composed of in-class **Investigations** that provide a mix of worked-out examples, direct modeling through cartoons, and interactive problem sets. During Investigations, the mathematics, not an artificial format, determines the approach and the day's activity. Each Investigation is designed to last about 45 minutes or one class period. Positioned at logical breaking points, Investigations help teachers determine pacing and help make multiday lessons manageable.

The **Share & Summarize** questions signal the end of each Investigation. These questions offer students an opportunity to share what they did and what was learned. They also provide a summary of major points. For teachers they offer an important assessment opportunity. When used as part of class discussion, Share & Summarize questions serve as a checkpoint to make sure that appropriate learning has taken place and that students can move forward in the lesson successfully.

## Assign & Assess

Independent assignments and opportunities to assess what students have learned are a regular part of the curriculum. The **On Your Own Exercises** at the end of each lesson are an integral part of program instruction and are intended for individual work done primarily outside of class. You will find three types of problems in each set of On Your Own Exercises.

- *Practice & Apply* problems provide opportunities for students to reinforce and directly apply the skills and concepts they have learned in each of the Investigations.
- *Connect & Extend* problems relate student learning in the lesson to other mathematical topics and strands, and sometimes require students to stretch their thinking. Connections may reach back to ideas previously developed in the program or might offer a preview of how current topics are related to what's to come.
- *Mixed Review* problems are an important part of the instructional and assignment structure. Frequent review of previously learned skills helps students maintain mastery and replaces the need to reteach topics.

Assignment guides for each Investigation are provided in the Teacher's Guide.

# Assessment in *Impact Mathematics*

The assessment tools in *Impact Mathematics* are broader than those in traditional mathematics programs. They encompass the processes of problem solving, reasoning, communication, connections, concepts, applications, representational strategies, and procedures.

The flexibility and variety of assessment in *Impact Mathematics* addresses the various ability levels and learning styles of students, as well as the instructional needs of teachers.

## In the Student Edition

- **Share & Summarize** questions provide a forum for students to summarize and share their learning with the class.
- **On Your Own Exercises,** an integral part of daily instruction, are independent assignments intended for individual work outside of class.
- **Review & Self-Assessment** provides students with an opportunity to reflect on the important topics within the chapter and to prepare for formal assessment.

## In the Teacher's Guide

- **Problem Set Wrap-Ups** ensure students are making appropriate progress through an Investigation.
- **Troubleshooting** notes provide remedial work students might need in order to move on to the next Investigation successfully.
- **Additional Examples** can be used as on-the-run assessment tools.
- **Quick Checks** provide checklists of what students should be able to do at the end of each lesson.
- **Quick Quizzes** provide brief end-of-lesson assessment opportunities.

## In the Assessment Resources Book

- A **Pretest** determines whether students have the prerequisite skills for the course.
- **Refresher Worksheets** help students review prerequisite skills.
- **Chapter Tests** provide a comprehensive evaluation of chapter content.
- **Performance Assessments** provide open-ended opportunities to measure student achievement. They can be used to supplement or replace items on chapter and semester tests, as take-home assignments, as group assessments, or as challenge or extra-credit problems.
- **Semester Tests** provide cumulative midyear and end-of-year evaluations.

# Pacing

*Impact Mathematics* and the accompanying support materials allow you to create a mathematics course that meets the needs of your students. The chart shown on these two pages offers general suggestions for pacing your students through the book.

| Chapter | Lesson (Investigation) | Day(s) |
|---|---|---|
| 1 | Lesson 1.1(1) | 1–2 |
| | Lesson 1.2(1) | 3–4 |
| | Lesson 1.2(2) | 5 |
| | Quiz, Lessons 1.1–1.2 | 6 |
| | Lesson 1.3(1) | 7–8 |
| | Lesson 1.3(2) | 9 |
| | Lesson 1.4(1) | 10–11 |
| | Lesson 1.4(2) | 12 |
| | Lesson 1.4(3) | 13 |
| | Lesson 1.4(4) | 14 |
| | Lab (optional); Quiz, Lesson 1.3–1.4 | 15 |
| | Chapter 1 Review | 16 |
| | Chapter 1 Test | 17 |
| 2 | Lesson 2.1(1) | 18–19 |
| | Lesson 2.1(2) | 20 |
| | Lesson 2.1(3) | 21 |
| | Lesson 2.1(4) | 22 |
| | Lab (optional); Quiz, Lesson 2.1 | 23 |
| | Lesson 2.2(1) | 24–25 |
| | Lesson 2.2(2) | 26 |
| | Lesson 2.2(3) | 27 |
| | Lesson 2.2(4) | 28 |
| | Lesson 2.3(1) | 29–30 |
| | Lesson 2.3(2) | 31 |
| | Lesson 2.3(3) | 32 |
| | Quiz, Lessons 2.2–2.3 | 33 |
| | Lesson 2.4(1) | 34–35 |
| | Lesson 2.4(2) | 36 |
| | Lesson 2.4(3) | 37 |
| | Quiz, Lesson 2.4 | 38 |
| | Lesson 2.5(1) | 39–40 |
| | Chapter 2 Review | 41 |
| | Chapter 2 Test | 42 |
| 3 | Lesson 3.1(1) | 43–44 |
| | Lesson 3.1(2) | 45 |
| | Lesson 3.1(3) | 46 |
| | Lab (optional); Quiz, Lesson 3.1 | 47 |
| | Lesson 3.2(1) | 48–49 |
| | Lesson 3.2(2) | 50 |
| | Lesson 3.2(3) | 51 |
| | Lesson 3.2(4) | 52 |
| | Lesson 3.2(5) | 53 |
| | Quiz, Lesson 3.2 | 54 |
| | Lesson 3.3(1) | 55–56 |
| | Lesson 3.3(2) | 57 |
| | Lesson 3.3(3) | 58 |
| | Lesson 3.3(4) | 59 |
| | Lesson 3.3(5) | 60 |
| | Chapter 3 Review | 61 |
| | Chapter 3 Test | 62 |
| 4 | Lesson 4.1(1) | 63–64 |
| | Lesson 4.1(2) | 65 |
| | Lesson 4.1(3) | 66 |
| | Lesson 4.1(4) | 67 |
| | Lesson 4.2(1) | 68–69 |
| | Lesson 4.2(2) | 70 |
| | Quiz, Lessons 4.1–4.2 | 71 |
| | Lesson 4.3(1) | 72–73 |
| | Lesson 4.3(2) | 74 |
| | Lab (optional); Quiz, Lesson 4.3 | 75 |
| | Chapter 4 Review | 76 |
| | Chapter 4 Test | 77 |
| 5 | Lesson 5.1(1) | 78–79 |
| | Lesson 5.1(2) | 80 |
| | Lesson 5.1(3) | 81 |
| | Lesson 5.2(1) | 82–83 |

| Chapter | Lesson (Investigation) | Day(s) |
|---|---|---|
| 5 | Lesson 5.2(2) | 84 |
| | Lesson 5.2(3) | 85 |
| | Quiz, Lessons 5.1–5.2 | 86 |
| | Lesson 5.3(1) | 87–88 |
| | Lesson 5.3(2) | 89 |
| | Lesson 5.3(3) | 90 |
| | Lab (optional); Quiz, Lesson 5.3 | 91 |
| | Chapter 5 Review | 92 |
| | Chapter 5 Test | 93 |
| 6 | Lesson 6.1(1) | 94–95 |
| | Lesson 6.1(2) | 96 |
| | Lesson 6.1(3) | 97 |
| | Lesson 6.2(1) | 98–99 |
| | Lesson 6.2(2) | 100 |
| | Lesson 6.2(3) | 101 |
| | Quiz, Lessons 6.2–6.3(3) | 102 |
| | Lesson 6.2(4) | 103 |
| | Lesson 6.2(5) | 104 |
| | Lesson 6.3(1) | 105–106 |
| | Lesson 6.3(2) | 107 |
| | Lab (optional); Quiz, Lessons 6.2(4)–6.3 | 108 |
| | Chapter 6 Review | 109 |
| | Chapter 6 Test | 110 |
| 7 | Lesson 7.1(1) | 111–112 |
| | Lesson 7.1(2) | 113 |
| | Lesson 7.1(3) | 114 |
| | Lesson 7.2(1) | 115–116 |
| | Lesson 7.2(2) | 117 |
| | Lesson 7.2(3) | 118 |
| | Quiz, Lessons 7.1–7.2 | 119 |
| | Lesson 7.3(1) | 120–121 |
| | Lesson 7.3(2) | 122 |
| | Chapter 7 Review | 123 |
| | Chapter 7 Test | 124 |
| 8 | Lesson 8.1(1) | 125–126 |
| | Lesson 8.1(2) | 127 |
| | Lesson 8.2(1) | 128–129 |
| | Lesson 8.2(2) | 130 |

| Chapter | Lesson (Investigation) | Day(s) |
|---|---|---|
| 8 | Quiz, Lessons 8.1–8.2 | 131 |
| | Lesson 8.3(1) | 132–133 |
| | Lesson 8.3(2) | 134 |
| | Lesson 8.3(3) | 135 |
| | Lesson 8.3(4) | 136 |
| | Lesson 8.4(1) | 137–138 |
| | Lesson 8.4(2) | 139 |
| | Lesson 8.4(3) | 140 |
| | Lab (optional); Quiz, Lessons 8.3–8.4 | 141 |
| | Lesson 8.5(1) | 142–143 |
| | Lesson 8.5(2) | 144 |
| | Chapter 8 Review | 145 |
| | Chapter 8 Test | 146 |
| 9 | Lesson 9.1(1) | 147–148 |
| | Lesson 9.1(2) | 149 |
| | Lab (optional) | 150 |
| | Lesson 9.2(1) | 151–152 |
| | Lesson 9.2(2) | 153 |
| | Lesson 9.2(3) | 154 |
| | Quiz, Lessons 9.1–9.2 | 155 |
| | Lesson 9.3(1) | 156–157 |
| | Lesson 9.3(2) | 158 |
| | Lesson 9.3(3) | 159 |
| | Chapter 9 Review | 160 |
| | Chapter 9 Test | 161 |
| 10 | Lesson 10.1(1) | 162–163 |
| | Lesson 10.1(2) | 164 |
| | Lab (optional) | 165 |
| | Lesson 10.2(1) | 166–167 |
| | Lesson 10.2(2) | 168 |
| | Lesson 10.2(3) | 169 |
| | Lesson 10.2(4) | 170 |
| | Quiz, Lessons 10.1–10.2 | 171 |
| | Lesson 10.3(1) | 172–173 |
| | Lesson 10.3(2) | 174 |
| | Chapter 10 Review | 175 |
| | Chapter 10 Test | 176 |

# Problem-Solving Strategies and Estimation

## Problem Solving

Problem solving occurs when students are engaged in activities for which the method of solving the problem is not immediately known. Problem solving is an integral part of *Impact Mathematics*. Students have frequent opportunities to formulate problems, solve them, and reflect upon the process.

The following problem-solving strategies are embedded throughout *Impact Mathematics*. The lesson notes in the Teacher's Guide indicate places where these problem-solving strategies are used.

### Problem-Solving Strategies

- Act it out
- Choose the method of computation
- Determine reasonable answers
- Draw a picture or diagram
- Eliminate possibilities
- Guess-check-and-improve
- Identify irrelevant or missing information
- Look for a pattern
- Make a model
- Make a table or chart
- Make an organized list
- Make and test a conjecture
- Solve a simpler problem
- State problem in own words
- Use a graph
- Use a Venn diagram
- Use benchmarks
- Use logical reasoning
- Work backward
- Write an equation or rule

## Estimation

Students in the middle grades are learning to compute fluently with rational numbers—fractions, decimals, percents, and integers. One important part of the computation process involves estimation. In *Impact Mathematics*, students develop and use strategies to estimate the results of rational number computations and use the results to judge whether the result is reasonable.

The lesson notes in the Teacher's Guide indicate places where the students estimate with rational numbers. The notes also indicate where estimation is used in other aspects of mathematics, such as geometry, graph interpretation, and statistics.

# Course 1 CONTENTS

## Volume A

## Chapter One

## Chapter Two

# Chapter Three

## Working with Fractions and Decimals . . . . . . . . . . . . . . . . .152



# Chapter Four

## Making Sense of Percents . . . . . .224

# Chapter Five

## Exploring Graphs .............276



# Volume B

# Chapter Six

## Analyzing Data .............340

# Chapter Seven

## Variables and Rules . . . . . . . . . . .408

# Chapter Eight

## Geometry and Measurement . . . .464

## Chapter Nine



## Chapter Ten

# CHAPTER 6

# Analyzing Data

## Chapter Overview

In this chapter students work with mathematical tools for analyzing data to help them uncover and understand the story behind the data. In some investigations, students role-play as analysts in a data-consulting group named Data Inc. to use their developing knowledge about data to solve problems for clients.

The first lesson builds on many of the ideas about graphing from Chapter 5. Students work with familiar line graphs and bar graphs and are introduced to a new visual display, histograms. They consider the shape of the graph to help them describe data.

In the second lesson, students learn the measures of central tendency (mode, median, and mean) and dispersion (range) to help them describe what is typical in a data set. They consider how changing or adding data values affects these statistical measures. They also learn two new ways to display data, the line plot and the stem-and-leaf plot.

In the third lesson, students apply their analytical skills to carry out a research project. They survey classmates about what activities they participate in outside of school; use statistical measures, when appropriate, to analyze the resulting data; and consider which data displays, if any, they will use to support their findings. The chapter includes an optional lab that guides students through using spreadsheet software to calculate statistical measures.

## the Big Picture

| Chapter 6 Highlights | Links to the Past | Links to the Future |
|---|---|---|
| Interpreting and creating bar graphs and histograms (6.1, 6.3) | **Elementary Grades:** Creating and interpreting bar graphs | **Course 2:** Choosing appropriate displays for data sets<br>**Course 3:** Analyzing data using graphs<br>**Course 3 and High School:** Showing probability distributions |
| Interpreting and creating line plots and stem plots (6.2, 6.3) | **Elementary Grades:** Understanding number lines<br>**Elementary Grades:** Creating and interpreting pictographs | **Course 2:** Choosing appropriate displays for data sets<br>**Course 3:** Analyzing data using graphs |
| Finding and interpreting the mode, median, and mean (6.2, 6.3) | **Elementary Grades:** Finding greatest values<br>**Elementary Grades:** Finding averages (means) | **Course 2:** Creating and interpreting box-and-whisker plots<br>**Course 3:** Analyzing data in tables |
| Choosing the best average for a given situation (6.2, 6.3) | **Elementary Grades and Chapter 3:** Choosing the appropriate operation to solve a problem | **Course 2:** Choosing appropriate displays for data sets<br>**Course 3:** Analyzing data in tables |

Time-Saving Tool TeacherWorks™ All-In-One Planner and Resource Center

# Planning Guide

| Lesson Objectives | Pacing | Materials | NCTM Standards | Hot Topics |
|---|---|---|---|---|
| **6.1 Using Graphs to Understand Data** page 341b<br>• To interpret and compare line graphs and investigate reasons for patterns observed in line graphs<br>• To use bar graphs and histograms to display data, identify patterns, and draw conclusions<br>• To create bar graphs and histograms from sets of data<br>• To decide whether to use a bar graph or a histogram to display a given set of data | 4 class periods | • Masters 55 and 56 or transparencies of them<br>• Master 57 or graph paper<br>• Master 58 or graph paper<br>• Master 59 or graph paper<br>• Masters 60a and 60b (optional)<br>• Master 61 (optional)<br>• Master 62 (optional)<br>• Transparency of Master 59 (optional) | 2, 5, 6, 7, 8, 10 | |
| **6.2 What Is Typical?** page 361b<br>• To find the mode, median, mean, and range of a data set<br>• To make line plots and stem-and-leaf plots and use them to describe a data set<br>• To predict how changing a value in a data set impacts the mean and median<br>• To use various measures to interpret and compare data sets | 6 class periods | • Master 63 (optional)<br>• Rulers (optional)<br>• Transparency of homework data from Investigation 1 (optional)<br>• Measuring tape or yardstick<br>• Class Roster (optional) | 1, 5, 7, 8, 10 | pp. 190–201, 202–203 |
| **6.3 Collecting and Analyzing Data** page 389b<br>• To prepare a plan that includes selecting the appropriate statistical measures as part of a research project<br>• To select a way to display data<br>• To follow a plan to reach a conclusion | 3 class periods | • Master 64 (optional)<br>• Transparency of Master 64 (optional)<br>• Masters 65a and 65b (optional)<br>• Master 66 (optional)<br>• Computer with spreadsheet software (1 per group) | 1, 5, 6, 7, 8, 10 | pp. 184–189 |

**Key to NCTM Curriculum and Evaluation Standards:** 1=Number and Operations, 2=Algebra, 3=Geometry, 4=Measurement, 5=Data Analysis and Probability, 6=Problem Solving, 7=Reasoning and Proof, 8=Communication, 9=Connections, 10=Representation

# Assessment Opportunities

## Standard Assessment

*Impact Mathematics* offers three types of formal assessment. The Chapter 6 Review and Self-Assessment in the Student Edition serves as a self-assessment tool for students. In the Teacher's Guide, a Quick Quiz at the end of each lesson allows you to check students' understanding before moving to the next lesson. The Assessment Resources include blackline masters for chapter and quarterly tests.

- **Student Edition** Chapter 6 Review and Self-Assessment, pages 402–407
- **Teacher's Guide** Quick Quizzes, pages 361, 389, 401
- **Assessment Resources** Chapter 6 Test Form A, pages 151–154; Chapter 6 Test Form B, pages 155–158

## Ongoing Assessment

*Impact Mathematics* provides numerous opportunities for you to assess your students informally as they work through the investigations. Share & Summarize questions help you determine whether students understand the important ideas of an investigation. If students are struggling, Troubleshooting tips provide suggestions for helping them. On the Spot Assessment notes appear throughout the teaching notes. They give you suggestions for preventing or remedying common student errors. Assessment Checklists in the Assessment Resources provide convenient ways to record student progress.

- **Student Edition** Share & Summarize, pages 345, 349, 352, 365, 369, 372, 376, 379, 393, 395
- **Teacher's Guide** On the Spot Assessment, pages T346, T351, T364, T368, T369, T393, T397
  Troubleshooting, pages T345, T349, T352, T365, T369, T372, T376, T379, T393, T395
- **Assessment Resources** Chapter 6 Assessment Checklists, pages 264–265

## Alternative Assessment, Portfolios, and Journal Ideas

The alternative assessment items in *Impact Mathematics* are perfect for inclusion in student portfolios and journals. The In Your Own Words feature in the Student Edition gives students a chance to write about mathematical ideas. The Performance Assessment items in the Assessment Resources provide rich, open-ended problems, ideal for take-home or group assessment.

- **Student Edition** In Your Own Words, pages 359, 386, 401
- **Assessment Resources** Chapter 6 Performance Assessment, pages 159–161

# Assessment Resources

The Assessment Resources provide a chapter test in two equivalent forms, along with additional performance items. The performance items can be used in a variety of ways. They are ideal for take-home assessment or in-class group assessment.

- Chapter 6 Test Form A, pages 151–154
- Chapter 6 Test Form B, pages 155–158
- Chapter 6 Performance Assessment, pages 159–161
- Chapter 6 Assessment Solutions, pages 162–166

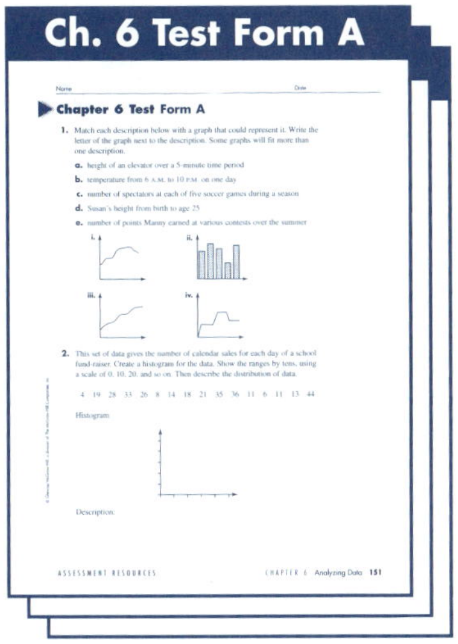


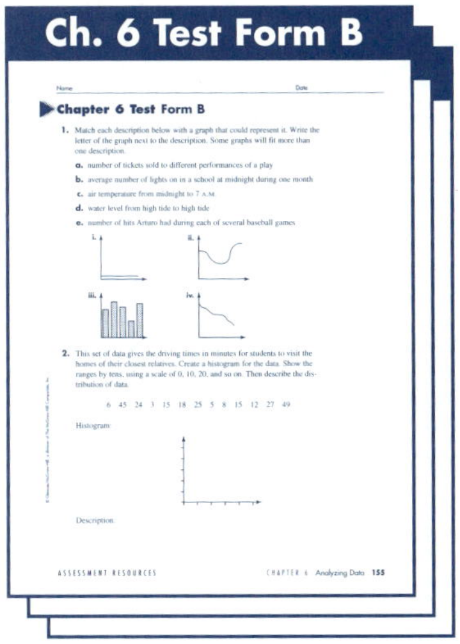


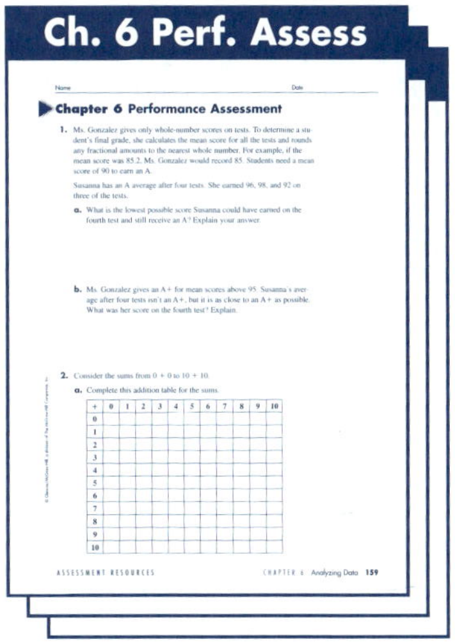


# Additional Resources

- **Math Skills Maintenance Workbook,** 2, 10
- **Investigations for the Special Education Student in the Mathematics Classroom,** 3, 4, 6
- **Virtual Activities CD-ROM,** Make a Table
- **What's Math Got To Do With It? Videos,** Level 1, Video 1
- **StudentWorks™ CD-ROM**
- **Reading and Writing in the Mathematics Classroom**
- **Using the Internet in the Mathematics Classroom**

## ExamView® Pro

Use ExamView® Pro Testmaker CD-ROM to:

- Create Multiple versions of tests.
- Create Modified tests for Inclusion students with one mouse click.
- Edit existing questions and Add your own questions.
- Build tests aligned with state standards using built-in State Curriculum Correlations.
- Change English tests to Spanish with one mouse click and vice versa.

# CHAPTER 6

# Analyzing Data

**Introduce**
Ask students what they think is meant by the phrase *average American*. Accept all reasonable answers. Be sure the discussion includes the mathematical definition of using a single number to describe a typical value in a set of data.

You might have students find reports in newspapers, magazines, or on the Internet that give information about the average American. Have students compare themselves to the average American and explain why these descriptions may not accurately describe them. They might think that adults or young children have different interests and affect the average. Most people are probably not typical because the average is determined from a large group of people.

**Think About It**
Discuss students' ideas. They may suggest collecting the heights of all the students in the school and determining which one is the most common. You might demonstrate by collecting data and finding the height of the "average" student in the class.

### Real-Life Math

**The Average American** You have probably read or heard reports that give information about the "average American." In these reports, a single number is often used to describe what is typical about the entire U.S. population. Usually, such a number is found by calculating an "average" value from a large set of data. As you will discover in this chapter, there are several types of averages—and for some sets of data, they may be very different from one another.

**Think About It** How might you find the height of the "average" student in your school?

# *Family Letter*

Dear Student and Family Members,

Did you know that the "average American" eats 24 quarts of ice cream and 68 quarts of popcorn every year? Did you ever wonder how people come up with numbers like these? Our next chapter is about analyzing data—numbers, facts, or other measurable information.

You'll explore three ways of measuring what is typical—the median, mode, and mean. Here's an example. Suppose you ask 15 of your friends to tell you how many movies they've seen in the past month. You organize the data into a list.

**0, 0, 0, 0, 0, 2, 2, 2, 2, 3, 3, 3, 3, 4, 6**

- The median is the middle number in the list, 2.
- The mode is the number that appears the most often, 0.
- The mean is the number you get if the total, 30 movies, is divided evenly among all your friends. The mean is $30 \div 15$ or 2.

So, you might conclude that your friends saw an average of 2 movies in the last month.

Along with analyzing data, it's important to display the data. You're already familiar with bar graphs and line graphs, but in this chapter you'll learn about histograms, stem-and-leaf plots, double bar graphs, and line plots too. Here's a line plot of the data above.

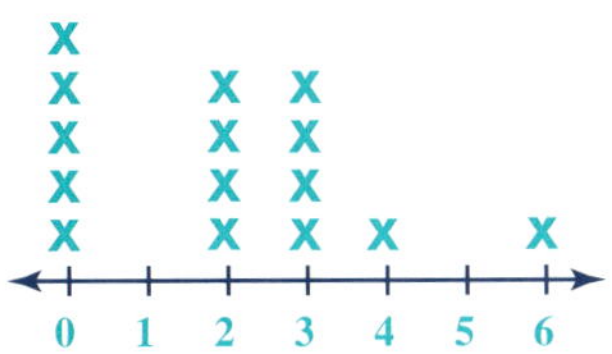


Graphs can make it easier to see patterns in data and to draw conclusions. What is obvious in the line plot that wasn't when the data was shown in the list?

**Vocabulary** Throughout the chapter you'll see these new vocabulary terms:

| | | |
|---|---|---|
| **distribution** | **mean** | **outlier** |
| **histogram** | **median** | **range** |
| **line plot** | **mode** | **stem-and-leaf plot** |

## What can you do at home?

During the next few weeks, your student may show an interest in data outside of the classroom. You might help him or her to think of different kinds of data people collect and why—purchases in the grocery story, telephone opinion surveys, political exit polls, or hits at a Web site.

**impactmath.com/family_letter**



Another version of the Family Letter, available in English and Spanish, is in the Teaching Resources. You may want to send a copy of this letter home with your students.

**Teaching Resources**

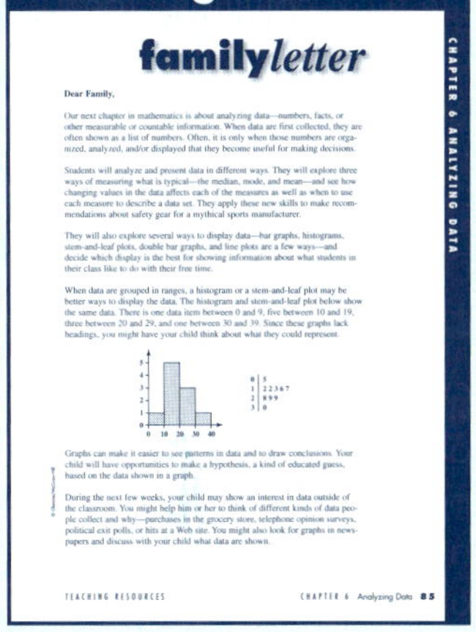

# Mathematical Background

Key characteristics are often used to make comparisons that help people to describe and make sense of the world. Any description of large groups based on particular characteristics results in large sets of data. Data analysis provides a set of mathematical tools to help people summarize data, see what is typical, and understand the shape of the data.

**Measures of central tendency** The three measures of central tendency, or measures of center, are the mean, median, and mode. In everyday language, the mean is often called the *average* but statisticians use the term *average* to refer to any measure of central tendency.

Consider the data set below, which shows the numbers of hours of television that some students watched during a week.

15, 12, 10, 5, 5, 5, 5, 5, 4, 3, 3, 3, 2, 2, 2, 2, 1, 1, 1, 0

The *mean* is the value that would result if all the values were combined and the sum were distributed evenly among all the members of the data set. The mean for this data set is calculated as

$$\frac{15 + 12 + 10 + 5 + 5 + 5 + 5 + 5 + 4 + 3 + 3 + 3 + 2 + 2 + 2 + 2 + 1 + 1 + 1 + 0}{20}, \text{ or } 4.3.$$

Another measure of what is typical in a data set is the median. The *median* is the middle value in a set of data. To find the median, put the items in a data set in order and then count in from either end to the middle. If a data set has an odd number of items, the median is the single middle value. If the data set has an even number of items, the median is the value halfway between the two middle values. The median of the TV data is 3.

A third measure of central tendency is the *mode,* the value or values that appear most frequently in a data set. The mode of the TV data above is 5.

**Choosing a typical value** The three measures of central tendency may give different ideas about a set of data because each measure reflects a different aspect of the data. In the television example, 8 members of the group watched more than 4.3 hours, while the other 12 members watched fewer than 4.3 hours. The median of 3 may be a better measure of average than the mean to describe the viewing habits of these students.

The values differ because the mean is sensitive to outliers, or isolated values at the extremes of a data set, whereas the median is not. If the student who watched 15 hours had actually watched 51 hours, the mean would change significantly while the median would not change at all.

The mode of the television data is greater than either the mean or the median. Some people may prefer to use another measure as the typical value since only 3 students watched more television than the mode and 12 students watched less.

For categorical data, the mode is the only statistical measure available. The mode is also different from the mean and median in that a data set can have no mode, one mode, or more than one mode but, if the data are numerical, the data set has exactly one mean and one median.

• *Teaching notes continued on page A656*

# 6.1

# Using Graphs to Understand Data

## Objectives

- To interpret and compare line graphs and investigate reasons for patterns observed in line graphs
- To use bar graphs and histograms to display data, identify patterns, and draw conclusions
- To create bar graphs and histograms from sets of data
- To decide whether to use a bar graph or a histogram to display a given set of data.

## Overview (pacing: about 4 class periods)

In this lesson, students use tables, line graphs, bar graphs, and histograms to analyze data. Double bar graphs are reviewed in the On Your Own Exercises. This lesson provides an introduction to data analysis. Students not only engage with real data and experience the power of graphical representations of data, but also think about and try to understand what stories the data tell in each situation.

In the first two investigations, students assume the roles of management consultants and write reports summarizing their findings. They also use reasoning to solve a mystery involving line graphs and construct bar graphs and histograms using data from tables.

## Advance Preparation

You will need to either provide students with copies of Masters 55 and 56 or make transparencies of them to display as students work through Problem Set A. These two masters contain additional information needed to explain Gerald Orkney's disappearance.

You may want to provide students with copies of Masters 57, 58, and 59 to help them organize their work and complete graphs as they work in class. These three masters reproduce the tables and graphs used in Problem Sets B, C, D, E, and F.

You may also want to provide copies of Masters 60a, 60b, 61, and 62 to students as they work through On Your Own Exercises 5, 6, and 9. These masters reproduce the tables and partially completed graphs shown in the exercises. Students may need graph paper or copies of Master 81 or Master 82 to make graphs in On Your Own Exercises 3, 7, 8, and 10. You may also substitute graph paper for any of the masters for this lesson except Masters 55 and 56.

## Lesson Planner

| | Summary | Materials | On Your Own Exercises | Assessment Opportunities |
|---|---|---|---|---|
| Investigation 1 page T343 | Students use unlabeled line graphs and other clues to solve a mystery. | • Masters 55 and 56 or transparencies of them | Practice & Apply: 1, 2, p. 353<br>Connect & Extend: 7, 8, pp. 357–358<br>Mixed Review: 11–23, p. 361 | Share & Summarize, pages T345, 345<br>Troubleshooting, page T345 |
| Investigation 2 page T346 | Students use information from multiple bar graphs to analyze trends in vehicle emissions. | • Master 57 or graph paper<br>• Master 58 or graph paper<br>• Master 62 (optional) | Practice & Apply: 3, 4, pp. 354–355<br>Connect & Extend: 9, p. 359<br>Mixed Review: 11–23, p. 361 | On the Spot Assessment, page T346<br>Share & Summarize, pages T349, 349<br>Troubleshooting, page T349 |
| Investigation 3 page T350 | Students make histograms to display and interpret frequency data. | • Master 59 or graph paper<br>• Transparency of Master 59 (optional)<br>• Masters 60a and 60b (optional)<br>• Master 61 (optional) | Practice & Apply: 5, 6, pp. 356–357<br>Connect & Extend: 10, p. 360<br>Mixed Review: 11–23, p. 361 | On the Spot Assessment, page T351<br>Share & Summarize, pages T352, 352<br>Troubleshooting, page T352<br>Informal Assessment, page 361<br>Quick Quiz, page 361 |

# Introduce

1 Ask students what kinds of tasks they think a data analyst performs. Continue the discussion to include the tools analysts might use and the skills they might need to complete their work. Inform students that they will use some of these tools as they take on the roles of data analysts for a company called Data Inc.

2 Remind students that in Chapter 5 they found patterns in line graphs. Point out that they can also find patterns and compare data found in tables and other types of graphs to help them make decisions.

3 **Explore**
Ask students to identify the two types of graphs shown. They should be familiar with both the line graph and the bar graph.

Then have groups or pairs of students work through the Explore, making sure they understand that there are multiple parts to consider for each situation. For example, when one graph could describe the situation, students must also decide what each axis would represent and what the graph reveals about the situation. When neither graph describes the situation, students must draw and label a graph for that situation.

4 As groups share their answers, you can point out that there may be more than one reasonable answer for each situation. For example, the bar graph may not be a good display for the fifth situation, the number of school days remaining, if the school is on a year-round schedule.

# 6.1 Using Graphs to Understand Data

People in many professions use data to help make decisions. When data are first collected, they are often just lists of numbers and other information. Before they can be understood, data must be organized and analyzed. In this chapter, you will investigate several tools for understanding data.

In many activities in this chapter, your class will play the role of a company called Data Inc.—a consulting group that specializes in organizing and analyzing data. Various people and organizations will come to Data Inc. for advice and suggestions.

In Chapter 5, you saw how graphs can help you discover patterns and trends. In this lesson, you will look at several types of graphs and compare the kinds of information each tells you about a set of data.

1 Discuss the types of tools data analysts use.

2 Remind students of patterns that can be found in line graphs.

3 You may have students work through the Explore in pairs or small groups.

4 Tell students that there may be more than one answer for each situation.

**Just the facts**

The word *data* is plural and means "bits of countable or measurable information." The singular form of data is *datum.* When you talk about data, you should use the plural forms of verbs—for example *are, were,* or *show.*

① Possible answer: Graph A; The horizontal axis would show days from January 1 through December 31, and the vertical axis would show the number of visitors. The graph shows that there are more visitors during the summer.

## Explore

These graphs have no labels or scale values. Tell whether one of these graphs could describe each situation below. If it could, tell what the axes would represent and what the graph would reveal about the situation. If neither graph could describe the situation, describe or sketch a graph that could.

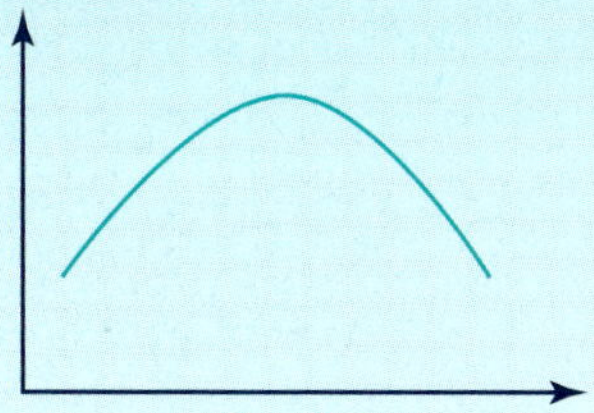

Graph A

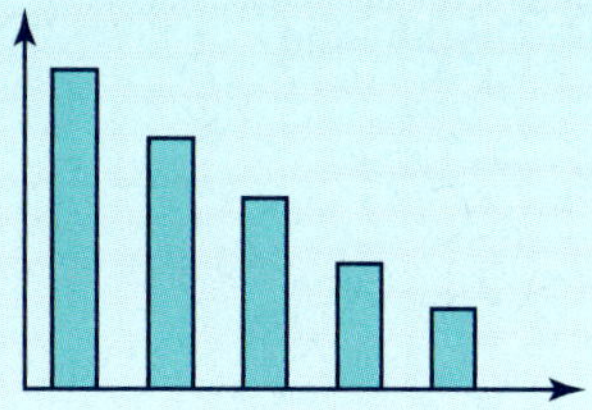

Graph B

- The number of visitors to a zoo over the past year See ①.
  ②–⑥. See Additional Answers.
- The weight of a young hippo from birth to age one See ②.
- The distance from a ball to the ground after the ball is dropped See ③.
- The number of minutes of daylight each day during a year See ④.
- The number of school days remaining on the first of each month, from February through June See ⑤.
- The number of children born each month in one year in Canada See ⑥.



**Additional Answers**

② Possible answer: Neither; this situation could be described by a line graph that increases from left to right.

③ Possible answer: Neither; this situation could be described by a line graph that decreases from left to right.

④ Possible answer: Graph A; the horizontal axis would show days from January 1 through December 31, and the vertical axis would show minutes of daylight. The graph shows there are more minutes of daylight during the summer.

⑤ Possible answer: Graph B; the bars could represent the months, with the height of each bar representing the number of school days left. The graph shows that the number of school days remaining decreases each month.

⑥ Possible answer: Neither; a bar graph with 12 bars would be most appropriate. Graph B has only 5 bars.

# Investigation 1

1 In this investigation, students act as analysts employed by Data Inc. to solve the mystery of the disappearance of Gerald Orkney and his pet iguana. Students make hypotheses about what the unlabeled graphs might represent to help them determine Gerald's whereabouts. During the investigation, more information is revealed so that students can check and revise their hypotheses.

Students will need graph paper or copies of Master 81 or Master 82 to complete On Your Own Exercises 7 and 8, if assigned.

Introduce the mystery situation by having students read the clues. Then have them look at the three graphs.
2 Although students analyze the graphs in more detail in the Think & Discuss on page 344, you might want to discuss the two-quadrant graph at the bottom of the page since one of its scales includes negative values. Ask students what the coordinates of the point closest to the vertical axis would be if both scales were in intervals of 1 and the origin was (0, 0). (1, −1). After this brief review of axes with negative numbers, proceed to the Think & Discuss on page 344.

## Investigation 1 Using Line Graphs to Solve a Mystery

**1** Discuss the mystery situation with the class before going on to the Think & Discuss.

The Smallville police are investigating the disappearance of a man named Gerald Orkney. Here is what they know so far:

- Mr. Orkney lives alone with his pet iguana, Agnes.
- Mr. Orkney didn't show up for work on December 15. When his friends came to check on him, he and Agnes were gone.
- An atlas and the graphs below, which have no scales or labels, were found in Mr. Orkney's apartment.

The police have asked Data Inc. to help them figure out what happened to Gerald Orkney.

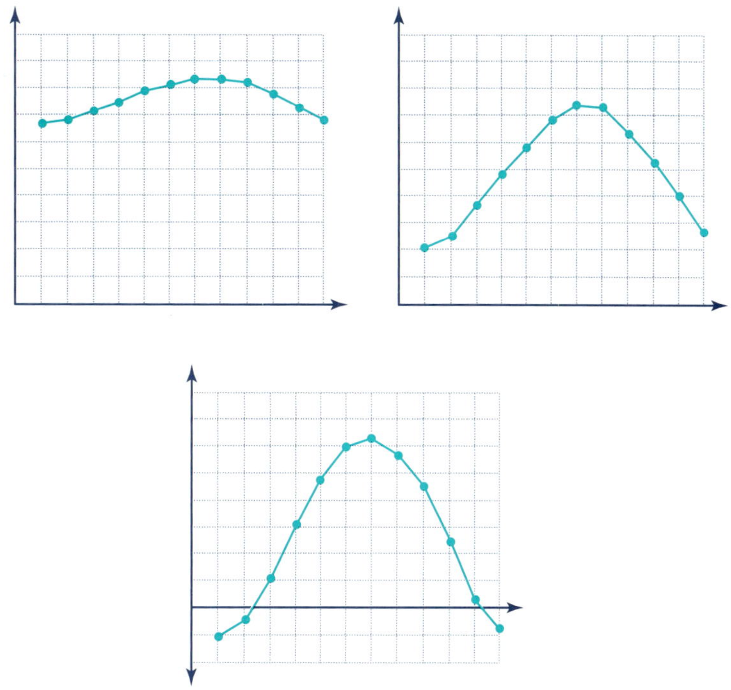

**2** Discuss the two-quadrant graph that contains negative values.



LESSON 6.1 Using Graphs to Understand Data 343

## Think & Discuss

1 Have students compare the graphs as part of a class discussion. Encourage them to look for patterns and consider how high and how low the graphs are when they describe the shape.

2 Discuss the definition of *hypothesis* as an educated guess. Then have students work in small groups to think about what the graphs might show and to develop their hypotheses of what happened to Gerald Orkney and his iguana.

3 As groups share their results, you may want to list facts and hypotheses on the board. Remind students that facts are information that is known and hypotheses are guesses that are supported by known information but are not proven or known. Your list may look like the one below.

| **Facts**<br>Things We Know from the Data | **Hypotheses**<br>Ideas Supported by Data but Not Stated |
|---|---|
| • One graph has negative values. | • The graphs are about three of the same kind of thing.<br>• The horizontal axis represents time. |

If students have difficulty distinguishing facts from hypotheses, encourage them to think about how they derived the information. For example, if someone says that the vertical axis represents height, ask what data supports that statement or ask for examples of something that would fit the graphs and have height on the vertical axis. Then ask for examples of what the graphs might represent if the vertical axis is not height.

Another approach is to have students first share observations and facts, and then consider hypotheses. If students begin to put forth such hypotheses as "This might be a graph of tides," tell them that this is a hypothesis and ask them to keep this idea in mind to present when hypotheses are discussed. Once there are a number of observations, collect hypotheses about what the graphs might represent and what might have happened to Gerald Orkney. Then have students work through Problem Set A.

## Tips from Teachers

"Sometimes my students are intimidated and don't readily offer ideas. I stress that students should respect one another and try to give them ways to politely disagree. When discussing facts and hypotheses, I encourage students to try to relate all hypotheses back to the data by asking questions such as 'What do you see in the graph that makes you think this way?' This approach often helps students make better hypotheses and models a way for them to offer constructive criticism without demeaning other students' ideas."

4 **Problem Set A** **Suggested Grouping: Small Groups**

In this problem set, students are gradually given more information about Gerald Orkney's disappearance and asked to refine their hypotheses. As part of this refinement process, students should see the importance of labels and titles on graphs.

You will either need to make transparencies of Masters 55 and 56 or provide students with copies of these masters. These masters contain more detailed versions of the graphs and provide additional clues to help students determine Gerald Orkney's location. They should be distributed as students work through the problem set. Each problem parallels many of the steps in the Think & Discuss as students make observations and offer hypotheses for the situation.

Students will need the information from Master 55 to work through **Problem 1.** Remind students to use the data at the top of page 343 along with the detailed graphs to make their hypotheses in **Problem 1c.** Ask them to present and explain how they derived their hypotheses.

**Teaching Resources**

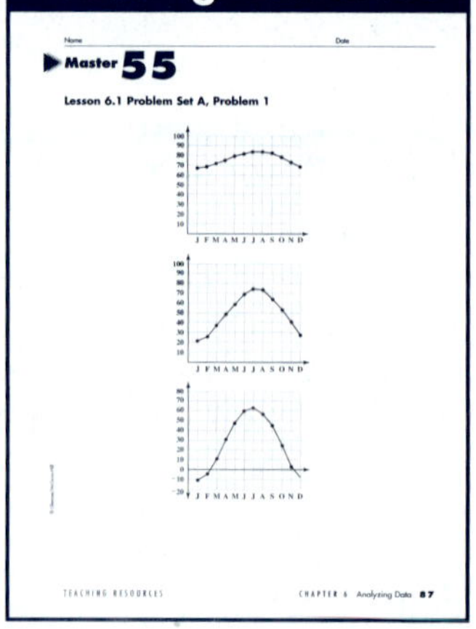

## Think & Discuss

Look carefully at the three graphs the police found. Think about the shape of each graph and about the axes. How are the graphs alike? How are they different? See ①.

Since the graphs don't have labels, it is impossible to know exactly what they represent or how they are related to Gerald Orkney's disappearance. However, you may be able to make a *hypothesis,* or an educated guess, based on the information you do have.

Try to think of some ideas about what the graphs might show. Consider both what the graphs look like and how they might be related to the other information the police found. See ② in Additional Answers.

Using the graphs and the other information given, try to come up with at least one hypothesis about what might have happened to Gerald Orkney. See ③.

① Possible answer: All three graphs have 12 divisions of the horizontal axis. The third graph has points that are below the horizontal axis. All three graphs go up and then come down again. The first graph is the "flattest" of the three. The third graph has the most dramatic increase and decrease.

③ Possible answers: He might have gone on a trip or moved somewhere. If the graphs show change over 12 months, he might have gone somewhere for a long time, rather than just a few days.

1 Problem-Solving Strategy
Look for a pattern

2 Introduce the term *hypothesis.*

3 List the facts and hypotheses on the board as the groups share their results.

4 You may have students work in small groups.

MATERIALS
copies of the "clues"

## Problem Set A

1. During a search of Gerald Orkney's office, the police found more detailed versions of the graphs. Your teacher will give you copies of the new graphs. Look at them closely.
   a. What new information do the graphs reveal? Does this information fit any of the ideas your class had in the Think & Discuss? Now what do you think the graphs might show? See below.
   b. Does the new information support any of the hypotheses your class made about what happened to Gerald Orkney? Explain.
   c. Make a new hypothesis, or make changes to an earlier hypothesis, to fit all the information you have so far.
   Possible answers: If the graphs show monthly temperatures of different places, Orkney may have been trying to figure out where to go. Since he left in December, maybe he went to a place with a warm climate in the winter (or all year long).

**1b.** Answers will vary.

**1a.** Possible answer: The horizontal axis represents months, so the graphs probably show how something changes over a year. The highest part of each graph is in the summer months and the lowest parts are in the winter months. The first graph has points with vertical-axis values from about 65 to 85, the second has values from about 20 to 75, and the third has values from about $^-10$ to 65. These could be temperatures. The graphs could show monthly temperatures of different places.



**Additional Answers**
**Think & Discuss**

② Possible answer: Since the horizontal axis has 12 intervals, the graphs might show some sort of change over 12 months—such as change in temperature, rainfall, or hours of daylight. Since an atlas was found by the graphs, they might have to do with different places.

1

## Access for all Learners

**Extra Challenge** Have students use the information available after **Problem 1c** or after **Problem 2c** to write stories that could explain what each graph represents. The stories should use logical reasoning to explain why Gerald Orkney made the graphs. One story for Problem 2c could be as follows: Gerald Orkney was trying to decide where to go for a vacation that would have a climate suitable for his pet iguana, Agnes. Agnes needs a warm climate to be healthy, and Orkney did not want to take the heat lamp with him.

After providing students with the list of cities and the veterinarian's letter, given on Master 56, have them work in their groups to complete **Problem 2.**

### Teaching Resources

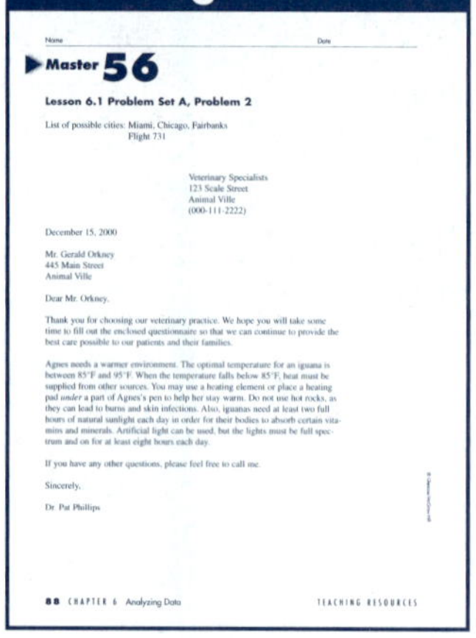
Name Date

**Master 56**

**Lesson 6.1 Problem Set A, Problem 2**

List of possible cities: Miami, Chicago, Fairbanks
Flight 731

Veterinary Specialists
123 Scale Street
Animal Ville
(000-111-2222)

December 15, 2000

Mr. Gerald Orkney
445 Main Street
Animal Ville

Dear Mr. Orkney,

Thank you for choosing our veterinary practice. We hope you will take some time to fill out the enclosed questionnaire so that we can continue to provide the best care possible to our patients and their families.

Agnes needs a warmer environment. The optimal temperature for an iguana is between 85°F and 95°F. When the temperature falls below 85°F, heat must be supplied from other sources. You may use a heating element or place a heating pad *under* a part of Agnes's pen to help her stay warm. Do not use hot rocks, as they can lead to burns and skin infections. Also, iguanas need at least two full hours of natural sunlight each day in order for their bodies to absorb certain vitamins and minerals. Artificial light can be used, but the lights must be full spectrum and on for at least eight hours each day.

If you have any other questions, please feel free to call me.

Sincerely,

Dr. Pat Phillips

88 CHAPTER 6 Analyzing Data TEACHING RESOURCES

## Access for all Learners

**Extra Help** Some students may need geographic or climate information for the cities on the list. You may want to have maps or atlases available so that students can locate each city. If students don't know that cities closer to the equator frequently have warmer temperatures than those farther from the equator, you may want to discuss these facts. Encourage students to use the eliminate possibilities problem-solving strategy.

You may also want to encourage students to consider the time of year and the data in the veterinarian's letter when making their hypotheses in **Problem 2c.**

**Problem Set Wrap-Up** As students share their hypotheses from **Problem 2c,** have each group tell what facts support the hypotheses. Students will summarize their investigation in written form in the Share & Summarize.

2

## Share & Summarize

You may want to briefly discuss or show a sample of a business letter, pointing out the various parts of the letter, such as the business address and salutation. Suggest that students give explanations to support any hypotheses that they state in the letter. Students could practice their word processing skills by composing the letter on a computer.

**Troubleshooting** If students have difficulty distinguishing between facts and hypotheses, have pairs choose a line graph in a newspaper and list some facts they know simply by reading the graph. To help them, suggest that they write the names of the axes and any information given by looking at one point in the graph. Then ask them to make hypotheses by looking for patterns in the graph to predict what might happen and to explain why the graph was made. Have them cut apart the hypotheses and the facts and trade the slips of paper and the graph with another pair of students. Have the second pair determine whether each statement is a hypotheses or a fact.

### On Your Own Exercises

Practice & Apply: 1, 2, p. 353
Connect & Extend: 7, 8, pp. 357–358
Mixed Review: 11–23, p. 361

**2c.** Possible answer: He went to Florida because the vet said that Agnes needed a warmer climate, and the note included a flight number.

**Share & Summarize Answer**

Possible letter:

Dear Smallville Police,

We think that Mr. Orkney and Agnes are in Miami, Florida. At first we didn't have much information, but since an atlas was found near the graphs, we thought maybe they had to do with population or temperatures or rainfall in different places. When we saw the scale labels, we were pretty confident that the graphs show temperatures over different months. We thought that maybe each graph represented a different place. When we saw the list and the note, we were able to match the cities with the graphs. The note indicated that Agnes needed a warmer climate. That information, and the fact that the note included a flight number, led us to the conclusion that Gerald Orkney flew to Miami to help Agnes. We hope you find this information useful.

Sincerely,

Data Inc.

**2.** The police have just discovered more clues: a list and a note. Your teacher will give you a copy of this new information.

**a.** What might the list have to do with the graphs? Do you have new ideas about what each graph might show? If so, add appropriate titles and axis labels to the graphs. See Additional Answers.

**b.** Does this new information support your hypothesis about what happened to Gerald Orkney? Explain. Answers will vary.

**c.** Now what do you think happened to Gerald Orkney? Make a new hypotheses, or make changes to an earlier hypothesis, to fit this new information.

**1** Problem-Solving Strategy
Use logical reasoning

## Share & Summarize

Write a letter to the Smallville police summarizing your group's investigation and presenting your hypothesis about what happened to Gerald Orkney.

**2** Have students give explanations to support their hypotheses.

• ***Additional Answers on page A656***

LESSON 6.1 Using Graphs to Understand Data **345**

# Investigation 2

In this investigation, students use data from tables to construct bar graphs and then use their graphs to explore how pollution emissions from cars have changed over many years. You may want to provide students with Masters 57, 58, 62, or 81 and 82 to complete the problems in this investigation and in On Your Own Exercises 3 and 9, if assigned. Master 57 reproduces the sets of axes in Problem Set B and Problem Set C. Master 58 reproduces the table in Problem Set D and provides an axis for graphing. Master 62 reproduces the data set and partially completed graph in On Your Own Exercise 9.

1 Introduce the investigation by reminding students that even though cars make many people's lives easier, this comfort comes at a cost to the environment. Ask students how cars contribute to pollution. Their responses may include the facts that exhaust from tailpipe emissions pollutes the air, oil or gas leaks pollute the ground and water, and waste from gasoline production can pollute air, ground, and water.

Then ask students how air pollution caused by cars could be reduced. Answers may include driving less, improving engines so they emit fewer pollutants, finding better fuels that cause less pollution, and improving public transportation systems.

2 Present the setting for the investigation by discussing the information in the chart, which gives the average emissions per vehicle, per mile driven. Point out that the figures for years 2000 and beyond are predictions. You might ask students to think about ways the Environmental Protection Agency (EPA) makes these predictions.

3 **Problem Set B** **Suggested Grouping: Small Groups**

In the problem set, students make bar graphs of the average per vehicle emissions data from the table in the introduction to this investigation. They analyze the data and compare the table and graph.

4 Some students may need a brief review of the different components of a bar graph before completing the graph in **Problem 2.** Connect the importance of a title on the graph and labels on the axes to students' experiences in Investigation 1. You may want to distribute Master 57 to help students create their graphs in this problem set and the next.

Review the guidelines for making bar graphs and have students use the graph on page 342 as a model when making their graphs. Some students may find it easier to use Master 57 since the axes are already labeled.

**Teaching Resources**

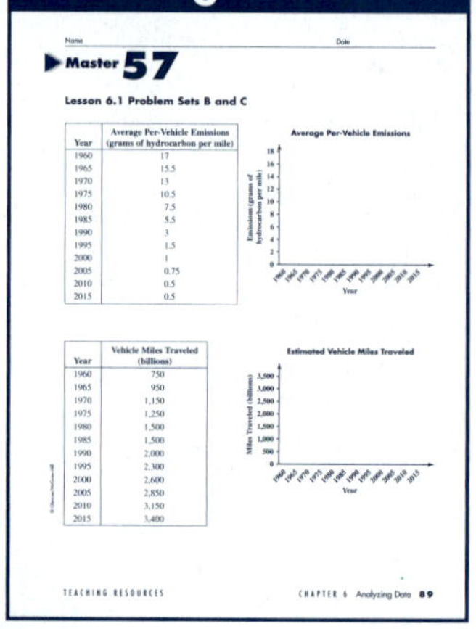

Name Date

**Master 57**

**Lesson 6.1 Problem Sets B and C**

| Year | Average Per-Vehicle Emissions (grams of hydrocarbon per mile) |
|---|---|
| 1960 | 17 |
| 1965 | 15.5 |
| 1970 | 13 |
| 1975 | 10.5 |
| 1980 | 7.5 |
| 1985 | 5.5 |
| 1990 | 3 |
| 1995 | 1.5 |
| 2000 | 1 |
| 2005 | 0.75 |
| 2010 | 0.5 |
| 2015 | 0.5 |

Average Per-Vehicle Emissions

| Year | Vehicle Miles Traveled (billions) |
|---|---|
| 1960 | 750 |
| 1965 | 950 |
| 1970 | 1,150 |
| 1975 | 1,250 |
| 1980 | 1,500 |
| 1985 | 1,500 |
| 1990 | 2,000 |
| 1995 | 2,300 |
| 2000 | 2,600 |
| 2005 | 2,850 |
| 2010 | 3,150 |
| 2015 | 3,400 |

Estimated Vehicle Miles Traveled

TEACHING RESOURCES CHAPTER 6 Analyzing Data 89

### On the Spot Assessment

Watch for students who incorrectly draw the bars when making bar graphs in Problem 2. Besides drawing the bars with incorrect heights, students may make bars with unequal, or varying, widths. These errors can distort the meaning of the graphs.

# Investigation 2 Using Bar Graphs to Analyze Data

The environmental group Citizens for Safe Air has asked Data Inc. to analyze some data about *hydrocarbons.* These compounds are part of the emissions from cars and other vehicles that pollute the air. The group wants to know how the total amount of hydrocarbon emitted by vehicles has changed over the past several decades—and how it might change in the future.

**Just the facts**

Hydrocarbons react with nitrogen oxides and sunlight to form ozone, a major component of smog. Ozone causes choking, coughing, and stinging eyes, and it damages lung tissue.

The table shows estimates of the typical amount of hydrocarbon emitted per vehicle for each mile driven in the United States for years from 1960 to 2015. The values from 2000 to 2015 are predictions.

| Year | Average Per-Vehicle Emissions (grams of hydrocarbon per mile) |
|---|---|
| 1960 | 17 |
| 1965 | 15.5 |
| 1970 | 13 |
| 1975 | 10.5 |
| 1980 | 7.5 |
| 1985 | 5.5 |
| 1990 | 3 |
| 1995 | 1.5 |
| 2000 | 1 |
| 2005 | 0.75 |
| 2010 | 0.5 |
| 2015 | 0.5 |

Source: "Automobiles and Ozone," Fact Sheet OMS-4 of the Office of Mobile Sources, the U.S. Environmental Protection Agency.

1 Discuss how cars pollute the environment and how we can reduce this pollution.

2 Remind students that figures for the years 2000 and beyond are predictions.

## Problem Set B

3 You may have students work in small groups.

**1.** For which 5-year period is the decrease in per-vehicle emissions greatest? For which 5-year period is it least?

**1.** 1975–1980, 2010–2015

**2.** On a set of axes like this one, draw a bar graph showing the typical per-vehicle emissions for each year given in the table.

4 Review the different components of a bar graph.

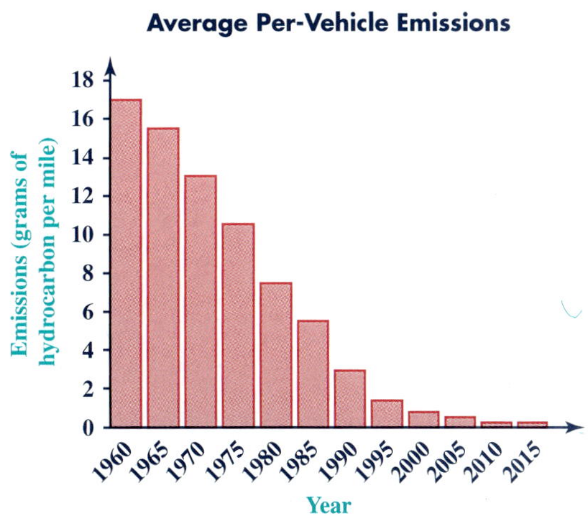

# Develop

1 **Problem Set Wrap-Up** Have students share their answers to **Problem 3.** Students should understand that the emissions from individual vehicles have tapered off over time. When discussing **Problem 4,** make sure that students mention that a bar graph makes it easier to see patterns and draw conclusions about trends while a table makes it easier to find exact values. ■

Read and discuss together the paragraphs and tables that present additional information needed to determine how the *total* hydrocarbon emissions have changed over the past several decades. The data in Problem Set B showed a decrease in the emissions each vehicle emits per mile traveled. These new data present the number of miles driven by all vehicles during the same time period.

**3.** Possible answer: The graph indicates that per-vehicle emissions have decreased over the years and are expected to level off in the future. The graph is at its highest point in 1960 and then the height of the bars decreases each year until 2010, when the bar height levels off. The greatest change occurs from 1975 to 1980. There is no change from 2010 to 2015.

**4.** Possible answer: It is easier to use the graph because the differences in emissions are clearly shown.

**3.** Describe what your graph indicates about the change in per-vehicle emissions over the years. Discuss high and low points, periods of greatest and least change, and any other patterns you see.

**4.** To determine when the greatest decrease in per-vehicle emissions occurred, is it easier to use the table or the graph? Explain.

**1** Problem-Solving Strategy
Use a graph

You have seen that the amount of hydrocarbon *each vehicle* emits *per mile* has decreased over the years, but this is not enough information to conclude that the *total amount* of hydrocarbon emitted by *all vehicles* is decreasing. You also need to consider the total number of miles driven by all vehicles.

This table shows estimates of the number of miles driven, or expected to be driven, by all vehicles in the United States for various years between 1960 and 2015.

| Year | Vehicle Miles Traveled (billions) |
|---|---|
| 1960 | 750 |
| 1965 | 950 |
| 1970 | 1,150 |
| 1975 | 1,250 |
| 1980 | 1,500 |
| 1985 | 1,500 |
| 1990 | 2,000 |
| 1995 | 2,300 |
| 2000 | 2,600 |
| 2005 | 2,850 |
| 2010 | 3,150 |
| 2015 | 3,400 |

Source: "Automobiles and Ozone," Fact Sheet OMS-4 of the Office of Mobile Sources, the U.S. Environmental Protection Agency, Jan 1993.

LESSON 6.1 Using Graphs to Understand Data 347

# Develop

1 **Problem Set C** **Suggested Grouping: Small Groups**

In this problem set, students make and interpret bar graphs about the total number of miles driven by all vehicles. Master 57 reproduces the data and axes for this problem set.

**Access for all Learners**

**For Early Finishers** Ask students to find the percents of increase in the total number of miles driven in 1960 and the total number of miles expected to be driven in 2015. The answer can be found as follows:

$$\frac{3400 - 750}{750} \approx 353\%$$

2 **Problem Set Wrap-Up** Briefly discuss **Problem 3.** Encourage students to share any patterns they found. Be sure they conclude that the number of miles driven has increased dramatically since 1960 and that the increase is *projected* through 2015, the last year given on the graph. ■

Recap the data previously given: the average per-vehicle emissions and the total vehicle miles traveled for the same years. Remind students that the objective is for Data Inc. to look at the change in *total* hydrocarbon emissions. They will look for ways to accomplish this goal in the Think & Discuss.

3 **Think & Discuss**

Be sure that all students have a reasonable way to use the tables shown before Problem Set B and Problem Set C to find the total amount of hydrocarbon emitted from all vehicles each year. They will use their method to complete Problem 1 in the next problem set.

## Problem Set C

1. 1985–1990, 1980–1985

3. Possible answer: The graph indicates that the total number of miles driven has increased over the years and will continue to increase. The graph is at its lowest point in 1960. The bar heights increase each year shown in the graph, increasing more slowly at first and then by greater amounts. The graph does not level off as the previous graph did.

**1.** Look at the table on page 347. During which 5-year period does the number of miles driven increase most? During which 5-year period does it increase least?

**2.** On a set of axes like the one below, draw a bar graph showing the billions of vehicle miles traveled for each year given.

**Estimated Vehicle Miles Traveled**

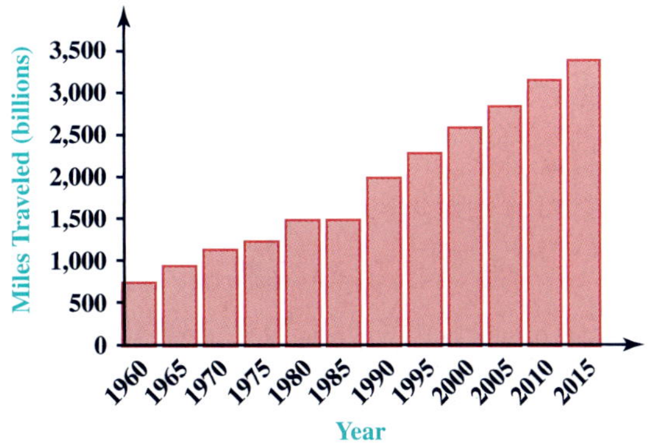


**3.** Describe what your graph indicates about the change in the total number of miles driven over the years. Discuss high and low points, periods of greatest and least change, and any other patterns you see.

You now know that, over time, the typical amount of hydrocarbon emitted per vehicle has decreased. You also know that more miles are driven each year. In the next problem set, you will combine this information to answer this question: Is the *total amount* of hydrocarbon emitted from vehicles increasing or decreasing?

### Think & Discuss

How could you use the data in the two previous tables to calculate estimates of the total amount of hydrocarbon emitted by all vehicles each year? Multiply the emissions per mile driven by the number of miles driven.

1 You may have students work in small groups.

2 Problem-Solving Strategy
Look for a pattern

3 Make sure students know they will use this method to complete the next problem set.

1

## Problem Set D **Suggested Grouping: Small Groups**

In this problem set, students apply the method from the Think & Discuss to create a table and a bar graph showing the estimated total emissions. Then they analyze the bar graph. You may want to distribute Master 58, which reproduces the table and includes axes for constructing the graph.

2

### Teaching Resources

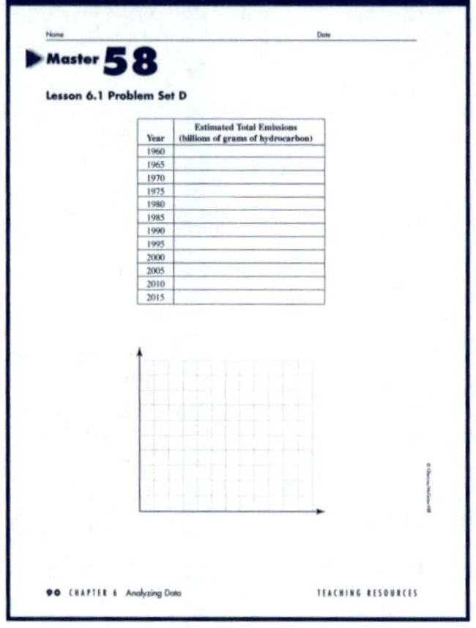

Master 58

Lesson 6.1 Problem Set D

| Year | Estimated Total Emissions (billions of grams of hydrocarbon) |
|---|---|
| 1960 | |
| 1965 | |
| 1970 | |
| 1975 | |
| 1980 | |
| 1985 | |
| 1990 | |
| 1995 | |
| 2000 | |
| 2005 | |
| 2010 | |
| 2015 | |

90 CHAPTER 6 Analyzing Data — TEACHING RESOURCES

### Access for all Learners

**Extra Help** To help students more easily make their calculations in **Problem 1** suggest that they use the make an organized list problem-solving strategy by inserting two columns in their tables: one showing the average per-vehicle emissions (from the table on page 346) and the other showing the vehicle miles traveled (from the table on page 347).

3

**Problem Set Wrap-Up** Discuss **Problem 3**, and ask students to relate their new knowledge to what they have already learned about hydrocarbon emissions. Students may note that the average per-vehicle emissions have decreased since 1960, which might indicate that air quality has improved. During the same time period, the number of miles driven has increased. The increasing number of miles driven only partially offsets the decreasing rate of emissions, so total emissions have declined.

4

## Share & Summarize

Students are asked to prepare a brief report organizing the data they have displayed and analyzed during this investigation, specifically the trends discovered in Problem Set D. Encourage them to include explanations of how they used the data to reach their conclusions, essentially summarizing the findings from Problem 3 in Problem Set D. Although Data Inc. is asked only to present the results of the analysis, you may also want students to provide possible ways to reduce emissions.

**Troubleshooting** Encourage students who are having difficulty describing the changes shown in a bar graph to choose a bar graph published in a newspaper or magazine. Have them answer the questions below and then write a summary of the graph based on their answers.

- What is happening at the highest point in the graph?
- What is happening at the lowest point in the graph?
- What is the greatest change?
- What is the least change?
- Do the changes show increases or decreases in values?
- Are there any patterns in the data?

### On Your Own Exercises

Practice & Apply: 3, 4, pp. 354–355
Connect & Extend: 9, p. 359
Mixed Review: 11–23, p. 361

## Problem Set D

1. Copy and complete the table to show the total amount of hydrocarbon emitted by U.S. vehicles each year.

| Year | Estimated Total Emissions (billions of grams of hydrocarbon) |
|---|---|
| 1960 | 12,750 |
| 1965 | 14,725 |
| 1970 | 14,950 |
| 1975 | 13,125 |
| 1980 | 11,250 |
| 1985 | 8,250 |
| 1990 | 6,000 |
| 1995 | 3,450 |
| 2000 | 2,600 |
| 2005 | 2,137.5 |
| 2010 | 1,575 |
| 2015 | 1,700 |

2. Make a bar graph of the data in the table. See Additional Answers.

3. Describe what your graph indicates about the change in the total hydrocarbon emissions over the years. Discuss high and low points, periods of greatest and least change, and any other patterns you see.

3. Possible answer: The graph indicates that, although the total emissions have decreased since 1970, they may level off or begin to increase again in the future. The graph goes up for the first three bars and then goes down, slowly at first, then more quickly, and then more slowly again. In 2015, it rises slightly. The greatest increase is between 1960 and 1965. The greatest decrease is between 1980 and 1985.

## Share & Summarize

Write a letter to Citizens for Safe Air. Describe Data Inc.'s investigation of hydrocarbon emissions, summarizing your findings about how total emissions have changed over the past few decades and how they might change in the future. See Additional Answers.

1 You may have students work in small groups.

2 You may wish to distribute Master 58 to help students complete the table and make the bar graph.

3 Discuss students' answers.

4 Have students list ways to reduce emissions.

• *Additional Answers on page A657*

LESSON 6.1 Using Graphs to Understand Data 349

# Investigation 3

This investigation introduces histograms. Histograms, like bar graphs, use bars to show categorical data. Unlike bar graphs, the data on the horizontal axis represent ranges of data, and the heights of the bars show the frequency of data. In this investigation, students are given a bar graph that shows the times of participants in the men's 10K cross-country skiing event at the 1998 Olympics. Students create frequency tables to sort the items into ranges and then make histograms to display the data. In the last problem set, students describe and interpret histograms with percents on the vertical axis.

You may want to provide students with copies of Master 59, which reproduces the bar graph, frequency table, and partially constructed histogram shown in Problem Set E and Problem Set F. Masters 60a and 60b and Master 61 reproduce the data in On Your Own Exercises 5 and 6. You may want to provide students with copies of these masters if the exercises are assigned. Students may need graph paper for On Your Own Exercise 10, if assigned, and for the other problems and exercises, if you elect not to distribute masters, for this investigation. Master 81 or Master 82 can be used for any of these exercises.

1 Introduce the investigation by having students name some of the types of graphs they have studied. Encourage them to describe some situations that can best be shown by each type of graph.

2 **Problem Set E** **Suggested Grouping: Pairs**
This problem set asks students to interpret a bar graph showing Olympic skiing data. They analyze the data in terms of ranges. This time-consuming task previews the idea of ranges that will be explored through histograms in the next two problem sets.

**Teaching Resources**

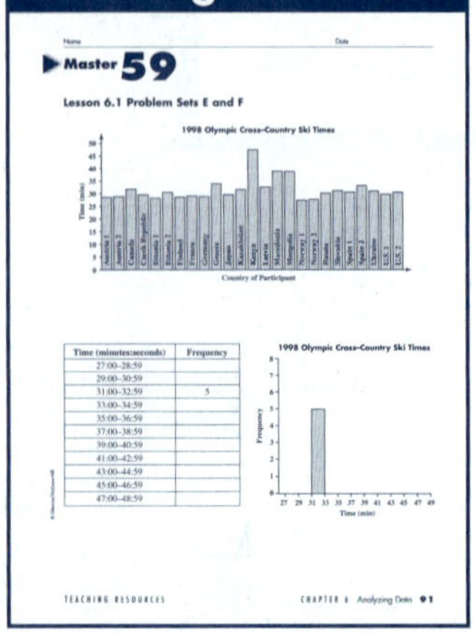


## Access for all Learners

**Extra Help** In **Problem 3,** it may be difficult for students to determine which skiers have times that fall in the desired range. Suggest that they estimate where the values for 31 minutes and 32 minutes 59 seconds are on the vertical scale and then use a straightedge to form horizontal lines for these values to help them find the relevant data.

3 It is likely that students will have different answers for **Problem 3.** If this occurs, you can either challenge the students to suggest a way to reconcile the differences or simply make a list on the board of the countries that students find and look for omissions or disagreements. You might point out that the bars allow readers to estimate the times for each skier. Since the times are recorded to the nearest hundredth and the graph shows intervals of 5, it is difficult to read the precise values on the graph.

The actual times for the skiers (rounded to the nearest second) are in the table below.

| Country | Time (min.: sec.) | Country | Time (min.: sec.) |
|---|---|---|---|
| Austria 1 | 28:25 | Latvia | 32:06 |
| Austria 2 | 28:59 | Macedonia | 39:39 |
| Canada | 31:56 | Mongolia | 39:39 |
| Czech Republic | 29:41 | Norway 1 | 27:25 |
| Estonia 1 | 28:01 | Norway 2 | 27:59 |
| Estonia 2 | 30:58 | Russia | 30:03 |
| Finland | 28:51 | Slovakia | 31:29 |
| France | 29:04 | Spain 1 | 30:45 |
| Germany | 28:48 | Spain 2 | 33:03 |
| Greece | 34:14 | Ukraine | 30:51 |
| Japan | 29:45 | U.S. 1 | 29:58 |
| Kazakhstan | 31:3 | U.S. 2 | 30:26 |
| Kenya | 47:26 | | |

Source: www.cbs.sportsline.com

• *Teaching notes continued on page A657*

# Investigation 3 Making Histograms

There are many types of graphs. The graph that is best for a given situation depends on the data you have and the information you want to convey.

1 Encourage students to name the different types of graphs they have learned and provide a situation that each graph describes.

## Problem Set E

2 You may have students work in pairs.

This bar graph shows the times of some of the participants in the men's 10-kilometer cross-country skiing event at the 1998 Winter Olympics:

2. Norway, about 27.5 min

**1998 Olympic Cross-Country Ski Times**

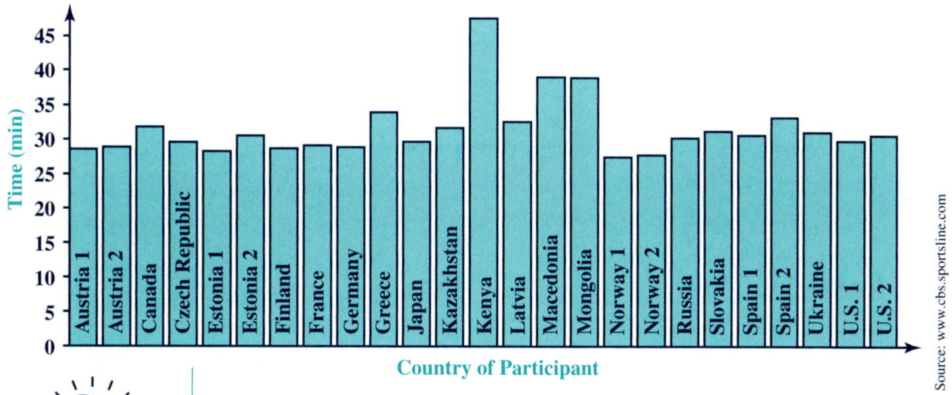


**Just the facts**

The oldest known skis are 4,000 to 5,000 years old. In the 10th century, Viking soldiers used skis for transportation. From the 15th to the 17th centuries, several northern European armies had companies of ski troops.

1. How many participants' times are shown in the graph? How many countries are represented? 25, 20
2. From which country was the gold medalist? What was his time?
3. How many participants completed the race with a time between 31 minutes and 32 minutes 59 seconds? 5
4. The bars in the graph are arranged alphabetically by country. Think of another way the bars could be ordered. What kinds of questions would be easier to answer if they were ordered that way? See Additional Answers.

3 Challenge pairs to reconcile their differences in Problem 3.

It probably took you some time to figure out the answer to Problem 3 of Problem Set E. Although it is easy to use the bar graph to find the time for each participant, it is not as easy to find the number of skiers who finished within a particular time interval.

VOCABULARY
**histogram**

You will now use a *histogram* to display the ski times. In a **histogram,** data are divided into equal intervals, with a bar for each interval. The height of each bar shows the number of data values in that interval. There are no gaps between intervals.

4 Introduce the term *histogram.*



**Additional Answers**

4. Possible answer: From least to greatest (or greatest to least) time. It would be easier to answer questions about how the skiers placed and about who finished within a particular period of time.

# Develop

1 **Problem Set F** **Suggested Grouping: Pairs**
In this problem set, students use data from the bar graph on page 350 to create a frequency table and a histogram. It may take some time for students to accurately correlate the data. You may want to make a transparency of Master 59 to show the histogram when discussing Problem 2.

2 Before students begin working through the problem set, you may want to introduce the term *frequency* and point out that the table in Problem 1 is called a frequency table since it shows the number of skiers (frequency) that had times in a given interval.

**Tips** from **Teachers**

"Have students complete their histograms on transparencies of Master 59 so that they can display their graphs for the class. This can facilitate discussion and help students internalize the characteristics of a histogram and distinguish the similarities and differences of histograms and bar graphs."

**Problem Set Wrap-Up** Have students share their histograms from Problem 2. Then discuss their answers to
3 Problems 3 and 4. Be sure students understand that the *distribution* of data means how the data are spread out over the axes.

**Teaching Resources**

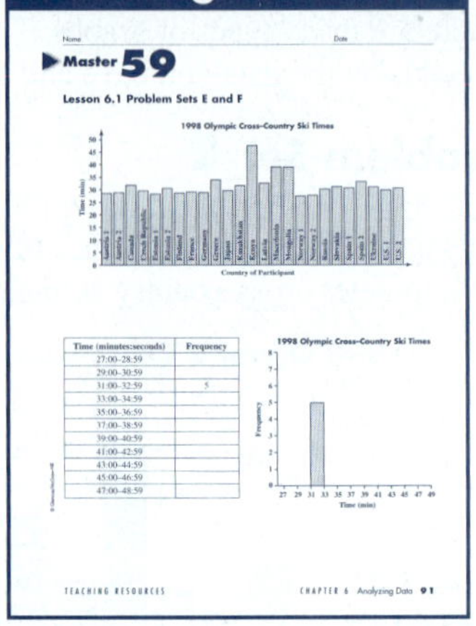

## Problem Set F

In this problem set, you will make a table of frequencies. *Frequencies* are counts of the number of data values in various intervals. You will use your frequency table to create a histogram.

1 You may have students work in pairs.

2 Introduce the term *frequencies.*

**1.** Copy this table. Use the bar graph in Problem Set E to count the number of participants who finished in each time interval. Record this information in the "Frequency" column.

| Time (minutes:seconds) | Frequency |
|---|---|
| 27:00–28:59 | 7 |
| 29:00–30:59 | 9 |
| 31:00–32:59 | 4 |
| 33:00–34:59 | 2 |
| 35:00–36:59 | 0 |
| 37:00–38:59 | 0 |
| 39:00–40:59 | 2 |
| 41:00–42:59 | 0 |
| 43:00–44:59 | 0 |
| 45:00–46:59 | 0 |
| 47:00–48:59 | 1 |

**2.** Copy the axes below. Create a histogram by drawing bars showing the number of participants who finished in each time interval. The bar for the interval 31:00–32:59 has been drawn for you.

**1998 Olympic Cross-Country Ski Times**

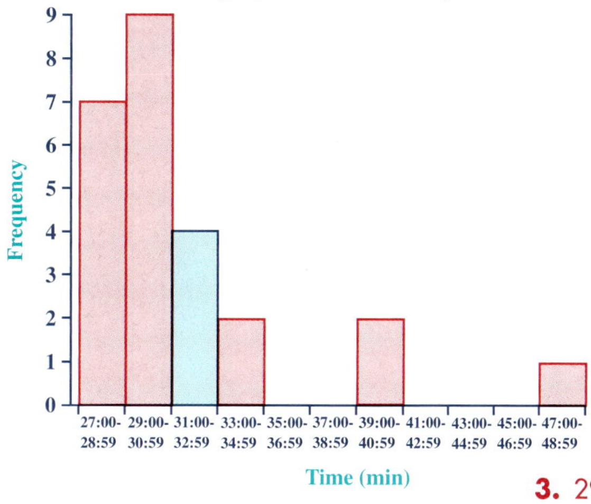


**3.** 29:00–30:59

**Just the facts**

*Relative frequency* is the ratio of the number of data in an interval to the total number of data in all intervals. For example, the relative frequency of the data in the 31:00–32:59 interval is $\frac{4}{25}$ or 0.16.

VOCABULARY
**distribution**

**3.** In which 2-minute interval did the greatest number of skiers finish?

**4.** The shape of a histogram reveals the **distribution** of the data values. In other words, it shows how the data are spread out, where there are gaps, where there are many values, and where there are only a few values. What can you say about the distribution of times for this event? See Additional Answers.

3 Introduce the idea of *a distribution of data.*



### Additional Answers

**4.** Possible answer: Most of the data are concentrated between 27 min and 33 min, with the greatest number of times (9) falling in the 29 to 33 interval. There are gaps from 35 to 39 and from 41 to 47 in which no times fall. Two skiers finished between 39 and 41 and one between 47 and 49.

1 Have students look at the histogram for the Exam Scores at the top of the page. Ask them how this histogram differs from those they made for the cross-country skiing times. Point out that the label for the vertical axis of the "Exam Scores" graph shows the *percent* of the class achieving the scores, if students do not make this observation on their own. Discuss how to read the graph, especially if you did not show a sample histogram when introducing histograms.

2 **Problem Set G** **Suggested Grouping: Pairs**
In this problem set, students interpret a histogram with percents on the vertical axis.

In **Problem 4,** students must use the histogram to find the percent of students that received a specific interval of scores and use that percentage to determine how many students in a class of 64 earned those scores. Since 14% of 64 is 8.96, students must round to find the number of students.

### Access for all Learners

**For Early Finishers** Ask students to approximate the number of students in a class of 75 that received a score less than 55. Challenge them to find two ways to the answer so they can check their work. Two possible ways are (6% + 8% + 10% + 11%) × 75 = 26.25, or 26 students, and (100% − 11% − 14% − 24% − 16%) × 75 = 26.25, or 26 students.

3 **Share & Summarize**
Students are asked to compare bar graphs and histograms to determine which would be the most appropriate way for displaying given data. Be sure students understand that histograms show frequency and ranges of data and are useful when looking at the distribution of data. Bar graphs show individual items of data and are best in situations when it is necessary to know specific values.

**Troubleshooting** If students have difficulty deciding when to make a bar graph and when to make a histogram, ask them to choose some context, such as football or rock music, and then think of two kinds of related questions. The first questions should ask how many members of a data set reached different levels of something; the second should focus on comparing different individuals or groups. Examples of the first type—which include such questions as "How many football players played in less than $\frac{1}{3}$ of the games? at least $\frac{1}{3}$ of the games? $\frac{2}{3}$ of the games?"—are best shown with histograms. Examples of the second type—which include such questions as "Who had the top-selling CD in 2000?" and "Whose CD was the least popular?"—compare individual members of a data set and are best shown with a bar graph. Since this is not a simple idea, it may take many examples for students to get a feel for the differences.

If students have difficulty interpreting data from histograms, have them create a frequency table for the data in the "Exam Scores" histogram on page 352; if this is still too complex, have them use the histogram they created in Problem Set F. In the latter case, students can compare their answers to the frequency table they created in Problem 1 of that problem set. The frequency table for the exam scores data is given below.

| Test Scores | Percent of Students |
|---|---|
| 36 to 40 | 6 |
| 41 to 45 | 8 |
| 46 to 50 | 10 |
| 51 to 55 | 11 |
| 56 to 60 | 16 |
| 61 to 65 | 24 |
| 66 to 70 | 14 |
| 71 to 75 | 11 |

### On Your Own Exercises

Practice & Apply: 5, 6, pp. 356–357
Connect & Extend: 10, p. 360
Mixed Review: 11–23, p. 361

Rather than showing the *number* of values in each interval, some histograms show the *percent* of values in each interval. For example, this histogram shows how the test scores for Mr. Lazo's math exam were distributed. The maximum possible score was 75 points.

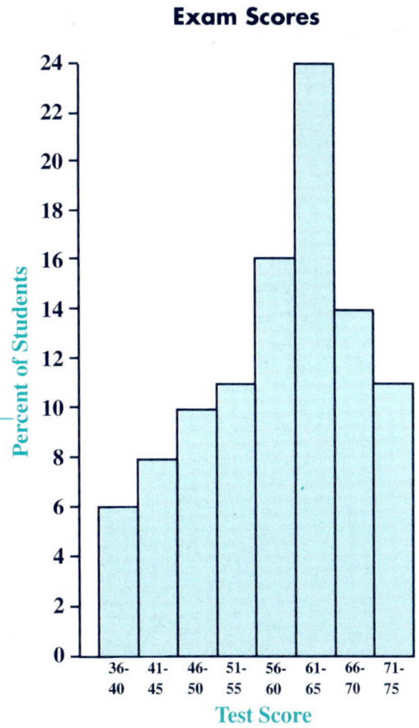


## Problem Set G

1. Describe the shape of the histogram. Tell what the shape indicates about the distribution of test scores.
2. Which interval includes the greatest percent of test scores? About what percent of scores are in this interval? 61–65, 24%
3. Which interval includes the least percent of test scores? About what percent of scores are in this interval? 36–40, 6%
4. Suppose 64 students took Mr. Lazo's test. How many of them received a score from 66 to 70? about 9
5. If you were to add the percents for all the bars, what should the total be? Why? 100%; the graph shows how all the scores (100%) are distributed among the various intervals.

1. See Additional Answers.

## Share & Summarize

1. What type of information does a histogram display? Give an example of a situation for which it would make sense to display data in a histogram. See ① in Additional Answers.
2. In this investigation, you looked at a bar graph and a histogram of Olympic ski data. What are some things the bar graph shows better than the histogram? What are some things the histogram shows better than the bar graph? See ② in Additional Answers.

### Just the facts

*Cumulative frequency* is the total number of all data values less than the upper limit of a certain interval. This is found by adding together the frequencies of the interval and all other intervals that come before it. For example, the cumulative frequency of the data less than 50 is 6 + 8 + 10 or 24.

*Cumulative relative frequency* is the ratio of the cumulative frequency for an interval to the total number of data in all intervals. For example, the cumulative relative frequency of the 46–50 interval is $\frac{24}{100}$ or 0.24.



1 Discuss the differences between this histogram and the one they created on page 351.

2 Problem-Solving Strategy
Use a graph

3 Have students discuss the differences between bar graphs and histograms.

### Additional Answers

1. Possible answer: Moving from left to right, the bar height increases, reaching its maximum height in the 61–65 interval, and then decreases again. The shape indicates that there are a few very low scores and a small number of very high scores, with most of the scores somewhere in between and the greatest number in the 61–65 interval.

• Additional Answers continued on page A658

# Teacher Notes

# On Your Own Exercises

## Practice & Apply

**1.** In Parts a–d, tell which graph could represent the situation.

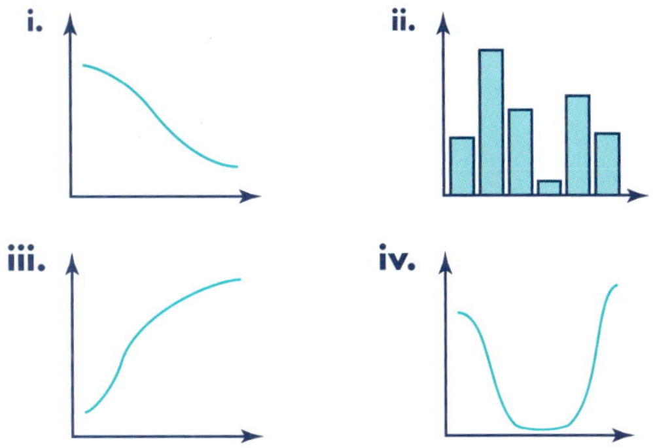


**a.** a child's activity level from before a nap until after a nap iv

**b.** the populations of six cities ii

**c.** the change in water level from high tide to low tide i

**d.** the change in the weight of a cat from birth until age 2 iii

**2.** **Earth Science** This line graph shows the monthly normal temperatures for a U.S. city. Compare this graph to the temperature graphs for Miami, Chicago, and Fairbanks from Problem Set A.

**Monthly Normal Temperatures**

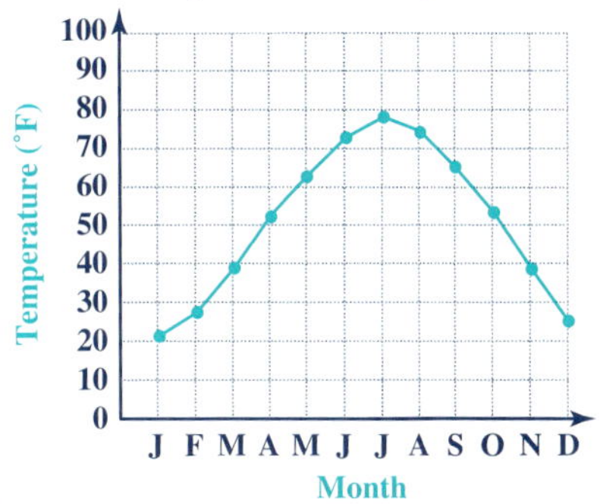


**a.** Which city's graph is this graph most similar to? Chicago

**b.** In which of the following regions do you think this city is located: North, South, East Coast, Midwest, or West? Midwest (Note: The graph shows temperatures for Omaha, Nebraska.)

Chicago, Illinois

impactmath.com/self_check_quiz



## On Your Own Exercises

Investigation 1, pp. 343–345
Practice & Apply: 1, 2
Connect & Extend: 7, 8

Investigation 2, pp. 346–349
Practice & Apply: 3, 4
Connect & Extend: 9

Investigation 3, pp. 350–352
Practice & Apply: 5, 6
Connect & Extend: 10

Assign Anytime
Mixed Review: 11–23

**Exercise 1:**
Encourage students to provide an explanation for the graph they chose in each situation.

**Exercise 2b:**
You may need to review the regions of the country before assigning this exercise.

LESSON 6.1 Using Graphs to Understand Data 353

**Exercise 3a Extension:**
Have students give the greatest and the least changes in terms of the number of people. The change from 1991 to 1992 is an increase of 4.6 million people and from 1993 to 1994 is a decrease of 1 million people.

**Exercise 3c:**
Students may need graph paper or copies of Master 81 or Master 82 to make their bar graphs.

**3.** **Social Studies** This table shows the number of people who visited the United States from other countries in the years from 1990 to 2001:

**Visitors to United States**

| Year | Visitors (millions) |
|---|---|
| 1990 | 39.4 |
| 1991 | 42.7 |
| 1992 | 47.3 |
| 1993 | 45.8 |
| 1994 | 44.8 |
| 1995 | 43.3 |
| 1996 | 46.5 |
| 1997 | 47.8 |
| 1998 | 46.4 |
| 1999 | 48.5 |
| 2000 | 50.9 |
| 2001 | 45.5 |

Source: *World Almanac and Book of Facts 2003*. Copyright © 2003 Primedia Reference Inc.

**a.** The change in the number of visitors is greatest from 2000 to 2001. Between which two years is the change in the number of visitors least? 1993 to 1994

**b.** By what percent did the number of visitors increase from 1990 to 2001? about 15.5%

**c.** Make a bar graph showing the number of visitors to the United States during the years shown in the table.

**d.** Describe what your graph indicates about the change in the number of visitors to the United States. Discuss high and low points, periods of greatest and least change, and any other patterns you see.

**3c.** See Additional Answers.

**3d.** The number of visitors increased from 1990 to 1992, decreased from 1992 to 1995, increased again from 1995 to 1997, dropped slightly between 1997 and 1998, increased steadily from 1998 to 2000, then dropped dramatically from 2000 to 2001. The greatest change occurs from 2000 to 2001, and the least change occurs from 1994 to 1995.

The Coliseum in Rome, Italy

354 CHAPTER 6 Analyzing Data

• *Additional Answers on page A658*

**4. Ecology** The bar graph on the left shows the number of farms in the United States from 1940 to 2000. The bar graph on the right shows the size of the average farm, in acres, for the same years.

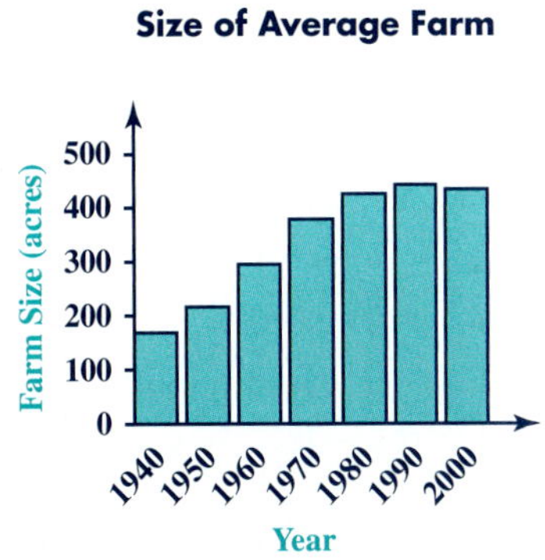


Source: U.S. Department of Agriculture.

**a.** Describe how the number of U.S. farms has changed over the years. See Additional Answers.

**b.** Describe how the size of the average U.S. farm has changed over the years.

**4b.** Possible answer: The size of the average farm increased for each 10-year interval from 1940 to 1990. From 1950 to 1990, the amount of increase got smaller with each 10-year interval. There was a very small decrease in farm size between 1990 and 2000. It appears that the size of the average farm has begun to level off.

**c.** Jing used the data from the two farm graphs to create the bar graph below. Her graph shows the total amount of U.S. land devoted to farms from 1940 to 2000. How do you think Jing calculated the values for her graph? See Additional Answers.

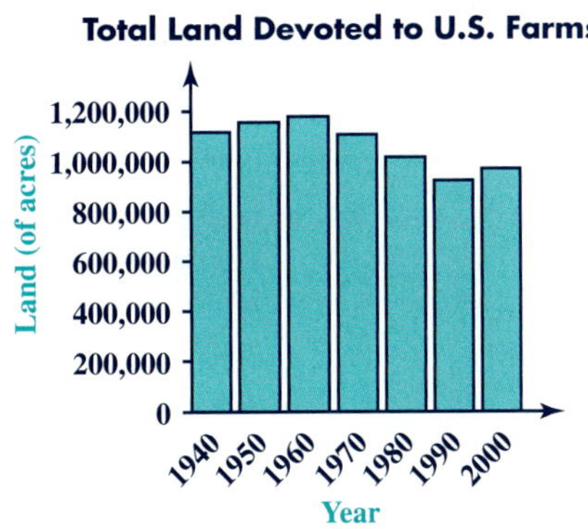

**d.** Describe how the total amount of land devoted to U.S. farms has changed over the years. Possible answer: The total number of acres increased from 1940 to 1960 and then decreased until 1990. From 1990 to 2000, the total amount of land decreased slightly. The change between 10-year intervals is not as drastic as in the other graphs.



## Additional Answers

**4a.** Possible answer: The number of farms decreased for each 10-year interval from 1940 to 1990. From 1950 to 1990, the amount of decrease got smaller with each 10-year interval. There was a very small increase in the number of farms between 1990 and 2000. It appears that the number of farms has begun to level off.

**4c.** For each year, she found the total number of farms from the first graph and the average size of each farm from the second graph, and multiplied the two numbers.

# Assign and Assess

**Exercise 5:**
You may want to provide students with copies of Master 60a and 60b. As an alternative, they could use graph paper or Master 81 or Master 82 to make their histograms but would not have the preprinted table.

For **Part C,** some students may find it easier to organize their data if they add a column for tallies to the table. Others may prefer to shade products in each range a different color and then count the number of products shaded in each color.

**5.** Think about the multiplication facts from $0 \times 0$ to $12 \times 12$. You can group the products into intervals of 10. For example, a product can be between 0 and 9, between 10 and 19, between 20 and 29, and so on.

**a.** Do you think the products are evenly distributed among the intervals of 10, or do you think some intervals contain more products than others? Answers will vary.

**b.** Copy and complete this multiplication table:

| × | 0 | 1 | 2 | 3 | 4 | 5 | 6 | 7 | 8 | 9 | 10 | 11 | 12 |
|---|---|---|---|---|---|---|---|---|---|---|---|---|---|
| **0** | 0 | 0 | 0 | 0 | 0 | 0 | 0 | 0 | 0 | 0 | 0 | 0 | 0 |
| **1** | 0 | 1 | 2 | 3 | 4 | 5 | 6 | 7 | 8 | 9 | 10 | 11 | 12 |
| **2** | 0 | 2 | 4 | 6 | 8 | 10 | 12 | 14 | 16 | 18 | 20 | 22 | 24 |
| **3** | 0 | 3 | 6 | 9 | 12 | 15 | 18 | 21 | 24 | 27 | 30 | 33 | 36 |
| **4** | 0 | 4 | 8 | 12 | 16 | 20 | 24 | 28 | 32 | 36 | 40 | 44 | 48 |
| **5** | 0 | 5 | 10 | 15 | 20 | 25 | 30 | 35 | 40 | 45 | 50 | 55 | 60 |
| **6** | 0 | 6 | 12 | 18 | 24 | 30 | 36 | 42 | 48 | 54 | 60 | 66 | 72 |
| **7** | 0 | 7 | 14 | 21 | 28 | 35 | 42 | 49 | 56 | 63 | 70 | 77 | 84 |
| **8** | 0 | 8 | 16 | 24 | 32 | 40 | 48 | 56 | 64 | 72 | 80 | 88 | 96 |
| **9** | 0 | 9 | 18 | 27 | 36 | 45 | 54 | 63 | 72 | 81 | 90 | 99 | 108 |
| **10** | 0 | 10 | 20 | 30 | 40 | 50 | 60 | 70 | 80 | 90 | 100 | 110 | 120 |
| **11** | 0 | 11 | 22 | 33 | 44 | 55 | 66 | 77 | 88 | 99 | 110 | 121 | 132 |
| **12** | 0 | 12 | 24 | 36 | 48 | 60 | 72 | 84 | 96 | 108 | 120 | 132 | 144 |

**c.** Make a table, like that at right, showing the number of products that fall in each interval of 10.

| Product | Frequency |
|---|---|
| 0–9 | 48 |
| 10–19 | 23 |
| 20–29 | 19 |
| 30–39 | 15 |
| 40–49 | 15 |
| 50–59 | 8 |
| 60–69 | 9 |
| 70–79 | 8 |
| 80–89 | 7 |
| 90–99 | 6 |
| 100–109 | 3 |
| 110–119 | 2 |
| 120–129 | 3 |
| 130–139 | 2 |
| 140–149 | 1 |

**d.** Make a histogram that shows the number of products in each interval of 10. Be sure to include axes labels and scale values.

**e.** What does the shape of your histogram reveal about the distribution of the products?

**f.** Now make another histogram showing the number of products that fall into intervals of 20—that is, 0–19, 20–39, 40–59, and so on.

**g.** Describe the similarities and differences in the two histograms. See Additional Answers.

**5d.** See Additional Answers.

**5e.** The first interval has by far the most products, and the number of products in each interval decreases from left to right, with an occasional "bump."

**5f.** See Additional Answers.



**Additional Answers**

**5d.**

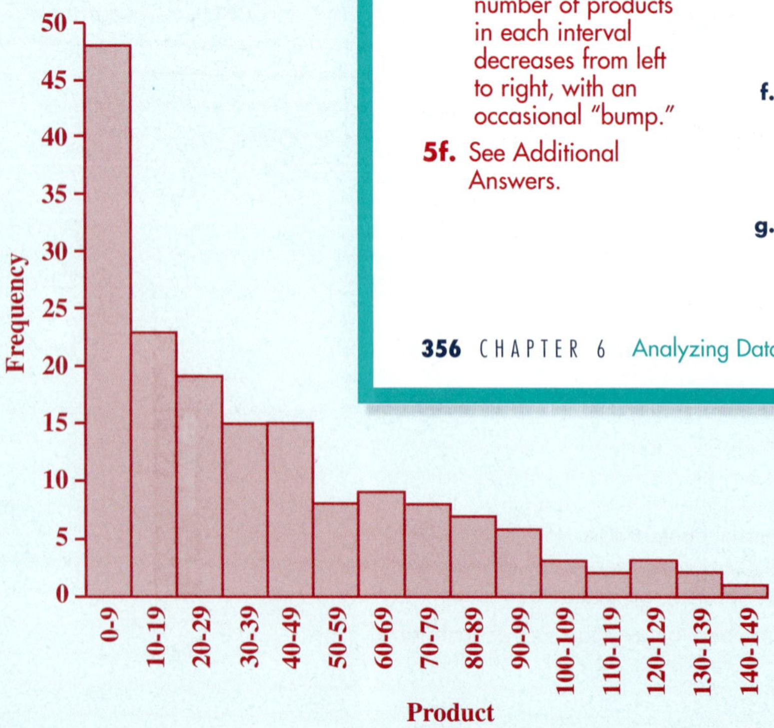


**5f.** See page A658.

**5g.** Possible answer: In both graphs, the bar for the first interval is much higher than the others, indicating that most products fall in the first interval (0–9 in the first graph and 0–19 in the second). In the second histogram, the height of the bar decreases with each interval. In the first histogram, the heights fluctuate a bit. For example, the bar for 50–59 is shorter than the bar for 60–69.

• ***Additional Answers continued on page A658***

**Just the facts**

The U.S. women's soccer team won the 1999 World Cup competition, defeating China by a score of 5 to 4. The game was tied after two overtime periods and had to be decided on penalty kicks.

6. **Sports** This table shows the number of U.S. girls of various ages who played soccer in leagues recognized by the American Youth Soccer Organization in a recent year:

**Soccer Players**

| Ages | Girls |
|---|---|
| 5 and 6 | 23,805 |
| 7 and 8 | 45,181 |
| 9 and 10 | 46,758 |
| 11 and 12 | 39,939 |
| 13 and 14 | 26,147 |
| 15 and 16 | 11,518 |
| 17 and 18 | 4,430 |

Source: American Youth Soccer Organization

a. Create a histogram showing these data. The first bar, which includes 5- and 6-year-old girls, has been drawn for you.

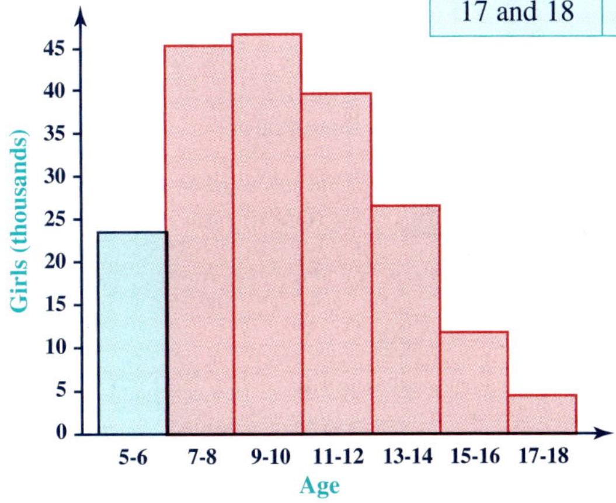


b. Describe the shape of the histogram. Tell what the shape indicates about the distribution of ages. See Additional Answers.

**Connect & Extend**

7. **Economics** Here are data about the number of motor vehicles manufactured in the United States, Europe, and Japan from 1993 to 2001:

**Motor Vehicles Manufactured (millions)**

| Year | U.S. | Europe | Japan |
|---|---|---|---|
| 1993 | 10.9 | 15.2 | 11.2 |
| 1994 | 12.3 | 16.2 | 10.6 |
| 1995 | 12.0 | 17.0 | 10.2 |
| 1996 | 11.8 | 17.6 | 10.3 |
| 1997 | 12.1 | 17.8 | 11.0 |
| 1998 | 12.0 | 16.3 | 10.1 |
| 1999 | 13.1 | 17.6 | 9.9 |
| 2000 | 12.8 | 17.7 | 10.1 |
| 2001 | 11.5 | 17.7 | 9.8 |

Source: *World Almanac and Book of Facts 2003*. Copyright © 2003 Primedia Reference Inc.

a. On the same set of axes, make a line graph of the data for each group. Use a different point shape or line color for each group.

**7a.** See Additional Answers.

**Remember**

To make a line graph, plot the data points and connect them with line segments.



**Exercise 6:**
You may want to provide students with copies of Master 61 to complete the table and histogram. As an alternative, students could use graph paper or copies of Master 81 or Master 82 to make their histograms.

**Teaching Resources**

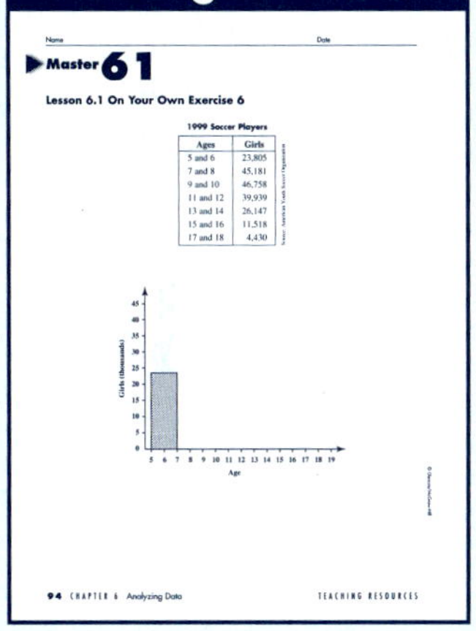

**Exercise 7:**
Students will need graph paper or copies of Master 81 or Master 82 to make their graphs. You may want to point out that while Europe is a continent, not a country, the name also describes a group of countries.

**Additional Answers**

**6b.** The heights of the bars increase, reaching their highest point for the "9 and 10" age group, and then decrease again, reaching their lowest point for the "17 and 18" age group. The shape indicates that most of the girls who play soccer are in the middle of the age range shown on the graph.

• ***Additional Answers continued on page A658***

**Exercise 7:**
You may want to point out that the conclusions for **Parts b and c** are facts while the prediction in **Part d** is a hypothesis.

**Exercise 8:**
Students will need graph paper or copies of Master 81 or Master 82. You may want to suggest that students graph latitude on the horizontal axis and temperature on the vertical axis for **Part a.**

**b.** Is there one group that consistently produces more motor vehicles than the others? If so, which group is it? yes, Europe

**c.** Write two or three sentences comparing the number of vehicles manufactured in the United States to the number manufactured in Japan for the years from 1993 to 2001.

**d.** Given the trends in these data, which group do you think produced the most motor vehicles in 2002? Which group do you think produced the fewest? Give reasons for your answers.

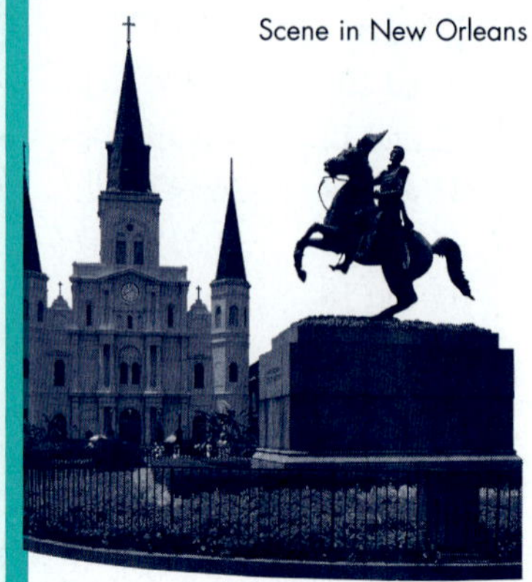
Scene in New Orleans

**7c.** Possible answer: Japan initially produced more vehicles, but from 1994 on, the United States produced more. 1999 shows the greatest difference between the number of vehicles produced in the two countries.

**7d.** See Additional Answers.

**8. Earth Science** The *latitude* of a location indicates how far it is from the equator, which has latitude 0°. The farther from the equator a place is, the greater its latitude. The latitude measure for a location includes the letter N or S to indicate whether it is north or south of the equator.

This table gives the lowest average monthly temperature and the latitude of nine cities:

| City | Latitude | Lowest Average Monthly Temp. (°F) |
|---|---|---|
| Albuquerque, New Mexico, U.S.A. | 35° N | 34 |
| Georgetown, Guyana | 7° N | 79 |
| New Orleans, Louisiana, U.S.A. | 30° N | 51 |
| Portland, Maine, U.S.A. | 44° N | 22 |
| Porto Alegre, Brazil | 30° S | 58 |
| Recife, Brazil | 8° S | 75 |
| San Juan, Puerto Rico | 18° N | 72 |
| St. John's, Newfoundland, Canada | 48° N | 23 |
| Stanley, Falkland Islands | 52° S | 36 |

Source: www.worldclimate.com

**a.** Make a line graph of the latitude and temperature data. When you graph the latitude values, ignore the N and S, and just graph the numbers. This way you will be graphing each city's distance from the equator. See Additional Answers.

**b.** Does there appear to be an overall relationship between the latitude of a city and its lowest average monthly temperature? If so, describe the relationship.

**8b.** Possible answer: In general, as latitude increases, lowest average monthly temperature decreases.

**c.** The island of Nassau in the Bahamas has a latitude of about 25° N. Predict Nassau's lowest average monthly temperature. Explain how you made your prediction.
See Additional Answers.



### Additional Answers

**7d.** Possible answer: I would expect that Europe produced the most because they produced the most for all the years shown. I would guess that Japan produced the fewest in 2002, because they produced between 9,000 and 11,000 for each of the previous 5 years while the United States produced between 11,000 and 13,000 in each of the previous 5 years.

**8a.**

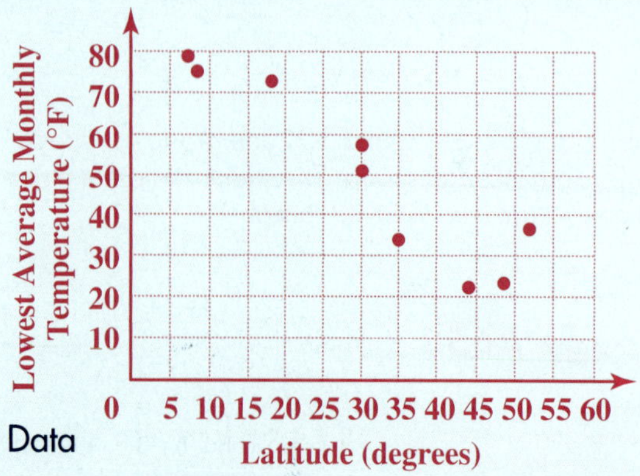


**8c.** Possible answer: About 63°F. I looked at the pattern in the graph. There are points on the graph for cities with latitudes 18° and 30°. The latitude for Nassau is between these, so I predicted that its temperature would be between the temperatures for those two cities (72°F and 54.5°F, the average of 51°F and 58°F). (Note: Nassau's actual lowest monthly temperature is 69°F.)

**In** your **own words**

Describe two types of graphs you have used to display data. For each type of graph, give an example of a set of data you might display with that type of graph.

**9. Preview** A survey asked middle school students how much time they spend with their parents or guardians on a typical weekend. Here are the results:

| Time Spent with Parents | Boys (percent) | Girls (percent) |
|---|---|---|
| Almost all | 39.6 | 49.6 |
| One full day | 18.6 | 21.8 |
| Half a day | 17.5 | 17.1 |
| A few hours | 24.3 | 11.5 |

If you wanted to compare the boys' responses with the girls' responses, you could display these data in two circle graphs.

**Time Spent with Parents**

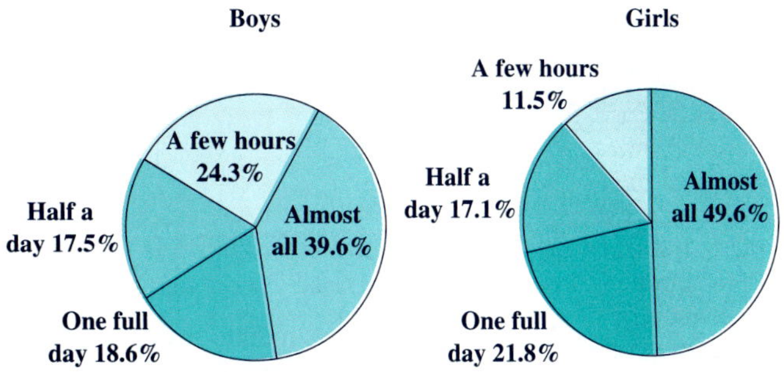


**a.** You could also show the data in a *double bar graph.* For each time category, the graph will have two bars—one showing the percent of boys in that category and the other showing the percent of girls. Copy and complete the graph below.

**Time Spent with Parents**

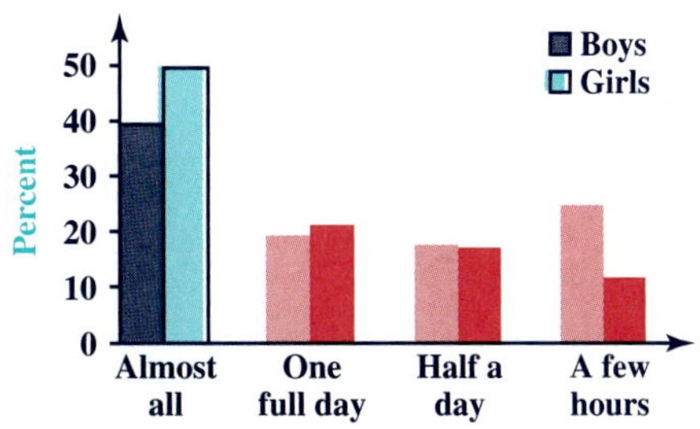


**b.** Which display do you think makes it easier to compare the two categories of data? Give reasons for your choice.
Possible answer: The double bar graph; it lets you see the percent of boys and girls in each category side by side.



**Exercise 9:**
This exercise introduces students to double bar graphs. You may want to briefly discuss the partial graphs, pointing out that the bars showing the time boys spend and the time girls spend are adjacent to one another but in different colors. Be sure students understand that they should use two adjacent bars, and that all the bars for boys should be the same color and all the bars for girls should be the same color.

You may want to provide students with copies of Master 62, which reproduces the data set and the partially constructed graph. Alternatively, students could use graph paper or copies of Master 81 or Master 82 to make their graphs.

**Teaching Resources**

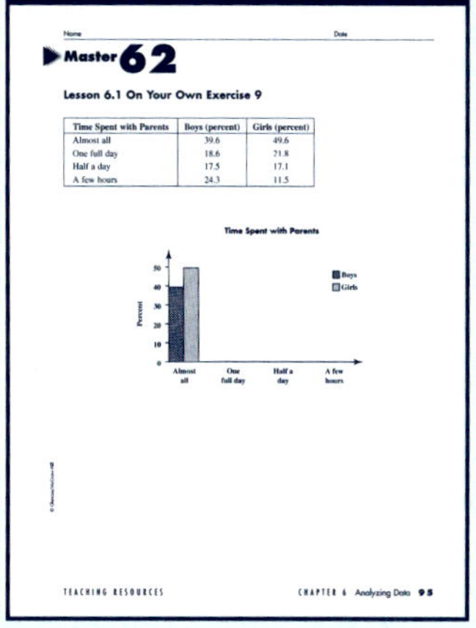


LESSON 6.1 Using Graphs to Understand Data **359**

**Exercise 10:**
This exercise has students consider some aspects that make a bar graph misleading. Since this may be a new idea for many students, you may want to work through the exercise as a class activity or, if assigned as independent study, have students share their answers. Students may want to use graph paper or Master 81 or Master 82 to make their graphs in Part c.

**10.** Drake's mother told him he could not play video games after school until his performance in math class improved significantly. Drake's math teacher gives a 20-point quiz each week. Drake made this graph to show his mother how much his scores had improved over the past five weeks:

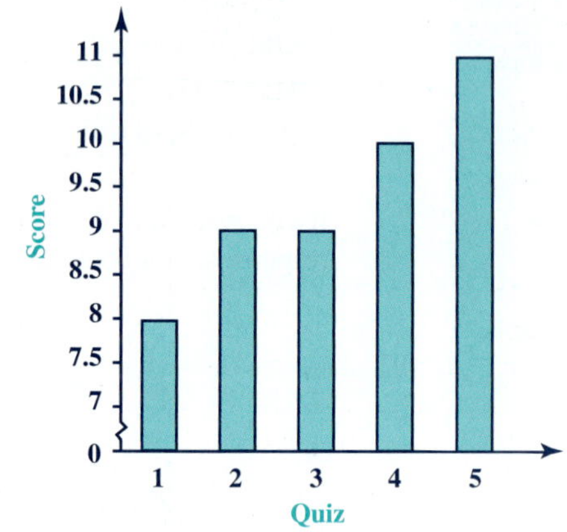


**a.** The bar for Quiz 5 is four times the height of the bar for Quiz 1. Is Drake's score on Quiz 5 four times his score on Quiz 1? no

**b.** Drake's mother says his graph is misleading because it makes his improvement look more dramatic than it really is. What features of the graph make it misleading?

**c.** Make a new bar graph that you feel gives a more accurate view of Drake's performance on the weekly quizzes.

**10b.** The scale on the vertical axis starts at 7 rather than 0. Each grid line on the vertical axis represents only 0.5, but because the space between grid lines is so great, an improvement of 1 point looks very large.

**10c.** Possible graph:

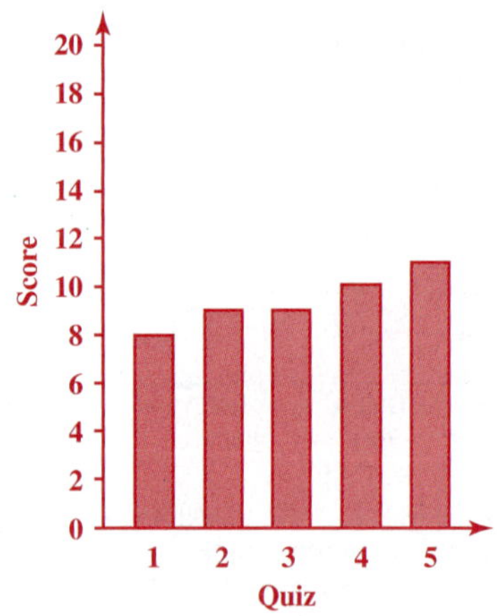

360 CHAPTER 6 Analyzing Data

## Mixed Review

Find each product or quotient.

**11.** $\frac{3}{4} \cdot \frac{4}{3}$ 1 **12.** $\frac{3}{4} \div \frac{4}{3}$ $\frac{9}{16}$ **13.** $\frac{12}{21} \cdot \frac{7}{16}$ $\frac{1}{4}$

**14.** $\frac{27}{32} \cdot \frac{24}{45}$ $\frac{9}{20}$ **15.** $2\frac{2}{5} \cdot \frac{1}{3}$ $\frac{4}{5}$ **16.** $3\frac{5}{8} \div \frac{1}{4}$ $14\frac{1}{2}$

**17.** $1\frac{3}{8} \cdot 4\frac{1}{2}$ $6\frac{3}{16}$ **18.** $4\frac{4}{7} \div 1\frac{1}{2}$ $3\frac{1}{21}$ **19.** $5 \div \frac{1}{9}$ 45

**Geometry** Find each missing angle measure.

**20.**

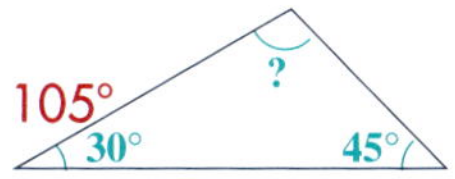


**21.**

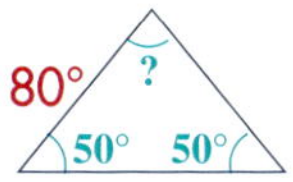


**22.**

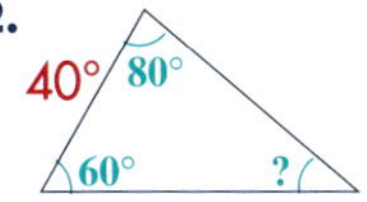


**23.** The 180 sixth-grade girls at Wright Middle School were asked to name their favorite activity in gym class. The results are shown in this circle graph:

**Favorite Gym Activities**

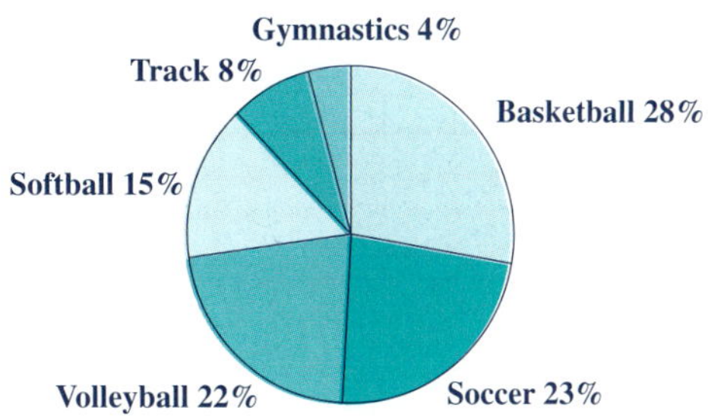


**a.** Which activity is most popular? About how many girls chose that activity? basketball, 50

**b.** Which activity is least popular? About how many girls chose that activity? gymnastics, 7

**c.** What is the difference in the *percent* of girls who chose volleyball and the percent who chose track? What is the difference in the *number* of girls who chose these sports? 14%, about 25



**Exercises 11–19:** Encourage students to use division to check their answers when multiplying and to use multiplication to check their answers when dividing.

## Quick Check

**Informal Assessment**
Students should be able to:

- ✔ interpret and compare line graphs and investigate reasons for patterns observed in line graphs
- ✔ use bar graphs and histograms to display data, identify patterns, and draw conclusions
- ✔ create bar graphs and histograms from sets of data
- ✔ decide whether to use a bar graph or a histogram to display a given set of data

**Quick Quiz**

1. What does it mean when a line graph is a horizontal line? Give an example of a graph that could have this shape. ***Possible answer: that the values do not change; a car's speed when it is on cruise control***

• ***Quick Quiz continued on page A658***

# Teacher Notes

# 6.2 What Is Typical?

## Objectives

- To find the mode, median, mean, and range of a data set
- To make line plots and stem-and-leaf plots and use them to describe a data set
- To predict how changing a value in a data set impacts the mean and median
- To use various measures to interpret and compare data sets

## Overview (pacing: about 6 class periods)

In this lesson, students use line plots and stem-and-leaf plots to display sets of data and practice differentiating between numerical and categorical data. They use range and measures of central tendency to analyze data sets displayed in tables, line plots, and stem-and-leaf plots. Mode and median (measures of central tendency) and range (a measure of dispersion) are introduced first. Mean is introduced in Investigation 3 since research has shown that many students develop a better conceptual understanding of mean when this concept is introduced after they have worked with median, mode, and range. Students then further develop their analytical skills by looking at how median, mean, and mode can be used to compare groups and how changing a value in a data set affects these measures.

## Advance Preparation

You may want to make copies of Master 63, which reproduces the table and partial line plot from Problem Set A. Otherwise, students may want to use rulers to make their line plots. At the beginning of the lesson, students estimate the time they spent on homework the previous day; you may want to make a transparency of the data for students to refer to in Investigation 2. You may want to provide students with a copy of the class roster for Investigation 5, making sure there is sufficient room to record each student's forearm measurement.

## Lesson Planner

| | Summary | Materials | On Your Own Exercises | Assessment Opportunities |
|---|---|---|---|---|
| Investigation 1<br>page T362 | Students make line plots. They find the mode, median, and range of data sets displayed in tables and line plots. | • Master 63 (optional)<br>• Rulers (optional) | Practice & Apply: 1–3, p. 380<br>Connect & Extend: 13, 14, pp. 384–385<br>Mixed Review: 20–39, p. 389 | On the Spot Assessment, page T364<br>Share & Summarize, pages T365, 365<br>Troubleshooting, page T365 |
| Investigation 2<br>page T366 | Students make stem-and-leaf plots and use them to analyze data. | • Transparency of homework data from Investigation 1 (optional) | Practice & Apply: 4, 5, p. 381<br>Connect & Extend: 15, p. 386<br>Mixed Review: 20–39, p. 389 | On the Spot Assessment, pages T368, T369<br>Share & Summarize, pages T369, 369<br>Troubleshooting, page T369 |
| Investigation 3<br>page T370 | Students compute the mean. | | Practice & Apply: 6, 7, p. 382<br>Connect & Extend: 16, 17, p. 387<br>Mixed Review: 20–39, p. 389 | Share & Summarize, pages T372, 372<br>Troubleshooting, page T372 |
| Investigation 4<br>page T373 | Students compare mean and median as measures of typicality and decide which measure better describes a set of data. | | Practice & Apply: 8–10, pp. 382–383<br>Connect & Extend: 18, p. 387<br>Mixed Review: 20–39, p. 389 | Share & Summarize, pages T376, 376<br>Troubleshooting, page T376 |
| Investigation 5<br>page T377 | Students analyze data, decide on ways to display the data, and make recommendations. | • Measuring tape or yardstick<br>• Class roster (optional) | Practice & Apply: 11, 12, p. 384<br>Connect & Extend: 19, p. 388<br>Mixed Review: 20–39, p. 389 | Share & Summarize, pages T379, 379<br>Troubleshooting, page T379<br>Informal Assessment, page 389<br>Quick Quiz, page A660 |

# Introduce

Have a volunteer explain what factors they considered when they described the graphs in Lesson 6.1. They may mention looking at the maximum and minimum values of the data set and at how fast the values change (the change in heights of the bars of a bar graph or how steep the various segments of a line graph appear). Explain that they will now consider other ways to describe data, such as using the typical value, and other ways to display data.

1 **Think & Discuss**
Write the time students spent doing homework on the board or a transparency as each student reports his or her estimate. Students should copy the list to use with Problem Set D in Investigation 2. Be sure to have a student copy the list for your use as well. You could also save the transparency to display the results for students.

2 Ask volunteers to share their descriptions of the data. Some students may be familiar with some measures of central tendency and mention them at this time. Students will find the median, mode, and range of the data set in Investigation 2.

# Investigation 1

In this investigation, students find the mode, median, and range of various data sets. They analyze data from tables and line plots.

Have students look at the tables at the bottom of the page showing basketball shoe sales. Inform students that they will analyze these data in Problem Set A and Problem Set B.

# 6.2 What Is Typical?

To help people understand a set of data, it is useful to give them an idea of what is *typical,* or average, about the data. In this lesson, you will learn three ways to describe the typical value in a data set. You will also learn about two simple types of graphs that are useful for showing the distribution of values in a set of data.

## Think & Discuss See Additional Answers.

Have each student in your class estimate how many minutes he or she spent doing homework yesterday. Your teacher should record the data on the board. How would you describe to someone who is not in your class what is typical about your class's data?

1 Have students copy the data as others share them with the class.

2 Encourage students to describe the data.

## Investigation 1 Mode and Median

The Jump Shot shoe store sells basketball shoes to college players. The table shows the brand and size of each pair the store sold one Saturday.

| Brand | Size |
|---|---|
| Swish | 13 |
| Dunkers | 14.5 |
| Hang Time | 8.5 |
| Swish | 13 |
| Hang Time | 13 |
| Airborne! | 15 |
| Swish | 12.5 |
| Swish | 8 |
| Dunkers | 13 |
| Big J | 13.5 |
| Hang Time | 14 |
| Big J | 14.5 |
| Dunkers | 14 |

| Brand | Size |
|---|---|
| Swish | 14 |
| Hang Time | 10 |
| Airborne! | 11 |
| Dunkers | 12 |
| Airborne! | 14 |
| Big J | 12.5 |
| Swish | 14 |
| Swish | 14.5 |
| Hang Time | 10.5 |
| Swish | 10.5 |
| Swish | 14 |
| Hang Time | 13.5 |



**Additional Answers**

Answers will vary. Students may suggest reporting the value that occurs most often or adding the values and dividing by the number of values (finding the mean). Students may also suggest reporting the entire range or the part of the range in which most values occur (for example, most students spent between 30 and 40 minutes doing homework).

# Develop

**1**

**Problem Set A** **Suggested Grouping: Pairs**
Students are introduced to the terms *line plot, median, mode,* and *range* as they work through this problem set.

**2**

You may want to provide students with copies of Master 63 to complete the line plot in **Problem 1a.** If you elect not to use the master, you might have rulers available for students to draw the number line.

**Teaching Resources**

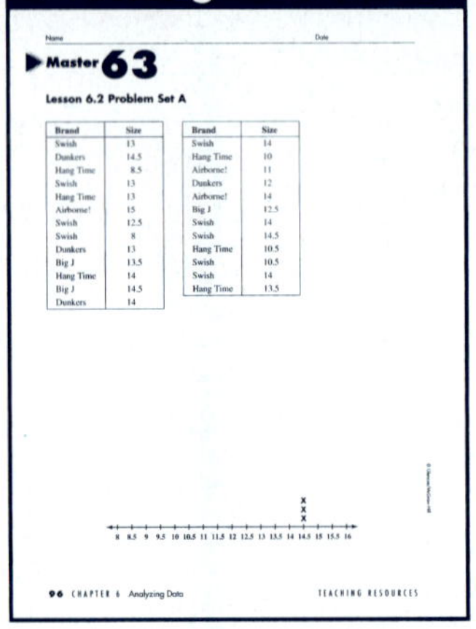
Name Date

**Master 63**

**Lesson 6.2 Problem Set A**

| Brand | Size |
|---|---|
| Swish | 13 |
| Dunkers | 14.5 |
| Hang Time | 8.5 |
| Swish | 13 |
| Hang Time | 13 |
| Airborne! | 15 |
| Swish | 12.5 |
| Swish | 8 |
| Dunkers | 13 |
| Big J | 13.5 |
| Hang Time | 14 |
| Big J | 14.5 |
| Dunkers | 14 |

| Brand | Size |
|---|---|
| Swish | 14 |
| Hang Time | 10 |
| Airborne! | 11 |
| Dunkers | 12 |
| Airborne! | 14 |
| Big J | 12.5 |
| Swish | 14 |
| Swish | 14.5 |
| Hang Time | 10.5 |
| Swish | 10.5 |
| Swish | 14 |
| Hang Time | 13.5 |

Encourage students to confirm that all data have been recorded on their line plots by making sure that the number of values in the table is the same as the number of X's on the line plot.

## Tips from Teachers

"Some students find it helpful to use the word association to help remember the definitions of *mode* and *median.* For example, students could think of the 'mō' sound in *mode* and *most* to remember that mode is the value that appears the most often. A median is also the name for the middle (often grassy) strip that separates lanes of traffic on a multi-lane highway; a median in data analysis is the middle number in an ordered set of data."

In the next two problem sets, you will learn some ways to summarize the shoe store's data.

## Problem Set A

VOCABULARY
**line plot**

1. You can create a line plot to show the sizes of the shoes sold on Saturday. A **line plot** is a number line with X's indicating the number of times each data value occurs.

   **a.** To make the plot, copy the number line below. Mark an X above a shoe size each time it appears in the data set. For example, 14.5 appears three times, so put three X's above 14.5.

   **b.** Describe the shape of your line plot. Tell what the shape indicates about the distribution of shoe sizes.

2. When you describe a data set, it is helpful to give the *minimum* (least) and *maximum* (greatest) values.

   **a.** Give the minimum and maximum shoe sizes in the data set.

   **b.** How can you find the minimum and maximum values by looking at a line plot?

VOCABULARY
**median**
**mode**
**range**

3. The **range** of a data set is the difference between the minimum and maximum values. Give the range of the shoe-size data. 7

4. The **mode** of a data set is the value that occurs most often.

   **a.** Give the mode of the shoe-size data. 14

   **b.** How can you find the mode of a data set by looking at a line plot?

5. The **median** is the middle value when all the values in a data set are ordered from least to greatest.

   **a.** List the shoe-size data in order from least to greatest, and then find the median size.

   **b.** How can you find the median of a data set by looking at a line plot? Possible answer: Start at one end and count the X's until you reach the middle value. In this data set, there are 25 values, so count until you get to the 13th X. The number this X is over is the median.

**1.** See Additional Answers.

**2a.** minimum: 8; maximum: 15

**2b.** Find the least and greatest values with X's over them.

**4b.** Look for the tallest stack of X's.

**5a.** 8, 8.5, 10, 10.5, 10.5, 11, 12, 12.5, 12.5, 13, 13, 13, 13, 13.5, 13.5, 14, 14, 14, 14, 14, 14, 14.5, 14.5, 14.5, 15; median: 13

**1** You may have students work in pairs.

**2** Introduce the terms *line plot, range, mode,* and *median.*



**Additional Answers**

**1a.**

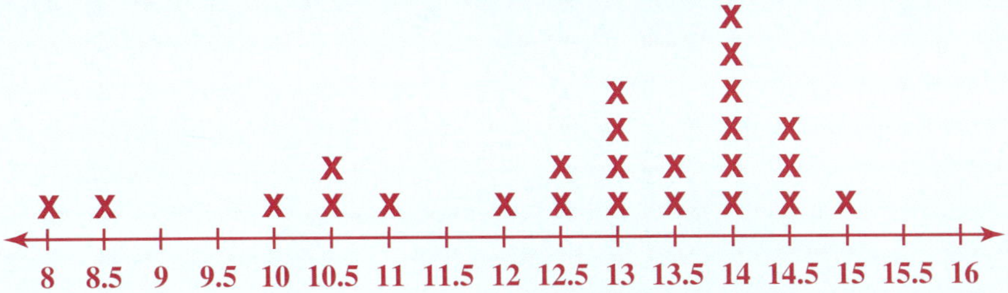


**1b.** Possible answer: There are a few scattered values on the left side of the plot, but most of the values are concentrated at the right side between 12 and 15. This indicates that many more large sizes than small sizes were sold. Size 14 has the tallest stack of X's, so it was the most popular size. There is a gap between 8.5 and 10 and between 11 and 12, indicating that no shoes with sizes 9, 9.5, or 11.5 were sold. There are only two values of 13.5, although several shoes of the next smaller and next larger size were sold.

# Develop

1 Students examine a data set with two modes in **Problem 6b.** Be sure they understand that the two values are both modes and that they should not find their average or median. You may want to mention that a data set can have more than two modes, and point out that when all the values in a data set appear the same number of times, the data set is considered to have no mode.

If students are concerned that the median in **Problem 6c** is not an actual shoe size, point out that the median is a mathematical concept that, in this case, reveals that half the shoes sold are in sizes greater than 13.75 and half the shoes sold are in sizes less than 13.75. It does not mean that any shoes sold are in that particular size.

## Access for all Learners

**Extra Challenge** Challenge students to find a data set of shoe sizes for Problem 6c in which the median would be a shoe size. Ask them to try several data sets to see if there are any patterns. Encourage them to find an explanation for the patterns.

They may note that data sets with an odd number of values always have a median that is an actual shoe size while data sets with an even number of values have a median that may or may not be an actual shoe size. This is because the median for a data set with an odd number of values is a number in the data set and therefore must be an actual shoe size. When there are an even number of data items, the median is an average of the two middle values. In this case, whether or not the median is an actual size depends upon which sizes are the middle values. If the middle values were 13 and 14, the median would be 13.5, which is a shoe size.

**Problem Set Wrap-Up** Discuss the vocabulary terms, *line plot, median, mode,* and *range* if not done previously to make sure students understand them. Have them share their answers to **Problems 3–6.** ■

2 Introduce the terms *numerical data* and *categorical data.* Point out that students used numerical data in Problem Set A.

3 **Problem Set B** **Suggested Grouping: Pairs** Students look at ways to analyze categorical data in this problem set. In the process, they discover that some of the techniques they used in Problem Set A do not apply to this type of data.

Remind students to use the data from the table on page 362 as they work through the problem set.

## On the Spot Assessment

Watch for students who think that a line plot can show the categorical data in **Problem 1** if the brand names are placed along a horizontal axis and an X is placed above the brand name each time it is mentioned. Point out that while data can be shown in this way, the graph would not have the characteristics of a line plot. In a line plot, the horizontal axis is a number line and must have numerical values. Writing the brand names along the axis is a form of a pictograph.

4 **Problem Set Wrap-Up** Be sure students understand that categorical data cannot be shown on a line plot and do not have a range or a median. ■

Point out that both the median and the mode describe an average value in a data set and that these measures are often not the same number. Ask students which value they think would be a better way to describe a data set. Accept any ideas, and tell students that sometimes one measure provides a better description than the other and that they will explore this idea in the next problem set.

5 **Problem Set C** **Suggested Grouping: Individuals** In this problem set, students decide whether to use the mode or the median to describe specific sets of data. They also are given facts about a data set and asked to make a data set that fits the description.

## On the Spot Assessment

Watch for students who do not find the correct median in **Problems 1a, 2a, and 3a.** Have them write the values in numerical order and cross off the greatest and least numbers, continuing the process until only one number remains. You might point out that when there is an even number of values in a data set, there will be two numbers left, and students should find the number halfway between them to determine the median.

6. Suppose the store discovered five Saturday sales that weren't recorded: 14.5, 14.5, 15, 14.5, and 16. Add these to your line plot.

a. What is the range of the data now? 8

b. What is the mode now? There are two modes, 14 and 14.5.

c. When a data set has an even number of values, there is no single middle value. In such cases, the median is the number halfway between the two middle values. Find the median of the new data set. 13.75

1 Reinforce the fact that there can be more than one mode in a set of data.

In Problem Set A, you looked at ways to summarize *numerical data*—that is, data that are numbers. You will now look at the brand-name data, which are not numbers. Non-numerical data are sometimes called *categorical data* because they can be thought of as names of *categories*, or groups.

2 Introduce the terms numerical data and categorical data.

**Problem Set B Answers**

1. No; the data are not numbers, so you can't plot them on a number line.
2. No; there is no least or greatest value.
4. No; since the data are not numbers, they can't be listed from least to greatest, so there is no middle value.
5. Possible answer: You could make a table or bar graph showing each brand name and the number of times it occurs. You could make a graph like a line plot, but with brand names listed along the horizontal axis. You could list all the brands that were sold.

## Problem Set B

3 You may have students work in pairs.

1. Is it possible to make a line plot to show the distribution of the brand-name data? Explain.
2. Can you find the range of the brand-name data? If so, find it. If not, explain why it is not possible to find the range.
3. Do the brand-name data have a mode? If so, find it. If not, explain why it is not possible to find the mode. yes, Swish
4. Do the brand-name data have a median? If so, find it. If not, explain why it is not possible to find the median.
5. What are some other ways you might summarize the brand-name data?

4 Discuss how categorical data cannot have a median or a range.

The mode and the median are two measures of the typical, or average, value of a data set. In some cases, one of these measures describes the data better than the other.

**Problem Set C Answer**

1a. range: 8; mode: 10; median: 7

## Problem Set C

5 You may have students work individually.

1. Ms. Washington gave her class a 10-point quiz. Here are her students' scores:

7 9 10 5 5 8 6 10 6 7 10 2

7 5 8 8 4 9 10 4 10 7 6

a. Find the range, mode, and median of the quiz scores.

b. Do you think the mode or the median is a better measure of what is typical in this data set? Explain.
Possible answer: The median; in this case, the mode is the highest score. Since the values in the data set are spread out between 2 and 10, the median of 7 gives a better idea of what is typical.

## Access for all Learners

**Extra Challenge** Have students decide whether they would choose the mode or the median to describe their data sets from **Problem 4** and to explain why they made the choice they did. Since the values are close together, the measures may be equally good.

Challenge students to write another data set that fits the description in Problem 4 and has a different measure that might better describe the typical value in the data set.

1 **Problem Set Wrap-Up** Discuss students' answers and reasoning for **Problems 1b, 2b, and 3b.** If time permits, have students share and compare their data sets from **Problem 4.**

### Share & Summarize

2 Have students work in pairs to review the vocabulary from this investigation and compare two different ways to display data. When discussing **Question 3,** be sure students realize they are considering categorical data.

**Troubleshooting** If students are having difficulty understanding and finding median and mode, try this class activity:

Have each student write the number of siblings he or she has on a card. Then ask students to line up in order from those having the fewest siblings to those having the most siblings. Once students are in order, either have them count off from either end to find the median value or have the students at either end of the line sit down, leaving two new students as the "ends." Repeat the process until one or two students remain standing. One student standing represents the median value. Two students standing indicate that the median is halfway between the two values represented.

If you have an even number of students, you might want to add your own piece of data to the set.

### On Your Own Exercises

Practice & Apply: 1–3, p. 380
Connect & Extend: 13, 14, pp. 384–385
Mixed Review: 20–39, p. 389

**2a.** range: 4; modes: 0 and 4; median: 1

**2b.** Possible answer: The modes; all but one person has 0 pets or 4 pets, so the modes give a very good indication of what is typical.

**Just the facts**

The sport of javelin throwing evolved from ancient spear-throwing contests introduced as part of the pentathlon in the Olympics in 708 B.C.

**2.** Hannah asked nine of her classmates how many pets they have. Here are the results:

0 4 1 0 0 4 4 0 4

**a.** Find the range, mode, and median of these data.

**b.** Do you think the mode or the median is a better measure of what is typical in this data set? Explain.

**3.** During one afternoon practice, an athlete threw a javelin 13 times. Here are the distances for each throw, rounded to the nearest foot:

257 210 210 255 210 220

275 253 210 255 250 252 200

**a.** Find the range, mode, and median of the data. range: 75; mode: 210; median: 250

**b.** If you had to use only one type of average—the mode or the median—to summarize this athlete's performance, which would you choose? Give reasons for your choice. See Additional Answers.

**4.** When you are given summary information about a set of data, you can sometimes get an overall picture of how the values are distributed. For example, suppose you know these facts about a data set:

- It has 15 values.
- The minimum value is 50, and the maximum value is 100.
- The mode is 55.
- The median is 57.

**a.** What do you know about how the data values are distributed?

**b.** Make up a data set that fits this description.

**4a.** Possible answer: Many values are bunched at the beginning of the range. Half of the values are less than 57, but greater than 50, and the other half are between 57 and 100.

**4b.** Possible answer: 50, 51, 51, 53, 55, 55, 55, 57, 60, 65, 65, 72, 83, 90, 100

**1** Discuss students' answers.

## Share & Summarize

See Additional Answers.

**1.** How is a line plot similar to a histogram? How is it different?

**2.** Describe what the range, mode, and median tell you about a set of numerical data.

**3.** Which measure—the range, the mode, or the median—can be used to describe a set of categorical data? Explain.

**2** Have students share the similarities and differences.



### Additional Answers

**3b.** Possible answer: The median; since half of the values are above the median and half are below, I think the median gives a good picture of the overall performance. Since only 1 of the 13 throws is shorter than the mode, I don't think the mode reflects what is typical.

### Share & Summarize

**1.** Possible answer: Both graphs show a distribution of data values. The histogram uses bars, while the line plot uses X's. The line plot tells you how many times each individual data value occurs, while the histogram tells you the number of data values in particular intervals.

**2.** The range indicates the difference between the least and greatest values. The mode tells you the value that occurs most often. The median tells you the middle value.

**3.** The mode; since categorical data are not ordered, they don't have a range or a middle value. However, there may be a value that occurs most often.

# Investigation 2

Students will now be introduced to stem-and-leaf plots. They begin by reading and interpreting stem plots showing the time various students spent on homework. Then they use the homework data from the Think & Discuss on page 362 to make a stem plot for their class data. Students interpret the data and compare the two data sets. They reinforce these skills as they expand their knowledge to create stem-and-leaf plots with two-digit numbers in the stems.

You may want to use a transparency of the homework data from Investigation 1 to provide data for students who were absent or lost their notes. You could also make copies of the list for these students or pair them with students that have the data.

1 Before students open their books, ask them how line plots can be used to help analyze data. They may mention that line plots show data distribution as well as individual data values so that the median, mode, and range can be determined.

Remind students that they have studied different ways to display data and that some ways show certain types of data better than other ways. Suggest that this may also be true for line plots.

### Think & Discuss

Have pairs of students think about how a line plot of the data in the table would look and whether it would be a good way to display the data. Then have students share their ideas.

2 Once students understand that there are some disadvantages to displaying the data set as a line plot because of the wide range of data values and the few values that are repeated, introduce stem-and-leaf plots. Detailed guidelines for making a stem plot are given in the Example on page 367.

## Investigation 2 Stem-and-Leaf Plots

You have seen that line plots are useful for showing the distribution of numerical data and for locating the mode and median. However, for some data sets, creating a line plot may not be practical.

1 Ask students how line plots can be used to help analyze data.

### Think & Discuss

The students in Ms. Washington's class kept track of how many minutes they spent doing homework one evening. Here are their results:

| Student's Initials | Time Spent Doing Homework (minutes) |
|---|---|
| AF | 42 |
| JB | 5 |
| RC | 60 |
| HE | 30 |
| JL | 45 |
| MM | 47 |
| DL | 0 |
| SK | 25 |
| FR | 67 |
| CO | 51 |
| DW | 56 |
| PG | 20 |

| Student's Initials | Time Spent Doing Homework (minutes) |
|---|---|
| RL | 90 |
| DD | 39 |
| AG | 30 |
| RT | 49 |
| CB | 58 |
| MC | 55 |
| FB | 75 |
| JM | 45 |
| TK | 44 |
| MG | 37 |
| LK | 62 |
| EL | 65 |

Think about what a line plot of these data would look like. Do you think a line plot is a good way to display these data? Why or why not?

VOCABULARY
**stem-and-leaf plot**

When a data set contains many different values, or when the values are spread out, a **stem-and-leaf plot** (also called a *stem plot*) may be more helpful than a line plot.

2 Introduce the term *stem-and-leaf plot.*

**Think & Discuss Answer**

Possible answer: No; the number line would have to go from 0 to 90. If I included a mark and a label for every whole number, it would be a very long number line. There are also many different values in this data set, and most values occur only once, so the line plot would have at most one or two X's over a number—it really wouldn't be very interesting or give me a good picture of the distribution.

# Develop

1 **Example**
As you discuss the example, you might model how to make the stem-and-leaf plot on the board or on a transparency. You can have students follow along at their desks. Students may initially resist the idea that these numerals must be separated into tens and ones before plotting the values, but most will understand the concept after entering a few data values.

For the first step, writing the stems, ask students why 0 is included as a stem. If no one mentions that 0 is needed to record values less than 10, point this out.

After discussing the leaves in the second step, have students look at the completed stem-and-leaf plot. Ask them such questions as these:

How are the leaves for each stem similar? They are in numerical order from least to greatest.

Why are there two 0s as leaves for stem 3? To show that two students, HE and AG, spent 30 minutes working on homework.

How do you know that all the data values have been recorded? The number of leaves and the number of values in the table are the same. ■

2 Discuss the similarities and differences between stem plots and histograms, noting that both displays show numerical data in intervals, but the stem plot also shows individual values.

Have students retrieve their copies of the homework data that they recorded for the Think & Discuss on page 362 before they begin working through Problem Set D.

**EXAMPLE**

To make a stem-and-leaf plot of the homework-time data, think of each data value as being made up of two parts: a tens digit and a ones digit.

Write the tens digits, from least to greatest, in a column. Draw a vertical line to the right of the digits. The values in this column are the *stems.*

| Stem | Leaf |
|---|---|
| 0 | |
| 1 | |
| 2 | |
| 3 | |
| 4 | |
| 5 | |
| 6 | |
| 7 | |
| 8 | |
| 9 | |

To add the *leaves,* write the ones digit for each data value to the right of the appropriate tens digit. For example, to plot the first time of 42 minutes, write a 2 next to the stem value 4. The first five data values—42, 5, 60, 30, and 45—have been plotted here.

| Stem | Leaf |
|---|---|
| 0 | 5 |
| 1 | |
| 2 | |
| 3 | 0 |
| 4 | 2 5 |
| 5 | |
| 6 | 0 |
| 7 | |
| 8 | |
| 9 | |

After you have plotted all the data values, redraw the plot, listing the ones digits for each stem in order from least to greatest. Give the plot a title and a key.

**Time Spent Doing Homework (min)**

| Stem | Leaf |
|---|---|
| 0 | 0 5 |
| 1 | |
| 2 | 0 5 |
| 3 | 0 0 7 9 |
| 4 | 2 4 5 5 7 9 |
| 5 | 1 5 6 8 |
| 6 | 0 2 5 7 |
| 7 | 5 |
| 8 | |
| 9 | 0 |

**Key:** 4|2=42

Stem plots and histograms both group data into intervals. However, unlike a histogram, a stem plot allows you to read individual values.

For example, if you made a histogram that grouped the data above into the intervals 0–9, 10–19, and so on, it would show that there are four values between 50 and 59, but it would not show that these values are 51, 55, 56, and 58.



**1** Go through this example with the class.

**2** Discuss differences between stem plots and histograms.

LESSON 6.2 What Is Typical? 367

1 **Problem Set D** **Suggested Grouping: Pairs**
In this problem set, students use the stem-and-leaf plot on page 367 to find the range, mode, and median of the homework data. Then they make a stem plot of the homework data for their class and analyze the data.

> On the **Spot Assessment**
>
> Watch for students who look only at the leaves when finding the mode in **Problems 1b and 3.** For example, they may think that 4 is the mode in Problem 1b because the greatest number of leaves have a value of 4. Remind them that each value is composed of a stem *and* a leaf.

2 It may take students some time to think through how to find the median in **Problem 1c.** Not only are they becoming familiar with the stem plot, but the data set has an even number of values. You may want to point out that students will need to use the stem plot and some mathematical calculations to find the median of this data set.

Data sets with more than two-digit values are used in Problem Set E. If your class has data values greater than or equal to 100 in **Problem 2,** you can have pairs use all the data, model that aspect of the stem-and-leaf plot for the class, or eliminate those data pieces until students are more comfortable making stem plots.

**Problem-Solving Strategies** Students may use one of the strategies below to make their stem-and-leaf plots in **Problem 2.**

- Order the values in the data set. Then use the ordered values to make the stem plot.
- Plot all the values as they appear in the table. Then revise the stem plot to put each set of leaves in numerical order.
- Plot the values as they appear in the table, but leave space between the leaves to insert other leaves. This may result in unevenly spaced leaves.

**Problem Set Wrap-Up** Have students share their stem-and-leaf plots from Problem 2. Then have volunteers explain how to use a stem plot to find the median and mode of a data set. ■

3 Discuss how to decide which numbers to use as the stems in a stem-and-leaf plot. Point out that there will be at least one leaf for a stem with the least value in the data set and for a stem with the greatest value in the data set, but that the stems between these values may or may not have leaves.

4 **Problem Set E** **Suggested Grouping: Pairs**
In this problem set, students make stem-and-leaf plots for data values in the thousandths and for data values in the tens and hundreds. They analyze the data in each data set.

MATERIALS
homework-time data for your class

## Problem Set D

1. Look at the completed stem-and-leaf plot in the Example on page 367.
   a. Explain how you can use this graph to find the range of the homework times.
   b. Use the stem plot to find the mode homework time. How did you find it? See Additional Answers.
   c. Use the stem plot to find the median homework time, and explain how you found it. See Additional Answers.
   d. Describe the shape of the graph. Explain what the shape tells you about how the data are distributed. See Additional Answers.
2. In the Think & Discuss at the beginning of this lesson, you collected homework times for your class. Make a stem plot of these data. (If your data include values of 100 minutes or more, your plot will need to include two-digit stem values. For example, a value of 112 would have a stem of 11 and a leaf of 2.) Plots will vary.
3. Find the range, median, and mode of your class data.
4. Describe the distribution of your class data. Answers will vary.
5. Write a few sentences comparing your class data to the data from Ms. Washington's class. Answers will vary.

**1a.** The minimum time is the first number in the first row (in this case, 0). The maximum time is the last number in the last row (in this case, 90). To find the range, subtract the minimum from the maximum. The range is 90 min.

**3.** Answers will vary.

In the plots in Problem Set D, the tens digits are used as stem values and the ones digits are used as leaves. The stem and leaf values you choose for a given situation depend on the minimum and maximum values in the data set.

**Just the facts**

A player's batting average is the number of hits divided by the number of times the player is officially at bat.

## Problem Set E

1. This table shows the batting averages for the 2002 Cleveland Indians baseball team:

**2002 Cleveland Indians**

| Batter | Average | Batter | Average |
|---|---|---|---|
| Bradley | .249 | Lawton | .236 |
| Branyan | .205 | Magruder | .217 |
| Burkes | .301 | McDonald | .250 |
| Diaz | .206 | Selby | .214 |
| Fryman | .217 | Stevens | .222 |
| Garcia | .297 | Thome | .304 |
| Gutierrez | .275 | Vizquel | .275 |

Source: *World Almanac and Book of Facts 2003*. Copyright © 2003 Primedia Reference Inc.

**1** You may have students work in pairs.

**2** Problem-Solving Strategy
Use a graph

**3** Encourage students to discuss which numbers to use as the stems in a stem-and-leaf plot.

**4** You may have students work in pairs.



### Additional Answers

**1b.** 30 min and 45 min; I looked for the digit that appears the most times in one stem. The digit 0 appears twice in the 3 stem, and the digit 5 appears twice in the 4 stem. No other digit appears more than once in the same stem.

**1c.** 46 min; Possible explanation: I put one finger on the first value and one on the last value and moved my fingers in, one value at a time, until they met in the middle. They met with one finger on 45 and the other on 47, so the median is 46 min.

**1d.** Possible answer: There are two values in the 0 stem, indicating that two students spent less than 10 min on their homework. The number of values in each stem increases until the 4 stem and then decreases until the 7 stem. The 4 stem has the most values, indicating that more homework times were in the 40–49 interval than in any other 10-minute interval. There are no values in the 8 stem, indicating that no one spent from 80 to 89 min on homework. One student spent 90 min on homework.

1

On the **Spot**
**Assessment**

Watch for students who do not use hundreds and tens in the stems of their stem-and-leaf plots for **Problem 2.** Remind them that a leaf can only have one digit, but a stem can have as many digits as needed to show the data values.

2

**Problem Set Wrap-Up** Have students describe the distributions of values from **Problems 1b and 2b.**

You might extend the discussion by quickly sketching a line graph of the data in **Problem 2,** and helping students compare the information shown in the line plot to that shown in the stem plot that students made. Students may realize that the line graph shows the values in a time continuum, making it easier to discover patterns, but the stem plot shows all the data values and can make it easier to find the median and mode of the data.

3

**Share & Summarize**
Have students work in pairs to answer these questions about when to use a stem-and-leaf plot and how stem plots and histograms differ.

**Troubleshooting** Have students who are having difficulty reading and interpreting stem-and-leaf plots write the values represented in each row of a plot to the right of the leaves in that row.

Encourage students having difficulty making stem-and-leaf plots to write the values of the data set vertically, in order from least to greatest, aligning all place values. Have them then draw a vertical line so that one place value is to the right of the line. The numbers to the left of the line will be the stems in the stem plot. Then have students consolidate the leaves for each stem as they make the stem-and-leaf plot.

**On Your Own Exercises**

Practice & Apply: 4, 5, p. 381
Connect & Extend: 15, p. 386
Mixed Review: 20–39, p. 389

**Just the facts**

*Runs batted in, or RBI, is the number of people who score a run as a result of a player's hitting the ball.*

**a.** The batting averages range from .205 to .304. In this case, you can use the tenths and hundredths digits as stem values and the thousandths digits as leaves.

Make a stem-and-leaf plot using the stem values shown at right. See Additional Answers.

**b.** Describe the distribution of data values.

**c.** Find the mode and median batting average.

**2.** This table shows the American League RBI leaders for each season from 1973 to 2002:

**2002 Cleveland Indians Batting Averages**

| Stem | Leaf |
|---|---|
| .20 | |
| .21 | |
| .22 | |
| .23 | |
| .24 | |
| .25 | |
| .26 | |
| .27 | |
| .28 | |
| .29 | |
| .30 | |

Key: .27|5=.275

**American League RBI Leaders**

| Year | Player | RBI |
|---|---|---|
| 1973 | Reggie Jackson | 117 |
| 1974 | Jeff Burroughs | 118 |
| 1975 | George Scott | 109 |
| 1976 | Lee May | 109 |
| 1977 | Larry Hisle | 119 |
| 1978 | Jim Rice | 139 |
| 1979 | Don Baylor | 139 |
| 1980 | Cecil Cooper | 122 |
| 1981 | Eddie Murray | 78 |
| 1982 | Hal McRae | 133 |
| 1983 | C. Cooper / J. Rice | 126 |
| 1984 | Tony Armas | 123 |
| 1985 | Don Mattingly | 145 |
| 1986 | Joe Carter | 121 |
| 1987 | George Bell | 134 |
| 1988 | Jose Canseco | 124 |
| 1989 | Ruben Sierra | 119 |
| 1990 | Cecil Fielder | 132 |
| 1991 | Cecil Fielder | 133 |
| 1992 | Cecil Fielder | 124 |
| 1993 | Albert Belle | 129 |
| 1994 | Kirby Puckett | 112 |
| 1995 | A. Belle / M. Vaughn | 126 |
| 1996 | Albert Belle | 148 |
| 1997 | Ken Griffey, Jr. | 147 |
| 1998 | Juan Gonzalez | 157 |
| 1999 | Manny Ramirez | 165 |
| 2000 | Edgar Martinez | 145 |
| 2001 | Brett Boone | 141 |
| 2002 | Alex Rodriguez | 142 |



**a.** Make a stem-and-leaf plot of the RBI data.

**b.** Describe the distribution of data values.

**c.** Find the minimum, maximum, mode, and median of the RBI data.

**1b.** Possible answer: There are two values in the .270s, one in the .290s, and two in the .300s. The rest are in the .200s through the .250s.

**1c.** modes: .217 and .275; median: .243

**2.** See Additional Answers.

## Share & Summarize

**1.** In what types of situations would you use a stem-and-leaf plot, rather than a line plot, to display a set of data?

**1.** Possible answer: when there are many different data values or when the data values are very spread out

**2.** What information can you get from a stem plot that you can't get from a histogram? individual data values



**1** Remind students that a leaf can only have one digit while a stem can have many digits.

**2** Help students realize that a stem plot makes it easier to find the median and mode of a data set.

**3** Have students discuss their answers.

**Additional Answers**

**2a.**

**American League RBI Leaders**

| Stem | Leaf |
|---|---|
| 7 | 8 |
| 8 | |
| 9 | |
| 10 | 9 9 |
| 11 | 2 7 8 9 9 |
| 12 | 1 2 3 4 4 6 6 9 |
| 13 | 2 3 3 4 9 9 |
| 14 | 1 2 5 5 7 8 |
| 15 | 7 |
| 16 | 5 |

Key: 11|2=112

**2b.** Possible answer: There is one data value in the 70s and then no data until the 100s. The number of leaves increases from the 100s to the 110s to the 120s and then decreases again. Most of the values clump around the 120s. The 120s have the greatest number of leaf values.

**2c.** minimum: 78; maximum: 165; modes: 109, 119, 124, 126, 133, 139, and 145; median: 127.5

# Investigation 3

In this investigation, students find the mean for various sets of data, beginning with a pictorial exercise using whole numbers and progressing to calculating means that may contain decimals. They explore how changing the values or the number of values in a data set affects the measures of central tendency. They also create data sets that meet given parameters.

You may want to let students use calculators when finding the mean of a data set, even though most students will be able to do the necessary calculations mentally or using paper and pencil.

Tell students that mode and median are two measures of central tendency. Explain that they will learn about a third measure, the mean, in this investigation.

1 **Problem Set F** **Suggested Grouping: Pairs**
Students explore mean in a pictorial situation in this problem set. This approach helps students understand the concept of mean as the idea of equally sharing a given amount.

2 Students may want to discuss whether this plan for dividing the berries is fair. Some may think that it is not fair since a person may not get to keep all of the strawberries he or she picked. Others may think it is fair because it accounts for physical differences like a sore back.

3 **Problem Set Wrap-Up** Have students discuss how they found their answers to **Problem 3.** Some students may draw pictures or use objects to help them understand the problem. Be sure they understand the mathematical basis for finding the answer. The definition for finding a mean is on page 371.

# Investigation 3 The Meaning of *Mean*

The median and the mode are two ways to describe what is typical, or average, about a set of data. These values are sometimes referred to as *measures of central tendency*, or simply *measures of center*, because they give an idea of where the data values are centered. In this investigation, you will explore a third measure of center: the *mean*.

1 You may have students work in pairs.

## Problem Set F

**1.** Althea's scouting troop went strawberry picking. They decided to divide the strawberries they picked equally, so each girl would take home the same amount. The drawing shows how many quarts each girl picked.

How many quarts did each girl take home? How did you find your answer?

**1.** 6; Possible explanations: I found the total number of quarts and divided by 6. *Or,* I redistributed the quarts from the larger piles to the smaller piles until all the piles had the same number.

2 Discuss whether the plan for dividing berries is fair.

**2.** Another group of friends went blueberry picking. If they divided their berries equally, how many quarts did each friend get? 8.5

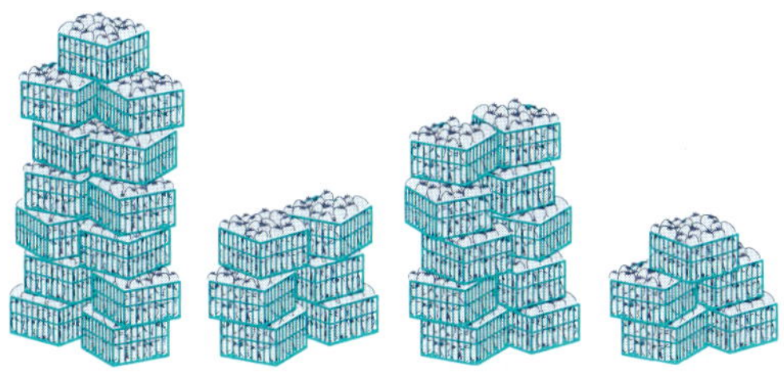


**3.** A group of 10 friends picked the following numbers of quarts of blackberries:

5 10 4 5 7 9 9 6 8 7

If they divided the blackberries equally, how many quarts did each friend get? 7

3 **Problem-Solving Strategies**

- Draw a picture
- Make a model

**Just the facts**

Strawberries were originally known as "strewberries" because they appeared to be strewn among the leaves of the strawberry plant.

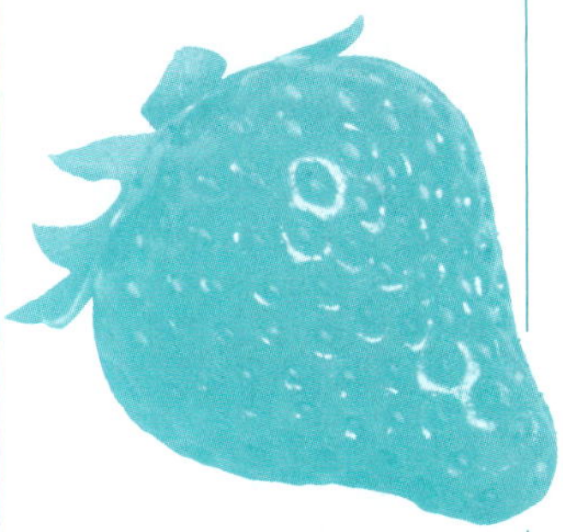

# Develop

1 Introduce the term *mean,* relating it to the redistribution, or sharing, of quarts of strawberries in Problem Set F. Emphasize that the mean is computed by adding the values and dividing the sum by the number of values, if students have not made this observation. You may want to provide students with another example. If so, have them find the mean for this data set: 2, 4, 6, 8. 5.

2 Stress that the word *average* often refers to the mean when used in everyday language but that, mathematically, the term can refer to any of the measures of central tendency. This can be a difficult distinction for students to make.

3 **Problem Set G** **Suggested Grouping: Pairs**
This problem set focuses on how changing values in a data set can affect the different measures of central tendency.

In **Problem 1b,** students should see that *adding* new data at the high end of the range results in a new data set with a greater median and a greater mean than the median and mean of the original data set. You may want to point out that students can use their prior calculations to find the new mean by adding the values to the prior sum and then dividing the new sum by the new number of values.

In **Problem 1c,** students see that *substituting* values at the high end of the range will not affect the median but does increase the mean.

**Problem 2** asks students to determine what value is added to a data set when they know all the other values in the data set and either the mean or the median. They should find the exact value of a missing data item when given the mean but not when given the median.

4 **Problem Set Wrap-Up** Have students generalize about how adding a great value to a data set affects the mean and the median. Then discuss their answers to Problems 2a and 2b. You may want to have students predict the effect of a value substituted at the low end of the range. They should see that the median would remain the same but the mean would decrease.

VOCABULARY
**mean**

In each situation in Problem Set F, you redistributed the quarts to give each person the same number. The result was the *mean* of the number of quarts picked. The **mean** of a set of values is the number you get by distributing the total evenly among the members of the data set. You can compute the mean by adding the values and dividing the total by the number of values.

The mean is another measure of the typical, or average, value of a data set. In everyday language, the word *average* is often used to mean *mean*. However, it is important to remember that the mean, median, and mode are *all* types of averages.

1 Introduce the term mean.

2 Remind students that average can refer to any of the measures of central tendency.

## Problem Set G

3 You may have students work in pairs.

**1.** The astronomy club is selling calendars to raise money to purchase a telescope. The 10 club members sold the following numbers of calendars:

3 5 7 10 5 3 4 6 9 8

**a.** Find the mean, median, and mode of the numbers of calendars the club members sold. mean: 6; median: 5.5; modes: 3 and 5

**b.** Suppose two very motivated students join the club. One sells 20 calendars and the other sells 22. Find the new mean, median, and mode. mean: 8.5; median: 6.5; modes: 3 and 5

**c.** In Part b, suppose that instead of 22 calendars, the 12th club member had sold 100 calendars. What would the new mean, median, and mode be? mean: 15; median: 6.5; modes: 3 and 5

**d.** How does the median in Part c compare to the median in Part b? How do the two means compare? Explain why your answers make sense.

**1d.** The medians are the same. The mean in Part c is greater. The median is the middle value when the numbers are listed in order. In both Part b and Part c, two numbers are added to the end of the list; the size of the numbers does not affect which number is in the middle of the list. The mean is calculated by adding the numbers and dividing by 12. The greater the numbers added, the greater the total and therefore the greater the mean.

**2.** Luke asked 12 students in his class how many books they had read (other than school books) in the past six months. Here are the responses given by 11 of the students:

3 5 7 10 5 3 4 6 9 8 20

**a.** Suppose you know that the mean number of books read by the 12 students is 10. Is it possible to find the number of books the 12th student read? If so, explain how. If not, explain why not. See below.

**b.** Suppose you know that the median number of books read by the 12 students is 5.5. Is it possible to find the number of books the 12th student read? If so, explain how. If not, explain why not. No; any value of 5 or less would give a median of 5.5.

4 Have students generalize how adding high or low numbers to the list affects the mean and median.

**2a.** yes; Possible explanation: The sum of the 11 numbers is 80. To get a mean of 10, the total divided by 12 must be 10. Since 120 ÷ 12 = 10, the total must be 120, so the 12th person must have read 40 books.



### Additional Answers

**1d.** The medians are the same. The mean in Part c is greater. The median is the middle value when the numbers are listed in order. In both Part b and Part c, two numbers are added to the end of the list; the size of the numbers does not affect which number is in the middle of the list. The mean is calculated by adding the numbers and dividing by 12. The greater the numbers added, the greater the total and therefore the greater the mean.

LESSON 6.2 What Is Typical? 371

# Develop

1 **Problem Set H** **Suggested Grouping: Pairs**
This problem set provides more practice in thinking about how adding values to a data set affects the mean and the median of the set. Students also create data sets that have given characteristics.

All but the first problems have many possible answers. To emphasize this fact, you could have volunteers write their solutions on the board and have the rest of the class check the solutions. When it is clear that there are many correct solutions, challenge the class to explain why this is true.

### Access for all Learners

**For Early Finishers** Challenge students to choose a description from **Problems 2–5** and create three data sets that fit the description. One data set should have no mode, one should have one mode, and one should have two or more modes.

## Share & Summarize

2 **Question 1** asks students to explain what each measure tells about a data set. This is a chance to check on students' understanding of the meaning of the terms and the differences between the mode, the median, and the mean.

3 **Question 2** may be too abstract for some students. Encourage them to create a data set to check their answers.

**Troubleshooting** If students have difficulty finding the mean of a data set, have them use the make a model problem-solving strategy by using manipulatives such as cubes or counters to model the situation. For example, they could use counters to show each number in the data set. Then they could either move the counters from one stack to another until all the piles are equal, or they could put all the counters in one pile and distribute them into equal stacks so that the number of stacks is the same as the number of pieces of data. Of course, this works best when the mean is a whole number. They can also use the problem-solving strategy of drawing and using a picture as was done in Problem Set F.

If students seem to have difficulty articulating the differences between the median and the mean, the next investigation may clarify the issue, as it focuses in depth on using the mean and the median to describe and compare groups.

### On Your Own Exercises

Practice & Apply: 6, 7, p. 382
Connect & Extend: 16, 17, p. 387
Mixed Review: 20–39, p. 389

## Problem Set H

1. At her party, Ines had a contest to see who could pick up the most jelly beans in a handful. The nine people at her party reached into a bag and pulled out these numbers of jelly beans:

   22 23 24 28 32 32 35 37 37

   **a.** Find the mean and median of the data set. mean: 30; median: 32

   **b.** Add two values to the data set so the median remains the same but the mean decreases. Give the new mean.

   **c.** Start with the original data set. Add two values to the set so the median remains the same but the mean increases. Give the new mean. Possible answer: 32, 60; mean: $32.\overline{90}$

   **d.** Start with the original data set. Add two values to the set so the mean remains the same but the median changes. Give the new median. Possible answer: 30, 30; median: 30

Create a data set with 10 values that fits each description below.

2. The minimum is 45, the maximum is 55, and the median and mean are both 50.
3. The minimum is 10, the maximum is 90, and the median and mean are both 50.
4. The range is 85, the mean is 50, and the median is 40.
5. The range is 55, the mean is 40, and the median is 50.

**1b.** Possible answer: 10, 33; mean: $28.\overline{45}$

**2.** Possible answer: 45, 46, 47, 48, 49, 51, 52, 53, 54, 55

**3.** Possible answer: 10, 20, 30, 40, 50, 50, 60, 70, 80, 90

**4.** Possible answer: 10, 15, 20, 25, 40, 40, 80, 85, 90, 95

**5.** Possible answer: 10, 15, 20, 25, 50, 50, 50, 55, 60, 65

1 You may have students work in pairs.

## Share & Summarize

1. Jing said that the students in her class have an average of three pets each.

   **a.** If Jing is referring to the mode, explain what her statement means. More students have 3 pets than any other number.

   **b.** If Jing is referring to the median, explain what her statement means. Half of the students have 3 or more pets and half have 3 or fewer.

   **c.** If Jing is referring to the mean, explain what her statement means. See below.

2. Suppose you have a data set for which the mean and median are the same. If you add a value to the set that is much greater than the other values in the set, would you expect the median or the mean to change more? Explain. See Additional Answers.

2 Check students' understanding of the differences of mode, median, and mean.

3 Challenge students to check their answers by creating a set of data.



### Additional Answers
### Share & Summarize

**1c.** Possible answer: If you took all the pets and distributed them evenly among the students, each student would get 3 pets.

**2.** The mean; the median is the middle number in an ordered list (or the value halfway between the two middle numbers). When you add a number to the end of the list, it doesn't matter how large the number is, the median may change slightly. However, adding a number much greater than the others would cause a big change in the sum of the data values, so it would also cause a big change in the mean.

# Investigation 4

In this investigation, students look at measures of central tendency and determine what facts can be ascertained from a data set. They explore the impact of outliers on each of the measures. They also consider how different measures can be used to influence decision making.

Introduce students to the title of the investigation, and read aloud the opening paragraph.

1

## Problem Set I Suggested Grouping: Small Groups

This problem set may be challenging for many students. These kinds of problems are often best approached by having students discuss the problems in small groups and then share their findings with the whole class.

2 Encourage students to make arguments and give examples to explain why the claims might be true or false. Making these kinds of arguments may not be easy for some students. Two important ideas are that one counterexample will disprove a claim that something is always true and one positive example proves that something can sometimes be true. Proving that something is always true cannot be done by example unless the examples are exhaustive. Thus it needs to be shown using logical reasoning.

While it is not expected that all students will be experts at this kind of argumentation, this is an opportunity for students to practice this kind of thinking. It is often useful in these kinds of contexts to continually challenge students who are not participating by asking questions about the evidence. For example, you might ask "Does Susan's example prove that this claim is valid?"

**Problem 1** focuses on conclusions that can be derived about the two data sets when given their medians.

### Access for all Learners

**Language Diversity** Students who think that the statement in **Problem 1b** can be true or false may be misinterpreting the word *must*. Point out that *must* means the action will always happen, not that the action may or may not happen. Thus, by definition, this statement is definitely false because the shortest student can be in either class.

**Problem 2** involves conclusions that can be drawn about the two data sets when given their means.

# Investigation 4 Mean or Median?

This investigation will help you better understand what the mean and median reveal about a set of data.

## Problem Set I

1 You may have students work in small groups.

2 Problem-Solving Strategy
Use logical reasoning

**1.** Lee and Arturo collected the heights of the students in their math classes. They found that the median height of students in Arturo's class is greater than the median height of students in Lee's class.

Tell whether each statement below is *definitely true,* is *definitely false,* or *could be true or false* depending on the data. In each case, explain why your answer is correct. (Hint: For some statements, it may help to create data sets for two small classes, with three or four students each.)

**a.** The tallest person is in Arturo's class. See Additional Answers.

**b.** Lee's class must have the shortest person. See Additional Answers.

**c.** If you line up the students in each class from shortest to tallest, each person in Arturo's class will be taller than the corresponding person in Lee's class. (Assume the classes have the same number of students.) See Additional Answers.

**d.** If you line up the students in each class from shortest to tallest, the middle person in Arturo's class would be taller than the middle person in Lee's class. (Assume the classes have the same odd number of students.)

**1d.** Definitely true; this is the definition of *median.*

**2.** Marta and Grace collected height data for their math classes. They found that Marta's class has a greater mean height than Grace's class.

Tell whether each statement below is *definitely true,* is *definitely false,* or *could be true or false* depending on the data. In each case, explain how you know your answer is correct.

**a.** The tallest person is in Marta's class. See Additional Answers.

**b.** Grace's class must have the shortest person.

**c.** If you line up the students in each class from shortest to tallest, each person in Marta's class will be taller than the corresponding person in Grace's class. (Assume the classes have the same number of students.) See Additional Answers.

**d.** If you line up the students in each class from shortest to tallest, the middle person in Marta's class would be taller than the middle person in Grace's class. (Assume the classes have the same odd number of students.) See Additional Answers.

**2b.** definitely false; Possible explanation: Marta's class could have heights 55, 65, 65, 65 and Grace's class could have heights 57, 57, 57, 57. Marta's class would have the greater mean and the shortest person.



### Additional Answers

**1a.** could be true or false; Possible explanation: The median is simply the middle value. It tells you nothing about the greatest value. Consider these two cases, for heights in inches:

- Case 1: Arturo's class: 50, 51, 52; Lee's class: 48, 50, 58
- Case 2: Arturo's class: 50, 51, 58; Lee's class: 48, 50, 52

In both cases, Arturo's class has the greater median. In Case 1, Lee's class has the tallest person. In Case 2, Arturo's class has the tallest person.

**1b.** definitely false; Possible explanation: Arturo's class could have heights 57, 60, 64 and Lee's class could have heights 58, 59, 61. Arturo's class would have the greater median and the shortest person.

• Additional Answers continued on page A659

# Develop

1 Have students read the statistics in the illustration. Remind them that an average can be any of the measures of central tendency and that the average that is reported depends on other factors, such as the desire to support or emphasize a particular outcome or opinion. Sometimes one measure can be more "typical" than another and provide a better description of the data set. Tell students they will consider these kinds of other factors as they make decisions in Problem Set J.

2 **Problem Set J** **Suggested Grouping: Small Groups or Pairs**

In this problem set, students look at how a large outlier can have a greater impact on the mean than on the median or the mode of a data set. They also consider which measure to use to describe a data set.

The salaries include one graduate making $1,000,000 a year. Although most graduates do not have starting salaries of this magnitude, this graduate could possibly be a sports figure, an entertainer, or a software developer. If your students interpret this as meaning that 1 in 20 graduates command this starting salary, you might want to discuss the rarity of this event.

Another grouping alternative is to have students work in the same groups they were in for the other problem sets involving analysis for Data Inc.

3 Students' answers to **Problem 1b** should consider only the data, not the purpose for which the average will be used. The purpose should be considered in Problem 3 when students make their recommendations.

Books, news reports, and advertisements often mention average values.

You have learned about three types of average: the mode, the median, and the mean. The average reported in a particular situation depends on many factors. Sometimes, one measure is "more typical" than the others. Other times, a measure is selected to give a particular impression or to support a particular opinion.

1 Remind students that average can refer to any of the measures of central tendency.

## Problem Set J

2 You may have students work in small groups or pairs.

Career Connections is a small company that helps college graduates find jobs. They are creating a brochure to attract new clients and would like to include the average starting salary of their recent clients. They have asked Data Inc. to help them determine which type of average to use.

Listed below are the starting salaries of the clients they have helped in the past three months.

| | | | | |
|---|---|---|---|---|
| \$30,000 | \$25,000 | \$60,000 | \$40,000 | \$25,000 |
| \$50,000 | \$70,000 | \$50,000 | \$25,000 | \$60,000 |
| \$25,000 | \$1,000,000 | \$60,000 | \$25,000 | \$40,000 |
| \$50,000 | \$25,000 | \$50,000 | \$25,000 | \$25,000 |

**1.** You know how to compute three types of averages: the mode, the median, and the mean.

3 Have students only consider the data.

**a.** Find the mode, median, and mean for these data.

**b.** Which average do you think best describes a typical value in this data set? Explain. Possible answer: The median; most of the data cluster around the median. The mode is the lowest salary, and since there are many greater values, it does not seem very typical. The very high salary of \$1,000,000 makes the mean much higher than most of the other values.

**1a.** mode: \$25,000; median: \$40,000; mean \$88,000

# Develop

1 The term *outlier* is introduced in **Problem 2.** You may want to reiterate that the outlier in this problem is much greater than the other values, but an outlier can also be a value that is much less than the other values in a data set.

> **Access for all Learners**
>
> **For Early Finishers** Ask students which measure—mode, median, mean—is used to describe the typical income in the United States. They may use their results from this problem set to conclude that the median is a better average to use because a few high or low salaries will not distort the description as may be the case if the mean is used.

2 **Problem Set Wrap-Up** Discuss students' conclusions from **Problem 3.** The variety of possible answers and the reasoning behind students' choices can help others broaden their thinking.

Use the experience to point out that different points of view can be supported by the same set of data.

3 **Problem Set K** **Suggested Grouping: Pairs**
In this problem set, student analysts must use their knowledge of statistics to support two opposing contentions.

4 Some students may question whether the Wolves' player actually played in the game in which she scored 0 points. If they think she did not play, they may choose to disregard this score when finding the measures of central tendency to support their arguments.

5 **Problem Set Wrap-Up** Have students share their arguments for each problem. You might extend the discussion to include the attributes that make up an MVP (most valuable player) and how people's differing opinions influence the measures used to identify and select an MVP.

VOCABULARY
**outlier**

**2a.** mode: $25,000; median: $40,000; mean: $40,000

**2b.** The mode and the median stay the same, but the mean drops by $48,000.

**2.** One of the salaries, $1,000,000, is much greater than the rest. A value that is much greater than or much less than most of the other values in a data set is called an **outlier.** One Data Inc. analyst suggested that this outlier should not be included when determining the average salary.

**a.** Remove $1,000,000 from the data set, and recompute the mode, median, and mean.

**b.** How does removing the outlier affect the three measures of center?

**3.** Write a brief letter to Career Connections telling them what value you recommend they report as the average starting salary of their clients. Give reasons for your choice. Consider all the averages you have computed for the salary data, both including and not including the $1,000,000 salary. See Additional Answers.

**1** *Introduce the term outlier.*

**2** *Have students share their letters with the class since they may help other students.*

In the next problem set, you will use a single set of data to support two very different points of view.

## Problem Set K

**3** *You may have students work in pairs.*

The Hillsdale School District will hold its annual girls' basketball banquet next Friday night. The head of athletics will present an award to the best scorer in the two high schools. Below are the points per game earned for each school's best offensive player.

Points per game for Westside Wolves' best player: 30, 61, 10, 0, 28, 48, 55, 12, 23, 55, 6, 25, 39, 18, 55, 31, 30

Points per game for Eastside Eagles' best player: 22, 35, 12, 37, 19, 36, 39, 13, 13, 36, 11, 37, 13, 38, 21, 37, 35

**4** *Students may choose not to include the 0 since it may support their arguments.*

Each coach wants to be able to argue that his player deserves the award. Both coaches have come to Data Inc. for help.

**1.** Use your knowledge of statistics to argue that the Westside Wolves' player deserves the award.

**2.** Use your knowledge of statistics to argue that the Eastside Eagles' player deserves the award.

Possible answer: The median points per game for the Eagles' player is 5 points greater than the median points per game for the Wolves' player (Eagles: 35; Wolves: 30). The Eagles' player is also more consistent; most of her scores are clumped around the median.

**5** *Extend the discussion on the types of attributes that make up an MVP.*

### Just the facts

The first women's intercollegiate basketball game was played in 1896 in San Francisco. The game pitted Stanford University against the University of California at Berkeley. Stanford won the game by a score of 2 to 1. Male spectators were not allowed at the game.

**1.** Possible answer: The mean points per game for the Wolves' player is more than 4 points greater (Wolves: 30.9; Eagles: 26.7). The Wolves' player also has the greatest number of points scored in a single game.



### Additional Answers
### Problem Set J

**3.** Possible answers:

- I recommend reporting the median because it seems "most typical" and is not influenced by the outlier.
- I recommend reporting $40,000, the mean when the outlier is removed. This value seems to summarize the data quite well, and since it ignores the $1,000,000 value, it will not mislead potential clients.
- I recommend reporting $88,000, the mean with the outlier. Since this is the greatest measure of center, it will give the best impression to potential clients.

1 **Problem Set L** **Suggested Grouping: Individuals**
In this problem set, students construct data sets to fit certain criteria. If time is limited, you may want to have three or four students work on each part of **Problem 1** and share their measures with the class. This will allow students to quickly see that there are many ways to create a data set that fulfills a set of criteria.

### Access for all Learners

**For Early Finishers** Have students change a single value of the data set in **Problem 2** so that the mean changes from 3 to 4, from 3 to 5, and then from 3 to 6. Ask them to look for and describe the pattern in the increases and explain why the pattern occurs.

The values change by 5, 10, and 15 respectively. The pattern is that for each increase of 5 in the sum of the data values, the mean increases by 1. This is true because there are five data values, so the sum is divided by 5. If the sum increases by 5, the mean will increase by 1.

Then challenge students to find what the increase would be if one value were changed so that the mean changed from 3 to 13. Students should find that one value would have to increase by 50.

### Share & Summarize

Students' ability to describe these data sets in abstract terms is an indication of their mastery of the concepts of the measures of central tendency. For example, some students may conclude that when the mean is higher than the median, the data set *may* have some very large outliers.

2 **Troubleshooting** If students have difficulty working in the abstract, have them refer to data sets from prior problem sets or create data sets with a few values that meet the conditions in each question. For example, if the median and mean are the same, students can change one value in the set to see the effect on the measures.

### On Your Own Exercises

Practice & Apply: 8–10, pp. 382–383
Connect & Extend: 18, p. 387
Mixed Review: 20–39, p. 389

The next problem set will help you better understand what the mean and median tell you about the distribution of a data set.

## Problem Set L

1. Create a data set with eight values from 1 to 20 that fits each description. Give the median and mean of each data set you create.
   a. The median is greater than the mean.
   b. The mean is greater than the median.
   c. The mean and median are equal.
   d. The mean is 3 more than the median.
   e. The median is 3 more than the mean.
2. The data set 1, 2, 3, 4, 5 has a mean of 3. Change two values so the new data set has a mean of 4. Possible answer: 1, 2, 3, 5, 9

## Share & Summarize

1. Reports of the typical income of a city, state, or country often use the median rather than the mean. Why do you think this is so?
2. What might cause the mean of a data set to be much greater than the median?
3. What might cause the median of a data set to be much greater than the mean? See Additional Answers.
4. What might cause the median of a data set to be equal to the mean? Possible answer: The data values might be fairly evenly distributed and have no outliers.

**Problem Set L Answers**

**1a.** Possible answer: 2, 3, 4, 5, 10, 10, 10, 10; median: 7.5; mean: 6.75

**1b.** Possible answer: 2, 3, 4, 5, 6, 11, 12, 13; median: 5.5; mean: 7

**1c.** Possible answer: 7, 8, 9, 10, 10, 11, 12, 13; median: 10; mean: 10

**1d.** Possible answer: 10, 10, 10, 10, 12, 20, 20, 20; median: 11; mean: 14

**1e.** Possible answer: 1, 2, 3, 10, 12, 12, 12, 12; median: 11; mean: 8

**Share & Summarize Answers**

**1.** Possible answer: A few very large or small incomes can pull the mean up or down, resulting in a misleading "typical" income. The median is influenced very little by such values.

**2.** Possible answer: There might be some outliers that are much greater than the other values. It is also possible that the values below the median are fairly close to the median while the values above the median are not as close to the median.



**1** You may have students work individually.

**2** Have students use or create and test smaller data sets if they are having trouble.

**Additional Answers**
**Share & Summarize**

**3.** Possible answer: There might be some outliers that are much less than the other values. It is also possible that the values above the median are fairly close to the median while the values below the median are not as close to the median.

# Investigation 5

In this lesson, students work as analysts for Data Inc. to analyze data regarding safety armbands for skaters. Students measure their own forearms, collect and analyze the class data, add data from another class that is provided in the text, and finally analyze the combined data to determine an average.

1 One way to introduce the lesson is to have students read the initial letter to Data Inc. from SportSafe. Then explain that the analysts need to collect data, make conclusions about the data, and present their findings to SportSafe. Reassure students that they will be guided through the process as they work through the problem sets.

Ask students how they think a user would put on the armband. Be sure they understand that it would slip over the thumb and up the arm, ending just above the elbow.

2 **Problem Set M** **Suggested Grouping: Small Groups**

In this problem set, students measure their forearms and record the measures in a class table. They analyze the class's results and prepare a short report for SportSafe.

3 It is important that everyone understand what part of the arm they should be measuring so that all results are consistent. You can use the drawing in the problem set to point out that the forearm measurement extends from just above the elbow to just below the top set of knuckles. You may want to model how to measure the arm by either having a student measure your forearm or measuring a volunteer's forearm.

One way to be sure students have a copy of the class data is to provide each student with a class roster. Write students' names on the board or on a transparency in the order they are listed on the class roster. It will also help students determine gender as requested in On Your Own Exercise 11, if assigned. If you do not want to use student names, have students write B or G next to their measure to note the gender, or make two tables—one for boys' forearm lengths and one for girls'.

Another alternative is to have students write their forearm measures and genders on sticky notes. The sticky notes can be arranged on the board to help students find the mode, median, and range or organized as a line plot. This makes some of the problem set a class activity, but groups can still prepare a short report analyzing the data.

**Tips from Teachers**

"My students find it easiest to measure their forearm lengths by putting their arms on the table with their thumbs pointing up. Partners put the ruler down on the table next to the forearm with one end just past the elbow. They read the measurement at the point just below the knuckles."

The report in **Problem 2** can simply be a rough outline of the information that would be included in a formal presentation.

4 **Problem Set Wrap-Up** Discuss the data and the answers to Problem 2, focusing on what information will be most useful in determining the size or sizes of armbands that should be manufactured and how Data Inc. should present the information. Encourage students to think about exactly what the typical measure will tell SportSafe and whether that measure will provide enough information for them to make their decision.

Students may have fewer ideas about displaying the information, particularly if they are uncertain about what information will be most helpful. Discuss various ways to display data and the strengths and weaknesses of each. For example, a line plot would show all the different forearm lengths and the number of students with each so SportSafe would have an idea of the distribution of the data. A histogram would show how many students have forearms with lengths that fall within particular ranges. A bar graph may be less helpful if there are numerous bars making it difficult to assess the data.

You may also want students to identify any outliers.

If students suggest separating the boys' results from the girls' results, confirm that this is a good idea that is addressed in On Your Own Exercise 11.

# Investigation 5 Analyzing Data

Data Inc. has just received a letter:

Dear Data Inc.,

We are a sports company specializing in protective gear for kids. We have an idea for a new product called the ForArm—an armband that would protect the area from the wrist to the elbow. Skaters and skateboarders would find this protective gear very useful. We would like to make a few samples to test with middle school students, but we need your help to determine how long middle school students' forearms are.

We would like you to start by analyzing the forearm lengths for students in your class. We are also gathering data from another middle school class and will send it to you as soon as it is available.

Thank you for your help.

Sincerely,
SportSafe

1 Introduce the lesson by having the students read the letter to Data Inc.

MATERIALS
*measuring tape or yardstick*

## Problem Set M Answers will vary.

2 You may have students work in small groups.

1. Measure the forearm of each member of your group to the nearest quarter inch. Measure from just below the knuckles to just above the elbow. Check your results by having two people do each measurement. Record the data for your group in a class table.

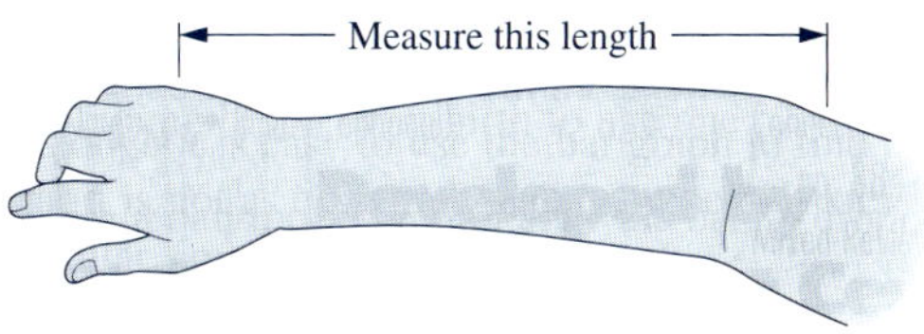


3 Make sure students are measuring from just above the elbow to just below the knuckles.

2. Work with your group to organize, analyze, and summarize your data. Prepare a short report of your findings. Include the range, mode, median, and mean, and any other information or graphs you think SportSafe would find useful.

Save your class data. You will need it for the On Your Own Exercises.

4 Have students think about whether the typical measure will provide enough information for SportSafe to make a decision.



LESSON 6.2 What Is Typical? 377

# Develop

1 Have students look at the new letter from SportSafe supplying data from another middle school class. Review the three analyses requested by the company: analysis of the new data, a comparison of the two data sets, and an analysis of the combined data.

Have students prepare their responses by working through Problem Set N.

**Just the facts**

A *vambrace* is a protective armband worn as part of a suit of armor in the 14th century.

Another letter just arrived from SportSafe.

Dear Data Inc.,

We just received the forearm-length data from the other middle school class.

Please send us a report that includes the following:

- An analysis of the data for this class
- A comparison of these data with the data for your class
- An analysis of the combined data for the two classes

Thank you for all your hard work.

Sincerely,
SportSafe

| Name | Forearm Length (inches) |
|---|---|
| Albert | $15\frac{3}{4}$ |
| Ali | $13\frac{1}{2}$ |
| Anna | 13 |
| Bianca | $13\frac{1}{2}$ |
| Jackie | $12\frac{3}{4}$ |
| Mariel | $12\frac{3}{4}$ |
| Michael | $13\frac{3}{4}$ |
| Nick | 14 |
| Olivia | 13 |
| Pablo | $13\frac{1}{2}$ |
| Powell | $13\frac{1}{2}$ |
| Sumi | $13\frac{1}{2}$ |
| Susan | $13\frac{3}{4}$ |
| Tomas | $13\frac{1}{2}$ |

1 Discuss the new letter with the class.

# Develop

1 **Problem Set N** **Suggested Grouping: Small Groups**

In this problem set, students analyze the additional data provided by SportSafe. They combine the two data sets and compare findings from the original and combined data sets. Then they decide which data set provides the most useful information to determine armband sizes.

In **Problem 1,** you may want to allow students to create any type of display or specify that *plot* refers to a line plot.

2 Encourage students to use the same tools and methods to analyze the combined data in **Problem 3** as they used to analyze the new class data in Problem 1.

**Problem 4** previews the idea that a larger sample usually provides more accurate data. The greater number of values will reduce the influence of outliers or other unusual occurrences that may skew the results of a smaller sample.

**Problem Set Wrap-Up** Discuss how students compared the data sets in Problems 2 and 3 and which results Data Inc. will report in Problem 4.

3 **Share & Summarize**

This is a good activity for pairs or individuals. Students will have to use their understanding of statistical measures to create the requested data set. Remind them that all values should be in multiples of $\frac{1}{4}$. You may want to have students share their strategies for creating their data sets.

**Troubleshooting** If students are still having difficulty creating data sets that meet specific statistical parameters, have them use one of the strategies presented when students discussed the Share & Summarize.

To provide more practice, you could ask students to create a data set with six values that has a median between 4 and 5. If necessary, demonstrate how to choose two middle values whose halfway value is a number between 4 and 5. Then have them add two lesser values and two greater values to complete the data set.

## On Your Own Exercises

Practice & Apply: 11, 12, p. 384
Connect & Extend: 19, p. 388
Mixed Review: 20–39, p. 389

## Problem Set N

1. Analyze the data SportSafe has sent. Find the mean, median, mode, and range, and create a plot of the data. See below.
2. Write a few sentences comparing data from your class with the data SportSafe collected. Answers will vary.
3. Combine your class data with the data sent by SportSafe. Analyze the combined data set, and describe how it is similar to and different from your class data. Answers will vary.
4. You have now analyzed three sets of forearm data: your class data, SportSafe's data, and the combined set of data. Which results do you think SportSafe should use to determine the sizes of their sample products? Why?

**4.** Possible answer: The results from the combined data; since the results are based on more data, they are more likely to represent what is typical of middle school students than the results based on either smaller data set.

## Share & Summarize

Answers will vary.

Create a set of forearm lengths containing at least 10 measurements, a range from 12 to 15, and a median between the medians you found for your class and for the SportSafe data. (If the medians of the two sets were the same, make the median of your new data set 1 inch greater.)

**Problem Set N Answer**

**1.** mean: about 13.55 in.; median: 13.5 in.; mode: 13.5 in.; range: 3 in. Possible plot:

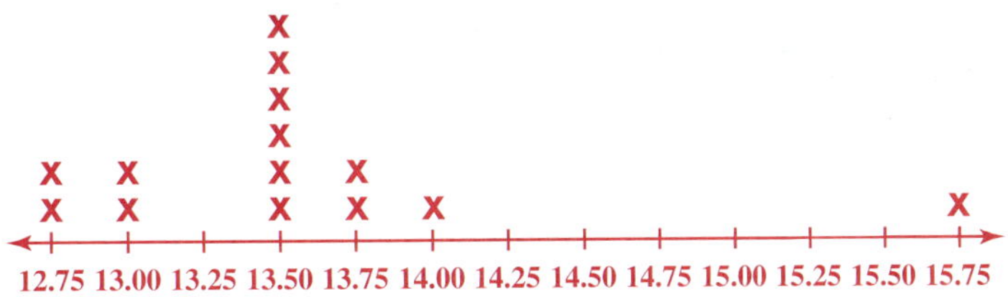




**1** You may have students work in small groups.

**2** Encourage students to use the same tools and methods to analyze the combined data.

**3** You may have students work in pairs or individually on the Share & Summarize.

LESSON 6.2 What Is Typical? 379

# On Your Own Exercises

## On Your Own Exercises

Investigation 1, pp. 362–365
Practice & Apply: 1–3
Connect & Extend: 13, 14

Investigation 2, pp. 366–369
Practice & Apply: 4, 5
Connect & Extend: 15

Investigation 3, pp. 370–372
Practice & Apply: 6, 7
Connect & Extend: 16, 17

Investigation 4, pp. 373–376
Practice & Apply: 8–10
Connect & Extend: 18

Investigation 5, pp. 377–379
Practice & Apply: 11, 12
Connect & Extend: 19

Assign Anytime
Mixed Review: 20–39

**Exercise 1:**
Students will need to recognize that **Parts a–c** use numerical data while **Parts d and e** use categorical data, although they are not asked to identify these classifications.

**Exercise 2:**
The data set has an even number of values, so you might want to have students explain how they found the median. Some students may simply state that the two middle values are 3, so the median must also be 3. Others may give a more detailed

## Practice & Apply

**1a, b.** See Additional Answers.

**1c.** range: $1\frac{1}{8}$; mode: $7\frac{1}{4}$; median: $7\frac{1}{4}$

**1.** The table shows the style and size of all the hats sold at the Put a Lid on It! hat shop last Thursday.

**a.** Make a line plot of the hat-size data.

**b.** Describe the shape of your line plot. Tell what the shape indicates about the distribution of the hat sizes sold.

**c.** Find the range, mode, and median of the hat-size data.

**d.** Find the mode of the hat-style data. cap and fedora

**e.** Is it possible to find the median and range of the hat-style data? If so, find them. If not, explain why it is not possible.
No; the data cannot be ordered, so there can be no least value, greatest value, or middle value.

| Style | Size |
|---|---|
| Cap | $6\frac{5}{8}$ |
| Beret | $7\frac{3}{8}$ |
| Fedora | $7\frac{1}{4}$ |
| Sombrero | 7 |
| Cap | $7\frac{1}{4}$ |
| Cap | $7\frac{3}{8}$ |
| Fedora | $7\frac{3}{8}$ |
| Cap | $7\frac{1}{4}$ |
| Beret | $6\frac{7}{8}$ |
| Panama hat | $7\frac{5}{8}$ |
| Fedora | $6\frac{7}{8}$ |
| Cap | $7\frac{1}{4}$ |
| Sombrero | $7\frac{1}{2}$ |
| Fedora | $7\frac{1}{2}$ |
| Fedora | $7\frac{1}{4}$ |
| Chef's hat | $7\frac{1}{8}$ |
| Beret | 7 |
| Derby | $7\frac{1}{8}$ |
| Beret | $7\frac{1}{8}$ |
| Top hat | $7\frac{3}{4}$ |
| Panama hat | $7\frac{3}{8}$ |

**2b.** Possible answer: The median; the mode is 0, the smallest value, and in most games Jing did get two or more hits. The median, 3, gives a much better idea of the typical number of hits.

**2. Sports** This list shows the number of hits Jing got in each softball game this season:

0 3 2 7 4 2 3 0 4 0 6 5 5 2 4 0

**a.** Find the mode and median of these data. mode: 0; median: 3

**b.** Do you think the mode or the median is a better measure of what is typical in this data set? Explain.

**3.** Create a data set with 13 values that satisfies these conditions: the minimum value is 3, the maximum value is 13, the mode is 4, and the median is 8. Possible answer: 3, 4, 4, 4, 4, 5, 8, 9 ,9, 10, 10, 12, 13



impactmath.com/self_check_quiz

## Additional Answers

**1a.**

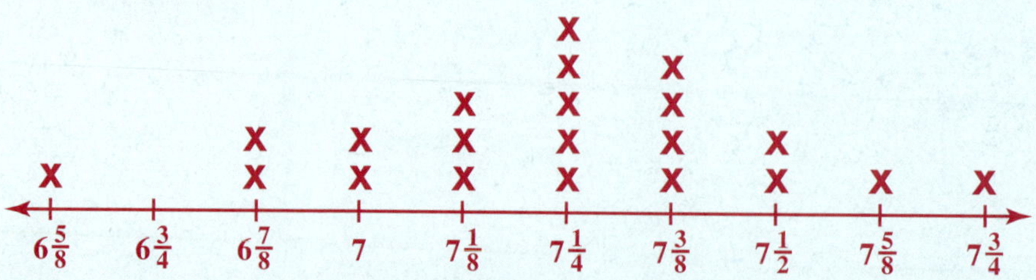


• Additional Answers continued on page A659

**4.** **Sports** Caroline had a bowling party for her birthday. Each person at the party bowled three games. The stem-and-leaf plot shows their scores.

**a.** What is the lowest score? What is the highest score? 52, 185

**b.** Find the mode and median scores.

**c.** Describe the distribution of scores.

**Bowling Party Scores**

| Stem | Leaf |
|---|---|
| 5 | 2 |
| 6 | |
| 7 | 3 7 |
| 8 | 1 6 |
| 9 | 2 9 |
| 10 | 0 3 |
| 11 | 0 1 4 4 4 9 |
| 12 | 3 3 |
| 13 | 2 6 6 |
| 14 | |
| 15 | |
| 16 | |
| 17 | |
| 18 | 5 |

**Key:** 9|2=92

**4b.** mode: 114; median: 111

**4c.** Possible answer: Although there is one low score of 52 and one high score of 185, most of the scores are in the 70s through the 130s. And, with the exception of the 11 stem (the 110s), the scores are fairly evenly distributed within this range. The 11 stem has the most values.

**5.** **Measurement** Two classes of elementary school students measured their heights in centimeters. Here are the results:

Ms. Cho's class: 117, 117, 119, 122, 127, 127, 114, 137, 99, 107, 114, 127, 122, 114, 120, 125, 119

Mr. Diaz's class: 130, 147, 137, 142, 140, 135, 135, 142, 142, 137, 135, 132, 135, 120, 119, 125, 142

**a.** For each class, make a stem-and-leaf plot of the height data. See below.

**b.** Find the range, mode, and median for each class.

**c.** The two classes are at two different grade levels. Which class do you think is the higher grade? Mr. Diaz's class

**d.** What percent of the students in Mr. Diaz's class are as tall or taller than the median height of the students in Ms. Cho's class? 100%

**5a.** **Ms. Cho's Class**

| Stem | Leaf |
|---|---|
| 9 | 9 |
| 10 | 7 |
| 11 | 4 4 4 7 7 9 9 |
| 12 | 0 2 2 5 7 7 7 |
| 13 | 7 |

**Key:** 11|4=114

**Mr. Diaz's Class**

| Stem | Leaf |
|---|---|
| 11 | 9 |
| 12 | 0 5 |
| 13 | 0 2 5 5 5 5 7 7 |
| 14 | 0 2 2 2 2 7 |

**Key:** 12|0=120

**5b.** Ms. Cho's class: range: 38; modes: 114 and 127; median: 119
Mr. Diaz's class: range: 28; modes: 135 and 142; median: 135

**Just the facts**

In the United States, bowling involves 10 pins. In Canada, 5-pin bowling is popular. In the Canadian game, the pins are smaller and the ball weighs only 3.5 pounds.



mathematical explanation such as $\frac{3+3}{2} = 3$.

**Exercise 5a:**
Point out that students should make two stem plots—one for each class. The plot for the data from Ms. Cho's class will have one- and two-digit stems.

*Quick Review* **Math Handbook**

hot words hot topics

Hot Topics pp. 190–201, 202–203

LESSON 6.2 What Is Typical? 381

# Assign and Assess

**Exercise 6c:**
Remind students that in mathematics, both the mean and the median are considered averages.

**Exercise 8:**
You might remind students that 0 is an outlier in the data set since it is much less than the other scores.

**6.** A scientist in a science fiction story finds the mass of alien creatures from four areas of the planet Xenon.

| Area of Xenon | Mass (kilograms) |
|---|---|
| Alpha | 6, 21, 12, 36, 15, 12, 27, 12 |
| Beta | 18, 36, 36, 27, 21, 48, 36, 33, 21 |
| Gamma | 12, 18, 12, 21, 18, 12, 21, 12 |
| Delta | 30, 36, 30, 39, 36, 39, 36 |

**a.** Find the range, mean, median, and mode of the masses for each area of Xenon. Round to the nearest tenth. See Additional Answers.

**b.** The scientist realizes he made a mistake. One of the 39-kilogram creatures in the Delta area actually has a mass of 93 kilograms. Compute the new mean and median for the Delta area.

**c.** Compare the original Delta mean and median to the mean and median you computed in Part b. Which average changed more? Explain why this makes sense. See Additional Answers.

**d.** The scientist realized that one value for the Alpha area is missing from the table. He doesn't remember what the value is, but he remembers that the complete data set has a mean of 19. What value is missing? 30

**6b.** mean: 42.9; median: 36

**7b.** Possible answer: 4, 40; mean: $12.\overline{18}$

**7c.** Possible answer: 10, 10; median: 10

**7.** Listed here are the number of unusual birds spotted by each member of a bird-watching club on a weekend excursion:

4 4 6 10 11 11 11 14 19

**a.** Find the mean and median of the data. mean: 10; median: 11

**b.** Add two values to the data set so the median remains the same but the mean increases. Give the new mean.

**c.** Add two values to the original data set so the median decreases but the mean remains the same. Give the new median.

**d.** Add two values to the original data set so both the mean and median stay the same. Possible answer: 5, 15

**8.** Lonnie earns money by tutoring students in algebra. Here are the scores his students received on their most recent algebra tests:

0 60 78 79 90 95 95

**a.** Lonnie claims that the students he tutored received an average score of 95 on their tests. Which measure of center is he referring to? Do you think 95 is a good measure of what is typical about these tests scores? Explain. See Additional Answers.



**Additional Answers**

**6a.** See below right.

**6c.** The mean; this makes sense because the new total is much greater than the original total, but you are dividing by the same number of values. The median didn't change because 39 and 93 are in the same position when the numbers are listed in order, so the middle value does not change.

**8a.** the mode; no; Possible explanation: 95 is the highest score. Since some of the test scores were much lower, and one of the scores was 0, reporting 95 as the average does not reflect how low some of the scores were.

**6a.**

| Area | Range | Mean | Median | Mode |
|---|---|---|---|---|
| Alpha | 30 | 17.6 | 13.5 | 12 |
| Beta | 30 | 30.7 | 33 | 36 |
| Gamma | 9 | 15.8 | 15 | 12 |
| Delta | 9 | 35.1 | 36 | 36 |

**8b.** mean: 71; median: 79

**8c.** mean: $82.8\overline{3}$ median: 84.5

**8d.** Possible answer: $82.8\overline{3}$ (the mean with the 0 dropped); this score takes into account both the high and low scores and gives a good idea about what is average.

**9a.** mean: 80; median: 80

**9b.** The median; no matter how Elsa does on her last test, she will get a grade of at least 79.

**9c, 10c.** See Additional Answers.

**b.** Find the mean and median of the test scores.

**c.** Lonnie said the score of 0 should not be counted when finding the average because the student didn't even show up for the test. Delete the 0, and find the new mean and median.

**d.** Which of the averages you computed in Parts b and c do you think best represents the typical test scores for the students Lonnie tutored?

**9.** Elsa received the following scores on her first four math tests this semester: 81, 79, 90, 70. There is one more test left. Elsa's teacher has told her she may choose to use her mean or her median test score as her final grade, but she must decide *before* she takes the final test.

**a.** Calculate Elsa's current mean and median test scores.

**b.** If Elsa is not confident she will do well on the final test, should she choose the mean or the median? Explain.

**c.** If Elsa is confident she will do well on the final test, should she choose the mean or the median? Explain.

**10.** **Sports** Alano and Kate are swimming instructors at the local recreation center. One day, both instructors asked their students to swim as many laps as they could. The results are shown in the table.

| Student | Instructor | Laps Swum |
|---|---|---|
| Lucinda | Kate | 7 |
| Jay | Alano | 15 |
| Guto | Kate | 9 |
| Deb | Alano | 11 |
| Ebony | Kate | 6 |
| Darius | Kate | 7 |
| Carlos | Alano | 9 |
| Carmen | Alano | 4.5 |
| Avi | Kate | 8 |
| Toku | Kate | 7 |
| Gil | Alano | 4 |
| Lana | Alano | 4 |

**a.** Find the range, mean, median, and mode for all 12 swimmers.

**b.** Find the range, mean, median, and mode for each instructor's students.

**c.** Alano said his students were stronger swimmers. Kate argued that her students were stronger. Use your knowledge of statistics to write two arguments—one to support Alano's position and one to support Kate's position.

**10a.** range: 11; mean: 7.625; median: 7; mode: 7

**10b.** Kate's students: range: 3; mean: $7.\overline{3}$; median: 7; mode: 7 Alano's students: range: 11; mean: $7.91\overline{6}$; median: 6.75; mode: 4

**Just the facts**

The first public swimming pool in the United States was built in Brookline, Massachusetts, in 1887. There are now more than 200,000 public swimming pools in the United States.



**Exercise 9:**
Some students may consider the grading scale when choosing which way to calculate the final grade and decide that the median is the best option. If 70 to 80 is a C on the grading scale, either option would give Elsa a C as long as she receives a score of at least 30 and less than 80. If her score is 80 or above, both the mean and the median will be at least 80 but less than 90, so she will receive a B. If her score is less than 30, she would receive a C if she selects the median but a D if she selects the mean.

## Additional Answers

**9c.** The mean; even if Elsa gets 100, her median score will be only 81. If she chooses the mean, a score of 85 or above will give her a final grade of 81 or higher.

**10c.** Possible answer:

Argument for Kate: Kate's group has a greater median than Alano's. It also has a smaller range, so Kate's group is more consistent, with most scores clumping near the median.

Argument for Alano: Alano's group has a greater mean. It also includes the two strongest swimmers. The strongest swimmer swam 15 laps, 6 more than Kate's strongest swimmer.

LESSON 6.2 What Is Typical? **383**

# Assign and Assess

**Exercise 11:** Students are asked to use the data collected in Problem Set M. As an alternative, you could have them analyze the data provided by SportSafe in the letter on page 378 (though their may be disagreement about the gender of some of the names).

**Exercise 13:** In **Part a,** students use data from a line plot to calculate the total number of people in 27 families. As they analyze the data, they may see that more can be learned from a graph of a data set than may be apparent at first glance. Watch for students who count the number of families (27) instead of the total number of people.

**Part b** focuses on the distinction between the size of a data item and its frequency. Students commonly confuse these two concepts.

**12.** Possible answer: She should buy four. For the six classrooms in the table, the mean is 3.7 and the median is 3.5, and only two classes have more than 4 left-handed students.

**11.** Divide your class's forearm data into two sets—one with the girls' measurements and one with the boys' measurements.

**a.** Organize, analyze, and summarize each set of data. Include a line plot of each set. Answers will vary.

**b.** Write a few sentences comparing the data for the girls with the data for the boys. Answers will vary.

**c.** If SportSafe wants to make different armbands for boys and for girls, what size recommendations would you make based on your class data? Answers will vary.

**12.** The table shows the number of left-handed and right-handed students in each homeroom class at Martin Middle School.

| Room Number | Left-handed Students | Right-handed Students |
|---|---|---|
| 101 | 3 | 27 |
| 102 | 4 | 26 |
| 103 | 2 | 28 |
| 104 | 5 | 25 |
| 105 | 2 | 29 |
| 106 | 6 | 23 |

The principal is buying new desks for a 30-student classroom. How many left-handed desks do you think she should buy? Use statistics to defend your answer.

## Connect & Extend

**13.** Zeke's class made a line plot showing the number of people in each student's family.

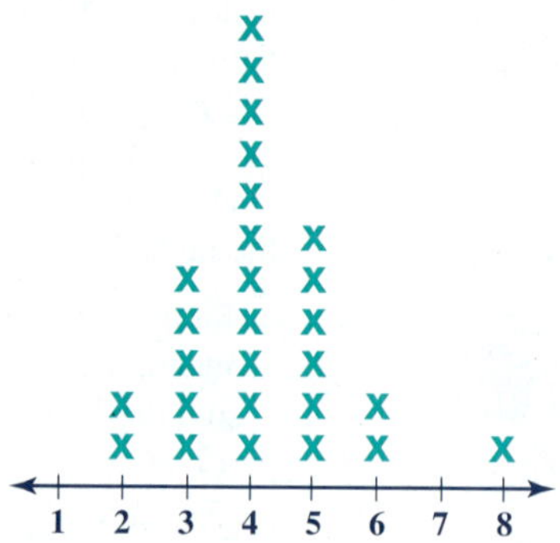


**13a.** 113

**a.** What is the total number of people in all the students' families?

**b.** Zeke said, "The plot can't be right! My family has 8 people. If I have the largest family, why is the stack of X's over the 8 the shortest one on the graph?" Answer Zeke's question.
An X over a number indicates that one person in the class has a family of that size. So, the single X over the 8 indicates that only one family (Zeke's) has eight people.

**14. Sports** The students in Consuela's gym class recorded how many times they could jump rope without missing. The results are shown in the table.

| Team | Name | Gender | Jumps |
|---|---|---|---|
| Group 1 | Jorge | male | 1 |
| | Felise | female | 1 |
| | Lana | female | 5 |
| | Sean | male | 7 |
| | Matt | male | 8 |
| | David | male | 11 |
| | Aaron | male | 16 |
| | Karen | female | 26 |
| | Brandon | male | 26 |
| | Enrique | male | 26 |
| | Emma | female | 40 |
| | Nicholas | male | 50 |
| | Shondra | female | 95 |
| | Selena | female | 300 |
| Group 2 | Lucas | male | 4 |
| | Colin | male | 4 |
| | Olivia | female | 4 |
| | Trent | male | 23 |
| | Lauren | female | 35 |
| | Tyrone | male | 48 |
| | Francisca | female | 68 |
| | Elsa | female | 83 |
| | Shari | female | 89 |
| | Kiran | male | 96 |
| | Meela | female | 110 |
| | Consuela | female | 138 |
| | Tino | male | 151 |

**a.** Group 1 claims they did better. Use what you have learned about statistics, along with any other information you think is useful, to write an argument Group 1 could use to support their claim.

**b.** Group 2 says they did better. Use what you have learned about statistics, along with any other information you think is useful, to write an argument Group 2 could use to support their claim.

**14a.** Possible answer: Group 1 had the best jumper. This person jumped 300 times—149 times more than the best jumper in Group 2! The mode for Group 1 is 26 while the mode for Group 2 is only 4. Clearly, Group 1 did better.

**14b.** Possible answer: It's obvious Group 2 did better. Three in Group 2 jumped more than 100 times, while only one in Group 1 jumped more than 100 times. The median value for Group 2 is 68, so more than half of them jumped 68 or more times. The median for Group 1 is 21, so half of them jumped 21 times or fewer. The mean for Group 2 is 65.6 while that for Group 1 is only 43.7. The range for Group 2 is 147 while the range for Group 1 is 299. This indicates that Group 2 was much more consistent.

**Just the facts**

*Double Dutch* jump roping involves skipping two ropes as they are swung in opposite directions. Since 1973, the American Double Dutch League has held an annual rope-jumping competition.

**Exercise 14:**
Students should recognize that both the mode and the median can be used as a measure of typicality. Sometimes one measure may provide users with a better understanding of what is happening in a data set. In this exercise, there are a number of arguments to support that Group 1 did better and a number to support that Group 2 did better. The maximum value in the data set supports Group 1's contention, but looking at the number of participants who jumped more than 100 times indicates that Group 2 did better. Not all of these mathematical ideas were introduced in the investigation, but creative students might make these kinds of arguments.

The sample answers give the mean for each data set. Some students may be familiar with this measure from prior classes. It is formally introduced in Investigation 3 of this lesson.



LESSON 6.2 What Is Typical? 385

**Exercise 15:** Back-to-back stem plots are introduced in this exercise. You might want to review this type of display with students, noting that there are places for leaves on both sides of the stem values. Point out that the leaves to the left of the stem are for the Yankees and are arranged in decreasing order from left to right while the leaves for the Mets, located to the right of the stem, are in increasing order.

**15a.** Possible answer: The Mets have a greater range of averages than the Yankees. The Mets have the lowest average and the Yankees have the highest. The Yankees have a greater median batting average (.260 versus .259). The Yankees have a greater mean batting average (.271 versus .257). The Yankees' batting average has a greater mode (.247 versus .254).

**15. Sports** You can create a *back-to-back stem plot* to display and compare two data sets. This plot shows the batting averages for the 2000 New York Yankees and the 2000 New York Mets.*

Leaf values for the Yankees are givento the left of the stem, and leaf values for the Mets are given to the right of the stem. For example, the leaves 7 and 5 in the eleventh row indicate a Yankee batting average of .297 and a Met batting average of .295.

**Yankees and Mets 2000 Batting Averages**

| Yankees | Stem | Mets |
|---|---|---|
| | .19 | 9 |
| | .20 | |
| | .21 | 5 |
| | .22 | |
| 2 | .23 | |
| 7 7 3 0 | .24 | 0 |
| | .25 | 4 4 9 |
| 8 0 | .26 | 0 6 |
| | .27 | |
| | .28 | 0 |
| 7 | .29 | 5 |
| 0 | .30 | 8 |
| 4 | .31 | |
| | .32 | |
| 3 | .33 | |

**Key:** .21|5=.215

**a.** Write a few sentences comparing the two teams' batting averages. Be sure to discuss the distribution of values and the mean and mode.

**b.** These data give the total number of home runs hit by American League and National League teams in 2002. Make a back-to-back stem plot comparing the number of home runs for the two leagues.* See Additional Answers.

**National League**

| Team | Home Runs |
|---|---|
| Arizona | 165 |
| Atlanta | 164 |
| Chicago | 200 |
| Cincinnati | 169 |
| Colorado | 152 |
| Florida | 146 |
| Houston | 167 |
| Los Angeles | 155 |
| Milwaukee | 139 |
| Montreal | 162 |
| New York | 160 |
| Philadelphia | 165 |
| Pittsburgh | 142 |
| San Diego | 136 |
| San Francisco | 198 |
| St. Louis | 175 |

**American League**

| Team | Home Runs |
|---|---|
| Anaheim | 152 |
| Baltimore | 165 |
| Boston | 177 |
| Chicago | 217 |
| Cleveland | 192 |
| Detroit | 124 |
| Kansas City | 140 |
| Minnesota | 167 |
| New York | 223 |
| Oakland | 205 |
| Seattle | 152 |
| Tampa Bay | 133 |
| Texas | 230 |
| Toronto | 187 |

*Source: *World Almanac and Book of Facts 2003.* Copyright © 2003 Primedia Reference Inc.

**c.** Write a few sentences comparing the numbers of home runs for the two leagues. Be sure to discuss the distribution of values and the mean and mode. See Additional Answers.

**In your own words**

Describe what a line plot and a stem plot are. Tell how you can find the median and mode of a data set from each type of graph.



**Additional Answers**

**15b.**
**Home Run Hits By League**

| National | Stem | American |
|---|---|---|
| | 10 | |
| | 11 | |
| | 12 | 4 |
| 9 6 | 13 | 3 |
| 6 2 | 14 | 0 |
| 5 2 | 15 | 2 2 |
| 9 7 5 5 4 2 0 | 16 | 5 7 |
| 5 | 17 | 7 |
| | 18 | 7 |
| 8 | 19 | 2 |
| 0 | 20 | 5 |
| | 21 | 7 |
| | 22 | 3 |
| | 23 | 0 |

Key: 19|2=192

**15c.** Possible answer: The mean for the National League is 162; the mean for the American League is 176. The median for the National League is 163; the median for the American League is 172. The mode for the American League is 152; the mode for the National League is 165. The American League has a much greater range of values, and it has both the least and greatest values. There is only one gap in the values for the National League; there are no gaps in the American League values. The National League has seven values in the 160s, while the American League has no more than two leaves for any stem.

17a. Possible answer: About half of the numbers are at the mean or above and about half are at the mean or below.

17b. It must be much greater than the other values. That would make the mean go up by a lot and the median by a little or not at all.

**Just the facts**

It's difficult to think of *The Wizard of Oz* without picturing Dorothy's ruby slippers. However, in the book on which the movie was based, Dorothy wore silver shoes, not ruby slippers.

18a, c, d. See Additional Answers.

18b. The mean and median are both 3 for all three movies.

**16.** If you have a data set that includes only whole numbers, which measures of center—mode, median, or mean—will *definitely be* whole numbers? Which measures of center *may or may not be* whole numbers? Explain your answers. See Additional Answers.

**17.** Suppose a data set has a mean and a median that are equal.

**a.** What must be true about the distribution of the data values?

**b.** Suppose one value is added to the set, and the new mean is much greater than the median. What must be true about the new value? Explain.

**c.** Now suppose you start with a new data set in which the mean and median are equal. You add one value to the set, and the new median is much greater than the mean. What must be true about the new value? See Additional Answers.

**18.** Emelia asked her friends to rate three movies on a scale from 1 to 5, with 5 being terrific and 1 being terrible.

**Movie Ratings**

| Friend | *Star Wars* | *The Sound of Music* | *The Wizard of Oz* |
|---|---|---|---|
| Adam | 2 | 1 | 4 |
| Ashley | 5 | 1 | 5 |
| Corey | 2 | 4 | 3 |
| Emelia | 4 | 4 | 2 |
| Eric | 4 | 4 | 4 |
| Hector | 3 | 5 | 5 |
| Ilene | 2 | 5 | 3 |
| Jay | 3 | 2 | 1 |
| Jose | 5 | 1 | 5 |
| Kareem | 3 | 3 | 2 |
| Karen | 2 | 5 | 3 |
| Lauren | 4 | 5 | 4 |
| Letonya | 3 | 5 | 4 |
| Lynn | 1 | 2 | 2 |
| Mai Lei | 4 | 3 | 3 |
| Maria | 3 | 2 | 1 |
| Michael | 3 | 1 | 1 |
| Peter | 1 | 1 | 2 |

**a.** For each movie, make a line plot of the friends' ratings.

**b.** Compute the mean and median rating for each movie.

**c.** Do you think reporting the means and medians is a good way to summarize the ratings for the three movies? Explain.

**d.** How would you summarize these data if you wanted to emphasize the differences in the ratings among the three movies? Explain why you would summarize the data this way.



**Exercises 16 and 17:** Encourage students to either refer to their work from the problem sets or create a data set to check that their observations are accurate. Students will need to be familiar with these kinds of generalizations as they work through Investigation 4.

**Exercise 17:** Students construct data sets to fit median, mode, and range criteria in this exercise. This is a preview to Investigation 4.

**Additional Answers**

**16.** The mode will be a whole number; the median and mean may not be. The mode is a value in the data set, so it will be a whole number. If the data set has an even number of values, the median will be halfway between the two middle values, so it is possible that it will not be a whole number (for example, if the two middle values are 10 and 13, the median will be 11.5). If the number of values does not divide evenly into the total of the values, the mean will not be a whole number.

**17c.** It must be much less than the other values. That would make the mean go down by a lot and the median by a little or not at all.

• *Additional Answers continued on page A660*

**Exercise 19c:**
You might review the data displays that students have studied this year, such as bar graphs, line graphs, circle graphs, line plots, histograms, and stem-and-leaf plots. Remind students that some displays better represent certain data sets than other displays.

**19a.** Possible answer: morning glories with peat: 0; zinnias with peat: 5 and 12; morning glories with topsoil: 4 and 15; zinnias with topsoil: 0

**19. Life Science** As their science project, Althea and Luke compared how two soil mixtures affected the growth of morning glories and zinnias. They planted some seeds of each type in a mixture of peat moss and sand and some in a mixture of top soil, compost material, and sand. They put all the plants by the same window and watered them at the same time with the same amount of water. After 20 days, they measured the height of each plant.

| Peat Moss and Sand | | Topsoil, Compost, and Sand | |
|---|---|---|---|
| Morning Glory Height (mm) | Zinnia Height (mm) | Morning Glory Height (mm) | Zinnia Height (mm) |
| 145 | 47 | 157 | 27 |
| 156 | 49 | 155 | 42 |
| 139 | 52 | 167 | 50 |
| 142 | 42 | 149 | 54 |
| 154 | 45 | 87 | 51 |
| 0 | 43 | 127 | 38 |
| 151 | 47 | 116 | 36 |
| 147 | 50 | 145 | 0 |
| 143 | 56 | 4 | 39 |
| 145 | 12 | 15 | 32 |
| 168 | 47 | 143 | 35 |
| 129 | 52 | 105 | 23 |
| 148 | 5 | 132 | 33 |

**19b, c.** See Additional Answers.

**a.** Each column contains data for one type of plant in one type of soil mixture. Althea and Luke decided to ignore the outliers in each column, reasoning that these seeds probably wouldn't grow well in any type of soil. For each column, tell which values you think are outliers.

**b.** Compute the mean, median, and mode for each column of data. Ignore the outliers.

**c.** For each column of data, make a display that shows how the values are distributed. Do not include outliers in your display.

**d.** Do you think the different types of soil affected the growth of the plants? Use your answers from Parts b and c, along with any other information you think is useful, to support your answer.
Possible answer: Yes; peat moss and sand seem to be better for both. For both plants, the height values for the topsoil mixture were more spread out than with the peat-moss mixture. With the peat moss, the heights were consistently high values. For both plants, both the mean and the median were greater with the peat-moss mixture than with the topsoil mixture.



**Additional Answers**

**19b.** Answers will depend on the values students considered to be outliers. For the outliers given in Part a, the statistics are as follows:
morning glories with peat: mean: 147.25; median: 146; mode: 145
zinnias with peat: mean: $48.\overline{18}$; median: 47; mode: 47
morning glories with topsoil: mean: $134.\overline{81}$; median: 143; mode: none
zinnias with topsoil mean: $38.\overline{3}$; median: 37; mode: none

• ***Additional Answers continued on page A660***

388 CHAPTER 6 Analyzing Data

## Mixed Review

Fill in the blanks.

**20.** 25% of 56 = 14

**21.** $33\frac{1}{3}$ % of 75 = 25

**22.** 80% of 200 = 160

**23.** 125% of 240 = 300

**24.** 85 % of 280 = 238

**25.** 1% of 30 = 0.3

Fill in each ● with <, >, or = to make a true statement.

**26.** $\frac{6}{5}$ ● 1.2 =

**27.** 0.37 ● $\frac{7}{20}$ >

**28.** $\frac{13}{18}$ ● $\frac{7}{10}$ >

**29.** 1.5 ● $\frac{25}{19}$ >

**30.** 0.0375 ● $\frac{3}{80}$ =

**31.** $\frac{23}{24}$ ● $\frac{24}{25}$ <

**Geometry** In Exercises 32–34, tell whether the given lengths could be side lengths of a triangle.

**32.** 3 cm, 3 cm, 3 cm yes

**33.** 3 in., 4 in., 7 in. no

**34.** 5 ft, 12 ft, 14 ft yes

**35.** Possible graph:

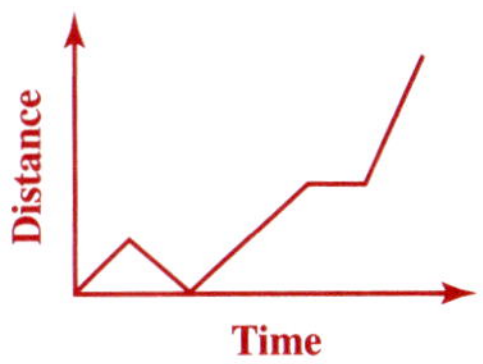


**35.** Jahmal left his apartment and started walking toward school. After walking a couple of blocks, he realized he had forgotten his homework. He walked back home to get it, and then started walking toward school again. He stopped to meet his friend Miguel and had to wait a few minutes while Miguel finished his breakfast. The two boys thought they might be late, so they ran from Miguel's house to the school.

Sketch a graph showing how Jahmal's distance from home might have changed from the time he first left his apartment.

Find the next three terms in each sequence.

**36.** 99, 98, 96, 93, 89, 84, . . . 78, 71, 63

**37.** 729, 243, 81, 27, 9, . . . 3, 1, $\frac{1}{3}$

**38.** 1, 1, 2, 6, 24, 120, 720, . . . 5,040; 40,320; 362,880

**39.** ❀, ▲, ❀, ▲, ▲, ❀, ▲, ▲, ▲, ❀, ▲, ▲, . . . ▲, ▲, ❀



**Exercises 32–34 Extension:**
Have students identify the types of triangles that would have sides with these lengths. The triangle in **Exercise 32** is acute and equilateral; the triangle in **Exercise 34** is obtuse and scalene.

### Quick Check

**Informal Assessment**
Students should be able to:

- ✔ find the mode, median, mean, and range of a data set
- ✔ make line plots and stem-and-leaf plots and use them to describe a data set
- ✔ predict how changing a value in a data set impacts the mean and median
- ✔ use various measures to interpret and compare data sets

• ***Quick Quiz on page A660***

LESSON 6.2 What Is Typical? 389

# Teacher Notes

# 6.3

# Collecting and Analyzing Data

## Objectives

- To prepare a plan that includes selecting the appropriate statistical measures as part of a research project
- To select a way to display data
- To follow a plan to reach a conclusion

## Overview (pacing: about 3 class periods)

In this lesson, students apply data analysis tools to complete a project about which activities middle school students participate in outside of school. The first investigation focuses on planning as students consider what information they need to help them answer six questions. They decide which statistical measures, mathematical calculations, and data displays they would use when reporting their findings. They also complete a form and collect data. Students implement their plans in Investigation 2.

These investigations require a great deal of work and may require more than the suggested number of days to complete.

The optional lab investigation guides students through using spreadsheet statistical functions to find the measures of central tendency. If you choose to have students complete the lab investigation, you might consider teaching it first and then have students use spreadsheets to record and analyze data from the project.

## Advance Preparation

You may want to provide students with copies of Masters 64, 65a, and 65b to record their nonschool activities and to analyze the combined class data. A transparency of Master 64 can help ensure that all students record the same changes to the form. Masters 65a and 65b can be taped together to make a form for all students in the class to record their data from the survey on Master 64. You can make copies of Masters 65a and 65b once students have entered in their results or you can display the form and have students copy the data.

If you elect to have students complete the lab investigation it is preferable that they all use the same spreadsheet software. If students will not use Microsoft® Excel, you may need to revise the directions given in the text. If possible, work through the investigation in a computer lab, although it can be done in other ways.

## Lesson Planner

| | Summary | Materials | On Your Own Exercises | Assessment Opportunities |
|---|---|---|---|---|
| **Investigation 1** page T391 | Students plan how they will collect, analyze, and present data for a research project. | • Master 64 (optional)<br>• Transparency of Master 64 (optional)<br>• Master 65a (optional)<br>• Master 65b (optional) | Practice & Apply: 1, p. 399<br>Connect & Extend: 3, p. 400<br>Mixed Review: 5–12, p. 401 | On the Spot Assessment, page T393<br>Share & Summarize, pages T393, 393<br>Troubleshooting, page 393 |
| **Investigation 2** page T394 | Students analyze data and present their findings. | • Master 66 (optional) | Practice & Apply: 2, p. 400<br>Connect & Extend: 4, p. 401<br>Mixed Review: 5–12, p. 401 | Share & Summarize, pages T395, 395<br>Troubleshooting, page T395<br>Informal Assessment, page 401<br>Quick Quiz, page 401 |
| **Lab Investigation** page T396 | Students use spreadsheet functions to analyze a set of data. | • Computer with spreadsheet software (1 per group) | | On the Spot Assessment, page T397 |

# Introduce

1 Introduce the lesson by asking students to name some of the ways they have learned to analyze data. They may mention measures of central tendency and various types of graphs. Then introduce the context of this lesson, pointing out that students will use some of the methods they mentioned to collect and analyze data for a magazine article about activities that middle school students participate in and enjoy.

Remind students that data must be collected before they can be analyzed. Have them read the six questions that the magazine editors would like to have answered.

Point out the form supplied by the magazine editors that is printed at the bottom of the page. Students will discuss this form in more detail in the Think & Discuss on page 391.

Lab Investigation

# Statistics and Spreadsheets

Analyzing data can be time-consuming, especially if you are working with a large data set. In this investigation, you will see how using a spreadsheet program can make analyzing a data set easier.

MATERIALS
computer with spreadsheet software (1 per group)

## The Situation

Conor read an article about how difficult it is to hold a book at arm's length with one hand for very long. He tried it himself and was amazed how hard it was.

Conor hypothesized that students who work out in the school weight room would be able hold a book longer than students who don't. To test his hypothesis, he recruited 20 students: 5 girls and 5 boys who work out and 5 girls and 5 boys who don't. He had each student hold a 7-pound book at arm's length, using the weaker arm, for as long as he or she could. He summarized the data in a table.

| Name | Work Out? | Time (seconds) |
|---|---|---|
| Abby | yes | 44 |
| Barbara | yes | 40 |
| Ben | no | 30 |
| Cassie | yes | 52 |
| Chris | yes | 49 |
| Eileen | yes | 58 |
| Grason | no | 20 |
| Jackie | no | 38 |
| John | no | 36 |
| Jon | yes | 45 |

| Name | Work Out? | Time (seconds) |
|---|---|---|
| Ken | yes | 18 |
| Kiran | no | 34 |
| Lee | yes | 55 |
| Liz | no | 25 |
| Michele | no | 30 |
| Roland | yes | 55 |
| Sara | no | 29 |
| Sydney | yes | 38 |
| Tia | no | 20 |
| Tomas | no | 28 |

**1.** What do you think Conor should do to determine whether students who work out did better than those who don't?

Possible answer: Separate the data into two groups—one group for students who work out and one for those who don't. Find the mode, median, and range for each group.



**1** You may have students work in pairs.

**2** You may want to review relevant computer skills.

**3** Problem-Solving Strategy
Make and test a conjecture

# Develop

1

### Set Up the Spreadsheet

Students who are familiar with spreadsheet software should have no trouble with this material, but you may want to do a brief review of spreadsheet basics before students begin. If you can project your computer screen on an overhead, you can model the process for students.

This first section focuses on entering the data and then sorting it into two groups—those who work out are designated with a Y and those who do not are designated with an N.

2

#### On the Spot Assessment

Watch for students who select too many cells when sorting data. They may include labels, select more than one column, or select too many rows. Remind students to check that they have the correct cells selected. Reassure them that this will be easier as they gain more practice on the computer.

#### Access for all Learners

**For Early Finishers** To help students strengthen their computer skills, suggest that they sort the data by time. This allows them to practice their sorting skills on a different column of data. Ask them to predict the order of the times before they actually perform the procedure. If students perform this sort before they finish the lab investigation, they will need to resort the data to group together students who worked out and students who did not work out.

### Find the Minimum and Maximum

3

These steps provide a straightforward description of how to calculate maximum and minimum values for a set of data. Students may need some support on the syntax of the commands. If they do make errors, this misunderstanding will probably be the source. You may want to stress that although students can easily pick out the maximum and minimum values for this data set by scanning the table on page 396, they might prefer using the spreadsheet when there are many pieces of data.

Some students may need more guidance on how to enter formulas. You may want to point out where students should enter the formula and review how to identify the cells.

If you are working in a lab, you can circulate around the room and check students' results as they work through the problems.

## Set Up the Spreadsheet

Conor set up a spreadsheet to help him analyze the data. He decided not to include the students' names. Here are the first few rows of his spreadsheet:

| | A | B | C | D |
|---|---|---|---|---|
| 1 | Work Out? | Time (s) | | |
| 2 | Y | 44 | | |
| 3 | Y | 40 | | |
| 4 | N | 30 | | |
| 5 | Y | 52 | | |

1 Review some spreadsheet basics before you start.

Set up your spreadsheet like Conor's. Enter the column heads shown, and then enter the data.

The results will be easier to analyze if you divide the data values into two groups—data for students who work out and data for students who don't. You can do this using the Sort command.

- Highlight all the cells containing data. (Don't select the column labels.)
- Choose the Sort command.
- Indicate that you want the data sorted according to the values in Column A (the Y or N values) and that you want the values sorted in ascending order. (*Ascending* sorts from least to greatest or, in this case, from A to Z.)

2 Have students make sure they have the correct cells selected before they sort the data.

**2.** Describe what happened to the data.

**3.** Choose Sort again, but this time choose to sort the data values in descending order. Describe what happened to the data.

**2.** The data are sorted so all the N's come first, followed by all the Y's.

**3.** The data are sorted so all the Y's come first, followed by all the N's.

## Find the Minimum and Maximum

You can use a spreadsheet to find the minimum and maximum values in a data set. For this data set, you can probably find these values fairly quickly by looking at the table. However, when you have a large data set, it is much more efficient to have a spreadsheet do it for you.

You can tell the spreadsheet to display the minimum value in any empty cell. In this example, you will put the result in Cell C2. First, enter the column head "Min. Time" in Cell C1. In Cell C2, type the following:

=MIN(B2:B21)

The = sign tells the spreadsheet that the entry is a formula to evaluate. MIN means "find the minimum value." B2:B21 tells the spreadsheet to search for the minimum value in Column B from Cell B2 to Cell B21.

| | A | B | C | D |
|---|---|---|---|---|
| 1 | Work Out? | Time (s) | Min. Time | |
| 2 | Y | 44 | =MIN(B2:B21) | |

3 Help students with the syntax of the formulas



LESSON 6.3 Collecting and Analyzing Data 397

For **Question 6,** point out that students need to use different cells in the formula when finding the maximum and minimum values for students who work out and for those who do not work out.

### Find the Median and Mean

If students are not using Excel software, they may not have a MEDIAN function.

1 Remind students that mathematicians use the word *average* to refer to any measure of central tendency. Point out that Excel uses the everyday meaning of *average* and it refers only to the mean.

Students may need assistance in determining the formulas needed in **Questions 8 and 10.**

### Back to the Situation

2 In **Questions 11 and 12,** it is vital that students recognize that although Conor's hypothesis is supported by the data from this group and makes sense, a small sample may not provide enough data to make an accurate generalization. You may want to use presidential elections as an example of small samples giving inaccurate information. Newspapers used the results of telephone interviews to incorrectly predict Wendell Willkies presidential win over Franklin D. Roosevelt in 1940. More recently, predictions regarding the 2000 presidential election first declared Al Gore as the winner of the Florida electoral votes, then George W. Bush, before finally deciding that the vote count was too close to call from samples.

### What Did You Learn?

3 You may want to consider whether you want students to provide additional data to test the hypothesis in **Question 13.** If some students may be uncomfortable participating, you could ask for volunteers. Groups could use their data to perform the computer calculations.

4. After you enter the formula and press Return, what value appears in Cell C2? 18

Enter the column head "Max. Time" in Cell D1. In Cell D2, use the MAX command to calculate and display the maximum time.

5. What is the maximum time for these data? 58 s

6. Use your spreadsheet to find the minimum and maximum values for each group—for the students who work out and for those who don't.

**6.** workout group: min = 18, max = 58; non-workout group: min = 20, max = 38

## Find the Median and Mean

For most spreadsheets, the command for computing the median is MEDIAN. So, to compute the median of all the time values, you would enter =MEDIAN(B2:B21).

7. Use your spreadsheet to find the median of all the time values. 37

8. Now find the median time value for each group.

**8.** workout median: 47; non-workout median: 29.5

For most spreadsheets, the command for computing the mean is AVERAGE.

**1** Remind students that Excel uses the regular meaning of average.

9. Use your spreadsheet to find the mean of all the time values. 37.2

10. Now find the mean time value for each group.

**10.** workout mean: 45.4; non-workout mean: 29

## Back to the Situation

11. Do the data Conor collected support his hypothesis? Use what you know about statistics to justify your answer.

**11.** Possible answer: Yes; the median and the mean of the group of students who work out are greater than the median and mean of the group of students who don't.

12. Does Conor's experiment *prove* his hypothesis? Explain.

**12.** no; Possible explanation: Conor tested his hypothesis for a very small group of students. We cannot generalize from such a small sample.

**2** Discuss the fact that small samples may not provide enough data to make a generalization.

## What Did You Learn?

13. Have each student in your group hold a heavy book at arm's length for as long as he or she can. Record the times for your group in a class table. Enter the times for your class into a spreadsheet, and calculate the minimum value, maximum value, mean, and median of the data. Answers will vary.

**3** You may want students to provide additional data to test the hypothesis.

# Teacher Notes

# On Your Own Exercises



## Practice & Apply

In Exercises 1 and 2, use this information:

The editors of *All about Kids!* would like to publish an article about teenage Internet users. Here are the questions they would like to answer:

**Question 1:** How much time do teen Internet users typically spend on the Internet each week?

**Question 2:** How much time do they typically spend on various Internet activities?

**Question 3:** What are these teenagers' favorite Internet activities?

The magazine editors have collected data from 15 students who are regular Internet users. The green entries indicate the students' favorite activities.

**Just the facts**

The average college student spends three hours per day on-line.

**Weekly Time Spent on the Net (minutes)**

| Student Initials | Chatting in a Chat Room | Playing Games | Doing Homework | Surfing the Web | E-mail |
|---|---|---|---|---|---|
| AB | 90 | 75 | 80 | 90 | 12 |
| BT | 75 | 150 | 0 | 150 | 15 |
| CP | 0 | 75 | 60 | 150 | 5 |
| CT | 0 | 240 | 0 | 0 | 0 |
| GO | 120 | 90 | 90 | 60 | 3 |
| KQ | 135 | 60 | 40 | 80 | 15 |
| LM | 75 | 160 | 30 | 45 | 6 |
| MC | 15 | 0 | 35 | 60 | 15 |
| MH | 80 | 180 | 30 | 90 | 6 |
| NM | 90 | 90 | 45 | 90 | 15 |
| PD | 100 | 150 | 60 | 90 | 0 |
| RL | 100 | 90 | 45 | 90 | 0 |
| SK | 90 | 135 | 60 | 240 | 10 |
| SM | 60 | 135 | 40 | 60 | 22 |
| YS | 120 | 30 | 60 | 45 | 15 |

1. See Additional Answers.

1. Complete Parts a and b for Questions 1, 2, and 3 above.
   **a.** Describe any statistical measures you will need to find or computations you will need to do to answer the question.
   **b.** Describe a graph that would be appropriate to include with the answer to the question.



## On Your Own Exercises

Investigation 1, pp. 391–393
Practice & Apply: 1
Connect & Extend: 3

Investigation 2, pp. 394–395
Practice & Apply: 2
Connect & Extend: 4

Assign Anytime
Mixed Review: 5–12

**Exercises 1 and 2:** In **Exercise 1**, students perform an analysis similar to that done in Investigation 1—deciding which data, which statistical measures, and which mathematical calculations will be needed to answer specific questions. Remind students that they are not asked to perform any calculations, but they may do so if it helps them create plans.

You may want to plan class time to discuss students' responses. Have them keep their plans to help them when they complete the analysis in **Exercise 2**.

**Quick Review Math Handbook**

hot words hot topics: Hot Topics pp. 184–189

## Additional Answers

**1a.** Possible answers:

Question 1: For each student, I will need to add the times for all the activities to find the total time spent on the Internet. Then I will need to find the mean, median, and mode of the totals.

Question 2: I will find the mean, median, and mode time for each activity.

Question 3: I will list the favorite activities and find the mode.

• ***Additional Answers continued on page A661***

# Assign and Assess

**Exercise 3:**
Students interpret a double bar graph. Gaining this familiarity will help students when they make graphs in Investigation 2.

**2.** See Additional Answers.

**2.** Refer to the information on page 399. Analyze the given data, and use your results to answer the three questions posed by the magazine editors. Provide the following information for each question:

**a.** The results of your computations and the measures you found

**b.** One or more sentences answering the question

**c.** A graph to help illustrate your answer

## Connect & Extend

**3. Social Studies** This double bar graph shows the number of licensed drivers per 1,000 people and the number of registered vehicles per 1,000 people in seven states:

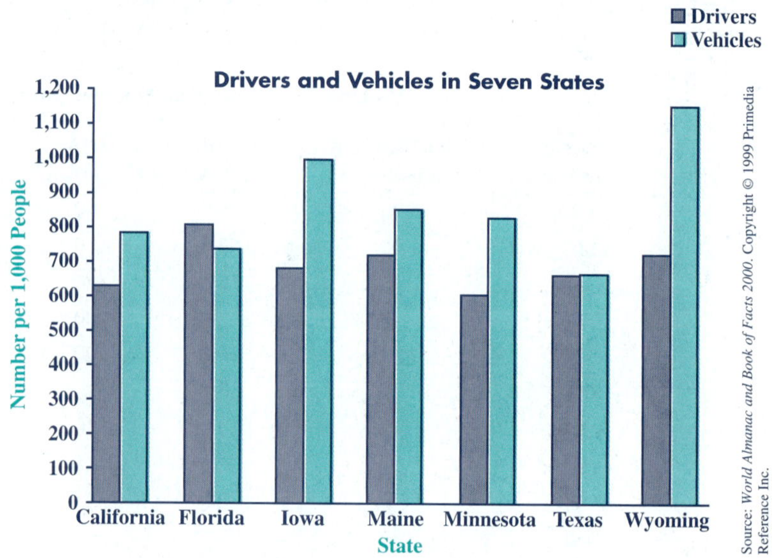


**a.** Which state has the greatest difference between the number of registered vehicles and the number of licensed drivers? Wyoming

**b.** Which state has about one registered vehicle per licensed driver? Texas

**c.** Which state has less than one registered vehicle per licensed driver? Florida

**d.** Which two states have about the same number of registered vehicles? Maine and Minnesota

**Just the facts**

Automobiles were commercially available in the United States beginning in 1896. In 1903, Missouri and Massachusetts adopted the first driver's license laws. In 1908, Rhode Island became the first state to require a driver's test.

### Additional Answers

**2.** Possible answers:

Question 1:

**a.** The total times for the 15 students are 347, 390, 290, 240, 363, 330, 316, 125, 386, 330, 400, 325, 535, 317, and 270. The median and mode are both 330, and the mean is about 331.

**b.** The typical total time is about 330 min, or $5\frac{1}{2}$ h. The mean, median, and mode are all close to this time.

**c.**

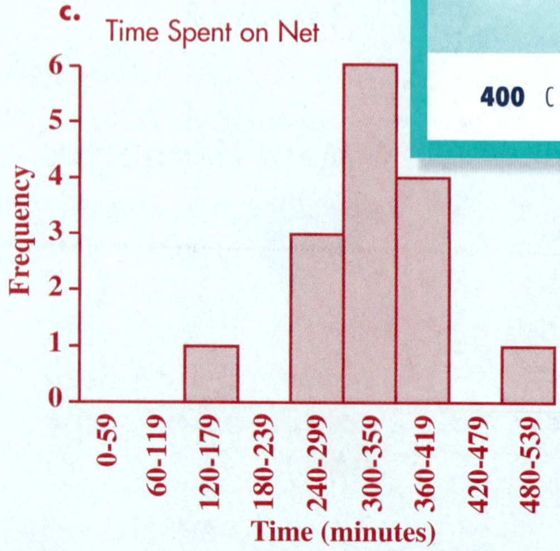


Question 2:

**a.** This table gives the mode, median, and mean (rounded to the nearest whole number) time for each activity:

| Activity | Mode | Median | Mean |
|---|---|---|---|
| Chatting | 90 | 90 | 77 |
| Playing games | 90 | 90 | 111 |
| Doing homework | 60 | 45 | 45 |
| Surfing the Web | 90 | 90 | 89 |
| Using e-mail | 15 | 10 | 9 |

• **Additional Answers continued on page A661**

## In your own words

Think of a question you could answer by collecting data from a particular group of people. Describe the group you would survey and the type of information you would collect. Then tell what statistical measures and graphs might be useful for summarizing the data.

4. **Social Studies** This stacked bar graph shows the number of bachelor's degrees awarded to men and women in the United States in various years.

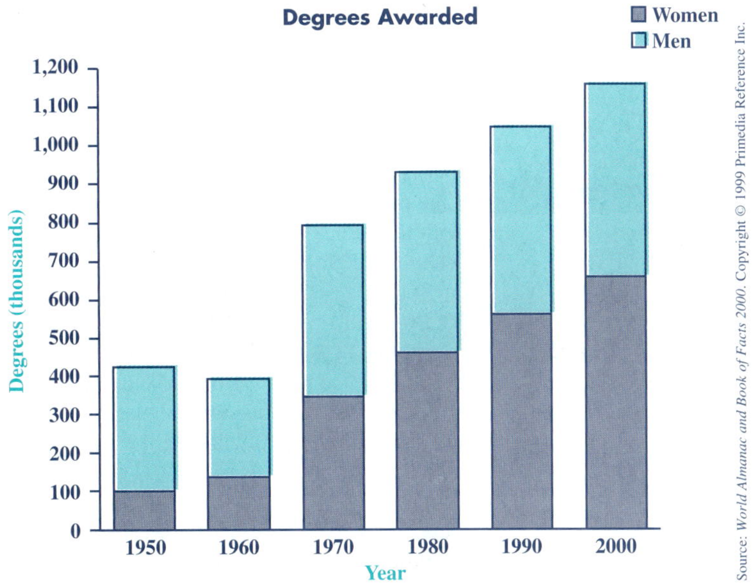


**4a, b.** See Additional Answers.

**a.** Describe how the total number of bachelor's degrees awarded changed over the years shown in the graph.

**b.** Describe how the total number of bachelor's degrees awarded to women has changed over the years shown in the graph.

**4c.** about 450,000; about 340,000

**c.** In 1970, about how many bachelor's degrees were received by men? About how many were received by women?

**4d.** about $\frac{1}{4}$, about $\frac{1}{2}$

**d.** In 1950, about what fraction of bachelor's degrees were received by women? In 2000, about what fraction were received by women?

## Mixed Review

**Number Sense** Order each set of numbers from least to greatest.

**5.** 4, $^-3$, 0, $^-1.5$, 0.5, 2, $^-4$ $^-4$, $^-3$, $^-1.5$, 0, 0.5, 2, 4

**6.** 0.108, 0.08, 0.1, 0.018, 0.081, 0.801

**6.** 0.018, 0.08, 0.081, 0.1, 0.108, 0.801

**7.** $\frac{3}{4}, \frac{8}{11}, \frac{21}{33}, \frac{15}{21}, \frac{209}{330}, \frac{41}{52}$ $\frac{209}{330}, \frac{21}{33}, \frac{15}{21}, \frac{8}{11}, \frac{3}{4}, \frac{41}{52}$

**8.** $\frac{4}{3}, ^-\frac{3}{4}, ^-0.999, ^-1, 0.009, \frac{2}{3}, ^-\frac{1}{3}$ $^-1, ^-0.999, ^-\frac{3}{4}, ^-\frac{1}{3}, 0.009, \frac{2}{3}, \frac{4}{3}$

List all the factors of each number.

**9.** 115 **10.** 92 **11.** 71 **12.** 90

**9.** 1, 5, 23, 115

**10.** 1, 2, 4, 23, 46, 92

**11.** 1, 71

**12.** 1, 2, 3, 5, 6, 9, 10, 15, 18, 30, 45, 90



**Exercise 4:**
The stacked bar graph was introduced in Investigation 1 on page 392. You might want to have students discuss how they found their answers.

## Quick Check

### Informal Assessment

Students should be able to:

- ✔ prepare a plan that includes selecting the appropriate statistical measures as part of a research project
- ✔ select a way to display data
- ✔ follow a plan to reach a conclusion

### Quick Quiz

1. Describe the graph you would use to support your answer about how long the typical student spends doing household chores each week. Explain your answer. ***Possible answer: Stem-and-leaf plot because it shows intervals of data to help summarize times and also provides the actual times.***

2. Describe the statistical measure or mathematical computation you would use to support your answer about the three most common chores done by students. Explain your answer. ***Possible answer: Choose the three chores that were named by the greatest numbers of students.***

### Additional Answers

**4a.** Possible answer: The number of degrees decreased from 1950 to 1960 but then increased dramatically between 1960 and 1970. Since 1970, the increase in the number of degrees has increased steadily.

**4b.** Possible answer: The number of degrees awarded to women has increased over the years, with the most dramatic increase between 1960 and 1970.

# Chapter 6 Review

**Chapter Summary**
This summary helps students recall the major topics of the chapter.

**Vocabulary**
Students should be able to explain each of the terms listed in the vocabulary section.

**Problem-Solving Strategies and Applications**
The questions in this section help students review and apply the important mathematical ideas and problem-solving strategies developed in the chapter. The questions are organized by mathematical highlights. The highlights correspond to those in "The Big Picture" chart on page 339a.

# Chapter 6 Review & Self-Assessment

VOCABULARY
**distribution**
**histogram**
**line plot**
**mean**
**median**
**mode**
**outlier**
**range**
**stem-and-leaf plot**

## Chapter Summary

In this chapter, you learned methods for organizing, analyzing, and displaying data. First you looked at *bar graphs* and *histograms.* You saw that bar graphs are used to show information about individuals or groups, while histograms are used to show the number or percent of data values that fall into various intervals.

You then learned about two more types of displays: *line plots* and *stem-and-leaf plots.* Both are helpful for showing a distribution of data values. Line plots show a stack of X's for each value. Stem plots group values with the same "stem" and are useful when the values are spread out or when there are many different values.

You were also introduced to some statistics used to summarize a data set. The *range* of a data set is the difference between the minimum and maximum values. The *mode* is the value that occurs most often. The *median* is the middle value. The *mean* is the value arrived at by dividing the sum of the data values equally among the data items. The mean, median, and mode are all measures of what is typical, or average, about a set of data.

Finally, you conducted a data investigation in which you had to decide what information, statistics, and graphs to use to answer a series of questions.

A snow-survey crew gathering data in Nevada



impactmath.com/chapter_test

## Strategies and Applications

The questions in this section will help you review and apply the important ideas and strategies developed in this chapter.

### Interpreting and creating bar graphs and histograms

**1.** The bar graph shows the number of cassettes and CDs sold in the years from 1992 to 2000.

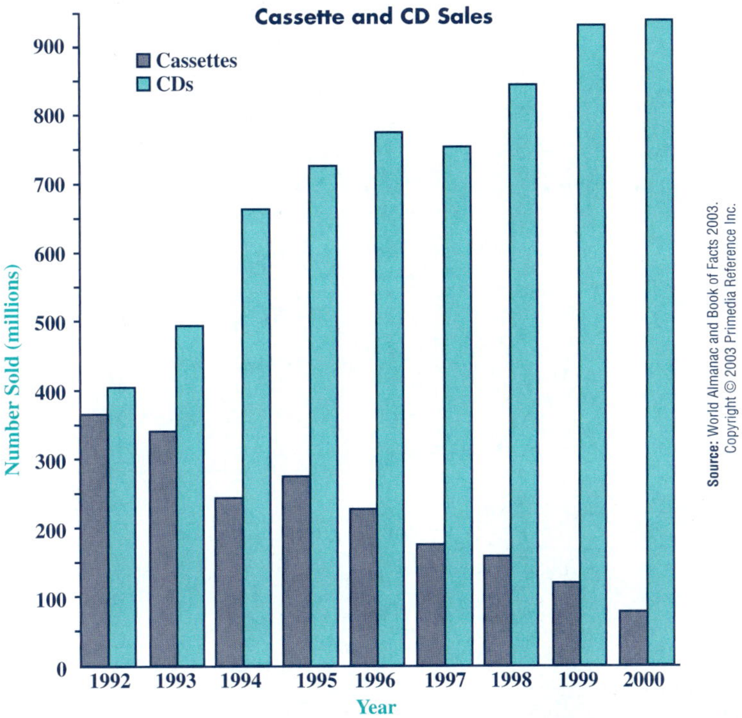


**a.** Describe what the graph indicates about the change in cassette sales over the years.

**b.** Describe what the graph indicates about the change in CD sales over the years.

**c.** Describe how the difference between CD sales and cassette sales has changed over the years.

**1a.** Possible answer: Cassette sales were highest in 1992 and then decreased fairly significantly from 1993 to 1994. Sales increased slightly in 1995 and then began to decrease gradually between 1996 and 1998. Cassette sales fell sharply in 2000.

**1b.** Possible answer: In general, CD sales have increased over the years, although there was a slight decrease from 1996 to 1997. The greatest increase in sales occurred from 1993 to 1994.

**1c.** Possible answer: In 1992, there were more CD sales than cassette sales, and the difference has steadily increased.



Review and Self-Assessment 403

# Chapter 6 Review

**Question 2a:**
Suggest that students add a column for tallies to their frequency tables. They can use the tallies to record data and then add them to find each frequency.

**Question 2b:**
You may want to provide students with copies of Master 81 or Master 82 or graph paper to draw their histograms.

**2.** The American Film Institute released a list of the best American movies of all time. Below are the years in which the top 50 movies were released.

| Movie Ranking | Year Released | Movie Ranking | Year Released | Movie Ranking | Year Released |
|---|---|---|---|---|---|
| 1 | 1941 | 18 | 1960 | 35 | 1934 |
| 2 | 1942 | 19 | 1974 | 36 | 1969 |
| 3 | 1972 | 20 | 1975 | 37 | 1946 |
| 4 | 1939 | 21 | 1940 | 38 | 1944 |
| 5 | 1962 | 22 | 1968 | 39 | 1965 |
| 6 | 1939 | 23 | 1941 | 40 | 1959 |
| 7 | 1967 | 24 | 1980 | 41 | 1961 |
| 8 | 1954 | 25 | 1982 | 42 | 1954 |
| 9 | 1993 | 26 | 1964 | 43 | 1933 |
| 10 | 1952 | 27 | 1967 | 44 | 1915 |
| 11 | 1946 | 28 | 1979 | 45 | 1951 |
| 12 | 1950 | 29 | 1939 | 46 | 1971 |
| 13 | 1957 | 30 | 1948 | 47 | 1976 |
| 14 | 1959 | 31 | 1977 | 48 | 1975 |
| 15 | 1977 | 32 | 1974 | 49 | 1937 |
| 16 | 1950 | 33 | 1952 | 50 | 1969 |
| 17 | 1951 | 34 | 1962 | | |

**a.** Copy and complete the frequency table to show the number of movies released during each decade.

| Decade | Frequency |
|---|---|
| 1910–1919 | 1 |
| 1920–1929 | 0 |
| 1930–1939 | 6 |
| 1940–1949 | 8 |
| 1950–1959 | 11 |
| 1960–1969 | 11 |
| 1970–1979 | 10 |
| 1980–1989 | 2 |
| 1990–1999 | 1 |

**2c.** Possible answer: Most of the best 50 movies were released between 1930 and 1980. More of the movies were released in the 1950s and the 1960s than any other decades. Only one of the movies was released before 1930, and only three were released after 1980.

**b.** Use your frequency table to help you make a histogram showing the number of values in each decade. See Additional Answers.

**c.** What does the shape of your histogram reveal about the distribution of years in which the best 50 movies were released? See above.



**Additional Answers**

**2b.**

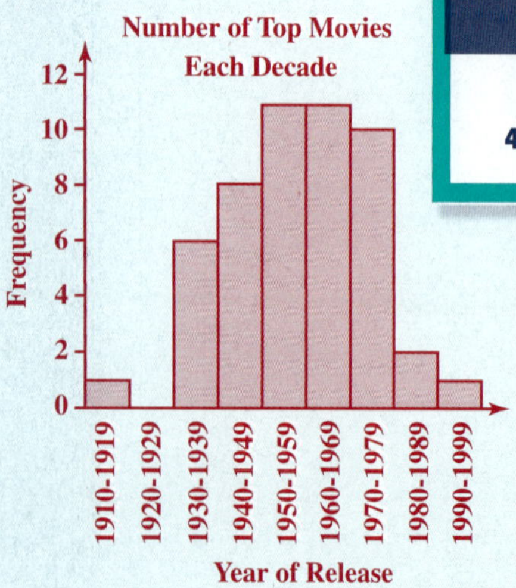

### Interpreting and creating line plots and stem plots

**3.** The table shows the number of cookies sold by the girls in a particular scouting troop.

| Initials | Boxes Sold |
|---|---|
| JJ | 12 |
| PK | 80 |
| TT | 29 |
| CR | 23 |
| HT | 90 |
| SI | 19 |
| FM | 87 |
| VY | 72 |

| Initials | Boxes Sold |
|---|---|
| PN | 21 |
| KT | 99 |
| FV | 84 |
| DY | 25 |
| HH | 33 |
| SE | 85 |
| MI | 79 |
| CK | 21 |

**a.** Do you think a line plot or a stem plot would be better for displaying the "Boxes Sold" values? Explain why.

**b.** Make the plot you suggested in Part a.

**c.** Use your plot to find the mode, median, and range of the "Boxes Sold" values. mode: 21; median: 52.5; range: 87

**d.** Describe the shape of the plot. Explain what the shape tells you about how the data are distributed. See Additional Answers.

**4.** How can you find the mode of a data set by looking at a line plot? How can you find the mode by looking at a stem plot?

### Finding and interpreting the mode, median, and mean

**5.** Marcus asked the students in his class how many first cousins they have. He summarized the data by reporting three averages.

**a.** Marcus said the mode number of cousins is 6. What does this average tell you about the data?

**b.** He reported that the median number of cousins is 4. What does this average tell you about the data?

**c.** Marcus found that the mean number of cousins is 5. What does this average tell you about the data?

**3a.** Possible answer: A stem plot; there are many different values and a large range. A line plot would have only one or two X's over each number. A stem plot would group the values and give a better picture of the distribution.

**3b.** Possible plot:

**Boxes of Cookies Sold**

| Stem | Leaf |
|---|---|
| 0 | |
| 1 | 2 9 |
| 2 | 1 1 3 5 9 |
| 3 | 3 |
| 4 | |
| 5 | |
| 6 | |
| 7 | 2 9 |
| 8 | 0 4 5 7 |
| 9 | 0 9 |

**Key:** 2|1=21

**4.** In a line plot, look for the number with the tallest stack of X's. In a stem plot, look for the leaf value that occurs most in a single stem.

**5a.** More students have 6 cousins than any other number.

**5b.** About half of the students have 4 cousins or fewer, and about half have 4 cousins or more.

**5c.** Possible answer: If the cousins were distributed equally among the students, each student would have 5.



**Question 5:** This question provides a review of the definitions of mean, median, and mode.

### Additional Answers

**3d.** Possible answer: The values are in two clumps, one near the top of the graph (for the 1, 2, and 3 stems) and one at the bottom of the graph (for the 7, 8, and 9 stems). There is a big gap between the clumps. The plot indicates that the data values can be divided into a group of low values and a group of high values, with no values in between. In the low group, most values are in the 20s; in the high group, the greatest number of values are in the 80s.

**Questions 6b and 6d:** In these questions, students must consider the effect on all three measures of central tendency when they add two values to the data set.

**Question 8d:** In this question, students consider how to use statistics to support the best outcome. If Hannah chooses the mean with the outlier, she has the best chance of getting a high allowance. However, students might want to consider whether this is really the best choice if Hannah wants to report a value that shows the typical allowance.

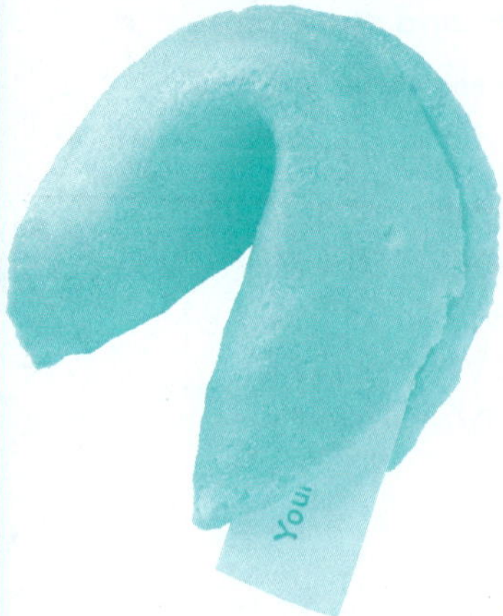

**6.** Angelia measured the mass of each of the fortune cookies her family brought home from dinner at a Chinese restaurant. Here are her measurements, in grams:

11 13 13 14 15 17 19 20 22

**a.** Find the mode, median, and mean of these data.

**b.** Add two values to the data set so the mean decreases but the median and mode remain the same. Possible answer: 10, 16

**c.** Start with the original data set. Add two values to the set so the mean remains the same but the median changes.

**d.** Start with the original data set. Add two values to the set so the median, mean, and mode all increase. Possible answer: 22, 22

**6a.** mean: 16; median: 15; mode: 13

**6c.** Possible answer: 16, 16

**7.** You learned that an *outlier* is a value that is much greater or much less than most of the other values in a data set. Which measure of center—the mean or the median—is more influenced by outliers? Explain your answer, and give an example to illustrate.
See Additional Answers.

### Choosing the best average for a given situation

**8.** Hannah's father wants to start giving Hannah a weekly allowance, but he isn't sure how much to give her. Hannah asked 13 of her friends how much weekly allowance they get. Here are the results:

\$0 \$2.50 \$3 \$2.50 \$5 \$15

\$7.50 \$0 \$10 \$0 \$10 \$75 \$5

**a.** Find the mean, median, and mode of the allowance data.

**b.** Which of the averages you found do you think best represents a typical value in this data set? Give reasons for your answer.

**c.** The \$75 allowance is an outlier, since it is much greater than the other data values. Remove this value and find the new mean, median, and mode. mean: \$5.04; median: \$4; mode: \$0

**d.** Hannah thinks telling her father the average allowance of her friends will help him decide how much to give her. Which average do you think she should report? Consider the averages for the data set with and without the outlier. Give reasons for your choice. Possible answers:

- The median or the mean with the outlier removed because these values best represent what is typical.
- The mean with the outlier because it is the greatest value and may convince Hannah's father to give her a high allowance.

**8a.** mean: \$10.42; median: \$5; mode: \$0

**8b.** Possible answer: The median; the mode is the least allowance and does not reflect the fact that most of the allowances are greater than \$0. The outlier of \$75 pulls up the mean. Only 2 of the 13 values are greater than the mean, so it is not typical. The median is the middle value and is not affected by the outlier.



### Additional Answers

**7.** the mean; Possible explanation: When finding the median, the only thing that matters is the order of the values. It does not matter whether the values less than the median are much less than the median or just a little less. Similarly, it doesn't matter whether the values greater than the median are much greater or just a little greater. The mean of a data set is the sum of the values divided by the number of data values. If a data set has a value that is much greater than the others, the sum will be greater than it would be if the value were closer to the others. And if the sum is greater, the mean will be greater. Similarly, a very small value gives a smaller sum and a smaller mean.

## Demonstrating Skills

Find the range, mean, median, and mode of each set of numbers.

**9.** 9, 3, 7, 3, 3, 2, 5, 5, 7, 7

**9.** range: 7, mean: 5.1; median: 5; modes: 3 and 7

**10.** 144, 120, 196, 95, 180, 5, 136, 175, 114

**10.** range: 191; mean: $129.\overline{4}$; median: 136; mode: none

**11.** Find the range, median, and mode of the data in this stem plot: range: 75; median: 545; mode: 562

| Stem | Leaf |
|---|---|
| 50 | 0 3 7 7 |
| 51 | 0 5 |
| 52 | |
| 53 | 3 5 7 |
| 54 | |
| 55 | 3 9 |
| 56 | 0 2 2 2 4 |
| 57 | 3 5 |

**Key:** 50|3=503

**12.** Make a line plot of the data below.

11 9 13 9 11 10 12 16

13 8 13 11 8 13 15 16 13 11

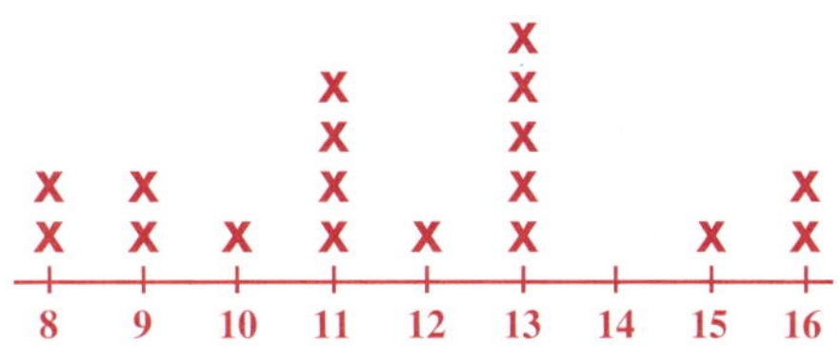


**Questions 9 and 10:** You might suggest that students put the data in order from least to greatest or greatest to least to help them find the statistical measures.



Review and Self-Assessment **407**

# CHAPTER 7

# Variables and Rules

## Chapter Overview

In this chapter, students examine variables and use them to describe rules and build mathematical expressions. Students begin by learning that a variable is a quantity that changes in value. They use letters to represent variable quantities and examine symbolic expressions that describe patterns in sequences and tables of numbers. They see that different symbolic expressions can sometimes describe the same pattern.

Students then use rules, expressed in symbolic form, to study real-life situations. They translate verbal descriptions of situations into symbolic form. Students see how these symbolic expressions can be used to clarify and study relationships between variable quantities.

Finally, students use what they know about operations to simplify expressions and write them in equivalent forms.

## the Big Picture

| Chapter 7 Highlights | Links to the Past | Links to the Future |
|---|---|---|
| Writing and interpreting rules for sequences and input/output tables (7.1) | **Elementary Grades:** Writing multiplication number models for rectangular arrays<br>**Chapter 1:** Determining a rule when given an input/output table. | **Course 2, Chapter 6:** Writing algebraic expressions for a variety of situations<br>**Course 3, Chapter 8:** Identifying functions in a variety of contexts and representations |
| Showing that two rules for a sequence are equivalent (7.1, 7.2) | **Chapter 1:** Writing rules that connect two quantities | **Course 2, Chapter 6:** Generating equivalent expressions to solve and create number puzzles<br>**Course 3, Chapter 8:** Understanding different ways of representing functions |
| Writing and interpreting rules for real-life situations (7.2) | **Elementary Grades:** Writing algebraic expressions to describe situations, using variables | **Course 2, Chapter 4:** Practicing using expressions to represent specific situations<br>**Course 3, Chapter 4:** Interpreting situations mathematically and creating equations to represent them |
| Using variables to explain number relationships (7.2, (7.3) | **Elementary Grades:** Writing algebraic expressions to represent situations described in words | **Course 2, Chapter 6:** Using variables to write algebraic expressions for a variety of situations and numerical relationships |

Time-Saving Tool **TeacherWorks**™ All-In-One Planner and Resource Center

# Planning Guide

| Lesson Objectives | Pacing | Materials | NCTM Standards | Hot Topics |
|---|---|---|---|---|
| **7.1 Patterns and Variables** page 409b<br>• To write symbolic rules for a toothpick or dot sequence and show how the number of toothpicks or dots in each term relates to the term number<br>• To use diagrams to explain why a rule for a toothpick or dot sequence is correct<br>• To recognize when rules for a pattern are equivalent<br>• To use symbols to write rules for tables made during games of *What's My Rule?* | 4 class periods | • Master 86 (optional) | 1, 2, 6, 7, 8, 9 | |
| **7.2 Rules in Real Life** page 429b<br>• To use letters for variables to write rules for real-life situations<br>• To analyze real-life situations, identify related variables, and write rules describing relationships between the variables<br>• To use related multiplication and division equations and related addition and subtraction equations to write rules in equivalent forms | 4 class periods | | 2, 6, 7, 8 | pp. 254–261 |
| **7.3 Explaining Number Relationships** page 449b<br>• To discover patterns describing how the starting number is related to the final number in "think of a number" tricks<br>• To use operations on symbolic expressions to explain why "think of a number" tricks work<br>• To use symbolic expressions that contain variables to represent consecutive whole numbers, consecutive even numbers, and consecutive odd numbers, and to explain patterns in sets of these numbers | 3 class periods | | 1, 2, 7, 8 | |

**Key to NCTM Curriculum and Evaluation Standards:** 1=Number and Operations, 2=Algebra, 3=Geometry, 4=Measurement, 5=Data Analysis and Probability, 6=Problem Solving, 7=Reasoning and Proof, 8=Communication, 9=Connections, 10=Representation

# Assessment Opportunities

## Standard Assessment

*Impact Mathematics* offers three types of formal assessment. The Chapter 7 Review and Self-Assessment in the Student Edition serves as a self-assessment tool for students. In the Teacher's Guide, a Quick Quiz at the end of each lesson allows you to check students' understanding before moving to the next lesson. The Assessment Resources include blackline masters for chapter and quarterly tests.

- **Student Edition** Chapter 7 Review and Self-Assessment, pages 460–463
- **Teacher's Guide** Quick Quizzes, pages A664, 449, 459
- **Assessment Resources** Chapter 7 Test Form A, pages 170–174; Chapter 7 Test Form B, pages 175–179

## Ongoing Assessment

*Impact Mathematics* provides numerous opportunities for you to assess your students informally as they work through the investigations. Share & Summarize questions help you determine whether students understand the important ideas of an investigation. If students are struggling, Troubleshooting tips provide suggestions for helping them. On the Spot Assessment notes appear throughout the teaching notes. They give you suggestions for preventing or remedying common student errors. Assessment Forms in the Assessment Resources provide convenient ways to record student progress.

- **Student Edition** Share & Summarize, pages 414, 418, 421, 432, 439, 442, 454, 456
- **Teacher's Guide** On the Spot Assessment, pages T410, T414, T431, T441, T451 Troubleshooting, pages T414, T418, T421, T432, T439, T442, T454, T456
- **Assessment Resources** Chapter 7 Assessment Checklists, pages 266–267

## Alternative Assessment, Portfolios, and Journal Ideas

The alternative assessment items in *Impact Mathematics* are perfect for inclusion in student portfolios and journals. The In Your Own Words feature in the Student Edition gives students a chance to write about mathematical ideas. The Performance Assessment items in the Assessment Resources provide rich, open-ended problems, ideal for take-home or group assessment.

- **Student Edition** In Your Own Words, pages 428, 448, 458
- **Assessment Resources** Chapter 7 Performance Assessment, pages 180–182

# Assessment Resources

The Assessment Resources provide a chapter test in two equivalent forms, along with additional performance items. The performance items can be used in a variety of ways. They are ideal for take-home assessment or in-class group assessment.

- Chapter 7 Test Form A, pages 170–174
- Chapter 7 Test Form B, pages 175–179
- Chapter 7 Performance Assessment, pages 180–182
- Chapter 7 Assessment Solutions, pages 183–187

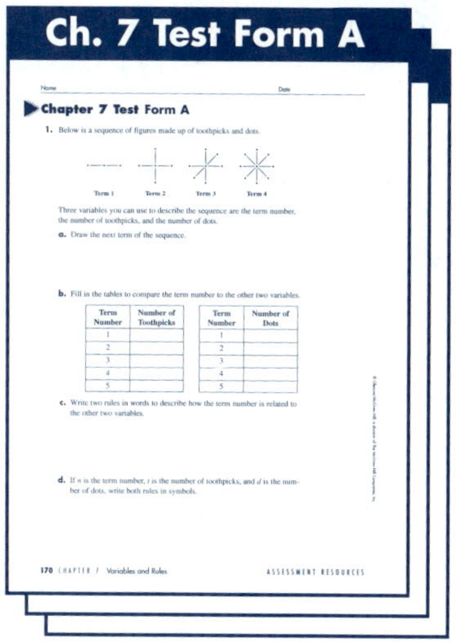


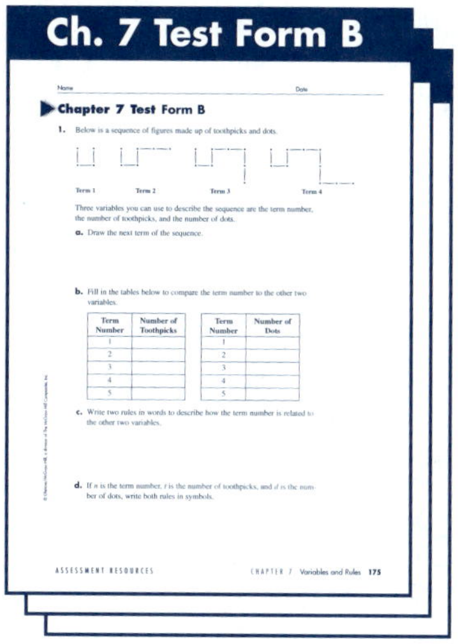


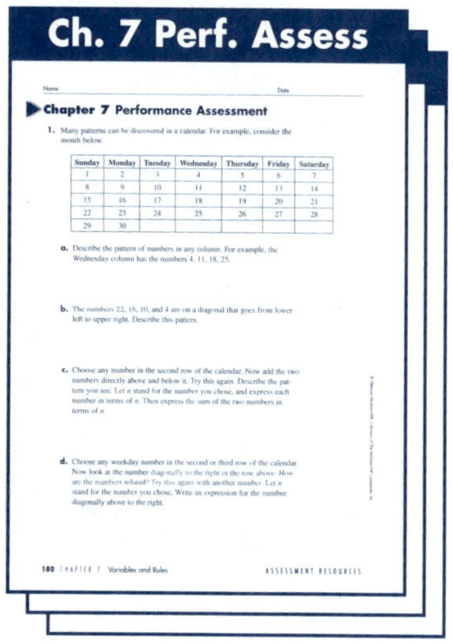


# Additional Resources

- **Math Skills Maintenance Workbook,** 10, 16, 17
- **Investigations for the Special Education Student in the Mathematics Classroom,** 10
- **What's Math Got To Do With It? Videos,** Level 1, Video 3
- **StudentWorks™ CD-ROM**
- **Reading and Writing in the Mathematics Classroom**
- **Using the Internet in the Mathematics Classroom**

## ExamView® Pro

Use ExamView® Pro Testmaker CD-ROM to:

- Create Multiple versions of tests.
- Create Modified tests for Inclusion students with one mouse click.
- Edit existing questions and Add your own questions.
- Build tests aligned with state standards using built-in State Curriculum Correlations.
- Change English tests to Spanish with one mouse click and vice versa.

**Introduce**

To introduce this chapter, you might ask students what they know about formulas.

Have students share and discuss any formulas they are already familiar with and how they are used. You may also introduce some geometric formulas (such as $A = l \times w$, and $P = 2 \times l + 2 \times w$) if students have not already done so.

Have students focus on the fact that formulas are actually rules written with variables represented by letters. Each variable has its own meaning and stands for a different value. They should understand that the rules they learned in Chapter 1 can be represented by variables.

**Think About It**

Use the formula $T = 120V \div d$. Substitute 2 for $V$ and 4 for $d$. So, $T = 120 \cdot 2 \div 4$, or 60. The diver can stay under water for 60 minutes.

# CHAPTER 7

# Variables and Rules

## *Real-Life Math*

**Under the Sea** A *formula* is an algebraic "recipe" for finding the value of one variable based on the values of one or more other variables. Letters are usually used to represent the variables. For example, a scuba diver can use this formula to figure out how long she can stay under water.

$$T = 120V \div d$$

In the formula, $T$ represents the time in minutes the diver can stay under water, $V$ represents the volume of air in the diver's tank in cubic meters, and $d$ represents the water's depth in meters.

**Think About It** Suppose a diver has 2 cubic meters of compressed air in her tank and is 4 meters underwater. How long can the diver stay under water?

# *Family Letter*

Dear Student and Family Members,

Earlier in the year, you worked with many patterns, finding rules to describe them. Here's a pattern from Chapter 1.

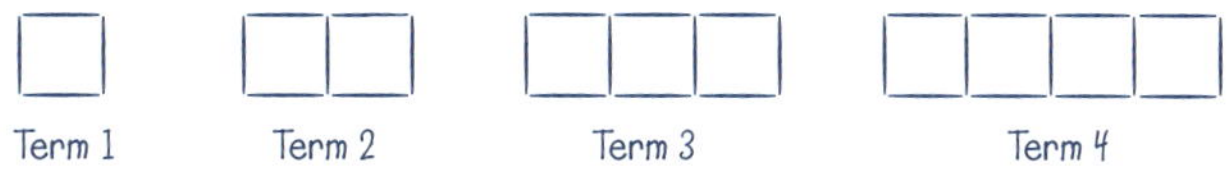


Each new term has one more square added on, made up of three more toothpicks. A rule that relates the term number to the number of toothpicks is:

$$\text{number of toothpicks} = 3 \cdot (\text{term number}) + 1.$$

Can you use the rule to find how many toothpicks would be in Term 5?

In this chapter, you will learn about shorter and more useful ways of writing rules to describe patterns—an important and fundamental part of algebra. Quantities that change are called *variables*. For example, the term number above is a variable because it has different values at different terms. If $t$ represents the number of toothpicks, and $n$ represents the term number, then the algebraic rule is $t = 3 \cdot n + 1$.

You will also look at rules that apply to everyday situations. For example, suppose a phone company charges $3.00 per month plus $0.07 per minute for each phone call. One way to understand the situation is by making a table.

| Month | June | July | August | September | October |
|---|---|---|---|---|---|
| Minutes | 120 | 90 | 95 | 150 | 80 |
| Cost | $11.40 | $9.30 | $9.65 | $13.50 | ? |

You can also write a rule to calculate the phone bill for any month using $m$ to represent the number of minutes and $c$ to represent the cost. Using the rule $c = \$0.07 \cdot m + \$3.00$, can you find the cost of October's phone bill?

### What can you do at home?

During the next few weeks, be on the lookout for rules in real life, such as finding the cost of $m$ movie tickets or $c$ CDs. Ask your student to write rules using variables for each situation.

impactmath.com/family_letter



Another version of the Family Letter, available in English and Spanish, is found in the Teaching Resources. You may want to send a copy of this letter home with your students.

**Teaching Resources**

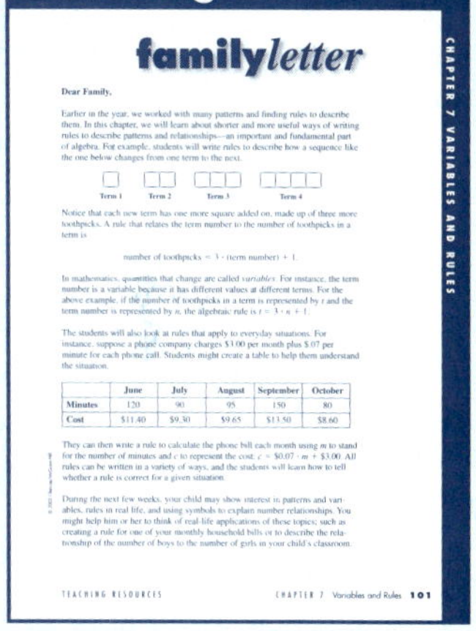


409

# Mathematical Background

Chapter 7 begins with students examining sequences of different toothpick arrangements. From a diagram, students are able to make a table showing the number of toothpicks in each term (stage). A rule can usually be stated, both verbally and mathematically, to show how the two variables are related. For example, consider this sequence of toothpicks:

Term 1 Term 2 Term 3 Term 4 Term 5

This table contains the term number and the number of toothpicks in each term.

| **Term Number** | 1 | 2 | 3 | 4 | 5 |
|---|---|---|---|---|---|
| **Number of Toothpicks** | 2 | 3 | 4 | 5 | 6 |

The rule can be stated as the following: number of toothpicks = term number + 1. Letters are used to represent variables in algebra. So, if the variable $t$ designates the number of toothpicks and the variable $n$ is used to show the term number, the symbolic rule is $t = n + 1$.

Two students will often look at a sequence of toothpicks, dots, or numbers and come up with different rules. Each student can show that his or her rule is correct by explaining why it will work for any term in the sequence. Thus, the rules will give the same result for every term. The rules are then said to be *equivalent*. Students learn to write symbolic expressions in equivalent forms by using properties and relationships between basic number operations.

A fun way for students to practice developing and writing rules is by playing the game *What's My Rule?* This adaptation of the game is slightly different from the one used in Chapter 1. Here, students guess the symbolic rule. In groups of at least three students, the first student develops a rule. When the other students provide inputs, the first student returns the output according to the rule. The other students compete to determine the rule.

Another way for students to practice writing rules is to show how they appear in the world around them. Introducing everyday situations of how rules are used allows students to see that important decisions are often made after those rules were analyzed. Determining which long-distance telephone service to use or which car rental company is more affordable are two of the many ways students can see how algebra is used in the real world. Students need to remember that creating a table and looking for a pattern can make finding the rule a little easier. To test whether a rule is correct, they will need to evaluate the equation for a value they know the answer for.

Determining and writing symbolic expressions can also help students describe number relationships. "Think of a number" tricks easily lend themselves to writing symbolic rules. These tricks have students start with a number and go through a series of (arithmetic) steps. The result of their calculations usually is the number they started with or a specific number. Letters and symbols show how a number trick works.

• *Teaching notes continued on page A662*

# 7.1 Patterns and Variables

## Objectives

- To write symbolic rules for a toothpick or dot sequence and show how the number of toothpicks or dots in each term relates to the term number
- To use diagrams to explain why a rule for a toothpick or dot sequence is correct
- To recognize when rules for a pattern are equivalent
- To use symbols to write rules for tables made during games of *What's My Rule?*

## Overview (pacing: about 4 class periods)

The concept of a *variable* is one of the fundamental concepts of algebra, and well before students begin their study of algebra, the concept has many uses. Although the term can have various meanings, it is introduced in this lesson with its most basic meaning, namely that of a quantity that changes.

Students are also introduced to the use of letters to represent variable quantities. They learn how variables can be used to describe and study patterns such as those in Chapter 1. They write rules for patterns in symbolic form, and they learn how to present convincing arguments that their rules will always work. They also learn to recognize equivalent rules—that is, different rules that generate the same pattern. These understandings will prove useful in problem solving and in later courses involving mathematical proof.

## Advance Preparation

For On Your Own Exercise 18, you may want to give students copies of Master 86.

## Lesson Planner

| | Summary | Materials | On Your Own Exercises | Assessment Opportunities |
|---|---|---|---|---|
| **Investigation 1** page T411 | Students learn that a sequence is an ordered list of items. They identify variables in various sequences and use symbols to write rules that describe how the variables are related to the term number. | • Master 86 (optional) | Practice & Apply: 1–5, pp. 422–423<br>Connect & Extend: 14–16, pp. 425–426<br>Mixed Review: 22–34, p. 429 | On the Spot Assessment, page T410<br>Share & Summarize, pages T414, 414<br>Troubleshooting, page T414 |
| **Investigation 2** page T414 | Equivalent rules are introduced as different rules that describe the same relationship or pattern. Students use diagrams and reasoning to explain how they know whether two rules are equivalent. | | Practice & Apply: 6–8, pp. 423–424<br>Connect & Extend: 17, 18, pp. 426–427<br>Mixed Review: 22–34, p. 429 | On the Spot Assessment, page T414<br>Share & Summarize, pages T418, 418<br>Troubleshooting, page T418 |
| **Investigation 3** page T419 | Students use symbols to describe rules for games of *What's My Rule?* They use tables to help decide whether different rules correctly describe the relationship between input and output values. | | Practice & Apply: 9–13, pp. 424–425<br>Connect & Extend: 19–21, pp. 428–429<br>Mixed Review: 22–34, p. 429 | Share & Summarize, pages T421, 421<br>Troubleshooting, page T421<br>Informal Assessment, page 429<br>Quick Quiz, page 429 |

## Introduce

1 Begin by discussing the meaning of the two new words in the opening paragraphs on page 410: *sequence* and *term*. Ask students to examine the two sequences that are given. Ask such questions as the following:

What is Term 4 of the first sequence? **13**

What is Term 4 of the second sequence? **a diamond**

Which sequence shows that the terms of a sequence do not all have to be different? **the second sequence**

What do the three dots in each sequence indicate? **the sequence goes on and on**

You may wish to have students describe any patterns they observe that would let them predict the next four or five terms in the second sequence. Then move on to the Explore.

### Explore

In this Explore, students examine figures in a toothpick sequence and look for quantities that vary from one term to the next. Read through the first two paragraphs with the class. Then discuss the meaning of the term *variable*. 2

3 Have students work in groups of 3 or 4 to examine the terms of the toothpick sequence, look for variables, and find patterns in accordance with the instructions. Allow five to ten minutes for students to work, and then conduct a class discussion of their results.

#### On the Spot Assessment

4 Watch for students who have difficulty coming up with things that change in the toothpick sequence. Help students by having them list some of the things they might count in the terms of the toothpick sequence. Have them count these things in each term and look at the resulting number sequence to see whether it matches any of the three given sequences.

#### Access for all Learners

**Extra Challenge** If students are adept at the tasks in the Explore, you may wish to challenge them to come up with additional problems that involve finding a variable that matches a given number sequence related to the toothpick sequence. They can ask a classmate to solve the problems that they devise.

# 7.1 Patterns and Variables

① Possible answer: The number of vertical toothpicks increases by 1 with each term: 2, 3, 4, 5, . . . . The number of toothpicks inside the outer rectangle increases by 1 with each term: 0, 1, 2, 3, . . . . The number of 90° angles in the figure increases by 4 with each term: 4, 8, 12, 16, . . . .

In Chapter 1 you explored several sequences. A *sequence* is an ordered list of items such as numbers, symbols, or figures. Here are two sequences:

4, 7, 10, 13, 16, . . .     ♦, ♥, ♦, ♦, ♥, ♦, ♦, ♦, ♥, . . .

Each item in a sequence is called a *term*. So, in the second sequence above, Term 1 is a diamond, Term 2 is a heart, and Term 3 is a diamond.

**1** Introduce the terms *sequence* and *term*.

## Explore

You looked at this sequence of toothpick figures in Chapter 1:

Term 1     Term 2     Term 3     Term 4

Write a rule that describes how to create each term of this sequence from the previous term. Possible rule: Add 3 toothpicks to the right side to create another square.

In this sequence, there are many *variables*—things that change, or vary. For example, the *number of squares* increases by 1 with each term. You can write a number sequence to show the values of this variable:

1, 2, 3, 4, . . .

**2** Introduce the term *variable*.

Look for a pattern in the way each variable given below changes from term to term. First describe the pattern in words. Then write a number sequence to show the values of the variable.

- the number of toothpicks in a term increases by 3 with each term; 4, 7, 10, 13, . . .
- the number of toothpicks along the top of each term increases by 1 with each term; 1, 2, 3, 4, . . .

**3** Problem-Solving Strategy: Look for a pattern

Look for at least three more variables in the toothpick sequence above. Describe how each variable changes from term to term, and give the first four terms of the matching number sequence. See ① in margin.

**4** Have students list the things that they can count in the terms of the sequence.

Try to find a variable in the toothpick sequence that changes according to each of these number sequences:

2, 3, 4, 5, . . . See ②.     2, 4, 6, 8, . . . See ③.     1, 3, 6, 10, . . . See ④.

**Just the facts**

A numeric sequence created by adding the same number to each term to get the next term is called an *arithmetic sequence.* Here are two examples:

4, 9, 14, 19, 24, . . .

$\frac{3}{4}, 1\frac{1}{2}, 2\frac{1}{4}, 3, \ldots$

② Possible variable: number of vertical toothpicks

③ Possible variable: number of horizontal toothpicks

④ Possible variable: number of rectangles (including squares)

# Investigation 1

The purpose of this investigation is to introduce letters to represent variables and describe relationships between variable quantities. Using letters for variables to express general relationships and patterns is one of the basic differences between algebra and ordinary arithmetic. Some students may already have used letters as a shorthand for numbers, but for other students the idea may be quite new.

1 Before students begin Problem Set A, copy the table from the top of page 411 onto the board. Point out how this table is obtained from the toothpick sequence in the Explore.

Ask students if they can find a rule that relates the term number to the number of toothpicks used to make that term of the sequence. Another way to put the question is this: If you are told a term number, how can you figure out the number of toothpicks in that term? Encourage students to describe their rules in ordinary language. Lead them to see that one rule that seems to work is the rule presented in the paragraph just above Problem Set A.

2 **Problem Set A** **Suggested Grouping: Pairs**
This problem set provides opportunities for students to discover rules that describe how the values of two variables are related. Students test their rules and consider whether the rules will always work. Working in pairs allows students to discuss their thinking as they arrive at their answers.

3 In **Problem 1,** students make tables showing the values of two variables from the toothpick sequence in the Explore. You may want to point out that the top row of each table should contain the term numbers 1 through 4.

## Investigation 1 Sequences, Rules, and Variables

In the Explore, you looked at variables in a sequence of toothpicks, including the number of squares, the number of toothpicks, and the number of toothpicks along the top of each term. The term number is also a variable.

Making a table is a good way to compare the values of variables. This table shows the term number and the number of toothpicks for the first five terms:

| Term Number | 1 | 2 | 3 | 4 | 5 |
|---|---|---|---|---|---|
| Number of Toothpicks | 4 | 7 | 10 | 13 | 16 |

**1** Problem-Solving Strategies

- Make a table
- Write an equation or rule

You can sometimes write a rule to show how two variables are related. In Chapter 1, you wrote a rule relating the number of toothpicks to the term number. Here is one possible rule:

$$\text{number of toothpicks} = 3 \cdot \text{term number} + 1$$

In the next problem set, you will make tables and find rules for some of the other variables in the toothpick sequence.

### Problem Set A

**2** You may have students work in pairs.

**Just the facts**

A numeric sequence created by multiplying each term by the same number to get the next term is called a *geometric sequence.* Here are two geometric sequences:

1, 4, 16, 64, 256, . . .

72, 36, 18, 9, 4. 5, . . .

**1.** In Parts a–d, make a table showing the values of the given variable for the first four terms of the sequence. The first table has been started for you.

Term 1 Term 2 Term 3 Term 4

**a.** number of squares

| Term Number | 1 | 2 | 3 | 4 |
|---|---|---|---|---|
| Number of Squares | 1 | 2 | 3 | 4 |

**3** Tell students that the top row of each table should contain the term numbers.

**b.** number of vertical toothpicks See Additional Answers.

**c.** number of horizontal toothpicks See Additional Answers.

**d.** number of rectangles (Hint: In each term, count the squares, the rectangles made from two squares, the rectangles made from three squares, and so on.)

| Term Number | 1 | 2 | 3 | 4 |
|---|---|---|---|---|
| Number of Rectangles | 1 | 3 | 6 | 10 |



**Additional Answers**

**1b.**

| Term Number | 1 | 2 | 3 | 4 |
|---|---|---|---|---|
| Number of Vertical Toothpicks | 2 | 3 | 4 | 5 |

**1c.**

| Term Number | 1 | 2 | 3 | 4 |
|---|---|---|---|---|
| Number of Horizontal Toothpicks | 2 | 4 | 6 | 8 |

In **Problem 2,** students write rules to describe how the variables in their tables are related. The rules can be
1 stated using words alone or with a combination of
words and symbols. The rule in **Part d** may be difficult for many students. You might suggest that students consider how the values of the second variable change from one term to the next. Discovering the pattern +1, +2, +3, may lead students to state the rule as follows

$$\text{number of rectangles} = \text{sum of the whole numbers from 1 through the term number}$$

Some students may arrive at the additional insight that this sum can be found by calculating

$$\text{term number} \cdot \frac{\text{term number} + 1}{2}.$$

**Problem 3** asks students to check their rules from Problem 2 by using them to predict the value of Term 5. You may want to call attention to the Remember at the top of the page. Remind students that checking a rule in this way lets them see whether the rule works once again, for the next term in the sequence. It does not prove for all instances that the rule will *always* work. If students have come up with a rule that works for the first four terms but not for Term 5, they will need to revise their rule.

**Problem 4** has students explain why their first three rules for Problem 3 will always work. This assumes that they have indeed arrived at rules that work for the first five terms for each situation. Putting their reasoning into words may be difficult, but being able to do so is the main way for students to be confident that their thinking is correct. It is also important in communicating their thinking to others.

2 **Problem Set Wrap-Up** After students have finished work on the problems, you may want the whole class to discuss and compare results. ■

Order-of-operations issues may have surfaced in Problems 3 and 4 as students attempted to describe and
3 defend their rules. Even if this has not occurred, you will
probably want to discuss the paragraphs following Problem 4 with the whole class. This material introduces the use of letters to represent variable quantities. Students will henceforth encounter more and more compact ways of writing rules. The synopsis of order of
4 operations in the Remember at the bottom of the page is
an appropriate part of the discussion.

**2.** For each table from Problem 1, try to find a relationship between the term number and the other variable. Then write a rule to describe the relationship.

**a.** number of squares = term number

**b.** number of vertical toothpicks =

**c.** number of horizontal toothpicks =

**d.** **Challenge** number of rectangles =

**2b.** Possible rules: 1 + term number, 2 + (term number − 1)

**2c.** Possible rules: 2 · term number, term number + term number

**2d.** Possible rules: term number · $\frac{\text{term number} + 1}{2}$, $\frac{\text{term number} \cdot (\text{term number} + 1)}{2}$

**3.** To check your rules, you can test them for a particular term number. Although this won't tell you for certain that a rule is correct, it's a good way to find mistakes. For each part of Problem 2, use your rule to predict the value of the variable for Term 5. Then draw Term 5 and check your predictions. See Additional Answers.

**4.** Explain how you know that the rules you wrote in Parts a–c of Problem 2 will work for every term. See Additional Answers.

**Remember**

It is not enough to show that a rule works in a few specific cases. Try to explain why it works based on how the terms are built.

In algebra, letters are often used to represent variables. For example, consider this rule:

$$\text{number of toothpicks} = 3 \cdot \text{term number} + 1$$

If you use the letter $n$ to represent the term number and the letter $t$ to represent the number of toothpicks, you can write the rule like this:

$$t = 3 \cdot n + 1$$

This rule is much shorter and easier to write than the original rule.

When a number is multiplied by a variable, the multiplication symbol is often left out. So, you can write the rule above in an even shorter form:

$$t = 3n + 1$$

You can use any letter to represent a variable, as long as you say what the letter represents. For example, you could let $w$ represent the term number and $z$ represent the number of toothpicks, and write the rule as $z = 3w + 1$.

A single rule can usually be written in many ways. Here are six ways to write the rule for the number of toothpicks in a term:

$t = n \cdot 3 + 1$ $\quad t = (n * 3) + 1$ $\quad t = 1 + 3n$

$t = 1 + (3 \cdot n)$ $\quad t = 1 + n \cdot 3$ $\quad t = 3 \times n + 1$

None of the rules above need parentheses, because order of operations tells you to multiply before you add. However, it is not incorrect to include them. Some rules do need parentheses, so be careful when you write your rules.

**Remember**

Order of operations:

- Evaluate expressions inside parentheses and above and below fraction bars.
- Do multiplications and divisions from left to right.
- Do additions and subtractions from left to right.

**1** Have students explore how the values of the second variable change from one term to the next.

**2** Discuss students' answers.

**3** Remind students that when a number is multiplied by a variable, the multiplication symbol is usually left out.

**4** Have students follow the order of operation rules when evaluating expressions.



**Additional Answers**
**Problem Set A**

**3.**

**Term 5**

**a.** number of squares = term number = 5

**b.** number of vertical toothpicks = term number + 1 = 5 + 1 = 6

**c.** number of horizontal toothpicks = 2 · term number = 2 · 5 = 10

**d.** number of rectangles = term number · $\frac{\text{term number} + 1}{2}$ = $5 \cdot \frac{6}{2} = 15$

• *Additional Answers continued on page A662*

## Develop

1 **Problem Set B** **Suggested Grouping: Pairs**
Problem Set B gives more practice in identifying variables in toothpick sequences and finding rules that relate the values of the variables to the term numbers. In these problems, students work with rules that are written in symbolic form.

In **Problem 1,** students are asked to write their rules from Problem Set A in shorter, symbolic forms by using letters for the variables. They use *n* to represent the first variable (the term number), but they may choose any other letter they please for the other variable. A helpful
2 convention is to choose a letter that will serve as a reminder of what the letter represents. For example, *s* (the first letter of "square") might be used if the number of squares is the second variable. Any letter (other than *n*) can be used for the second variable provided the student states clearly what the letter represents.

In **Problems 2–4,** students are given new toothpick sequences and a dot sequence. They identify a second variable suggested by the figures in each sequence and make a table that shows the values of the second variable. They then find a rule that relates the term number to the value of the second variable and write the rule in symbolic form.

**Problems 5 and 6** take the opposite approach: students use a rule expressed in symbolic form and use it to list the first four terms in the number sequence. They then try to discover and draw a toothpick sequence or dot sequence that fits the rule and the number sequence. Students may find this a bit challenging. In Problem 5, a helpful approach may be to imagine a toothpick sequence that starts with 6 toothpicks and then imagine how to generate subsequent figures by adding on 4 new toothpicks each time. A similar hint applies to Problem 6.

3 **Problem Set Wrap-Up** You might ask students to share their variables from Problems 2–4 and their sequences from Problems 5 and 6. The variety of answers may surprise them.

## Problem Set B

1. See Additional Answers.

**1.** Rewrite your rules from Problem Set A in a shorter form by using $n$ for the term number and a different letter for the other variable. Make sure to state what variable each letter represents.

**2.** Consider this toothpick sequence:

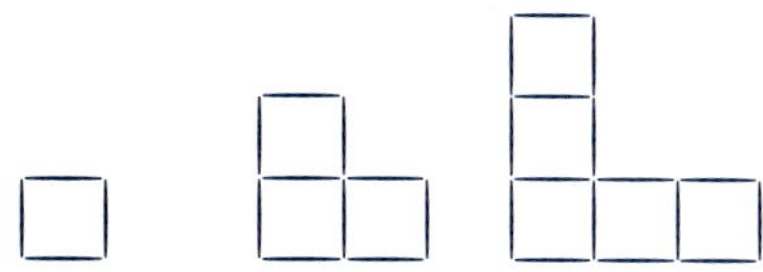

Term 1 Term 2 Term 3

**a.** Choose a variable other than the term number. See margin.

**b.** Create a table showing the value of your variable for each term.

| Term Number, $n$ | 1 | 2 | 3 | 4 |
|---|---|---|---|---|
| Your Variable | | | | |

**2b, c.** See Additional Answers.

**c.** Try to find a rule that connects the term number and your variable. Write the rule as simply as you can, using $n$ to represent the term number and a different letter to represent your variable.

**3.** Complete Parts a–c of Problem 2 for this sequence of toothpicks:

Term 1 Term 2 Term 3 Term 4

**4.** Complete Parts a–c of Problem 2 for this sequence of dots:

Term 1 Term 2 Term 3 Term 4

**5.** Consider the rule $t = 4 \cdot n + 2$, where $n$ represents the term number and $t$ represents the number of toothpicks in a sequence.

**a.** Write the first four numbers in the sequence. 6, 10, 14, 18

**b.** Draw a toothpick sequence that fits the rule. See Additional Answers.

**6.** Consider the rule $d = 3 \cdot (n + 1)$, where $n$ represents the term number and $d$ represents the number of dots in a sequence.

**a.** Write the first four numbers in the sequence. 6, 9, 12, 15

**b.** Draw a dot sequence that fits the rule.

Possible sequence:

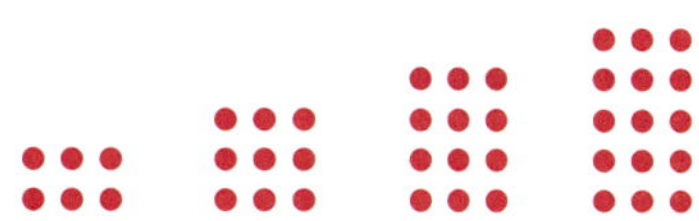

Term 1 Term 2 Term 3 Term 4

**3, 4.** See Additional Answers.

### Remember

When a number is multiplied by a quantity in parentheses, the multiplication symbol is often left out. So, $3 \cdot (n + 1)$ can be written $3(n + 1)$.

**2a.** Possible variables: number of toothpicks, number of squares, number of outer toothpicks, number of inner toothpicks

**1** You may have students work in pairs.

**2** Have students choose a letter that will serve as a reminder of what the letter represents.

**3** Have students share the variables they chose.



### Additional Answers

**1.** Possible answer:

**a.** $s = n$, where $s$ represents the number of squares

**b.** $v = n + 1$, where $v$ represents the number of vertical toothpicks

**c.** $h = n \cdot 2$, where $h$ represents the number of horizontal toothpicks

**d.** $r = n \cdot \frac{n+1}{2}$, where $r$ represents the total number of rectangles

• ***Additional Answers continued on page A662***

# Develop

**1** **Share & Summarize**
Give students time to write out their answers to these questions. For **Question 1,** you may want to suggest that students show at least four terms in their toothpick or dot sequence. In **Question 2,** the variables should be described clearly in words. Each description should make it possible for the student or another person to easily determine the value of the variable for any desired term of the sequence. In **Question 3,** students should express their rules in clear and accurate symbolic form. They should check that the rules are written in agreement with order of operations rules.

**2** **Troubleshooting** Students may have difficulty writing symbolic rules that relate values of variables. If so, they may find it helpful first to write the rule in words and then to translate the verbal rule into symbolic form. You might offer them the example below.

**Additional Example** Lorraine studied the following dot sequence:

Term 1 Term 2 Term 3 Term 4

She wrote the following rule relating the term number to the number of "outside" dots in the term: *the number of outside dots equals 4 times the term number.*

In all, how many dots will there be in Term 5? *36*

How many dots in Term 5 are outside dots? *20*

Does Lorraine's rule correctly predict the number of outside dots in Term 5? Explain. *yes;* $4 \cdot 5 = 20$

Write Lorraine's rule using $n$ as the term number and $d$ as the number of outside dots in the term. $d = 4 \cdot n$

### On Your Own Exercises

Practice & Apply: 1–5, pp. 422–423
Connect & Extend: 14–16, pp. 425–426
Mixed Review: 22–34, p. 429

# Investigation 2

In this investigation, students see that they can use different symbolic expressions to describe the same sequence. Their explanations of the rules should show that the rules must work for any term in the sequence. Their roles will typically describe how to build the same toothpick or dot sequence in different ways. Such rules are said to be *equivalent.* This work will give students a good foundation for expanding, rearranging, and simplifying expressions that contain variables.

Look at the given toothpick sequence with the class, and read the two rules written to describe the sequence.

**3** **Think & Discuss**
Have students work in pairs to answer the Think & Discuss questions. Students need to understand that stating three instances in which a rule works is not enough to prove that the rule *always* works.

### On the Spot Assessment

Some students may have difficulty performing the calculations in the Think & Discuss. Encourage students to show all their work. This will help you determine the source of any errors students have made. If there are errors involving order of operations, you may need to remind students of the order-of-operations rules.

2. Possible variables: total number of dots, number of dots in bottom row

3. Possible answer: $t = n \cdot n + 1$, where $t$ is the total number of dots and $n$ is the term number; $b = n$, where $b$ is the number of dots in the bottom row and $n$ is the term number

## Share & Summarize

1. Possible sequence:

Term 1 Term 2 Term 3 Term 4

1. Draw a toothpick or dot sequence. Make sure your sequence changes in a predictable way.
2. Name two variables in your sequence.
3. For each variable you named, try to write a rule relating the term number to the variable. Use letters to represent the variables, and tell what each letter represents.

1 Allow students enough time to write out their answers.

2 You might suggest students write the rule in words, then translate it into symbolic form.

# Investigation 2 Are These Rules the Same?

Sometimes two people can look at the same pattern and write rules that look very different. This may have happened when you and your classmates wrote rules in the last investigation.

**Remember**
The multiplication symbol is often left out when a number is multiplied by a variable. So, $2n$ is the same as $2 \cdot n$.

Consider this toothpick sequence:

Term 1 Term 2 Term 3 Term 4

Rosita and Conor wrote rules for the number of toothpicks in each term. Both students used $n$ to represent the term number and $t$ to represent the number of toothpicks.

Rosita's rule: $t = 3 + 2 \cdot (n - 1)$ Conor's rule: $t = 1 + 2n$

## Think & Discuss

Use the two rules to find the number of toothpicks in Term 10. Check your results by drawing Term 10 and counting toothpicks. See ① in Additional Answers.

Show that both rules give the same result for Term 20 and for Term 100. See ② below.

Do you think the rules will give the same result for every term? Answers will vary.

3 Encourage students to show all of their work.

One way to show that the two rules will give the same result for every term is to explain why both rules must work for any term in the sequence.

② Rosita's rule: $t = 3 + 2 \cdot (20 - 1) = 41$, $t = 3 + 2 \cdot (100 - 1) = 201$
Conor's rule: $t = 1 + 2 \cdot 20 = 41$, $t = 1 + 2 \cdot 100 = 201$



**Additional Answers**
**Think & Discuss**

①

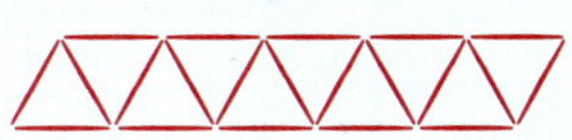

Term 10

There are 21 toothpicks in Term 10. Rosita's rule gives $t = 3 + 2 \cdot (10 - 1) = 21$, and Conor's rule gives $t = 1 + 2 \cdot 10 = 21$. Both rules give the correct result.

# Develop

1 **Example**
Discuss with the class the Example, presented in the cartoon. This Example demonstrates clearly that the toothpick sequence on page 414 can be constructed in two different ways. Rosita's rule describes one of these ways, and Conor's rule describes the other. Since the same sequence results no matter which rule is used, the rules are equivalent.

**EXAMPLE**

Rosita explains why her rule works.

Conor explains why his rule works.

Rosita's and Conor's rules both correctly describe the toothpick sequence, so they *will* give the same result for every term. Two rules that look different, but that describe the same relationship, are said to be *equivalent.*

**1** Discuss the different rules presented and how they are equivalent.



LESSON 7.1 Patterns and Variables **415**

# Develop

## 1 Problem Set C Suggested Grouping: Small Groups

In this problem set, students use toothpick sequences and
dot sequences to consider equivalent rules. You should
2 allow plenty of time for these problems, since many stu-
dents may find them challenging.

In **Problem 1,** students are given a toothpick sequence
and two equivalent rules that describe the sequence. A
diagram is provided to show why one of the rules cor-
rectly describes the number of toothpicks in Term *n*.
Students are asked to use diagrams to explain why the
3 other rule also works. They may need to make sketches
and try several ideas before they hit on an explanation
for the second rule.

1. See Additional Answers.

## Problem Set C

1. Consider this sequence:

Term 1 Term 2 Term 3

Althea and Marcus wrote equivalent rules for this sequence. Both students used $n$ to represent the term number and $t$ to represent the number of toothpicks.

Althea's rule: $t = 2 \cdot n + 4$ Marcus' rule: $t = 2 \cdot (n + 2)$

Althea used diagrams to explain why her rule is correct.

The ⟨ and ⟩ at the ends of each term make a total of four toothpicks. The number of toothpicks along the top of a term is equal to the term number. So is the number of toothpicks along the bottom.

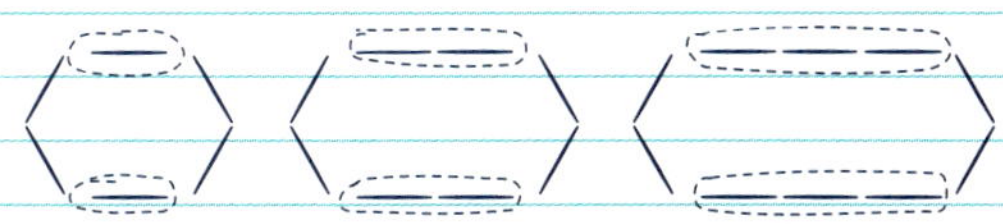

| Term 1 | Term 2 | Term 3 |
|---|---|---|
| 2 groups of 1, plus 4 | 2 groups of 2, plus 4 | 2 groups of 3, plus 4 |

So, Term n has 2 • n + 4 toothpicks.

Use diagrams to help explain why Marcus' rule is correct.

2. Consider this sequence:

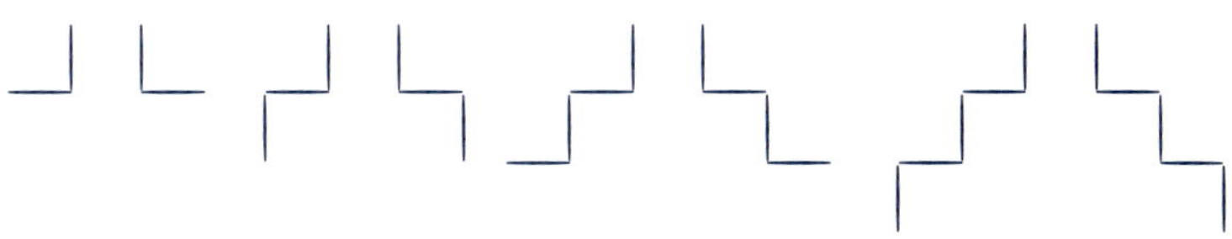

Term 1 Term 2 Term 3 Term 4

Caroline and Jing wrote equivalent rules for this sequence. Both girls used $n$ to represent the term number and $t$ to represent the number of toothpicks.

Caroline's rule: $t = 2n + 2$ Jing's rule: $t = 2 \cdot (n + 1)$

The Mayan temple of Kukulcan on the Yucatan Peninsula, Mexico



1 You may have students work in small groups.

2 Allow plenty of time for students to complete these problems.

3 Problem-Solving Strategy
Draw a diagram

### Additional Answers

1. Possible explanation: Each term has two halves, and each half has a number of toothpicks equal to the term number plus 2. So, Term $n$ has $2 \cdot (n + 2)$ toothpicks.

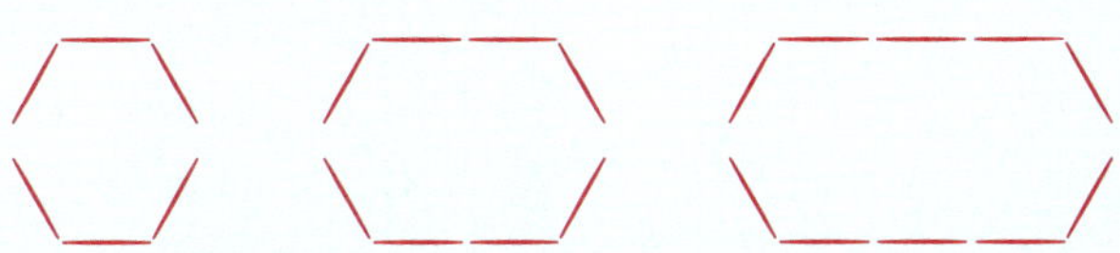

| Term 1 | Term 2 | Term 3 |
|---|---|---|
| 2 groups of 3 | 2 groups of 4 | 2 groups of 5 |

# Develop

1 In **Problems 3 and 4,** students are given sequences and are asked first to write two equivalent rules for the sequence and then to explain why each rule is correct. There is no sure-fire method for coming up with rules or explanations. To find a rule for a sequence, it is often helpful to make a table of values. In these problems, tables that show the term number in the top row and the number of toothpicks or dots in the second row may suggest a rule for the sequence. A good approach is to observe how the numbers in the second row change from one term to the next. One can also inspect the toothpick or dot sequence itself to look for a pattern. It may be helpful to use both approaches in conjunction.

**2b, c.** See Additional Answers.

**3b.** See Additional Answers.

**4a.** Possible rules: $d = 2 + 4n$, $d = 2 \cdot (2n + 1)$

**4b.** See Additional Answers.

**Just the facts**

Morse code is a system of communication in which letters, numbers, and punctuation marks are represented by combinations of dots, dashes, and spaces. Morse code messages are transmitted as electrical pulses of various lengths—a short pulse represents a dot, and a long pulse represents a dash.

**a.** Copy and complete the table to show that both rules work for the first five terms of the sequence.

| Term Number, $n$ | 1 | 2 | 3 | 4 | 5 |
|---|---|---|---|---|---|
| Number of Toothpicks, $t$ | 4 | 6 | 8 | 10 | 12 |
| $2n + 2$ | 4 | 6 | 8 | 10 | 12 |
| $2 \cdot (n + 1)$ | 4 | 6 | 8 | 10 | 12 |

**b.** Use words and diagrams to explain why Caroline's rule is correct.

**c.** Use words and diagrams to explain why Jing's rule is correct.

**3.** Consider this sequence:

Term 1 Term 2 Term 3

**a.** Write two equivalent rules for the number of toothpicks in each term. Possible rules: $t = 2 \cdot (n + 2)$, $t = 4 + 2n$

**b.** Use words and diagrams to explain why each rule is correct.

**1** Problem-Solving Strategy
Make a table

**4.** Consider this sequence:

Term 1 Term 2 Term 3

**a.** Write two equivalent rules for the number of dots in each term.

**b.** Use words and diagrams to explain why each rule is correct.

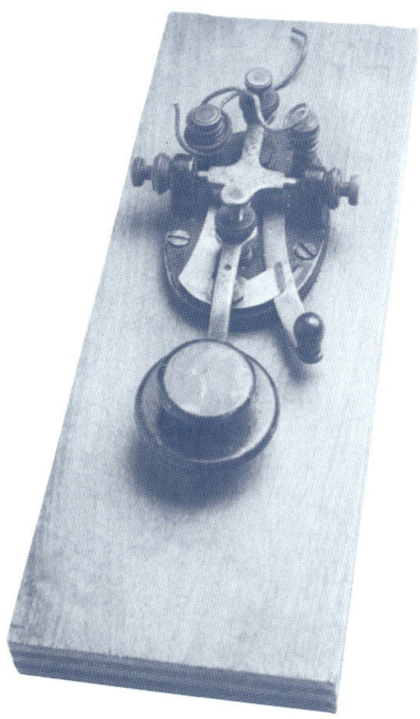



**Additional Answers**
**Problem Set C**

**2b.** Possible answer: Each term has two vertical toothpicks at the top, and two equal groups of toothpicks. The number of toothpicks in each group is equal to the term number. So, the number of toothpicks in Term $n$ is $2n + 2$.

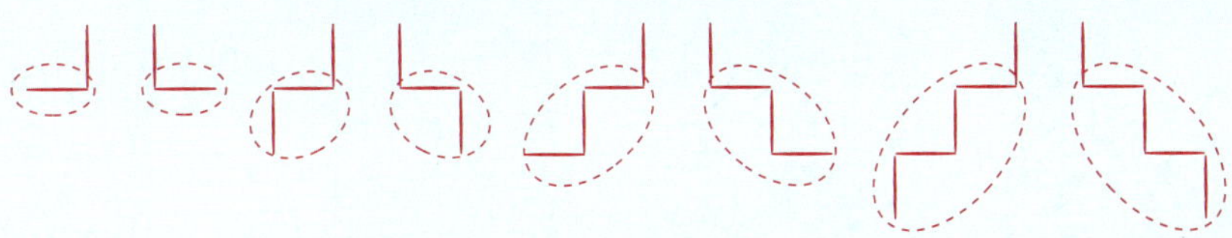

| Term 1 | Term 2 | Term 3 | Term 4 |
|---|---|---|---|
| 2 groups of 1, plus 2 | 2 groups of 2, plus 2 | 2 groups of 3, plus 2 | 2 groups of 4, plus 2 |

• *Additional Answers continued on page A663*

## Develop

1 In **Problem 5,** students should note that Rosita's rule uses the number of hexagons and not the term number. However, the toothpick diagrams make it clear that the number of hexagons in a term is the same as the term number.

2 Students can quickly rule out Jahmal's rule. If they make tables of values for each rule, they can see that Jahmal's rule does not correctly predict the number of toothpicks in Terms 2 and 3.

Students may want to test whether Conor's and Rosita's rules correctly predict the number of toothpicks in Terms 4 and 5. Next they can examine the sequence more closely to come up with a reasoned explanation of why these two rules always work.

3 **Problem Set Wrap-Up** Having students share some of their rules and explanations for Problems 3 and 4 might be very helpful. Coming up with rules and explanations of why the rules work can take time and patience. Students must be willing to try different ideas. With practice, they can improve their ability to solve problems of this kind.

### Share & Summarize

4 These questions will help students focus on what it takes to be confident that two rules are equivalent. You may want to have students work on these questions in small groups. Then have the groups present their answers in a discussion with the whole class.

**Troubleshooting** Thinking about rules that can be directly linked to a visual pattern can be helpful for many visual learners. You may want to discuss the following example before you go to Investigation 3. In Investigation 3, students will continue their study of rules, but without the emphasis on visual models.

**Additional Example** Use diagrams to show the first four terms of a dot sequence described by the rule $d = 3 + 5n$, where $n$ is the term number and $d$ is the number of dots in the term.

*Possible dot sequence:*

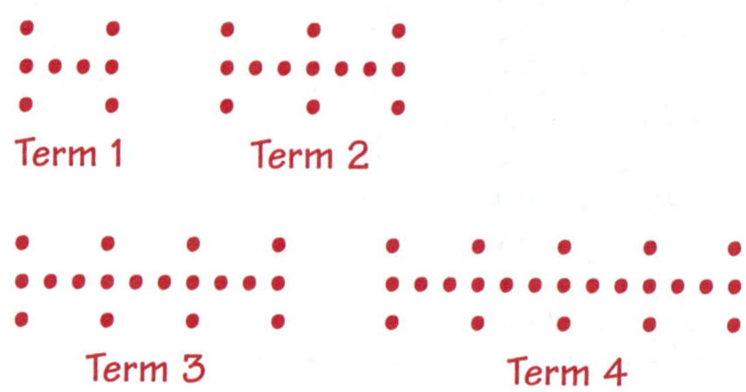

### On Your Own Exercises

Practice & Apply: 6–8, pp. 423–424
Connect & Extend: 17, 18, pp. 426–427
Mixed Review: 22–34, p. 429

5. Conor, Jahmal, and Rosita each wrote a rule for this sequence:

Term 1 Term 2 Term 3

Conor's rule: $t = 5n + 1$, where $t$ is the number of toothpicks and $n$ is the term number

Jahmal's rule: $t = 3 \cdot n + 3$, where $t$ is the number of toothpicks and $n$ is the term number

Rosita's rule: $t = 6h - (h - 1)$, where $t$ is the number of toothpicks and $h$ is the number of hexagons

Determine whether each rule correctly describes the sequence. If it does, explain how you know it is correct. If it does not, explain why it isn't correct. See Additional Answers.

6. Consider strips of T-shapes like this one, in which a strip can have any number of T-shapes.

a. Find a rule that connects the number of toothpicks in a strip to the number of T-shapes.

b. Explain why your rule is correct. Use diagrams if they help you to explain. See Additional Answers.

6a. Possible rule: $t = 3 + 7 \cdot k$, where $k$ is the number of T-shapes and $t$ is the number of toothpicks

1 Remind students that Rosita's rule uses the number of hexagons as a variable.

2 Have students explain why Conor's and Rosita's rules always work.

3 It might be helpful for students to share their rules and explanations with the class.

4 You may have students work in small groups.

## Share & Summarize

1. How can you show that two rules for a sequence are equivalent?

2. Suppose two different rules give the same number of dots for Term 4 of a sequence. Can you conclude that the two rules are equivalent? Give an example to support your answer.

no; Possible example: Suppose the rules are $d = n + 4$ and $d = 2 \cdot n$, where $n$ is the term number and $d$ is the number of dots. Both rules give eight dots for Term 4, but the first rule gives five dots for Term 1 while the second gives two.

1. Possible answer: Show that both rules correctly describe the sequence. If both rules are correct, they must be equivalent.



**Additional Answers**

5. See page A663.

6b. Possible explanation: You can think of starting with three toothpicks and adding seven more to form each T-shape. So, a strip with $k$ T-shapes would have $3 + 7 \cdot k$ toothpicks.

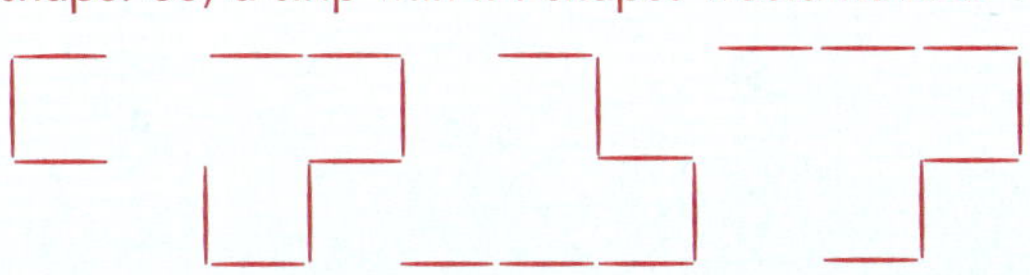

3 +7 +7 +7

• **Additional Answers continued on page A663**

# Investigation 3

In this investigation, students take another look at the *What's My Rule?* game from Chapter 1. Now, however, they are asked to state their rules by using letters to represent input and output variables.

Students are reminded that rules that describe the same relationship are equivalent. Knowing that two rules give the same outputs for a limited number of inputs is not sufficient to prove that the rules are equivalent. On the other hand, a single instance of rules that give different outputs for the same input is enough to show that the rules are *not* equivalent.

1 Review the directions for playing *What's My Rule?* with the class. Discuss the idea of writing rules for the game by using letters to represent the input and output variables.

### Think & Discuss

This introduction focuses on the idea that symbolically written rules can be used to find outputs for given inputs in games of *What's My Rule?* Use a class discussion to
2 draw attention to the similarities among the rules that Rosita, Jahmal, and Althea are using. You may want to put input-output tables for the rules on the board. Label these *Rosita's Table, Jahmal's Table,* and *Althea's Table.* Use the same input values in each table. Guide the discussion to help students observe that the order in which they perform the operations is the only significant difference between Rosita's and Jahmal's rules. In the case of Rosita's and Althea's rules, there is a superficial difference in the notation.

3 **Access for all Learners**

**Extra Help** Although the use of different letters for the input and output is only a superficial difference, some students may not see it that way. Auditory learners may find it helpful to describe to you in their own words what each rule says should be done to find the output for a given input. Check their explanations for clarity, and help them check whether the verbal description matches the symbolic rule.

## Investigation 3 What's My Rule?

In Chapter 1, you played the game *What's My Rule?* Here's how you play:

- One player, the rule-maker, thinks of a secret rule for calculating an output number from a given input number. For example:

  *To find the output, add 3 to the input and multiply by 4.*

- The other players take turns giving the rule-maker input numbers. For each input, the rule-maker calculates the output and says the result out loud.
- By comparing each input to its output, the players try to guess the secret rule. The first player to guess the rule correctly wins.

In *What's My Rule?* the input and output are variables. In this investigation, you will play *What's My Rule?* using letters to represent these variables. For example, if you let $i$ represent the input and $o$ represent the output, you can write the rule above as

$$o = (i + 3) \cdot 4$$

**1** Review the directions for playing *What's My Rule?* with the class.

### Think & Discuss

Rosita, Jahmal, and Althea are playing *What's My Rule?* Rosita's secret rule is $a = 3b + 4$, where $a$ is the output and $b$ is the input.

- Jahmal guesses that the rule is $a = 4 + 3b$, where $a$ is the output and $b$ is the input.
- Althea guesses that the rule is $x = 3y + 4$, where $x$ is the output and $y$ is the input.

Rosita is not sure whether the rules Jahmal and Althea wrote are the same as her secret rule. Tell whether each rule is correct, and explain how you know.

**2** **Problem-Solving Strategy** Make a table

**3** Encourage students to choose a letter that will serve as a reminder of what the letter represents.

**Remember**

The multiplication symbol is often left out when a number is multiplied by a variable. So, $3b$ is the same as $3 \cdot b$.

**Think & Discuss Answer**

Both rules are correct. In Jahmal's rule, the addition is written before the multiplication, but since order of operations says that multiplication is done first, his and Rosita's rules will always give the same result. Althea's rule simply uses different letters to represent the variables.



LESSON 7.1 Patterns and Variables 419

# Develop

1

## Problem Set D **Suggested Grouping: Small Groups**

In this problem set, students play several rounds of the *What's My Rule?* game. The players may want to agree each time on what letters to use for the input and output variables.

The first of the additional rules for these games requires the rule-maker to write the secret rule in symbols. This rule is intended, in part, to limit the rules that are easily expressed in terms of fundamental arithmetical operations.

2

It is quite possible that students will limit their inputs to whole numbers, though there is nothing in the game that says they must do so.

3

**Problem Set Wrap-Up** The group's observations about strategies may vary widely. Therefore, you will probably find it helpful to have a class discussion of results. You might ask the groups to keep track of the rules that were easiest and hardest to discover and to comment on what makes a rule easy or hard to guess.

4

## Problem Set E **Suggested Grouping: Small Groups**

In this problem set, students use given tables of inputs and outputs and test proposed rules to see whether they could be correct.

In **Problem 1,** some students may observe that the first rule they test is equivalent to $p = 3 \cdot q + 4$. This would probably be enough for them to see that the first and second rules cannot both be correct.

In **Problem 3,** you may want to help students observe that the rules are equivalent because they say the same thing in a slightly different form. Therefore, as soon as they determine that the first rule does not work, they know that the second rule will not work either.

**Problem Set D Answer**

Possible answer: I found that guessing numbers in order, starting with 0, helped me guess the rule. After the first guess, I wrote down several rules that might work and then checked to see if any of them worked for the second guess. Rules that have only one operation are easiest to guess. Rules with a lot of operations or parentheses are harder. To see if two rules were the same, we first checked that both gave the same outputs for all of the inputs we used. If they did, we tried to explain why they were the same, like we did in the last lesson. Sometimes we couldn't show they were the same, but we decided that if they gave the same output for four or five different inputs, we would say they were the same.

## Problem Set D

Play *What's My Rule?* with your group, using these added rules:

- The rule-maker should write the secret rule with symbols, using letters for the variables.
- The other players should make a table to keep track of the inputs and outputs.
- When a player guesses the rule, he or she should write it with symbols, using letters for the variables.

Take turns being the rule-maker so everyone has a chance. As you play, you may have to decide whether a guessed rule is equivalent to the secret rule even though it looks different.

After your group has played several games, write a paragraph describing what you learned while playing. In your paragraph, you might discuss the following:

- Strategies you used to help you guess the rule
- A description of what makes a rule easy to guess and what makes a rule difficult to guess
- Strategies you used to decide whether two rules were equivalent even when they looked different

**1** You may have students work in small groups.

**2** Challenge students to use inputs other than whole numbers.

**3** Discuss groups' results.

## Problem Set E

These tables were made during games of *What's My Rule?* Two rules are given for each table. Determine whether each rule could be correct, and explain how you know. **1–3.** See below.

**4** You may have students work in small groups.

1.

| $q$ | 1 | 2 | 3 | 10 |
|---|---|---|---|---|
| $p$ | 7 | 11 | 15 | 43 |

$p = q + q + q + 4$

$p = 4 \cdot q + 3$

2.

| $s$ | 1 | 2 | 5 | 10 |
|---|---|---|---|---|
| $t$ | 10 | 20 | 50 | 100 |

$t = 4 \cdot s + 6 \cdot s$

$t = 10s$

3.

| $k$ | 0 | 2 | 5 | 10 |
|---|---|---|---|---|
| $j$ | 1 | 17 | 101 | 402 |

$j = 5 \cdot k^2 + 1$

$j = 5 \cdot k \cdot k + 1$

**Remember**

An exponent tells how many times a number is multiplied together. For example, $4^2 = 4 \cdot 4$ and $t^4 = t \cdot t \cdot t \cdot t$.

In the next problem set, you will try to figure out the rules for some *What's My Rule?* games.

1. Possible explanation: $p = q + q + q + 4$ is not correct because it works only for $q = 1$ and $p = 7$. $p = 4 \cdot q + 3$ could be correct because it works for all the values in the table.
2. Possible explanation: Both rules could be correct because they work for all the values in the table.
3. Neither rule is correct. Possible explanation: Both rules give an output of 21 for an input of 2, but the table lists 17 as the output for an input of 2.

# Develop

1 **Problem Set F** **Suggested Grouping: Pairs**
In this problem set, students must discover a rule that relates the inputs and outputs that are presented in each of several tables. Students should test each rule they pro-
2 pose to be sure it gives the correct output for each input shown. The inputs in the tables have been selected more or less at random. As a result, it may take some time for students to come up with a rule for some of the tables.

3 In **Problem 1,** it may help students to look at what happens if the inputs are arranged in order from least to greatest. It then appears that increasing the input by 1 causes the output to increase by 4.

In **Problem 2,** the first three inputs are whole numbers that increase by 1 each time. When the input increases by 1, the output increases by 5. This may help students get started in their search for a rule.

4 In **Problems 3 and 4,** the fractions and decimals may suggest that it would be a good idea to consider rules that involve division.

5 **Problem Set Wrap-Up** For each problem, have students share the various rules they wrote. Remind them that rules that look different but describe the same relationship are said to be equivalent.

## Share & Summarize

Give students time to write out their answers to the questions. Then have volunteers read their answers. Discuss the answers with the class. For **Question 1,** be sure students make appropriate use of parentheses when they write Abby's rule in symbols.

6 **Troubleshooting** If students are having difficulty generating rules for tables, review with them specific strategies that might help. For example, suggest putting inputs in order in the input-output tables. Remind students to think about multiples and squares. You might offer the following example.

**Additional Example** A student played *What's My Rule?* and came up with the following input-output table:

| $a$ | 5 | 6 | 7 | 1 | 2 |
|---|---|---|---|---|---|
| $b$ | 36 | 43 | 50 | 8 | 15 |

If the input increases by 1, what seems to happen to the output? **It seems to increase by 7.**

What does this suggest about a rule for the table? **The rule might involve multiplication by 7.**

What rule seems to work for this table? $b = 7a + 1$

### On Your Own Exercises

Practice & Apply: 9–13, pp. 424–425
Connect & Extend: 19–21, pp. 428–429
Mixed Review: 22–34, p. 429

## Problem Set F

These tables were made during games of *What's My Rule?* In each table, the values in the top row are the inputs and the values in the bottom row are the outputs.

Write a rule for each table, using the given letters to represent the variables.

1.

| $a$ | 2 | 5 | 3 | 6 | 1 |
|---|---|---|---|---|---|
| $b$ | 9 | 21 | 13 | 25 | 5 |

2.

| $y$ | 4 | 5 | 6 | 1 | $\frac{3}{5}$ |
|---|---|---|---|---|---|
| $z$ | 18 | 23 | 28 | 3 | 1 |

3.

| $w$ | $\frac{12}{7}$ | 11 | 19 | 4 | 7 |
|---|---|---|---|---|---|
| $g$ | $\frac{2}{7}$ | $1\frac{5}{6}$ | $3\frac{1}{6}$ | $\frac{2}{3}$ | $1\frac{1}{6}$ |

4.

| $q$ | 10 | 5.5 | 1 | 2 | 3 |
|---|---|---|---|---|---|
| $p$ | 6 | 3.75 | 1.5 | 2 | 2.5 |

5.

| $c$ | 100 | 42 | 17 | 1 | 0.3 |
|---|---|---|---|---|---|
| $d$ | 10,000 | 1,764 | 289 | 1 | 0.09 |

6.

| $s$ | 1 | 3.1 | 10 | 5 | 6.5 |
|---|---|---|---|---|---|
| $t$ | 3 | 11.61 | 102 | 27 | 44.25 |

**1.** Possible rule: $b = 4a + 1$

**2.** Possible rule: $z = 5y - 2$

**3.** Possible rule: $g = w \div 6$

**4.** Possible rule: $p = (q \div 2) + 1$

**5.** Possible rules: $d = c^2$, or $d = c \cdot c$

**6.** Possible rules: $t = s^2 + 2$, or $t = s \cdot s + 2$

## Share & Summarize

Abby and Ji-Young were playing *What's My Rule?* Ji-Young's secret rule was "To get the output, multiply the input by itself and subtract 1." Abby guessed that the rule was "Subtract 1 from the input, and multiply the result by itself."

1. Write both rules with symbols. Use $m$ to represent the output and $n$ to represent the input.
2. Is Abby's rule equivalent to Ji-Young's rule? Explain.

**1.** Ji-Young's rule: $m = n^2 - 1$, or $m = n \cdot n - 1$; Abby's rule: $m = (n - 1)^2$, or $m = (n - 1) \cdot (n - 1)$

**2.** No; the two rules give the same output only when the input is 1.



**1** You may have students work in pairs.

**2** Problem-Solving Strategy
Make and test a conjecture

**3** It may be helpful for students to arrange the inputs from least to greatest.

**4** Stress that fractions and decimals usually suggest rules that involve division.

**5** Wrap-up the activity by having students discuss their various rules.

**6** It may be helpful for students to look for multiples and squares when looking for rules.

LESSON 7.1 Patterns and Variables 421

# On Your Own Exercises

## On Your Own Exercises

Investigation 1, pp. 411–414
Practice & Apply: 1–5
Connect & Extend: 14–16

Investigation 2, pp. 414–418
Practice & Apply: 6–8
Connect & Extend: 17, 18

Investigation 3, pp. 419–421
Practice & Apply: 9–13
Connect & Extend: 19–21

Assign Anytime
Mixed Review: 22–34

## Practice & Apply

**1a.** 2, 3, 4, 5, . . .

**2a.** 3, 5, 7, 9, . . .

**2b.** Possible rule: $d = 2n + 1$, where $d$ represents the number of dots and $n$ represents the term number

**1.** Consider this sequence of toothpick “houses”:

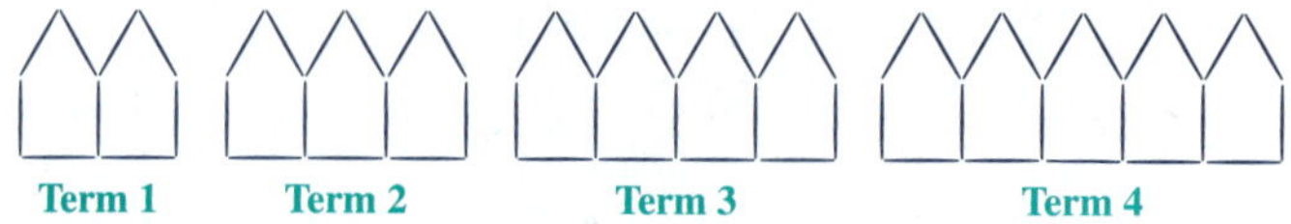


**a.** Write a number sequence for the number of houses in each term.

**b.** Write a rule that connects the number of houses to the term number. Use letters to represent the variables, and tell what each letter represents. **1b.** Possible rule: $h = n + 1$, where $h$ represents the number of houses and $n$ represents the term number

**2.** Consider this dot sequence:

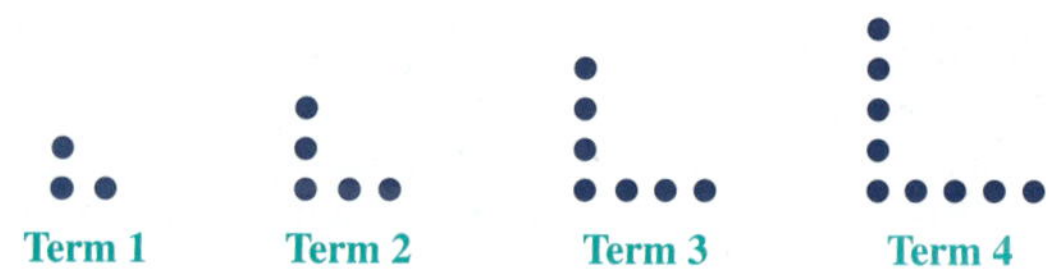


**a.** Write a number sequence for the number of dots in each term.

**b.** Write a rule that connects the number of dots to the term number. Use letters to represent the variables, and tell what each letter represents.

In Exercises 3 and 4, give the first four numbers in the sequence. Then draw a sequence of toothpicks or dots that fits the rule.

**3.** $t = 4 \cdot (k + 1)$, where $t$ represents the number of toothpicks and $k$ represents the term number See Additional Answers.

**4.** $d = 3 \cdot p - 2$, where $d$ represents the number of dots and $p$ represents the term number 1, 4, 7, 10, . . .

Possible sequence:

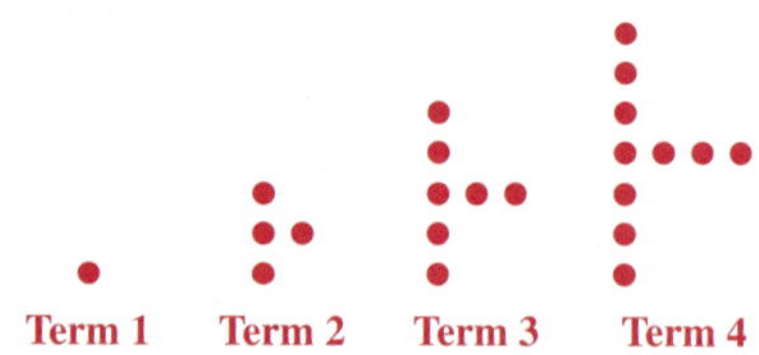


### Just the facts

Braille is a system of writing used by the blind. Each Braille character is made from 1 to 6 raised dots. The system was invented in 1824 by Louis Braille while he was a student at the National Institute for Blind Youth in Paris.

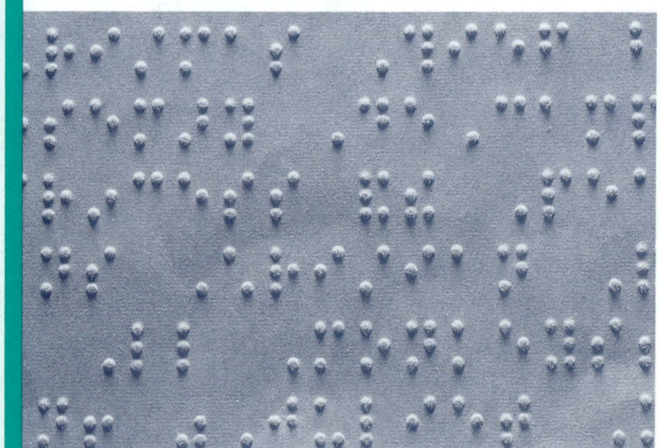



impactmath.com/self_check_quiz

## Additional Answers

**3.** 8, 12, 16, 20, . . .

Possible sequence:

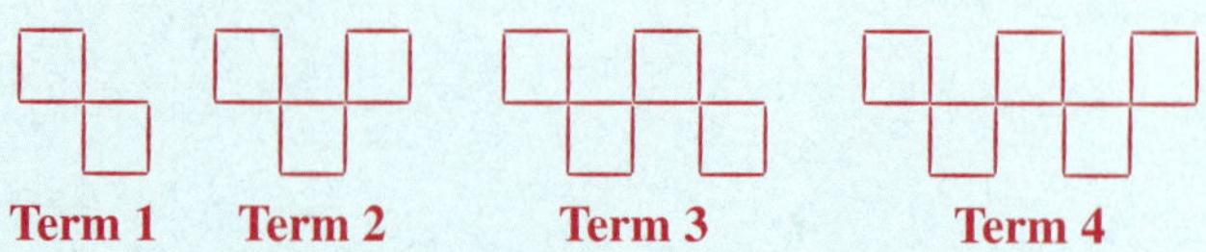

**5.** Consider this sequence:

Term 1 Term 2 Term 3 Term 4

**a.** Write two equivalent rules for the number of toothpicks in each term.

**b.** Use words and diagrams to explain why each of your rules is correct. See Additional Answers.

**6.** Three students wrote rules for this dot sequence. Determine whether each rule correctly describes the pattern. If it does, explain how you know it is correct. If it does not, explain why it is not correct.

Term 1 Term 2 Term 3 Term 4

**a.** Ilsa's rule: $d = 5 \cdot m$, where $d$ is the number of dots and $m$ is the term number

**b.** Mattie's rule: $d = 1 + 4k$, where $d$ is the number of dots and $k$ is the term number See Additional Answers.

**c.** Mateo's rule: $d = 4 \cdot (j - 1) + 5$, where $d$ is the number of dots and $j$ is the term number See Additional Answers.

**7.** When tiling a walkway, a particular contractor surrounds each pair of blue tiles with white tiles as shown at right. Four copies of this design are put together below.

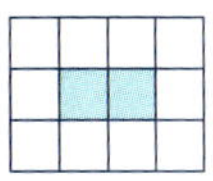

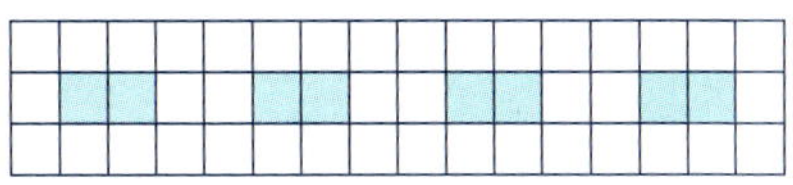

**a.** Copy and complete the table to show the number of white tiles needed for each given number of blue tiles.

| Blue Tiles | 2 | 4 | 6 | 8 | 20 | 100 |
|---|---|---|---|---|---|---|
| White Tiles | 10 | 20 | 30 | 40 | 100 | 500 |

**b.** Find a rule that describes the connection between the number of white tiles and the number of blue tiles. Use letters for the variables in your rule, and tell what each letter represents.

**c.** Explain how you know your rule is correct. Use diagrams if you need to.

**5a.** Possible rules: $t = 2 \cdot n + 1$, $t = 3 + 2 \cdot (n - 1)$, where $t$ is the number of toothpicks and $n$ is the term number

**6a.** incorrect; Possible explanation: Her rule gives $d = 5 \cdot 2 = 10$ dots in Term 2, but there are only 9.

**7b.** Possible rule: $w = 5 \cdot b$, where $b$ is the number of blue tiles and $w$ is the number of white tiles

**7c.** Possible explanation: Each section of the design has five times as many white tiles as blue tiles (2 blue and 10 white). When the sections are put together, there will always be five times as many white tiles as blue tiles.



**Exercise 5:**
Students may find it helpful to make a table showing term numbers and term values as they search for rules. Their rules may suggest how diagrams can be used to argue that the rules always work.

**Exercise 6:**
To eliminate the rule in **Part a,** students need only find one case in which it fails to predict the term value correctly. However, students should confirm that the rules in **Parts b and c** give correct values for all four terms shown in the dot sequence. They must then explain why the rules will continue to give correct values.

**Exercise 7:**
This exercise is somewhat different from the preceding exercises in which students were asked to find a rule that gives the value of a term in terms of the term number. Here, the value of $b$ (the number of blue tiles) is always an even number. Nevertheless, students should notice fairly quickly that each number in the second row of the table for **Part a** is 5 times the number of blue tiles.

**Additional Answers**

**6b.** correct; Possible explanation: Each term has one center dot and four "arms." The number of dots in each arm is equal to the term number. So, the number of dots in Term $k$ is $1 + 4k$.

| Term 1 | Term 2 | Term 3 | Term 4 |
|---|---|---|---|
| 4 groups of 1, plus 1 | 4 groups of 2, plus 1 | 4 groups of 3, plus 1 | 4 groups of 4, plus 1 |

• ***Additional Answers continued on page A664***

LESSON 7.1 Patterns and Variables 423

**Exercise 8a:** Another rule that students may give is $c = (n + 2)^2 - n^2$.

**8a.** Possible rule: $c = 4 + 4 \cdot n$, where $c$ is the number of cubes and $n$ is the term number

**9.** Possible rule: $g = 10 + 3f$

**10.** Possible rule: $k = j \cdot \frac{1}{4} + 3$ or $k = j \div 4 + 3$

**8.** Consider this sequence of cubes:

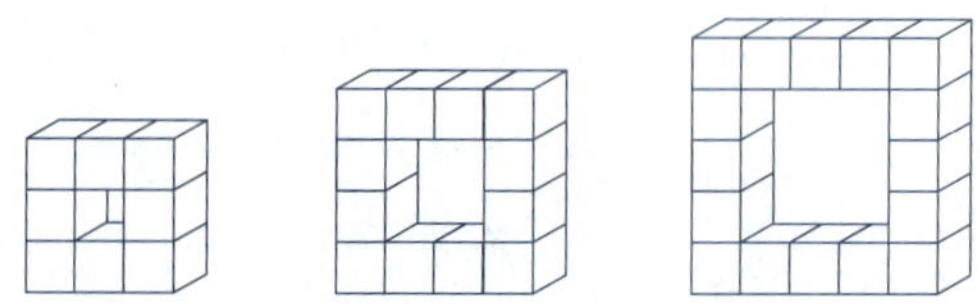

Term 1 Term 2 Term 3

**a.** Find a rule for the number of cubes in each term. You may want to make a table first. Use letters for the variables in your rule, and tell what each letter represents.

**b.** Explain how you know your rule is correct. Use diagrams if you need to. See Additional Answers.

The tables below were made during games of *What's My Rule?* The values in the top row are inputs, and the values in the bottom row are outputs. Write a rule for each table, using the given letters to represent the variables.

**9.**

| $f$ | 11 | 4 | 1 | 7 | 2 |
|---|---|---|---|---|---|
| $g$ | 43 | 22 | 13 | 31 | 16 |

**10.**

| $j$ | 12 | 9 | 2 | 16 | 23 |
|---|---|---|---|---|---|
| $k$ | 6 | 5.25 | 3.5 | 7 | 8.75 |

**Remember**

An exponent tells you how many times a number is multiplied together. For example, $4^2 = 4 \cdot 4$ and $t^4 = t \cdot t \cdot t \cdot t$.

**11.** In a game of *What's My Rule?* Hiam's secret rule was $a = b^2 \cdot 3$, where $b$ is the input and $a$ is the output. Complete the table to show the outputs for the given inputs.

| $b$ | 8 | 0 | 15 | 7 | 4 |
|---|---|---|---|---|---|
| $a$ | 192 | 0 | 675 | 147 | 48 |



## Additional Answers

**8b.** Possible explanation for $c = 4 + 4 \cdot n$: Each term has four corner cubes. If you remove these cubes, each side has a number of cubes equal to the term number. So, Term $n$ has $4 + 4 \cdot n$ cubes.

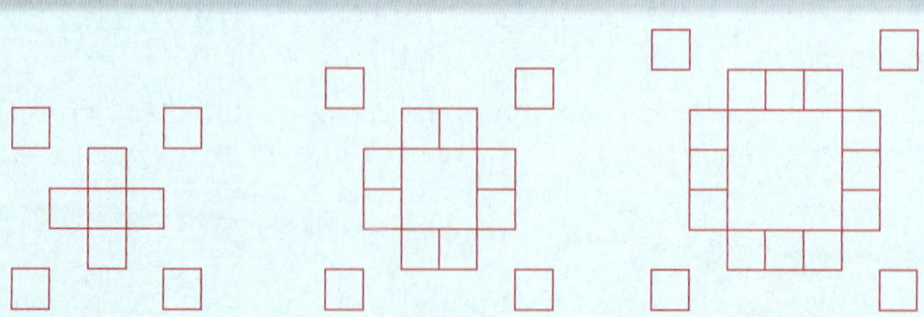

**Term 1** 4 groups of 1, plus 4

**Term 2** 4 groups of 2, plus 4

**Term 3** 4 groups of 3, plus 4

The tables in Exercises 12 and 13 were made during games of *What's My Rule?* Two rules are given for each table. Determine whether each rule could be correct, and explain how you know.

**12.**

| $s$ | 2 | 5 | 11 | 7 | 10 |
|---|---|---|---|---|---|
| $t$ | 5 | 14 | 32 | 20 | 29 |

$s = (t - 1) * 3$

$t = 2s + (s - 1)$

**12.** See Additional Answers.

**13.**

| $p$ | 2 | 5 | 1 | 4 | 3 |
|---|---|---|---|---|---|
| $m$ | 10 | 127 | 3 | 66 | 29 |

$m = 2 + p \cdot p \cdot p$

$m = p^3 + 2$

## Connect & Extend

**13.** Possible explanation: Both rules could be correct because they work for all the values in the table.

**14.** See Additional Answers.

**14.** Here are the first and fifth terms of a toothpick sequence:

Term 1 Term 5

**a.** What might Terms 2, 3, and 4 look like?

**b.** Write a rule that connects the term number and the number of toothpicks in your sequence. Use letters to represent the variables.

**15.** Each term of this sequence is made from a 1-inch straw cut into equal-sized pieces:

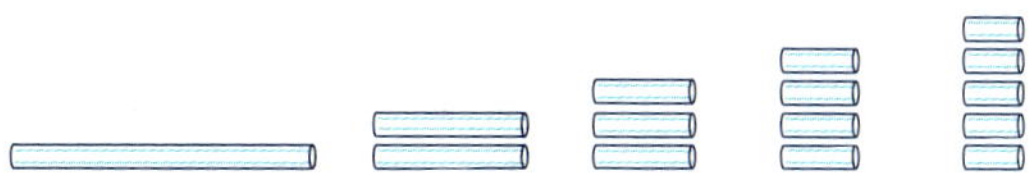

Term 1 Term 2 Term 3 Term 4 Term 5

**a.** How many pieces of straw will be in Term 10? What fraction of an inch will the length of each piece be? 10, $\frac{1}{10}$

**b.** Write a rule that connects the number of straw pieces to the term number.

**c.** Write a rule that connects the length of each straw piece in a term to the term number.

**15b.** $p = n$, where $p$ represents the number of pieces and $n$ represents the term number

**15c.** $L = \frac{1}{n}$, where $L$ represents the length of each piece and $n$ represents the term number

**Exercises 12 and 13:** These exercises make the point that a valid rule must work in every case.

**Exercise 14:** Encourage students to make a table of term numbers and number of toothpicks. This will help them draw the middle three terms and subsequently, find the rule.

### Additional Answers

**12.** Possible explanation: $s = (t - 1) * 3$ is not correct because it gives an $s$ value of 12 for a $t$ value of 5. Possible explanation: $t = 2s + (s - 1)$ could be correct because it works for all the values in the table.

**14a.** Possible answer:

Term 2 Term 3 Term 4

**14b.** Possible rule: $t = 4 + 6 \cdot (n - 1)$, where $t$ represents the number of toothpicks and $n$ represents the term number

## Assign and Assess

**Exercise 17:**
This exercise is unlike preceding exercises in that the rules involve three rather than two variables. In **Part b,** students may realize that $2 \cdot L \cdot W$ does not represent the perimeter of the rectangle but rather twice the *area* of the rectangle.

**16.** The Briggs made a patio using square bricks. They worked in stages, completing one stage before moving on to the next.

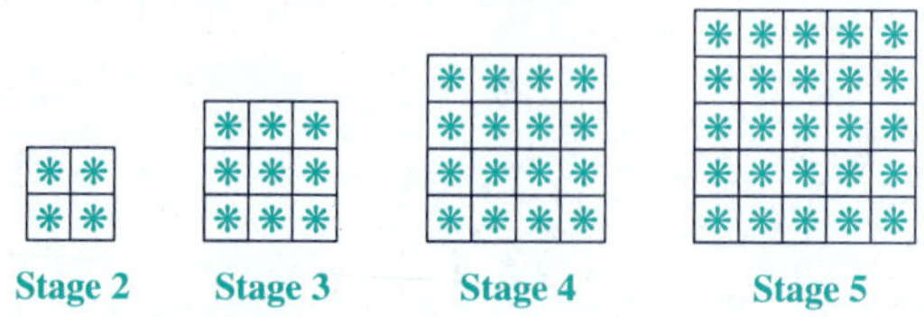


**a.** Write a number sequence showing the number of bricks in the patio in each stage, starting with Stage 2. 4, 9, 16, 25, . . .

**b.** Write a rule that connects the number of bricks in the patio to the stage number. Use letters to represent the variables, and tell what each letter represents.

**c.** Use your rule to figure out how many bricks were in the patio in Stage 1. 1

**d.** Use your rule to predict how many bricks will be in the patio at Stage 14. 196

**Remember**
The *perimeter* of a figure is the distance around it.

**16b.** $b = s^2$, or $b = s \cdot s$, where $b$ is the number of bricks and $s$ is the stage number

**17. Geometry** Gage and Tara wrote formulas for the perimeter of a rectangle. Both students used $P$ for the perimeter, $L$ for the length, and $W$ for the width.

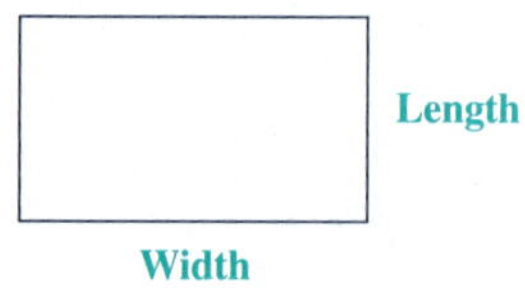


Gage's formula: $P = 2 \cdot L + 2 \cdot W$

Tara's formula: $P = 2 \cdot (L + W)$

**a.** Tell whether each formula is correct. If it is correct, draw diagrams showing why it is correct. If it is not correct, explain what is wrong. See Additional Answers.

**b.** Gage said, "I wonder whether I could just write my formula as $P = 2 \cdot L \cdot W$." Does this formula give the correct perimeter for a rectangle? Explain how you know. no; Possible explanation: A rectangle with width 2 cm and length 5 cm has perimeter 14 cm, but this formula gives a perimeter of 20 cm.



**Additional Answers**

**17a.** Gage's formula is correct. Possible explanation: The rectangle can be broken into two lengths and two widths, so the perimeter is $2L + 2W$.

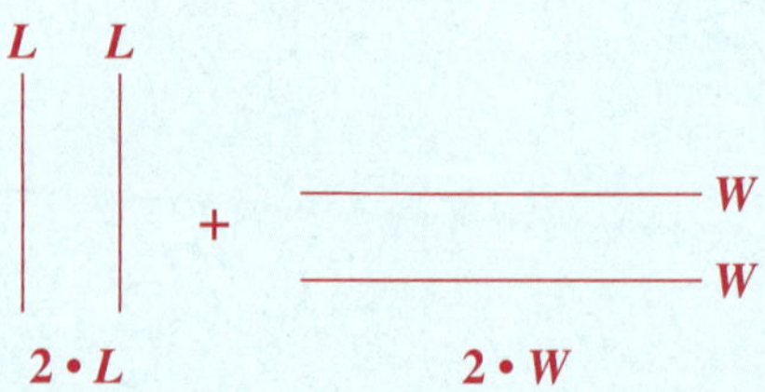


Tara's formula is correct. Possible explanation: The rectangle can be broken into two parts made of one length and one width. So, the perimeter is $2 \cdot (L + W)$.

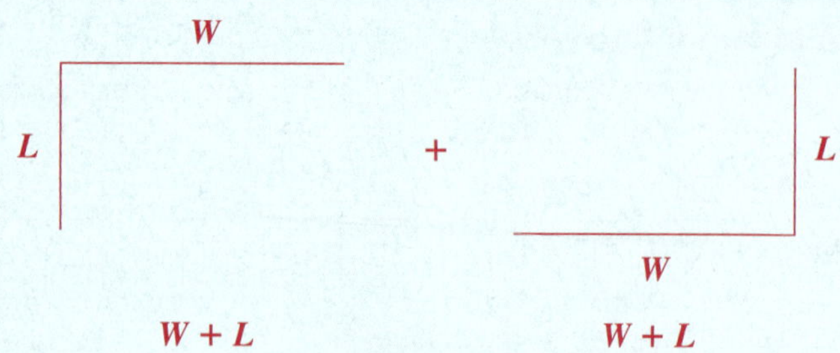


426 CHAPTER 7 Variables and Rules

**18.** Ben made this sequence from watermelon seeds:

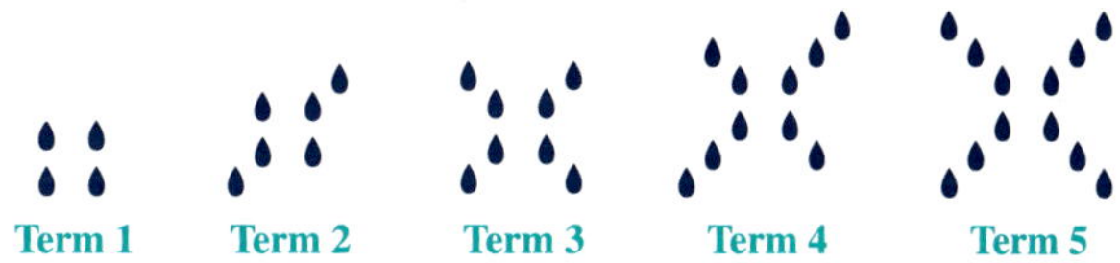


**a.** Complete the table to show the number of seeds in each term.

| Term Number | 1 | 2 | 3 | 4 | 5 |
|---|---|---|---|---|---|
| Number of Seeds | 4 | 6 | 8 | 10 | 12 |

**b.** On a set of axes like the one below, graph the data in your table.

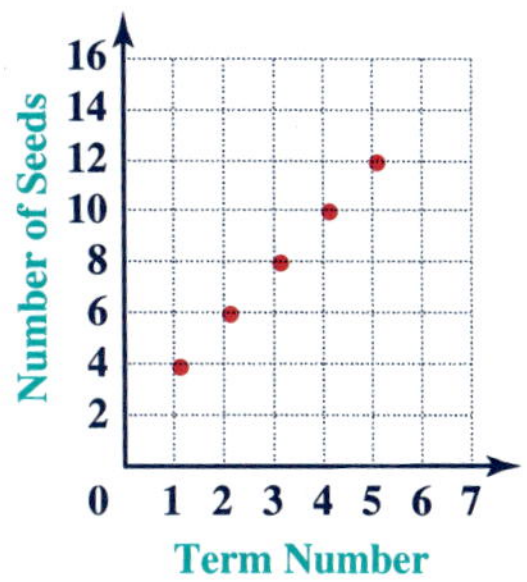


**18c.** Possible answer: The points fall on a straight line that slants upward from left to right.

**c.** Describe the pattern of points in your graph.

**d.** Use your graph to predict the number of seeds in Terms 6 and 7. Check your prediction by drawing these terms and counting the seeds. 14, 16

**Exercise 18:**
You may want to distribute copies of Master 86 to students for use with this exercise.

In **Part b,** students should note that the scale on the vertical axis is not the same as the scale on the horizontal axis. In **Part d,** you may want students to put dots or small x's at the points on the graph they think will correspond to Terms 6 and 7.

LESSON 7.1 Patterns and Variables **427**

# Assign and Assess

**Exercise 19c:**
Other rules that students may give are $s = n^2 + 2n$ and $s = n \cdot (n + 1) + n$.

**Exercise 20:**
This exercise should help students recall how multiplication and division are related. If no one mentions the possibility of using fractions, you may want to discuss the fact that the rule in **Part b** can also be written as $m = \frac{1}{13} \cdot n$.

**19.** Jacob made the graph below to show the number of stars in each term of this sequence:

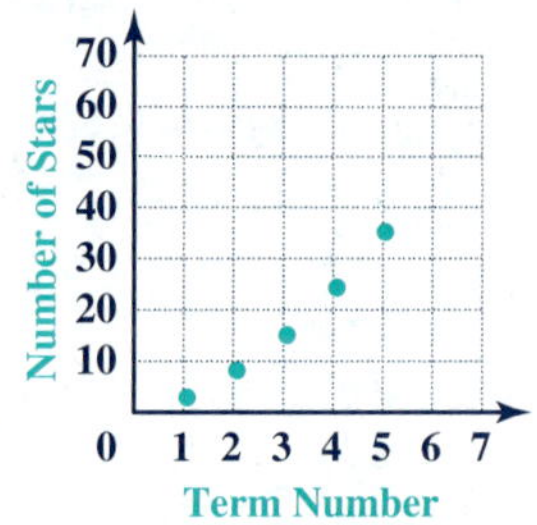


**19a.** Possible answer: The points seem to fall on a curve that rises slowly at first and then more quickly.

**19b.** Answers will vary. Students may say that it is difficult to make an accurate prediction because the points do not fall in a straight line. However, a reasonable prediction might be 50 and 65.

**19c.** Possible rule: $s = (n + 1) \cdot (n + 1) - 1$, where $n$ is the term number and $s$ is the number of stars. For Term 6, this rule gives $s = (6 + 1) \cdot (6 + 1) - 1 = 48$. For Term 7, it gives $s = (7 + 1) \cdot (7 + 1) - 1 = 63$.

**a.** Describe the pattern of points in Jacob's graph.

**b.** Use the graph to predict the number of stars in Terms 6 and 7. Do you think your predictions are accurate? Explain.

**c.** Write a rule for finding the number of stars in any term of the sequence. Use your rule to check your predictions from Part b.

**20.** This table shows values of the variables $m$ and $n$:

| $m$ | 6 | 0 | 11 | 7 | 3 |
|---|---|---|---|---|---|
| $n$ | 78 | 0 | 143 | 91 | 39 |

**a.** Complete this rule for the relationship between $m$ and $n$:

$n =$ <u>13$m$</u>

**b.** Complete this rule for the relationship between $m$ and $n$:

$m =$ <u>$n \div 13$</u>

**c.** Explain how the two rules you wrote could describe the same relationship.

**20c.** Possible answer: The first rule says that $n$ is equal to 13 times $m$. If this is true, then $m$ must equal $n$ divided by 13, which is what the second rule says. The two rules are just different ways of expressing the same relationship.

**In your own words**

*Describe how you can determine whether two rules for a sequence are equivalent.*

**Just the facts**

In race walking, a competitor's leading leg must be straight from the time her heel hits the ground until her leg is under her hip, and one foot must be in contact with the ground at all times.

**21. Sports** Danilo and his brother Adan are having a walking race. Since Adan is younger, Danilo gives him a head start. The table shows the number of minutes after the start of the race that each boy has reached the given distance. The boys are walking at a steady pace, and all the blocks are about the same length.

| Distance (blocks) | 1 | 4 | 6 | 10 |
|---|---|---|---|---|
| Adan's Time (min) | 3 | 12 | 18 | 30 |
| Danilo's Time (min) | 22 | 28 | 32 | 40 |

**a.** Write a rule for finding the number of minutes $M$ that it takes Adan to reach block $N$. $M = 3 \cdot N$

**b.** Write a rule for finding the number of minutes $M$, after the start of the race, that it takes Danilo to reach block $N$. See margin.

**c.** If both boys stop walking an hour after the race began, will Danilo catch up to Adan? Explain. See margin.

## Mixed Review

Find each product without using a calculator.

**22.** 44 ∗ 781 34,364 **23.** 4.4 · 0.781 3.4364 **24.** 440 ∗ 781,000 343,640,000

**25.** 0.44 × 7.81 3.4364 **26.** 0.044 · 0.0781 0.0034364 **27.** 440 × 78.1 34,364

Fill in the blanks.

**28.** <u>75</u> % of 152 = 114 **29.** 0.5% of 200 = <u>1</u>

**30.** $33\frac{1}{3}$% of <u>45</u> = 15 **31.** <u>300</u> % of 20 = 60

**32.** 45% of 45 = <u>20.25</u> **33.** 80% of <u>400</u> = 320

**34. Statistics** Jaleesa received the following scores on her 20-point spelling quizzes:

20 15 18 19 20 0 14 17 16

**a.** Find the mean, median, and mode of her scores.

**b.** Which measure of center do you think best represents Jaleesa's typical quiz score? Give reasons for your choice. See below.

**c.** Jaleesa's teacher has agreed to drop each student's lowest score. Drop the lowest score, and compute the new mean, median, and mode of Jaleesa's scores. mean: 17.375; median: 17.5; mode: 20

**34b.** Possible answer: The median; the mode is the highest score, and it does not reflect the lower scores. The mean seems a little low; it was influenced by the score of 0. The median seems like the best average. Except for the 0, most of the scores are close to 17.

**21b.** Possible rule: $M = 20 + 2 \cdot N$

**21c.** Yes; after 60 min, Adan will have walked $N = \frac{60}{3}$, or 20, blocks, and Danilo will have walked $N = \frac{60 - 20}{2} = \frac{40}{2}$, or 20, blocks. The boys will reach the 20-block point at the same time.

**34a.** mean: $15.\overline{4}$; median: 17; mode: 20



## Quick Check

### Informal Assessment

Students should be able to:

- ✔ write symbolic rules for a toothpick or dot sequence and show how the number of toothpicks or dots in each term relates to the term number
- ✔ use diagrams to explain why a rule for a toothpick or dot sequence is correct
- ✔ recognize when rules for a pattern are equivalent
- ✔ use symbols to write rules for tables made during games of *What's My Rule?*

• ***Quick Quiz on page A664***

LESSON 7.1 Patterns and Variables 429

## Teacher Notes

# 7.2 Rules in Real Life

## Objectives

- To use letters for variables to write rules for real-life situations
- To analyze real-life situations, identify related variables, and write rules describing relationships between the variables
- To use related multiplication and division equations and related addition and subtraction equations to write rules in equivalent forms

## Overview (pacing: about 4 class periods)

In this lesson, students will see how rules expressed in symbolic form can be used to examine real-life situations. Most of the patterns in Lesson 7.1 came from sequences. By contrast, real-life situations are often about variables that involve continuous quantities. In this lesson, students will consider rules that apply to both types of quantities.

In the lab investigation, students explore how to get people across a small bridge in the fewest number of trips. This situation involves not two but three variables. It leads naturally into the investigation that focuses on expressing relationships described in words in a more compact, symbolic form.

The lesson concludes with another look at equivalent rules, especially those based on related operations.

## Advance Preparation

No advance preparation is required.

## Lesson Planner

| | Summary | Materials | On Your Own Exercises | Assessment Opportunities |
|---|---|---|---|---|
| **Investigation 1** page T431 | Students use letters for variables to write rules that allow them to compare costs for long-distance telephone plans and auto-rental plans. | | Practice & Apply: 1–3, pp. 443–444<br>Connect & Extend: 14, 15, p. 446<br>Mixed Review: 22–36, p. 449 | On the Spot Assessment, page T431<br>Share & Summarize, pages T432, 432<br>Troubleshooting, page T432 |
| **Lab Investigation** page T433 | Students discover and write rules that relate three variables in a real-life situation. | | | |
| **Investigation 2** page T436 | Students analyze situations described in words and use letters to write rules describing how the variables in the situations are related. | | Practice & Apply: 4–8, pp. 444–445<br>Connect & Extend: 16, 17, pp. 446–447<br>Mixed Review: 22–36, p. 449 | Share & Summarize, pages T439, 439<br>Troubleshooting, page T439 |
| **Investigation 3** page T439 | Students use their knowledge of the relationship between addition and subtraction, and the relationship between multiplication and division to write a variety of rules in equivalent forms. | | Practice & Apply: 9–13, p. 445<br>Connect & Extend: 18–21, pp. 447–448<br>Mixed Review: 22–36, p. 449 | On the Spot Assessment, page T441<br>Share & Summarize, pages T442, 442<br>Troubleshooting, page T442<br>Informal Assessment, page 449<br>Quick Quiz, page 449 |

# Introduce

1 Introduce the lesson by briefly discussing two real-life situations involving variable quantities that are related. You might ask students to consider, for example, the height of a bean plant several days after the seed was planted. Ask the students to name some things that might affect the height of the plant. **amount of water the plant gets, temperature, type of soil, amount of sunlight, and so on** Help students see that because there are so many things that can change in this situation, it would probably be very difficult to predict exactly how tall the plant would be after a given number of days.

Next, you might have them consider the weight of an aquarium. Suppose the weight of the tank is known. Ask students what they would need to know to say how much the aquarium will weigh when full. **weight of water, fish, plants, and other contents** Discuss the fact that the water is usually the heaviest thing in an aquarium, and point out that you could probably get a good idea about how much the aquarium will weigh if you know the number of gallons of water it will hold.

2

## Explore

In this Explore, students compare two long-distance plans. They complete a table for each plan to show the monthly charges over a 5-month period. The work students do to complete the tables will point them toward rules that allow them to calculate the monthly charges for an arbitrary number of minutes of long-distance calls. Students write the rule in symbolic form and consider which of the two long-distance plans is more economical.

3 You might want to discuss how to find the number of minutes of usage that would result in equal charges under the two plans. Help students understand that they can substitute the same number for $m$ in each rule to see how much it would cost. Help them see how their tables provide clues about which numbers to test.

# 7.2 Rules in Real Life

You have seen that using letters to stand for variables lets you write rules more simply. In Lesson 7.1, the rules you worked with came from sequences. In this lesson, you will look at rules for everyday situations.

**Just the facts**

Long-distance service between London and Paris was introduced in 1891.

## Explore

Last summer, Jing's best friend moved to another state. Since then, Jing has been making lots of long-distance calls. Jing's mother has asked her to use her allowance to help pay the phone bill.

- Jing's family uses TalkCo long-distance service. The service charges \$3.00 per month plus \$0.07 per minute for each phone call. Use this information to complete the table below.

**TalkCo Charges**

| | Jun | Jul | Aug | Sep | Oct |
|---|---|---|---|---|---|
| **Minutes** | 120 | 90 | 95 | 150 | 80 |
| **Cost** | \$11.40 | \$9.30 | \$9.65 | \$13.50 | \$8.60 |

- Write a rule Jing's family could use to calculate their monthly long-distance bill. Use $m$ to stand for the number of minutes they talked and $c$ to stand for the cost in dollars. $c = 0.07 \cdot m + 3$

- Complete this table to show what Jing's family's monthly charges would have been if they had used Chatterbox long-distance service instead.

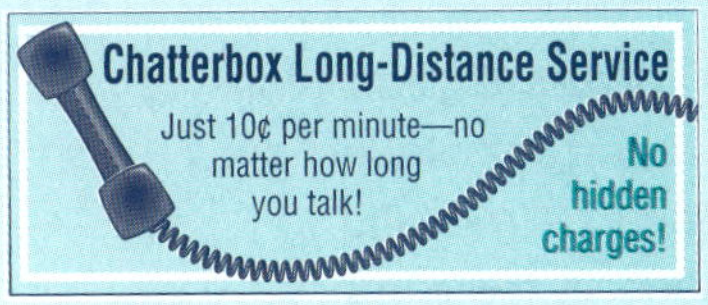


**Chatterbox Charges**

| | Jun | Jul | Aug | Sep | Oct |
|---|---|---|---|---|---|
| **Minutes** | 120 | 90 | 95 | 150 | 80 |
| **Cost** | \$12.00 | \$9.00 | \$9.50 | \$15.00 | \$8.00 |

- Write a rule Chatterbox customers could use to calculate their monthly bills. Use $m$ to stand for the minutes and $c$ for the cost. $c = 0.1 \cdot m$

- Do you think Jing's family should switch to Chatterbox? Explain. See ①.

① Possible answer: No; over the 5 months they have averaged 107 min, and if they talk more than 100 min per month, TalkCo is cheaper.

1 Introduce this lesson by discussing two real-life situations.

2 Have students compare the two long-distance plans.

3 Show students how the tables provide clues as to which number to test.

# Investigation 1

In this investigation, students see how rules can be used to describe relationships between variables in real-life situations.

1 **Problem Set A** **Suggested Grouping: Pairs**
Problem Set A asks students to consider the cost of renting a sedan or an SUV from a company that uses two different rental plans for the two types of vehicles. Students identify the variables for each plan. They make a table and write a rule for each plan. Then, they use their rules to decide which type of vehicle is less expensive for certain trips. Finally, they consider how the cost of renting a sedan is affected if the fixed cost is eliminated and the per-mile charge is increased.

2 In **Problem 1,** be sure students understand that for calculating the rental cost of the SUV in **Part a,** the number of miles driven is not relevant. In **Part b,** the number of days the sedan is rented is not relevant. In **Part c,** students should focus on the relevant variables for each type of vehicle. The fixed costs need not be thought of as variables even though they are different in the two situations.

3 In **Problem 2,** check that students' verbal explanations in **Part a** provide general directions for calculating the rental cost. In **Part b,** urge students to be realistic about the number of miles they include in their tables. For example, it would hardly be realistic for a customer to rent a sedan to drive a distance of only 1 mile. In **Part c,** the symbolic rule should be in agreement with the explanation in Part a.

### On the Spot Assessment

Use Problem 2 to check that students recall the correct procedures for multiplying and adding decimals. If students have errors that result from applying these procedures incorrectly, you may want to review the procedures on an individual basis.

## Investigation 1 Rules in Context

Just like sequences and patterns, real situations often involve variables you can represent with letters. In the telephone-service problem, the variables are the number of long-distance minutes $m$ and the cost $c$. In this lesson, you will write lots of rules to represent real situations you might encounter.

**Remember**
A *variable* is a quantity that can change.

### Problem Set A

1 You may have students work in pairs.

Haley's car rental company rents sedans and sports utility vehicles (SUVs).

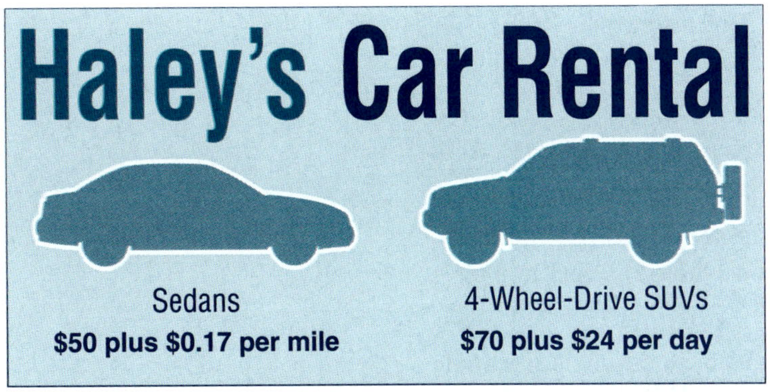


**Just the facts**
The sale of sports utility vehicles grew from 960,852 in 1988 to 2,796,310 in 1998—a 191% increase!

2 Make sure students understand how to calculate the rental cost.

**1.** The ad shows Haley's car rental rates.

**a.** If a customer rents an SUV for 14 days and drives it 2,000 miles, what will the cost be? $406

**b.** If a customer rents a sedan for 14 days and drives it 2,000 miles, what will the cost be? $390

**c.** What are the variables in this situation?

**1c.** cost of renting a sedan, cost of renting an SUV, number of miles the sedan is driven, number of days the SUV is rented

**2a, b.** See Additional Answers.

**2.** Consider the cost of renting a sedan.

**a.** Explain in words how to calculate the cost of renting a sedan.

**b.** Create a table that shows the cost of renting a sedan and driving it for various numbers of miles.

3 Encourage students to be realistic when filling out the table.

**c.** Write a rule for calculating the cost of renting a sedan. Use $m$ to represent the number of miles driven and $c$ to represent the cost.
$c = m \cdot 0.17 + 50$

**d.** A customer rented a sedan and drove it 567.6 miles. Use your rule to determine how much the customer should be charged.
$146.49



**Additional Answers**

**2a.** Multiply the number of miles driven by $0.17 and add $50.

**2b.** Possible table:

| Miles Driven | 10 | 20 | 30 | 100 | 200 |
|---|---|---|---|---|---|
| Cost | $51.70 | $53.40 | $55.10 | $67.00 | $84.00 |

LESSON 7.2 Rules in Real Life 431

# Develop

1 In **Problem 4,** students need to be careful to note whether they are substituting the number of miles or the number of days when they use their rules to calculate the costs. They should not forget to say which vehicle will be less expensive in each case.

In **Problem 5,** students write a new rule for the cost of renting a sedan.

**Problem Set Wrap-Up** You may want to discuss Problem 5 with the entire class. If students had difficulty with **Part b,** you may want to suggest that they make a table for the new rule and compare it to the table they made in Problem 2.

2 **Share & Summarize**

This question asks students to think about the advantages of symbolic rules over rules expressed in words. You may want to have some volunteers read their answers to the class.

**Troubleshooting** Some students may be unclear about how to identify the variables in a real-life situation. For these students, it may be helpful to describe some additional situations and ask what the variables are.

**Additional Example** Margarita is trying to decide which of several books to buy. She wants to mail copies to several friends.

What variables would you take into account in writing a rule for the total cost of the books? *Possible answer: the cost of the book she decides to buy, the number of friends she plans to buy a copy for, the total cost*

What are some of the variables involved in writing a rule for the total mailing cost? *Possible answer: the weight of the book, the number of books she plans to send*

### On Your Own Exercises

Practice & Apply: 1–3, pp. 443–444
Connect & Extend: 14, 15, p. 446
Mixed Review: 22–36, p. 449

★ indicates multi-step problem

**3a.** Multiply the number of days by 24 and add 70. This is different from the sedan rule because it depends on the number of days instead of the number of miles.

**3c.** $c = 70 + 24d$

**4a.** sedan cost: \$160.50, SUV cost: \$142.00; The SUV is less expensive.

**4b.** sedan cost: \$279.50, SUV cost: \$310.00; The sedan is less expensive.

**4c.** sedan cost: \$71.25, SUV cost: \$406.00; The sedan is less expensive.

**5b.** It depends on how far you drive. For example, if you drive the sedan 100 miles, it would cost \$67 with the old plan and \$85 with the new plan. If you drive only 50 miles, it would cost \$58.50 with the old plan and \$42.50 with the new plan.

**3.** Consider the cost of renting an SUV.

**a.** Explain in words how to calculate the cost of renting an SUV. How is this rule different from the rule for the sedan?

**b.** Create a table that shows the cost of renting an SUV for various numbers of days. See Additional Answers.

**c.** Write a rule for calculating the cost of renting an SUV. Use $d$ to represent the number of days and $c$ to represent the cost.

**d.** A customer rented an SUV for 27 days. Use your rule to determine how much the customer should be charged. \$718

**4.** Each customer described below wants to rent a vehicle. Use the rules you wrote in Problems 2 and 3 to decide which type of vehicle would be less expensive. Explain your answers.

★**a.** Mr. Houlihan is taking a business trip to Denver, Colorado. He will be driving 650 miles and will be gone for 3 days.

★**b.** Ms. Basil is traveling to Orlando, Florida, for 10 days. She will be driving 1,350 miles round-trip.

★**c.** Mr. and Mrs. Iso are planning to visit their grandchildren in Portland, Maine. They will be driving 125 miles and will be gone for 2 weeks.

**5.** Haley's has changed its rate for renting a sedan. It now charges \$.85 per mile, with no fixed amount.

**a.** Write a new rule for calculating the rental cost for a sedan. Use $c$ to represent the cost and $m$ to represent the number of miles driven. $c = 0.85 \cdot m$

**b.** Does this change make it more or less expensive to rent a sedan? Explain.

## Share & Summarize

Possible answer: Symbolic rules are shorter and easier to write, and they make it easier to do calculations. If rules are written with symbols, it is easier to compare them.

In Chapter 1, you worked with real-life rules written in words. In this investigation, you wrote and used symbolic rules, with letters for the variables. What are some advantages of using rules written with letters and symbols?

**1** Be sure students are careful using the correct rules when calculating the costs.

**2** Have students read their answers to the class.



### Additional Answers

**3b.** Possible table:

| Days | 1 | 2 | 3 | 4 | 5 |
|---|---|---|---|---|---|
| Cost | \$94 | \$118 | \$142 | \$166 | \$190 |

# Lab Investigation

1 **Suggested Grouping: Small Groups**
In this lab, students explore a situation that involves getting varying numbers of adults and children across a bridge. There are restrictions on how many people can safely cross the bridge at one time. The object is to find the fewest number of trips necessary if you know how many adults and how many children are in the group. The puzzle-like character of the problem will make the problem intriguing for students.

2 To introduce this investigation, you may want to have a group of students act out the first few moves. Have eight students (two designated as the children) stand on one side of the room. Use a small object to represent the flashlight, which must be carried on each trip. Have the class tell the group who should cross for the first few stages. If one adult walks over with the flashlight, what happens then? How does the flashlight get back to the group that needs it? Do not take time to get the whole group to the other side of the room. Have the class demonstrate that they understand the procedure needed to get everyone across.

Make sure students understand that each crossing counts as one trip. Then have students work in groups to answer the questions in the investigation.

### Try It Out

For **Question 1,** students can answer **Part a** by describing the procedure used when the class considered how to act out the problem. **Part b** may prove to be more difficult. Students who quickly see the pattern that is unfolding will probably be able to arrive at the answer rather fast. Others may feel it necessary to devise a way to record what happens for each trip and to count the number of trips. Another approach is to see what would happen if there were only one adult and two children, and then to increase the number of adults one step at a time.

The following scheme (illustrated for two adults and two children) makes it easy to keep track of the trips needed to get everyone across the bridge. Each row in the body of the table shows where the adults and children are after the trip number for that row. For example, after trip 4, there will be one adult and two children on the first side of the bridge, and there will be one adult on the second side.

| Trip Number | First Side | Second Side |
|---|---|---|
| 1 | 2A | 2C |
| 2 | 2A, 1C | 1C |
| 3 | 1A, 1C | 1A, 1C |
| 4 | 1A, 2C | 1A |
| 5 | 1A | 1A, 2C |
| 6 | 1A, 1C | 1A, 1C |
| 7 | 1C | 2A, 1C |
| 8 | 2C | 2A |
| 9 | | 2A, 2C |

**Question 2** can be approached in much the same way as Question 1.

3 **Access for all Learners**

**Problem-Solving Strategy**
Make a model

**Extra Help** Tactile learners may find it helpful to explore the bridge-crossing problem by using small pieces of paper to represent the people. Have students tear a sheet of paper into 8 or 10 pieces. Next, have them write the letter C on two pieces of paper and the letter A on the other pieces. The pieces can represent the children and adults, respectively.

# Lab Investigation ▶ Crossing a Bridge

1 You may have students work in small groups.

While walking at night, a group of eight hikers—six adults and two children—arrives at a rickety wooden bridge. A sign says the bridge can hold a maximum of 200 pounds. The group estimates that this means

- one child can cross alone, *or*
- one adult can cross alone, *or*
- two children can cross together.

2 Problem-Solving Strategy
Act it out

Anyone crossing the bridge will need to use a flashlight. Unfortunately, the group has only one flashlight.

**1a.** Possible explanation: Repeat the following steps until all the adults are across:

- The two children cross the bridge.
- One child brings back the flashlight, and one stays on the other side.
- The child that returned gives an adult the flashlight, and the adult crosses.
- The adult gives the flashlight to the child on the other side, and that child crosses back.

After all the adults are across, the two children cross together.

**2a.** Yes; use the same process as the first group (2 children cross, 1 child returns with the flashlight, 1 adult crosses, the other child brings back the flashlight).

## Try It Out

3 Problem-Solving Strategy
Make a model

**1.** Find a way to get all the hikers across the bridge in the fewest number of trips. Count one trip each time one or two people walk across the bridge.

**a.** Describe your plan using words, drawings, or both.

**b.** How many trips will it take for everyone in the group to get across the bridge? 25

**2.** A second group of hikers approaches the bridge. This group has 10 adults, 2 children, and one flashlight.

**a.** Can everyone in this group get across the bridge? If so, describe how.

**b.** What is the least number of trips it will take for this group to cross the bridge? 41



LESSON 7.2 Rules in Real Life 433

**Question 3** should be easy for students who have been successful in seeing the pattern to which Questions 1 and 2 are pointing.

1 **Question 4** asks students to write, in symbols, the rule they have discovered in their work on Questions 1–3. Students will probably benefit from a discussion of why the rule works. If students have any difficulty explaining this, you may want to use tables of trips (like the one shown on page T433) to make the point that each additional adult requires four more trips.

**Question 5** is important because it shows that the bridge problem has no solution if there is only one child in the group.

### Try It Again

Before students work on Questions 6–9, take time to have them observe that they have so far varied only the number of adults in the group. Tell them that they will now consider what happens when the number of children changes.

2 In **Question 6,** students consider the number of trips needed to get various numbers of children across the bridge. They may use methods similar to those used for earlier questions. Students should note that this question assumes that no adults are in the group crossing the bridge.

For **Question 7,** students look for a pattern that generalizes what they discovered in Question 6.

3 For **Question 8,** students may need to think a bit about how to handle variation in the number of both adults and children. After some thought and experimentation, they should realize that they can get all of the children across and then address the problem of getting all the adults across.

For **Question 9,** students may reason that getting $c$ children from one side to the other will require $2c - 3$ trips (assuming that $c$ is 2 or greater). If there are adults in the group, each adult will require another 4 trips.

3. In Parts a–c, find the least number of trips it would take for the group to get across the bridge. Assume each group has only a single flashlight.

   a. 8 adults and 2 children 33

   b. 1 adult and 2 children 5

   c. 100 adults and 2 children 401

4. Suppose a group has two children, $a$ adults, and one flashlight. Write a rule that relates these two variables: the least number of trips $t$ needed to get everyone across, and the number of adults.

5. Could a group with 15 adults, one child, and one flashlight cross the bridge? Explain.

4. $t = 4a + 1$

5. No; a group needs at least two children to get across so there can be one child on the other side to return the flashlight when an adult brings it across.

**1** Encourage students to discuss why the rule works in their groups.

## Try It Again

Now you will explore how the number of children in a group affects the number of trips needed for the group to cross the bridge. You will start by thinking about groups with no adults at all.

**2** Be sure students use methods similar to those used in earlier questions.

6. Tell how many trips it would take to get each of these groups across the bridge. (Hint: First figure out a method for getting everyone across, and then think about how many trips it would take.)

   a. 2 children 1  b. 3 children 3

   c. 6 children 9  d. 10 children 17

7. Look for a pattern relating the number of children to the number of trips.

   a. Write rule that relates the number of trips $t$ to the number of children in the group $c$. Assume the group has no adults, more than one child, and just one flashlight. $t = 2c - 3$

   b. Part a states that the group must have more than one child. How many trips would it take a single child to get across the bridge? Does your rule give the correct result for one child? 1, no

8. Tell how many trips it would take to get each of these groups across the bridge:

   a. 6 adults and 2 children 25  b. 6 adults and 3 children 27

   c. 6 adults and 6 children 33  d. 8 adults and 10 children 49

**3** Make sure students realize they would need to get all of the children across first.

9. **Challenge** Write a rule that relates the number of trips to the number of children and adults in the group. Assume the group has at least two children and exactly one flashlight.

   $t = 4a + 2c - 3$, where $a$ is the number of adults and $c$ is the number of children

### What Did You Learn?

1 These questions allow students to apply their methods from earlier questions to a situation involving different restrictions on the number of people who can cross a bridge at one time. You may want to take time to discuss the new restrictions with the class. Students should note that if only one person needs to cross the second bridge, then one trip will do. But the problem is really about trying to get a *group* of people across the bridge. Use a discussion to help students see that to get everyone in a group of two or more across the bridge, there must be at least one child.

Methods similar to those used for Questions 1–3 can be helpful with **Questions 10–12.** Encourage students to make a table to help them develop a rule. Some students may discover that if there are $p$ people in the group (where $p$ is at least 2 and the group includes at least one child), then $2p - 3$ trips are required to get everyone across.

2 For **Question 13,** students can make a table to help them develop the rule $t = 2a - 1$. You may want to discuss the fact that this rule assumes at least one adult is in the group.

## What Did You Learn?

Another bridge has a different weight restriction. For this bridge,

- one adult can cross alone, *or*
- one child can cross alone, *or*
- one adult and one child can cross together, *or*
- two children can cross together.

**10.** With these new rules, a group might be able to use a different method to get across the bridge.

**a.** Find a way to get a group of eight adults and two children across the bridge in the fewest trips. Describe your method using words, pictures, or both.

**b.** How many trips does it take for everyone to get across? 17

**c.** How does the number of trips needed to get the group across this bridge compare to the number of trips needed for the first bridge? (See your answer for Part a of Problem 3.) See margin.

**11.** A group of 15 adults and 1 child could not get across the first bridge. Could this group cross the second bridge? If so, explain how, and tell how many trips it would take.

**12.** Tell how many trips it would take each of these groups to get across the second bridge:

**a.** 3 adults and 1 child 5

**b.** 5 adults and 1 child 9

**c.** 100 adults and 1 child 199

**13.** Write a rule that relates the number of trips needed to cross the second bridge to the number of adults in the group. Assume the group has one child and one flashlight. $t = 2a - 1$, where $t$ is the number of trips and $a$ is the number of adults

**10a.** Possible answer: An adult crosses with a child. The adult stays behind while the child crosses back over. This is repeated until all the adults have crossed. Finally, the two children cross together.

**Just the facts**

The world's longest overwater bridge is the Lake Pontchartrain Causeway, which connects Mandeville and Metairie, Louisiana. The bridge is almost 24 miles long.

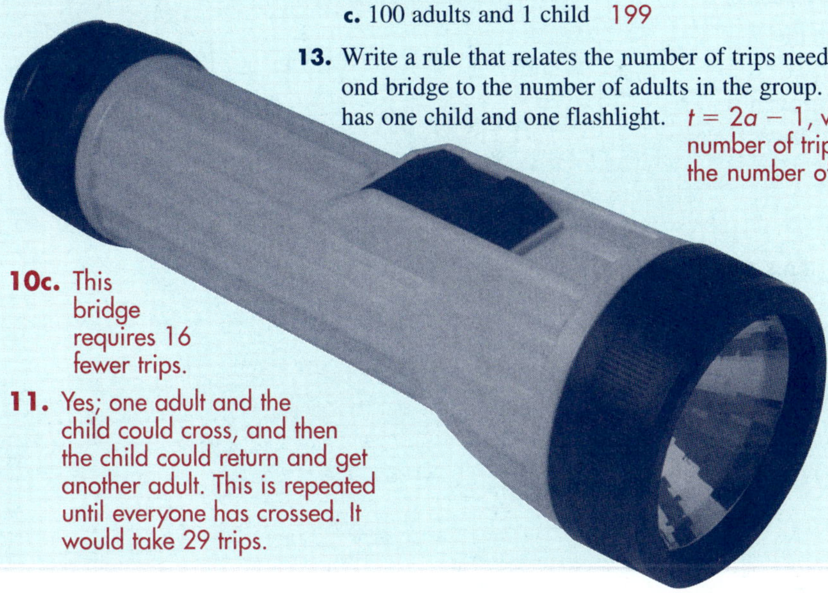

**10c.** This bridge requires 16 fewer trips.

**11.** Yes; one adult and the child could cross, and then the child could return and get another adult. This is repeated until everyone has crossed. It would take 29 trips.

**1** Problem-Solving Strategy
Make a table

**2** Discuss the new restrictions with the class.



LESSON 7.2 Rules in Real Life 435

# Investigation 2

Students who are adept at *What's My Rule?* games may not be equally adept at translating relationships expressed in words into symbolic form. This investigation addresses some of the difficulties students have when they try to translate quantitative relationships from every-day language into the more compact symbolic language of mathematics.

1 **Think & Discuss**
Read the question aloud to the class. After students have had a short time to think about the question, ask for a show of hands of those who selected each of the rules. Then have volunteers from each group explain the thinking that led them to the answer they selected.

2 Ask students to identify the variables in the situation. Be sure they understand that the variables are not spiders and legs but the *number of spiders (S)* and the *number of legs (L)*. Watch for students who say that since there are eight times as many legs as spiders, the rule is $S = 8 \cdot L$. These students are probably replacing words for the things (legs, spiders) with the symbols for the variables. They are focusing narrowly on the word order rather than thinking about the mathematical relationship between the variables. Point out that if $S = 8 \cdot L$ were correct, there would be more spiders than legs.

3 **Example**
The table in this Example will probably help students see why $L = 8 \cdot S$ is the correct rule. Discuss why testing the

4 rule is a good idea. Ask questions to achieve closure.

> In any spider collection, will there be more spiders or more legs? more legs
>
> The table tells you that three spiders will have how many legs? 24 legs
>
> How many times 3 is 24? 8

# Investigation 2 Translating Words into Symbols

**Just the facts**

The body and legs of a tarantula are covered with hairs. Each hair has a tiny barb on the end. When a tarantula is being attacked, it can rub its hairs on its attacker, causing itching and even temporary blindness.

Writing a rule for a real-life situation can be difficult, even when the situation is fairly simple. It's easy to make a mistake if you don't pay close attention to the details.

## Think & Discuss

$L = 8 \cdot S$; Possible explanation: There are 8 legs per spider, so the number of legs is 8 times the number of spiders. This is what the second rule says.

A spider has eight legs. If $S$ represents the number of spiders and $L$ represents the number of legs, which of the following rules is correct? How do you know?

$$S = 8 \cdot L \qquad L = 8 \cdot S$$

1 Discuss the question with the class.

2 Be sure students are focusing on the mathematical relationship between the variables, and not the word order.

In the spider situation, it is easy to confuse the two rules. The Example shows how Luke thought about the situation.

3 Discuss Luke's explanation in the cartoon.

**EXAMPLE**

Creating a table and looking for a pattern can make finding a rule a little easier. Notice that, after Luke wrote his rule, he checked it by testing a value he knew the answer for. It is a good idea to test a value whenever you write a rule. Although this won't guarantee that your rule is correct, it's a helpful way to uncover mistakes.

4 **Problem-Solving Strategy**
Make and test a conjecture

1

**Problem Set B** **Suggested Grouping: Pairs**
For each problem in this problem set, students are to identify the two variables in the situation, make a table of values, describe how to calculate the values of one variable from the values of the other, and write a rule in symbolic form. By working in pairs, students will be able to discuss and check their thinking at each stage.

In **Problem 1,** students should be careful in **Part a** to identify the variables as the *number* of blue tiles and the *number* of white tiles (as opposed to simply blue tiles and white tiles).

In **Part b,** some students may think that only whole tiles should be used. This may influence the values they choose for variables in their tables but is not likely to present any difficulty when they get to **Parts c and d.**

2

In **Problem 3,** the situation requires that students use only whole-number values for the variables when they make the table for **Part b.** (One does not want a fraction of a hamster.)

For the table in **Problem 5,** students must use a price greater than $3 for an adult's ticket.

## Access for all Learners

**Extra Challenge** Students may have noted that for some of the situations in Problems 1–6, it is possible to identify more than two variables. It is usually clear which variables to focus on. Nevertheless, other choices are possible.

For example, in Problem 2, a student could let $m$ be the total number of mints in a package, $g$ be the number of green mints, and $p$ be the number of pink mints. A rule relating $m$ to $g$ would then be $m = p + g$. A rule relating $m$ to $p$ is $m = p + 4p$ (since $g = 4p$). A rule relating $m$ to $g$ is $m = \frac{1}{4}g + g$ (since $p = \frac{1}{4}g$). In the case of this last rule, it should be understood that $g$ needs to be a multiple of 4 to ensure that whole mints are contained in each package.

3

**Problem Set Wrap-Up** Have students share the various rules they wrote for each situation. Review the idea that even though two equations may use different variables, they are still equivalent if they describe the same relationship.

## Problem Set B

1. To tile his bathroom, Mr. Drury needs twice as many blue tiles as white tiles.
   a. What are the two variables in this situation?
   b. Make a table of values for the two variables. See Additional Answers.
   c. Look for a pattern in your table. Describe how to calculate the values of one variable from the values of the other.
   d. Write a rule for the relationship between the two variables. Use letters to represent the variables, and tell what each letter represents. Be sure to check your rule by testing a value.

In Problem 2–6, complete Parts a–d of Problem 1.

2. In packages of Cool Breeze mints, there are four green mints for every pink mint.
3. A pet store always carries six times as many fish as hamsters.
4. In a factory, each assembly worker earns one seventh as much money as his or her manager.
5. A community theater charges \$3 less for a child's ticket than for an adult's ticket.
6. In a toothpick sequence, the total number of toothpicks in a term is 4 more than twice the number of vertical toothpicks in the term.

**1a.** number of white tiles, number of blue tiles

**1c.** Possible answers: To find the number of blue tiles, multiply the number of white tiles by 2. *Or,* to find the number of white tiles, divide the number of blue tiles by 2.

**1d.** $b = 2w$, or $w = \frac{1}{2} \cdot b$, or $w = b \div 2$, where $b$ represents the number of blue tiles and $w$ represents the number of white tiles

**2–6.** See Additional Answers.

**1** You may have students work in pairs.

**2** Be sure students use whole numbers for the variables.

**3** Review with students that two equations can be equivalent even though they use different variables.



### Additional Answers

**1b.** Possible table:

| White Tiles | 1 | 2 | 3 | 4 | 5 |
|---|---|---|---|---|---|
| Blue Tiles | 2 | 4 | 6 | 8 | 10 |

**2a.** number of pink mints, number of green mints

**2b.** Possible table:

| Pink Mints | 1 | 2 | 3 | 4 | 5 |
|---|---|---|---|---|---|
| Green Mints | 4 | 8 | 12 | 16 | 20 |

• ***Additional Answers continued on page A665***

LESSON 7.2 Rules in Real Life **437**

# Develop

1 **Problem Set C** **Suggested Grouping: Pairs**
In this problem set, students consider situations that involve several related variables. They identify the variables and write rules that tell how pairs of variables are related.

2 In **Problem 1,** students who make a table to help them write the rule may write a correct rule but have tables that are not fully in accord with the conditions of the problem. For example, a table that shows a value of 3 for $n$ and $1\frac{1}{2}$ for $a$ does not make sense, since Andrea cannot have $1\frac{1}{2}$ prizes. You may want to discuss this with students, but commend students who nevertheless arrive at the correct rule in **Part d.**

In **Problem 2,** students may need help with **Part d.** One approach is to make a table of values for $n$ and $g$. However, it would be good for students to see that they can use the rule from Problem 1b, $j = 2n$, and the rule from Part a of this problem, $g = j + 2$, to get $g = 2n + 2$. Simply replace $j$ with $2n$ in $g = j + 2$.

3 **Problem Set Wrap-Up** Ask students to explain how they arrived at the rule for Christa's weight in **Part b** of **Problem 3.** One approach is to write the rule $c = m + 3$ to show the relationship between Christa's weight and Michael's weight. Since $m = \frac{s}{2}$, $\frac{s}{2}$ can be substituted for $m$ in the rule $c = m + 3$ to obtain $c = \frac{s}{2} + 3$.

## Problem Set C

Below are more situations involving rules. Make a table whenever you feel it will help you understand the problem. Also, be sure to test all your rules.

**1.** Nick and his friends collect the prizes hidden in boxes of Flako cereal. Joel has twice as many prizes as Nick. Rafael has 3 more prizes than Nick. Andrea has half as many prizes as Nick.

**a.** If Nick has 6 prizes, how many prizes do the other friends have?

**b.** Write a rule for the relationship between the number of prizes Joel has $j$ and the number of prizes Nick has $n$. $j = 2n$

**c.** Write a rule for the relationship between the number of prizes Rafael has $r$ and the number Nick has $n$. $r = 3 + n$

**d.** Write a rule for the relationship between the number of prizes Andrea has $a$ and the number Nick has $n$. $a = \frac{1}{2} \cdot n$ or $a = n \div 2$

**2.** Suppose Gish has 2 more prizes than Joel.

**a.** Write a rule for the relationship between the number of prizes Gish has $g$ and the number Joel has $j$. $g = j + 2$

**b.** If Nick has 19 prizes, how many prizes does Joel have? How many does Gish have?

**c.** Describe in words the relationship between the number of prizes Gish has and the number Nick has.

**d.** Write a rule for the relationship between the number of prizes Gish has $g$ and the number Nick has $n$.

**3.** Spiro Papadopoulos weighs $s$ pounds.

- His wife, Toula, is 20 pounds lighter than Spiro.
- His son, Michael, weighs half as much as Spiro.
- His daughter, Christa, weighs 3 pounds more than his son.

**a.** If Spiro weighs $155\frac{1}{2}$ pounds, how much do each of the others weigh? Toula: $135\frac{1}{2}$ lb; Michael: $77\frac{3}{4}$ lb; Christa: $80\frac{3}{4}$ lb

**b.** Write a rule for calculating the weight of each member of Spiro's family based on Spiro's weight $s$. Tell what variable each letter represents.

**Just the facts**

The first breakfast cereal, known as Granula, was created in the 1860s. The cereal consisted of heavy bran nuggets and had to be soaked overnight before it could be chewed.

**1a.** Joel has 12. Rafael has 9. Andrea has 3.

**2b.** Joel has 38. Gish has 40.

**2c.** Gish has twice as many as Nick, plus 2.

**2d.** $g = 2 \cdot n + 2$

**3b.** $t = s - 20$, where $t$ represents Toula's weight

$m = \frac{s}{2}$, where $m$ represents Michael's weight

$c = \frac{s}{2} + 3$, where $c$ represents Christa's weight

**1** You may have students work in pairs.

**2** Make sure students have realistic values in the table that match the rule and the problem.

**3** Have students discuss the rule they found for Christa's weight.

### 1 Share & Summarize

After students have had time to write their answers to these questions, have some volunteers read their answers. You may also want to ask students to give some examples to illustrate what they say. They can refer to the problems in Problem Set B.

**Troubleshooting** If students are having difficulty writing rules for situations given in words, it may help to provide more practice with this skill before beginning work on the next investigation.

**Additional Example** There are five times as many boys as girls in a class. Let $b$ stand for the number of boys, $g$ for the number of girls, and $t$ for the total number of students in the class.

What rule describes how the variable $b$ is related to the variable $g$? **Possible rule: $b = 5g$**

What rule describes how $t$ is related to $b$ and $g$? **Possible rule: $t = b + g$**

What rule describes how $t$ is related to $g$? **Possible rule: $t = 5g + g$, or $t = 6g$**

### On Your Own Exercises

Practice & Apply: 4–8, pp. 444–445
Connect & Extend: 16, 17, pp. 446–447
Mixed Review: 22–36, p. 449

# Investigation 3

In this investigation, students use what they know about the relationship between multiplication and division, and between addition and subtraction, to decide whether two rules are equivalent. They also use these relationships to write rules in equivalent forms. Finally, students construct examples to show that a rule written in symbols may describe many different situations.

### 2 Think & Discuss

This introduction will get students to think about related multiplication and division equations. Read through the Think & Discuss with the class. Give students a minute or two to think about the questions. Then have students discuss which rules they think are correct. Encourage them to use the diagrams to help guide their thinking. Once students are confident that all the rules except Caroline's are correct, discuss the information at the top of page 440.

## Share & Summarize

**1.** How can making a table help you find a rule for a situation?

**2.** Once you have written a rule, how can you test it to check for mistakes?

**1.** Possible answer: If I make a table, I can look for patterns and relationships in the values. I can think of the values as inputs and outputs and try to find a rule that fits them, as I did when I played *What's My Rule?*

**2.** Possible answer: Test the rule for a case for which you know the result.

**1** Discuss students' answers and have them provide some examples.

# Investigation 3 Equivalent Rules

Every rule can be written in a variety of ways. As you discovered when working with toothpick sequences, when two rules look different, it is sometimes difficult to tell whether they represent the same relationship.

### Think & Discuss See Additional Answers.

Each of these rectangles is five times as long as it is wide:

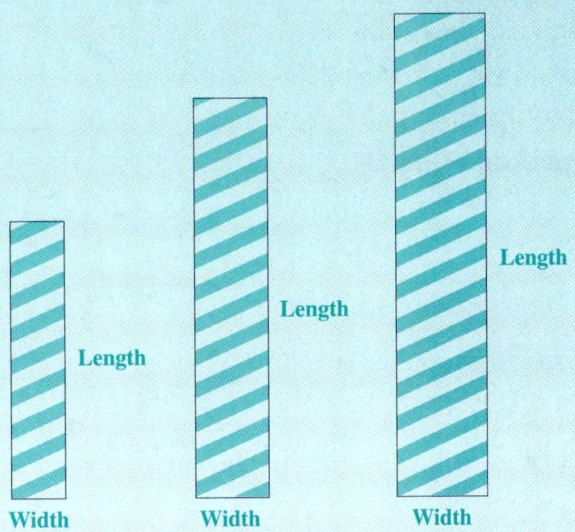


Four students wrote rules to describe the relationship between the length and width of these rectangles. In each rule, $W$ represents the width and $L$ represents the length.

Caroline's rule: $W = 5 \cdot L$

Jahmal's rule: $L = 5W$

Jing's rule: $W = L \div 5$

Rosita's rule: $\frac{L}{W} = 5$

Which of these rules are correct? How do you know?

**2** Encourage students to use the diagrams to help them guide their thinking.

**Remember**
The multiplication symbol is often left out when a number is multiplied by a variable. So, $5W$ is the same as $5 \cdot W$.



**Additional Answers**
**Think & Discuss**

Jahmal's, Jing's, and Rosita's rules are correct. Jahmal's rule says the length is 5 times the width, which is correct. If the length is 5 times the width, then the length divided by 5 equals the width and the length divided by the width equals 5. This is what Jing's and Rosita's rules say. Caroline's rule says that the width is 5 times the length, which is not true.

LESSON 7.2 Rules in Real Life 439

# Develop

1 Read through the information about related multiplication and division equations with the class.

2 **Problem Set D** **Suggested Grouping: Pairs**
In this problem set, students use reasoning about related operations to decide whether given rules are equivalent.

3 In **Problem 1,** students may find it helpful in **Part a** to first try their hand at writing a rule relating $c$ and $t$. This should help them see that the relation involves multiplication or division, but not addition or subtraction. After testing their rules, they can use related equations to see which multiplication and division equations match the rules they wrote.

In **Part b,** students can replace $t$ with 1 and $c$ with 4 in each equation to see whether the resulting equation is true or false. The rules that give false equations must be incorrect.

4 For **Problem 2,** students need to look at the diagram and note how the number of nuts in a package is related to the number of bolts.

**1a.** $t = c \div 4$ and $c = 4t$

**1b.** Possible answer: $c \cdot 4 = t$ and $c = t \div 4$ are incorrect because they say there are 4 tables for 1 chair (when really it is the other way around). $c + 4 = t$ is incorrect because if there are 4 chairs, there should be 1 table, but this rule says there are 8 tables. $t + 4 = c$ is incorrect because when there are 2 tables, it says there should be 6 chairs, when there really should be 8 chairs.

You can sometimes use what you know about the relationship between multiplication and division, or between addition and subtraction, to rewrite a rule in a different way. For example, you know that for every multiplication equation, there are two related division equations:

| Multiplication Equation | Division Equations | |
|---|---|---|
| $3 \cdot 5 = 15$ | $15 \div 5 = 3$ | $15 \div 3 = 5$ |

You can use this idea to rewrite a rule involving multiplication, such as $2 \cdot q = p$:

| Multiplication Equation | Division Equations | |
|---|---|---|
| $2 \cdot q = p$ | $p \div q = 2$ | $p \div 2 = q$ |

All three of the rules above represent the same relationship—that is, they are all *equivalent.*

In the next problem set, you will consider different rules that represent the same situation, and you will rewrite rules in different ways.

## Problem Set D

**1.** In the Aster banquet hall, there are four chairs at every table. Let $c$ represent the number of chairs and $t$ represent the number of tables.

**a.** Which of these rules correctly describes this situation?

$c \cdot 4 = t$ $\quad t = c \div 4$ $\quad c = t \div 4$

$c = 4t$ $\quad c + 4 = t$ $\quad t + 4 = c$

**b.** Explain how you know each of the other rules is incorrect.

**2.** A hardware store sells nuts and bolts in three package sizes.

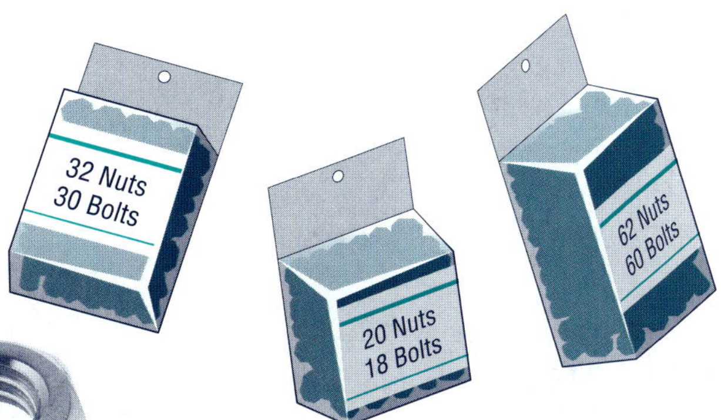


**Just the facts**

*The phrase* nuts and bolts *means the practical or basic elements of something: "He is interested in the nuts and bolts of how a television operates."*

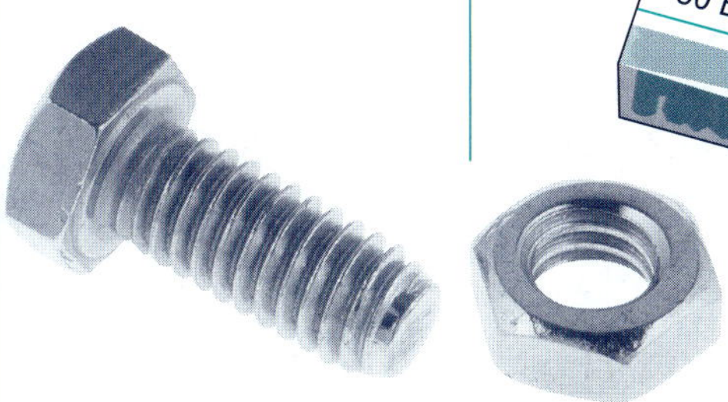

**1** You may want to discuss how different rules can describe the same relationship.

**2** You may have students work in pairs.

**3** Have students write a rule relating *c* and *t*.

**4** Point out that students should look at the diagram to find how the number of nuts and bolts are related.

# Develop

1 In **Part b,** students can easily confirm that both rules work for special values of $b$ and $n$. But to be confident that the rules are equivalent, students should use reasoning to argue that the rules are different ways of saying the same thing.

In **Problem 3,** some students may propose answers such as $p = 5e + 2e$ for **Part b.** This rule is indeed equivalent to $p = 7e$. Make sure, however, that students are able to give at least one equivalent rule that involves division.

2 On the **Spot Assessment**

**Problem-Solving Strategy**
Solve a simpler problem

Watch for students who have difficulty using equivalent multiplication and division equations (or equivalent addition and subtraction equations). If the problem seems to occur mainly when they are working with equations that have letters for variables, provide practice with equations that involve only specific numbers. Then move to such equations as $5 \cdot \triangle = \square$. After practicing with equations of this type, students can move on to equations that use letters for the variables.

In **Problem 6,** students can use what they know about related multiplication and division equations to decide which variables are represented by which letters. First, however, they need to be clear about what the variable quantities are and how they are related.

3 **Problem Set Wrap-Up** Ask students to share their rules for **Problem 4.** If no one offers the rules $d \div h = \frac{1}{24}$ or $d \div \frac{1}{24} = h$, you may want to point out that these are additional possibilities.

**2a.** Possible answer: $n = b + 2$ says the number of nuts is 2 more than the number of bolts. $b = n - 2$ says the number of bolts is 2 fewer than the number of nuts.

**2b.** yes; Possible explanation: Saying there are two more nuts than bolts is the same as saying there are two fewer bolts than nuts.

**3a.** The number of sheets of paper is 7 times the number of envelopes.

**3b.** Possible answer: $\frac{p}{7} = e$, $7 = \frac{p}{e}$

**4a.** Possible answer: The number of days is equal to $\frac{1}{24}$ of the number of hours.

**4b.** Possible answer: $24 \cdot d = h$, $24 = \frac{h}{d}$

**5c.** Possible answer: $g = \frac{5}{2} \cdot b$, $\frac{b}{g} = \frac{2}{5}$

**6.** For Lee's rule, *s* is the number of tires and *t* is the number of cars.

For Gavin's rule, *a* is the number of tires and *b* is the number of cars.

For Chitra's rule, *w* is the number of tires and *c* is the number of cars.

Althea and Luke wrote rules to describe the relationship between the number of nuts $n$ and the number of bolts $b$:

Althea's rule: $n = b + 2$ Luke's rule: $n - 2 = b$

**a.** Explain in words what each rule says.

**b.** Do the rules represent the same relationship? In other words, are the two rules equivalent? Explain.

**3.** The rule $p = 7e$ expresses the relationship between the number of envelopes $e$ and the number of sheets of paper $p$ in a package of stationery.

**a.** What does the rule tell you about relationship between paper and envelopes in the package?

**b.** Rewrite this rule in at least two different ways.

**4.** The rule $\frac{1}{24} \cdot h = d$ expresses the relationship between the number of days $d$ and the number of hours $h$.

**a.** What does the rule tell you about the relationship between days and hours?

**b.** Rewrite this rule in at least two different ways.

**5.** A jar contains black and green jelly beans. There are $\frac{2}{5}$ as many black jelly beans as green jelly beans.

**a.** Draw two jars that fit this description. See Additional Answers.

**b.** Write a rule that relates the number of black jelly beans $b$ to the number of green jelly beans $g$. Possible rule: $b = \frac{2}{5} \cdot g$

**c.** Rewrite your rule in at least two different ways.

**6.** Three students wrote rules connecting the number of tires (including the spare) to the number of cars in a parking lot. If all the rules below are correct, tell what variable each letter must stand for.

Lee's rule: $s = 5 \cdot t$

Gavin's rule: $5 \times b = a$

Chitra's rule: $c = w \div 5$

**1** Challenge students to reason that the two rules are equivalent.

**2** Help students who have trouble with equivalent equations.

**3** Discuss students' answers.



## Additional Answers

**5a.** Possible drawing:

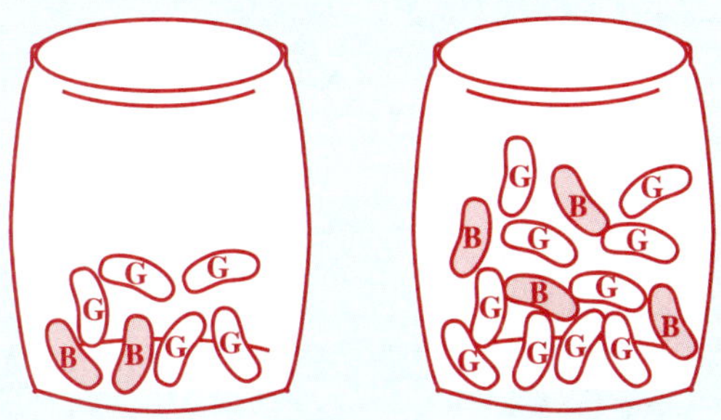

LESSON 7.2 Rules in Real Life 441

# Develop

1 **Problem Set E** **Suggested Grouping: Pairs or Small Groups**

The problems in this problem set make explicit something that some students may have already noticed: different situations may lead to the same rule. To be sure, the variables may represent quite different quantities, but the forms of the rules are the same.

### Share & Summarize

These questions will provide insight into how well students grasp the connection between descriptions in words and descriptions in symbols. They also help summarize how ideas about related multiplication and division equations, and related addition and subtraction equations, can be used to write equivalent rules.

2 **Troubleshooting** If students are having difficulty devising situations that can be described by a given rule, you might provide clues to help them get started. One way to do this is to suggest what variables the letters might represent.

**Additional Examples**

- Describe a situation that can be represented by the rule $e = f + 2$, where $e$ represents Elsa's age and $f$ represents Francisco's age. **Possible answer: Elsa is 2 years older than Francisco.**
- Describe a situation that can be represented by $T = 35L$, where $T$ is time in minutes and $L$ is the number of loads of laundry Lewis has to do. **Possible answer: Lewis has several loads of laundry to do. Each load will take 35 minutes.**

## On Your Own Exercises

Practice & Apply: 9–13, p. 445
Connect & Extend: 18–21, pp. 447–448
Mixed Review: 22–36, p. 449

So far in this chapter, you have looked at and written rules that fit specific situations. Now you will describe situations that match given rules.

## Problem Set E

**1.** Think about this rule: $t = 3 \cdot c$.

Describe three different situations this rule could represent. Tell what the variables stand for in each situation.

**2.** Sandra and Tim collect old cameras. In the rules below, $s$ represents the number of cameras Sandra has and $t$ represents the number of cameras Tim has. Explain what each rule tells you.

**a.** $s = t + 1$

**b.** $s = 2 \times t$

**c.** $s = 2t + 1$

**d.** $t = s + 3$

## Share & Summarize

Think about these rules:

$$\frac{1}{5} \cdot f = b \qquad g = a - 7$$

**1.** Describe a situation each rule could represent. Tell what variable each letter stands for. See below.

**2.** Rewrite each rule in a different way so that it still describes the situation. Possible answers: $f = 5 \cdot b$, $\frac{b}{f} = \frac{1}{5}$, $\frac{f}{b} = 5$; $g + 7 = a$, $a - g = 7$, $a = g + 7$

**Share & Summarize Answer**

**1.** Possible answer:

$\frac{1}{5} \cdot f = b$: If $b$ is the number of boys in a class and $f$ is the number of girls, this rule says the number of boys is $\frac{1}{5}$ the number of girls.

$g = a - 7$: If $a$ is my brother's allowance and $g$ is my allowance, this rule says my allowance is $7 less than my brother's.

**1.** Possible answer:

- The total number of quiz questions a teacher has to grade if there are three questions on each quiz. $t$ represents the total number of questions and $c$ represents the number of students who took the quiz.
- The number of wheels in a group of tricycles. $t$ represents the number of wheels and $c$ represents the number of tricycles.
- The amount of money a baby-sitter earns if she gets $3 per hour. $t$ represents the number of dollars earned and $c$ represents the number of hours of baby-sitting.

**2a.** Sandra has one more camera than Tim.

**2b.** Sandra has twice as many cameras as Tim.

**2c.** Sandra has one more camera than twice the number Tim has.

**2d.** Tim has three more cameras than Sandra.

**1** You may have students work in pairs or small groups.

**2** Provide clues to students to help them devise situations.

# Teacher Notes

# On Your Own Exercises

★ indicates multi-step problem

## Practice & Apply

1. When he grades a test, Mr. Bonilla awards 2 points for each correct answer, plus 6 points just for taking the test.

   a. Copy and complete the table to show the score a student would receive for the given numbers of correct answers.

| Correct Answers | 30 | 35 | 40 | 45 |
|---|---|---|---|---|
| Test Score | 66 | 76 | 86 | 96 |

   b. Write a rule you could use to calculate a test score if you know the number of questions a student answered correctly. Use $c$ to represent the number of correct answers and $s$ to represent the score. $s = 6 + 2c$

2. **Sports** The Get Wet beach shop rents sailboards for windsurfing. The cost to rent a sailboard is \$10 plus \$5 per hour. Get Wet also teaches people how to windsurf for \$15 per lesson.

   a. What are the variables in this situation? See margin.

   b. Complete the table to show the cost of renting a sailboard for various numbers of hours.

| Hours | 3 | 4 | 4.5 | 5 | 5.25 |
|---|---|---|---|---|---|
| Cost | \$25.00 | \$30.00 | \$32.50 | \$35.00 | \$36.25 |

   c. Write a rule you could use to calculate the cost of renting a sailboard for any number of hours. Use $c$ to represent the cost and $h$ to represent the number of hours. $c = 5h + 10$

   d. Create a table to show the costs for different numbers of lessons.

   e. Use letters and symbols to write a rule for calculating the cost for any number of lessons. Tell what variable each letter represents.

   ★f. Get Wet is running an early-summer special. If you take two lessons, you get 50% off the cost of your next rental. How much would it cost to take two lessons and then rent a sailboard for 2 hours? Show how you found your answer.
   \$40; \$30 for two lessons + \$10 (half the usual cost) for a 2-hour rental

**2a.** number of hours, cost, number of lessons

### Just the facts

The first windsurfing board was patented in 1968. The board was developed by Hoyle Schweitzer (a surfer) and Jim Drake (a sailor), who wanted to combine surfing with sailing.

**2d.** Possible table:

| Number of Lessons | 1 | 2 | 3 | 4 | 5 |
|---|---|---|---|---|---|
| Cost | \$15 | \$30 | \$45 | \$60 | \$75 |

**2e.** $D = 15 \cdot L$, where $D$ is the cost and $L$ is the number of lessons

impactmath.com/self_check_quiz



## Assign and Assess

### On Your Own Exercises

Investigation 1, pp. 431–432
Practice & Apply: 1–3
Connect & Extend: 14, 15

Investigation 2, pp. 436–439
Practice & Apply: 4–8
Connect & Extend: 16, 17

Investigation 3, pp. 439–442
Practice & Apply: 9–13
Connect & Extend: 18–21

Assign Anytime
Mixed Review: 22–36

**Exercise 2:**
In **Part f,** students will need to recall how to take 50% off a given cost.

*Quick Review* Math Handbook

hot words hot topics

Hot Topics pp. 254–261

**Exercise 3:**
You may want to ask students which is the greater distance, 1 kilometer or 1 mile. It is not uncommon for students to read such a statement as "there are approximately 1.6 kilometers in a mile" and to think that 1 kilometer is greater than 1 mile simply because the number preceding the word *kilometer* is greater than 1.

**Exercise 5a:**
Check that students are using a correct procedure to change 236 min to hours and minutes. You may want to discuss why a cook would find 3 h 56 min a more useful result than either 236 min or 3.93 h.

**Exercise 6e:**
You may wish to call on volunteers to explain how they found the distance Ana ran.

**3b.** $k = 1.6 \cdot m$, where $k$ is the number of kilometers and $m$ is the number of miles

**6a.** $j = 2a$, where $j$ represents the number of miles Juan ran

**6b.** $k = a + 3$, where $k$ represents the number of miles Kai ran

**6c.** $t = \frac{2}{3} \cdot a$, where $t$ represents the number of miles Toshio ran

**Just the facts**

On April 14, 2002, Khalid Khannouchi of Morocco completed the Flora London Marathon in 2 hours 5 minutes 38 seconds, setting a new world record.

3. **Measurement** There are approximately 1.6 kilometers in a mile.
   a. Create a table comparing miles to kilometers. See Additional Answers.
   b. Write a rule to represent the relationship between the number of miles and the number of kilometers. Tell what each letter in your rule represents.
   c. The driving distance from New York to Boston is about 200 miles. Use your rule to calculate how many kilometers this is. 320 km

4. **Economics** Tickets for a school play cost $3.75 each. Write a rule connecting the total cost in dollars $c$ and the number of tickets bought $t$. Make a table if you need to, and check your rule by testing a value. $c = 3.75 \cdot t$

5. The cooking time for a turkey is 18 minutes for every pound plus an extra 20 minutes.
   a. How long will it take to cook a 12-pound turkey? 236 min, or 3 h 56 min
   b. Write a rule for this situation, using $m$ for the number of minutes and $p$ for the number of pounds. Make a table if you need to, and check your rule by testing a value. $m = 18p + 20$

6. **Sports** Ana and her friends are training for a marathon. Today Ana ran $a$ miles. In Parts a–d, write a rule expressing the relationship between the number of miles the person ran and the number of miles Ana ran. Be sure to state what the letters in your rules represent. Make a table if you need to, and check your rule by testing a value.
   a. Juan ran twice as far as Ana.
   b. Kai ran 3 miles more than Ana.
   c. Toshio ran $\frac{2}{3}$ as far as Ana.
   d. **Challenge** Melissa ran three times as far as Toshio. (Remember, your rule should relate Melissa's distance to Ana's distance.)
   e. If Toshio ran 2 miles, how far did Ana, Juan, Kai, and Melissa run? Ana ran 3 mi. Everyone else ran 6 mi.

**6d.** $m = 2 \cdot a$, where $m$ represents the number of miles Melissa ran

**Additional Answers**

**3a.** Possible table:

| Miles | 1 | 2 | 3 | 4 | 10 |
|---|---|---|---|---|---|
| Kilometers | 1.6 | 3.2 | 4.8 | 6.4 | 16 |

**7. Sports** In football, a team receives 6 points for a touchdown, 1 point for making the kick after a touchdown, and 3 points for a field goal.

**a.** If a team gets three touchdowns, makes two of the kicks after the touchdowns, and scores two field goals, what is the team's total score? 26 points

**b.** Write a rule for a team's total score $S$ if the team gets $t$ touchdowns, makes $p$ kicks after touchdowns, and scores $g$ field goals. Be sure to check your rule by testing it for a specific case.

**8.** Three students wrote rules for the relationship between the number of eyes and the number of noses in a group of people. Each student used $e$ to represent the number of eyes and $n$ to represent the number of noses. Which of the rules are correct? Explain how you decided. See Additional Answers.

Miguel's rule: $2 \cdot e = n$

Althea's rule: $n \times 2 = e$

Hannah's rule: $n = e \div 2$

**9a.** The length of the room is 1 ft more than the width.

**9. Measurement** The rule $L = W + 1$ expresses the relationship between the length $L$ and width $W$ of a room. Both dimensions are in feet.

**a.** What does this rule tell you about the dimensions of the room? See above.

**b.** Rewrite this rule in at least two different ways. Explain how you know your rules are correct. See Additional Answers.

**10.** The rule $s = 3{,}600h$ gives the relationship between the number of hours and the number of seconds. See Additional Answers.

**a.** What does this rule tell you about the relationship between hours and seconds?

**b.** Rewrite this rule in at least two different ways.

**11.** The rule $m = \frac{3}{7} \times b$ describes the relationship between the number of mountain bikes $m$ in a cycling shop and the total number of bikes $b$.

**a.** What does this rule tell you about the relationship between the number of mountain bikes and the total number of bikes?

**b.** Rewrite this rule in at least two different ways.

In Exercises 12 and 13, describe a situation the rule could represent. Tell what variables the letters stand for.

**12.** $n = m - 8$

**13.** $p = 3r + 4$

**12, 13.** See Additional Answers.

**7b.** $S = 6 \cdot t + p + 3 \cdot g$

**11a.** $\frac{3}{7}$ of the bikes are mountain bikes.

**11b.** Possible answer: $b = \frac{7}{3} \cdot m$, $\frac{m}{b} = \frac{3}{7}$, $\frac{7}{3} = \frac{b}{m}$

**Just the facts**

*Penny farthing bicycles, introduced in England in the early 1870s, had front wheels with diameters up to 5 feet.*



**Exercise 8:**
Students who have difficulty selecting the correct rules may find it helpful to ask themselves, "Which rule says that there are more eyes than noses?"

**Exercise 9:**
Some students may find it helpful to sketch a diagram.

**Exercise 10:**
You may want to discuss why the rule $s = 3{,}600h$ is correct.

**Exercise 11b:**
You may want to count how many equivalent rules students can devise.

**Additional Answers**

**8.** Althea's and Hannah's rules are correct. Possible explanation: According to Miguel's rule, there would be four noses if there are two eyes. This is not correct. Althea's rule says there are twice as many eyes as noses, which is correct. Hannah's rule says there are half as many noses as eyes, which is correct.

**9b.** Possible answer: $W = L - 1$, $L - W = 1$; If the length is 1 ft longer than the width, then the width is 1 ft shorter than the length and the difference between the length and the width is 1 ft.

**10a.** The number of seconds is 3,600 times the number of hours.

**10b.** Possible answer: $h = \frac{s}{3{,}600}$; $\frac{s}{h} = 3{,}600$

**12.** Possible answer: Nellie is 8 years younger than Marie. $n$ represents Nellie's age, and $m$ represents Marie's age.

**13.** Possible answer: The price of renting a boogie board is \$4.00 plus \$3.00 per hour. $p$ represents the price of the rental, and $r$ represents the number of hours you use the board.

**Exercise 14:**
Check that students use the correct units when they give their answers.

**Exercise 15:**
Students need to understand that in **Part a** they are simply replacing the words for the variables with the letters for the variables.

You may want to discuss the details of the calculations for **Part b.**

## Connect & Extend

**14.** **Physics** Suppose you throw a ball straight up into the air and catch it again. The relationship between the time the ball is in the air $t$ and the approximate speed $s$ at which you threw the ball, in meters per second, is given by this rule:

$$s = 5 \cdot t$$

**a.** If the ball was in the air for 4.5 seconds, about how fast did you throw it? 22.5 m/s

**b.** If you throw the ball at a speed of about 25 meters per second, how long will it stay in the air? about 5 s

**15.** **Science** This rule is used to calculate the amount of a particular medicine a child should be given:

$$\text{child's dose} = \text{adult's dose} \times \frac{\text{child's age}}{\text{child's age} + 12}$$

**15a.** $c = d \cdot \frac{y}{y + 12}$

**a.** Rewrite this rule using $c$ to represent the child's dose, $d$ to represent the adult's dose, and $y$ to represent the child's age.

**b.** If the adult's dose is 50 milliliters, how much medicine should a 3-year-old child be given? 10 ml

**16.** In a children's story, peacocks and rabbits lived in a king's garden. A peacock has two legs, and a rabbit has four legs.

**a.** Complete the table to show the total number of legs in the garden for the given numbers of peacocks and rabbits.

| Peacocks | 2 | 4 | 6 | 8 | 10 |
|---|---|---|---|---|---|
| Rabbits | 3 | 6 | 9 | 12 | 15 |
| Legs | 16 | 32 | 48 | 64 | 80 |

**b.** Describe how you calculated the total number of legs for each group of animals. See below.

**c.** Use letters and symbols to write a rule to calculate the total number of legs in the garden if you know the number of rabbits and the number of peacocks. Tell what variable each letter represents.
$L = 4 \cdot r + 2 \cdot p$, where $L$ is the total number of legs, $r$ is the number of rabbits, and $p$ is the number of peacocks

**16b.** I multiplied the number of peacocks by 2 and the number of rabbits by 4 and added the results.



446 CHAPTER 7 Variables and Rules

★ indicates multi-step problem

**17d.** Possible answer: $d = 2$ and $q = 20$, $d = 7$ and $q = 18$, $d = 12$ and $q = 16$

**17.** Lina has $q$ quarters in her pocket. Alan has $d$ dimes in his pocket.

**a.** Write a rule to express the value of Lina's quarters *in dollars.* Use $v$ to represent the value. $v = 0.25 \cdot q$

**b.** Write a rule to express the value of Alan's dimes *in dollars.* Use $v$ to represent the value. $v = 0.10 \cdot d$

**c.** Alan combines his dimes with Lina's quarters. Write a rule to express the total value of the coins. $v = 0.25 \cdot q + 0.10 \cdot d$

**d.** Together, Lina and Alan have $5.20. Find three possibilities for the values of $d$ and $q$.

**18.** **Physical Science** Water is made up of hydrogen atoms and oxygen atoms. A sample of water contains twice as many hydrogen atoms as oxygen atoms.

*A molecule of water*

**a.** Write a rule to express the relationship between the number of hydrogen atoms and the number of oxygen atoms in a sample of water. Tell what the letters in your rule represent. See below.

**b.** A particular sample of water contains 476 hydrogen atoms. How many oxygen atoms does it contain? 952

**19.** **Science** A pharmacist is filling bottles with capsules. Let $n$ represent the number of bottles she has. a–c. Possible answers given.

**a.** She takes half of the bottles and puts 40 capsules in each. Write a rule for the total number of capsules $p$ in these bottles. (Your rule should include the letter $n$.) $p = \frac{1}{2} \cdot n \cdot 40$

**b.** She puts 50 capsules in each of the remaining bottles. Write a rule for the total number of capsules $q$ in these bottles. (Your rule should include the letter $n$.) $q = \frac{1}{2} \cdot n \cdot 50$

**c.** Use your answers to Parts a and b to help you write a rule for the total number of capsules $t$ the pharmacist has put into all the bottles. $t = \frac{1}{2} \cdot n \cdot 40 + \frac{1}{2} \cdot n \cdot 50$

★**d.** If the pharmacist started with 4 bottles, how many capsules did she use? If she started with 100 bottles, how many capsules did she use? 180; 4,500

**18a.** Possible answer: $o = 2h$, where $h$ is the number of hydrogen atoms and $o$ is the number of oxygen atoms



**Exercise 19:**
For **Part d,** some students may do the calculations without substituting 4 or 100 into the rule from **Part c.** You may want to suggest that they check their results by using the substitution method.

LESSON 7.2 Rules in Real Life 447

## Assign and Assess

**Exercise 20e:**
Students should be careful how they write the rule. You may want to discuss with the class why $r = 1 \div p^2$ is acceptable while $r = 1 \div p \cdot p$ is unacceptable. They may need to be reminded what the order of operations dictates in this situation.

**Exercise 21:**
You may want to discuss how the rules in **Parts a and b** can be written using fractions. Possible rules are $c = \frac{5}{9} \cdot (f - 32)$ and $f = \frac{9}{5} \cdot c + 32$.

★ indicates multi-step problem

### In your own words

Write a paragraph explaining what you have learned about writing rules for real-life situations. Be sure to discuss

- strategies for making rule-writing easier
- how to tell whether a rule is correct

**20a.** The [$x^2$] key multiplies a number by itself.

**20c.** It divides the number into 1 (or divides 1 by the number).

**21a.** $c = (f - 32) \cdot 5 \div 9$

**21b.** $f = c \cdot 9 \div 5 + 32$

**20. Technology** In this exercise, you will investigate some keys on your calculator.

**a.** Describe in words what the [$x^2$] key does. If you are not sure, enter a number, press the key, then press [ENTER]. Try this a few times until you are certain you know what the key does.

**b.** Suppose you enter a number into your calculator and press [$x^2$] [ENTER]. Write a rule to describe the relationship between the number you enter *n* and the result *r*. $r = n \cdot n$, or $r = n^2$

**c.** Your calculator has the command $x^{-1}$, accessed by pressing [2nd] [$x^{-1}$]. Use this command a few times with different numbers until you think you know what it does. Describe in words what this command does.

**d.** Suppose you enter a number into your calculator and press [2nd] [$x^{-1}$] [ENTER]. Write a rule to describe the relationship between the number you enter *m* and the result *r*. $r = 1 \div m$, or $r = \frac{1}{m}$

**e. Challenge** Suppose you enter a number, press [$x^2$], and then press [2nd] [$x^{-1}$] [ENTER]. Write a rule to describe the relationship between the number you enter *p* and the result *r*. Check that your rule gives the same result as the calculator for at least three values of *p*. $r = 1 \div p^2$

**21. Physical Science** To convert a temperature from degrees Fahrenheit (°F) to degrees Celsius (°C), use this rule: *Subtract 32, multiply the result by 5, and then divide by 9.*

**a.** Use symbols to write the rule for converting degrees Fahrenheit *f* to degrees Celsius *c*. (Hint: Make sure you think about order of operations.)

**b.** To convert a temperature from degrees Celsius to degrees Fahrenheit, just "undo" the operations in the reverse order. In other words, use this rule: *Multiply the Celsius temperature by 9, divide the result by 5, and then add 32.*

Use this information to write a rule for converting degrees Celsius to degrees Fahrenheit.

★**c.** Use the rule you wrote in Part a to convert 212°F to degrees Celsius. Then convert your answer back to degrees Fahrenheit using the rule you wrote in Part b. Do you end with 212°F?
212°F = 100°C = 212°F, yes

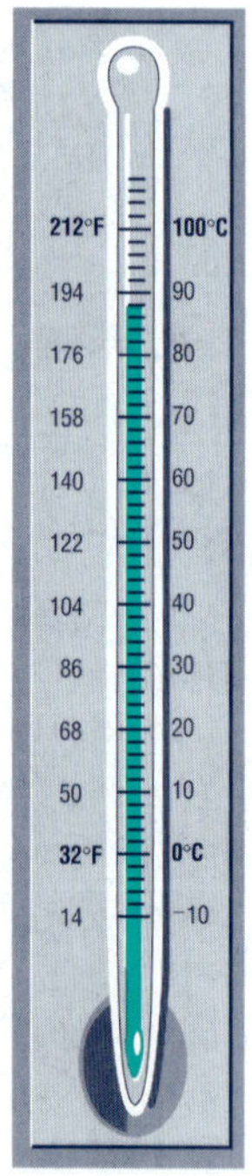




448 CHAPTER 7 Variables and Rules

## Mixed Review

Write each decimal as a fraction in lowest terms.

**22.** 0.05 $\frac{1}{20}$ **23.** $0.\overline{6}$ $\frac{2}{3}$ **24.** 0.6 $\frac{3}{5}$

**25.** 0.0075 $\frac{3}{400}$ **26.** 0.4545 $\frac{909}{2,000}$ **27.** 0.99 $\frac{99}{100}$

Write each fraction as a decimal.

**28.** $\frac{1}{8}$ 0.125 **29.** $\frac{3}{5}$ 0.6 **30.** $\frac{1}{11}$ $0.\overline{09}$

**31.** $\frac{10}{12}$ $0.8\overline{3}$ **32.** $\frac{228}{475}$ 0.48 **33.** $\frac{8}{9}$ $0.\overline{8}$

**34.** Stewart answered 14 out of 18 questions correctly on his math test. What percent did he answer correctly? Round your answer to the nearest whole percent. 78%

**35. Economics** Mr. Rosen's car cost $13,500 when he bought it. Since then, the value of the car has decreased by 45%. What is Mr. Rosen's car worth now? $7,425

**36. Statistics** Althea asked 20 of her classmates to keep track of the number of movies they watched on video or at the theater over a two-month period. She made a stem-and-leaf plot of the results.

**Number of Movies Watched**

| Stem | Leaf |
|---|---|
| 0 | 0 2 3 5 6 8 9 9 |
| 1 | 2 2 4 5 6 |
| 2 | 0 4 4 4 7 |
| 3 | 2 |
| 4 | |
| 5 | 5 |

**Key:** 2|4=24

**a.** What was the least number of movies watched? What was the greatest number? 0, 55

**b.** Describe the overall distribution of data values.

**c.** Find the mode, median, and mean of the data.
mode: 24; median: 13; mean: 15.85

**36b.** Possible answer: The interval from 0 to 9 has the greatest number of data values. The number of values decreases as the stems increase. Only two students watched more than 27 movies. One student watched 55 movies—23 more than the student with the next greatest value.



## Quick Check

**Informal Assessment**

Students should be able to:

- ✔ use letters for variables to write rules for real-life situations
- ✔ analyze real-life situations, identify related variables, and write rules describing relationships between the variables
- ✔ use related multiplication and division equations and related addition and subtraction equations to write rules in equivalent forms

**Quick Quiz**

1. A library charges $1.00 when a book is late and 10 cents per day for each additional day the book is late.

   a. What are the variables in this situation? ***number of days the book is late, cost of returning the book late***

   b. Create a table showing the late fee for four different numbers of days.

| Days Late | Fee |
|---|---|
| 1 | $1.00 |
| 2 | $1.10 |
| 3 | $1.20 |
| 4 | $1.30 |

c. Write a rule for calculating the late fee for any number of days. ***$f = 1 + 0.1 \cdot a$, where $f$ is the late fee and $a$ is the additional days the book is late (Note: Students may write the rules in words.)***

• ***Quick Quiz continued on page A665***

LESSON 7.2 Rules in Real Life 449

# Teacher Notes

7.3

# Explaining Number Relationships

## Objectives

- To discover patterns describing how the starting number is related to the final number in "think of a number" tricks
- To use operations on symbolic expressions to explain why "think of a number" tricks work
- To use symbolic expressions that contain variables to represent consecutive whole numbers, consecutive even numbers, and consecutive odd numbers, and to explain patterns in sets of these numbers

## Overview (pacing: about 3 class periods)

In this lesson, students continue to use letters to represent variables. Here the emphasis is less on writing rules to show relationships between variables and more on the numbers represented by symbolic expressions such as $n + 2$ and $n^2$. The idea that $n + 2$ can be treated as a number may seem novel to many students. One of the key features of this lesson is having students explore how to use symbolic expressions to make convincing arguments that certain calculations will always yield predictable results and that certain statements will always be true.

## Advance Preparation

No advance preparation is required.

## Lesson Planner

| | Summary | Materials | On Your Own Exercises | Assessment Opportunities |
|---|---|---|---|---|
| Investigation 1<br>page T451 | Students learn to use variables to predict and explain the outcome of "think of a number" tricks. | | Practice & Apply: 1–3, p. 457<br>Connect & Extend: 6, 7, p. 458<br>Mixed Review: 11–30, p. 459 | On the Spot Assessment, page T451<br>Share & Summarize, pages T454, 454<br>Troubleshooting, page T454 |
| Investigation 2<br>page T455 | Students use variables to represent and to explain properties of consecutive whole numbers, consecutive even numbers, and consecutive odd numbers. | | Practice & Apply: 4, 5, pp. 457–458<br>Connect & Extend: 8–10, pp. 458–459<br>Mixed Review: 11–30, p. 459 | Share & Summarize, pages T456, 456<br>Troubleshooting, page T456<br>Informal Assessment, page 459<br>Quick Quiz, page 459 |

# Introduce

Before you begin the Think & Discuss, ask students to have a piece of scratch paper handy in case they need it.

1

### Think & Discuss

Tell students that they are going to try two number tricks. Use each of the number tricks in the Think & Discuss, asking students for their results each time. After you have gone through both tricks, tell students that this lesson will help them understand why such tricks work. Then have students begin work on Investigation 1.

# 7.3 Explaining Number Relationships

You have used letters and symbols to describe sequences and real-life situations. In this lesson, you will use symbols to help describe and explain some number relationships.

## Think & Discuss

Try this number trick with your class. Each student should start with his or her own number.

- Think of a number.
- Add 5.
- Double the result.
- Subtract twice the number you started with.

What number do you end with? How does your result compare with other students' results? Everyone should get 10.

Now try this number trick:

- Think of a number.
- Add 5.
- Multiply the result by 5.
- Subtract 25.
- Divide the result by 5.

What number do you end with? How does your result compare with other students' results?

Everyone should get the number he or she started with.

1 Discuss both number tricks.

# Investigation 1

In this investigation, students look at more "think of a number" tricks. They then consider how to explain why each trick works.

1 **Problem Set A** **Suggested Grouping: Pairs**
The tricks in this problem set are similar to those in the
Think & Discuss on page 450. Students are asked to
select four starting numbers for each trick. In each prob-
lem, one of the starting numbers should be a fraction
and one a decimal. When students try "think of a num-
2 ber" tricks, they tend to use only whole numbers. But the
tricks presented here will work for *all* numbers, and it is
good for students to be aware of this.

In **Problem 2,** students may use the same starting numbers they used in Problem 1 or they may select entirely different numbers.

On the **Spot Assessment**

3 It would be a good idea to move around the room and watch for students who have computational errors that make it impossible for them to discover patterns in the results. Students often try to "beat the game" by picking what they consider to be "oddball" numbers that they think will make the trick fail. Numbers that are chosen in this way are often difficult to use in computations and can make errors more likely.

If you notice computational errors, have both students check their work until their results clearly reveal the pattern the problem intends to display.

## Investigation 1 Think of a Number

In this investigation, you will try some more number tricks. You will then use variables to help you understand how the tricks work.

1 You may have students work in pairs.

### Problem Set A

**1.** The first column of this table gives the steps for a number trick:

| Step | Result | | | |
|---|---|---|---|---|
| Think of a number. | 10 | 0.6 | $\frac{1}{4}$ | 7 |
| Add 2. | 12 | 2.6 | $2\frac{1}{4}$ | 9 |
| Multiply by 3. | 36 | 7.8 | $\frac{27}{4}$ | 27 |
| Subtract 3 times the number you started with. | 6 | 6.0 | $\frac{24}{4} = 6$ | 6 |

2 Remind students that these tricks work for all types of numbers.

**a.** Choose four different starting numbers, and record the result of each step in your table. One of your starting numbers should be a fraction and one should be a decimal. See the table for a possible answer.

**b.** Describe your results.

**1b.** The result is 6 no matter what number you start with.

**2.** This table gives the steps for another number trick:

| Step | Result | | | |
|---|---|---|---|---|
| Think of a number. | 10 | 0.6 | $\frac{1}{4}$ | 7 |
| Multiply by 10. | 100 | 6 | $\frac{10}{4}$ | 70 |
| Add 10. | 110 | 16 | $\frac{50}{4}$ | 80 |
| Subtract 9 times the number you started with. | 20 | 10.6 | $\frac{41}{4}$ | 17 |
| Subtract 10. | 10 | 0.6 | $\frac{1}{4}$ | 7 |

3 Walk around the room to check that students are not making computational errors.

**a.** Choose four different starting numbers, and record the result of each step in the table. One of your starting numbers should be a fraction and one should be a decimal. See the table for a possible answer.

**b.** Describe your results.
The result is always your original number.



LESSON 7.3 Explaining Number Relationships 451

1 **Example**
This Example presents a detailed, step-by-step account of why the first trick in the Think & Discuss on page 450 will always work.

Be sure to discuss the Example carefully with the class. If students have difficulty working with the symbolic expressions, it may help to write the expressions on the board and then point to the relevant parts as you use words to help students understand what is going on. For instance, in the equations at the end of the Example, you can point to the two places where $2 \cdot n$ occurs and say, "We're adding 2 times a number and then subtracting 2 times the very same number. That's going to give us 0. When we add 10 to 0, we'll get 10 as the final answer."

2 A key step in the Example is the one in which $2 \cdot (n + 5)$ is transformed into $2 \cdot n + 10$. Be sure students understand that in rewriting $2 \cdot (n + 5)$ as $(n + 5) + (n + 5)$, we are simply using the meaning of multiplication by 2.

Because you can change the starting number each time you try a number trick, the starting number is a variable. Using a letter to represent this variable can help you understand how the trick works. The Example explains the first trick in the Think & Discuss on page 450.

**EXAMPLE**

- Think of a number.

Use the letter $n$ to represent the starting number.

- Add 5.

Adding 5 to $n$ gives $n + 5$.

- Double the result.

Doubling the result of the previous step gives $2 \cdot (n + 5)$.

Multiplying a number by 2 is the same as adding the number to itself. You can use this idea to rewrite $2 \cdot (n + 5)$:

$$\begin{aligned} 2 \cdot (n + 5) &= (n + 5) + (n + 5) \\ &= n + 5 + n + 5 \end{aligned}$$

When you add a string of numbers, you can always rearrange them:

$$\begin{aligned} n + 5 + n + 5 &= n + n + 5 + 5 \\ &= 2 \cdot n + 10 \end{aligned}$$

So, the result of this step is $2 \cdot n + 10$.

- Subtract twice the number you started with.

The number you started with is $n$. Twice this number is $2 \cdot n$. Subtract this from the result of the last step:

$$2 \cdot n + 10 - 2 \cdot n$$

$2 \cdot n$ represents a number, and if you subtract a number from itself, the result is 0. In other words, $2 \cdot n - 2 \cdot n = 0$. You can use this fact to simplify the expression above:

$$\begin{aligned} 2 \cdot n + 10 - 2 \cdot n &= 2 \cdot n - 2 \cdot n + 10 \\ &= 0 + 10 \\ &= 10 \end{aligned}$$

So, no matter what number you start with, the result will be 10!

**Just the facts**

Coins with heads on both sides, and cards with back designs on both sides, are common magician's accessories.

1 Discuss the Example with the class.

2 Be sure students understand how $2 \cdot (n + 5)$ is rewritten.

# Develop

1 The table at the top of page 453 summarizes what is happening at each stage of the "think of a number" trick from the Example. It provides a convincing argument that the result is always 10 no matter what number $n$ you start with.

### Think & Discuss

In this Think & Discuss, students practice multiplying a sum or difference that contains a variable by a whole number greater than 1. Allow a few minutes for students to work on the expressions given. Then discuss each
2 example writing the appropriate steps on the board. Be sure students understand that the first step is to use the idea of multiplication as repeated addition. For the first expression, the work will look like this:

$3 \cdot (n + 2) = (n + 2) + (n + 2) + (n + 2)$

Apply the meaning of multiplication by 3.

$= n + 2 + n + 2 + n + 2$

$= n + n + n + 2 + 2 + 2$

Group the $n$'s and the 2's.

$= 3 \cdot n + 3 \cdot 2$

Repeated addition of the same number can be written as multiplication.

$= 3 \cdot n + 6$

### About the Mathematics

Each problem in the Think & Discuss is an instance of the *distributive property* of multiplication over addition or subtraction. The distributive properties are valid for all numbers. Thus, for all numbers $a$, $b$, and $c$, the following equations are true:

$$a \cdot (b + c) = (a \cdot b) + (a \cdot c)$$

$$a \cdot (b - c) = (a \cdot b) - (a \cdot c)$$

Students will study the distributive property in Course 2. At this point in time, it is enough for them to deal with special cases in which the first factor $a$ is a whole number greater than 1.

Once students have grasped the idea of how to rewrite expressions such as those in the Think & Discuss, have them work on Problem Set B.

The table summarizes the steps in the Example.

| Step in Words | Step in Symbols |
|---|---|
| Think of a number. | $n$ |
| Add 5. | $n + 5$ |
| Double the result. | $2 \cdot (n + 5) = n + 5 + n + 5$<br>$= 2 \cdot n + 10$ |
| Subtract twice the number you started with. | $2 \cdot n + 10 - 2 \cdot n = 10$ |

To understand the steps in the Example, you need to think about and work with expressions as if they were numbers. For instance, you need to think about doubling $n + 5$. It takes practice to become comfortable thinking in this way.

1 Have students review the table.

## Think & Discuss

You can think of multiplication as repeated addition. For example, $3 \cdot 5$ can be thought of as $5 + 5 + 5$.

2 Be sure students understand that multiplication is the same as repeated addition.

Rewrite each expression below by thinking about the multiplication as repeated addition. Then rearrange and rewrite the result to make it as simple as you can.

| $3 \cdot (n + 2)$ | $5 \cdot (n - 1)$ | $4 \cdot (3 + n)$ | $2 \cdot (10 + n)$ |
|---|---|---|---|
| $3 \cdot n + 6$ | $5 \cdot n - 5$ | $12 + 4 \cdot n$ | $20 + 2 \cdot n$ |

Compare the original expressions above and the final expressions you wrote. Find a pattern or shortcut you could use for multiplying an expression, such as $(n + 4)$ or $(n - 3)$, by a number without writing out all the steps. See ①.

Use your shortcut to find $3 \cdot (6 + n)$. Check your answer by writing out all the steps. See ②.

① Possible answer: Multiply the number part of the expression by the number, and multiply the variable part by the number. Then add or subtract the results (depending on the sign in the original expression).

② $18 + 3 \cdot n$; Check: $3 \cdot (6 + n) = 6 + n + 6 + n + 6 + n = 6 + 6 + 6 + n + n + n = 18 + 3n$.

In Problem Set B, you will use symbols to try to explain how some of the number tricks you have tried in this investigation work.



LESSON 7.3 Explaining Number Relationships 453

# Develop

1 **Problem Set B** **Suggested Grouping: Small Groups**

In this problem set, students use ideas they have worked with in Problem Set A to study new "think of a number" tricks.

2 In **Problem 1,** if students are unsure how to start, suggest they use the table on p. 453 as a guide.

In **Problem 3,** some students may be unsure what to do with $10 \cdot n + 10 - 9 \cdot n$. You might point out that this expression can be rewritten as $10 + 10 \cdot n - 9 \cdot n$. Think of $10 \cdot n$ as the sum of $n$ and 9 more $n$'s to get $10 + n + 9 \cdot n - 9 \cdot n$, or $10 + n + 0$, or $10 + n$.

3 **Share & Summarize**

You may want to call on volunteers to read their answers to the class. Discuss the answers carefully. Hopefully, students will have realized by now that nothing magical is happening in the number tricks. All the tricks lead to perfectly predictable results that can be determined on the basis of sound mathematical principles.

**Troubleshooting** By analyzing students' work in Problem Set B, you may be able to pinpoint specific types of difficulties students are having. This can help you come up with tailor-made, simplified examples. For example, if there are students who have trouble working with symbolic expressions, they may find it helpful to work with some simplified examples in which the principle behind the trick is close to the surface.

**Additional Example** Suppose you start with a number, add 5, and then subtract 5. How will your result compare with the number you started with? *It will be the same as the starting number.*

## On Your Own Exercises

Practice & Apply: 1–3, p. 457
Connect & Extend: 6, 7, p. 458
Mixed Review: 11–30, p. 459

## Problem Set B

**1.** The second number trick on page 450 always gives back the number you start with. You can use symbols to understand why this happens.

**a.** Copy and complete the table to show each step in symbols.

| Step in Words | Step in Symbols |
|---|---|
| Think of a number. | $n$ |
| Add 5. | $n + 5$ |
| Multiply by 5. | $5 \cdot (n + 5) = 5 \cdot n + 25$ |
| Subtract 25. | $5 \cdot n + 25 - 25 = 5 \cdot n$ |
| Divide by 5. | $5 \cdot n \div 5 = n$ |

**b.** In the second step, you add 5 to your number. By the end of the trick, the 5 has "disappeared." What happens to the 5?

**2.** Here is a trick you worked with in Problem Set A. For this trick, you end up with 6 no matter what number you start with.

**a.** Complete the table to show each step in symbols.

| Step in Words | Step in Symbols |
|---|---|
| Think of a number. | $n$ |
| Add 2. | $n + 2$ |
| Multiply by 3. | $3 \cdot (n + 2) = 3 \cdot n + 6$ |
| Subtract 3 times the number you started with. | $3 \cdot n + 6 - 3 \cdot n = 6$ |

**b.** In the first step, you start with $n$. What happens to $n$ by the end?

**3.** Here is the second trick from Problem Set A. For this trick, you always end up with your original number. Complete the table.

| Step in Words | Step in Symbols |
|---|---|
| Think of a number. | $n$ |
| Multiply by 10. | $10 \cdot n$ |
| Add 10. | $10 \cdot n + 10$ |
| Subtract 9 times the number you started with. | $10 \cdot n + 10 - 9 \cdot n = n + 10$ |
| Subtract 10. | $n + 10 - 10 = n$ |

**1b, 2b.** See Additional Answers.

### Share & Summarize

See Additional Answers.

Explain how using letters and symbols can help you understand how a number trick works.

1 You may have students work in small groups.

2 Have students use the table on page 453 to help them.

3 Discuss students' answers.



### Additional Answers

**1b.** After you add 5, you multiply by 5 to get 25. Then you subtract 25, which makes the 5 "disappear."

**2b.** In the third step, you multiply $n$ by 3 to get $3 \cdot n$, but then you subtract $3 \cdot n$ in the last step.

### Share & Summarize

Possible answer: Symbols let you follow the steps and track what is happening without worrying about particular numbers. You can see exactly why the result is always a particular number or always the number you started with.

# Investigation 2

This investigation provides additional opportunities for students to perform operations on expressions that use letters to represent variables. Students will see how they can use expressions with variables to discover and justify new properties of whole numbers, including special properties of even numbers and odd numbers.

1 Before students work on the Think & Discuss, read the definition of consecutive numbers at the top of the page. Discuss the examples of consecutive whole numbers, consecutive odd numbers, and consecutive even numbers.

2 **Think & Discuss**
You may want to have students work on these questions in small groups. Then have a class discussion of the results. The key idea is that no matter what whole number you choose, you can get the next consecutive numbers by adding 1 successively. To generate consecutive even (or odd) numbers, start with an even (or odd) number and add 2 successively.

### About the Mathematics

If you use successive *subtraction* to generate consecutive numbers, you may produce some negative numbers. In this case, you will have consecutive *integers* rather than consecutive whole numbers. For example, if you start with 3 and subtract 1 successively you get 2, 1, 0, $^{-}1$, $^{-}2$. Similar remarks apply for consecutive even numbers and consecutive odd numbers.

In this investigation, it is assumed that the numbers are whole numbers. However, the rules and patterns that students discover in this investigation are true in general, even when negative numbers are included.

3 **Problem Set C** **Suggested Grouping: Pairs**
In this problem set, students consider several sets of three consecutive whole numbers and look for a pattern that relates the sum of the numbers to the middle number.

4 In **Problem 2,** students may need some time to come up with the pattern. The amount of time they need will depend on the numbers they use in their tables.

In **Problem 4,** it is assumed that the number $m$ is greater than zero. This ensures that $m - 1$ will not be negative.

## Investigation 2 Consecutive Numbers

*Consecutive numbers* are numbers that follow one after the other.

- 199, 200, and 201 are consecutive whole numbers.
- 5, 7, 9, and 11 are consecutive odd numbers.
- 52, 54, 56, 58, and 60 are consecutive even numbers.

### Think & Discuss

If $n$ is some whole number, how would you write the next three consecutive whole numbers? $n + 1, n + 2, n + 3$

If $n$ is some whole number, what whole number comes just before $n$? What whole number comes before that? $n - 1, n - 2$

If $e$ is an even number, what are the next two consecutive even numbers? What two even numbers come just before $e$? $e + 2, e + 4; e - 2, e - 4$

### Problem Set C

Now you will think about sums of three consecutive whole numbers.

1. Write down three consecutive whole numbers and find their sum. Repeat this with at least four more sets of numbers. Record your results in a table.

| Three Consecutive Numbers | Sum |
|---|---|
| 5, 6, 7 | 18 |
| 20, 21, 22 | 63 |
| 0, 1, 2 | 3 |
| 40, 41, 42 | 123 |
| 100, 101, 102 | 303 |

2. Look for a pattern in your table. See if you can find a connection between the sum and the middle consecutive number. Write a rule in words that you could use to find the sum of three consecutive whole numbers without actually adding the numbers. Possible answer: The sum is equal to 3 times the middle number.
3. Test your rule on three more sets of consecutive numbers. Does it work for all the numbers you tried?
4. To show that your rule will *always* work, you need to show that it works for *any* three numbers, not just a few specific examples. You can do this by using letters to represent the variables. Use $m$ to stand for any whole number. How would you represent the number just before $m$ and the number just after $m$? $m - 1, m + 1$

1. See the table for a possible answer.

3. If students have written a correct rule, it will work for any three consecutive numbers.

1 Introduce the term *consecutive numbers.*

2 Have students work on these questions in small groups.

3 You may have students work in pairs.

4 **Problem-Solving Strategy** Look for a pattern

# Develop

1 In **Problem 5,** check that students understand how to simplify the sum. You may need to remind them that the numbers can be added in any order.

2 **Problem Set D** **Suggested Grouping: Pairs**
In this problem set, students examine still more properties of consecutive whole numbers. They also consider a property of consecutive even numbers.

3 In **Problem 3,** many students will follow the lead of earlier problems by, in **Part c,** choosing a letter to represent the middle number. It is possible, however, to let the letter represent the first number. If $e$ represents the first of three consecutive even numbers, then the other two numbers are $e + 2$ and $e + 4$. The sum of the three numbers is $e + (e + 2) + (e + 4)$. This can be rewritten as $3e + 6$. Notice that now the middle number is $e + 2$. Three times this number is $3 \cdot (e + 2)$. Since $3 \cdot (e + 2) = (e + 2) + (e + 2) + (e + 2) = 3e + 6$, the sum of the three numbers is again seen to be equal to 3 times the middle number.

4 **Share & Summarize**
This question will help you ascertain whether students understand how verifying a result in particular cases is different from presenting a general argument that shows that the result is true in *all possible* cases. Ask a few volunteers to present their answers for class discussion.

**Troubleshooting** If students are having difficulty understanding how to represent consecutive whole numbers using variables, offer them a few more examples to help them get the idea.

**Additional Examples** After offering these examples, you might ask students to come up with a few of their own.

- If $n$ is some whole number, how would you write the whole number just before $n$? $n - 1$ The whole number just after $n$? $n + 1$ What are the three whole numbers, in order? $n - 1, n, n + 1$
- If $j$ is an odd number, what is the next odd number? $j + 2$ The odd number after that? $j + 4$ What are the three odd numbers, in order? $j, j + 2, j + 4$
- Suppose $k$ represents the greatest of three consecutive even numbers. What are the three numbers, in order? $k - 4, k - 2, k$

## On Your Own Exercises

Practice & Apply: 4, 5, pp. 457–458
Connect & Extend: 8–10, pp. 458–459
Mixed Review: 11–30, p. 459

5. $m + m - 1 + m + 1$; $3 \cdot m$

6. Possible answer: Yes; when I added the three consecutive numbers $m - 1$, $m$, and $m + 1$, I got $3 \cdot m$, which is 3 times the middle number. Since $m$ can be any whole number, this shows the rule works for any three consecutive numbers.

**Problem Set D Answers**

1a. Possible answer:

3, 4, 5: $3 + 5 = 8$, $2 \cdot 4 = 8$

10, 11, 12: $10 + 12 = 22$, $2 \cdot 11 = 22$

61, 62, 63: $61 + 63 = 124$, $2 \cdot 62 = 124$

The rule works.

1b. Possible answer: If $m$ is the middle number, then $m - 1$ is the first number and $m + 1$ is the last number. The sum of the first and last numbers is $m - 1 + m + 1 = m + m + 1 - 1 = 2 \cdot m$, which is 2 times the middle number.

2, 3. See Additional Answers.

**5.** Now write the sum of three consecutive numbers—that is, the sum of $m$, the number just before $m$, and the number just after $m$. Then rearrange and rewrite the sum to make it as simple as possible.

**6.** Does your result show that your rule always works? Explain.

Since it is impossible to test a rule for *every* number, you can't prove that a rule is always true by giving lots of examples. However, if you use a letter to represent "any number," as you did in Problem Set C, you can often use symbols to show that a rule is always true.

## Problem Set D

**1.** Rosita thinks she has found another rule involving three consecutive whole numbers. She says that if you add the first number and the last number, the result will be twice the middle number.

**a.** Test Rosita's rule for at least three sets of numbers. Does it work for all the numbers you tried? If not, find a correct rule relating the sum of the first and last numbers to the middle number.

**b.** Using $m$ to stand for the middle number, explain why the correct rule (either Rosita's rule or the rule you wrote) always works.

**2.** Marcus wonders whether the sum of four consecutive whole numbers is always 4 times the second number.

**a.** Test Marcus' rule for at least three sets of numbers. Do you think it is correct? If not, find a correct rule relating the sum of four consecutive numbers to the second number.

**b.** Using $n$ to represent the second number, show that the correct rule (either Marcus' rule or the rule you wrote) always works.

**3.** Consider the sum of three consecutive even numbers—such as 6, 8, 10 or 42, 44, 46.

**a.** Find the sum of at least three sets of three consecutive even numbers.

**b.** Write a rule in words that you could use to find the sum of three consecutive even numbers without actually adding the numbers.

**c.** Use symbols to help explain why your rule is always true.

### Share & Summarize

See Additional Answers.

Suppose you want to convince someone that a relationship involving consecutive numbers is always true. Explain why it is better to use letters and symbols than to test lots of examples.



1 Make sure students know how to simplify the sum.

2 You may have students work in pairs.

3 Challenge students to let the variable represent the first number.

4 Have some students volunteer their answers to the class.

**Additional Answers**

2a. Possible answer:

1, 2, 3, 4: $1 + 2 + 3 + 4 = 10$, $4 \cdot 2 = 8$

10, 11, 12, 13: $10 + 11 + 12 + 13 = 46$, $4 \cdot 11 = 44$

7, 8, 9, 10: $7 + 8 + 9 + 10 = 34$; $4 \cdot 8 = 32$

Marcus' rule is not correct; the sum of four consecutive numbers is 4 times the second number, plus 2.

2b. Possible answer: If $n$ is the second number, then the four consecutive numbers are $n - 1$, $n$, $n + 1$, and $n + 2$. The sum is $n - 1 + n + n + 1 + n + 2 = 4 \cdot n + 2$, which is 4 times the second number, plus 2. $4 \cdot n + 2$

• Additional Answers continued on page A666

# Teacher Notes

# On Your Own Exercises

**Assign and Assess**

**Practice & Apply**

In Exercises 1–3, the first column of the table gives the steps for a number trick. Complete Parts a–c for each table.

**a.** Choose at least three different starting numbers, and record the result of each step in the table.

**b.** Describe your results.

**c.** Use symbols to explain how the trick works.

**1–3.** See Additional Answers.

**1.**

| Step | Result | | |
|---|---|---|---|
| Think of a number. | 10 | 0.5 | 0 |
| Add 6. | 16 | 6.5 | 6 |
| Multiply by 2. | 32 | 13 | 12 |
| Add your original number. | 42 | 13.5 | 12 |
| Subtract 3 times your original number. | 12 | 12 | 12 |

**2.**

| Step | Result | | |
|---|---|---|---|
| Think of a number. | 10 | 0.5 | 0 |
| Divide by 2. | 5 | 0.25 | 0 |
| Add 1. | 6 | 1.25 | 1 |
| Multiply by 4. | 24 | 5 | 4 |
| Subtract your original number. | 14 | 4.5 | 4 |
| Subtract 3. | 11 | 1.5 | 1 |

**3.**

| Step | Result | | |
|---|---|---|---|
| Think of a number. | 10 | 0.5 | 0 |
| Add 2. | 12 | 2.5 | 2 |
| Multiply by 5. | 60 | 12.5 | 10 |
| Subtract 10. | 50 | 2.5 | 0 |
| Subtract 2 times your original number. | 30 | 1.5 | 0 |
| Divide by 6. | 5 | 0.25 | 0 |

**4.** Jahmal said that for any four consecutive whole numbers, the sum of the first and last numbers is equal to the sum of the two middle numbers.

**a.** Test Jahmal's statement for at least three sets of numbers. Is it true for all the numbers you tried?

**b.** Using $f$ to represent the first number, use symbols to help explain why Jahmal's statement is true for any four consecutive numbers.

**4a.** Possible answer:
1, 2, 3, 4:
$1 + 4 = 5$,
$2 + 3 = 5$
4, 5, 6, 7:
$4 + 7 = 11$,
$5 + 6 = 11$
12, 13, 14, 15:
$12 + 15 = 27$,
$13 + 14 = 27$
The statement is true for all the numbers I tried.

**4b.** Possible answer: The four numbers are $f$, $f + 1$, $f + 2$, and $f + 3$. The sum of the first and last numbers is $f + f + 3 = 2 \cdot f + 3$. The sum of the middle numbers is $f + 1 + f + 2 = 2 \cdot f + 3$. So, the sum of the first and last numbers is equal to the sum of the two middle numbers.

impactmath.com/self_check_quiz



## On Your Own Exercises

Investigation 1, pp. 451–454
Practice & Apply: 1–3
Connect & Extend: 6, 7

Investigation 2, pp. 455–456
Practice & Apply: 4, 5
Connect & Extend: 8–10

Assign Anytime
Mixed Review: 11–30

**Exercise 2c:**
Some students may feel uncomfortable working with expressions that contain variables in fractions. If they prefer, they can use $\frac{1}{2} \cdot n$ or $0.5n$ instead of $\frac{n}{2}$.

**Additional Answers**

**1a.** See the table for a possible answer.

**1b.** The result is always 12.

**1c.**

| Step in Words | Step in Symbols |
|---|---|
| Think of a number. | $n$ |
| Add 6. | $n + 6$ |
| Multiply by 2. | $2 \times (n + 6) = 2 \times n + 12$ |
| Add your original number. | $2 \times n + 12 + n = 3 \times n + 12$ |
| Subtract 3 times your original number. | $3 \times n + 12 - 3 \times n = 12$ |

• *Additional Answers continued on page A666*

# Assign and Assess

**Exercise 5c:**
Students may prefer to let $n$ represent the first number. The sum of the five consecutive numbers is then $5n + 10$, which is equal to 5 times $(n + 2)$.

**Exercise 6:**
You can have an interesting class discussion by asking how students devised their number tricks. A good general approach is to start with a variable and then "build" and "unbuild" an expression to get a result of 0. For example, start with $n$, build the expression $5n + 5$, which equals $5(n + 1)$, and then subtract 5 times $n + 1$. The trick might read, "Think of a number. Multiply by 5. Add 5. From this, subtract 5 times 1 more than the original number."

**Exercise 7:**
This exercise calls on students to use logical reasoning to solve number puzzles.

**5b.** The sum of any five consecutive whole numbers is equal to 5 times the middle number.

**5c.** See Additional Answers.

## Connect & Extend

**5a.** Possible answer:
$1 + 2 + 3 + 4 + 5 = 15$
$10 + 11 + 12 + 13 + 14 = 60$
$45 + 46 + 47 + 48 + 49 = 235$
$7 + 8 + 9 + 10 + 11 = 45$

**6.** See Additional Answers.

**7b.** A = 5, C = 4, B = 3

**8c.** 33; I substituted 10 for $A$: $P = 3 \cdot (A + 1) = 3 \cdot (10 + 1) = 3 \cdot 11 = 33$.

### In your own words

Describe how using letters and symbols can help you
- show how number tricks work
- explain why relationships involving consecutive numbers are always true

**5.** Consider the sum of five consecutive whole numbers.

**a.** Find the sum of at least four sets of five consecutive whole numbers. See margin.

**b.** Write a rule in words relating the sum of five consecutive whole numbers to the middle number. See margin.

**c.** Use symbols to help show that your rule must always be true.

**6.** Make up a number trick with at least three steps that always gives a result of 0. Use symbols to show why your trick works.

**7.** In the addition puzzles below, each letter stands for a single digit. The same letter stands for the same digit, and different letters stand for different digits.

**a.** Look at this subtraction problem:

$$\begin{array}{r} 7\ 2\ 8 \\ -\ 2\ 4\ 5 \\ \hline X\ Y\ Z \end{array}$$

Z must be equal to 3, since 8 − 5 = 3.

$$\begin{array}{r} 7\ 2\ 8 \\ -\ 2\ 4\ 5 \\ \hline X\ Y\ 3 \end{array}$$

Figure out what digits X and Y must represent. X = 4, Y = 8

**b.** Solve this puzzle. In other words, figure out what digits A, B, and C represent.

$$\begin{array}{r} A\ 1\ 2 \\ +\ C\ B\ A \\ \hline 9\ C\ 7 \end{array}$$

**c.** Find at least three different solutions to this puzzle:

$$\begin{array}{r} U \\ +\ U \\ \hline H\ E \end{array}$$

**7c–e.** See Additional Answers.

**d.** In Part c, are there any digits U cannot be? Explain your answer.

**e.** In Part c, are there any digits H cannot be? Explain your answer.

**f.** Solve this puzzle:

$$\begin{array}{r} BBE \\ +\ HE \\ \hline HEEE \end{array}$$

B = 9, E = 0, H = 1

**8.** Paul is 3 times as old as Ann will be next year.

**a.** Suppose $A$ stands for Ann's age. How old will Ann be in a year? $A + 1$

**b.** Write a rule you can use to find Paul's age $P$ if you know Ann's age. $P = 3 \cdot (A + 1)$

**c.** If Ann is now 10 years old, how old is Paul? Explain how you found your answer. See margin.

**d.** If Ann is 15, how old is Paul? 48

**e.** If you know that Paul is 24, explain how you can find Ann's age. See Additional Answers.



## Additional Answers

**5c.** Possible answer: If $m$ is the middle number, the five numbers are $m - 2$, $m - 1$, $m$, $m + 1$, and $m + 2$. The sum of the numbers is $m - 2 + m - 1 + m + m + 1 + m + 2 = 5 \cdot m$. So, the sum is 5 times the middle number.

• Additional Answers continued on page A666

9. Jonas is twice as old as Eva was last year.

a. If $J$ represents Eva's age this year, how old was she last year? $J - 1$

b. Write a rule you can use to find Jonas' age $C$ if you know Eva's age. $C = 2 \cdot (J - 1)$

c. Suppose Eva is 15 *this* year. How old is Jonas this year? 28

d. Suppose Eva was 12 *last* year. How old is Jonas this year? 24

e. Suppose Jonas is 30 this year. How old is Eva this year? 16

10. Natalie is 3 times as old as Terry was last year.

a. Write a rule for this situation, using letters and symbols. Tell what the letters in your rule stand for.

b. If Terry is now 7, how old is Natalie? 18

## Mixed Review

**10a.** Possible answer: $N = 3 \cdot (T - 1)$, where $N$ is Natalie's age and $T$ is Terry's age

**Number Sense** Fill in each ● with >, <, or =.

11. $33\frac{1}{3}\%$ ● $\frac{1}{3}$ =

12. $\frac{7}{8}$ ● 85% >

13. 0.398 ● $\frac{2}{5}$ <

14. $^{-}5$ ● $^{-}1$ <

15. $0.\overline{5}$ ● $\frac{5}{9}$ =

16. $\frac{31}{40}$ ● 75% >

17. $\frac{347}{899}$ ● $\frac{347}{900}$ >

18. $\frac{6}{7}$ ● $\frac{7}{8}$ <

19. 80% ● $\frac{45}{60}$ >

20. 0.01 ● 0.1% >

**Measurement** Fill in the blanks.

21. 356 cm = 3.56 m

22. 356 cm = 3,560 mm

23. 44 m = 44,000 mm

24. 5 mm = 0.005 m

25. 5 mm = 0.5 cm

26. 89,000 mm = 89 m

**Geometry** Estimate the measure of each angle. **27-30.** Possible answers given.

27.

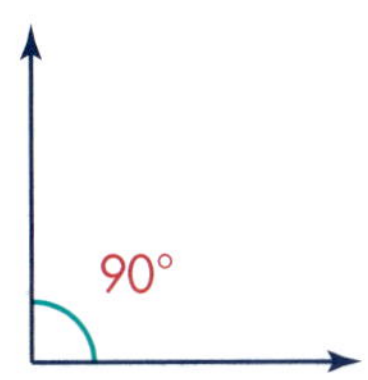


28.

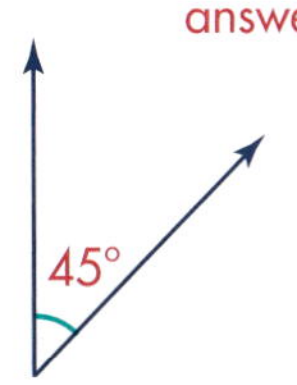


29.

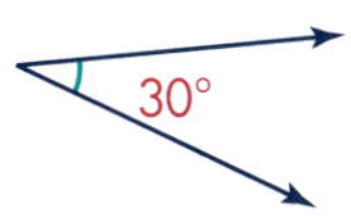


30.

## Quick Check

### Informal Assessment

Students should be able to:

✔ discover patterns describing how the starting number is related to the final number in "think of a number" tricks

✔ use operations on symbolic expressions to explain why "think of a number" tricks work

✔ use symbolic expressions that contain variables to represent consecutive whole numbers, consecutive even numbers, and consecutive odd numbers, and to explain patterns in sets of these numbers

### Quick Quiz

1. Consider this number trick: Start with a number. Multiply by 2. Add 4 times the original number. Divide the result by 6.

   a. Try the trick with four different starting numbers. Describe how the result is related to the starting number. **It is equal to the starting number.**

   b. Use expressions containing variables to explain why your pattern always works. **Let $n$ represent the starting number. Multiplying the starting number by 2 gives $2n$. Adding 4 times the original number gives $2n + 4n$, or $6n$. Dividing this by 6 gives $n$, the original number.**

2. Suppose $m$ is the first of five consecutive odd numbers.

   a. Use the variable $m$ to represent the other four odd numbers. **$m + 2$, $m + 4$, $m + 6$, $m + 8$**

   b. Show why the sum of the first and last numbers is equal to the sum of the second and fourth numbers. **$m + (m + 8) = 2m + 8$ and $(m + 2) + (m + 6) = 2m + 8$**

## Chapter Summary

This summary helps students recall the major topics of the chapter.

## Problem-Solving Strategies and Applications

The questions in this section help students review and apply the important mathematical ideas and problem-solving strategies developed in the chapter. The questions are organized by mathematical highlights. The highlights correspond to those in "The Big Picture" chart on page 407a.

# Chapter 7 Review & Self-Assessment

## Chapter Summary

In this chapter, you worked with *symbolic rules,* in which letters are used to represent variables. You started by writing rules to relate variables in sequences of dots and toothpicks. You found ways to explain why a rule works for every term in a sequence and to show that two different rules are equivalent. Then you played the game *What's My Rule?* and expressed input/output rules by using symbols.

You saw that real-life situations often involve variables, and you wrote symbolic rules to represent many such situations. You found that the same relationship can often be expressed in several ways, and you practiced rewriting rules to form equivalent rules.

Finally, you saw how using letters to represent variables can help you show how number tricks work and explain why certain relationships involving consecutive numbers are always true.

## Strategies and Applications

The questions in this section will help you review and apply the important ideas and strategies developed in this chapter.

### Writing and interpreting rules for sequences and input/output tables

**1.** Consider this toothpick pattern:

Term 1 Term 2 Term 3

**a.** Choose a variable other than the term number.

**b.** Create a table showing the value of your variable for each term.

**c.** Try to find a rule that connects the term number and your variable. Write the rule as simply as you can, using $n$ to represent the term number and another letter to represent your variable.

**1a.** Possible answers: total number of toothpicks, number of top and bottom toothpicks

**1b.** Possible table:

| Term Number | 1 | 2 | 3 |
|---|---|---|---|
| Number of Top and Bottom Toothpicks | 3 | 6 | 9 |

**1c.** Possible rules:

$t = 4n + 1$, where $t$ is the number of toothpicks

$b = 3 \cdot n$, where $b$ is the number of top and bottom toothpicks



impactmath.com/chapter_test

**4a.** Possible rules: $t = 2 + 4n$ and $t = 2 \cdot (1 + 2n)$, where $n$ is the term number and $t$ is the number of toothpicks

**5.** no; Possible example: Suppose the rules are $d = 2 \cdot n$ and $d = 2 + n$. These rules both give $d = 4$ when $n$ is 2, but they are not equivalent. For example, when $n$ is 1, the first rule gives $d = 2$ while the second rule gives $d = 3$.

**6d.** Possible answer: If I planned on skating for less than 3 hours, I would rent from the Skate Shop because it would be cheaper. If I wanted to skate for more than 3 hours, Rolling Along would be cheaper. If I planned on skating exactly 3 hours, it wouldn't matter where I rented my skates.

**2.** Stacey drew a dot sequence. The rule $d = 5 \cdot n - 2$, where $n$ represents the term number and $d$ represents the number of dots in a term, describes her sequence.

**a.** Find the number of dots in each of the first four terms of Stacey's sequence. 3, 8, 13, 18

**b.** Draw a dot sequence that fits the rule. See Additional Answers.

**3.** Write two different rules to express the relationship between $a$ and $b$.

| $a$ | 0 | 1 | 2 | 3 | 4 |
|---|---|---|---|---|---|
| $b$ | 3 | 6 | 9 | 12 | 15 |

Possible answer: $b = 3a + 3$, $b = 3 \cdot (a + 1)$

## Showing that two rules for a sequence are equivalent

**4.** Consider this toothpick sequence:

Term 1 Term 2 Term 3

**a.** Write two equivalent rules for the number of toothpicks in a term.

**b.** Use words and diagrams to explain why each of your rules is correct. See Additional Answers.

**5.** Suppose two different rules give the same number of dots for Term 2 of a sequence. Can you conclude that the rules are equivalent? Give an example to support your answer.

## Writing and interpreting rules for real-life situations

**6.** At the entrance to Fairmont Park, two stores rent in-line skates. Rolling Along charges $4.50, plus $2.50 per hour. The Skate Shop has a flat rate of $4 per hour.

**a.** How much would it cost to rent skates from Rolling Along for 2 hours? How much would it cost to rent skates from the Skate Shop for 2 hours? $9.50, $8

**b.** Write a rule for the cost $c$ of renting skates from Rolling Along for $h$ hours. $c = 4.5 + 2.5 \cdot h$

**c.** Write a rule for the cost $c$ of renting skates from the Skate Shop for $h$ hours. $c = 4 \cdot h$

**d.** Suppose you want to rent in-line skates to use in Fairmont Park. Assuming the stores have skates of the same quality, how would you decide where to rent your skates? See margin.



## Additional Answers

**2b.** Possible answer:

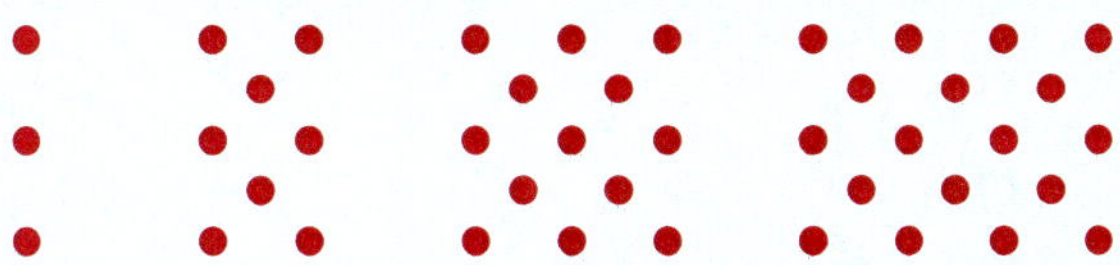

Term 1 Term 2 Term 3 Term 4

• Additional Answers continued on page A667

7a. Possible answer: The total cost $t$ to order a book is $d$ dollars, plus a shipping charge of \$3.50.

7b. Possible answer:

$t - 3.5 = d$; If $t$ is 3.5 more than $d$, then $d$ is 3.5 less than $t$.

$t - d = 3.5$; If $t$ is 3.5 more than $d$, then the difference between $t$ and $d$ is 3.5.

8a. See the table for a possible answer.

8c. See Additional Answers.

9c. If $f$ is the first number, then $f + 1$ is the second number and $f + 2$ is the third number. Their sum is $f + f + 1 + f + 2$, which equals $3 \cdot f + 3$. This is 3 times the first number, plus 3.

**7.** Consider this rule:

$$t = 3.5 + d$$

**a.** Describe a situation this rule could represent. Tell what variables the letters stand for.

**b.** Write the rule in two different ways. Explain how you know your rules are correct.

## Using variables to explain number relationships

**8.** The first column of this table gives the steps for a number trick:

| Step | Result | | |
|---|---|---|---|
| Think of a number. | 10 | 0.5 | 5 |
| Add 3. | 13 | 3.5 | 8 |
| Multiply by 4. | 52 | 14 | 32 |
| Subtract 12. | 40 | 2 | 20 |
| Subtract your original number. | 30 | 1.5 | 15 |
| Divide by 3. | 10 | 0.5 | 5 |

**a.** Choose at least three different starting numbers, and record the result of each step in the table.

**b.** Describe your results. I always get the number I started with.

**c.** Use symbols to explain how the trick works.

**9.** Consider sums of three consecutive whole numbers.

**a.** Write down three consecutive whole numbers and find their sum. Repeat this with at least three more sets of numbers. Record your results in a table. Possible answer:

| Three Consecutive Numbers | Sum |
|---|---|
| 5, 6, 7 | 18 |
| 20, 21, 22 | 63 |
| 40, 41, 42 | 123 |
| 100, 101, 102 | 303 |

**b.** Look for a connection between the first number and the sum. Write a rule in words that you could use to find the sum of three consecutive whole numbers based only on the value of the first number. Possible answer: The sum is equal to 3 times the first number, plus 3.

**c.** Using $f$ to stand for the first number, explain why your rule always works.



• Additional Answers continued on page A667

## Demonstrating Skills

Find the value of $t$ when $n$ is 7.

**12.** 15

**10.** $t = 3n - 13$ 8 **11.** $t = 2.5 + 1.5 \cdot n$ 13 **12.** $t = 5 \cdot (n - 4)$

Find the value of $y$ when $x$ is $\frac{2}{5}$.

**13.** $y = 20 \cdot x$ 8 **14.** $y = 4 \cdot x - x$ $\frac{6}{5}$, or $1\frac{1}{5}$ **15.** $y = x \div \frac{7}{10}$ $\frac{4}{7}$

Write a rule for the relationship between $x$ and $y$.

**16.**

| $x$ | 0 | 1 | 2 | 3 | 4 |
|---|---|---|---|---|---|
| $y$ | 1 | 2 | 5 | 10 | 17 |

Possible rule: $y = x^2 + 1$

**17.**

| $x$ | 9 | 0.75 | 1 | 7 | 5.5 |
|---|---|---|---|---|---|
| $y$ | 34 | 1 | 2 | 26 | 20 |

Possible rule: $y = 4 \cdot x - 2$

**18.** Possible rules: $q = \frac{v}{157}$, $v \div q = 157$

**19.** Possible rules: $n = m - 9$, $m - n = 9$

**20.** Possible answer: $c = \frac{1}{3} \cdot p$, where $c$ is the number of cantaloupes and $p$ is the number of pumpkins

**21.** $t = 7.5 \cdot a + 3.75 \cdot c$

**18.** Write two rules that are equivalent to $v = 157 \cdot q$.

**19.** Write two rules that are equivalent to $m = n + 9$.

**20.** The number of cantaloupes Miguel has growing in his garden is $\frac{1}{3}$ the number of pumpkins he has. Write a rule to express this relationship. Be sure to tell what the letters in your rule represent.

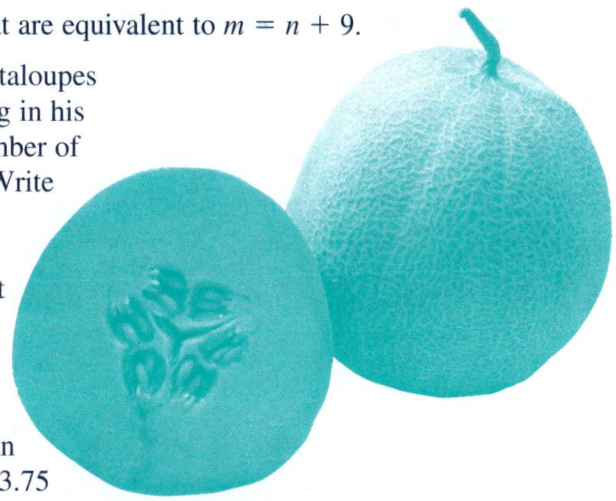

**21.** A movie theater charges \$7.50 for an adult's ticket and \$3.75 for a child's ticket. Write a rule for calculating the total cost $t$ if $a$ adults and $c$ children see a movie.

**22.** Members of the student council sold tickets to an after-school concert. Carlos sold 3 times as many tickets as Angie. Yoshi sold 5 fewer tickets than Angie. Mel sold 12 more tickets than Carlos.

**a.** Write a rule for the relationship between the number of tickets Carlos sold $c$ and the number of tickets Angie sold $a$. $c = 3 \cdot a$

**b.** Write a rule for the relationship between the number of tickets Yoshi sold $y$ and the number of tickets Angie sold $a$. $y = a - 5$

**c.** Write a rule for the relationship between the number of tickets Mel sold $m$ and the number of tickets Angie sold $a$. $m = 3 \cdot a + 12$



Review and Self-Assessment **463**

# CHAPTER 8

# Geometry and Measurement

## Chapter Overview

Students were introduced to the basic ideas of angles in Chapter 1. They will now extend their knowledge by learning how to measure, draw, and classify angles. They will investigate relationships between angles formed by intersecting lines as well as the measures (and the sum of the measures) of the interior angles of any polygon.

Students will also learn how to measure other aspects of polygons, including perimeter and area. They will be introduced to the Greek constant $\pi$ in the context of finding circumferences of circles. Students will learn to use formulas to help them find these measurements.

An extension to studying areas of squares is the squaring and taking the square root of a number. These two inverse operations will enable students to find side lengths of right triangles, work with the Pythagorean Theorem, and apply the Pythagorean Theorem to real-world situations.

## the Big Picture

| Chapter 8 Highlights | Links to the Past | Links to the Future |
|---|---|---|
| Measuring angles and drawing angles with given measures (8.1) | **Elementary Grades:** Classifying, estimating, and measuring angles<br>**Chapter 1:** Estimating angle measures; Finding measures of identical angles around a vertex summing to 90°, 180°, or 360° | **Course 2, Chapter 7:** Identifying congruent angles by matching or measuring<br>**Course 3, Chapter 5:** Determining the angle of rotation and the center of rotation given a figure and its rotated image |
| Finding and estimating perimeters (8.2) | **Elementary Grades:** Finding perimeter of a rectangle; Using formulas to find perimeter | **Course 2, Chapter 7:** Calculating the perimeter of a figure given the perimeter of a similar figure and the scale factor |
| Understanding $\pi$ and the formula for the circumference of a circle (8.2) | **Elementary Grades:** Reviewing circumference and radius; Measuring diameter and circumference of round objects | **Course 2:** Calculating the circumference of a circle<br>**Course 3, Chapter 3:** Understanding that irrational numbers cannot be expressed as the ratio of two integers |
| Finding and estimating areas (8.3, 8.4) | **Elementary Grades:** Reviewing area concepts and units of area; Developing and applying formulas for the areas of rectangles, triangles, parallelograms, and circles | **Course 2, Chapter 7:** Calculating the area of a figure given the area of a similar figure and the scale factor |
| Understanding and applying the ideas of squaring and taking the square root (8.3) | **Elementary Grades:** Exploring relationships between squaring a number and finding its square root | **Course 2, Chapter 4:** Recognizing graphs of square relationships<br>**Course 3, Chapter 7:** Backtracking to undo taking the square root of a number |
| Understanding and applying the Pythagorean Theorem (8.5) | **Elementary Grades:** Practicing measurement skills | **Course 2, Course 3:** Applying the Pythagorean Theorem when finding lengths of segments on a graph |

Time-Saving Tool TeacherWorks™ All-In-One Planner and Resource Center

# Planning Guide

| Lesson Objectives | Pacing | Materials | NCTM Standards | Hot Topics |
|---|---|---|---|---|
| **8.1 Angles** page 465b<br>• To measure angles by using a protractor<br>• To classify angles whose measures are greater than 0° and less than 180° as acute, right, or obtuse<br>• To find the sum of the angles of any polygon given the number of sides of the polygon | 3 class periods | • Board protractor (optional)<br>• Protractors<br>• Rulers<br>• Master 67<br>• Master 68 | 2, 3, 4, 6, 7, 8, 10 | pp. 296–303, 314–319 |
| **8.2 Measuring Around** page 481b<br>• To use the side lengths of a polygon to calculate the perimeter of the polygon<br>• To use polygons to approximate the perimeter of a shape with curved sides<br>• To understand and use formulas for the circumference of a circle | 3 class periods | • Metric rulers<br>• Master 69<br>• 5 objects with circular faces<br>• String or measuring tape<br>• Scissors | 1, 2, 3, 4, 6, 7, 8, 9, 10 | pp. 320–323 |
| **8.3 Areas and Squares** page 493b<br>• To find areas by counting unit squares<br>• To find the formula for the area of a rectangle<br>• To understand how squaring is used in the order of operations rules<br>• To evaluate expressions that involve squaring<br>• To find square roots of numbers that are squares of whole numbers or squares of quotients of whole numbers<br>• To estimate square roots of whole numbers that are not perfect squares | 5 class periods | • Master 70<br>• Scissors<br>*• 1-inch tiles<br>• Master 84<br>• Master 85<br>• Pages from a newspaper<br>• Metric rulers | 1, 2, 3, 4, 6, 7, 8, 9, 10 | pp. 174–177 |
| **8.4 Calculating Areas** page 513b<br>• To derive and apply a formula for the area of a parallelogram<br>• To derive and apply a formula for the area of a triangle<br>• To estimate areas of circles drawn on grids and derive a formula for the area of a circle | 4 class periods | • Master 82<br>• Master 83<br>• Master 81<br>• Master 84<br>• Master 71<br>• Metric rulers<br>• Scissors<br>• Tape<br>• Protractors<br>• Master 72<br>• Master 73<br>• Computers with spreadsheet software (1 per group) | 1, 2, 3, 4, 5, 6, 7, 8, 9, 10 | pp. 324–329, 340–345 |
| **8.5 The Pythagorean Theorem** page 535b<br>• To use right triangles and squares drawn on dot paper to arrive at the Pythagorean Theorem<br>• To prove the Pythagorean Theorem<br>• To use the Pythagorean Theorem to find missing lengths of sides of right triangles | 3 class periods | • Master 74<br>• Master 75<br>• Master 76<br>• Master 84<br>• Scissors | 2, 3, 4, 6, 7, 8, 9, 10 | |

* Included in Impact Mathematics Manipulative Kit

**Key to NCTM Curriculum and Evaluation Standards:** 1=Number and Operations, 2=Algebra, 3=Geometry, 4=Measurement, 5=Data Analysis and Probability, 6=Problem Solving, 7=Reasoning and Proof, 8=Communication, 9=Connections, 10=Representation

# Assessment Opportunities

## Standard Assessment

*Impact Mathematics* offers three types of formal assessment. The Chapter 8 Review and Self-Assessment in the Student Edition serves as a self-assessment tool for students. In the Teacher's Guide, a Quick Quiz at the end of each lesson allows you to check students' understanding before moving to the next lesson. The Assessment Resources include blackline masters for chapter and quarterly tests.

- **Student Edition** Chapter 8 Review and Self-Assessment, pages 551–555
- **Teacher's Guide** Quick Quizzes, pages 481, 493, 513, 535, 550
- **Assessment Resources** Chapter 8 Test Form A, pages 196–198; Chapter 8 Test Form B, pages 199–201

## Ongoing Assessment

*Impact Mathematics* provides numerous opportunities for you to assess your students informally as they work through the investigations. Share & Summarize questions help you determine whether students understand the important ideas of an investigation. If students are struggling, Troubleshooting tips provide suggestions for helping them. On the Spot Assessment notes appear throughout the teaching notes. They give you suggestions for preventing or remedying common student errors. Assessment Forms in the Assessment Resources provide convenient ways to record student progress.

- **Student Edition** Share & Summarize, pages 471, 476, 486, 489, 498, 501, 503, 507, 517, 521, 524, 540, 543
- **Teacher's Guide** On the Spot Assessment, pages T473, T483, T487, T517, T520 Troubleshooting, pages T471, T476, T486, T489, T498, T501, T503, T507, T517, T521, T524, T540, T543
- **Assessment Resources** Chapter 8 Assessment Checklists, pages 268–270

## Alternative Assessment, Portfolios, and Journal Ideas

The alternative assessment items in *Impact Mathematics* are perfect for inclusion in student portfolios and journals. The In Your Own Words feature in the Student Edition gives students a chance to write about mathematical ideas. The Performance Assessment items in the Assessment Resources provide rich, open-ended problems, ideal for take-home or group assessment.

- **Student Edition** In Your Own Words, pages 480, 493, 512, 535, 548
- **Assessment Resources** Chapter 8 Performance Assessment, pages 202–204

# Assessment Resources

The Assessment Resources provide a chapter test in two equivalent forms, along with additional performance items. The performance items can be used in a variety of ways. They are ideal for take-home assessment or in-class group assessment.

- Chapter 8 Test Form A, pages 196–198
- Chapter 8 Test Form B, pages 199–201
- Chapter 8 Performance Assessment, pages 202–204
- Chapter 8 Assessment Solutions, pages 205–207

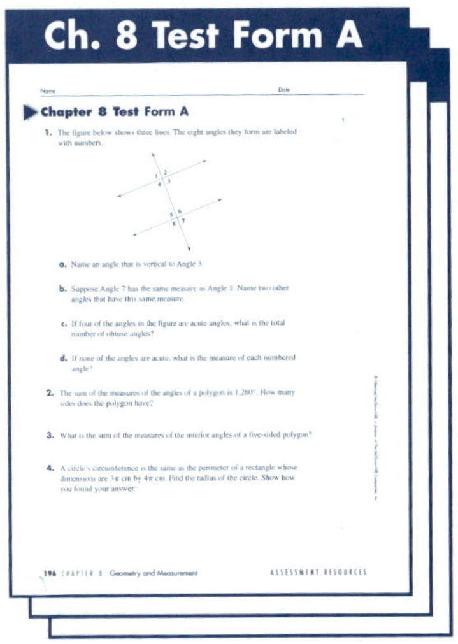


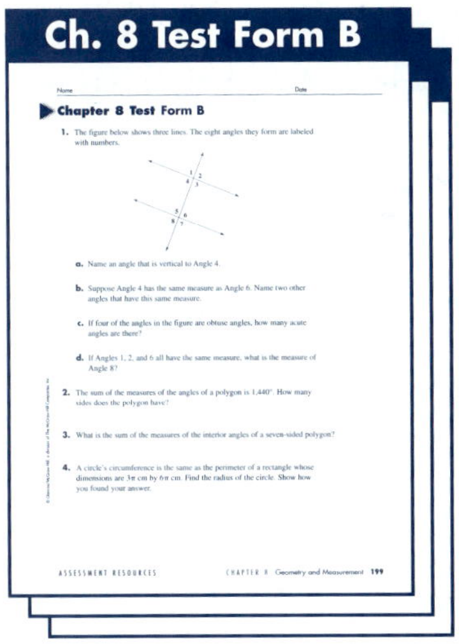


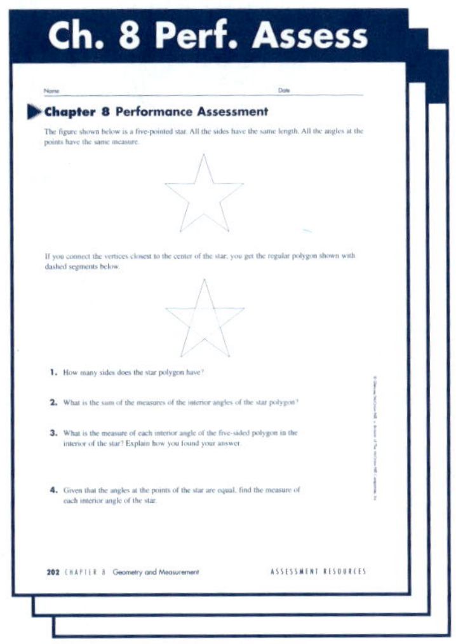


# Additional Resources

- **Math Skills Maintenance Workbook,** 2, 7, 8, 9, 10, 16, 17, 18, 32
- **Investigations for the Special Education Student in the Mathematics Classroom,** 2, 18, 19, 21, 23
- **Virtual Activities CD-ROM,** Area of Rectangles and Squares, Area of Triangles
- **What's Math Got To Do With It? Videos,** Level 1, Video 5
- **StudentWorks™ CD-ROM**
- **Reading and Writing in the Mathematics Classroom**
- **Using the Internet in the Mathematics Classroom**

## ExamView® Pro

Use ExamView® Pro Testmaker CD-ROM to:

- Create Multiple versions of tests.
- Create Modified tests for Inclusion students with one mouse click.
- Edit existing questions and Add your own questions.
- Build tests aligned with state standards using built-in State Curriculum Correlations.
- Change English tests to Spanish with one mouse click and vice versa.

## Introduce

Ideas of geometry and measurement are evident in the world around us. One place where measurement takes center stage is at the Olympics. The 2000 Summer Olympic Games held in Sydney, Australia featured 199 nations competing in 35 sports across 16 days. The size of the fields or courts on which the Olympians competed and the equipment they used are very important in their sports.

Have students look over the table on this page. Ask them if they know (and then explain) the difference between perimeter and area as they study the dimensions of different types of fields.

## Think About It

Students will need to measure the length and width of the classroom in meters. Have them find the perimeter and area of the classroom and compare their results to the perimeter and area of the volleyball court.

# CHAPTER 8

# Geometry and Measurement

### Real-Life Math

**Olympic Proportions** The 2000 Olympic Games in Sydney, Australia, featured 35 sports. The table below lists the dimensions, perimeter, and area of the rectangular courts, fields, mats, and pools where some of these sports were played.

| Sport *(meters)* | Length *(meters)* | Width *(meters)* | Perimeter *(meters)* | Area *(square meters)* |
|---|---|---|---|---|
| Football (soccer) field | 100 | 70 | 340 | 7,000 |
| Field hockey field | 91.4 | 55 | 292.8 | 5,027 |
| Swimming pool | 50 | 25 | 150 | 1,250 |
| Handball court | 40 | 20 | 120 | 800 |
| Water polo pool | 30 | 20 | 100 | 600 |
| Volleyball court | 18 | 9 | 54 | 162 |
| Badminton court (singles) | 13.4 | 5.18 | 37.16 | 69.41 |
| Fencing piste | 14 | 1.5 | 31 | 21 |

**Think About It** How do the dimensions of your classroom compare to the dimensions of the volleyball court?

# *Family Letter*

Dear Student and Family Members,

It's time for a little change of pace. Our next chapter in mathematics is about measurement in geometry. In this chapter, you will learn about angles, angle measurement, and perimeters and areas of squares, rectangles, and irregular shapes. Finally, you will learn about the famous Pythagorean Theorem, $a^2 + b^2 = c^2$, that is used with right triangles.

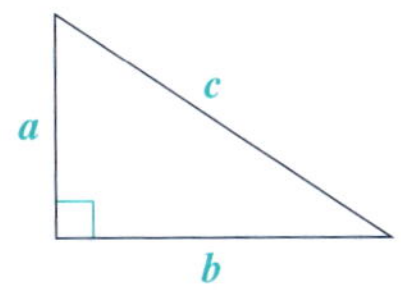


Perimeter is a measurement of the distance around a figure. When a figure is a circle, the distance around it is called the *circumference*. To find the circumference of a circle, you will discover that you can multiply the diameter by pi. Pi ($\pi$) is approximately equal to 3.14.

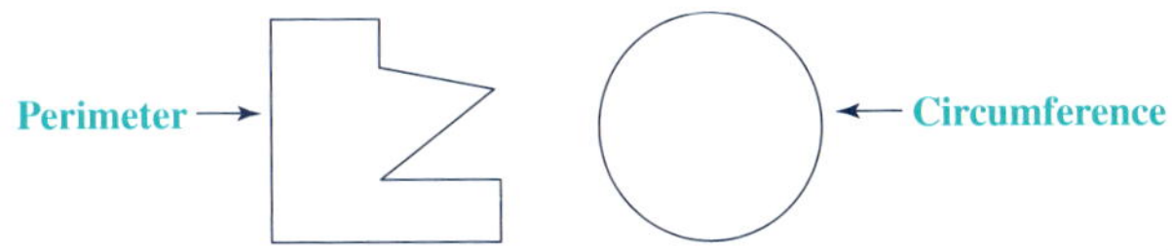


**Vocabulary** There are many vocabulary terms in this chapter. Some of them may be familiar to you.

| | |
|---|---|
| **acute angle** | **parallelogram** |
| **area** | **perfect square** |
| **chord** | **perimeter** |
| **circumference** | **perpendicular** |
| **diameter** | **radius** |
| **hypotenuse** | **right angle** |
| **inverse operations** | **right triangle** |
| **leg** | **square root** |
| **obtuse angle** | |

## What can you do at home?

Geometry is everywhere! Ask your student for examples of perimeter and area in his or her daily life.

impactmath.com/family_letter



Ask students if they watched the most recent summer or winter Olympics. Have them choose their favorite Olympic event and the field on which it is played, and ask them to compare the dimensions of the field with those listed here. Ask them whether perimeter or area of the field is more important in their favorite event.

Another version of the Family Letter, available in English and Spanish, is found in the Teaching Resources. You may want to send a copy of this letter home with your students.

**Teaching Resources**

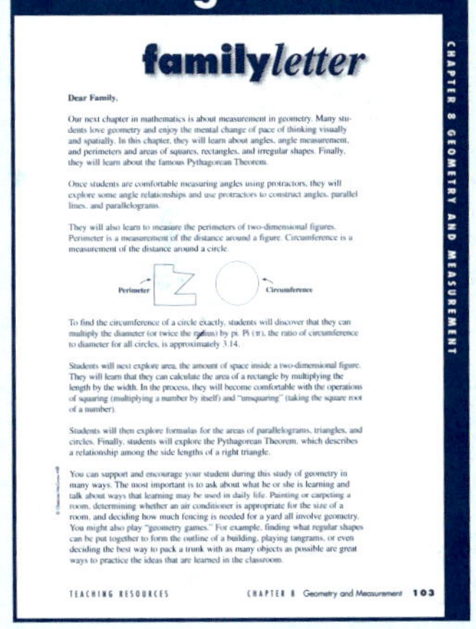

# Mathematical Background

Measurement is important not only in geometry but also in innumerable real-world applications of geometry. Students are already acquainted with some basic ideas about measuring angles, line segments, and areas.

**Angles** This chapter reviews benchmark angles, which students studied in Chapter 1. Students now learn to use a protractor to measure angles and to draw an angle of a given size. They go on to study angle relationships between angles formed by intersecting lines and angle sum relationships in polygons. Students learned in Chapter 1 that the sum of the measures of the angles of a triangle is 180°. In this chapter, they see how to draw diagonals of a polygon to divide the polygon into non-overlapping triangles, whose angle measures can be added to find the sum of the angle measures of the polygon. This gives a simple method for finding the sum of the angle measures of any polygon.

**Perimeter and Area** In this chapter, students will study perimeters and areas of polygons. For simple figures such as squares, rectangles, parallelograms, and triangles, they will develop methods and formulas for calculating perimeter and area. They will also see that, contrary to what their intuition might tell them, there is no hard-and-fast way to predict the area of a figure from its perimeter, and no way to predict the perimeter from the area. Such predictions are sometimes possible for certain basic figures such as squares, but as a general rule, the relationship between perimeter and area is not simple.

**Approximating Perimeters of Figures with Curved Sides** Polygons are useful for studying figures with curved sides. Closed, two-dimensional figures with curved sides can be approximated by selecting several points evenly spaced along the curved boundary and joining them to create a polygon. Think of a circle. If you draw a regular 10-sided polygon by connecting evenly spaced points around the circle, the polygon will somewhat resemble a circle. If you draw a polygon with many more vertices (and sides), the polygon looks even more like a circle. If you used a regular polygon with 10,000 vertices, the naked eye would not be able to distinguish the circle from the polygon.

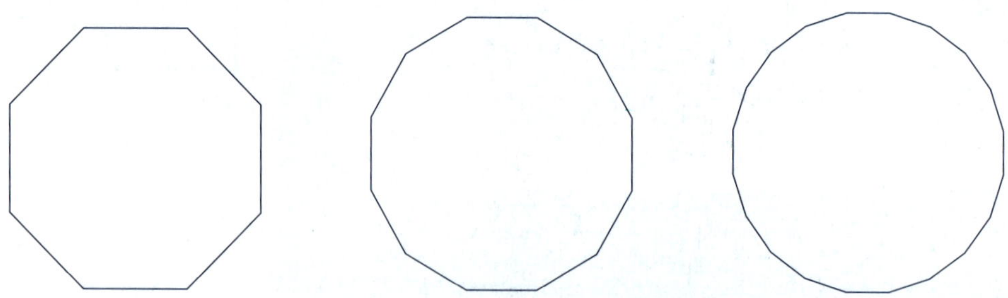

**Circles** Students will use the idea of approximating closed-curved figures by means of polygons when they examine the circumference and area of circles. Students will measure the diameters and circumferences of circular objects to see that the circumference divided by the diameter always seems to be the same, no matter what size the circular object they measure. This leads to a discussion of pi ($\pi$), which is the key concept for understanding the circumference and area of a circle.

• *Teaching notes continued on page A667*

# 8.1 Angles

## Objectives

- To measure angles by using a protractor
- To classify angles whose measures are greater than 0° and less than 180° as acute, right, or obtuse
- To find the sum of the angles of any polygon given the number of sides of the polygon

## Overview (pacing: about 3 class periods)

In this lesson, students get a brief review of the benchmark angles they studied in Chapter 1. They next learn how to use a protractor to measure angles. They learn how to classify angles with measures greater than 0° and less than 180° as acute, right, or obtuse angles. Perpendicular lines and segments are defined as pairs of lines or segments that intersect to form right angles.

Students learn what is meant by vertical angles and learn that vertical angles have the same measure. Next, students measure angles of triangles and compute their angle sums. They confirm that the angle sum for any triangle is 180°. They also discover how to find the angle sums for polygons with more than three sides by dividing the polygons into triangles.

## Advance Preparation

For Investigation 1, you will probably want to give students copies of Master 67. Students will also need Master 68 for On Your Own Exercise 24.

## Lesson Planner

| | Summary | Materials | On Your Own Exercises | Assessment Opportunities |
|---|---|---|---|---|
| **Investigation 1** page T467 | Students learn to use a protractor to measure angles and to classify angles whose measures are greater than 0° and less than 180° as acute, right, or obtuse. | • Board protractor (optional)<br>• Protractors<br>• Rulers<br>• Master 67<br>• Master 68 | Practice & Apply: 1–14, p. 477<br>Connect & Extend: 22, 23, pp. 478–479<br>Mixed Review: 26–38, p. 481 | Share & Summarize, pages T471, 471<br>Troubleshooting, page T471 |
| **Investigation 2** page T472 | Students measure angles and use reasoning to discover how measures of vertical angles are related; they then discover how to find the sum of the angle measures of any polygon, given the number of sides of the polygon. | • Rulers<br>• Protractors | Practice & Apply: 15–21, pp. 477–478<br>Connect & Extend: 24, 25, pp. 479–480<br>Mixed Review: 26–38, p. 481 | On the Spot Assessment, page T473<br>Share & Summarize, pages T476, 476<br>Troubleshooting, page T476<br>Informal Assessment, page 481<br>Quick Quiz, page 481 |

# Introduce

1 Take time to review the concept of an angle with the class. Using a yardstick, draw a long, horizontal segment on the board. Place the yardstick directly on top of the segment so that the left ends coincide. Then rotate the yardstick counterclockwise, keeping the two endpoints in contact with the board. Ask the class to call out "stop" when you reach the 90° position. Then continue the rotation. Ask the class to call out "stop" when you reach the 180° position.

2 **Think & Discuss**
Give students a short time to work on the Think & Discuss. Have volunteers present their answers, and discuss the answers with the class. If necessary, help students relate each diagram to one of the three benchmark angles discussed before the Think & Discuss.

Once you are confident that all students understand how to get the Think & Discuss answers, you may want to analyze the diagrams a bit more. You can put each diagram on the board and ask additional questions. For example, ask:

In the middle diagram, can you find a 60° angle? **yes**

How many of them do you see? **two**

Then call on a volunteer to go to the board and point out the two 60° angles contained in the middle diagram. Next, ask: How many 90° angles can you find in the third diagram? **three**

Call on a volunteer to point them out. Then ask for angles of other sizes that can be found in the third diagram, and have volunteers point them out. **some possibilities: 135°, 180°, 225°, 270°, 315°, and 360°**

# 8.1 Angles

## Remember

You can think of an angle as a rotation. A 360° angle is a rotation around a complete circle. A 180° angle is a rotation $\frac{1}{2}$ of the way around a circle. A 90° angle is a rotation $\frac{1}{4}$ of the way around a circle.

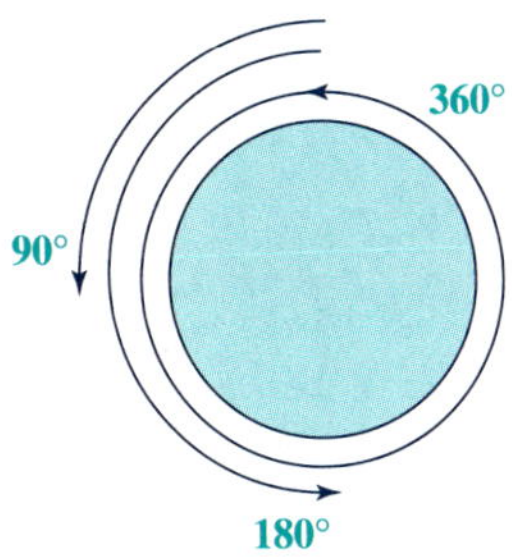

In Chapter 1, you investigated angles. You learned that an angle is defined as two rays with a common endpoint, called the *vertex.*

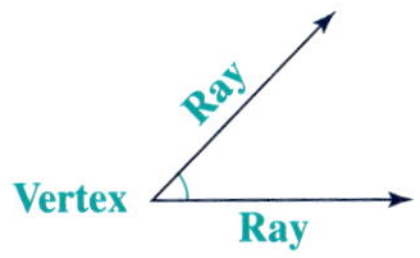


Angles are measured in degrees. In Chapter 1, you used 90°, 180°, and 360° angles as benchmarks to help estimate the measures of other angles.

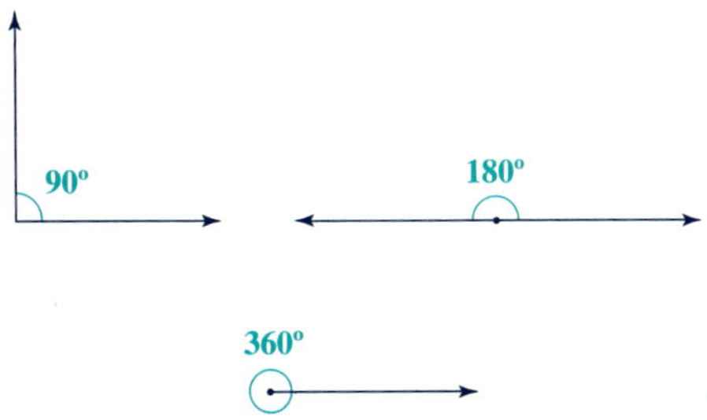


**1** Review the concept of an angle.

## Think & Discuss

Each diagram is constructed from angles of the same size. Estimate the measure of each marked angle, and explain how you found it.

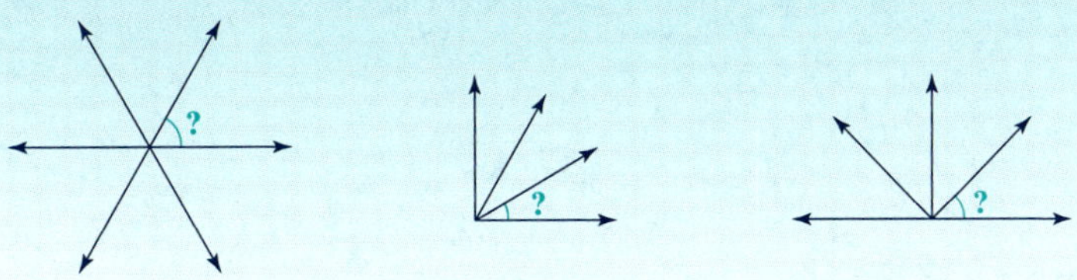


60°; Possible explanation: Together, the six equal angles make a 360° angle, so each must have measure 360 ÷ 6, or 60°.

30°; Possible explanation: Together, the three equal angles make a 90° angle, so each must have measure 90 ÷ 3, or 30°.

45°; Possible explanation: Together, the four equal angles make a 180° angle, so each must have measure 180 ÷ 4, or 45°.

**2** Have students discuss their answers with the class.

# Investigation 1

In this investigation, students practice using a protractor to measure angles.

1 It is helpful to have a board protractor that you can use to demonstrate to the class the steps involved in measuring an angle. Draw two angles with measures between 0° and 90° and two angles with measures between 90° and 180°. Before you demonstrate how to measure the angles, hold up the board protractor and point out the center point and reference line. Then measure each angle. Each time, explain the steps you take so that students can relate them to the three steps on page 467.

2 Discuss the given steps for measuring an angle. Have students confirm that the diagram correctly shows how to place the protractor. When you get to the third step—that of actually reading the angle measure—discuss the fact that the protractor has two scales. Ask students for any thoughts they have about which scale they should use when they read the measure for the angle. Tell them that they will take a good look at this question on page 468.

# Investigation 1 Measuring Angles

A *protractor* is a tool for measuring angles. A protractor has two sets of degree labels around the edge of a half circle (or sometimes a full circle). The line that goes through 0° is called the *reference line.*

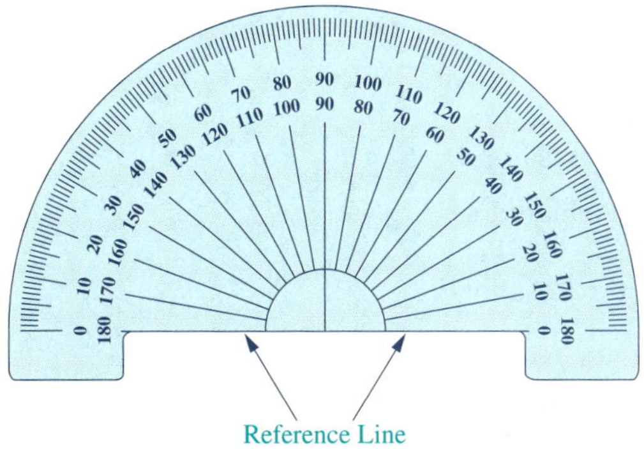


**Just the facts**

In in-line skating, 360s (full turns) and 540s (one and a half turns) are two of the more difficult stunts.

To measure an angle, follow these steps:

- Place the bottom center of the protractor at the vertex of the angle.
- Line up the reference line with one ray of the angle. Make sure the other ray can be seen through the protractor. (You may need to extend this ray so that you can see where it meets the tick marks along the edge of the protractor.)
- Read the angle measurement.

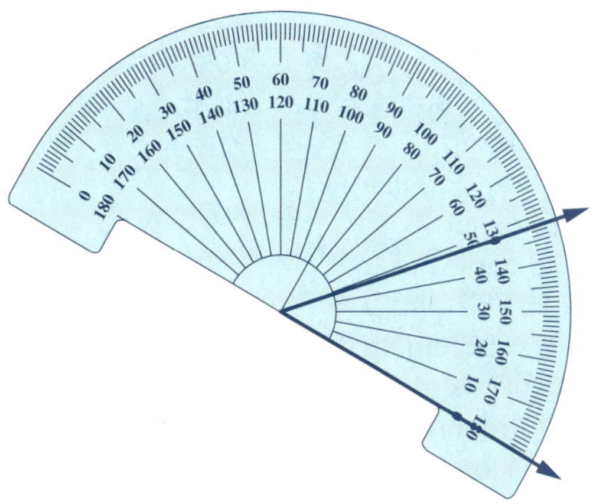

1 Demonstrate to the class how to measure an angle.

2 Students should discuss which scale to use when measuring angles.



LESSON 8.1 Angles 467

# Develop

### Think & Discuss

You will want to discuss carefully the decisions involved
in measuring the angles in the Think & Discuss. Students
1 should understand that positioning the protractor correct-
ly is crucial. There are always two correct ways to do
that, depending on which ray of the angle is lined up
with the reference line of the protractor.

Assuming that the protractor is in a correct position for measuring the angle, the question then becomes, "Which scale do I use?" Review the Think & Discuss questions to be sure students are able to address this question.

**Problem-Solving Strategies** Here are two ways to decide on the appropriate scale.

2 • Use benchmark angles to determine which scale is correct.

• Make sure the ray is lined up with the reference line
3 of the protractor and then check whether the ray pass-
es through the 0° mark of the inside or outside scale.
Once you have identified this scale, read the angle
measure from the place where the other ray crosses
this same scale. This procedure can be illustrated by
using the two diagrams in the Think & Discuss.

4 To measure the three angles shown at the bottom of the
page, students will need to extend the rays to see
where the rays cross the protractor scales. They can use
the straight side along the bottom of the protractor to
extend the sides. You may want to distribute copies of
Master 67, which contains the three triangles. Be sure
to discuss both methods given in the answers for finding
the measure of the last angle on the page. If the stu-
5 dents are using protractors marked to 360°, ask how
they would measure this angle using a protractor
marked to only 180°.

**Teaching Resources**

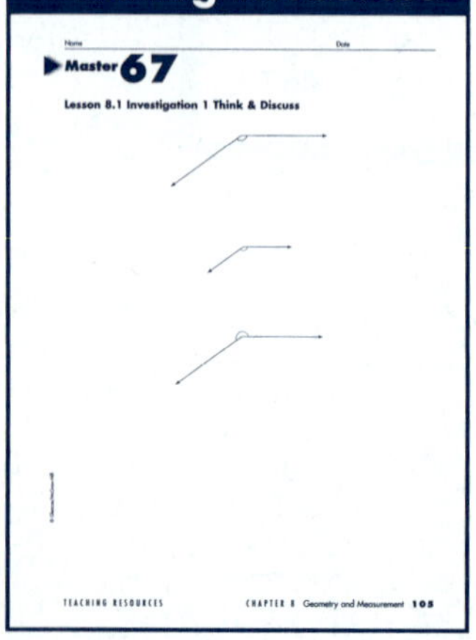

MATERIALS
- protractor
- copies of the angles

① a little more than 90°; Possible explanation: The angle is larger than the angle formed by a corner of a sheet of paper.

**Just the facts**

A *mariner's astrolabe* was a navigational instrument of the 15th and 16th centuries. Sailors used it to measure the angle of elevation of the sun or other star. This measurement could help them determine their ship's latitude.

## Think & Discuss

48°; Possible explanation: The angle is clearly less than 90°.

The angle below measures about 48°. Or is it about 132°? How do you know which number to use?

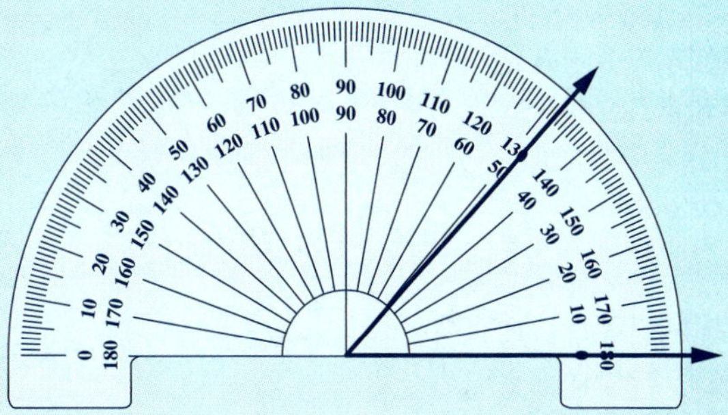

Is the measure of the angle below a little more than 90° or a little less than 90°? How do you know? See ①.

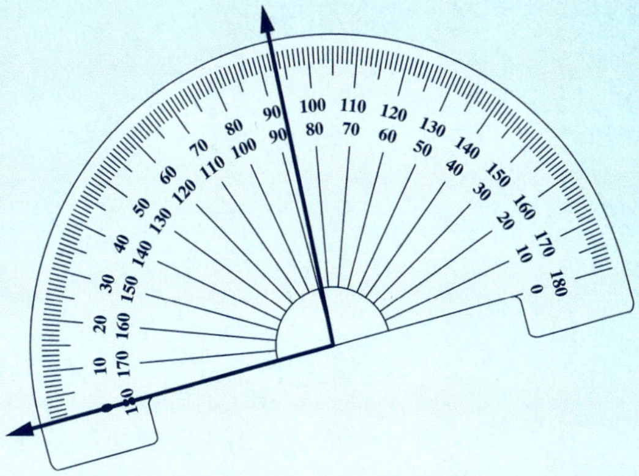

Measure these two angles. How do the measures compare? Both measure 145°.

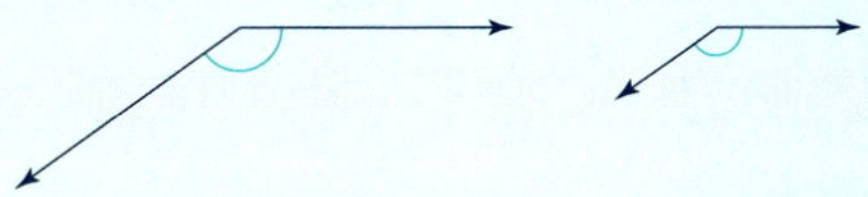

Find the measure of the angle below. Describe the method you used.

See ② in Additional Answers.

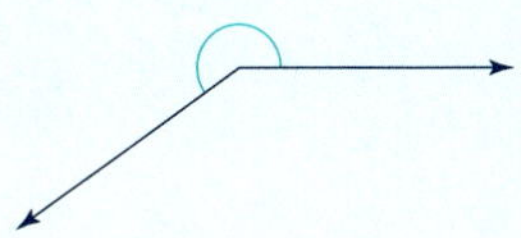

1 Remind students to position the protractor correctly.

2 **Problem-Solving Strategy**
Use benchmarks

3 Have students line up the ray with the reference line of the protractor.

4 Have students extend the rays to see where they cross the protractor scales.

5 Challenge students to measure the angle using a protractor marked only to 180°.



### Additional Answers

② 215°; Possible methods:

- Measure the smaller angle (the unmarked angle) and subtract the result from 360°.
- Extend one of the rays backward, dividing the angle into a 180° angle and a smaller angle. Measure the smaller angle, and add the result to 180°.

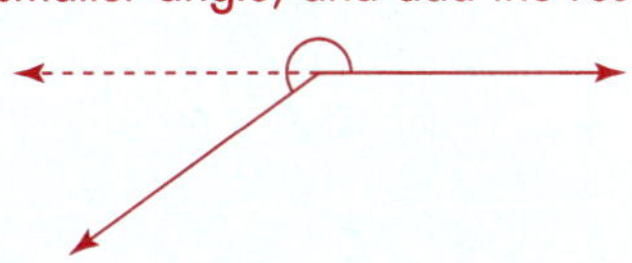

## Develop

1 When you discuss the new vocabulary terms introduced on this page, be sure students understand that the terms *acute, right,* and *obtuse* are used only for angles with measures greater than 0° and less than 180°. Acute and obtuse angles can often be identified by simply looking at the angle. A corner of a sheet of paper or a protractor can be used to classify angles that are difficult to classify by simple visual inspection.

2 If students ask whether there are special names for angles that are greater than obtuse angles, you may wish to mention that 180° angles are sometimes called *straight angles* because their sides form a straight line. Angles with measures greater than 180° but less than 360° are called *reflex angles.*

You may want to have students use a corner of a sheet of paper to confirm that all the angles formed by a pair of perpendicular lines or segments are right angles.

### About the Mathematics

The terms *acute, right,* and *obtuse* are important for classifying triangles as well as angles. In Investigation 2, students will gather evidence that the sum of the angle measures of any triangle is 180°. Knowing this fact, it is easy to deduce that any triangle falls into one of three categories: (1) three acute angles, or (2) two acute angles and one right angle, or (3) two acute angles and one obtuse angle. A triangle with three acute angles is called an *acute triangle.* A triangle with one right angle is called a *right triangle.* A triangle with one obtuse angle is called an *obtuse triangle.* Right triangles play a very important role in mathematics. Students will study an important fact about right triangles, the result known as the Pythagorean Theorem, in Lesson 8.5.

You have seen that when you measure an angle with a protractor, you must determine which of two measurements is correct. One way to decide is to compare the angle to a 90° angle. Because 90° is such an important benchmark, angles are sometimes classified by how their measures compare to 90°.

VOCABULARY
**acute angle**
**obtuse angle**
**perpendicular**
**right angle**

**1** Emphasize that acute, right, and obtuse angles are terms used only for angles greater than 0° and less than 180°.

**Acute angles** measure less than 90°.

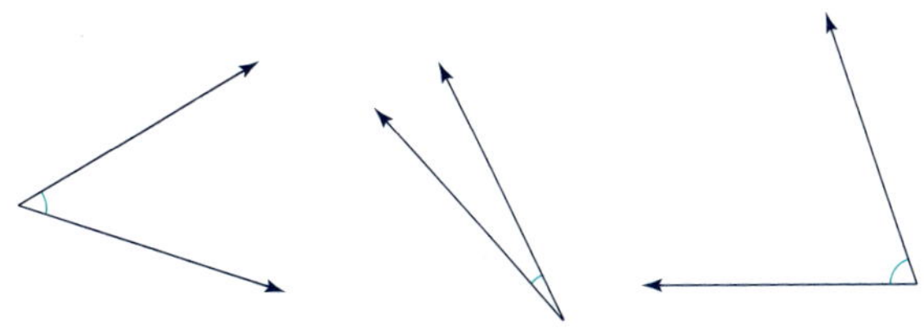

**Obtuse angles** measure more than 90° and less than 180°.

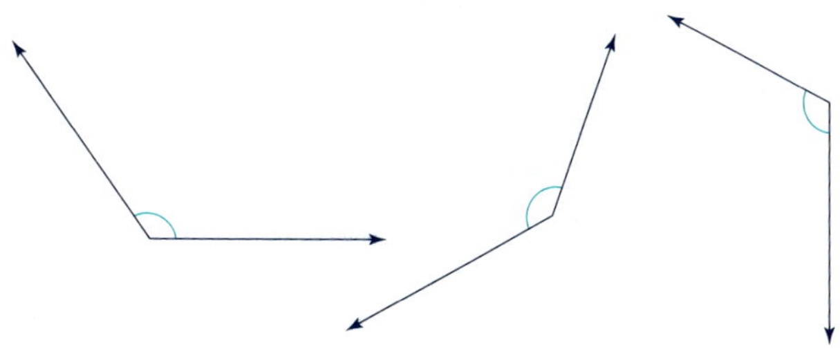

**Right angles** measure exactly 90°. Right angles are often marked with a small square at the vertex.

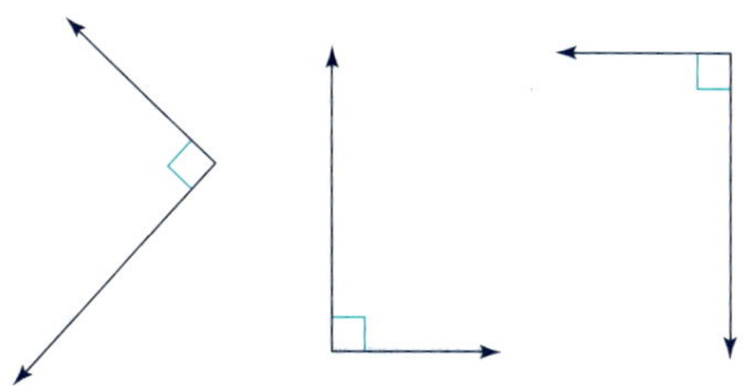

Two lines or segments that form a right angle are said to be **perpendicular.**

**2** You may want to tell students that angles measuring 180° are called *straight angles.*

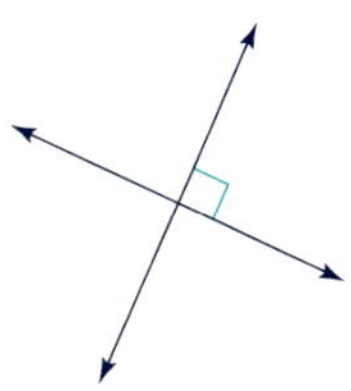

Perpendicular Lines

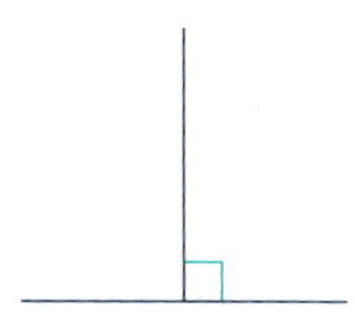

Perpendicular Segments



LESSON 8.1 Angles 469

1

### Problem Set A **Suggested Grouping: Pairs**

This problem set asks students to classify given angles as acute or obtuse. Indirectly, it also deals with right angles, since classifying an angle without a protractor implies that you have compared it to a right angle. Students should be able to decide what kind of angle is shown before measuring the angle. Often this can be done simply by looking. They can also use a corner of a sheet of paper to help them decide.

2

### Example

Discuss the Example, which shows how to draw an angle with a given measure. Be sure students understand that the scale to use is determined by which endpoint of the segment is used as the vertex of the angle. There are two 25° labels, the one to use is the one that will result in an acute angle. When the center point of the protractor is placed as shown in the diagram, the 25° label on the outside scale is the correct one.

**MATERIALS**
protractor

## Problem Set A

Tell whether each angle is acute or obtuse. Then find its measure.

**1.**

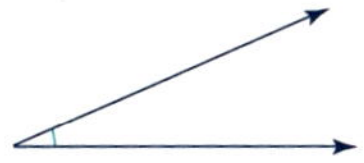

acute, 23°

**2.**

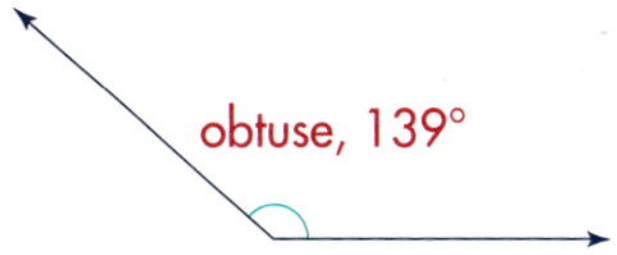

obtuse, 139°

**3.**

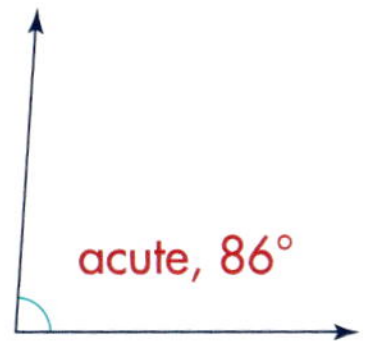

acute, 86°

**4.**

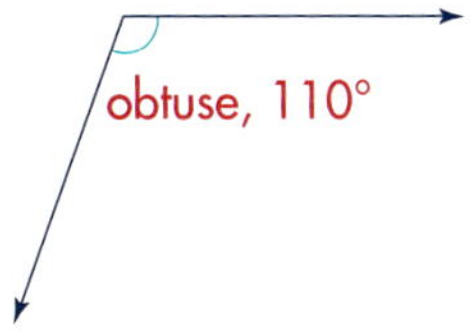

obtuse, 110°

**5.**

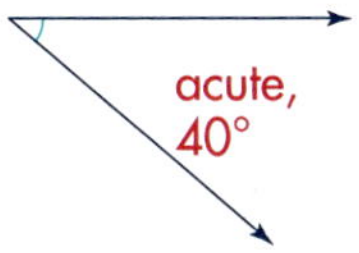

acute, 40°

**6.**

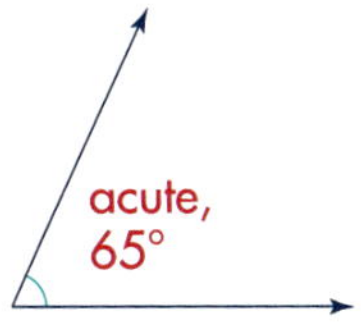

acute, 65°

You have seen that a protractor is a useful tool for measuring angles. You can also use a protractor to draw angles with given measures.

**EXAMPLE**

To create a 25° angle, start by drawing a line segment. This segment will be one side of the angle.

Line up the reference line of the protractor with the segment, with the center of the protractor at one endpoint of the segment. This endpoint will be the vertex of the angle.

Draw a mark next to the 25° label on the protractor. (Be sure to choose the correct 25° label!)

Remove the protractor, and draw a segment from the vertex (the endpoint that was at the center of the protractor) through the mark.

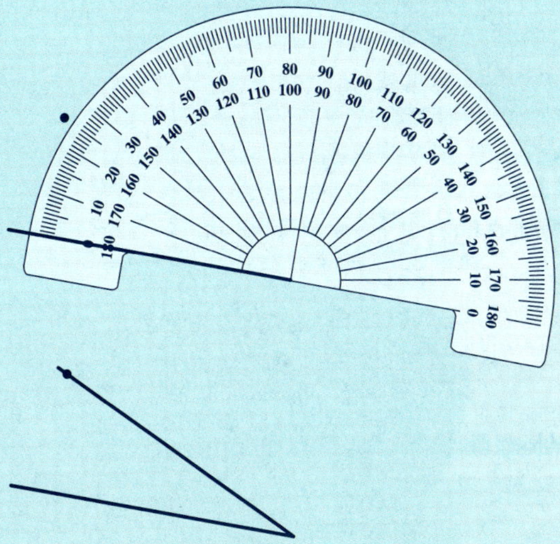

**1** You may have students work in pairs.

**2** Demonstrate how to draw an angle.

## Develop

1

### Problem Set B Suggested Grouping: Pairs

This problem set gives students practice drawing angles that have a given measure. It also provides additional practice in measuring angles.

**Sample Strategies** Students may have various approaches for **Problem 2.**

- One approach is to draw a 150° angle and mark the "outside" of this angle.
- Another approach is to draw a line and mark a point on it to serve as the vertex of a 180° angle. Next, draw a 30° angle that has this same vertex and one of the rays of the 180° angle as the other side. Mark the sum of the 180° angle and the 30° angle.

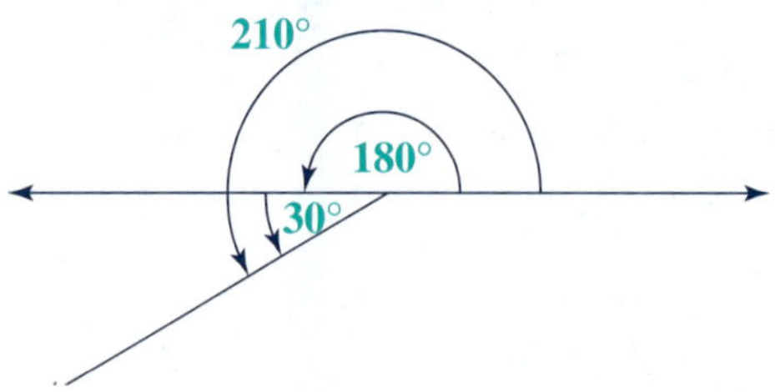


2

For **Problem 3,** you may need to remind students that perpendicular segments must meet to form right angles.

For **Problem 4,** students may need to draw a few triangles before they get one that works. Since the problem requires a 50° angle, students should draw that angle first.

3

After students have drawn a triangle for **Problem 6,** you may want to ask them to find the measure of the third angle of the triangle. If they have drawn the desired triangle accurately, the third angle will also have a measure of 60°.

4

### Share & Summarize

After students have had time to write their answers to the Share & Summarize questions, have volunteers read their answers to the class. If you have a board protractor, you may want to have the volunteers go to the board to demonstrate procedures.

**Troubleshooting** If students are having difficulty measuring angles with a protractor, work with them one-on-one. If you do not have time to do this yourself, ask a student who understands the procedure well to provide the help.

### On Your Own Exercises

Practice & Apply: 1–14, p. 477
Connect & Extend: 22, 23, pp. 478–479
Mixed Review: 26–38, p. 481

MATERIALS
- protractor
- ruler

## Problem Set B

1. Draw a 160° angle. Include a curved angle mark to show which angle is 160°.
2. Draw a 210° angle. Include an angle mark to show which angle is 210°.
3. Draw two perpendicular segments. See Additional Answers.
4. Draw a triangle in which one angle measures 50° and the other angles have measures greater than 50°. Label each angle with its measure.
5. Draw a triangle with one obtuse angle. Label each angle with its measure. See margin.
6. Draw a triangle with two 60° angles. See Additional Answers.
7. Measure the sides of the triangle you drew in Problem 6. What do you notice? The sides all have the same length.
8. Draw a square. Make sure all the sides are the same length and all the angles measure 90°. See Additional Answers.
9. Draw a polygon with any number of sides and one angle that has a measure greater than 180°. Mark that angle. Possible polygon:

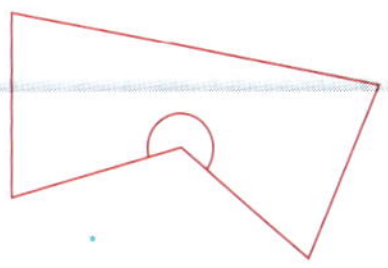

**1.**

**2.**

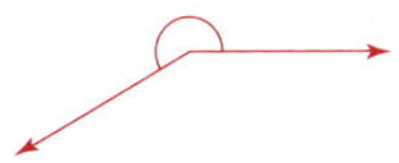

**4.** Possible triangle:

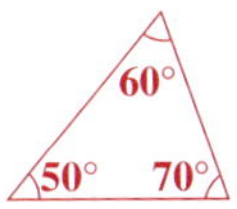


**Just the facts**

A *theodolite* is an angle-measuring instrument used in navigation, meteorology, and surveying.

**5.** Possible triangle:

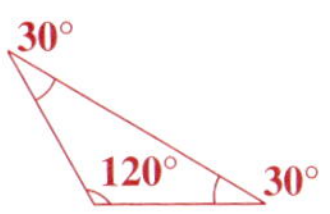


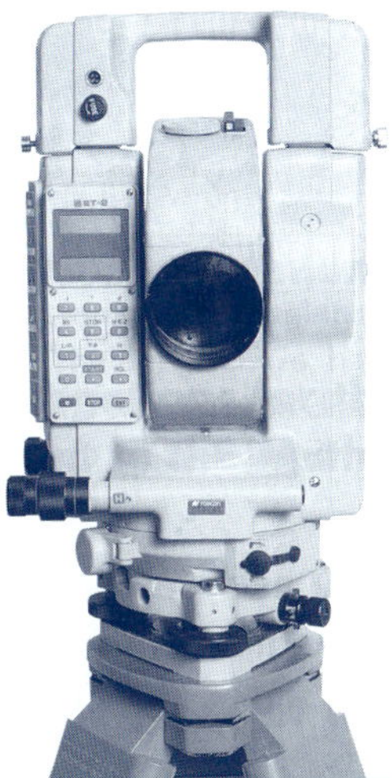

1 You may have students work in pairs.

2 Remind students that perpendicular segments must meet to form right angles.

3 Have students verify that the third angle of the triangle is 60°.

## Share & Summarize

1. When you measure an angle with a protractor, how do you know which of the two possible numbers to choose?
2. The protractor on page 470 has a scale up to 180°. Describe how you would use such a protractor to draw an angle with a measure greater than 180°. Give an example if it helps you to explain your thinking.

4 Discuss students' answers with the class.

**Share & Summarize Answers**

1. Possible answer: Determine whether the angle is greater than 90° or less than 90°. Then choose the appropriate measure.
2. Possible answer: I would draw an angle with measure equal to 360° − (the angle measure). The correct angle would be the "outside" of this angle. For example, to draw a 230° angle, I would draw a 130° angle.

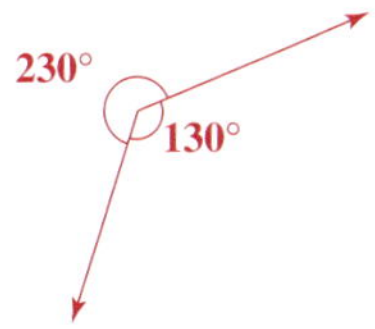




**Additional Answers**

3. Possible answers:

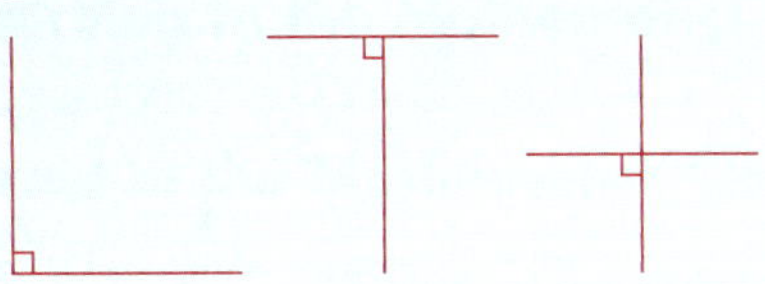

6. Note: The size of the triangles will vary, but the triangle must be equilateral.

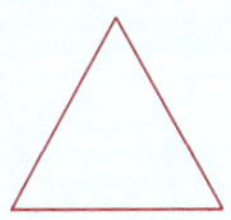

8. Note: Sizes will vary.

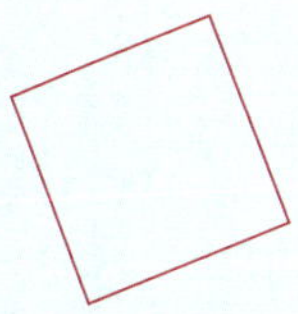

# Investigation 2

In this investigation, students study how angles are related in figures formed by lines and segments. First they examine the angles formed by two intersecting lines. Then they examine interior angles of polygons.

1 Read through the paragraphs at the top of page 472 with the class to be sure students understand the shorthand notation that is used to express angle measures.

2 **Problem Set C** **Suggested Grouping: Pairs**
In this problem set, students examine angles formed by two intersecting lines. They measure the angles formed and look for relationships between the measures.

3 In **Problem 1,** students measure the angles formed by the intersecting lines at the top of page 472. Encourage them to use the shorthand notation they have just learned when they record the angle measures.

In **Problem 2,** students observe that two of the angle pairs in **Problem 1** have the same measure. Students should note the relative positions of the angles in these pairs.

In **Problems 3 and 4,** students draw their own pairs of lines and examine the measures of the angles formed.

In **Problem 5,** students make a conjecture about how the angles formed by any two intersecting lines will be related.

4 **Problem Set Wrap-Up** Call on volunteers to read their answers for Problem 5. Students may have focused their attention only on pairs of angles with the same measure. You may want to ask whether they find any other relationships. They should be able to see without too much trouble that other pairs of angles in each figure have measures that add to 180°. ■

5 After you have discussed Problem 5 with the class, read through the paragraph at the bottom of page 472, where the term *vertical angles* is explained.

# Investigation 2 Investigating Angle Relationships

You can refer to the angles in a drawing more easily if you label them with numbers or letters.

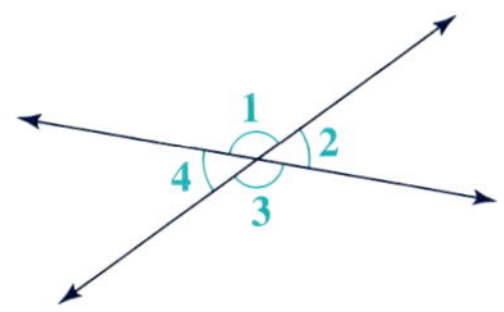


In the drawing above, the measure of Angle 1 is 135°. You can write this in symbols as $m\angle 1 = 135°$. The "*m*" stands for "measure," and $\angle$ is the symbol for "angle."

MATERIALS
- protractor
- ruler

## Problem Set C

Two lines that cross each other—like the lines in the drawing above—are said to *intersect.*

1. Measure angles 1, 2, 3, and 4 in the drawing above.
2. Which angles have the same measure? 1 and 3, 2 and 4
3. Use a ruler to draw another pair of intersecting lines. Measure each of the four angles formed, and label each with its measure.
4. Draw one more pair of intersecting lines, and label each angle with its measure. Answers will vary.
5. What patterns do you see relating the measure of the angles formed by two intersecting lines?

**1.** $m\angle 1 = 135°$, $m\angle 2 = 45°$, $m\angle 3 = 135°$, $m\angle 4 = 45°$

**3.** Answers will vary.

**5.** Possible answer: There are two pairs of equal angles. The angles that are not directly next to each other are equal. The measures of any two angles right next to each other add to 180°.

When two lines intersect, two angles that are not directly next to each other are called *vertical angles.* In the drawing below, $\angle a$ and $\angle c$ are vertical angles, and $\angle b$ and $\angle d$ are vertical angles.

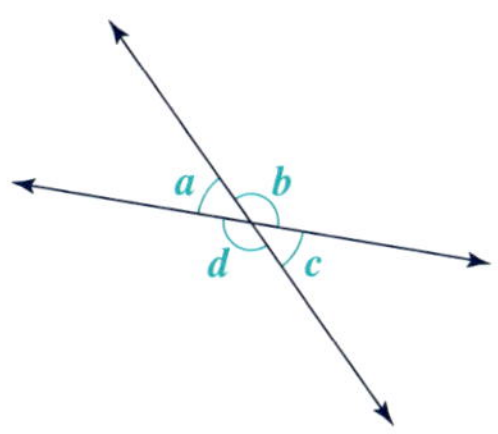


1 Be sure students understand how to label angle measurements.

2 You may have students work in pairs.

3 Encourage students to use the short-hand notation when they record the angle measures.

4 Discuss students' answers.

5 Introduce the term *vertical angles.*

# Develop

The cartoon at the top of page 473 shows how reasoning can be used to explain why two vertical angles have the same measure. In Problem Set C, this conclusion was obtained by first using a protractor to measure the angles and then observing that vertical angles have the same measure. In the cartoon, Conor is using logic to reach the same conclusion. He does this without making measurements.

1 **Think & Discuss**
Discuss with the class how an argument similar to Conor's argument shows that the other pair of vertical angles in the cartoon have equal measures. You may also use the discussion to help students see why the side-by-side angles in the diagram have measures that add to 180°. ■

2 After the Think & Discuss, have students consider the quadrilateral shown just above Problem Set D. Review the fact from Chapter 1 that the sum of the measures of the interior angles of any triangle is 180°. You may want to take time to have students draw some triangles, measure their angles, and add the measures to confirm that this is so. Then tell students they will be looking for similar rules for other polygons.

3 **Problem Set D** **Suggested Grouping: Pairs**
In this problem set, students draw three kinds of polygons, measure the interior angles, and find the sum of the interior angles.

Encourage students to draw polygons that are large enough to make it easy to measure the angles. Suggest that they write the measure of each interior angle inside the polygon, close to the vertex of the angle.

**On the Spot Assessment**

4 You can use the results of Problem Set D to identify particular difficulties students are having. If students get a sum that is not close to 360°, 540°, or 720°, check their work to see whether they made a computational error in computing the angle sum. If there is no computational error, then there was an error in measuring one or more of the angles. In this case, have students re-measure the angles. If possible, have them do this while you observe to be sure they are positioning and reading the protractor correctly. If you do not have time to do this yourself, have the students work with other students to correct the mistakes.

Below, Conor explains the relationship he discovered in Problem Set C.

**Think & Discuss Answer**

Possible answer: $\angle 2$ and $\angle 1$ form a straight line, so their measures add to 180°. $\angle 4$ and $\angle 1$ form a straight line, so their measures add to 180°. So, $m\angle 2 + m\angle 1 = 180°$ and $m\angle 4 + m\angle 1 = 180°$. Since adding $m\angle 2$ to $m\angle 1$ gives 180°, and adding $m\angle 4$ to $m\angle 1$ also gives 180°, $m\angle 2$ must be equal to $m\angle 4$.

## Think & Discuss

In the cartoon, Conor showed that $m\angle 1 = m\angle 3$. Explain why $m\angle 2 = m\angle 4$.

1 Discuss how the other pair of vertical angles have equal measures.

The *interior angles* of a polygon are the angles inside the polygon. In this quadrilateral, the interior angles are marked.

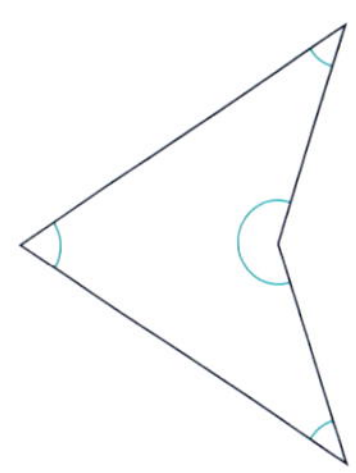

2 Introduce the term *interior angles.*

In Chapter 1, you discovered that the sum of the measures of the interior angles of any triangle is 180°. In the next problem set, you will look for similar rules about the angle sums of other polygons.

MATERIALS

- ruler
- protractor

## Problem Set D

3 You may have students work in pairs.

1. Use a ruler and pencil to draw a quadrilateral. Measure each interior angle, and then find the sum of the four angles.
   Measurements will vary. The sum should be very close or equal to 360°.
2. Now draw a pentagon. Measure each interior angle, and find the sum of the five angles.
   Measurements will vary. The sum should be very close or equal to 540°.
3. Finally, draw a hexagon. Measure each interior angle, and find the sum of the six angles. Measurements will vary. The sum should be very close or equal to 720°.

4 Check students' work for any computational or measuring errors.



LESSON 8.1 Angles 473

# Develop

1 **Think & Discuss**
For this Think & Discuss, students will use the look for a pattern, write a rule, and make and test a conjecture or prediction problem-solving strategies. Urge pairs of students to compare their results with those of at least two or three other pairs of students. When students have had time to do this, have a class discussion of any conclusions they are ready to offer. In particular, ask for their predictions of the angle sums of the two polygons shown at the end of the Think & Discuss. The class should be able to agree on a prediction of 360° for the angle sum of the quadrilateral and 720° for the hexagon. ■

2 When students are ready to consider the cartoon at the bottom of page 474, call their attention to the Remember note in the margin. Ask students which segments in the first panel of the cartoon are diagonals. *the dashed segments*

3 Be sure students understand that each of the dashed diagonals divides two interior angles of the polygon into two smaller angles and that the sum of the angle measures of the smaller angles at each vertex equals the measure of the interior angle at that vertex.

MATERIALS
protractor

## Think & Discuss

For each problem in Problem Set D, you and your classmates probably all drew different polygons. Compare the angle sums you found with the sums found by your classmates. What patterns do you see?
See ① below.

Describe a rule you could use to predict the sum of the interior-angle measures of a polygon when you know only the number of angles.
See ② in Additional Answers.

Use your rule to predict the interior-angle sums for each concave polygon below. Check your predictions by measuring the angles. Be sure to measure the *interior* angles.
360°, 720° (Note: The angle sums found by measuring may vary slightly.)

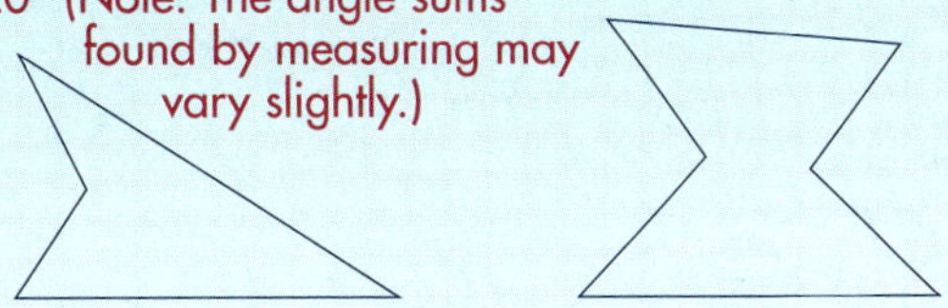

**Remember**
A *concave polygon* has an interior angle with measure greater than 180°. Concave polygons look "dented."

**1 Problem-Solving Strategies**
- Look for a pattern
- Write a rule
- Make and test a conjecture

By now, you have probably concluded that the sum of the angle measures in a polygon depends only on the number of angles (or the number of sides). You may have also discovered a rule for predicting the angle sum of any polygon when you know the number of angles.

**2** Remind students of the definition of a diagonal.

Hannah and Jahmal wondered whether they could use what they know about the angle sum for triangles to think about the angle sums for other polygons.

**Remember**
A *diagonal* is a segment that connects two vertices of a polygon but is not a side of the polygon.

**Think & Discuss Answer**
① The angle sums for quadrilaterals are all about 360°. The angle sums for pentagons are all about 540°. The angle sums for hexagons are all about 720°.

**3** Discuss that the dashed diagonals divide the interior angles into two smaller angles.



### Additional Answers

② Possible rules:
- A triangle has an angle sum of 180°. Each time you add an angle to the polygon, you add 180° to the angle sum.
- angle sum = 180° · (number of angles − 2)

# Develop

1 **Problem Set E** **Suggested Grouping: Pairs**
In this problem set, students investigate how to find the sum of the interior angles of any polygon by using diagonals to divide the polygon into nonoverlapping triangles. The students find the sum of the angle measures of the original polygon by using the fact that each of these triangles has an angle sum of 180°.

2 Students need to understand that when they use diagonals to divide a polygon into triangles, the diagonals should not cross each other and should lie *inside* the polygon. Also, students should not use a diagonal that overlaps a side of a polygon. These last two cautions are especially important for polygons that are concave (that is, polygons that have "dents").

For **Problem 1a,** students may think that the diagonals can be drawn from any vertex. For some pentagons, this is not the case. For example, in pentagon *ABCDE* below, you would not want to use the diagonals from point *A*. The diagonal from *A* to *D* lies outside the pentagon, and the diagonal from *A* to *C* overlaps a side of the pentagon.

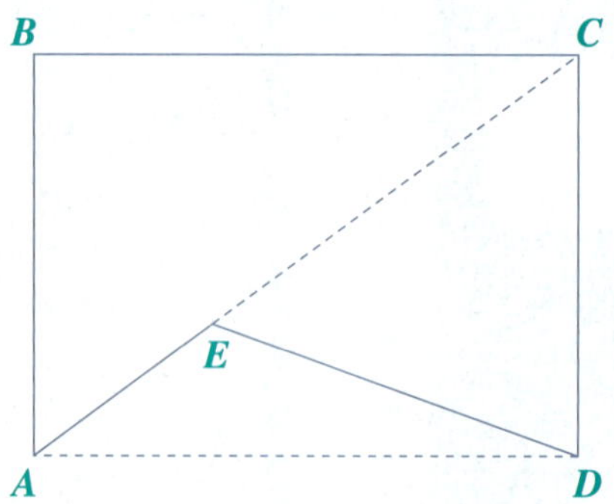


The difficulty is easily overcome by using the diagonals from point *E*, as shown below.

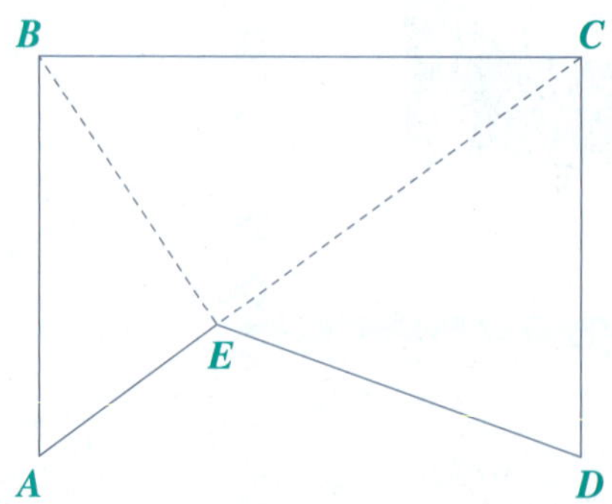


3 In **Problem 1c,** students need to recall that the sum of the angle measures in each triangle is 180°.

In **Problem 2c,** it is possible to draw a hexagon that cannot be divided into suitable triangles by drawing diagonals from a *single* vertex. Hexagon *ABCDEF*, shown below, is an example of such a hexagon. No matter which vertex is selected, one or more of the diagonals from that vertex will overlap a side or will fail to lie entirely inside the polygon. In the unlikely event that a student draws such a hexagon, simply relax the requirement that all the diagonals be drawn from the same vertex.

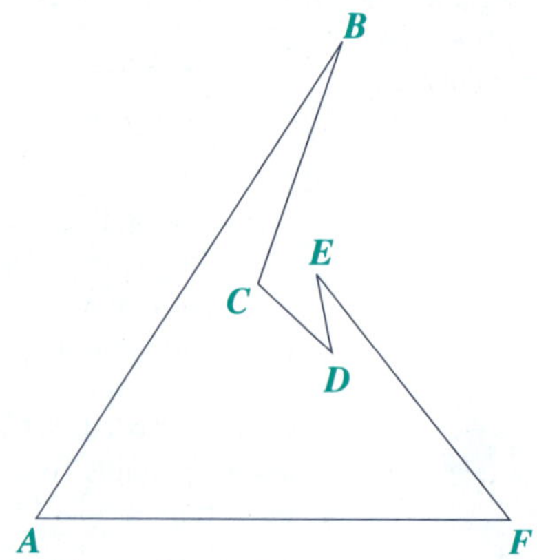


The diagram below, shows how hexagon *ABCDEF* can be divided into four triangles by drawing the diagonals from *A* to *C*, from *A* to *D*, and from *D* to *F*.

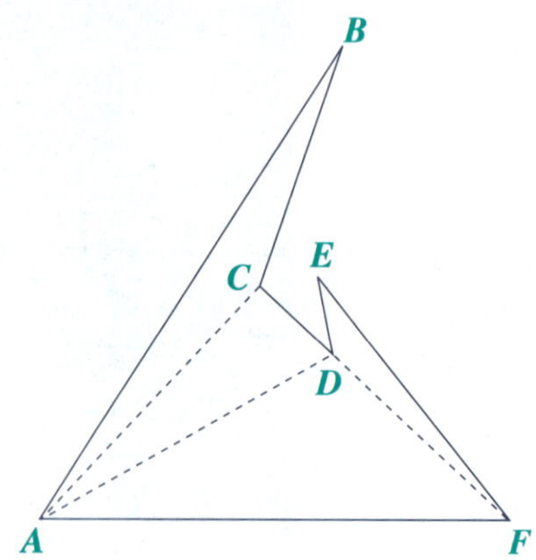

In the next problem set, you will investigate whether Hannah and Jahmal's strategy applies to other polygons. You will also see how their strategy leads to a rule for calculating angle sums.

MATERIALS
ruler

## Problem Set E

1. First consider pentagons.

   a. Draw two pentagons. Make one of the pentagons concave. Divide each pentagon into triangles by drawing diagonals from one of the vertices. Drawings will vary.

   b. Into how many triangles did you divide each pentagon? 3

   c. Use your answer to Part b to find the sum of the interior angles in a pentagon. 540°

2. Now consider hexagons.

   a. Draw two hexagons. Make one of the hexagons concave. Divide each hexagon into triangles by drawing diagonals from one of the vertices. Drawings will vary.

   b. Into how many triangles did you divide each hexagon? 4

   c. Use your answer to Part b to find the sum of the interior angles in a hexagon. 720°

**Just the facts**

*Crystallographers*—people who study the geometric properties and internal structures of crystals—use *reflecting goniometers* to measure the angles between the faces of a crystal.

1 You may have students work in pairs.

2 Be sure students know that diagonals should not cross each other or overlap a side when dividing polygons.

3 Remind students that the sum of the angle measures in a triangle is 180°.



LESSON 8.1 Angles 475

# Develop

1 In **Problem 3a,** students should not be too concerned if they sketch a concave octagon that cannot be divided into suitable triangles by drawing diagonals from a single vertex. Relax the requirement that all the diagonals be drawn from a single vertex.

In **Problem 6,** some students may draw a quadrilateral with three 90° angles. They may then measure the fourth angle to answer **Problem 6a** and use the appearance of the diagram to answer **Problem 6b.** While this approach can be used to obtain correct answers, it overlooks the fact that the problem can be solved by reasoning. Encourage students to examine the use of reasoning.

### Share & Summarize

2 You can use these questions to check that students understand what kinds of triangles are needed in order to use division of a polygon by diagonals to find the sum of the angle measures of a polygon. You may want to call on volunteers to read their answers to these questions. Discuss their answers with the class.

**Troubleshooting** If students have difficulty with **Question 1** in the Share & Summarize, sketch the quadrilaterals on the board. In each diagram, mark one of the angles where the diagonals intersect and ask students whether it is part of an interior angle of the quadrilateral. Then point out angles that are part of an interior angle of the quadrilateral.

If students have difficulty with **Question 2,** have them sketch a nonagon. Ask them to divide the polygon into appropriate triangles by drawing diagonals. Help them see that they can find the sum of the angle measures of the polygon by counting the triangles and multiplying by 180°.

**Additional Example** What is the sum of the angle measures of a decagon (a 10-sided polygon)? 1,440°

### On Your Own Exercises

Practice & Apply: 15–21, pp. 477–478
Connect & Extend: 24, 25, pp. 479–480
Mixed Review: 26–38, p. 481

**3a.** Possible answer: 6; so far, I have been able to divide each polygon into the number of triangles equal to 2 less than the number of sides.

**Just the facts**

*A 15-sided polygon is called a pentadecagon.*

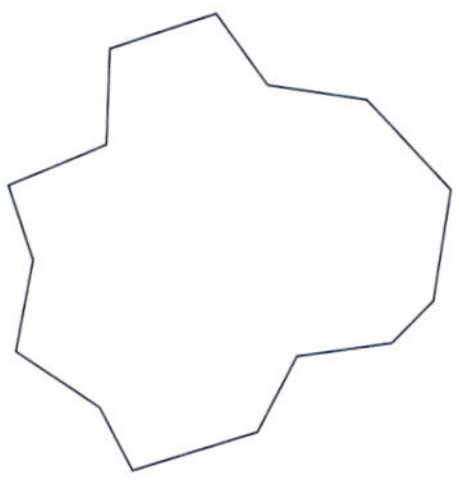

**3b.** Drawings will vary, but all should yield 6 triangles.

**6a.** 90°; the angles must add to 360°.

**3.** Now think about octagons, which are 8-sided polygons.

**a.** Without making a drawing, predict how many triangles you would divide an octagon into if you drew all the diagonals from one of the vertices. Explain how you made your prediction.

**b.** Draw an octagon, and check your prediction.

**c.** Use your answer to find the interior-angle sum for an octagon. 1,080°

**4.** Suppose you drew a 15-sided polygon and divided it into triangles by drawing diagonals from one of the vertices.

**a.** How many triangles would you make? 13

**b.** Use your answer to find the interior-angle sum for a 15-sided polygon. 2,340°

**5.** Suppose a polygon has $n$ angles.

**a.** How many triangles would you make if you divided the polygon into triangles by drawing diagonals from one of the vertices? $n - 2$

**b.** Use your answer to write a rule for calculating the sum of the angle measures $s$ in a polygon with $n$ angles. $s = (n - 2) \cdot 180°$

**6.** A particular quadrilateral has three 90° angles.

**a.** What is the measure of the fourth angle? How do you know?

**b.** What kind of quadrilateral is it? a rectangle

## Share & Summarize

**1.** Possible explanation: Some of the angles Marcus is counting are in the center of the quadrilateral; they are not part of the angles of the quadrilateral.

**1.** Marcus said the sum of the angle measures for a quadrilateral must be 720° because a quadrilateral can be split into four triangles by drawing both diagonals.

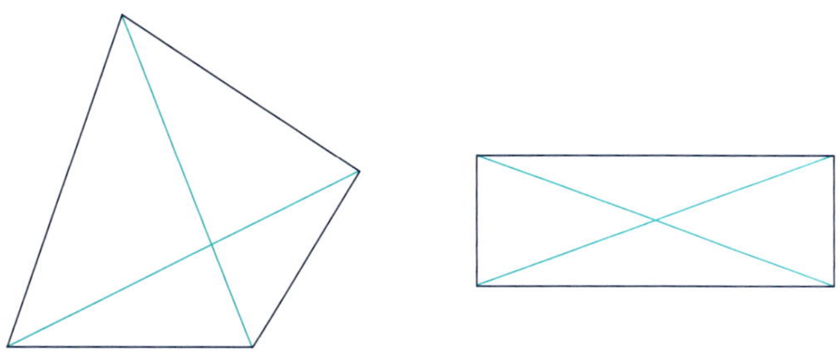

Explain what is wrong with Marcus' argument.

**2.** What is the sum of the angle measures of a nonagon (a 9-sided polygon)? 1,260°

**1** Have students draw the diagonals from a single vertex.

**2** Discuss students' answers with the class.

# Teacher Notes

# On Your Own Exercises



## Practice & Apply

Find the measure of each angle.

**1.**

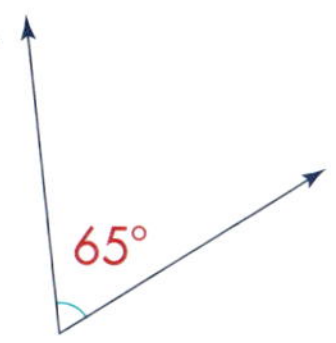


**2.**

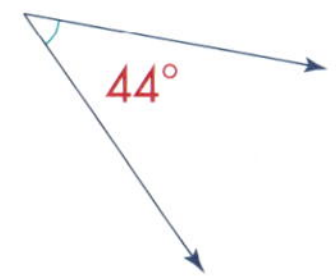


**3.**

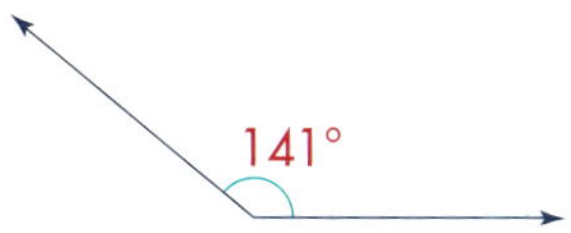


**4.**

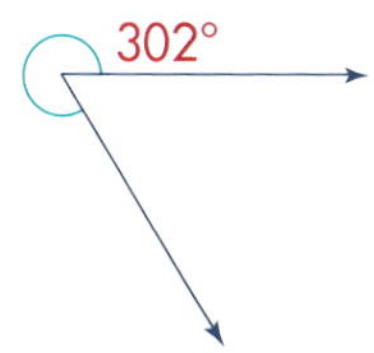


Without measuring, find the missing angle measures.

**5.**

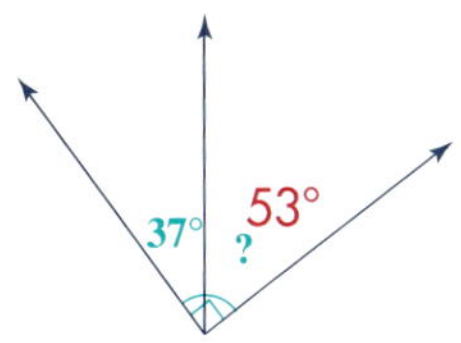


**6.**

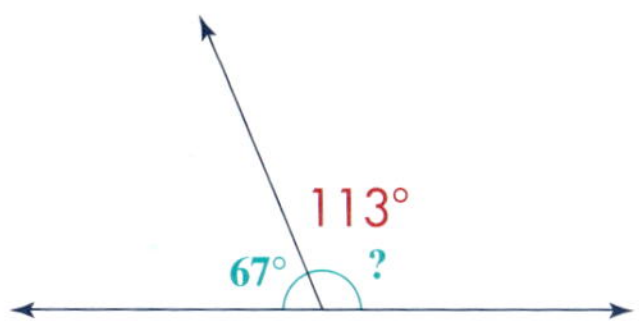


**7.**

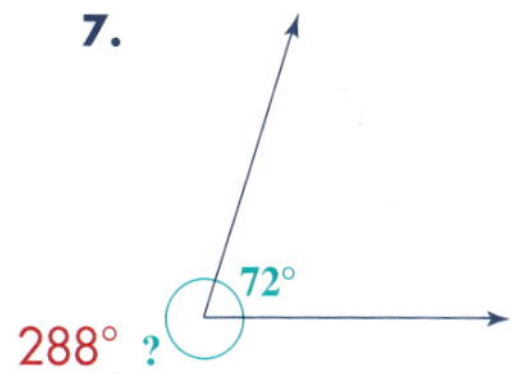


Draw an angle with the given measure.

**8.** 17° **9.** 75° **10.** 164° **11.** 290°

**8.**

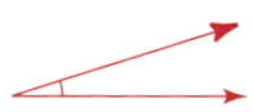

**9.**

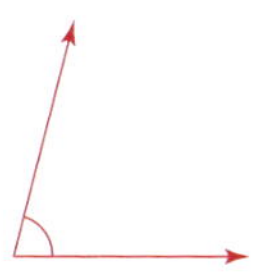

**10.**

Draw the figure described. Label every angle in the figure with its measure.

**12.** a quadrilateral with two 60° angles

**13.** a pentagon with two 90° angles

**14.** a quadrilateral with one 200° angle

**11–14.** See Additional Answers.

Without measuring, find the measure of each lettered angle.

**15.**

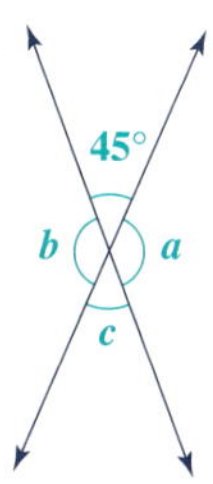


**15.** $m\angle a = 135°$, $m\angle b = 135°$, $m\angle c = 45°$

**16.**

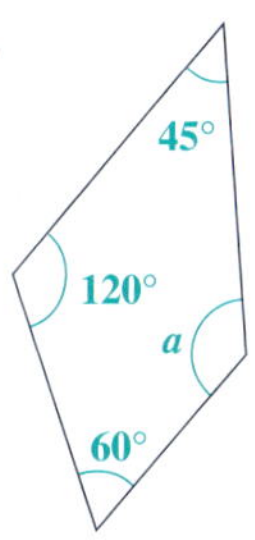


$m\angle a = 135°$

**17.**

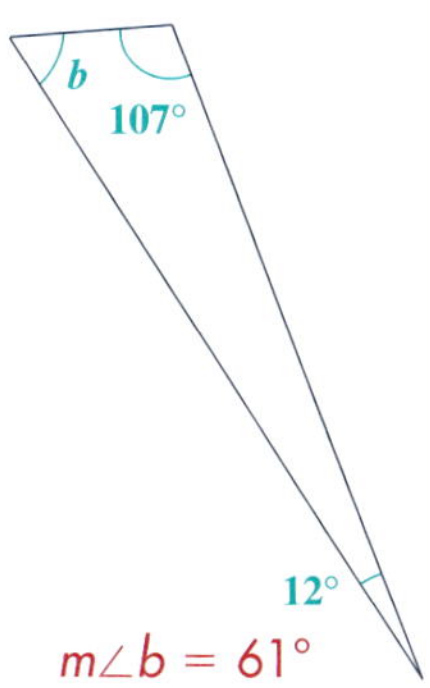


$m\angle b = 61°$

**18.**

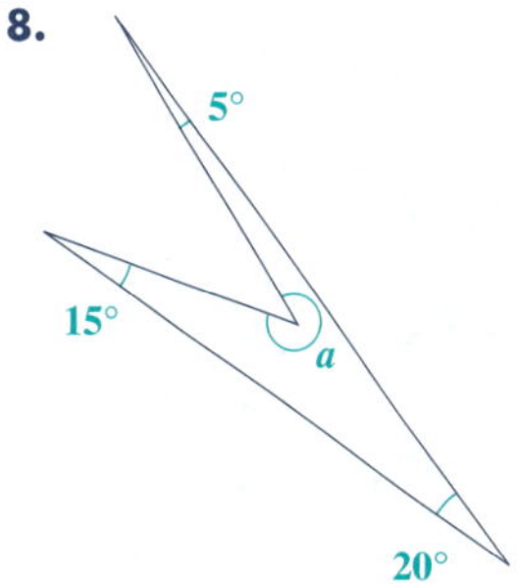


$m\angle a = 320°$

impactmath.com/self_check_quiz

**On Your Own Exercises**

Investigation 1, pp. 467–471
Practice & Apply: 1–14
Connect & Extend: 22, 23

Investigation 2, pp. 472–476
Practice & Apply: 15–21
Connect & Extend: 24, 25

Assign Anytime
Mixed Review: 26–38

**Exercises 15–18:** These exercises are intended to check that students understand the angle relationships for each figure. Encourage students to show their calculations so you can assess whether they used valid methods to find the angle measures.

*Quick Review* **Math Handbook**

hot words hot topics

Hot Topics pp. 296–303, 314–319

## Additional Answers

**11.**

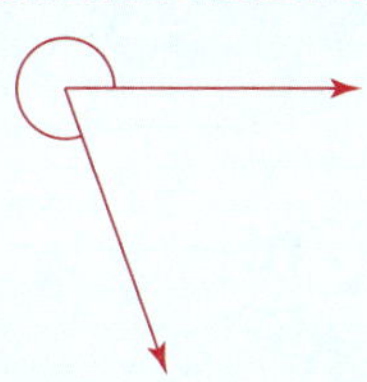

**12.** Possible quadrilateral:

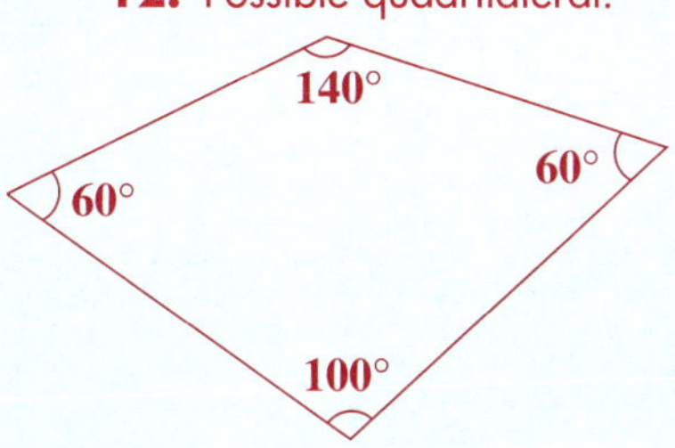


**13.** Possible pentagon:

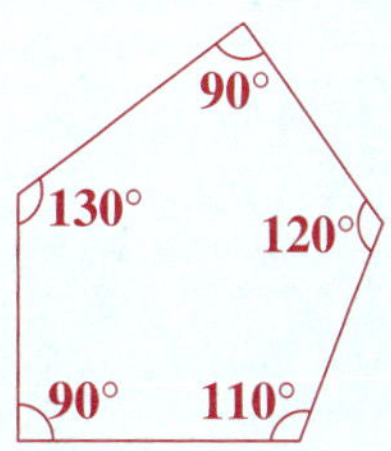


**14.** Possible quadrilateral:

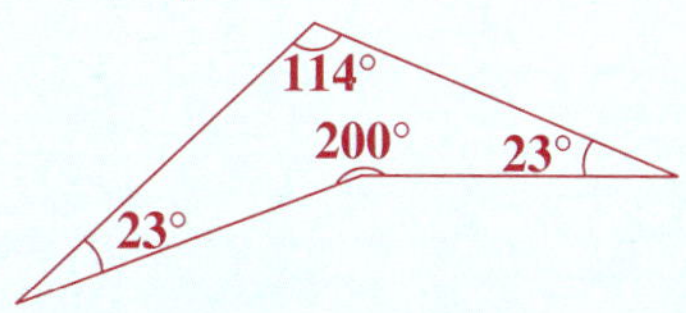

# Assign and Assess

**Exercise 19:**
Some students may be unsure how to approach this exercise. You might help by asking such questions as the following: How many sides does this polygon have? From what you know about the number of sides, what can you say about the total of all of the measures of the interior angles? If you subtracted the sum of the known angle measures from that total, how many degrees would that leave for $\angle a$ and $\angle b$ combined? Since $\angle a$ and $\angle b$ have the same measure, what is the measure of each angle?

**Exercise 22:**
When students trace the diagrams for this exercise, it would be good for them to use dots for the players rather than tracing each stick figure. Suggest that they mark the dots in a systematic way. They might, for example, place a dot at the head of each player or at the right foot of each player.

## Remember

*In a regular polygon, all sides are the same length and all angles the same measure.*

## Connect & Extend

## Just the facts

*Soccer is the national sport of most Latin American and European countries.*

**19.** In this polygon, $\angle a$ and $\angle b$ have the same measure. What is it? 160°

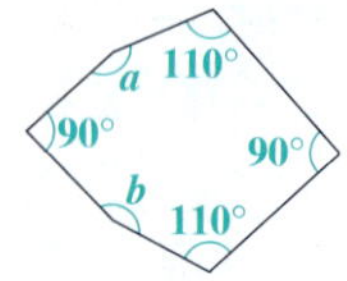


**20.** What is the measure of each angle of a regular pentagon? 108°

**21.** What is the measure of each angle of a regular hexagon? 120°

**22. Sports** The drawings below show angles formed by a soccer player and the goalposts. The greater the angle, the better chance the player has of scoring a goal. For example, the player has a better chance of scoring from Position A than from Position B.

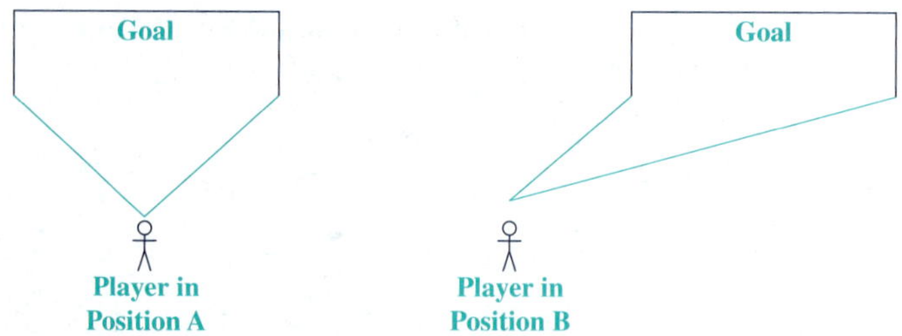


In Parts a and b, it may help to trace the diagrams and draw and measure angles.

**a.** Seven soccer players are practicing their kicks. They are lined up in a straight line in front of the goalposts. Which player has the best (the greatest) kicking angle? Player 4

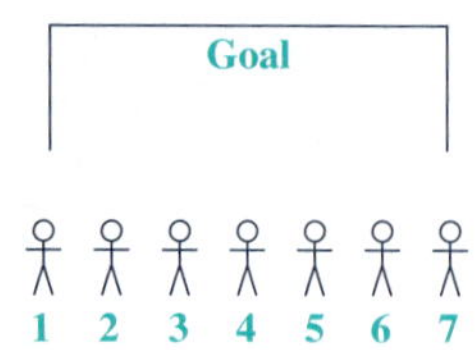


**b.** Now the players are lined up as shown. Which player has the best kicking angle? Player 3

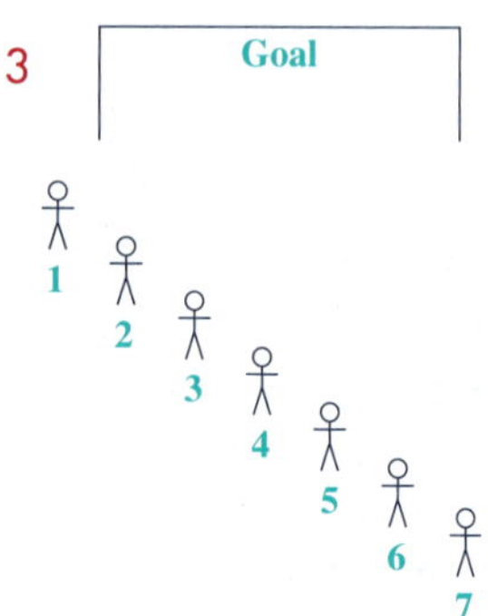




478 CHAPTER 8 Geometry and Measurement

**24a.** $m\angle 1 = 112°$, $m\angle 2 = 107°$, $m\angle 3 = 141°$; sum = 360°

**24b.** $m\angle 1 = 75°$, $m\angle 2 = 143°$, $m\angle 3 = 142°$; sum = 360°

**24c.** $m\angle 1 = 130°$, $m\angle 2 = 77°$, $m\angle 3 = 80°$, $m\angle 4 = 73°$; sum = 360°

**24d.** $m\angle 1 = 45°$, $m\angle 2 = 79°$, $m\angle 3 = 51°$, $m\angle 4 = 81°$, $m\angle 5 = 104°$; sum = 360°

**24e.** $m\angle 1 = 50°$, $m\angle 2 = 55°$, $m\angle 3 = 63°$, $m\angle 4 = 65°$, $m\angle 5 = 45°$, $m\angle 6 = 82°$; sum = 360°

**24f.** The exterior-angle sum is always 360°.

**23.** The *diameter* of a circle is a segment that passes through the center of the circle and has both its endpoints on the circle.

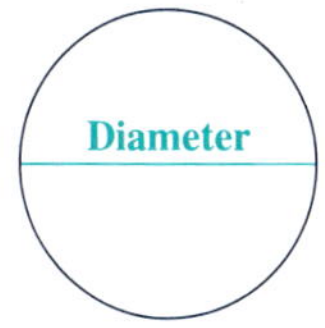


The four triangles below have all three vertices on a circle and the diameter as one side.

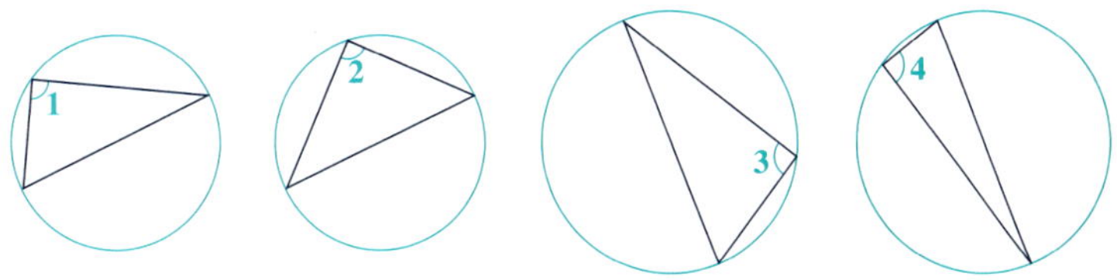

Measure each numbered angle. What do the measures have in common? All four angles measure 90°.

**24.** You discovered a rule about the sums of the interior angles of polygons. Polygons also have *exterior* angles, which can be found by extending their sides. In the drawings below, the exterior angles are marked.

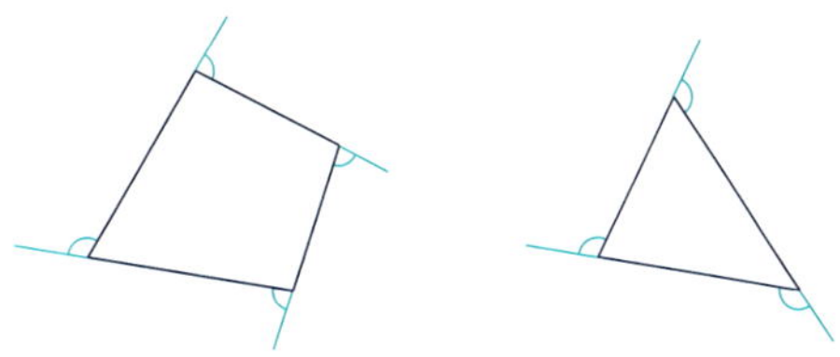

In Parts a–e, find the measure of each exterior angle. Then find the sum of the measures.

**a.**

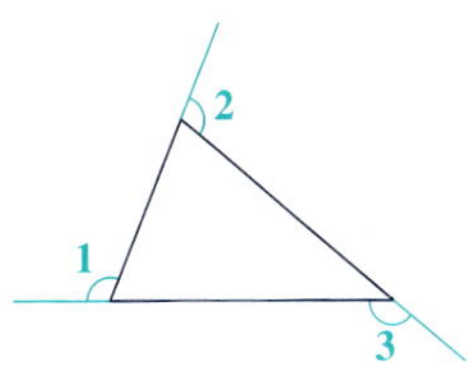

**b.**

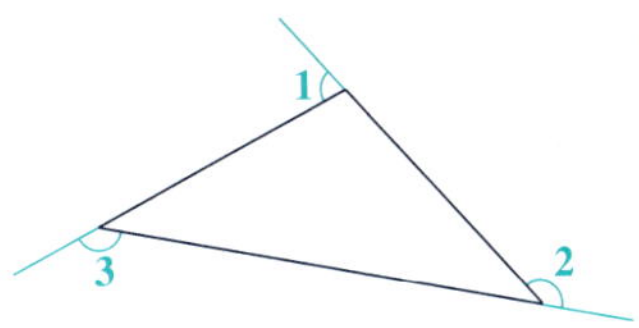

**c.**

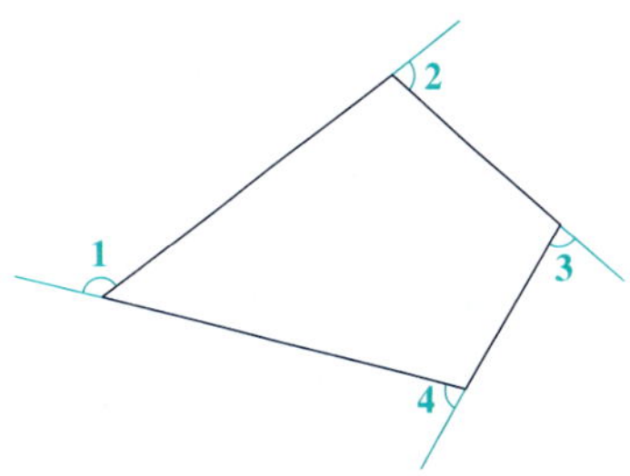

**d.**

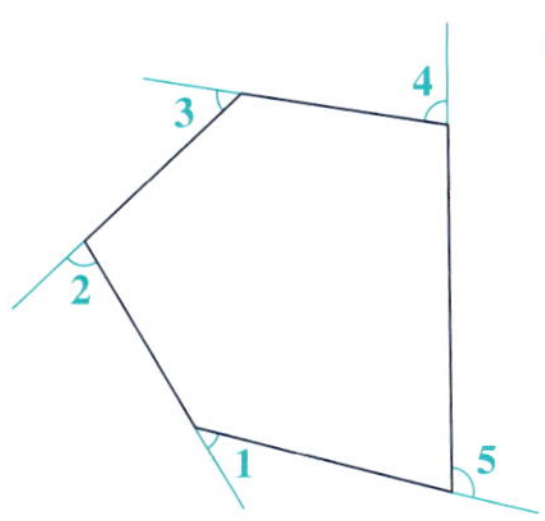

**e.**

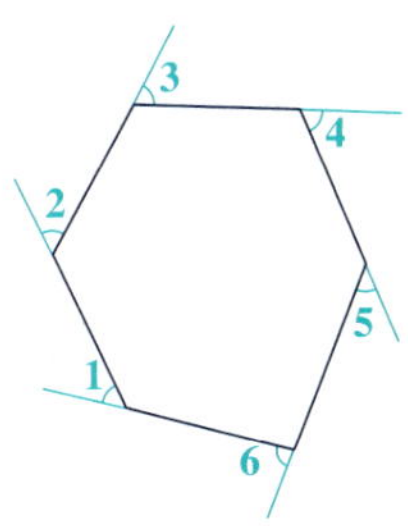

**f.** Describe any pattern you find in the measures in Parts a–e.



**Exercise 24:**
This exercise gives practice in measuring angles. It also leads students to discover an interesting property of convex polygons (polygons without "dents").

When students calculate the sum of the angle measures for the figures, their answers may not be equal to exactly 360°. This can easily be due to slight measurement errors. Nevertheless, students will probably be able to see the pattern. You may want to ask students if they can give a logical explanation for the pattern in **Part f.**

The figures in Exercise 24 are reproduced on Master 68.

**Teaching Resources**

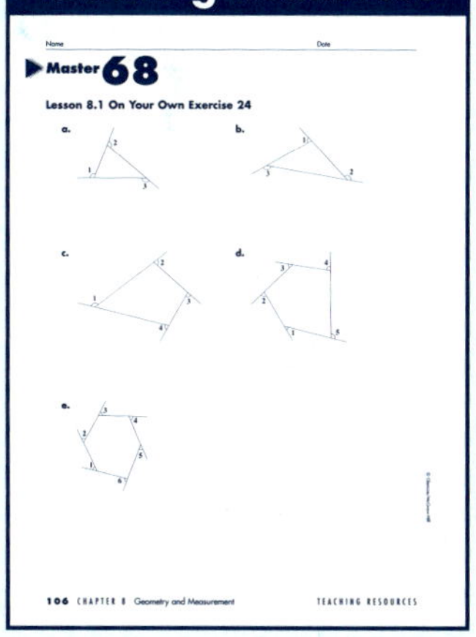



LESSON 8.1 Angles **479**

**Exercise 25:** Most students are likely to use a trial-and-error approach to solve **Part d.** Later, when students know more geometry, they will be able to solve the problem by using reasoning.

## Just the facts

When light hits a mirror, it behaves in the same way as a pool ball hitting the side of a table. If lights hits a mirror at an angle, it bounces off at the same angle. In physics, this law is often stated as "the angle of incidence = the angle of reflection."

## In your own words

Describe how you can find the interior-angle sum of any polygon without measuring any angles. Then explain how you know that your method works.

**25. Sports** The angle at which a pool ball hits the side of a table has the same measure as the angle at which it bounces off the side. This is shown in the drawing at right. The marked angles have the same measure, and the arrow shows the ball's path.

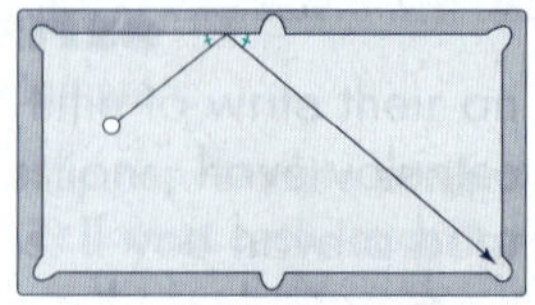

In Parts a–c, trace the drawing. Then use your protractor to find the path the ball will take when it bounces off the side. Tell whether the ball will go into a pocket or hit another side. (Draw just one bounce.)

**a.**

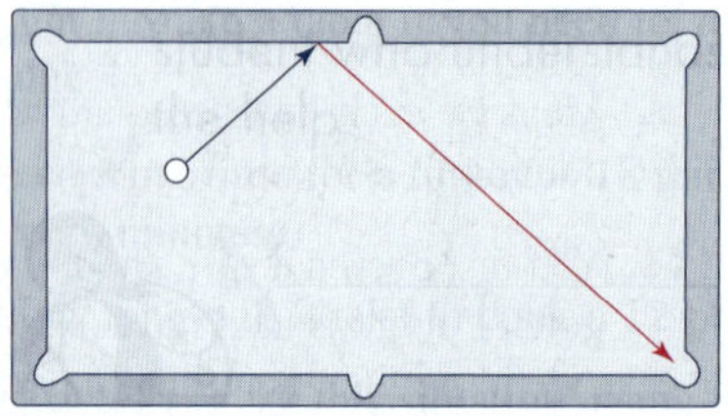

It will land in the corner pocket.

**b.**

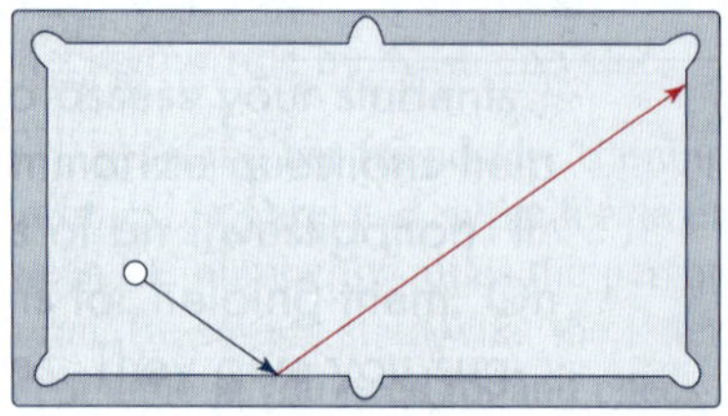

It will hit another side.

**c.**

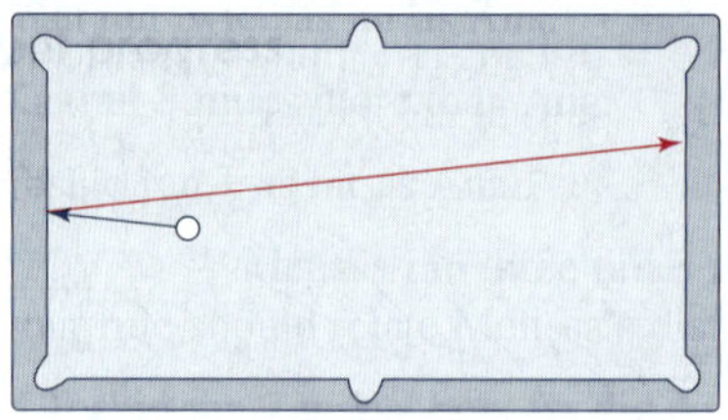

It will hit another side.

**d. Challenge** Trace this drawing. Draw a path for which the ball will bounce off a side and land in the lower-right pocket.

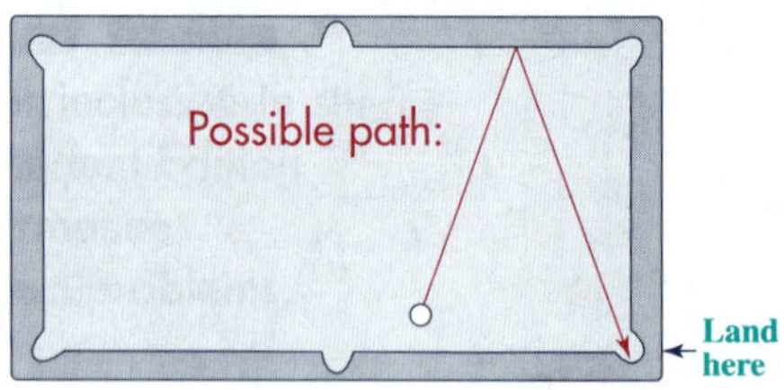




480 CHAPTER 8 Geometry and Measurement

## Mixed Review

Find the least common multiple of each pair of numbers.

**26.** 14 and 21 42 **27.** 17 and 51 51 **28.** 54 and 24 216

**29.** 13 and 5 65 **30.** 100 and 75 300 **31.** 32 and 36 288

Evaluate each expression.

**32.** $\frac{1}{3} + \frac{3}{4} \cdot \frac{4}{9}$ $\frac{2}{3}$ **33.** $\frac{5}{14} \div \frac{6}{7} - \frac{2}{7}$ $\frac{11}{84}$

**34.** $2\frac{1}{5} \div \frac{11}{15} \cdot \frac{7}{10}$ $2\frac{1}{10}$ **35.** $3\frac{5}{8} - 1\frac{1}{2} + \frac{3}{4}$ $2\frac{7}{8}$

**36.** $\frac{21}{34} \cdot \left(\frac{7}{12} + \frac{1}{8}\right)$ $\frac{7}{16}$ **37.** $\left(\frac{5}{6} \div \frac{5}{12}\right) + \left(1\frac{5}{6} \cdot 6\right)$ 13

**38. Technology** The graphs give information about two video games. Use the graphs to determine whether each statement is true or false, and explain how you decided.

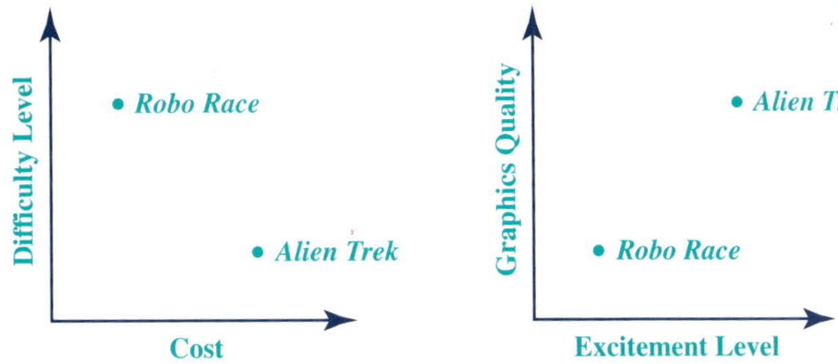


**a.** The more exciting game is less expensive.

**b.** The more difficult game has lower-quality graphics.

**c.** The less difficult game is more exciting.

**d.** The less expensive game has better graphics.

**e.** The game with better graphics is less exciting.
False; the graph on the right indicates that *Alien Trek* has better graphics and is more exciting.

**38a.** False; the graph on the right indicates that *Alien Trek* is more exciting. The graph on the left indicates that it is more expensive.

**38b.** True; the graph on the left indicates that *Robo Race* is more difficult. The graph on the right indicates that it has lower-quality graphics.

**38c.** True; the graph on the left indicates that *Alien Trek* is less difficult. The graph on the right indicates that it is more exciting.

**38d.** False; the graph on the left indicates that *Robo Race* is less expensive. The graph on the right indicates that it has lower-quality graphics.



## Quick Check

### Informal Assessment

Students should be able to:

- ✔ measure angles by using a protractor
- ✔ classify angles whose measures are greater than 0° and less than 180° as acute, right, or obtuse
- ✔ find the sum of the angles of any polygon given the number of sides of the polygon

### Quick Quiz

1. Measure each angle. Tell whether the angle is *acute, right,* or *obtuse.*

a.

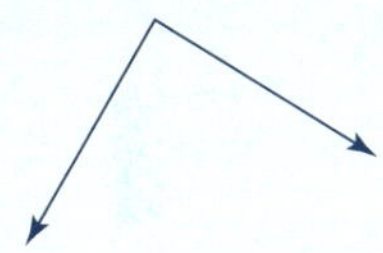

90°; right

b.

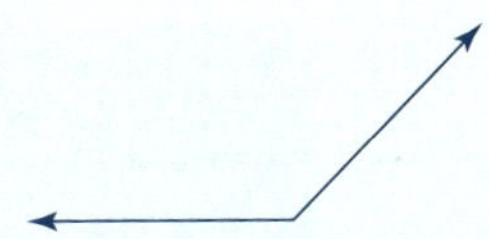

135°; obtuse

c.

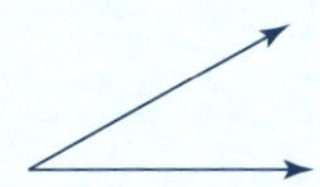

28°; acute

2. Draw an angle whose measure is 280°.

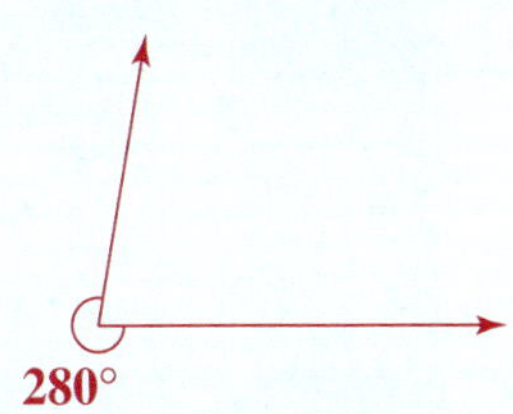


3. Without using a protractor, find the measure of each lettered angle.

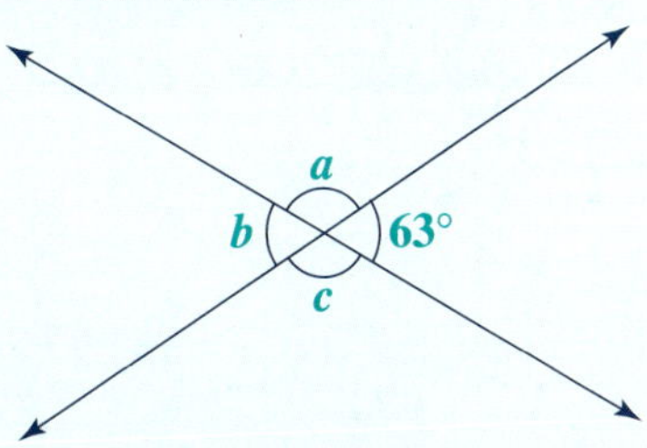


$m\angle a = 117°$, $m\angle b = 63°$, $m\angle c = 117°$

4. A quadrilateral has two 120° angles and an angle of 40°. What is the measure of the fourth angle? 80°

5. What is the sum of the angle measures of a polygon that has 11 sides? 1,620°

• See Questions 3–5 at left

# Teacher Notes

# 8.2 Measuring Around

## Objectives

- To use the side lengths of a polygon to calculate the perimeter of the polygon
- To use polygons to approximate the perimeter of a shape with curved sides
- To understand and use formulas for the circumference of a circle

## Overview (pacing: about 3 class periods)

In this lesson, students learn to calculate perimeters of polygons. They use their knowledge of rectangles to write a formula for the perimeter of a rectangle. Students investigate how perimeters of polygons can be used to approximate perimeters of figures with curved sides.

Students use measuring tape to measure the diameter and circumference of circular objects. They use these measurements to discover that dividing the circumference by the diameter gives the same number for all circles. Students learn the meaning of $\pi$ and learn to express the circumference and diameter of a circle in terms of $\pi$.

## Advance Preparation

For Investigation 1, you will need to provide students with copies of Master 69. For Investigation 2, you will need five objects with circular faces (for example, a CD, a coffee can, a roll of tape, a plate, and a nickel) that students can easily measure with string or measuring tape.

## Lesson Planner

| | Summary | Materials | On Your Own Exercises | Assessment Opportunities |
|---|---|---|---|---|
| **Investigation 1** page T482 | Students find perimeters of polygons and use perimeters of polygons to approximate perimeters of figures with curved sides. | • Metric rulers<br>• Master 69<br>• String | Practice & Apply: 1–5, pp. 490–491<br>Connect & Extend: 10, 11, pp. 491–492<br>Mixed Review: 15–24, p. 493 | On the Spot Assessment, pages T483, <br>Share & Summarize, pages T486, 486<br>Troubleshooting, page T486 |
| **Investigation 2** page T486 | Students use measurements of circles to approximate $\pi$; they use the definition of $\pi$ to write circumference formulas and then apply these formulas to solve problems that involve circles. | • 5 objects with circular faces<br>• String or measuring tape<br>• Rulers<br>• Scissors | Practice & Apply: 6–9, p. 491<br>Connect & Extend: 12–14, pp. 492–493<br>Mixed Review: 15–24, p. 493 | On the Spot Assessment, page T487<br>Share & Summarize, pages T489, 489<br>Troubleshooting, page T489<br>Informal Assessment, page 493<br>Quick Quiz, page A668 |

1 Read the opening paragraph with the class. Discuss the meaning of the term *perimeter* and how to calculate the *perimeter* of the picture of the house by using the information in the diagram.

2 **Think & Discuss**
Have a class discussion of the questions. In the discussion, consider any difficulties that might arise in measuring the perimeter of the floor of the classroom. For example, are there heavy bookcases that get in the way? If so, how could the perimeter be found without moving those objects? Would a string or yardstick be more helpful, or would you need both? After a brief discussion of these matters, go to Investigation 1.

# Investigation 1

In this investigation, students find perimeters of polygons. They develop a formula for the perimeter of a rectangle. They then use perimeters of polygons to approximate perimeters of figures with curved sides.

3 **Problem Set A** **Suggested Grouping: Pairs**
In this problem set, students use a metric ruler to find the perimeters of several rooms in a floor plan.

4 Have students examine the floor plan. Point out that the diagram is a *scale drawing.* This means that the shapes of the actual classrooms are just like those in the floor plan, only larger. Call attention to the scale for the floor plan, which is just below the diagram. Ask students how they could use the scale to estimate the actual length of Mrs. Stratton's classroom. Students will probably easily see that they could measure the length of the classroom in the drawing in centimeters and multiply by 2 to estimate the actual length in meters.

# 8.2 Measuring Around

**VOCABULARY**
**perimeter**

The **perimeter** of a two-dimensional shape is the distance around the shape. The perimeter of the shape at right is 10.8 cm.

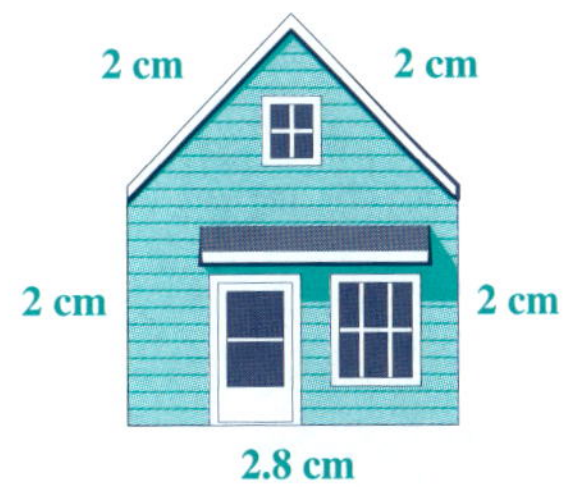


① Possible answer:
- Use a ruler, meterstick, or yardstick to measure each side, and add the measures.
- Wrap a string around the edges of the floor, and cut the string so its length equals the perimeter. Then measure the string with a ruler or yardstick.
- Wrap a long tape measure around the edges of the floor.
- Measure the pace of a student. Then count how many paces the student takes as he walks around the edge of the room, and multiply the result by the length of a pace.

1 Introduce the term *perimeter*.

## Think & Discuss

Describe as many methods as you can for measuring the perimeter of the floor of your classroom. See ①.

Which of your methods do you think will give the most accurate measurement? Answers will vary.

Which of your methods do you think is the most practical? Answers will vary.

2 Discuss the different methods students have used.

## Investigation 1 Finding Perimeter

To find the perimeter of a polygon, you simply add the lengths of its sides.

**MATERIALS**
metric ruler

### Problem Set A

3 You may have students work in pairs.

This is the floor plan of the second floor of Millbury Middle School. On the drawing, each centimeter equals 2 meters.

4 Introduce the term *scale drawing*.

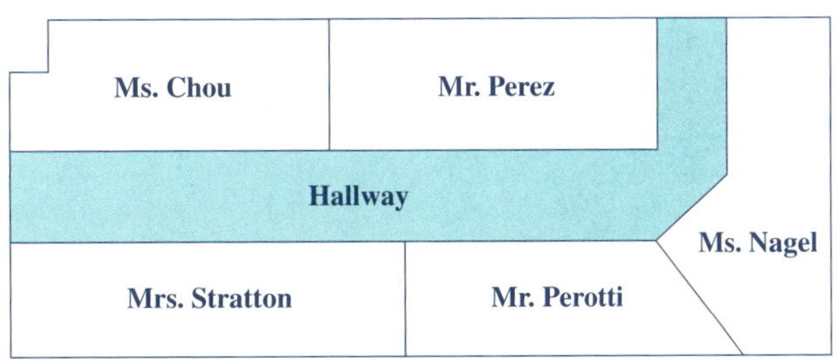


Scale: 1 cm = 2 m

# Develop

1 In **Problem 1,** students are asked to tell which teacher's classroom they think has the greatest perimeter. Encourage students to come up with a thoughtful reply. Use students' answers to discover which features of the shapes guided them toward their answer.

2 In **Problems 2–6,** encourage students to measure carefully.

In **Problem 3,** some students may assume that the room is rectangular and measure only two sides. Ask whether it is safe to make that assumption, or whether it is safer to measure each side. (Remember, the instructions ask for a very accurate perimeter.)

For **Problem 4,** students are using the idea that a quadrilateral having four right angles is a rectangle.

3 In **Problem 5,** students discover that the right-angle "dent" in the upper left-hand corner of Ms. Chou's room does not affect the perimeter. A rectangular room with the same *overall* length and width will have the same perimeter.

### On the Spot Assessment

Watch for students who have incorrect answers for Problem 5. Students who simply added the four lengths shown in the diagram need to see that they have not accounted for the lengths of all parts of the polygon. Have these students count the number of sides in the polygon and the number of numbers that they added. Other students may have failed to align digits correctly when they added lengths. Remind these students that it is important to align digits carefully when adding decimals.

★ indicates multi-step problem

**1.** Without measuring, tell whose classroom you think has the greatest perimeter, and explain why you think so. Answers will vary.

**2.** Look at the floor plan for Ms. Nagel's room.

**a.** What type of polygon is the floor of Ms. Nagel's room? hexagon

**b.** Find the perimeter of Ms. Nagel's floor plan to the nearest tenth of a centimeter. Then calculate the perimeter of the actual floor in meters. 11.4 cm, 22.8 m

**3.** Find the perimeter of Mrs. Stratton's floor plan to the nearest tenth of a centimeter. Then calculate the perimeter of the actual floor in meters. 12.6 cm, 25.2 m

**4.** Look at the floor plan for Mr. Perez's room.

**a.** Is Mr. Perez's floor a rectangle? How do you know?

**b.** Describe how to find the perimeter of Mr. Perez's floor plan by making only two measurements.

**c.** Measure the perimeter of Mr. Perez's floor plan to the nearest tenth of a centimeter. Then calculate the perimeter of the actual floor in meters. 11.4 cm, 22.8 m

**5.** To find the perimeter of Ms. Chou's floor plan, Althea made the measurements labeled below. She claims these are the only measurements she needs to make.

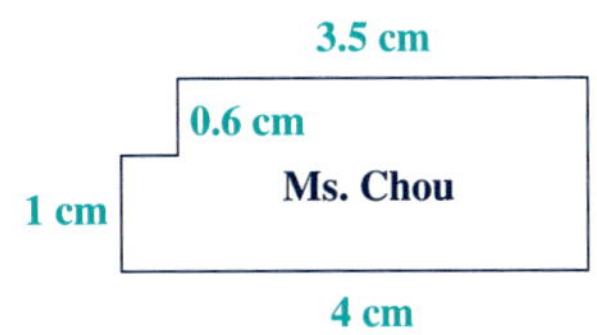


**a.** Is Althea correct? If so, explain how to find the perimeter of Ms. Chou's room using only these measurements. If not, tell what other measurements you would need. See Additional Answers.

★ **b.** Find the perimeter of Ms. Chou's floor plan to the nearest tenth of a centimeter. Then calculate the perimeter of the actual floor in meters. 11.2 cm, 22.4 m

★ **6.** Which teacher's classroom floor has the greatest perimeter? What is the perimeter? Mrs. Stratton's, 25.2 m

**4a.** Yes; it has four right angles and its opposite sides have the same length.

**4b.** Possible answers: Measure the length and the width, double each, and add the results. *Or,* measure the length and the width, add them, and double the result.

**Just the facts**

Many colleges and universities offer classes over the Internet, allowing students to earn college credit—and even a college degree—without setting foot in a classroom!

**1** Have students discuss which classroom has the greatest perimeter.

**2** Encourage students to measure carefully.

**3** Discuss whether the "dent" in the upper left-hand corner affects the perimeter of the room.



## Additional Answers

**5a.** yes; Possible explanations:

- You can imagine "pushing out" the indented upper-left corner so the floor plan becomes a rectangle. The length of the rectangle would be 4 cm, and the width would be 1 cm + 0.6 cm, or 1.6 cm. Find the perimeter by adding twice the length to twice the width.
- You can find the length of the right side by adding the two vertical lengths on the left; 0.6 cm + 1 cm, or 1.6 cm. Find the other missing length by subtracting the length of the top horizontal segment from the length of the bottom horizontal segment: 4 cm − 3.5 cm, or 0.5 cm. Then just add the side lengths.

LESSON 8.2 Measuring Around 483

## Develop

1 Discuss the explanation of the perimeter formulas for rectangles that is given at the top of the page. Students should not find it difficult to see why the two formulas are equivalent.

2 **Problem Set B** **Suggested Grouping: Pairs**
The problems in this problem set are straightforward applications of the perimeter formulas for rectangles.

**Sample Strategies** In **Problem 2,** students may arrive at their answers in different ways.

- Some may guess a number for the length or width, double that number, subtract from 42, and then divide by 2 to get the other dimension.
- Some may reason that if $2(L + W) = 42$, then $L + W = 21$. They can then look for pairs of numbers that have a sum of 21.

3 In **Problem 3,** students will find it easy to get the answer once they remember that a square is a special rectangle in which the length and width are equal.

**Problem-Solving Strategies** Students may have different solution strategies for **Problem 4.**

- Some students may come up with the answer by noting that a square is a rectangle with four sides of equal length $s$.
- Other students may use one of the formulas for the perimeter of a rectangle, replace $L$ and $W$ with $s$, and simplify. Both approaches are valid.

### Access for all Learners

**Extra Challenge** Refer students to Problem 2 of Problem Set B. Ask them how many dimensions are possible for a rectangle that has a perimeter of 42 feet Ask whether there are any limits on how large or how small the dimensions can be. If so, what are they? **An unlimited number of dimensions are possible. Both dimensions must be greater than 0, and the greater dimension must be less than 24 feet.**

The floor plan at the bottom of page 484 is used to focus students' attention on how to approximate the perimeter of a figure with curved sides. The Example on the next page uses the polygon approximation method.

In Problem Set A, you probably realized you could find the perimeter of a rectangle without measuring every side. This is because the opposite sides of a rectangle are the same length. If you measure the length and the width of a rectangle, you can find the perimeter using either of two rules:

*Add the length and the width, and double the result.*

*Double the length, double the width, and add the results.*

If you use $P$ to represent the perimeter and $L$ and $W$ to represent the length and width, you can write these rules in symbols.

Geometric rules expressed using symbols, like those above, are often called *formulas.*

> **Perimeter of a Rectangle**
>
> $P = 2 \cdot (L + W)$ $\qquad$ $P = 2L + 2W$
>
> In these formulas, $P$ represents the perimeter and $L$ and $W$ represent the length and width.

1 Discuss the perimeter formulas with the class.

## Problem Set B

2 You may have students work in pairs.

1. Use one of the perimeter formulas to find the perimeter of a rectangle with length 5.7 meters and width 2.9 meters. 17.2 m
2. The floor of a rectangular room has a perimeter of 42 feet. What are three possibilities for the dimensions of the floor?
3. A square floor has a perimeter of 32.4 meters. How long are the sides of the floor? 8.1 m
4. Write a formula for the perimeter of a square, using $P$ to represent the perimeter and $s$ to represent the length of a side. Explain why your formula works.

3 Remind students of the definition of a square.

**2.** Possible answer: 10 ft × 11 ft, 10.5 ft × 10.5 ft, and 20 ft × 1 ft

**4.** $P = 4s$; This works because the four sides of a square are the same length, so to find the perimeter, you can multiply the side length of the square by 4.

This floor plan is of the auditorium at Marshville Middle School.

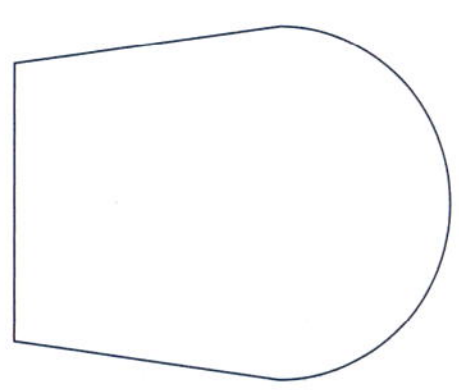

Scale: 1 cm = 5 m

Since part of the floor is curved, it is difficult to find the perimeter using just a ruler. You could use a measuring tape or a piece of string to find the length of the curved part. Another method is to use a polygon to *approximate* the shape of the floor.

# Develop

1 **Example**
Discuss the Example with the class. Then have students work on Problem Set C.

2 **Problem Set C** **Suggested Grouping: Pairs**
In this problem set, students explore how increasing the number of sides of a polygon need to approximate a shape's perimeter can affect the approximation. Distribute a copy of Master 69.

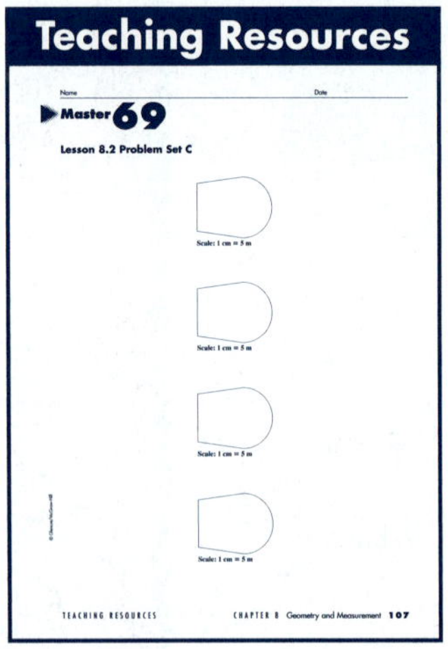


3 In **Problem 1a,** students will probably get a variety of estimates, depending on where they select the two vertices that are on the curved side. You may want to have pairs of students compare their polygons and their estimates. Spacing the points evenly along the curve will give a good approximation.

From **Problem 1b,** students will realize that no matter where they place the two vertices on the curved side, the perimeter of the polygon will be only an approximation of the perimeter of the floor plan.

In **Problem 1c,** the locations of the vertices on the curved side will again affect how good the approximation is. You may want to discuss with the class whether they think a heptagon can give a better approximation than a pentagon or a hexagon.

In **Problem 2,** students wrap a piece of string around the floor plan and measure how much string it took to go around.

4 **Problem Set Wrap-Up** You may want to discuss how the estimation procedures used for the floor plan would work if students had to measure the actual auditorium instead of the floor plan. Again discuss which procedure would give the best estimate and why.

## About the Mathematics

Approximating perimeters of shapes with curved sides by using polygons was an idea familiar to mathematicians in ancient times. When the curved sides can be described mathematically with great exactness, their lengths can be calculated with extraordinary accuracy without resorting to string or tape measures. The Greek mathematician and inventor Archimedes (about 287 B.C. to 212 B.C.) was an early master of these methods.

**1a.** Possible answer: The perimeter is about 11.2 cm.

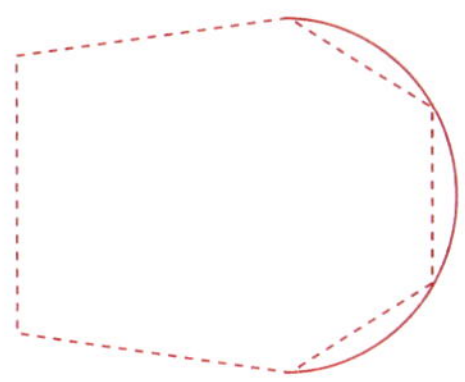

**1b.** Greater; each side of the polygon along the curved part of the floor plan is shorter than the corresponding length of the floor plan.

**EXAMPLE**

Luke drew a pentagon to approximate the shape of the floor:

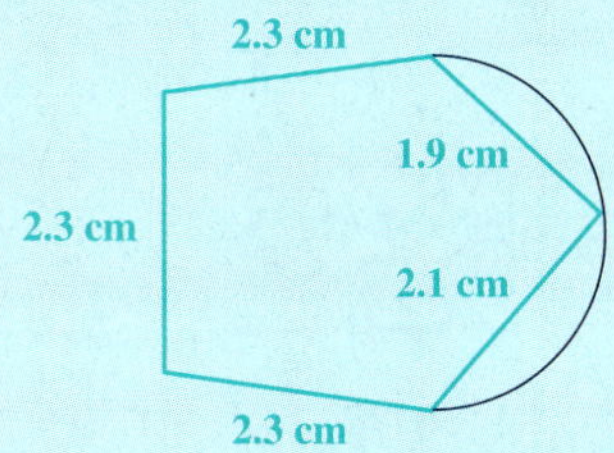


Scale: 1 cm = 5 m

Then he found the pentagon's perimeter:

$$2.3 + 2.3 + 2.3 + 1.9 + 2.1 = 10.9 \text{ cm}$$

**1** Discuss the Example with the class.

MATERIALS

- copies of the auditorium floor
- string
- ruler

## Problem Set C

**2** You may have students work in pairs.

**1.** You can get a closer approximation than Luke's by using a polygon with more sides.

**a.** Try using a hexagon—a polygon with six sides—to approximate the shape of the floor plan. What perimeter estimate do you get using a hexagon?

**b.** Is the actual perimeter greater than or less than your estimate? Explain.

**3** **Estimation** Have pairs discuss their estimates with other pairs.

**c.** Now try a heptagon—a polygon with seven sides—to approximate the shape of the floor plan. What perimeter estimate do you get using a heptagon? See Additional Answers.

**d.** Is the actual perimeter greater than or less than your estimate? Explain. See Additional Answers.

**e.** Using a polygon with more than seven sides, make another estimate. What is your estimate? See Additional Answers.

**2.** Wrap a piece of string around the floor plan. Try to keep the string as close to the sides of the floor plan as possible. Then mark the string to indicate the length of the perimeter, and measure the string's length up to the mark. What is your perimeter estimate?

**2.** Estimates will vary.

**3.** Which of your estimates do you think is most accurate? Explain.
Possible answer: The estimate I made with the string because I was able to get closer to the sides of the floor plan.

**4** Have students discuss how they would measure the actual auditorium.



**Additional Answers**

**1c.** Possible answer: The perimeter is about 11.3 cm.

**1d.** Greater; each side of the polygon along the curved part of the floor plan is shorter than the corresponding length of the floor plan.

**1e.** Possible answer: The perimeter is about 11.4 cm.

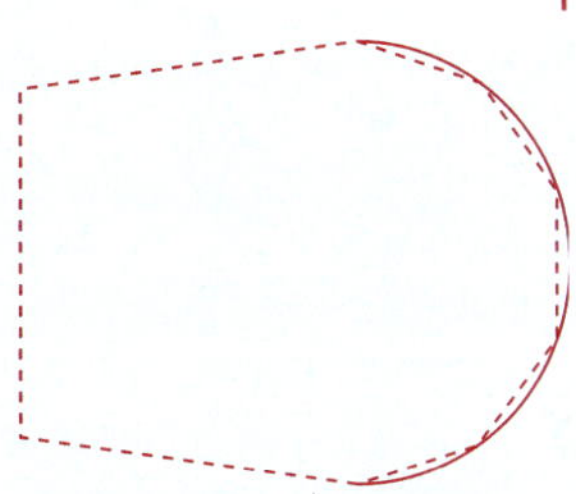

1 **Share & Summarize**
After students have written their paragraphs for the Share & Summarize, you may want to call on volunteers to read them to the class. If students' paragraphs make use of diagrams, have the students sketch them on the board.

**Troubleshooting** The discussion of students' work can give you valuable information about how well students have grasped the main ideas of Investigation 1. If they need more practice, you might have them create their own shapes with curved sides to measure. They could trade shapes with a partner to check their perimeter estimates.

### On Your Own Exercises

Practice & Apply: 1–5, pp. 490–491
Connect & Extend: 10, 11, pp. 491–492
Mixed Review: 15–24, p. 493

# Investigation 2

In this investigation, students learn several basic vocabulary terms related to circles. They measure the circumference and diameter of several circular objects and use their results to confirm that the circumference divided by the diameter is approximately equal to 3, regardless of the diameter of the object. They learn that the value of this quotient, assuming exact measures for the circumference and diameter, is called pi ($\pi$). They use this idea to write formulas for the circumference in terms of the diameter and radius. Finally, they apply these formulas to solve problems.

2 Before the Think & Discuss on page 486, discuss the new vocabulary terms. You may wish to tell students that the word *perimeter* can be used for the distance around any flat figure. The word *circumference* usually signals that the figure is a circle.

3 **Think & Discuss**
When you go over these questions with the class, be sure students understand that all the points on a circle are the same distance from the center. Refer to the diagram on page 486 as you discuss the questions. When you discuss whether all diameters are the same length, use the diagram to point out that every diameter is made up of two radii. This idea makes it easy to write a rule that relates the radius to the diameter.

## Share & Summarize

See Additional Answers.

Write a paragraph discussing what you know about finding the perimeter of two-dimensional shapes. Be sure to discuss

- polygons and nonpolygons
- ruler measurements and string measurements
- formulas

1 Call on students to read their paragraphs to the class.

# Investigation 2 Approximating $\pi$

In the last investigation, you found perimeters of polygons, and you estimated perimeters of a shape with curved sides. In this investigation, you will focus on circles.

**VOCABULARY**
**chord**
**circumference**
**diameter**
**radius**

The perimeter of a circle is called its **circumference.** Although you can estimate the circumference of a circle by using string or by approximating with polygons, there is a formula for finding the exact circumference. Before you begin thinking about circumference, you need to learn some useful words for describing circles.

A **chord** is a segment connecting two points on a circle. The **diameter** is a chord that passes through the center of the circle. *Diameter* also refers to the distance across a circle through its center. The **radius** is a segment from the center to a point on the circle. *Radius* also refers to the distance from the center to a point on the circle. The plural of *radius* is *radii.*

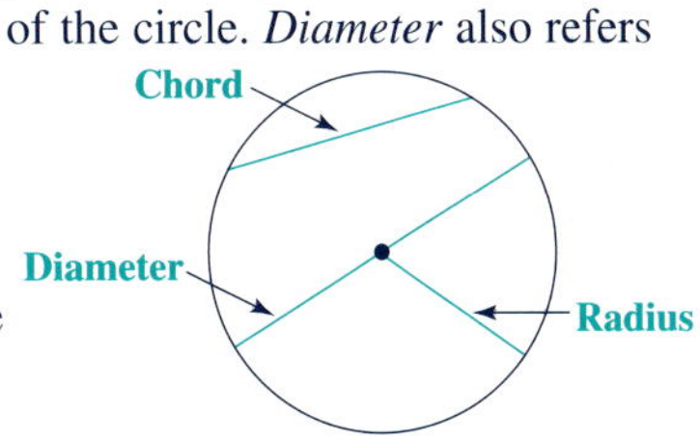


2 Introduce the terms *circumference, chord, diameter,* and *radius.*

## Think & Discuss

Are all the chords of a circle the same length? If not, which are the longest? No; the longest chords are those that pass through the center (the diameters).

Are all the diameters of a circle the same length? Are all the radii the same length? yes, yes

Write a rule for the relationship between the radius of a circle $r$ and its diameter $d$. $r = \frac{1}{2}d$, or $d = 2r$

3 Make sure students know that all points on a circle are the same distance from the center.



### Additional Answers
### Share & Summarize

Possible answer: Perimeter is the distance around a two-dimensional shape. You can find the perimeter of a polygon by measuring the lengths of its sides and adding them. For a rectangle, you can use the formula $P = 2 \cdot L + 2 \cdot W$, where $L$ is the length and $W$ is the width. You can approximate the perimeter of shapes with curved sides by using string or by drawing a polygon that closely matches the shape and finding its perimeter.

# Develop

1 **Problem Set D** **Suggested Grouping: Groups of 3 or 4**

In this problem set, students measure the circumference $C$ and diameter $d$ of five circular objects. They record the data in a table and then enlarge the table to include the values of $C \div d$. They use their results to make and test the conjecture that $C \div d$ is the same for all circles.

2 For **Problem 1,** you may want to check with each group to be sure students are measuring the circumference and diameter correctly. You may want to suggest measuring to the nearest tenth of a centimeter or to the nearest eighth of an inch.

### Access for all Learners

**Tactile Learners** Some students may find the following approach helpful in measuring the diameter of a circular object. Have students trace the object on a sheet of paper, cut out the circle, and fold the cutout so that the two halves match. Students can then measure the straight edge of the cutout to find the diameter.

Prepare your own table for **Problems 1 and 3.** The measurements and quotients that students record in their tables should be close to yours, though they need not be exactly the same as yours. Ask students to express the quotients as decimals rounded to the nearest tenth.

### On the Spot Assessment

Students may have made computational errors in Problem 3. Students who are using measurements made in inches may have had difficulty with division of fractions. Students who used measurements made in centimeters may be having trouble placing decimal points in the correct position in quotients. You may need to provide some individual help to address these difficulties.

For **Problem 4,** you can have groups record the quotients from Problem 3 on the board. This will facilitate class discussion of the results.

3 **Problem Set Wrap-Up** Be sure to discuss the results from Problems 4 and 5 with the class. The key observation is that all the quotients are close to 3.

This quote from the novel *Contact* by Carl Sagan mentions a relationship between the circumference and diameter of any circle:

> *In whatever galaxy you happen to find yourself,*
> *you take the circumference, divide it by its diameter,*
> *measure closely enough, and uncover a miracle.*

In the next problem set, you will examine the relationship that Sagan is describing.

**MATERIALS**

- 5 objects with circular faces (for example, a CD, a coffee can, a roll of tape, a plate, and a nickel)
- string or measuring tape
- ruler
- scissors

## Problem Set D

**1** Problem-Solving Strategy
Make and test a conjecture

For this problem set, your group will need five objects with circular faces.

**1.** Follow these steps for each object:

- Use string or a measuring tape to approximate the circumference of the object.
- Trace the circular face of the object. Cut out the tracing, and fold it in half to form a crease along the circle's diameter. Measure the diameter.

Record your measurements in a table like this one. Tables will vary.

**2** Have students measure the circumference and diameter to the nearest tenth of a centimeter, or to the nearest one-eighth of an inch.

| Object | Circumference, $C$ | Diameter, $d$ |
|---|---|---|
| | | |
| | | |
| | | |
| | | |
| | | |

**2.** Do you see a relationship between the circumference and the diameter of each circle? If so, describe it.

**3.** The quotation from *Contact* mentions dividing the circumference by the diameter. Add a column to your table showing the quotient $C \div d$ for each object. Describe any patterns you see.

**4.** On the board, record the values of $C \div d$ that your group found. How do your results compare with those of other groups?

**5.** Does the $C \div d$ value depend on the size of the circle? Explain. No; the $C \div d$ value is about the same for all the objects, regardless of their size.

**3** Make sure the quotients students find are close to 3.

**2.** Possible answer: The circumference is about 3 times the diameter.

**3.** Students should find that the quotient is about 3 for each object.

**4.** All groups should get values close to 3 for all the objects.



LESSON 8.2 Measuring Around 487

# Develop

1 Discuss the material on page 488 with the class. Students may be interested to know that the value of $\pi$ has been calculated to many millions of decimal places. Mathematicians have proved that the decimal for $\pi$ never comes to an end. The digits in this decimal never repeat in groups the way they do in decimals for fractions such as $\frac{1}{7}$ or $\frac{5}{9}$. This is why we use a special symbol when we want to talk about the exact value of the number.

2 The formulas for the circumference of a circle are easy to obtain if one accepts the fact that $C \div d$ is the same for all circles.

## About the Mathematics

Ancient Greek mathematicians knew that $C \div d$ is the same for all circles. Archimedes was able to prove that $\pi$ is between $3\frac{10}{71}$ and $3\frac{10}{70}$. Although we use the Greek letter $\pi$ to stand for the exact value of $C \div d$, the Greeks themselves never used $\pi$ to stand for this number. This use of the symbol $\pi$ became standard practice in mathematics thanks to Swiss mathematician and physicist Leonhard Euler (1707–1783).

It has long been known that the decimal for $\pi$ does not terminate and is not a repeating decimal. However, infinitely long decimals need not be repeating decimals for their digits to occur in *some* kind of pattern. Is there any kind of regular pattern in the decimal for $\pi$? No one knows!

**Just the facts**

Decimal numbers that never end or repeat are called *irrational numbers.* Whole numbers, fractions, and the decimals you have worked with to this point are called *rational numbers.*

No matter what size a circle is, the circumference divided by the diameter is always the same value. You probably discovered that this quotient is a little more than 3. The exact value is a decimal number whose digits never end or repeat. This value has been given the special name "pi" and is represented by the Greek letter $\pi$.

1 Discuss the history and concept of $\pi$.

$\pi = C \div d$, where $C$ is the circumference of a circle and $d$ is the diameter

Since the digits of $\pi$ never end or repeat, it is impossible to write its exact numeric value. The number 3.14 is often used as an approximation of $\pi$. You can press the $\pi$ key on your calculator to get a closer approximation.

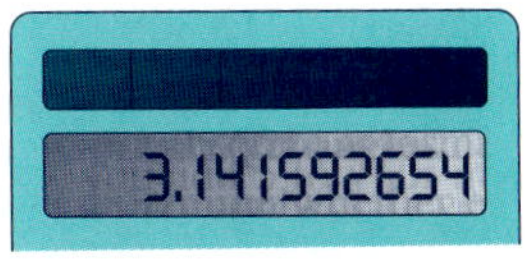


If you start with the division equation $\pi = C \div d$, you can write the related multiplication equation: $C = \pi \cdot d$. This is the formula for computing the circumference $C$ of a circle when you know its diameter $d$.

2 Discuss the formula for the circumference of a circle.

**Circumference of a Circle**

$$C = \pi \cdot d$$

In this formula, $C$ is the circumference and $d$ is the diameter. Since the diameter of a circle is twice the radius $r$, you can also write the formula in these ways:

$$C = \pi \cdot 2 \cdot r \qquad C = 2 \cdot \pi \cdot r$$

**Just the facts**

When $\pi$ is multiplied by a number or a variable, you can write the formula without using a multiplication sign:

$C = \pi d$ or $C = 2\pi r$

Since the radius of this circle is 2.5 cm, the diameter is 5 cm. So,

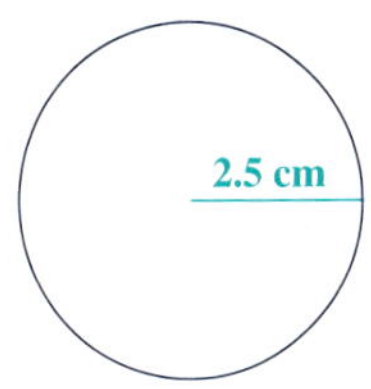


$$\begin{aligned} C &= \pi \cdot d \\ &= \pi \cdot 5 \end{aligned}$$

The exact circumference of the circle is $5 \cdot \pi$ cm. Although you can't write the circumference as an exact numeric value, you can use the $\pi$ key on your calculator to find an approximation.

$$C = \pi \cdot 5 \text{ cm} \approx 15.71 \text{ cm}$$

The symbol $\approx$ means "is approximately equal to."

### 1 Problem Set E Suggested Grouping: Individuals or Pairs

In this problem set, students use the formulas for the circumference of a circle to solve problems about circles.
2 Students are encouraged to use a calculator with a [π] key to do the calculations.

For **Problem 1,** students need only substitute 9 for $d$ in the formula $C = \pi \cdot d$ and evaluate the resulting expression.

In **Problem 2,** students use what they know about equivalent multiplication and division equations to solve $16 = \pi \cdot d$.

### 3 Share & Summarize

This Share & Summarize question is intended to help you assess how well students understand the meaning of $\pi$. You may want to add some related questions that ask students to find the circumference of a circle, given its diameter or radius.

**Troubleshooting** Students sometimes confuse the radius and diameter when they solve problems dealing with circles. Urge them to read problems and examine diagrams carefully. If students are having trouble recalling the correct meaning of the words, remind them that both words refer to segments that contain the center of a circle. A radius is shorter than a diameter, and the word *radius* is shorter than the word *diameter.*

### On Your Own Exercises

Practice & Apply: 6–9, p. 491
Connect & Extend: 12–14, pp. 492–493
Mixed Review: 15–24, p. 493

★ indicates multi-step problem

## Problem Set E

For this problem set, use your calculator's $\pi$ key to approximate $\pi$. (If your calculator does not have a $\pi$ key, use 3.14 as an approximation for $\pi$.) Give your answers to the nearest hundredth.

1. Find the circumference of a circle with diameter 9 centimeters. 28.27 cm

2. A circular pool has a circumference of about 16 meters. What is the pool's diameter? about 5.09 m

3. Find the circumference of this circle: 64.40 in.

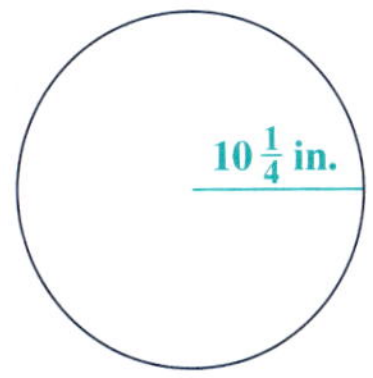


**Just the facts**

Earth is not perfectly round. The southern end bulges a bit, creating a slight pear shape. However, data collected by satellite indicate that Earth is gradually rounding itself out.

4. The radius of Earth at the equator is about 4,000 miles.

   a. Suppose you could wrap a string around Earth's equator. How long would the string have to be to reach all the way around? (Assume the equator is a perfect circle.) about 25,132.74 mi

   ★ b. Now suppose you could raise the string 1 mile above Earth's surface. How much string would you have to add to your piece from Part a to go all the way around? about 6.29 mi

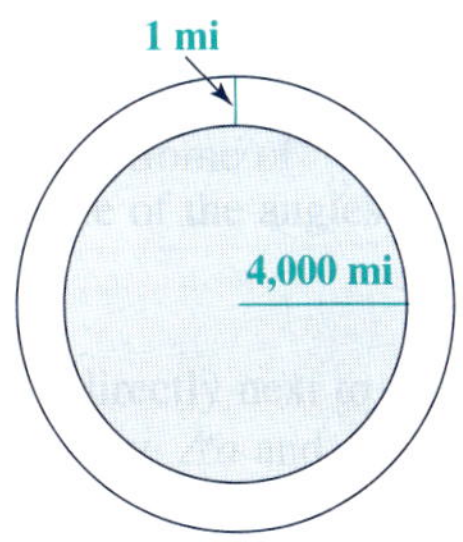


1 You may have students work individually or in pairs.

2 Have students practice using the $\pi$ key on their calculator when solving these problems.

## Share & Summarize

See Additional Answers.

Explain what $\pi$ is in your own words. Be sure to discuss

- how it is related to circles
- its approximate value

3 Discuss students' answers with the class.



### Additional Answers
### Share & Summarize

Possible answer: $\pi$ is the number you get when you divide a circle's circumference by its diameter. So, if you multiply the diameter of a circle by $\pi$, you get the circle's circumference. The value of $\pi$ is a little more than 3. The digits of $\pi$ never end or repeat, so it is impossible to give the exact value of $\pi$. My calculator gives the approximation 3.141592654.

## On Your Own Exercises

Investigation 1,
pp. 482–486
Practice & Apply: 1–5
Connect & Extend: 10, 11

Investigation 2,
pp. 486–489
Practice & Apply: 6–9
Connect & Extend: 12–14

Assign Anytime
Mixed Review: 15–24

**Exercise 2a:**
This is a good exercise for assessing how well students understand the use of polygons to estimate perimeters of figures with curved sides. All of the polygons students use for their approximations will have two or more vertices between the endpoints of the curved side of the infield. The closest approximations can be obtained by using vertices that are evenly spaced along the curved side.

**Exercise 3:**
In **Part a,** students can use string to measure the perimeter of the field. If they are using flexible plastic rulers, they can bend the ruler to measure the curved side and then add the lengths of the two straight sides.

In **Part b,** students can multiply the answer from Part a by 100 and then by 5 (for five laps) to find that the player will run slightly more than 5,560 ft. This is a distance of a little more than 1 mi.

# On Your Own Exercises

★ indicates multi-step problem

## Practice & Apply

**Sports** In Exercises 1–3, use this diagram of a baseball field.

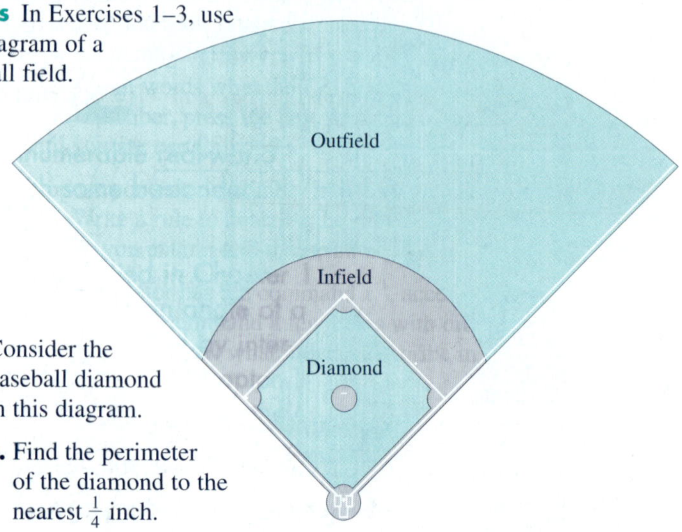


**1.** Consider the baseball diamond in this diagram.

**a.** Find the perimeter of the diamond to the nearest $\frac{1}{4}$ inch.

**b.** An actual baseball diamond is a square with sides 90 feet long. What is the perimeter of an actual baseball diamond? 360 ft

**c.** The perimeter of an actual baseball diamond is about how many times the perimeter of the baseball diamond in the diagram?

**2.** Rosita approximated the perimeter of the infield using a quadrilateral. She found a perimeter of about $5\frac{1}{8}$ inches.

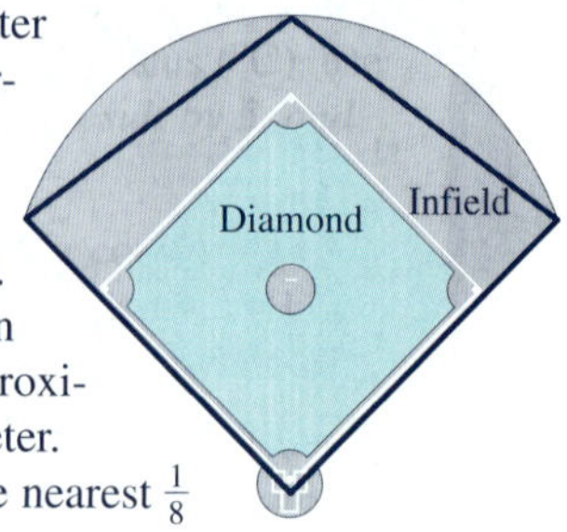


**a.** Trace the shape of the infield. Use a polygon with more than four sides to find a better approximation of the infield's perimeter. Make all measurements to the nearest $\frac{1}{8}$ of an inch.

**b.** How does your approximation compare to Rosita's?

**1a.** $3\frac{3}{4}$ in.
**1c.** about 1,152
**2a.** Possible answer: $5\frac{3}{8}$ in.
**2b.** It is more than hers, and it is closer to the actual perimeter.

**3.** Suppose the manager tells a player to run five laps around the entire baseball field (including the outfield), staying as close to the outer edge as possible.

**a.** Measure the perimeter of the field in the diagram at the top of this page to the nearest $\frac{1}{8}$ of an inch. about $11\frac{1}{8}$ in.

★ **b.** If 1 inch on the diagram represents approximately 100 feet on the actual field, about how many miles will the player run in his five laps around the field? about 1 mi

**Remember**
1 mile = 5,280 feet



impactmath.com/self_check_quiz

**Quick Review Math Handbook**
hot words hot topics
Hot Topics
pp. 320–323

490 CHAPTER 8 Geometry and Measurement

★ indicates multi-step problem

**Just the facts**

Founded in 1800 in Washington, DC, the Library of Congress is considered one of the greatest national libraries. In addition to 15,000,000 books, it houses impressive collections of manuscripts, music, prints, and maps.

**4.** This is the floor plan of the Harperstown Library. What is the perimeter of the floor? 130.6 ft

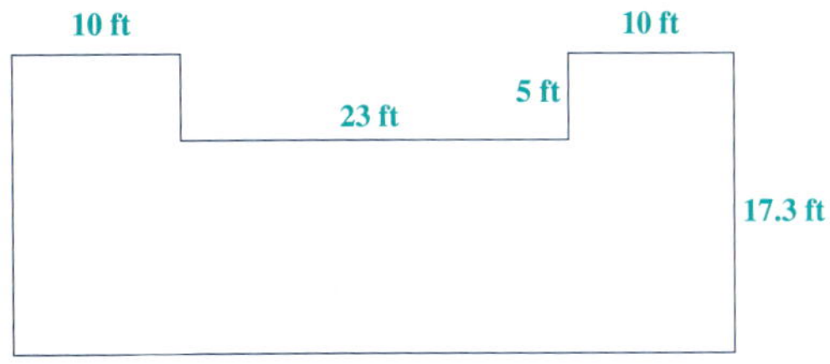


**5.** Give the dimensions of five rectangles that have a perimeter of 50 feet. See Additional Answers.

**6.** Find the circumference of a circle with diameter 7 meters. Round your answer to the nearest hundredth. 21.98 m

**7.** Find the circumference of a circle with radius 4.25 inches. Round your answer to the nearest hundredth. 26.70 in.

**8.** The circumference of a tire is 150 inches. What is the tire's radius? Round your answer to the nearest hundredth. 23.87 in.

**9. Challenge** The radius of the wheel on Jahmal's bike is 2 feet.

★ **a.** If he rides 18.9 feet, how many full turns will the wheel make?

★ **b.** If the wheel on Jahmal's bike turned 115 times, how many feet did Jahmal ride? About how many miles is this?

**9, 10.** See Additional Answers.

★ **c.** If Jahmal rides 20 miles, how many times will his wheel turn?

## Connect & Extend

**10.** Two shapes are *nested* when one is completely inside the other.

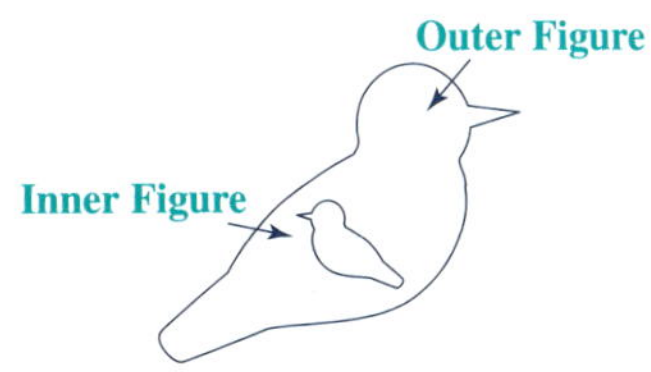


**a.** Draw two nested shapes so that the outer shape has a greater perimeter than the inner shape. Give the perimeters of both shapes.

**b.** Draw two nested shapes so that the inner shape has a greater perimeter than the outer shape. Give the perimeters of both shapes.

**c.** Draw two nested shapes so that the outer shape has the same perimeter as the inner shape. Give the perimeters of both shapes.

**d.** Look at your shapes from Parts a–c. In each case, which shape has more space inside: the inner shape or the outer shape? How do you know?



**Exercise 4:**
Students should be careful to note that there are two 5-ft sides in this figure.

**Exercises 6–9:**
Answers may vary slightly, depending on whether students are using the $\pi$ key on a calculator or are doing the calculations with pencil and paper. If they are using pencil and paper, suggest that they use 3.14 for $\pi$.

**Exercise 9:**
Ask students to show what expressions they evaluated to get their answers for each part of this exercise. This will help you check the validity of the approach they used to answer each question.

**Exercise 10:**
This exercise lends itself easily to a trial-and-error approach. It can also lead to some imaginative thinking about perimeter. The easiest figures to deal with are polygons. One highly systematic approach is to use polygons in which all intersecting sides are perpendicular. Other types of polygons are certainly possible, as the sample answers demonstrate.

### Additional Answers

**5.** Possible answer: 20 ft × 5 ft, 10 ft × 15 ft, 12.5 ft × 12.5 ft, 8 ft × 17 ft, 1 ft × 24 ft

**9a.** about $1\frac{1}{2}$

**9b.** $460\pi$ ft, or about 1,445 ft; about 0.27 mi

**9c.** about 8,403

• ***Additional Answers continued on page A667***

**Exercise 11:**
**Part c** of this exercise (and **Exercise 10d**) can be viewed as readiness-builders for Lessons 8.3 and 8.4, which deal with area.

**Exercise 12:**
Answers for this exercise may vary slightly, depending on whether students use the [π] key on a calculator or do the calculations with pencil and paper. The answer given here was obtained by using the calculator value of $\pi$.

**11. Fine Arts** The artist M. C. Escher often incorporated mathematics into his artwork. Many of his well-known works are tessellations. A *tessellation* is a design made of identical shapes that fit together without gaps or overlaps.

One way to make a shape that will tessellate is to cut a rectangle into two pieces and slide one piece to the other side.

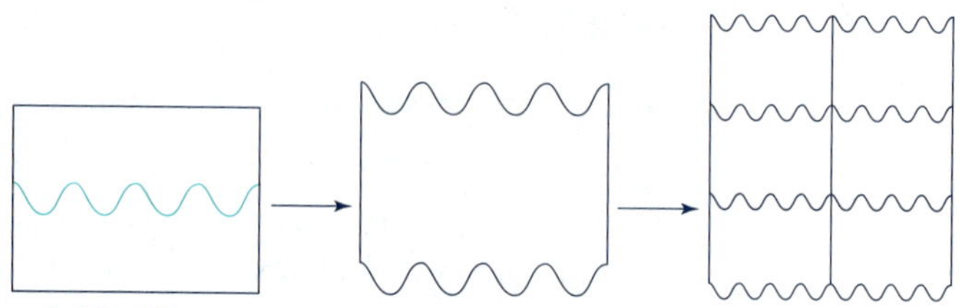


**11a.** 10 cm

**11b.** Possible answer: 10.6 cm

**11c.** Possible answer: The curvy side of the new shape is longer than the corresponding straight side of the original rectangle. The space inside is the same because everything that was cut off was added back.

**a.** Find the perimeter of the original shape above.

**b.** Trace the new shape, and estimate its perimeter by using a polygon approximation or a piece of string.

**c.** When the new shape is formed from the original, the space inside the shape—the area—stays the same, but the perimeter changes. Explain why this happens.

**12. Sports** This is a diagram of the outer lane of the track at Albright Middle School. The lane is made of two straight segments and two semicircles (half circles). If a student runs one lap around the track in this lane, how many yards will she run? about 438.76 yd

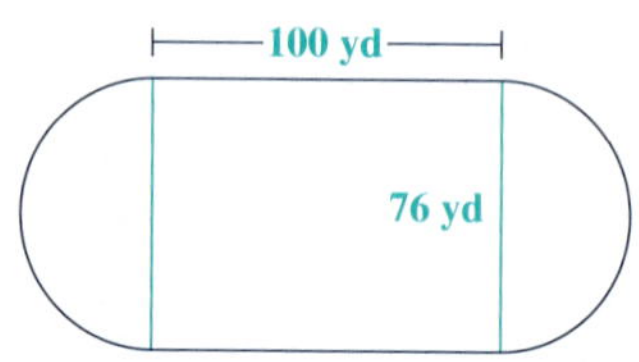




492 CHAPTER 8 Geometry and Measurement

## In your own words

Describe what perimeter is and how to find it for various shapes. Give an example of a situation in which finding a shape's perimeter would be useful.

## Mixed Review

13. Possible answer: 18.1 in. × 18 in., 30 in. × 6.1 in., 10 in. × 26.1 in.

24. See Additional Answers.

**13.** Caroline wrapped a piece of string around the circumference of a circle with a diameter of 23 inches. She cut the string to the length of the circumference and then formed a rectangle with the string. Give the approximate dimensions of three rectangles she could make. See margin.

**14.** A circle with radius 6.5 inches is cut into four wedges and rearranged to form another shape.

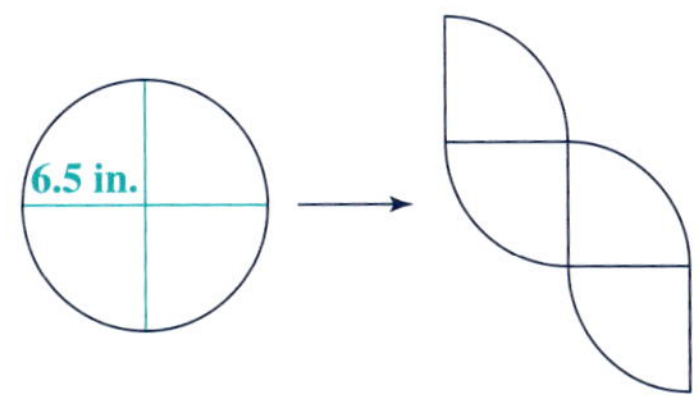


Does the perimeter change? How do you know? If it does change, by how much does it increase or decrease? See Additional Answers.

Find the greatest common factor of each pair of numbers

**15.** 14 and 21 7 **16.** 17 and 51 17 **17.** 54 and 24 6

**18.** 13 and 5 1 **19.** 100 and 75 25 **20.** 32 and 36 4

Write a rule for each relationship between $p$ and $q$.

**21.**

| $p$ | 2 | 3 | 4 | 5 | 6 | 7 |
|---|---|---|---|---|---|---|
| $q$ | 0.5 | 0.75 | 1 | 1.25 | 1.5 | 1.75 |

Possible rule: $q = 0.25p$

**22.**

| $p$ | 14 | 6 | 8.35 | 8 | 12 | 9 |
|---|---|---|---|---|---|---|
| $q$ | 9 | 1 | 3.35 | 3 | 7 | 4 |

Possible rule: $q = p - 5$

**23.**

| $p$ | 2 | $\frac{2}{3}$ | 7 | 3 | 10 | 1 |
|---|---|---|---|---|---|---|
| $q$ | 4 | 0 | 19 | 7 | 28 | 1 |

Possible rule: $q = 3p - 2$

**24. Statistics** Jing kept a record of the length (in minutes) of each phone call she made last week. Here are the results:

7 37 3 24 29 54 12 18 25

15 19 22 32 35 18 21 15 22

**a.** Make a histogram of the phone-call times. Use 10-minute intervals on the horizontal axis.

**b.** Describe the distribution of phone-call times.



## Quick Check

### Informal Assessment

Students should be able to:

- ✔ use the side lengths of a polygon to calculate the perimeter of the polygon
- ✔ use polygons to approximate the perimeter of a shape with curved sides
- ✔ understand and use formulas for the circumference of a circle

## Quick Quiz

1. Find the perimeter of each figure.

a.

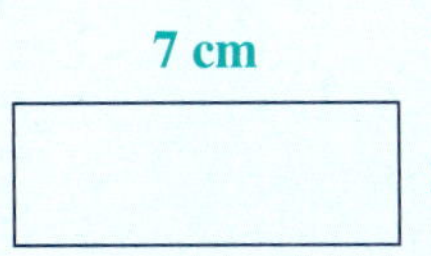


19 cm

b.

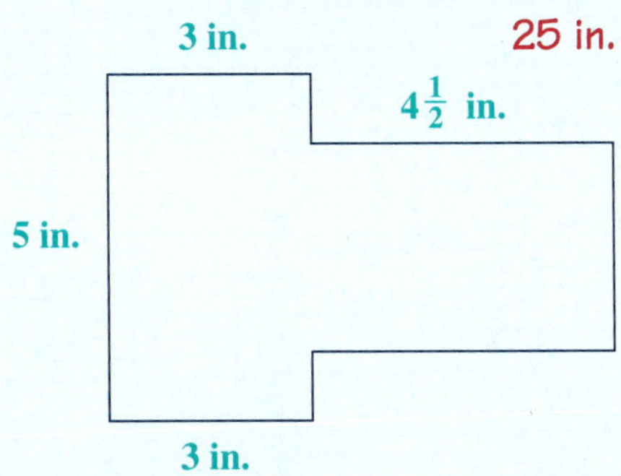


25 in.


• *Additional Answers on page A668*

• *Quick Quiz continued on page A668*

# Teacher Notes

# 8.3 Areas and Squares

## Objectives

- To find areas by counting unit squares
- To find the formula for the area of a rectangle
- To understand how squaring is used in the order of operations rules
- To evaluate expressions that involve squaring
- To find square roots of numbers that are squares of whole numbers or squares of quotients of whole numbers
- To estimate square roots of whole numbers that are not perfect squares

## Overview (pacing: about 5 class periods)

In this lesson, students examine basic area concepts. They count unit squares to find areas of simple polygons, and use their observations to arrive at the formula $A = L \times W$.

Next, students examine the concept of the square of a number and relate it to areas of squares. They examine the concept of a perfect square and see how it can be interpreted geometrically in terms of area.

Students go on to consider the squaring operation and see how squaring is incorporated with the order of operations rules. Finally, students examine the square root operation. They see that the square root operation and the squaring operation are inverses. They find square roots of numbers that are squares of whole numbers or squares of quotients of whole numbers. They also learn to find approximate square roots of whole numbers that are not perfect squares.

## Advance Preparation

For the Explore that precedes Investigation 1, students will need copies of Master 70. For Investigation 1, you will need to provide each group of students with a page from a newspaper. Be sure to select pages that have headlines, articles, photographs (or art), and advertisements. They will also need 1-centimeter (Master 84) and 1-inch (Master 85) dot paper.

## Lesson Planner

| | Summary | Materials | On Your Own Exercises | Assessment Opportunities |
|---|---|---|---|---|
| **Investigation 1** page T495 | Students use observations about areas of figures made from unit squares to arrive at a formula for the area of a rectangle. | • Master 70<br>• Scissors and metric rulers<br>*• 1-inch tiles (12 per group)<br>• Masters 84 and 85<br>• Pages from a newspaper | Practice & Apply: 1–10, p. 508<br>Connect & Extend: 44–46, p. 510<br>Mixed Review: 59–72, p. 513 | Share & Summarize, pages T498, 498<br>Troubleshooting, page T498 |
| **Investigation 2** page T498 | Students examine squares of numbers and the concept of a perfect square and relate these to areas of squares. | • Master 85<br>*• 1-inch tiles | Practice & Apply: 11–20, pp. 508–509<br>Connect & Extend: 47–50, p. 511<br>Mixed Review: 59–72, p. 513 | Share & Summarize, pages T501,501<br>Troubleshooting, page T501 |
| **Investigation 3** page T501 | Students see how the squaring operation is incorporated into the order of operations rules, and they evaluate expressions that involve squaring. | | Practice & Apply: 21–35, p. 509<br>Connect & Extend: 51–54, p. 511<br>Mixed Review: 59–72, p. 513 | Share & Summarize, pages T503, 503<br>Troubleshooting, page T503 |
| **Investigation 4** page T504 | Students find square roots of perfect squares and quotients of perfect squares, and they estimate square roots of whole numbers that are not perfect squares. | | Practice & Apply: 36–43, p. 509<br>Connect & Extend: 55–58, p. 512<br>Mixed Review: 59–72, p. 513 | Share & Summarize, pages T507, 507<br>Troubleshooting, page T507<br>Informal Assessment, page 513<br>Quick Quiz, page A668 |

* Included in Impact Mathematics Manipulative Kit

# Introduce

**1** Begin this lesson by reminding students that the perimeter of a figure tells us how far it is around a figure. Tell students that they will now use *area* to tell how much space is inside a figure.

**2** **Explore**
Have students work in pairs. Provide each pair of students with a copy of Master 70 and a pair of scissors.

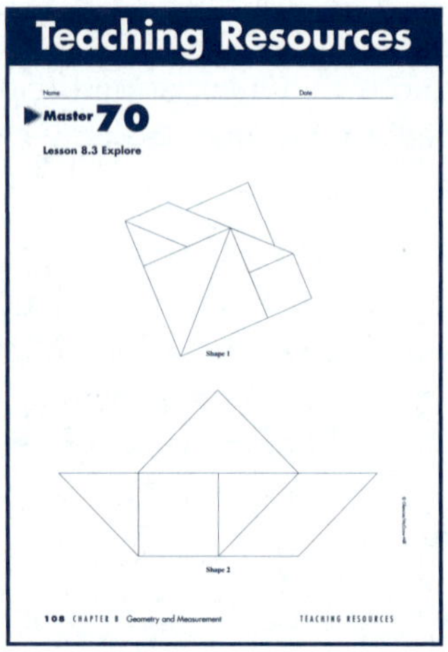


**3** Give students time to cut apart the two shapes and rearrange the pieces to make two squares. When they have done this, discuss the questions with the class. Tell the students that in this lesson they will use squares to get a numerical measure of the amount of space inside a two-dimensional figure.

# 8.3 Areas and Squares

**VOCABULARY**
**area**

**MATERIALS**
- copies of the two shapes
- scissors

You know that the perimeter of a two-dimensional shape is the distance around the shape. The **area** of a two-dimensional shape is the amount of space inside the shape.

1 Introduce the term *area*.

2 You may want to have students work in pairs for the Explore.

## Explore

Consider these shapes:

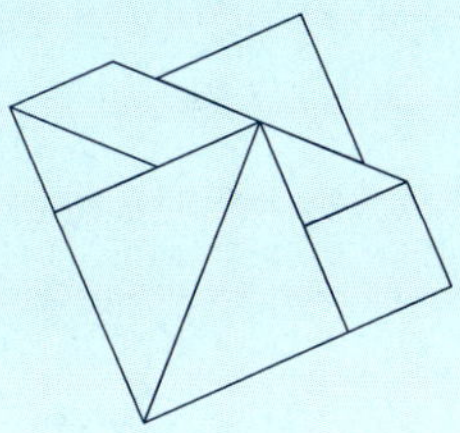

Shape 1

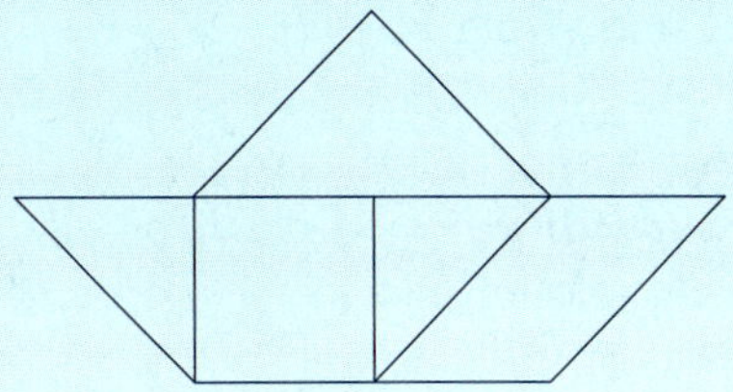

Shape 2

Which shape do you think is larger? That is, which shape do you think has the greater area? Answers will vary.

Cut out Shape 1 along the lines, and rearrange the pieces to make a square. Do the same for Shape 2. See ① in Additional Answers.

Of the two squares you made, which has the greater area? How can you tell? See ② in Additional Answers.

Do the original shapes have the same areas as the squares? Why or why not? Yes, because they are made of exactly the same pieces.

When determining which shape has the greater area, is it easier to compare the original shapes or the squares? Why? See ③ below.

3 Have students discuss their answers with the class.

**Just the facts**

*Shape 1 is made with tangram pieces. A tangram is a Chinese puzzle consisting of a square cut into five triangles, a square, and a rhombus (a quadrilateral with four equal sides) that can be put together to form various shapes.*

Squares are the basic unit used for measuring areas. In this lesson, you will look closely at areas of squares and at a special operation associated with the areas of squares.

③ Possible answer: The squares are easier to compare because they are the same shape.



**Additional Answers**

① Square for Shape 1

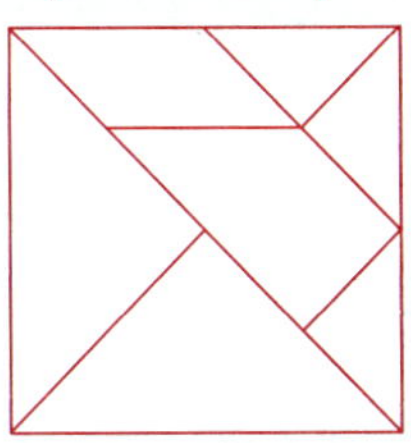

Square for Shape 2

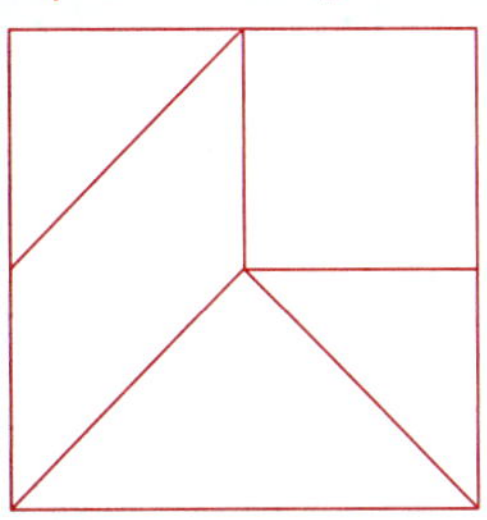

② the square for Shape 2; Possible explanations: You could put the square for Shape 1 on top of it, and it would fit inside. *Or,* the sides of the square for Shape 2 are longer.

# Investigation 1

1 In this investigation, students use tiles to explore the relationship between perimeter and area. They discover that two shapes with equal perimeters do not necessarily have equal areas. Likewise, shapes with equal areas do not necessarily have equal perimeters. Students will find areas of figures made from square tiles and areas of figures drawn on dot paper. They will use their results as evidence for the formula $A = L \times W$, which relates the area $A$ of a rectangle to the length $L$ and width $W$ of the rectangle.

2 **Problem Set A** **Suggested Grouping: Pairs**
In this problem set, students explore the relationship between area and perimeter. Each pair of students will need twelve 1-inch tiles for their work on the problems.

For **Problem 1,** three rectangles are possible and some students may mention this in their answers. Note that a $2 \times 4$ rectangle and a $4 \times 2$ rectangle are counted only once.

3 For **Problem 2,** there are only three rectangles that have an area of 12 square inches. They all have different perimeters.

For **Problem 3c,** you may wish to mention that shapes in which two squares have only a corner point in common are not allowed.

4 **Problem Set Wrap-Up** Discuss students' answers for these problems with the class. Ask for volunteers to summarize what they have learned from the problems about how area and perimeter are related. One conclusion is that knowing the perimeter of a rectangle does not let you predict its area. Likewise, knowing the area of a rectangle does not let you predict its perimeter.

# Investigation 1 Counting Square Units

**1.** Possible answer:

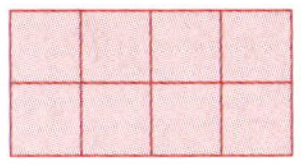

Area = 8 sq in.

Area = 5 sq in.

**2.** Possible answer:

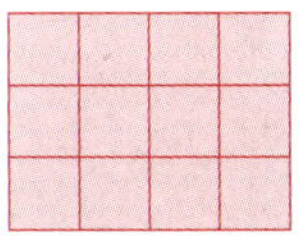

Perimeter = 14 in.

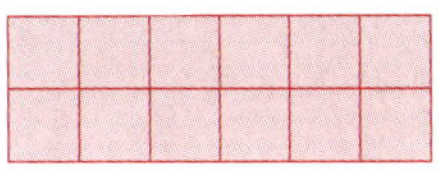

Perimeter = 16 in.

MATERIALS
1-inch tiles

**3a.** 18 in., 10 sq in.

**3b.** Possible answer: The new perimeter is 14 in. The area is the same.

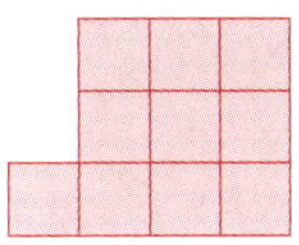

**3c.** Possible answer: The new perimeter is 22 in. The area is the same.

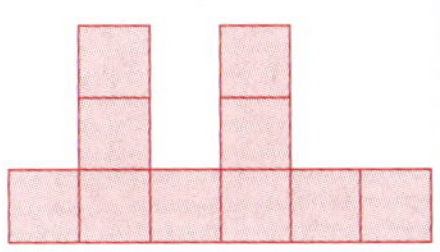

Area is measured in *square units,* such as square inches and square centimeters. A *square inch* is the area inside a square with sides 1 inch long. A *square centimeter* is the area inside a square with sides 1 centimeter long.

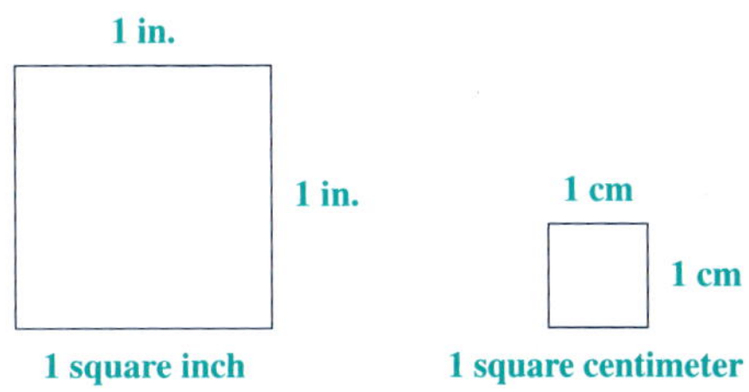


The area of a shape is the number of square units that fit inside it.

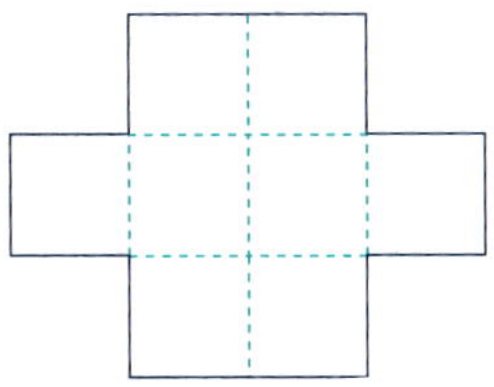

Area = 8 square centimeters

## Problem Set A

1. Use your tiles to create two rectangles with perimeters of 12 inches but different areas. Sketch your rectangles, and label them with their areas.
2. Now create two rectangles with areas of 12 square inches but different perimeters. Sketch your rectangles, and label them with their perimeters.
3. Now use your tiles to create this shape:

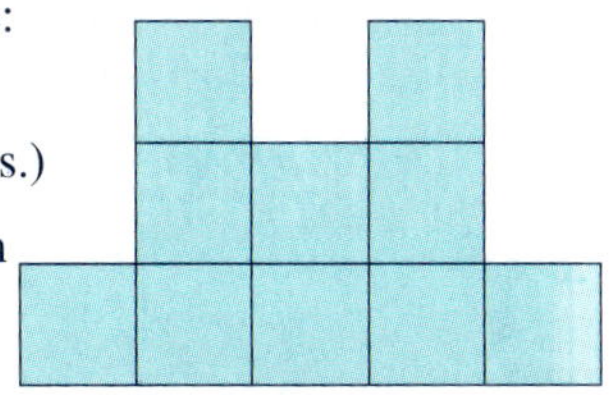

   a. Find the perimeter and area of the shape. (Don't forget to give the units.)
   b. Move one tile to create a shape with a smaller perimeter. Sketch the new shape, and give its perimeter. How does the new shape's area compare to the original area?
   c. Reconstruct the original shape. Move one tile to create a shape with a greater perimeter. Sketch the new shape, and give its perimeter. How does the new area compare to the original area?
4. Use your tiles to create two new shapes so that the shape with the smaller area has the greater perimeter. Sketch your shapes, and label them with their perimeters and areas. See Additional Answers.



**1** Introduce and differentiate the terms *square inch* and *square centimeter.*

**2** You may have students work in pairs.

**3** Make sure students understand that all three rectangles have different perimeters.

**4** Have students summarize their work on perimeter and area and state any conclusions.

### Additional Answers

**4.** Possible answer:

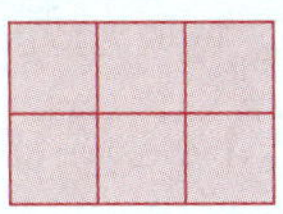

Perimeter = 10 in.
Area = 6 sq in.

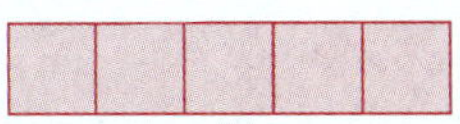

Perimeter = 12 in.
Area = 5 sq in.

LESSON 8.3 Areas and Squares 495

# Develop

1 **Problem Set B** **Suggested Grouping: Pairs**
In this problem set, students first consider areas of figures drawn on dot paper. They consider some figures that are not rectangles. In the final problems, they examine rectangles that do not lend themselves well to being drawn on dot paper. Students discover that they can find the area of any rectangle by simply multiplying the length by the width. Master 85 (1-inch dot paper) is available for students to use with this problem set.

For **Problem 2,** students can connect dots horizontally and vertically to show 6 unit squares. This allows them to think of the hexagon as being composed of 2 unit squares and 4 half-squares.

In **Problem 3,** students can think of the triangle as half of a $2 \times 4$ rectangle.

2 For **Problem 8,** students should recognize that the sides of the square must be drawn diagonally.

3 For **Problems 9–12,** students might recall models they have used to illustrate multiplication of whole numbers and to show multiplication with fractions.

Students should have little difficulty coming up with the generalization for **Problem 13.**

MATERIALS
1-inch dot paper

## Problem Set B

The shapes below are drawn on dot grids. Find the area of each shape. Consider the horizontal or vertical distance between two dots to be 1 unit.

**1.** 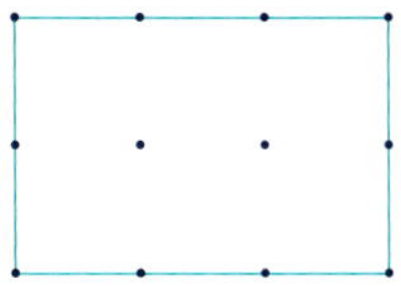 6 sq units

**2.** 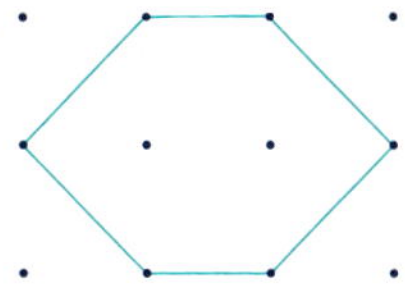 4 sq units

**3.** 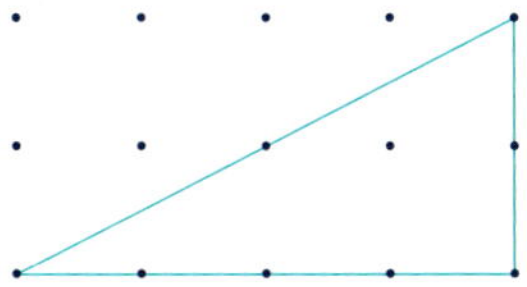 4 sq units

**4.** 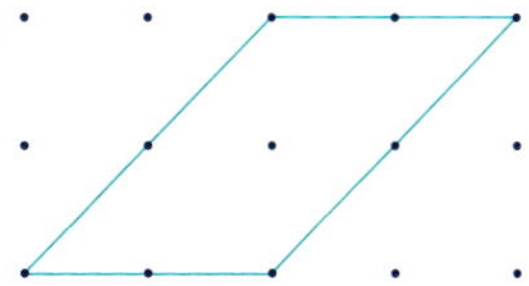 4 sq units

In Problems 5–8, draw the shape by connecting dots on a sheet of 1-inch dot paper.

**5.** a square with area 4 square inches

**6.** a rectangle with area 2 square inches

**7.** a shape with an area of at least 15 square inches and a perimeter of no more than 25 inches

**8.** **Challenge** a square with an area of 2 square inches

**5.** 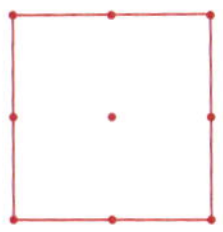

**6.** 

**7.** Possible shape:
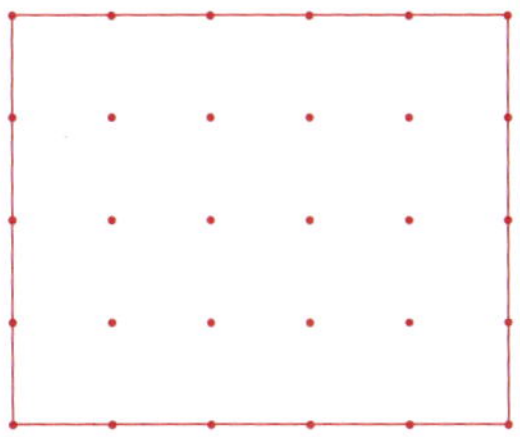

**8.** 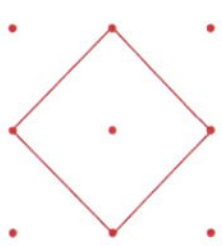

Find the area of each shape.

**9.** 7 in.  7 in.
49 sq in.

**10.** $\frac{1}{2}$ mi  $\frac{1}{2}$ mi
$\frac{1}{4}$ sq mi

**11.** 50 cm  70 cm
3,500 sq cm

**12.** 2 in. $\frac{1}{4}$ in. 
$\frac{1}{2}$ sq in.

**13.** If you know the length and width of a rectangle, how can you find the rectangle's area without counting squares?
Multiply the length by the width.

**1** You may have students work in pairs.

**2** Help students by drawing the sides of the square diagonally.

**3** Have students recall multiplication and division models of whole numbers to help them solve problems with fractions.

# Develop

1 Students should have little difficulty understanding, accepting, and applying the formula for the area of a rectangle. Emphasize that the formula works for all rectangles, regardless of the dimensions.

**Think & Discuss**
Students will need 1-centimeter dot paper (Master 84) to solve this Think & Discuss.

When you discuss this exercise, point out that the model makes it possible to picture the product. The area formula has the advantage that it can be used even when a model is not convenient.

2 **Problem Set C** **Suggested Grouping: Pairs**
In this problem set, students measure sections of a newspaper page and use the area formula for rectangles to help them find the areas used for certain types of features and for the entire page. They use their results to find the percent of the page that is used for articles, advertisements, and so on.

Each group of students will need a newspaper page that has the various elements listed in Problem 2: photographs and art, advertisements, articles, and headlines.

3 In **Problem 1,** encourage students to record their results systematically in a table or an organized list. They could, for example, number the rectangles, write the dimensions and area of each rectangle, and note what kind of item is in each rectangle.

You will probably find it a good idea to check with the groups as they get started. Students should not lump together types of features that are clearly different. Some judgment calls will be needed.

**Problem 2** requires students to use skills from Chapter 4. You may find it helpful to conduct a brief review with the whole class of how to find the percent of a number.

4 **Problem Set Wrap-Up** You might have groups put their results from Problem 2 on the board. The class can have an interesting discussion about which features predominate on each page and whether this seems to vary from one section of the newspaper to another.

Finding the area of a shape by counting squares is not always easy or convenient. Fortunately, there are shortcuts for some shapes.

To find the area of a rectangle, just multiply the length by the width.

**Area of a Rectangle**

$$A = L \cdot W$$

In this formula, $A$ represents the area of a rectangle, and $L$ and $W$ represent the length and width.

① 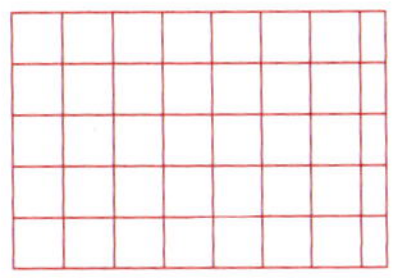

Emphasize that the formula for the area of the rectangle works for all rectangles regardless of their dimensions.

## Think & Discuss

MATERIALS
dot paper or grid paper

On dot or grid paper, draw a rectangle with side lengths 5 units and $7\frac{1}{2}$ units. See ①.

Use the formula above to find the area of your rectangle. Check that your answer is correct by counting the squares. $37\frac{1}{2}$ sq units

## Problem Set C

Answers will vary.

MATERIALS
- page from a newspaper
- metric ruler

You may have students work in pairs.

**1.** On your newspaper page, draw rectangles around the major items (photographs and art, advertisements, articles, and headlines).

**a.** Measure the sides of each rectangle to the nearest tenth of a centimeter.

**b.** Calculate the area of each rectangle.

**c.** Calculate the area of the entire page.

**2.** What percent of your newspaper page is used for

**a.** photographs and art?

**b.** advertisements?

**c.** articles?

**d.** headlines?

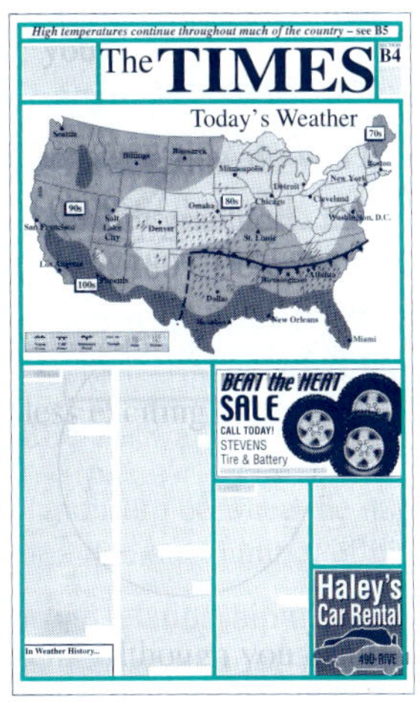


Problem-Solving Strategy
Make an organized list

Have the class combine all of the results and discuss which features predominate a newspaper page.

**Just the facts**

The first successful daily newspaper in the United States was the *Pennsylvania Packet & General Advertiser,* which was first printed on September 21, 1784.



LESSON 8.3 Areas and Squares 497

1 **Share & Summarize**
You may want to call on volunteers to read their answers to the class. Discuss the answers with the class.

2 **Troubleshooting** Many students find it difficult to remember and internalize the fact that greater area does not automatically imply greater perimeter. If students are having difficulty with this point, you can provide them with Master 84 and ask them to draw figures that illustrate how area and perimeter relationships may vary. You can model the work along the lines of Problems 1 and 2 in Problem Set A.

### On Your Own Exercises

Practice & Apply: 1–10, p. 508
Connect & Extend: 44–46, p. 510
Mixed Review: 59–72, p. 513

# Investigation 2

In this investigation, students examine the operation of squaring. They interpret the square of a number greater than zero geometrically. They also learn about numbers that are perfect squares.

3 Remind students that writing a number with an exponent means that the number is to be multiplied by itself. The exponent indicates the number of times the other number is to be used as a factor. Tell students that an expression such as $5^2$ can be read as "5 squared."

**Think & Discuss**
Students should not need much time to answer the questions in this Think & Discuss. If there is any hesitation about what square they should draw on their 1-centimeter dot paper (Master 84) ask students how they would write 5 squared as a product. Briefly discuss their answers to the questions. ■

4 Ask students to get their calculators. Have them evaluate $5^2$, $10^2$, $24^2$, and $136^2$ by using the $x^2$ key. Be sure to use the word "squared" when you read the expressions. Call students' attention to the last paragraph on page 498. You might mention that for "in.$^2$", the period is used to avoid confusion with the word "in".

**2.** No; the first shape has the greater area but the smaller perimeter.

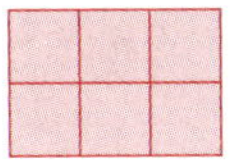

Perimeter = 10 in.
Area = 6 sq in.

Perimeter = 12 in.
Area = 5 sq in.

## Share & Summarize

**1.** Possible answer: the length of a pencil, the space on my desk

**1.** Give an example of something you would measure in inches and an example of something you would measure in square inches.

**2.** If one shape has a greater area than another, must it also have a greater perimeter? Explain or illustrate your answer.

**3.** Describe two ways to find the area of a rectangle. Count the number of unit squares inside the rectangle, or multiply the length by the width.

**1** Call on volunteers to read their answers to the class.

**2** Be sure students understand that greater area does not imply greater perimeter, and vice versa.

## Investigation 2 Squaring

**3** Review the meaning of an exponent.

Recall that an exponent tells you how many times a number is multiplied by itself. You can write the product of a number times itself using the exponent 2:

$$5 \cdot 5 = 5^2$$

Multiplying a number by itself is called *squaring* the number. The expression $5^2$ can be read "5 squared."

MATERIALS
dot paper

① $5^2 = 25$

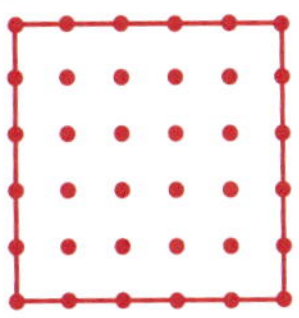

### Think & Discuss

Evaluate $5^2$. Then, on a sheet of dot paper, draw a square with an area equal to that many square units. See ①.

How long is each side of the square? 5 units

Why do you think $5^2$ is read "5 squared"? because $5^2$ is the area of a square with side length 5

You can use the [$x^2$] key on your calculator to square a number. To calculate $5^2$, press these keys:

[5] [$x^2$] [ENTER]

**4** Have students practice using the [$x^2$] key on the calculator.

The exponent 2 is often used to abbreviate square units of measurement. For example, *square centimeter* can be abbreviated $cm^2$, and *square inch* can be abbreviated $in.^2$.

# Develop

1 **Problem Set D** **Suggested Grouping: Individuals**
This problem set focuses on the geometric meaning of the square of a number.

2 In **Problem 3,** you may want to suggest that students write the number for the first blank in parentheses to avoid a result that makes it appear that only one of the numbers in the fraction is being squared.

In **Problems 5–7,** encourage students to use an exponent when they abbreviate the area units.

For **Problem 10,** students can reason that since $7^2 = 49$ and $8^2 = 64$, the answer will be a decimal between 7 and 8. Since a number is being multiplied by itself to give a product that ends in the digit 9, the number ought to be 7.3 or 7.7. Calculation shows that 7.3 is the number that works.

## Problem Set D

Fill in the blanks. Here is an example:

2 in.

2 in. Area = $\underline{2}^2$ in.$^2$ = $\underline{4}$ in.$^2$

**1.** 13 ft

13 ft Area = $\underline{13}^2$ ft$^2$ = $\underline{169}$ ft$^2$

**2.** 1.25 cm

1.25 cm Area = $\underline{1.25}^2$ cm$^2$ = $\underline{1.5625}$ cm$^2$

**3.** $\frac{7}{4}$ in.

$\frac{7}{4}$ in. Area = $\underline{\frac{7}{4}}^2$ in.$^2$ = $\underline{\frac{49}{16}}$ in.$^2$

**4.** Write a formula for finding the area $A$ of a square if you know the side length $s$. Use an exponent in your formula. $A = s^2$

Find the area of a square with the given side length.

**5.** 1 in. 1 in.$^2$ **6.** $\frac{1}{3}$ in. $\frac{1}{9}$ in.$^2$ **7.** 19 cm 361 cm$^2$

Find the side length of a square with the given area.

**8.** 144 ft$^2$ 12 ft **9.** 10,000 in.$^2$ 100 in. **10.** 53.29 cm$^2$ 7.3 cm

In the next problem set, you will look at the squares that can be made with square tiles.

**1** You may have students work individually.

**2** Suggest that students use parenthesis when squaring fractions.



LESSON 8.3 Areas and Squares 499

# Develop

1 **Problem Set E** **Suggested Grouping: Pairs**
The mathematical problem that is the focus of this problem set can be stated in a general form as follows: if $N$ is a whole number, is it possible to use unit tiles to make a square whose area is $N$ units$^2$? Students will need 1-inch tiles and 1-inch dot paper (Master 85) for these problems.

For **Problem 1,** a table is an excellent way to show all the possibilities.

For **Problem 2,** students can think about their answer for Problem 1. Since 20 square units was not among the areas listed, there is no way to use tiles to make a square with an area of 20 square units. Be sure students understand that this is *not* the same as saying that there is no square of area 20 square units, only that there is no such square made up entirely of *unit* (whole) tiles.

2 For **Problem 4,** some students may try the diagram shown in the answer and reject it. They may be deceived into thinking that the sides are 2 units long. It is true that each side joins two grid dots and has a single dot in the center, but this does not mean that the sides are 2 units long. The *vertical* and *horizontal* distances between neighboring dots represent 1 unit of length. The diagonal distance between neighboring dots is greater. ■

3 Discuss the information just below Problem Set E with the class to be sure students understand what is meant by a *perfect square.* It would be a good idea to take time to list several more perfect squares, continuing from where the list leaves off: $6^2 = 36$, $7^2 = 49$, $8^2 = 64$, and so on.

4 **Problem Set F** **Suggested Grouping: Pairs**
This problem set focuses on the concept of a perfect square. Students think about how to decide whether a number is a perfect square. They also think about whether the sums of perfect squares are perfect squares.

For **Problem 1,** students will readily see that it is easy to come up with perfect squares greater than 1,000. When you discuss this problem, you might ask where these numbers begin to occur. Since $30^2 = 900$, $31^2 = 961$, and $32^2 = 1{,}024$, students can see that the perfect squares greater than 1,000 are the squares of whole numbers greater than or equal to 32.

5 For **Problems 3–6,** discuss the strategies that students use. One strategy is to find a number whose square is less than the given number and one whose square is greater than the given number. Then try to narrow the gap. Also use the last digit of the given number to help limit the possibilities.

For **Problem 7,** one approach is to list the first few perfect squares starting with 1:1, 4, 9, 16, 25, 36, . . . .Then, scan the list for two numbers in the list that have a sum that is also in the list.

**Access for all Learners**

**Extra Challenge** There are infinitely many pairs of whole numbers that have squares whose sum is a perfect square. The smallest such pair is 3 and 4: $3^2 + 4^2 = 5^2$. Challenge students to find at least two other such pairs. **Possible answers: 5 and 12, ($5^2 + 12^2 = 13^2$); 7 and 24, ($7^2 + 24^2 = 25^2$)**

MATERIALS

- 1-inch tiles
- 1-inch dot paper

## Problem Set E

1. Find every square that can be made from 100 tiles or fewer. Give the side length and area of each square. See Additional Answers.
2. Is it possible to make a square with 20 tiles? If so, explain how. If not, explain why not.
3. Is it possible to make a square with 625 tiles? If so, explain how. If not, explain why not. Yes; make a square with side length 25.
4. **Challenge** Miguel tried to make a square with area 8 in$^2$ using tiles. After several tries, he said, "I don't think I can make this square using my tiles. But I know I can make it on dot paper."

   On dot paper, draw a square with area 8 in$^2$.
5. How can you tell whether a given number of tiles can be made into a square without actually making the square?

2. no; Possible explanation: A $4 \times 4$ square has 16 tiles, and a $5 \times 5$ square has 25 tiles. There are no whole numbers between 4 and 5, so there are no squares with between 16 and 25 tiles.

VOCABULARY

**perfect square**

A number is a **perfect square** if it is equal to a whole number multiplied by itself. In other words, a perfect square is the result of *squaring* a whole number.

| **Whole Number Squared** | $1^2$ | $2^2$ | $3^2$ | $4^4$ | $5^2$ |
|---|---|---|---|---|---|
| **Perfect Square** | 1 | 4 | 9 | 16 | 25 |

Geometrically, a perfect square is the area of a square with whole-number side lengths. In Problem Set E, the perfect squares were the numbers of tiles that could be formed into squares.

**Problem Set E Answers**

4.

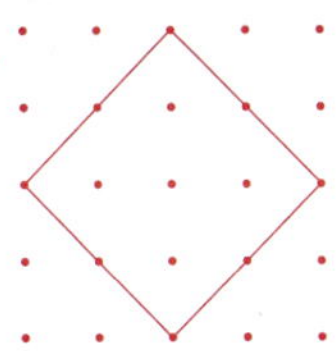

5. Possible answer: If the number of tiles is equal to a whole number times itself, you can make a square with them.

## Problem Set F

1. Find three perfect squares greater than 1,000. Possible answer: 1,024; 1,600; 5,625
2. Is 50 a perfect square? Why or why not? See Additional Answers.

Tell whether each number is a perfect square, and explain how you know.

3. 3,249
4. 9,196.81
5. 12,225
6. 184,041
7. Find two perfect squares whose sum is also a perfect square.
8. Find two perfect squares whose sum is not a perfect square.

   Possible answer: $16 + 4 = 20$, which is not a perfect square

**Problem Set F Answers**

3. yes; $3{,}249 = 57^2$
4. no; Possible explanation: You can't multiply a whole number by itself and get a decimal.
5. No; $110^2 = 12{,}100$ and $111^2 = 12{,}321$, so there is no whole number that can be squared to give 12,225.
6. yes; $184{,}041 = 429^2$
7. Possible answer: $9 + 16 = 25$



1 You may have students work in pairs.

2 Show students that diagonal distances between neighboring dots are greater than the horizontal and vertical distances.

3 Introduce the term *perfect square*.

4 You may have students work in pairs.

5 Discuss strategies that students used.

### Additional Answers

### Problem Set E

1.

| Side length (in.) | 1 | 2 | 3 | 4 | 5 | 6 | 7 | 8 | 9 | 10 |
|---|---|---|---|---|---|---|---|---|---|---|
| Area (in.$^2$) | 1 | 4 | 9 | 16 | 25 | 36 | 49 | 64 | 81 | 100 |

### Problem Set F

2. no; Possible explanation: $7^2 = 49$ and $8^2 = 64$, so you can't square a whole number and get 50.

1 **Share & Summarize**
After students have had time to answer these questions, ask for volunteers to share their answers with the class. As you discuss the answers, you may want to ask for examples that illustrate each correct answer.

**Troubleshooting** If students are having difficulty with the concept of a perfect square, have them do the following activity:

- Name two consecutive whole numbers. Suppose they name 12 and 13.
- Ask them to calculate $12^2$ and $13^2$. 144, 169
- Ask if any of the numbers between 144 and 169 can be perfect squares. Also ask how they know their answer is correct. No; 144 is a perfect square, and the very next perfect square is 169. There are no whole numbers between 12 and 13, so there are no perfect squares between 144 and 169.
- Use several more similar examples if necessary.

### On Your Own Exercises

Practice & Apply: 11–20, pp. 508–509
Connect & Extend: 47–50, p. 511
Mixed Review: 59–72, p. 513

# Investigation 3

In this investigation, students incorporate the squaring operation into the order of operations rules. They use the enlarged set of rules to evaluate expressions that involve squaring.

2 Discuss the order of operations rules shown in the box on page 501. Stress that evaluating the parts of an expression that involve exponents should come *before* doing multiplication, division, addition, and subtraction. Then go directly to the Think & Discuss.

**Think & Discuss**
When you discuss how to evaluate the two expressions, point out that the numbers that occur in the expressions are exactly the same. Go through the order of operations rules, in order, for each expression.

- Ask questions as you work on the first expression.
- Are there parentheses or fractions? *no*
- Are there exponents, and if so, where? *yes, in 11 squared*
- What is 11 squared? 121
- What multiplication do you need to do next? 2 times 121
- What is the product? 242

After you have arrived at the correct answer, point out that it would not be correct to find $2 \cdot 11$ first, because the order of operations rules require that the part with the exponent 2 be evaluated before multiplying by the factor 2.

For the second expression students should realize that the multiplication $2 \cdot 11$ should be done first because it is in parentheses. The order of operations rules require that expressions in parentheses come first. After multiplying 2 and 11 to get 22, we square the result. Students should note that the results for the two sets of calculations are quite different.

1. The square of a number is the area of a square whose side lengths are that number.

2. Yes; any number can be multiplied by itself.

## Share & Summarize

1. How is the idea of squaring a number related to the area of a square?
2. Can *any* number be squared? Why or why not?
3. Can *any* number be a perfect square? Why or why not?
No; only numbers that are the squares of whole numbers are perfect squares.

1 Discuss students' answers.

## Investigation 3 More about Squaring

Squaring is an operation, just like addition, subtraction, division, and multiplication. In Chapter 1, you learned about *order of operations*—a rule that specifies the order in which the operations in an expression should be performed. Below, the rule has been extended to include squares and other exponents.

**Order of Operations**

- Evaluate expressions inside parentheses and above and below fraction bars.
- Evaluate all exponents, including squares.
- Do multiplications and divisions from left to right.
- Do additions and subtractions from left to right.

2 Have students review the order of operations rules.

## Think & Discuss

Evaluate each expression.

$2 \cdot 11^2$ 242 $\qquad$ $(2 \cdot 11)^2$ 484

Explain how the order in which you performed the operations is different for the two expressions. For the first expression, I squared 11 and multiplied the result by 2. For the second expression, I multiplied 2 by 11 and squared the result.

## Develop

1

### Problem Set G Suggested Grouping: Pairs

This problem set gives students practice applying the order of operations rules that have been expanded to include exponents.

You will probably want to discuss Problems 1–3 with the class to be sure students are applying the order of operations rules correctly.

In **Problem 1,** watch how students evaluate $(3 + 5)^2$. Some may square 5 and add the result to 3 to get 28. Remind them that the expression in parentheses must be evaluated first.

In **Problem 2,** again check that students evaluate the expression in parentheses first. Some students may try to evaluate $(5 \cdot 3)^2$ by first squaring 3 and then multiplying by 5.

In **Problem 3,** students may be a bit unsure how to handle $(5 \cdot x)^2$. Remind them that raising a number to a power means using a factor repeatedly, as many times as the exponent indicates. So, $(5 \cdot x)^2 = (5 \cdot x) \cdot (5 \cdot x) = 5 \cdot x \cdot 5 \cdot x$. The factors can be rearranged to give $5 \cdot 5 \cdot x \cdot x$, which equals $5^2 \cdot x^2$.

For **Problem 4b,** students should see that there is no way to get an expression that makes sense by inserting parentheses on the right side of the equation. From **Problem 4a,** we know that $11^2 - 23 = 98$. So in **Problem 4b,** the object is to insert parentheses on the left side of the original equation to get an expression whose value is 98. Students can use trial and error until they find a way to do this that works.

2

### Problem Set H Suggested Grouping: Pairs

In this problem set, students play a game that involves squaring numbers. Students must decide whether the current number being used has a square that equals or exceeds 1 million. Each pair of students should have a calculator.

Do not give any hints about winning strategies. Students will probably start thinking rather soon about whether there is an easy way to ensure a win. Realizing that $1{,}000^2 = 1{,}000{,}000$ is the key.

3

Before students get started, point out that the rules do not prohibit decimals.

For **Problem 2,** since $1{,}000^2$ is equal to 1 million, 999 is the greatest whole number whose square will not equal or exceed 1 million.

1. no; Possible explanation: $(3 + 5)^2 = 64$, but $3^2 + 5^2 = 34$.

3. See Additional Answers.

**Just the facts**

The convention of using raised exponents was introduced by French mathematician, philosopher, and scientist René Descartes, who also invented the Cartesian coordinate system.

## Problem Set G

1. Does $(3 + 5)^2$ have the same value as $3^2 + 5^2$? Explain.
2. Does $(5 \cdot 3)^2$ have the same value as $5^2 \cdot 3^2$? Explain. yes; Possible explanation: Both expressions are equal to 225.
3. Is $(5 \cdot x)^2$ equivalent to $5^2 \cdot x^2$? Explain.
4. This equation is *not* true:

$$2 \times 5 + 2^2 = 11^2 - 23$$

   a. Show that the equation above is not true by finding the value of each side. $2 \times 5 + 2^2 = 14$, $11^2 - 23 = 98$

   b. **Challenge** Place one pair of parentheses in the equation to make it true. Show that your equation is true by finding the value of each side. See Additional Answers.

5. Consider the four digits of the year you were born. Write at least three expressions using these four digits and any combination of parentheses, squaring, addition, subtraction, multiplication, and division. Use each digit only once in an expression. Evaluate each expression. See Additional Answers.

In the next problem set, you will compare squaring to doubling.

## Problem Set H

In this problem set, you and a partner will play the game *Square to a Million.* The object of the game is to get a number as close to 1 million as possible, without going over, using only the operation of squaring.

Here are the rules for the game:

- Player 1 enters a number greater than 1 into a calculator.
- Starting with Player 2, players take turns choosing to continue or to end the game. In either case, the player states his or her decision and then presses $x^2$ ENTER.

  —*If the player chooses to continue the game* and the result is greater than or equal to 1 million, the player loses the round. If the result is less than 1 million, it is the other player's turn.

  —*If the player chooses to end the game* and the result is greater than or equal to 1 million, the player wins. If it is less than 1 million, the player loses.

Play six games with your partner, switching roles for each round.

1. On your turn, how did you decide whether to continue or to end the game?
2. What is the greatest whole number whose square is less than 1 million? 999

1. If the number was greater than 1,000, I ended the game because the square would be over 1 million. If it was less than 1,000, I squared it.

1 You may have students work in pairs.

2 You may have students work in pairs.

3 Have students note that the rules do not prohibit decimals.



**Additional Answers**
**Problem Set G**

3. yes; Possible explanation: $(5 \cdot x)^2 = 5 \cdot x \cdot 5 \cdot x = 5 \cdot 5 \cdot x \cdot x = 5^2 \cdot x^2$.

4b. $2 \times (5 + 2)^2 = 11^2 - 23$; $2 \times (5 + 2)^2 = 2 \times 49 = 98$ and $11^2 - 23 = 98$

5. Possible answer for 1999: $1 + 9 + 9 + 9 = 28$, $1 \times 9 \times 9 \times 9 = 729$, $9 \div 9 \div 9 \div 1 = \frac{1}{9}$, $1 \div 9 \div 9 \div 9 \approx 0.00137$, $1^2 + 9^2 + 9^2 + 9^2 = 244$

For **Problem 3,** students can use their result from Problem 2 and look for the greatest whole number whose square is less than or equal to 999. They can likewise make successive use of previous answers to answer **Problem 4** and then **Problem 5.**

1 In **Problem 6,** students should observe that if they use 1 or a number between 0 and 1, the game will never end.

**Problem Set Wrap-Up** Review student answers to **Problems 7** and **8** as a class. Have them discuss the differences between squaring and doubling numbers.

### Share & Summarize

2 You may want to have students work in pairs or groups of three on these questions. Students can have others in their group evaluate the expressions they write for **Question 1.**

3 You will find it worthwhile to discuss the table in **Question 2** with the entire class. Students may base their answers in each row on only one or two numerical examples. Use the class discussion to assess where they are able to use reasoning to support their answers. For rows 1 and 2, they should be able to see that the answers in columns 2 and 3 force the answer to be "doubling" in column 4. The last three rows are more difficult to answer by using reasoning alone. If students are able to construct a graph that shows values of $2x$ and $x^2$, they may feel more confident of the answers for column 4.

**Troubleshooting** If students are still having trouble applying the order of operations rules, you might give them a few more expressions—containing parenthesis, fractions, and exponents—to evaluate. They may need to refer back to the list on page 501.

### On Your Own Exercises

Practice & Apply: 21–35, p. 509
Connect & Extend: 51–54, p. 511
Mixed Review: 59–72, p. 513

**6.** The game would go on forever because you could never get to 1,000,000. If you started with 1, the result will always be 1, no matter how many times you pressed $x^2$. If you started with a number less than 1, the number would get smaller every time you pressed $x^2$, getting further away from 1 million.

**3.** What is the greatest whole number you could start with, press $x^2$ twice, and get a number less than 1 million? 31

**4.** What is the greatest whole number you could start with, press $x^2$ three times, and get a number less than 1 million? 5

**5.** What is the greatest whole number you could start with, press $x^2$ four times, and get a number less than 1 million? 2

**6.** What would happen if you started the game with a positive number less than or equal to 1?

**7.** Imagine you are playing the game *Double to a Million,* in which you double the number in the calculator instead of squaring it. If you start with the given number, how many times will you have to double until you produce a number greater than or equal to 1 million?

**a.** 50 15 **b.** 5 18 **c.** 1 20 **d.** 0.5 21

**8.** For each part of Problem 7, describe what would happen if you repeatedly squared the result instead of doubling it. See Additional Answers.

**1** Students should observe that the game will never end if they use 0 or 1.

## Share & Summarize

**1.** Possible answer: $4 + 2 \cdot 3^2 - (6 - 4)$; To evaluate the expression, first do the subtraction in the parentheses: $4 + 2 \cdot 3^2 - 2$. Then do the squaring: $4 + 2 \cdot 9 - 2$. Then multiply: $4 + 18 - 2$. Finally, add and subtract from left to right: $22 - 2 = 20$.

**1.** Write an expression that involves parentheses, squaring, and at least two other operations. Explain how to use order of operations to evaluate your expression.

**2.** Copy the table, and fill in the missing information. The first row has been completed for you.

| Number | **Double It** Is the result greater than, less than, or equal to the original number? | **Square It** Is the result greater than, less than, or equal to the original number? | **Which gives the greater result: squaring or doubling?** |
|---|---|---|---|
| Between 0 and 1 | greater than | less than | doubling |
| 1 | greater than | equal to | doubling |
| Between 1 and 2 | greater than | greater than | doubling |
| 2 | greater than | greater than | results are the same |
| Greater than 2 | greater than | greater than | squaring |

**2** You may have students work in pairs or in groups of 3.

**3** Discuss the table with the class.



### Additional Answers

**8a.** It would take only 2 presses.

**8b.** It would take only 4 presses.

**8c.** I would never reach 1 million; the result will always be 1.

**8d.** The result will get smaller and smaller, so it will never reach 1 million.

# Investigation 4

1 In this investigation, students study inverse operations. They relate the concept to what they already know about the relationship between addition and subtraction and between multiplication and division. They then explore the square root operation, which is the inverse of squaring. They also explore the geometric meaning of square roots.

2 Discuss the examples at the top of page 504 with the class, and use them to explain the meaning of inverse operations. Give one or two more examples that involve addition and subtraction and one or two more that involve multiplication and division.

### Think & Discuss

3 Discuss the Think & Discuss with the class. Students should have little difficulty coming up with the answers for the first four calculator screens. For the last screen, it may be necessary to ask some leading questions to help students see that the number sought must be between 5 and 6. They should then have no difficulty seeing that the answer is a decimal that ends in 5 (since the square on the screen ends in 5). ■

4 After the Think & Discuss, go over the explanation of the term *square root.* Discuss the radical sign, used to show the square root of a number. Also discuss the geometric interpretation of square root based on area.

5 Use the numbers from the Think & Discuss to present five more examples that parallel the one at the bottom of the page. Show the equations $\sqrt{100} = 10$, $\sqrt{2,500} = 50$, $\sqrt{36} = 6$, $\sqrt{0.64} = 0.8$, and $\sqrt{30.25} = 5.5$. Discuss what these equations mean in terms of squares and areas of squares.

# Investigation 4 Taking Square Roots

**VOCABULARY**
**inverse operations**

Two operations that "undo" each other are called **inverse operations.** Addition and subtraction are inverse operations.

**Add 12 to 15 to get 27. To undo the addition, subtract 12 from 27 to get 15.**

15 → (+ 12) → 27; 27 → (− 12) → 15

**Subtract 12 from 27 to get 15. To undo the subtraction, add 12 to 15 to get 27.**

Similarly, multiplication and division are inverse operations.

**Multiply 7 by 5 to get 35. To undo the multiplication, divide 35 by 5 to get 7.**

7 → (× 5) → 35; 35 → (÷ 5) → 7

**Divide 35 by 5 to get 7. To undo the division, multiply 7 by 5 to get 35.**

In this lesson, you will explore the operation that undoes squaring.

## Think & Discuss

Luke squared some numbers on his calculator. His results are shown below. In each case, find the number he started with.

10 50 6 0.8 5.5

**VOCABULARY**
**square root**

In each part of the Think & Discuss, you found the number you need to square to get a given number. The number you found is the **square root** of the original number. For example, the square root of 36 is 6.

The square root is shown using a *radical sign,* $\sqrt{\phantom{x}}$. You can think of $\sqrt{36}$ in any of these ways:

- the number you multiply by itself to get 36: $6 \cdot 6 = 36$
- the number you square to get 36: $6^2 = 36$
- the side length of a square with area 36:

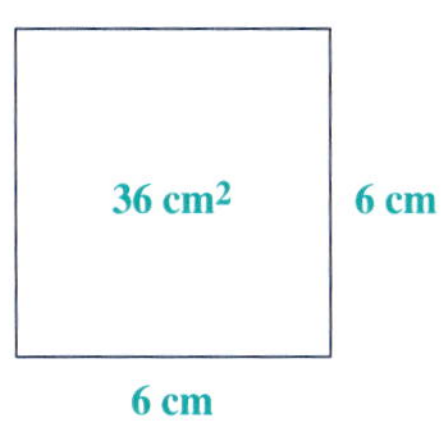


1 Introduce the term *inverse operations.*

2 Discuss the examples in terms of inverse operations.

3 Discuss students' answers.

4 Introduce the term *square root.*

5 Show the class a few more examples illustrating the square root function.

# Develop

1 Use the diagram at the top of page 505 to explain why squaring and taking the square root are considered inverse operations.

2 **Problem Set I** **Suggested Grouping: Pairs**
This problem set will help students think about the inverse operations of squaring and taking the square root. It will also help them make appropriate associations with areas of squares.

For **Problem 2,** students can recall that a square with side length 1 cm has an area of 1 $cm^2$ and that a square with side length 2 cm has an area of 4 $cm^2$. So it stands to reason that the side length will be between 1 cm and 2 cm. The students can then ask what decimal between 1 and 2 could be squared to get a product of 2.25. The final digit of 2.25 is 5, and this suggests trying 1.5. The number 1.5 does indeed work.

For **Problem 3,** if students are having difficulty, ask how they would square a fraction. This will probably be enough of a hint for most students.

For **Problem 4,** students must come up with an approximate value for $\sqrt{10}$. The sample answer suggests one possible line of thinking. Another approach is presented in the Example on page 506.

3 **Problem Set Wrap-Up** Remind students that squares which have whole-number areas, usually have decimal (and fraction) square roots. For these squares, students can only estimate the side lengths with decimal equivalents.

## Remember

Every positive number has both a positive and a negative square root. For example, the square roots of 36 are 6 and ⁻6. In this lesson, you will focus on positive square roots.

Squaring and taking the square root are inverse operations.

**Square 6 to get 36. To undo the squaring, take the square root of 36 to get 6.**

$6 \xrightarrow{\ \_\_^2\ } 36$, $36 \xrightarrow{\ \sqrt{\_\_}\ } 6$

**Take the square root of 36 to get 6. To undo taking the square root, square 6 to get 36.**

## Problem Set I

Fill in the blanks. Here is an example:

81 in.$^2$ — **Side length** $= \sqrt{81}$ **in.** $=$ 9 **in.**

**1.** 49 ft$^2$ — **Side length** $= \sqrt{49}$ **ft** $=$ 7 **ft**

**2.** 2.25 cm$^2$ — **Side length** $= \sqrt{2.25}$ **cm** $=$ 1.5 **cm**

**3.** $\frac{121}{4}$ yd$^2$ — **Side length** $= \sqrt{\_\_}$ **yd** $=$ $\frac{11}{2}$ **yd**

$\sqrt{\frac{121}{4}}$

**4.** Jing drew a square with area 10 in.$^2$ on a sheet of dot paper. She knew that the sides of the square must be $\sqrt{10}$ in. long. About how long is this? How do you know?

Area = 10 in.$^2$

Possible answer: A little more than 3 in. If you square 3, you get 9 in.$^2$, and if you square 3.5, you get 12.25 in.$^2$. 10 must be between 9 and 12.25.



**1** Describe that squaring and taking the square root are inverse operations.

**2** You may have students work in pairs.

**3** **Estimation** Remind students that whole-number areas usually have decimal and fraction square roots.

LESSON 8.3 Areas and Squares 505

When you discuss the paragraph at the top of page 506, you might remind students that fractions with whole-number numerators and denominators always have decimals that end or repeat. Since the decimals for $\sqrt{10}$ and $\sqrt{41}$ do not end or repeat, they are not equal to fractions with whole-number numerators and denominators.

1

### About the Mathematics

Decimals that neither end or repeat are called *irrational numbers.* Students have already encountered one irrational number, namely the number $\pi$. Now they have an unlimited number of irrational numbers: all the square roots of whole numbers that are not perfect squares.

The first proof of the irrationality of $\pi$ was given in 1761, but the irrationality of $\sqrt{2}$ was known to ancient Greek mathematicians. Students will learn about many more irrational numbers when they study algebra.

2

### Example

Go over the Example of estimating the decimal equivalent of a square root carefully. Some students may need more time to feel entirely comfortable with the thinking illustrated in the Example. It will probably help to call attention to the three squares in the first cartoon panel. The areas of the squares increase from left to right. Hand-in-hand with the areas, the side lengths also increase.

3

### Problem Set J Suggested Grouping: Pairs

In this problem set, students find the two whole numbers that the square root of a given nonperfect square lies between. They then find approximate decimal values for square roots of whole numbers that are not perfect squares.

4

For **Problems 1–3,** it may help students to write a list of the first 10 whole numbers and their squares. They can scan the list quickly and locate the given numbers (under the radical sign) between two consecutive squares in the list. From this, they can easily determine the two consecutive whole numbers the square root lies between.

For **Problem 4,** students may reason that since 26 is just a little greater than 25 ($5^2$) and quite a bit less than 36 ($6^2$), $\sqrt{26}$ should be just slightly more than 5. Since $5.1^2 = 26.01$, $\sqrt{26}$ is between 5.0 and 5.1. Students should be able to see that the closest approximation for the square root is 5.1.

Similar reasoning will work for **Problem 5.** Here, however, the thinking must be extended from one-place to two-place decimals.

Not all whole numbers have whole-number square roots. In fact, whole numbers that are not perfect squares have square roots that are decimals that never end or repeat. So, you can only estimate the decimal equivalents of numbers such as $\sqrt{10}$ and $\sqrt{41}$.

**Remember**

Decimal numbers that never end or repeat are called *irrational numbers.* In Lesson 8.2, you learned about the irrational number $\pi$:

$\pi = 3.14159265\ldots$

**EXAMPLE**

Luke estimated the decimal equivalent of $\sqrt{10}$.

## Problem Set J

Find the two whole numbers each given square root is between. Do not use your calculator.

1. $\sqrt{2}$ 1 and 2
2. $\sqrt{75}$ 8 and 9
3. $\sqrt{20}$ 4 and 5
4. To the nearest tenth, between which two numbers is $\sqrt{26}$? Do not use your calculator. 5.0 and 5.1
5. In the Example, Luke estimated $\sqrt{10}$ to one decimal place. Use Luke's method to estimate $\sqrt{10}$ to two decimal places. Explain each step in your work.

**5.** Possible answer: $\sqrt{10}$ is between 3.1 and 3.2, but it's closer to 3.2, so I'll try 3.17; $3.17^2 = 10.0489$, which is too large. I'll try 3.15: $3.15^2 = 9.9225$, which is too small. I'll try 3.16: $3.16^2 = 9.9856$, which is also too small, but closer than 3.15. So, $\sqrt{10}$ is between 3.16 and 3.17. Since 9.9856 is closer to 10 than 9.9225 is, $\sqrt{10}$ must be closer to 3.16.

**1** Explain that decimals that neither end nor repeat are called irrational numbers.

**2** **Estimation** Discuss the estimation Example with the class.

**3** You may have students work in pairs.

**4** **Problem-Solving Strategy** Make an organized list

# Develop

1 At this point, it is important to make sure students know how to use their calculators to find the square root of a number. If all students have the same kind of calculator, have them go through a few examples with you. Call out a perfect square, such as 576, and write the number on the board. Explain what calculator keys to use, and write the answer on the board in equation form. Read each equation aloud as you write it: $\sqrt{576} = 24$ (read "the square root of 576 is equal to 24").

Next, use some examples in which the whole number you call out is not a perfect square. Each time ask a volunteer to read the decimal displayed on the calculator screen. Ask students to round the decimal to the nearest hundredth.

2 **Problem Set K** **Suggested Grouping: Pairs**
In this problem set, students get more experience working with square roots.

**Problems 1–4** are straightforward and should present few difficulties.

In **Problem 5d,** students will only be able to speculate on why the answers in Parts c and d are different.

**Problem 7** is designed to emphasize that the squaring and square root operations are inverses.

3 **Share & Summarize**
You may want to have students work on these questions with a partner. Encourage students to illustrate their general descriptions with examples.

**Troubleshooting** If students are having difficulty with the concept of square root, you may want to provide additional practice. Have students complete tables such as the one shown below.

| $x$ | $x^2$ | $\sqrt{x^2} =$ |
|---|---|---|
| 5 | 25 | $\sqrt{25} = 5$ |
| 9 | | |
| 27 | | |
| 44 | | |
| 59 | | |
| 67 | | |
| 83 | | |

## On Your Own Exercises

Practice & Apply: 36–43, p. 509
Connect & Extend: 55–58, p. 512
Mixed Review: 59–72, p. 513

Your calculator has a command for finding square roots accessed by pressing [2nd] [$\sqrt{\ }$]. For example, to find $\sqrt{4}$, press [2nd] [$\sqrt{\ }$] 4 [ENTER].

## Problem Set K

Use your calculator to approximate each square root to the nearest hundredth. Compare your results with your answers in Problem Set J.

**1.** $\sqrt{2}$ 1.41 **2.** $\sqrt{75}$ 8.66 **3.** $\sqrt{20}$ 4.47 **4.** $\sqrt{26}$ 5.10

**5.** In this problem, you will look at the result your calculator gives for $\sqrt{1{,}000}$.

**a.** Use your calculator to approximate $\sqrt{1{,}000}$. Write down the exact result shown in the display. Do not clear the screen.

**b.** Square the number on your calculator by pressing [$x^2$] [ENTER]. What result does your calculator give? 1,000

**c.** Clear your calculator screen. Then enter your result from Part a and press [$x^2$] [ENTER]. What result does your calculator give?

**d.** Why do you think your results from Parts b and c are different?

**6.** Althea entered a number into her calculator. She then pressed [$x^2$] [ENTER] repeatedly until the calculator showed 43,046,721. What could her original number have been? List all the possibilities.

**7.** Suppose you enter a positive number less than 10 into your calculator and press [$x^2$] [ENTER] 5 times. Then you start with the final result and take the square root 5 times. What will happen? Explain why.

**5a.** 31.6227766

**5c.** 999.9999999

**5d.** Possible answer: The number shown in the display is just an approximation of $\sqrt{1{,}000}$ so its square is not equal to exactly 1,000. When I square the result without clearing the screen, the calculator squares the actual square root of 1,000, not just the digits shown in the display.

**6.** 3; 9; 81; or 6,561

**7.** You will get the original number. Each time you take the square root, you undo one of the squares.

## Share & Summarize

**1.** Describe the relationship between squaring a number and taking its square root.

**2.** Describe a method for approximating a square root *without* using a calculator. Possible answer: First find the two whole numbers that the square root is between. Then find the two tenths it is between, and so on, until your estimate is as close as you want it to be.

**1.** The operations undo each other. To find the number that was squared to make another number, take the square root.



**1** Have students practice using their square root key on their calculators.

**2** You may have students work in pairs.

**3** Encourage students to illustrate their descriptions with examples.

LESSON 8.3 Areas and Squares 507

# Assign and Assess

## On Your Own Exercises

Investigation 1, pp. 495–498
Practice & Apply: 1–10
Connect & Extend: 44–46

Investigation 2, pp. 498–501
Practice & Apply: 11–20
Connect & Extend: 47–50

Investigation 3, pp. 501–503
Practice & Apply: 21–35
Connect & Extend: 51–54

Investigation 4, pp. 504–507
Practice & Apply: 36–43
Connect & Extend: 55–58

Assign Anytime
Mixed Review: 59–72

**Exercise 1:**
Students can list all possible ways of writing 20 as a product of two whole numbers: $1 \times 20$, $2 \times 10$, $4 \times 5$ (with products such as $1 \times 20$ and $20 \times 1$ counted as the same). They can use the factors as the dimensions and calculate the perimeter of each rectangle to find the dimensions that give the greatest perimeter.

**Exercise 5:**
If students need a hint, suggest that they connect the four inside dots to show a $1 \times 1$ square, and then connect each vertex of the inside square to the nearest vertex of the large square.

# On Your Own Exercises

## Practice & Apply

**1.** See Additional Answers.

**2.**

Perimeter = 18 units

**10.** no; Possible explanation: A $10 \times 1$ rectangle has perimeter 22 and area 10. A $4 \times 4$ rectangle has perimeter 16 and area 16. So, the rectangle with the greater perimeter has the smaller area.

**15.** Possible answer: 121, 144, 169, 196, 225

**16.** See Additional Answers.

**17.** yes, $81 = 9^2$

**1.** On dot paper or grid paper, draw a rectangle with an area of 20 square units, whole-number side lengths, and the greatest possible perimeter. What is the perimeter of your rectangle?

**2.** On dot paper or grid paper, draw a rectangle with an area of 20 square units, whole-number side lengths, and the least possible perimeter. What is the perimeter of your rectangle?

These shapes are drawn on centimeter dot grids. Find the area of each shape.

**3.**

**4.**
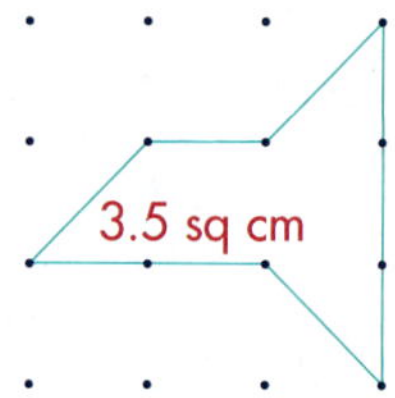


**5.**
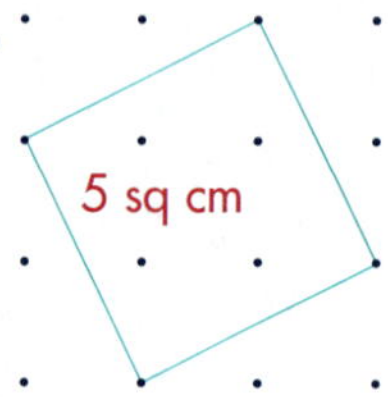


**6.** Find the area of a rectangle with length 7.5 feet and width 5.7 feet. 42.75 sq ft

**7.** Find the length of a rectangle with width 11 centimeters and the given area.

**a.** 165 square centimeters 15 cm **b.** 60.5 square centimeters 5.5 cm

**8.** Find the length of a rectangle with area 484 square inches and the given width.

**a.** 10 inches 48.4 in. **b.** 22 inches 22 in.

**9.** A square garden has area 289 square feet. How long is each side of the garden? 17 ft

**10.** If one rectangle has a greater perimeter than another, must it also have a greater area? Explain your answer.

Square each number.

**11.** 14 196 **12.** 21.5 462.25 **13.** $\frac{9}{10}$ $\frac{81}{100}$ **14.** 0.3 0.09

**15.** List five perfect squares between 100 and 500.

Tell whether each number is a perfect square, and explain how you know.

**16.** 40 **17.** 81 **18.** 125 **19.** 256

**18.** See Additional Answers.
**19.** yes, $16^2$



impactmath.com/self_check_quiz

## Additional Answers

**1.**

Perimeter = 42 units

**16.** no; Possible explanation: The factor pairs of 40 are $1 \times 40$, $2 \times 20$, $4 \times 10$, and $5 \times 8$. None of the pairs is a whole number squared.

**18.** no; Possible explanation: The factor pairs of 125 are $1 \times 125$ and $5 \times 25$. Neither of the pairs is a whole number squared.

**Just the facts**

Before the widespread use of portable calculators, slide rules were used to do complex calculations, including finding and estimating square roots.

**25–27.** See Additional Answers.

**20.** If a square has area 30.25 square feet, how long is each side? 5.5 ft

Find the value of each expression.

**21.** $5 \cdot 3^2 - 2$ 43 **22.** $2 \cdot (5^2 - 10)$ 30 **23.** $3^2 - 2^2$ 5 **24.** $7 + \frac{6^2}{3}$ 19

**25.** Does $(1 + 3)^2$ have the same value as $1^2 + 3^2$? Explain.

**26.** Does $(4 - 2)^2$ have the same value as $4 - 2^2$? Explain.

**27.** Does $(11 \cdot 7)^2$ have the same value as $11^2 \cdot 7^2$? Explain.

**28. Challenge** Place one pair of parentheses in the equation below to make it true. Show that it is true by computing the value of each side.

$$22 - (7 - 5)^2 \cdot 2 = 2 \cdot 3^2 - 4$$

**29.** Suppose you are playing *Square to a Million.* You chose the starting number 5, and your partner squared it. Now it is your turn. Should you continue or end the game? Explain. See Additional Answers.

**30.** Suppose you are playing *Square to a Million.* Your partner chose the starting number 1,001. Should you continue or end the game? Explain. End the game; 1,001 squared is 1,002,001, which is more than 1 million.

In Exercises 31–35, suppose you enter the number into a calculator and press $x^2$ three times. Without doing any calculations, tell whether the result will be *less than, greater than,* or *equal to* the original number.

**31.** 0.75 **32.** $\frac{2}{3}$ **33.** 1 **34.** 1.5 **35.** 5

**31–35.** See Additional Answers.

**36.** Luke squared a number and got 28,900. What number did he square? 170

**37.** Jing squared a number and got $\frac{121}{25}$. What number did she square? $\frac{11}{5}$

**38.** Find the side length of a square playground that has an area of 1,521 square yards. 39 yd

**39.** Without using your calculator, determine which two whole numbers $\sqrt{72}$ is between. 8 and 9

**40.** Without using your calculator, determine which two whole numbers $\sqrt{3}$ is between. 1 and 2

**41.** $\sqrt{53}$ is between 7 and 8. Without using your calculator, find five other whole numbers whose square roots are between 7 and 8.

**42.** Without using your calculator, determine whether $\sqrt{39}$ is closer to 6 or to 7. Explain how you know. **41, 42.** See Additional Answers.

**43.** Without using a calculator, approximate $\sqrt{75}$ to the nearest tenth. 8.7



**Exercises 21–24:**
If students have mistakes in these exercises, there is a good chance that either they are not observing the correct order of operations or they are using the exponent 2 as a factor. If students did not show all steps for the calculations, ask them to do so. This will help you pinpoint the source of the problem.

**Exercises 31–35:**
Watch that students are not thrown by the condition that the $x^2$ key is pressed three times. If they understand what happens (and why) when the $x^2$ key is pressed one time, they can deduce what will happen when the key is pressed two or three times.

**Quick Review Math Handbook**

hot words hot topics Hot Topics pp. 174–177

**Additional Answers**

**25.** no; Possible explanation: $(1 + 3)^2 = 16$ while $1^2 + 3^2 = 10$.

**26.** no; Possible explanation: $(4 - 2)^2 = 4$ while $4 - 2^2 = 0$.

**27.** yes; Possible explanation: $(11 \cdot 7)^2$ and $11^2 \cdot 7^2$ are both 5,929.

**29.** Continue the game; after my partner's turn, 25 is showing in the display. $25^2 = 625$, which is less than 1,000,000.

**31.** less than the original number

**32.** less than the original number

**33.** equal to the original number

**34.** greater than the original number

**35.** greater than the original number

**41.** Possible answer: 50, 55, 60, 61, 63

**42.** 6; Possible explanation: 39 is closer to $6^2$ than to $7^2$, so $\sqrt{39}$ must be closer to 6 than to 7.

# Assign and Assess

**Exercise 45:**
This exercise uses the idea that of all rectangles that have a given perimeter, the square having that perimeter has the greatest area.

**Exercise 46b:**
Some students may need a hint to get started. You might try the following questions: Can each of the whole-number areas from 2 square units to 30 square units be obtained by using a rectangle of width 1 unit? yes If that is the only rectangle possible for a certain area, what does that tell you about the factors of that area number? There are only two factors, 1 and the number itself.

## Connect & Extend

**44.** Ms. Johnson built this tile patio around a square fountain. The tiles measure 1 foot on each side. The patio is constructed of white, light green, dark green, and blue tiles.

**a.** What is the total perimeter of the patio? (Add the inner and outer perimeters.) 40 ft

**b.** What is the area of the patio? 60 sq ft

**c.** Express the portion of the patio that each color makes up as a fraction and as a percent.

**44c.** dark green: $\frac{1}{3}$, $33.\overline{3}\%$; white: $\frac{1}{3}$, $33.\overline{3}\%$; blue: $\frac{1}{6}$, $16.\overline{6}\%$; light green: $\frac{1}{6}$, $16.\overline{6}\%$

**45.** Hannah wants to build a fenced-in play area for her rabbit. She has 30 feet of fencing. Give the dimensions and area of the largest rectangular play area she can fence in. 7.5 ft × 7.5 ft, 56.25 sq ft

**Just the facts**

The length of a soccer field can vary from 100 yards to 130 yards. The width can vary from 50 yards to 100 yards. So, the least possible area is 100 · 50, or 5,000, square yards, and the greatest possible area is 130 · 100, or 13,000, square yards.

**46.** Each of these rectangles has whole-number side lengths and an area of 25 square units:

Below is the only rectangle with whole-number side lengths and an area of 5 square units:

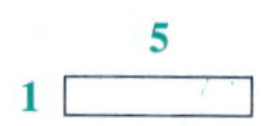


**46a.** 5 rectangles: 1 × 36, 2 × 18, 3 × 12, 4 × 9, 6 × 6

**a.** How many different rectangles are there with whole-number side lengths and an area of 36 square units? Give the dimensions of each rectangle. See above.

**b.** Consider every whole-number area from 2 square units to 30 square units. For which of these areas is there only one rectangle with whole-number side lengths? 2, 3, 5, 7, 11, 13, 17, 19, 23, 29

**46c.** They are prime numbers.

**c.** What do the areas you found in Part b have in common?

**d.** For which area from 2 square units to 30 square units can you make the greatest number of rectangles with whole-number side lengths? Give the dimensions of each rectangle you can make with this area. 24 sq units: 1 × 24, 2 × 12, 3 × 8, 4 × 6 or 30 sq units: 1 × 30, 2 × 15, 3 × 10, 5 × 6

**50b.** The areas should be 4 times the areas in Part a.

**50c.** No, the areas were multiplied by 4. Possible explanation: Since you multiply each side length by 2, and multiply the side lengths to get the area, the original area gets multiplied by 2 twice, which is the same as being multiplied by 4.

**50d.** Yes; you are still multiplying the length by 2 and the width by 2, so you are multiplying the area by 4.

**50e.** The area will be multiplied by 9.

**51–53.** See Additional Answers.

**47.** Althea squared a number, and the result was the same as the number she started with. What number might she have squared? Give all the possibilities. 0 or 1

**48.** Conor squared a number, and the result was 10 times the number he started with. What was his starting number? 10

**49.** Marcus squared a number, and the result was less than the number he started with. Give two possible numbers that Marcus might have started with. Possible answer: 0.3, 0.99

**50.** In this exercise, you will explore what happens to the area of a square when you double its side lengths.

**a.** Draw and label four squares of different sizes, and calculate their areas. Answers will vary.

**b.** For each square you drew, draw a square with sides twice as long. Calculate the areas of the four new squares.

**c.** When you doubled the side lengths of your squares, did the areas double as well? If not, how did the areas change? Why do you think this happened?

**d.** If you double the side lengths of a rectangle that is not a square, do you think the same pattern would hold? Why or why not?

**e.** If you triple the side lengths of a square, what do you think will happen to the area? Test your hypothesis on two or three squares.

**51.** Melissa receives \$1.00 a week as an allowance. Her older brother Owen gets \$1.50 each week. Her younger sister Simone receives \$0.75 each week. All three children have asked their parents for larger allowances. Their parents have given them these choices:

- Option I: Add \$0.50 to your current weekly allowance.
- Option II: Square your current weekly allowance.

Which option should each child choose? Why?

**52.** Miguel squared a number and got 390,625. Without using your calculator, find the possible ones digits of his original number. Explain.

**53.** Rosita squared a number and got 15,376. Without using your calculator, find the possible ones digits of her original number. Explain.

**54.** Caroline squared a number and got 284,089. Without using your calculator, find the possible ones digits of her original number. Explain. 3 or 7; Both numbers square to give a number that ends in 9.



**Exercise 49:**
If students have difficulty with this exercise, you may want to suggest that they look back at the table for Question 2 in the Share & Summarize on page 503.

**Exercise 50:**
Providing grid paper for this exercise will make it much easier for students to draw the squares and rectangles and make comparisons.

**Exercise 51:**
Students may find it helpful to make a table to show what each child would receive as his or her new allowance under each option.

**Additional Answers**

**51.** Melissa should choose Option I. It will give her \$1.50 a week, while squaring would not change her allowance at all. Nicholas should choose Option II. Squaring would give him \$2.25 a week, while adding \$0.50 would give him only \$2.00. Simone should choose Option I. It would give her \$1.25 a week, while squaring would decrease her allowance to about \$0.56.

**52.** 5; 5 is the only digit that can be squared to make a number that ends in 5.

**53.** 4 or 6; Both numbers square to give a number that ends in 6.

LESSON 8.3 Areas and Squares 511

**Exercises 55–57:**
In these exercises, students need to understand that the radical sign acts as an operation symbol and grouping symbol rolled into one. For example, in **Exercise 55,** instead of using parentheses and writing $\sqrt{}(36 - 25)$, we simply make the horizontal bar in the radical sign long enough to cover the entire expression (36 − 25). We have explicitly chosen not to incorporate the square root operation into the order of operations rules. It is easy to do so, and students will see just how to do so in later courses.

**Exercise 58:**
In **Parts a–e,** students can find or estimate the square root of the number on each price tag or package and compare the result to the number that is half the price in an organized list or table.

Some students may be observant enough to note that there is a connection between this exercise and Question 2 in the Share & Summarize on page 503.

**In** your **own words**

Explain how squaring and taking square roots is different for numbers greater than 1 than for numbers less than 1.

**55.** No; $\sqrt{36 - 25} = \sqrt{11}$ while $\sqrt{36} - \sqrt{25} = 1$.

**56.** Yes; both have value 30.

**57.** Yes; both have value $\frac{6}{5}$.

**55.** Does $\sqrt{36 - 25}$ have the same value as $\sqrt{36} - \sqrt{25}$? Explain.

**56.** Does $\sqrt{36 \cdot 25}$ have the same value as $\sqrt{36} \cdot \sqrt{25}$? Explain.

**57.** Does $\frac{\sqrt{36}}{\sqrt{25}}$ have the same value as $\sqrt{\frac{36}{25}}$? Explain.

**58. Economics** The school store is having a two-day sale. If you shop Thursday, the sale price of any item is the square root of the original price. If you shop Friday, the sale price of any item is half of the original price.

In Parts a–e, tell which day you should shop to get the item at the lowest price.

**a.** 

Thursday

**b.** 

Friday

**c.** 

Friday

**d.** 

Friday

**e.** 

Thursday

**f.** In general, for what prices do you save more by taking the square root? For what prices do you save more by taking half?
when the price is more than \$4.00, when the price is less than \$4.00



**512** CHAPTER 8 Geometry and Measurement

## Mixed Review

Find the prime factorization of each number.

**59.** 432 $2^4 \cdot 3^3$ **60.** 224 $2^5 \cdot 7$ **61.** 1,053 $3^4 \cdot 13$

**62.** 935 $5 \cdot 11 \cdot 17$ **63.** 198 $2 \cdot 3^2 \cdot 11$ **64.** 736 $2^5 \cdot 23$

**65.** In Art's Art Supply store, each box of colored pencils contains 12 pencils. Write a rule for the number of pencils $p$ in $b$ boxes.

**65.** Possible rule: $p = 12b$

**66.** **Economics** T. J. sells magazine subscriptions over the phone. He earns \$50 per day, plus \$1.25 for each subscription he sells. Write a rule for T. J.'s daily earnings $d$ if he sells $s$ subscriptions.

**66.** Possible rule: $d = 50 + 1.25s$

**67.** In the Spring Valley hot-air balloon race, there were 7 striped balloons entered for every 6 solid-color balloons. Write a rule for the relationship between the number of striped balloons $p$ and the number of solid-color balloons $s$.

**67.** Possible rule: $p = \frac{7}{6}s$

**68.** **Economics** A store is having a "25% off everything" sale. Write a rule for calculating the sale price $s$ of an item originally priced at $d$ dollars.

**68.** Possible rule: $s = 0.75d$

Find the measure of each angle.

**69.** 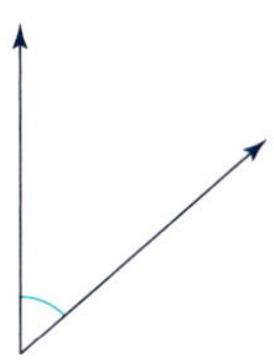 50°

**70.**  160°

**71.**  15°

**72.** 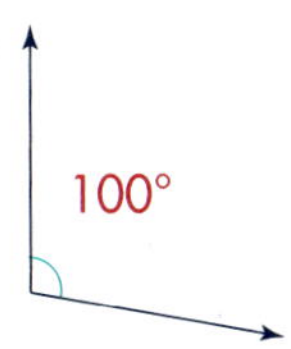 100°



### Quick Check

**Informal Assessment**

Students should be able to:

- ✔ find areas by counting unit squares
- ✔ find the formula for the area of a rectangle
- ✔ understand how squaring is used in the order of operations rules
- ✔ evaluate expressions that involve squaring
- ✔ find square roots of numbers that are squares of whole numbers or squares of quotients of whole numbers
- ✔ estimate square roots of whole numbers that are not perfect squares

• ***Quick Quiz on page A668***

# Teacher Notes

# 8.4 Calculating Areas

## Objectives

- To derive and apply a formula for the area of a parallelogram
- To derive and apply a formula for the area of a triangle
- To estimate areas of circles drawn on grids and derive a formula for the area of a circle

## Overview (pacing: about 4 class periods)

The lesson begins by having students draw an irregular figure (an outline of their own hand) on grid paper. Students estimate the area of the figure by counting grid squares. Students learn that formulas can be developed to find areas of many geometric figures. They cut apart parallelograms and reassemble the parts to make rectangles. This allows them to develop a formula for the area of a parallelogram. Students see that two identical triangles can be put together to form a parallelogram. This leads to a formula for the area of a triangle. Finally, students estimate areas of circles drawn on grids and use their results to arrive at a formula for the area of a circle.

## Advance Preparation

Masters have been provided that reproduce figures that students will need for the investigations. For the Explore that precedes Investigation 1, each student will need one copy of Masters 82 and 83. Each pair of students will need a copy of Masters 70, 71, 72, and 73. For the On Your Own Exercises, students should have 1-centimeter grid paper (Master 81), 1-centimeter dot paper (Master 84), a metric ruler, and a protractor.

## Lesson Planner

| | Summary | Materials | On Your Own Exercises | Assessment Opportunities |
|---|---|---|---|---|
| **Investigation 1** page T515 | Students develop a formula for the area of a parallelogram by using the fact that every parallelogram can be cut into two pieces and reassembled into a triangle. | • Master 82<br>• Master 83<br>• Master 81<br>• Master 84<br>• Master 71<br>• Metric rulers and protractors<br>• Scissors and tape | Practice & Apply: 1–5, p. 528<br>Connect & Extend: 15–18, pp. 530–531<br>Mixed Review: 25–36, p. 535 | On the Spot Assessment, page T517<br>Share & Summarize, pages T517, 517<br>Troubleshooting, page T517 |
| **Investigation 2** page T518 | Students develop a formula for the area of a triangle by using the fact that two identical triangles can be put together to form a parallelogram. | • Master 72<br>• Master 73<br>• Scissors<br>• Tape<br>• Protractors<br>• Metric rulers<br>• Master 84 | Practice & Apply: 6–10, pp. 528–529<br>Connect & Extend: 19–22, pp. 531–533<br>Mixed Review: 25–36, p. 535 | On the Spot Assessment, page T520<br>Share & Summarize, pages T521, 521<br>Troubleshooting, page T521 |
| **Investigation 3** page T522 | Students estimate areas of circles drawn on grids and use their results to arrive at the formula for the area of a circle. | | Practice & Apply: 11–14, p. 530<br>Connect & Extend: 23, 24, p. 534<br>Mixed Review: 25–36, p. 535 | Share & Summarize, pages T524, 524<br>Troubleshooting, page T524<br>Informal Assessment, page 535<br>Quick Quiz, page A669 |
| **Lab Investigation** page T525 | Students use spreadsheet software to discover that a square is the rectangle of greatest area for a given perimeter and that a circle with the same perimeter has an even greater area. | • Computers with spreadsheet software (1 per group) | | |

# Introduce

Remind students that the area of a two-dimensional shape is the number of square units that fit inside it. Tell them that in this lesson they will see how to estimate areas of irregular shapes with curved sides. Explain that they will then find formulas for areas of several geometric shapes that are not squares or rectangles. Then move on to the Explore.

## Explore

Conduct this Explore with the entire class. Each student will need a sheet of 1-inch grid paper (Master 83) and $\frac{1}{2}$-inch grid paper (Master 82). You may want to trace your own hand on a transparency of Master 83. You can project the transparency to discuss with the class how to get started on the Explore.

Tell students that they should draw a straight line at the bottom of the hand to show where the hand stops. Tell them that they will be counting the grid squares inside the tracing.

1 Students will easily see that if they count only whole squares entirely inside the hand outline, they will probably get a rather rough estimate of the area. To get a better estimate, they will need to use fractions for squares that are partially inside the outline.

After students have had time to estimate the area of their hands by using both the 1-inch and $\frac{1}{2}$-inch grids, discuss the results. Be sure students understand why the estimate with $\frac{1}{2}$-inch grid paper requires that the number of squares be multiplied by $\frac{1}{4}$ to get the area in square inches.

2 Discuss students' findings on both the 1-inch and $\frac{1}{2}$-inch grid paper. Ask them which grid was easier to use to estimate the area of their hand and why.

### Access for all Learners

**Early Finishers** If you have some students who are up to the challenge, have them trace the same hand they used earlier but with the fingers spread out. Have them count the grid squares inside the tracing and compare the area of their hand with the area they found previously. Discuss the following questions.

- Which area was harder to find (count)? Why?
- Are the two areas the same? Why or why not?
- What factors lead to the difference in areas?

When you have finished discussing the Explore activity, have students read the paragraph at the bottom of page 514. Then go to Investigation 1.

# 8.4 Calculating Areas

In the last lesson, you learned that the area of a shape is the number of square units that fit inside it.

MATERIALS
- 1-inch grid paper
- $\frac{1}{2}$-inch grid paper

## Explore

Place one hand on a sheet of 1-inch grid paper, with your fingers held together. Trace around your hand.

- Estimate the area of your hand tracing in square inches by counting grid squares. Estimates will vary.

Now trace your hand onto a sheet of $\frac{1}{2}$-inch grid paper.

- Estimate the number of squares inside the tracing. Estimates will vary.
- On $\frac{1}{2}$-inch grid paper, each small square has side length $\frac{1}{2}$ inch. What is the area of each small square in square inches? $\frac{1}{4}$ in.$^2$
- Use the previous two answers to estimate the area of your hand in square inches. Estimates will vary.

Which estimate do you think is more accurate: the estimate based on the 1-inch grid, or the estimate based on the $\frac{1}{2}$-inch grid? Why? See ①.

**1 Estimation** Encourage students to count fractions of squares when estimating the area of their hands.

**2** Discuss students' results.

**Just the facts**

The area of the palm of your hand is about 1% of the area of your skin. Doctors use this approximation to estimate the percent of a person's skin that is affected by a burn or other problem. It is known as the "rule of palms."

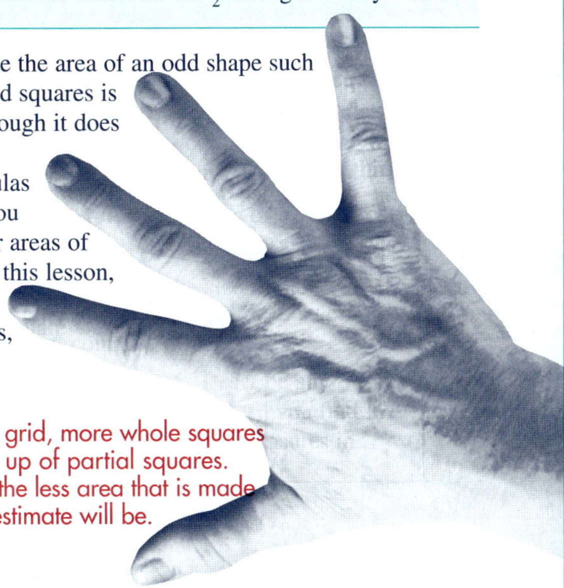

When you want to estimate the area of an odd shape such as your hand, counting grid squares is a fairly good method, although it does take time. For many other shapes, you can use formulas to find the area quickly. You already know formulas for areas of squares and rectangles. In this lesson, you will explore formulas for areas of parallelograms, triangles, and circles.

① Possible answer: The $\frac{1}{2}$-inch grid estimate; with this grid, more whole squares fit inside the tracing, and less of the area is made up of partial squares. With partial squares, you have to approximate, so the less area that is made up of partial squares, the more accurate the area estimate will be.

# Investigation 1

In this investigation, students learn what a parallelogram is. They find the areas of parallelograms drawn on grids. Next, they see that every parallelogram can be cut into two pieces and reassembled to form a rectangle whose length and width are the same as the base and height of the parallelogram. They see how this leads to a formula for the area of a parallelogram.

1 Have students read the opening paragraph in the investigation. Discuss the definition of the term *parallelogram*. Students should find it easy to accept the fact that if the opposite sides of a quadrilateral are parallel, the opposite sides have the same length. This means that if the opposite sides are parallel, the quadrilateral is a parallelogram.

2 Use the discussion to point out that rectangles are special kinds of parallelograms. However, not all parallelograms are rectangles. In a rectangle, all the angles are right angles. Have students look at the parallelograms shown in the opening paragraph. Ask which of these is a rectangle. **the third parallelogram** Ask why the other parallelograms are not rectangles. **They slant; their angles are not right angles.**

3 **Problem Set A** **Suggested Grouping: Pairs**
In this problem set, students find areas of parallelograms drawn on a 1-centimeter grid. Tell students that all the squares measure 1 cm on each side and that all the areas will be in square centimeters. The vertices of all three parallelograms are at points where grid lines intersect, so it is possible to find the exact area of each parallelogram. You may want to distribute Master 71, which reproduces the parallelograms from Problem 1.

**Teaching Resources**

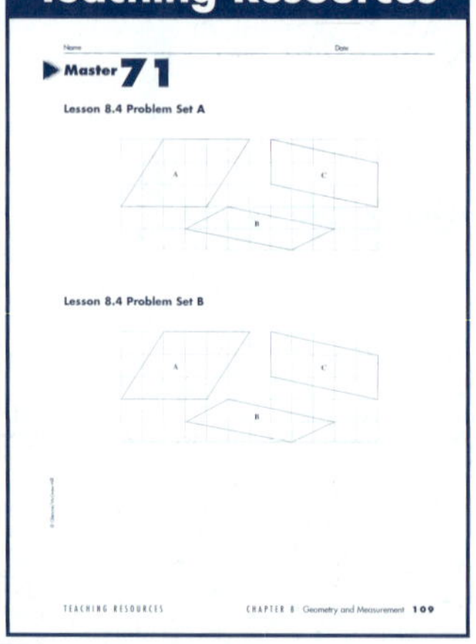


**Problem-Solving Strategies** In **Problem 1,** there are different ways to find the area of Parallelogram A.

- This diagram shows one approach. On the left, the triangle with gray shading is half of a $3 \times 2$ rectangle. So its area is $\frac{1}{2} \cdot 6$, or 3 $cm^2$. Likewise, the triangle on the right with gray shading has an area of 3 $cm^2$. The part with green shading is a rectangle made up of 6 squares. The total area of the shaded region is $3 + 6 + 3$, or 12 $cm^2$.

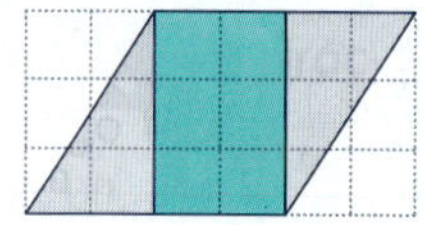

- Another approach is to imagine cutting off the gray triangle on the left and moving it to the right side to obtain the figure shown here. Pieces of the original parallelogram have been rearranged to form a rectangle. None of the original area has been lost. So, the area of the parallelogram is $4 \times 3$, or 12 $cm^2$.

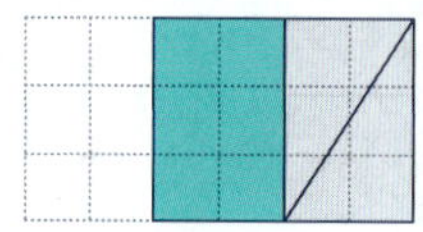

4 Discuss the method students used to find the area of Parallelogram B. One method is to think of the top part of the parallelogram as two triangles, as shown below.

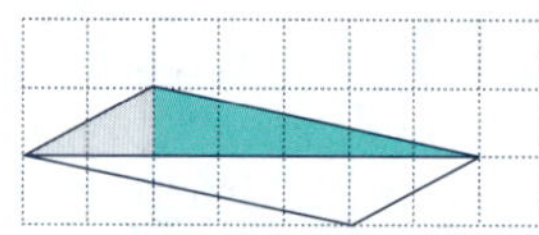

The small triangle, with gray shading, is half of a $1 \times 2$ rectangle, so its area is 1 $cm^2$. The large triangle, with green shading, is half of a $1 \times 5$ rectangle, so its area is 2.5 $cm^2$. The bottom part of the parallelogram can be treated the same way. The area of the parallelogram is therefore $1 + 2.5 + 2.5 + 1$, or 7 $cm^2$.

The area of Parallelogram C can be found by using the method that was used for Parallelogram A.

**Problem Set Wrap-Up** Discuss the results for **Problem 3.** Ask for which type of parallelogram it is possible to find the area by multiplying the lengths of two sides. **rectangles** ■

5 Discuss the vocabulary terms introduced in the paragraph just below Problem Set A. Students often get the idea that the term *base* can be used only if the side in question is in a horizontal position. This is not the case.

# Investigation 1 Areas of Parallelograms

VOCABULARY
**parallelogram**

A **parallelogram** is a quadrilateral with opposite sides that are the same length. The term *parallelogram* refers to the fact that the opposite sides are *parallel*—that is, they never meet no matter how far they are extended.

1 Introduce the term *parallelogram.*

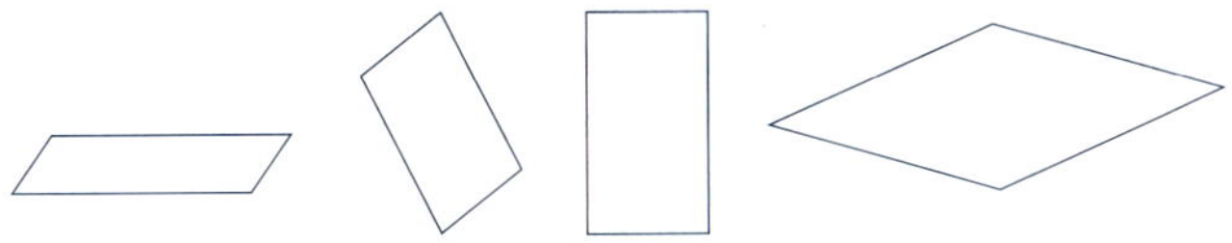

2 Show that rectangles are special kinds of parallelograms.

In this investigation, you will use what you know about finding areas of rectangles to develop a formula for the area of a parallelogram.

MATERIALS
- copies of the parallelograms
- metric ruler

## Problem Set A

3 You may have students work in pairs.

1. Find the area of each parallelogram below, and explain the method you used.

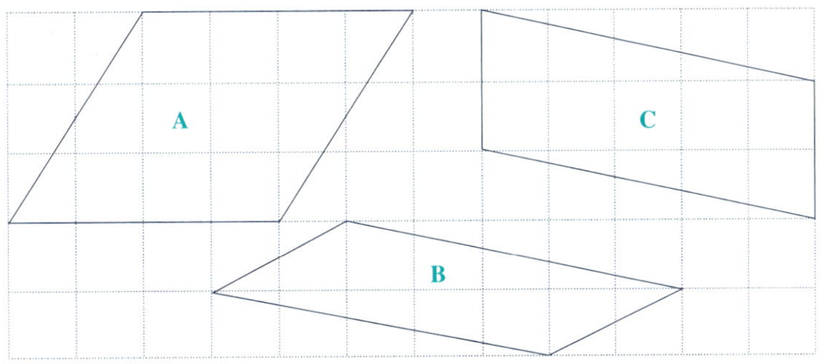


4 Discuss students' methods for finding the area of Parallelogram B.

2. Measure the lengths of the sides of each parallelogram to the nearest tenth of a centimeter.
3. Is the area of a parallelogram equal to the product of the lengths of its sides? no

**1.** Parallelogram A: 12 $cm^2$; Parallelogram B: 7 $cm^2$; Parallelogram C: 10 $cm^2$; See the teaching notes for possible strategies.

**2.** Parallelogram A: 4.0 cm and 3.6 cm; Parallelogram B: 5.1 cm and 2.2 cm; Parallelogram C: 5.1 cm and 2.0 cm

VOCABULARY
**base of a parallelogram**
**height of a parallelogram**

The **base of a parallelogram** can be any of its sides. The **height of a parallelogram** is the distance from the side opposite the base to the base. The height is always measured along a segment perpendicular to the base (or to the line containing the base).

5 Discuss the two different parts of a parallelogram.

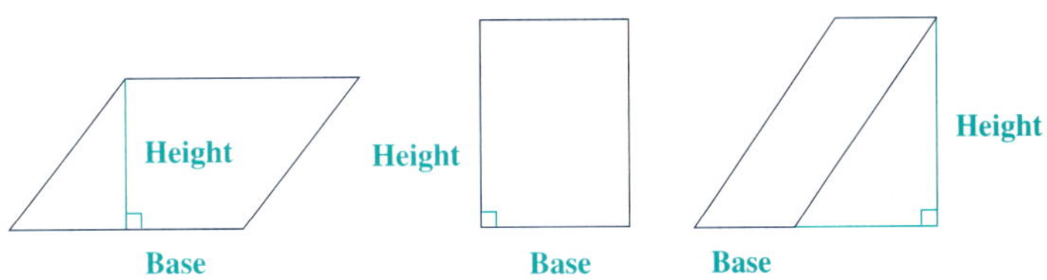




LESSON 8.4 Calculating Areas 515

1 **Problem Set B** **Suggested Grouping: Pairs**
In this problem set, students find the area of each parallelogram from Problem Set A (as reproduced on Master 71) by using a special method. Inside the parallelogram, they draw a segment whose length is a height of the parallelogram. They cut out the parallelogram and separate it into two pieces by cutting along the segment they have just drawn. They reassemble the pieces to form a rectangle. Students observe that the area of the original parallelogram equals the area of the rectangle. Also, the base and height of the parallelogram are equal to the length and width of the rectangle. Using the formula for the area of a rectangle, students can deduce that the area of the parallelogram is the product of the length of the base and the height.

2 You will probably need to allow about 15 minutes for this problem set. You should check on the students' work as they do Problem 1. You may want to discuss the results of Problems 1–3 with the class before students work on Problems 4 and 5.

In **Problem 1a,** the segment perpendicular to the base will correspond to the height of the parallelogram. In **Problem 1c,** students will cut along this segment and separate the parallelogram into the two pieces that will form a rectangle. Getting this rectangle requires that the segment drawn in Problem 1a lies completely inside the parallelogram. For Parallelograms A and C, this is quite easy. (Students can use segments that are parts of grid lines.) For parallelogram B, students must use one of the long sides of the parallelogram as the base. Otherwise, the segment for the height will not lie entirely inside the parallelogram. To get accurate results, students should use their protractors to draw the segment for the height.

In **Problem 1c,** students cut out each parallelogram, cut along the heights, and reassemble the pieces of the parallelogram to make a rectangle. They can use a piece of tape to hold the pieces together. (Note: Although the length and width values should match the base and height values, their order may be interchanged. In the given answers, for example base equals length and height equals length for Parallelogram C.)

For **Problem 2,** the rectangles students made for Problem 1 should make it clear that the base and height of each parallelogram are the same as the length and width of the corresponding rectangle. It is important that students understand this.

3 In **Problem 3,** students should recall that the area of each parallelogram is the amount of space *inside* the parallelogram. None of this space is lost in making the rectangles. Problem 3 does not ask the student to state the area of either figure, only to recognize that reassembling the cut-apart parallelogram does not change the total area.

**Problem 4** asks students to synthesize the results of Problems 1–3. The result will tell them how to find the area of any parallelogram if they know the length of the base and the corresponding height.

In **Problem 5,** students use the method they described in Problem 4 to find the area of a parallelogram. Suggest that students measure the lengths and heights in centimeters.

In Problem Set B, you will explore how the base and height of a parallelogram are related to its area.

MATERIALS
- copies of the parallelograms
- metric ruler
- scissors
- tape
- protractor

## Problem Set B

1. Complete Parts a–c for each parallelogram in Problem Set A.

   a. Choose a side of the parallelogram as the base. Draw a segment perpendicular to the base that extends to the side opposite the base. The segment should be completely inside the parallelogram. For Parallelogram A, you might draw the segment shown in green. See students' work.

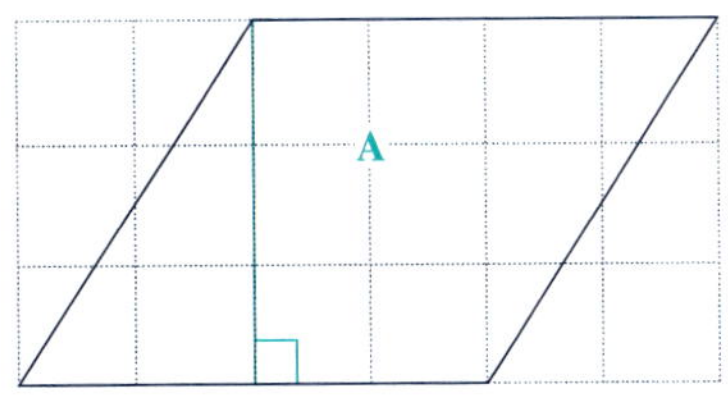


   b. Find the lengths of the base and the height. (The height is the length of the segment you drew in Part a.) Record these measurements in a table like the one below.

| | Parallelogram | | Rectangle | |
|---|---|---|---|---|
| | Base | Height | Length | Width |
| A | 4 cm | 3 cm | 4 cm | 3 cm |
| B | 5.1 cm | 1.4 cm | 5.1 cm | 1.4 cm |
| C | 2 cm | 5 cm | 5 cm | 2 cm |

   c. Divide the parallelogram into two pieces by cutting along the segment you drew in Part a. Then reassemble the pieces to form a rectangle. Record the length and width of the rectangle in your table.

2. How do the base and height of each parallelogram compare with the length and width of the rectangle formed from the parallelogram?

3. How does the area of each parallelogram compare with the area of the rectangle formed from the parallelogram? The areas are equal.

4. How can you find the area of a parallelogram if you know the length of a base and the corresponding height? Use what you have discovered in this problem set to explain why your method works.

5. Find the area of this parallelogram without forming it into a rectangle. Explain each step of your work.
   See Additional Answers.

**1b.** Possible answer: See the table. (Note: For each parallelogram, there are two possible sides that can be chosen as the base, so students may not get the values shown.)

**1c.** Possible answer: See the table. (Note: The length and width values should match the base and height values.)

**2.** The length and width are the same as the base and height.

**4.** See Additional Answers.

**1** You may have students work in pairs.

**2** Discuss the results of Problems 1–3 before going on to Problems 4 and 5.

**3** Remind students that reassembling the parallelogram into a rectangle does not change the total area.



### Additional Answers

**4.** Multiply the length of the base by the height. This works because you can form the parallelogram into a rectangle with length and width equal to the base and height of the parallelogram. The area of the rectangle is the length times the width, which is the same as the base times the height. Since the parallelogram's area is equal to the rectangle's area, the area of the parallelogram is also equal to the base times the height.

**5.** about 6 $cm^2$; Possible explanation: I chose a side to be the base and then used a protractor to help draw a segment from the opposite side perpendicular to the base. I measured the base and the height and multiplied these two measurements to get the area.

# Develop

1 **Problem Set Wrap-Up** Use the information at the top of page 517 as a wrap-up for Problem Set B. You may want to draw a parallelogram on the board when you discuss the area formula for parallelograms.

2 **Problem Set C** **Suggested Grouping: Pairs**
In this problem set, students calculate the areas of parallelograms using the formula for the area of a parallelogram. They also find the length of the base of a parallelogram, given the area and height of the parallelogram.

In **Problems 1–3,** students are given the side lengths and height of a parallelogram. The segment that is drawn perpendicular to a side of the parallelogram gives the height of the parallelogram and indicates which side is used as the base.

For **Problem 4,** students can find the value of $b$ by solving the equation $b \cdot 2.45 = 12.93$. Be sure they understand why this is the equation to use rather than $b \cdot 2.98 = 12.93$. To solve the equation, students can use related multiplication and division equations.

### On the Spot Assessment

Some students may multiply the wrong measures to find the area of the parallelograms in Problems 1–3. Use questions to help them understand which segment in each diagram indicates the height of the parallelogram. Ask them to show you the sides to which this segment is perpendicular. Then ask them the length of each of these sides. Finally, have them multiply this length by the height to find the area.

3 **Share & Summarize**
Call on volunteers to read their answers to the class. Discuss the answers. Pay special attention to whether the answers indicate a clear understanding of what is meant by the base and height of a parallelogram. You may want to put some diagrams on the board that students can refer to during the discussion.

**Troubleshooting** If students seem to have trouble determining how to identify the base and height of some parallelograms, you may want to give them a few more parallelograms to measure, and have them check their findings with a partner.

### On Your Own Exercises

Practice & Apply: 1–5, p. 528
Connect & Extend: 15–18, pp. 530–531
Mixed Review: 25–36, p. 535

You can find the area of a parallelogram by multiplying the length of the base by the height. This can be stated using a formula.

**Area of a Parallelogram**

$$A = b \cdot h$$

In this formula, $A$ represents the area, $b$ represents the base, and $h$ represents the height.

1 Review the formula for the area of a parallelogram with the class.

## Problem Set C

2 You may have students work in pairs.

Find the area of each parallelogram to the nearest hundredth of a square unit.

**1.** 1.32 in.$^2$

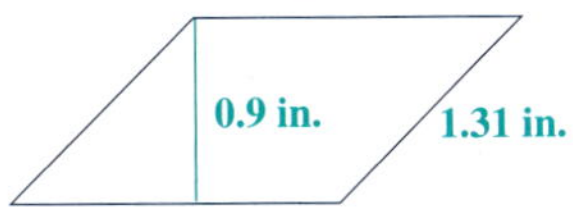


**2.** 8.20 cm$^2$

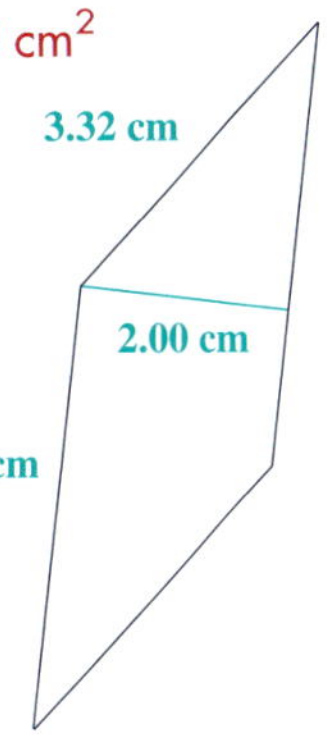


**3.** 5.55 cm$^2$

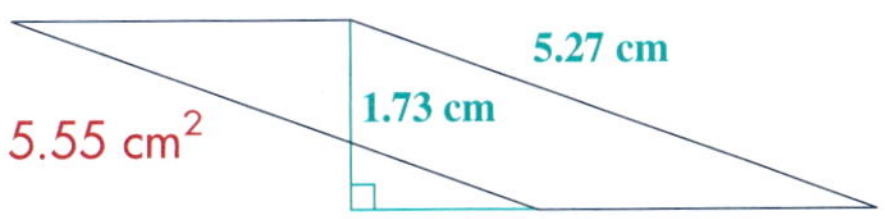


**4.** The area of the parallelogram below is 12.93 cm$^2$. Find the value of $b$ to the nearest hundredth. 5.28 cm

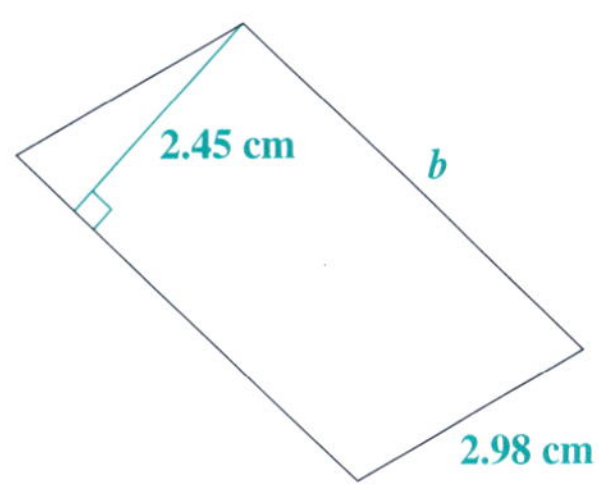


**Share & Summarize Answer**

Possible answer: In both cases, you multiply two measurements. For a rectangle, the measurements are both side lengths. For the parallelogram, one of the measurements (the length of the base) is a side length, but the other may not be. The height is the length of a segment from the side opposite the base to the base, perpendicular to the base. In both cases, the two segments whose lengths are multiplied are perpendicular.

## Share & Summarize

How is finding the area of a parallelogram similar to finding the area of a rectangle? How is it different?

3 Have students read their answers to the class.



LESSON 8.4 Calculating Areas 517

# Investigation 2

In this investigation, students apply what they have learned about the area of a parallelogram to develop a formula for the area of a triangle. After developing the formula, they apply it and explore some consequences.

1 **Problem Set D** **Suggested Grouping: Pairs**
The triangles for this problem set and the next can be found on Master 72. Distribute one copy to each pair.

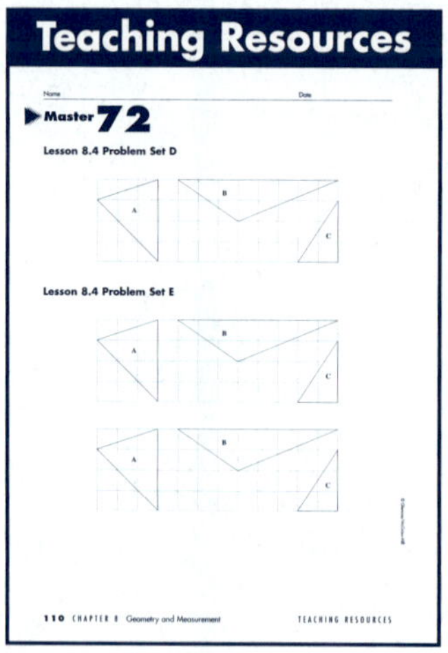
Teaching Resources

Master 72

Lesson 8.4 Problem Set D

Lesson 8.4 Problem Set E

Students can find the areas of these triangles by using ideas similar to those used for parallelograms drawn on grids.

Triangle A can be thought of as two triangles, each of which is half of a rectangle. Triangle B can be thought of in the same way. Triangle C is already half of a rectangle.

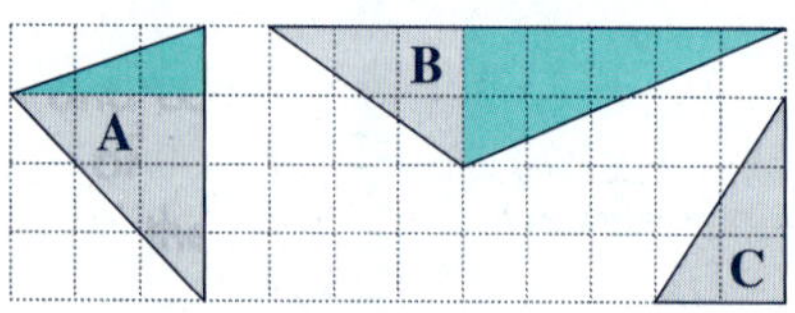


$$\text{Area of } \triangle A = \tfrac{1}{2} \cdot (1 \cdot 3) + \tfrac{1}{2} \cdot (3 \cdot 3)$$
$$= 1.5 + 4.5$$
$$= 6 \text{ cm}^2$$
$$\text{Area of } \triangle B = \tfrac{1}{2} \cdot (2 \cdot 3) + \tfrac{1}{2} \cdot (2 \cdot 5)$$
$$= 3 + 5$$
$$= 8 \text{ cm}^2$$
$$\text{Area of } \triangle C = \tfrac{1}{2} \cdot (3 \cdot 2)$$
$$= 3 \text{ cm}^2$$

Some students may see that triangle A can immediately be thought of as having half the area of a $4 \times 3$ rectangle, triangle B as having half the area of a $2 \times 8$ rectangle, and triangle C as having half the area of a $3 \times 2$ rectangle.

**Problem Set Wrap-Up** Have students share the methods they used in Problem Set A. ■

2 Discuss the new vocabulary terms. After their work with parallelograms, students may find it easy to grasp the idea that any side of a triangle can be used as the base. When one talks about the height of a triangle, it is presupposed that one has already decided which side is being considered the base. The base may or may not be in a horizontal position.

3 **Problem Set E** **Suggested Grouping: Pairs**
In this problem set, students explore ways to join two copies of a triangle to make a parallelogram. They use their observations to discover a formula for the area of a triangle in terms of the length of the base and the height.

Each pair of students will need two copies of Master 72. You should probably allow 15 minutes for this problem set.

4 For **Problem 1,** urge students to be careful when they cut out the triangles. This will make it possible for them to get a good fit when the copies are put together.

# Investigation 2 Areas of Triangles

You have looked at areas of rectangles and parallelograms. Now you will turn your attention to triangles.

MATERIALS
copies of the triangle

## Problem Set D

Find the area of each triangle, and explain the method you used.

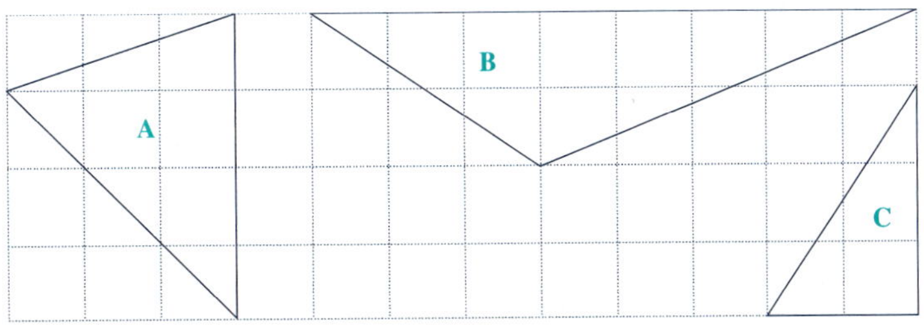


**Problem Set D Answer**

Triangle A: 6 $cm^2$; Triangle B: 8 $cm^2$; Triangle C: 3 $cm^2$; See the teaching notes for possible strategies.

1 You may have students work in pairs.

VOCABULARY
**base of a triangle**
**height of a triangle**

The **base of a triangle** can be any of its sides. The **height of a triangle** is the distance from the base to the vertex opposite the base. The height is always measured along a segment perpendicular to the base (or the line containing the base).

2 Discuss the two different parts of a triangle.

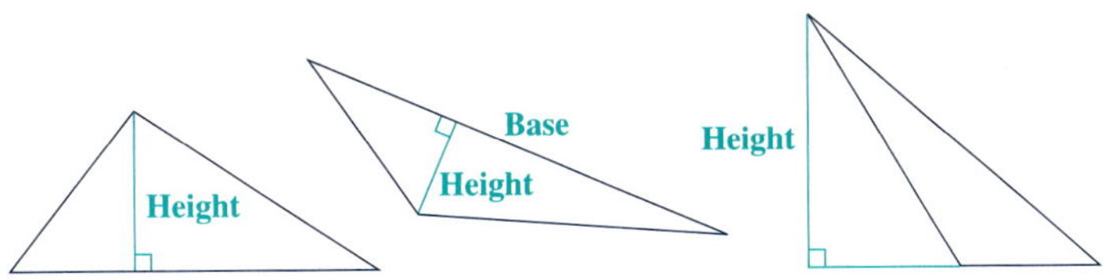


In Problem Set D, you may have used a variety of methods for finding the areas of the triangles. In the next problem set, you will see how you can find the area of a triangle by relating it to a parallelogram.

MATERIALS
- copies of the triangles
- scissors
- tape
- protractor
- ruler

## Problem Set E

3 You may have students work in pairs.

Cut out two copies of each triangle from Problem Set D.

1. Complete Parts a–d for each triangle.

   a. Make as many different parallelograms as you can by putting together the two copies of the triangle. (Do not tape them together.) Make a sketch of each parallelogram. See Additional Answers.

   b. How does the area of the triangle compare to the area of each parallelogram? The area of the triangle is half the area of each parallelogram.

4 Have students carefully cut out the triangles so they can be a good fit when they are put together.



**Additional Answers**

**1a.** For each triangle, you can make three different parallelograms. Here are the three parallelograms for Triangle B:

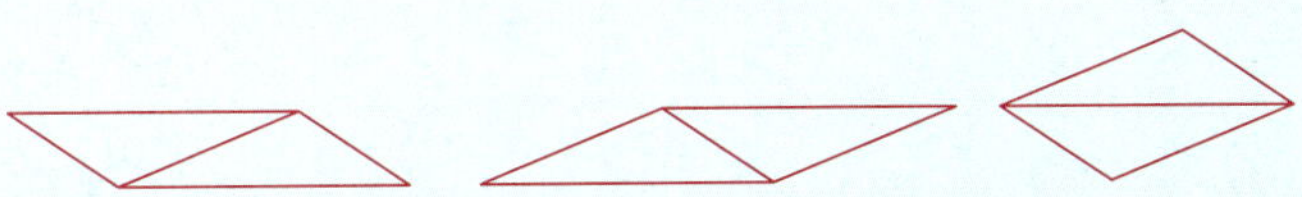

# Develop

In **Problem 1c,** students may try to draw over the tape that is holding the triangles together. If this is difficult, they can turn the parallelogram over and draw the segment for the height on the back.

1 After students have completed **Problem 2,** you should probably take time to discuss the results obtained so far. When you are confident students have understood everything up to this point, have them complete the problem set.

2 In **Problems 3–5,** students will need to identify the segments used for the base and height of each triangle. Once this is done, they can use the rule they found in Problem 2 to calculate the area of the triangle.

3 **Problem Set Wrap-Up** Discuss the area formula for triangles to wrap up the work for Problem Set E.

1c. Answers will depend on the parallelogram students chose and the side they designated as the base.

2. Possible answer: Multiply the base times the height and divide the result by 2.

**c.** Tape the two copies of the triangle together to form one of the parallelograms you sketched in Part a. Choose one side of the parallelogram as the base, and draw a segment perpendicular to the base extending to the opposite side.

**d.** Do the base and height of the parallelogram correspond to a base and height of the triangle? yes

**2.** Think about what you learned in Problem 1 about the relationship between triangles and parallelograms. How can you find the area of a triangle if you know the length of a base and the corresponding height?

Find the area of each triangle to the nearest hundredth.

**3.** 0.90 cm²

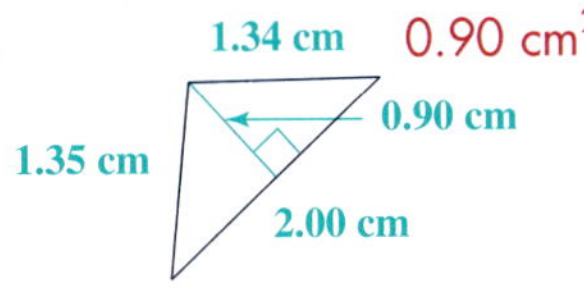


**4.** 0.60 cm²

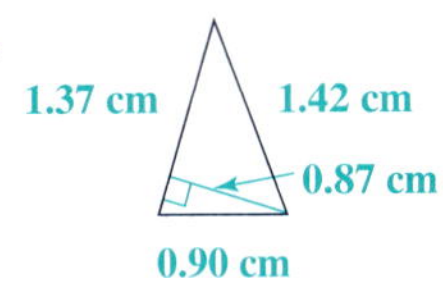


**5.** 1.03 cm²

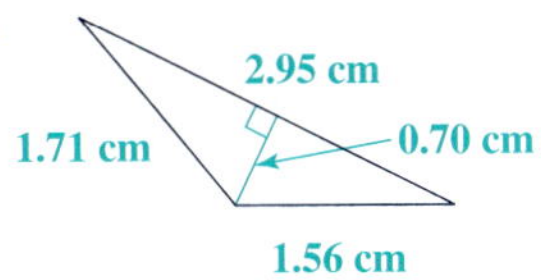


**Just the facts**

Triangles are rigid shapes. If you build a triangle out of a strong material, it will not collapse or change shape when you press on its sides or vertices. Because of this property, triangles are used frequently as supports for buildings, bridges, and other structures.

In Problem Set E, you probably discovered that the area of a triangle is half the length of the base times the height. You can state this using a formula.

**Area of a Triangle**

$$A = \frac{1}{2} \cdot b \cdot h$$

In this formula, $A$ represents the area, $b$ represents the base, and $h$ represents the height.

1 Discuss the results obtained thus far.

2 Make sure students identify the segments for the base and the height of each triangle.

3 Discuss the formula for the area of a triangle.



LESSON 8.4 Calculating Areas 519

# Develop

**1** **Problem Set F** **Suggested Grouping: Pairs**
In this problem set, students confirm that the area formula for triangles gives the same number of square units no matter which side is chosen as the base. Each pair of students will need three copies of Master 73.

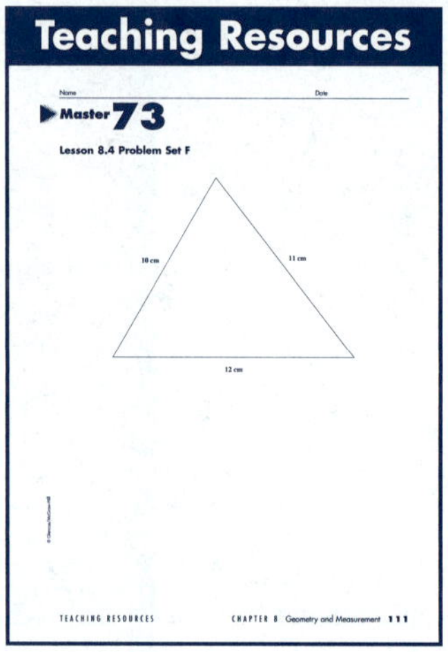
Teaching Resources

Name Date

Master 73

Lesson 8.4 Problem Set F



TEACHING RESOURCES CHAPTER 8 Geometry and Measurement 111

**2** For **Problem 2a,** you may want to demonstrate at the board how to draw the segment for the height accurately. Explain that the bottom line of the protractor (the line through the center point of the protractor) should be exactly on top of the base of the triangle (in this case, the side that measures 12 cm). Slide the protractor along this line until the opposite vertex (the top vertex of the triangle) is on the 90° mark. The center point of the triangle will be at a certain point of the base when this occurs. Mark this point. Then draw the segment from this point to the opposite vertex. Tell students to use the square corner symbol to show which segments are perpendicular.

Another method for locating the correct point on the base is to use a square corner of a sheet of paper. If students use this method, they can use their protractors to check that the base and the segment for the height are perpendicular.

In **Problems 2b and 2c,** if students have any trouble drawing the perpendicular segments when the base is not horizontal, tell them to turn the triangle to get the base in a horizontal position. When they have drawn the segment for the height, they should use the square corner symbol to show which segments are perpendicular.

### On the Spot Assessment

If students' answers for Problems 2–4 differ markedly from the sample answer, examine their work to see whether the students had difficulty drawing the segments for the heights. You may need to demonstrate the proper procedure again.

MATERIALS
- 3 copies of the triangle
- protractor
- ruler

## Problem Set F

Three students found the area of this triangle. Caroline used the 12-centimeter side as the base, Rosita used the 11-centimeter side, and Marcus used the 10-centimeter side.

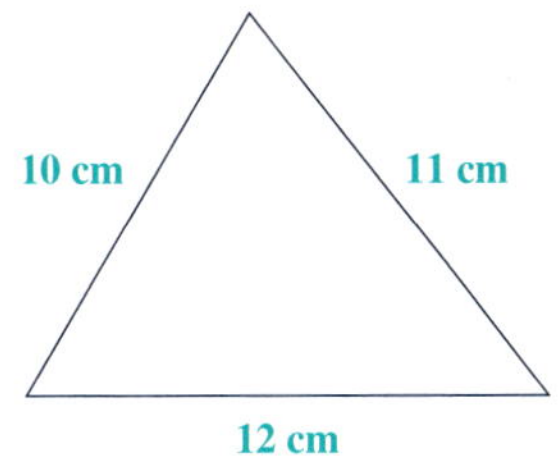


**1.** Answers will vary. The areas will be the same no matter which side is used as the base.

**1.** Assuming the students did the calculations correctly, do you think they found the same area or different areas? Explain.

**2.** Complete Parts a–c to find the area using the 12-centimeter side as the base.

**a.** Draw a segment perpendicular to the base from the vertex opposite the base. Use your protractor to make sure the base and the segment form a right angle.

**b.** Measure the height to the nearest tenth of a centimeter.

**c.** Use the base and height measurements to calculate the area of the triangle. Possible answer: 51.6 $cm^2$

**3.** Repeat Parts a–c of Problem 2 using the 11-centimeter side as the base.

**4.** Repeat Parts a–c of Problem 2 using the 10-centimeter side as the base.

**5.** Compare your results for Problems 2, 3, and 4. Did the area you calculated depend on the base you used? Explain.

No; in all three cases, I found an area of a little more than 51 $cm^2$. The small differences are probably due to measuring error.

**2a.**

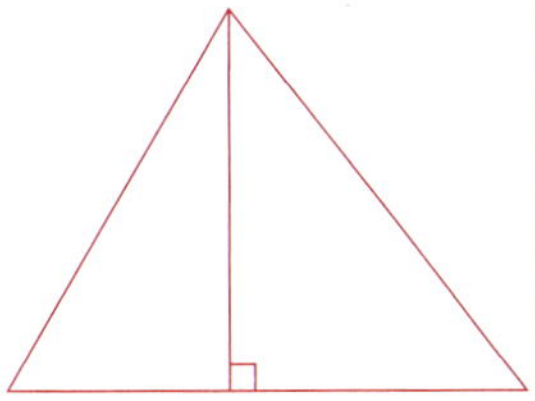

**2b.** Possible answer: 8.6 cm

**3a.**

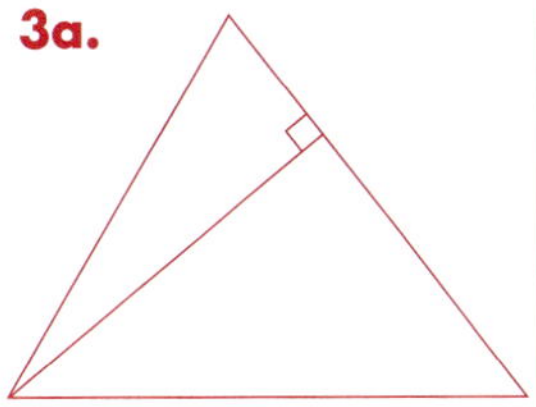

**3b.** Possible answer: 9.3 cm

**3c.** Possible answer: 51.2 $cm^2$

**4a.**

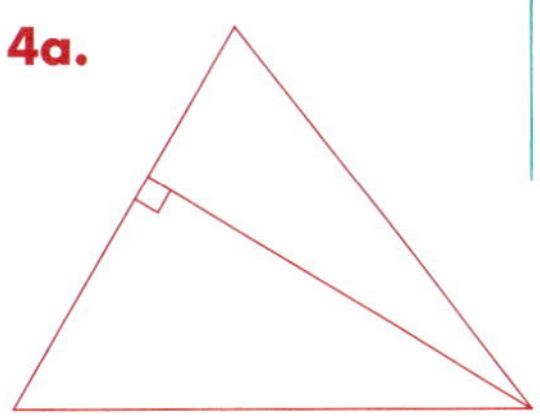

**4b.** Possible answer: 10.3 cm

**4c.** Possible answer: 51.5 $cm^2$

**1** You may have students work in pairs.

**2** Demonstrate how to draw the height accurately.

## Develop

1 **Problem Set G** **Suggested Grouping: Pairs**
In this problem set, students see that shearing a triangle does not change the area of the triangle but generally does change the perimeter.

Discuss the introductory paragraph to be sure students understand what is meant by *shearing* a triangle.

For **Problem 2,** have students copy the triangles in the diagram on their sheet of 1-centimeter dot paper (Master 84). They should label the triangles in the copy the same way as in the original.

For **Problem 3,** students should use the area formula for triangles to calculate the areas of the three original triangles and the two triangles they drew for Problem 2.

2 **Problem Set Wrap-Up** After students have finished the problems, discuss the results with the class. You may want to copy the original diagram on the board and have a volunteer show why the heights of the sheared triangles are all the same.

3 **Share & Summarize**
Have volunteers read their answers to the class. Encourage them to put diagrams on the board whenever they think they will be helpful in explanations.

4 **Troubleshooting** When students find areas of triangles, they may multiply the length of the base by the height but forget to multiply by $\frac{1}{2}$ (or divide by 2). If you find that students are making this mistake, suggest that they always write the formula they are using.

### On Your Own Exercises

Practice & Apply: 6–10, pp. 528–529
Connect & Extend: 19–22, pp. 531–533
Mixed Review: 25–36, p. 535

MATERIALS
dot paper

## Problem Set G

$\triangle ABD$ and $\triangle ABE$ were created by shearing $\triangle ABC$. *Shearing* a triangle means "sliding" one of its vertices along a line parallel to the opposite side. In this case, $\triangle ABC$ was sheared by sliding vertex $C$ to vertex $D$ and then to vertex $E$.

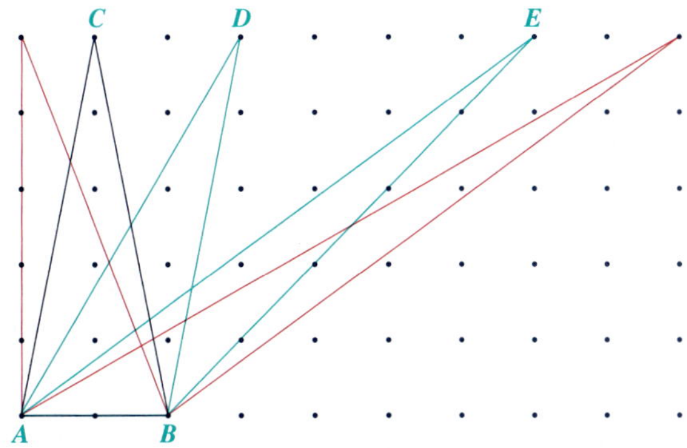


1. How are $\triangle ABC$, $\triangle ABD$, and $\triangle ABE$ alike? How are they different?
2. Draw two more triangles by shearing $\triangle ABC$.
3. Does shearing $\triangle ABC$ change its area? Explain.
4. Does shearing $\triangle ABC$ change its perimeter? Explain.

1. Possible answer: All share side $AB$, and all have vertices in the top row of dots. If side $AB$ is considered the base, then all three triangles have the same base and height. $\triangle ABE$ is longer and skinnier than $\triangle ABD$, which is longer and skinnier than $\triangle ABC$. $\triangle ABC$ has two sides of the same length. In the other triangles, all three sides have a different length.
2. See the figure for a possible answer.
3. No; the base and height stay the same, so the area is the same as well.
4. Yes; as the vertex moves farther from point $C$, the non-horizontal sides get longer and longer, so the perimeter increases.

**1** You may have students work in pairs.

**2** Have students discuss the results with the class.

**3** Have students discuss their answers with the class.

**4** Suggest that students always write the area formula when they are finding areas of triangles.

## Share & Summarize

Describe how finding the area of a triangle is related to finding the area of a parallelogram.

**Share & Summarize Answer**

Possible answer: The area of a triangle is half the area of a parallelogram formed by two copies of the triangle. When you form such a parallelogram, the parallelogram's base and height correspond to the triangle's base and height. Since the area of the parallelogram is the base times the height, the area of a triangle is half the base times the height.



LESSON 8.4 Calculating Areas 521

# Investigation 3

In this investigation, students approximate areas of circles drawn on a grid. They complete a table that will help them see how the area of a circle and the radius of the circle are related. They learn a formula for the area of a circle. They then apply the formula to solve mathematical and real-world problems.

1 **Problem Set H** **Suggested Grouping: Pairs**
In this problem set, students investigate how the area and the radius of a circle are related.

2 For **Problem 2,** encourage students to estimate the area of each circle as accurately as they can. More accurate estimates will give better results for the last column of the table.

# Investigation 3 Areas of Circles

Finding the area of a figure with curved sides often requires counting grid squares or using another estimation method. However, there is a surprisingly simple formula for calculating the area of a circle.

## Problem Set H

These circles are drawn on 1-centimeter grid paper:

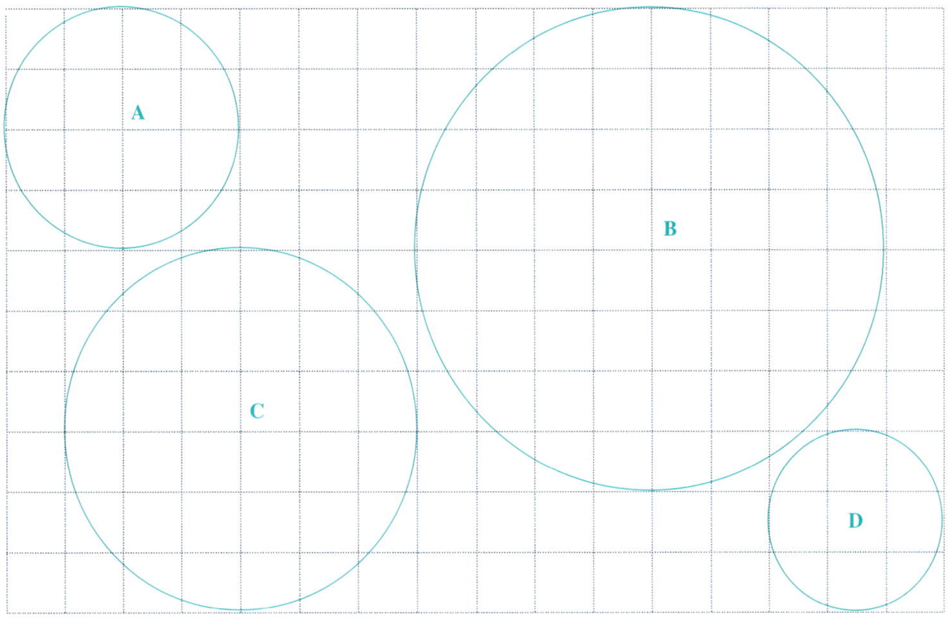


1. Copy the table. Find the radius of each circle, and record your results in the "Radius" column.

| Circle | Radius, $r$ (cm) | Estimated Area, $A$ ($cm^2$) | $A \div r$ | $A \div r^2$ |
|---|---|---|---|---|
| A | 2 | 12 | 6 | 3 |
| B | 4 | 48 | 12 | 3 |
| C | 3 | 28 | 9.3 | 3.1 |
| D | 1.5 | 7 | 4.7 | 3.1 |

**2, 3.** See the table.

2. Estimate the area of each circle by counting grid squares. Record your estimates in your table.
3. For each circle, divide the area by the radius, and record the results.

1 You may have students work in pairs.

2 **Estimation** Encourage students to accurately estimate the areas of the circles.

## Develop

1 For **Problem 5,** students are asked to look for a pattern and recall a similar pattern. You may want to suggest that students look back to Problem Set D in Lesson 8.2.

In **Problem 6a,** there are various ways to see that Jahmal's estimate is not reasonable. The sample answer points out that his estimate is out of line with the pattern in the table students completed on page 522. Here is another way to reason about this problem. Suppose you draw a quarter circle of radius 10 cm centered at the lower-right corner of the grid on page 522. The quarter circle will pass through the point at the upper-right corner of the grid and will cross the bottom side of the grid at the point 10 cm to the left of the bottom-right corner. If you connect these points, you get a triangle that lies entirely inside the quarter circle. By the formula for the area of a triangle, the area of this triangle is 50 $cm^2$, which is already greater than Jahmal's estimate.

2 **Problem Set Wrap-Up** Discuss the results from Problem Set H. Pay special attention to the table and to the answer to **Problem 5.** ■

3 Next, read with the class the information following Problem Set H. Discuss the formula for the area of a circle to be sure students understand the formula.

**5.** Possible answer: The $A \div r^2$ values show a pattern—all the results are around 3. This is similar to the pattern in the $C \div d$ values we saw in Lesson 8.2.

**4.** For each circle, divide the area by the radius squared, and record the results. See the table on page 522.

**5.** Look at the last two columns of the table. Do the values in either column show an obvious pattern? If so, does it remind you of other patterns you have seen in this chapter?

**6.** Jahmal estimated that the area of a circle with radius 10 cm is about 40 $cm^2$.

**a.** Explain why Jahmal's estimate is not reasonable.

**b.** What is a reasonable estimate for the area of a circle with radius 10 cm? about 300 $cm^2$

**c.** What is a reasonable estimate for the radius of a circle with area 40 $cm^2$? 3.5 cm to 4 cm

**Remember**

$\pi$ is a decimal number with digits that never end or repeat. It can be approximated as 3.14.

**6a.** Possible answer: $A \div r^2$ should be about 3, but for Jahmal's measurements, $A \div r^2$ is only about 0.4.

In Lesson 8.2, you learned about the number $\pi$ and how it is related to the circumference of a circle. You found that if $C$ is the circumference of a circle and $d$ is the diameter, the following is true:

$$\pi = C \div d$$

Amazingly, the number $\pi$ is also related to the area of a circle. If $A$ is the area of any circle and $r$ is the radius, the following is true:

$$\pi = A \div r^2$$

You can use this fact to develop the formula for the area of a circle.

**Area of a Circle**

$$A = \pi \cdot r^2$$

In this formula, $A$ is the area and $r$ is the radius.

**1** **Problem-Solving Strategies**

- Look for a pattern
- Determine reasonable answers

**2** Discuss students' results.

**3** Discuss the formula for the area of a circle.



LESSON 8.4 Calculating Areas 523

# Develop

1 **Problem Set I** **Suggested Grouping: Individuals**
In this problem set, students apply the formula for the area of a circle.

2 In **Problem 3,** students should note that they are told the *diameter* of the second circle. They should divide the diameter by 2 to find the radius. Some students may go on to calculate the approximate area of each circle. If they do this correctly, praise their good work, but point out that they could have answered the question in the problem without the calculations. Be sure they understand how.

For **Problem 4,** students can substitute 100 for $A$ in the formula $A = \pi \cdot r^2$ to get $100 = \pi \cdot r^2$. They can use a calculator to find the approximate value of $r^2$. They will then need to use the square root operation to get a value for $r$.

In **Problem 5,** some students may get off to a wrong start by trying to compare the perimeters of the pizzas. Help them understand that it makes more sense to use area. Once they know the area of each pizza, dividing price by area will give the cost per square inch.

For **Problem 6,** students should be careful to use the radius rather than the diameter when they calculate the total area of the semicircles.

3 **Share & Summarize**
These questions will help you ascertain whether students understand the formula for the area of a circle and how to use it. When you discuss the answers in class, review the meaning of the terms *radius* and *diameter.*

**Troubleshooting** If students are having trouble keeping the formulas straight, you might help them with memory devices. For example $C = \pi \cdot d$ contains the first letters of the words *circumference* and *diameter*. The area formula $A = \pi \cdot r^2$ contains the first two letters of the word *area*, and $r$ squared is a reminder that area is always given in square units.

## On Your Own Exercises

Practice & Apply: 11–14, p. 530
Connect & Extend: 23, 24, p. 534
Mixed Review: 25–36, p. 535

★ indicates multi-step problem

## Problem Set I

For this problem set, use the [π] key on your calculator to approximate π. (If your calculator doesn't have a [π] key, use 3.14 to approximate π.)

1. What is the area of a circle with radius 15 in.? Give your answer to the nearest hundredth of a square inch. 706.86 in.$^2$
2. What is the area of a circle with radius 10.15 cm? Give your answer to the nearest hundredth of a square centimeter. 323.65 cm$^2$
3. Which has the greater area: a circle with radius 7.2 cm, or a circle with diameter 12.75 cm? Explain your answer.
4. What is the radius of a circle with area 100 in.$^2$? Give your answer to the nearest hundredth of an inch. 5.64 in.

★ 5. A pizza parlor makes pizzas in two shapes. The circular pizza has a diameter of 10 inches. The rectangular pizza measures 16 inches by 10 inches. A circular cheese pizza costs $8, and a rectangular cheese pizza costs $14. Which shape gives you more pizza for your money? Explain how you found your answer.

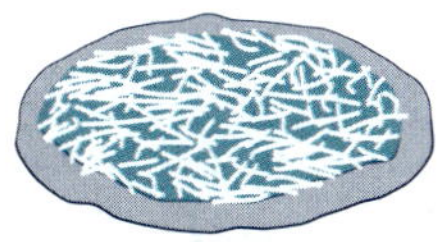

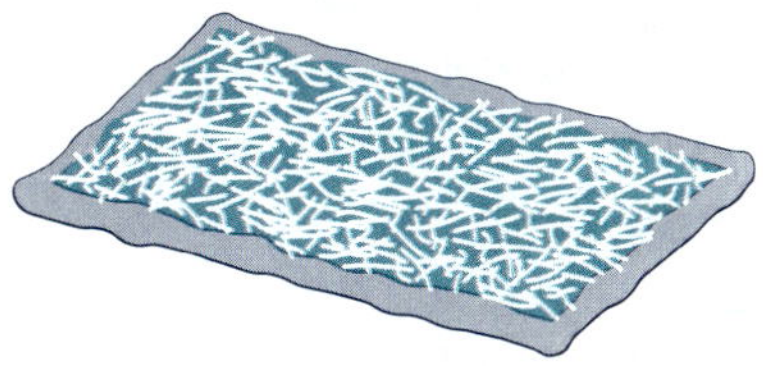

★ 6. This is a diagram of the inner lane of the track at Walker Middle School. The lane is made of two straight segments and two semicircles (half circles). The area inside the track is covered with grass. To the nearest tenth of a square yard, what is the area of the grass inside the track? Explain how you found your answer. See Additional Answers.

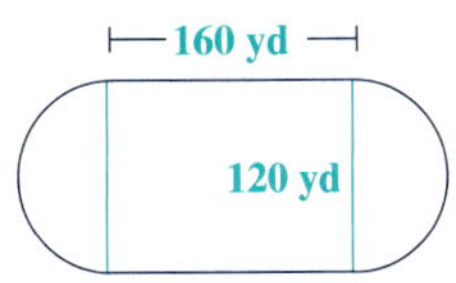


## Share & Summarize

1. $A = \pi \cdot r^2$, where $A$ is the area and $r$ is the radius
2, 3. See Additional Answers.

1. Give the formula for the area of a circle. Tell what the letters in the formula represent.
2. How can you calculate the area of a circle if you know only its diameter?
3. How can you calculate the area of a circle if you know only its circumference?

3. a circle with a radius of 7.2 cm; Possible explanation: A circle with diameter 12.75 cm has radius 6.375 cm. Since it has the smaller radius, it has less area.

**Just the facts**

There are more than 60,000 pizzerias in the United States, accounting for about 15% of all restaurants.

5. The rectangular pizza; I divided the area of each pizza by the price to find the amount per dollar. The circular pizza is 78.5 ÷ 8, or about 9.8 in.$^2$ per dollar. The rectangular pizza is 160 ÷ 14, or about 11.4 in.$^2$ per dollar.

1 You may have students work individually.

2 Be sure students know that they can answer the question without finding the area.

3 Review the meaning of radius and diameter.



### Additional Answers

6. 30,509.7 yd$^2$; The total area is the area of the rectangle plus the area of the two semicircles. Together, the two semicircles form a circle with radius 60 yd. So, the total area is $120 \cdot 160 + \pi \cdot 60^2$, or about 30,509.7 yd$^2$.

### Share & Summarize

2. Divide the diameter in half to get the radius. Then use the formula $A = \pi \cdot r^2$.
3. Divide the circumference by π to get the diameter, and divide the diameter in half to get the radius. Then use the formula $A = \pi \cdot r^2$.

# Lab Investigation

1 **Suggested Grouping: Groups of 2 or 3**
In this lab investigation, students use spreadsheet software to solve a problem about maximizing the area of a rectangular pen.

2
## Materials and Preparation

You will need a computer with spreadsheet software for each pair or group of three students. Many students may never have worked with spreadsheet software, so you will need to help them learn how to identify cells, enter formulas and data, edit formulas and data, and so on. Suggest that students take turns entering formulas and data into the spreadsheet. If you have the equipment to project a computer screen to the front of the room, you might use it to help students with the steps for creating their spreadsheets.

Read the problem at the top of page 525 with the class. Make sure students understand what information is given and what the problem is asking them to find.

3
## Think about the Problem

Discuss **Questions 1 and 2** with the class. Knowing how to answer these questions is crucial to understanding what will be happening in the spreadsheet.

## Create a Spreadsheet

Have students work in their groups, at the computers, as you explain how to set up the spreadsheet. Explain that they will be entering formulas and data for calculations. The computer will do the necessary calculations automatically.

Explain that the space in Column A Row 1 is called Cell A1, the space in Column B Row 2 is called Cell B2, and so on.

4 First, have students enter the labels shown in Row 1 of the spreadsheet diagram on page 525. Next, tell students that they are going to enter formulas that will tell the spreadsheet how to calculate the perimeter and area of a rectangle and the cost of fencing. Explain that each formula must begin with an equal sign (=). Help students enter =2*A2+2*B2 in Cell C2.

**Lab Investigation**

# Using a Spreadsheet to Maximize Area

MATERIALS
computer with spreadsheet software

1 You may have students work in groups of 2 or 3.

2 Help students learn how to use a spreadsheet.

Miguel's dog Max loves to play outside. When he is in the backyard, he is usually tied to a 10-foot leash in the center of the yard.

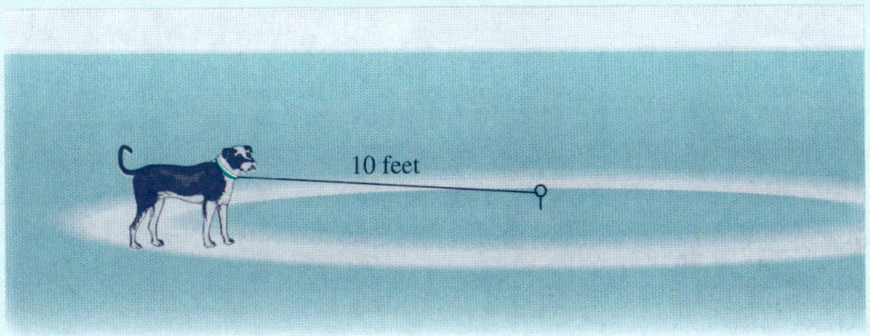

**Just the facts**

Dogs resembling modern-day greyhounds lived in Egypt as early as 4500 B.C. Hieroglyphs of dogs can be found on ancient Egyptian tombs.

Miguel has earned $400 mowing lawns, and he wants to use the money to build a pen for Max. The fencing Miguel has chosen costs $4.50 per foot. In this lab investigation, you will use a spreadsheet to figure out the largest rectangular pen Miguel can afford to build.

## Think about the Problem

3 Discuss Questions 1 and 2 with the class.

1. If you know the length and width of a rectangular pen, how can you figure out the area and perimeter?
   *1. To find the area, multiply the length and width. To find the perimeter, add twice the length to twice the width.*
2. How can you figure out how much the pen will cost?
   *Multiply the perimeter by $4.50.*

## Create a Spreadsheet

You will create a spreadsheet that automatically calculates the area, perimeter, and cost of the pen when you enter the length and the width. First enter column headings for length, width, perimeter, area, and cost.

4 Show students how to enter formulas into a spreadsheet.

| | A | B | C | D | E | F | G |
|---|---|---|---|---|---|---|---|
| 1 | Length | Width | Perimeter | Area | Cost | | |
| 2 | | | | | | | |

You want your spreadsheet to calculate perimeter, area, and cost when you enter length and width values in Cells A2 and B2. For this to work, you will enter *formulas* that tell the spreadsheet how to do the calculations.

The perimeter is 2 times the length value in Cell A2 plus 2 times the width value in Cell B2. So, in Cell C2, enter this formula:

=2*A2+2*B2



LESSON 8.4 Calculating Areas 525

# Develop

1 Give students time to think about and answer **Questions 3 and 4.** Then discuss the answers with the class. Next, help students fill in the appropriate formulas for Cells D2 and E2. If students make mistakes doing this and do not know how to correct them, ask them to raise a hand so you can help.

2 When students have correctly entered the formulas for area and cost, read the paragraph below **Question 4.** Tell students to enter the given numbers in the appropriate cells, as explained in the paragraph.

3 For **Question 7,** have students change the values in Cells A2 and B2. Provide help if they do not know or immediately discover how to do this.

4 After students have answered Question 7, read the information that follows Question 7. Have students highlight Cells C2 through C5. Show or tell them how to select the Fill Down command. Have them look at what the computer displays.

The = sign lets the spreadsheet know that the entry is a formula to evaluate. So, the formula you entered in Cell C2 tells the spreadsheet to add 2 times the value in Cell A2 to 2 times the value in Cell B2.

| | A | B | C | D | E | F | G |
|---|---|---|---|---|---|---|---|
| 1 | Length | Width | Perimeter | Area | Cost | | |
| 2 | | | =2*A2+2*B2 | | | | |
| 3 | | | | | | | |
| 4 | | | | | | | |
| 5 | | | | | | | |

3. In Cell D2, you need to enter a formula to calculate the area. What formula should you enter? (Your formula must begin with an = sign and should contain the cell names A2 and B2.) =A2*B2

4. In Cell E2, you need to enter a formula to calculate the cost. What formula should you enter? =4.5*C2, or =4.5*(2*A2+2*B2)

**1** Discuss students' answers to Questions 3 and 4.

Enter your area and cost formulas into the spreadsheet. Test the formulas by entering the length value 11 in Cell A2 and the width value 7 in Cell B2.

**2** Have students enter the numbers in the appropriate cells.

5. Does your spreadsheet give the correct perimeter, area, and cost values for a pen that is 11 feet by 7 feet? How do you know? If your formulas are not correct, adjust them.

**5.** Yes; the perimeter of an 11-by-7 rectangle is $2 \cdot 11 + 2 \cdot 7$, or 36; the area is $11 \cdot 7$, or 77; and the cost is $4.5 \cdot 36$, or \$162.

6. Can Miguel afford to build an 11-ft-by-7-ft pen? yes

7. Change the length and width values in Cells A2 and B2 to the length and width of your choice. What happens to the values in the other columns? They change as well.

**3** Help students change the values in Cells A2 and B2.

Now you will copy your formulas down to Row 25 of the spreadsheet. This will allow you to enter different length and width values in each row. Here are the steps for copying the perimeter formula:

- Highlight Cells C2 through C25.
- Select the Fill Down command.

**4** Explain how the Fill Down command works.

| | A | B | C | D | E | F | G |
|---|---|---|---|---|---|---|---|
| 1 | Length | Width | Perimeter | Area | Cost | | |
| 2 | | | =2*A2+2*B2 | | | | |
| 3 | | | | | | | |
| 4 | | | | | | | |
| 5 | | | | | | | |
| 6 | | | | | | | |
| 7 | | | | | | | |
| 8 | | | | | | | |
| 9 | | | | | | | |
| 10 | | | | | | | |

1

### Find the Largest Pen

**Question 9** may take students a while to complete. They will probably need to guess and check quite a few values for the length and width before they have values they find satisfactory. You may need to check on students' work to be sure they are keeping in mind that Miguel has no more than $400 to spend on fencing.

**Question 10** will be a bit difficult if students are basing their thinking on only the spreadsheet results.

### What Did You Learn?

**Question 12** calls on students to use their knowledge of circumference and area of circles. It also assesses whether they understand how to enter formulas into spreadsheets.

For **Question 13,** students need to use radius values that vary in a regular way. They must keep Miguel's cost restrictions in mind.

In **Question 14,** students will see that the area of the largest circular pen Miguel can afford exceeds that of the largest rectangular pen he can afford.

### Access for all Learners

2

**Extra Challenge** Students have probably observed that $\$400 \div \$4.50 \approx 88.888$ tells about how many feet of fencing Miguel can afford. Ask students if they can find other formulas that they could use in the spreadsheet to produce additional evidence that the rectangular pen of maximum area that Miguel can afford is a square pen.

8. Select Cell C3. What formula appears in this cell? Now select Cell C18. What formula appears here? What does the Fill Down command do?

Copy the formulas for area and cost down to Row 25. Now you can look at perimeter, area, and cost for several different pens at the same time.

## Find the Largest Pen

Test different length and width values to try to find the largest pen Miguel can afford. Enter a different pair of values in each row.

9. What are the dimensions and area of the largest rectangular pen Miguel can afford?

10. How do you know that the pen you found in Question 9 is the largest possible?

11. Will Max have more room to move in his new pen than he now has on his leash? Explain. See margin.

## What Did You Learn?

12. Create a spreadsheet that calculates the circumference, area, and cost of circular pens with different radii. Use 3.14 as an approximation for $\pi$. See the spreadsheet for possible formulas.

| | A | B | C | D |
|---|---|---|---|---|
| 1 | Radius | Circumference | Area | Cost |
| 2 | | =2*3.14*A2 | =3.14*A2*A2 | =B2*4.5 |

13. Find the approximate radius and area of the largest circular pen Miguel could build with the fencing. radius: 14.1 ft; area: 624.3 ft$^2$

14. How does the area of the largest circular pen compare to the area of the largest rectangular pen? The circular pen's area is greater.

8. =2*A3+2*B3, =2*A18+2*B18; Possible answer: It copies the formula but changes the numbers of the cells to use the length and width in that row.

9. a square with a side length of about 22.2 ft and an area of about 493 ft$^2$ (Note: Students may not be this precise. They may get 22 ft × 22 ft or 21 ft × 23 ft.)

10. Possible answer: After a lot of tries, I found that I needed to have dimensions that totaled about 89 feet in perimeter (400 ÷ 4.5). Of these, the square seems to have the greatest area.

11. Yes; now, he can move in a circle with radius 10 ft, so he has an area of $\pi \cdot 10^2$, or about 314 ft$^2$. With the new pen, he will have about 493 ft$^2$.

**Just the facts**

Dogs similar to modern-day mastiffs were used by ice-age hunters to help catch large game such as the woolly mammoth.

1 **Problem-Solving Strategy** Guess-check-and-improve

2 Challenge students to find a formula they can use in the spreadsheet that shows a rectangular pen of maximum area.



LESSON 8.4 Calculating Areas 527

# On Your Own Exercises

## On Your Own Exercises

Investigation 1, pp. 515–517
Practice & Apply: 1–5
Connect & Extend: 15–18

Investigation 2, pp. 518–521
Practice & Apply: 6–10
Connect & Extend: 19–22

Investigation 3, pp. 522–524
Practice & Apply: 11–14
Connect & Extend: 23, 24

Assign Anytime
Mixed Review: 25–36

**Exercise 2:**
In **Parts b and c,** there are no restrictions on the base or height except that both be greater than 0 and that the rectangle be sketchable.

**Exercise 5:**
Here, as in all exercises of this kind, the word *the* before *base* is not meant to imply that a parallelogram can have only one base.

**Exercise 6b:**
Students will probably use two copies of each triangle shown to make their parallelograms. However, the exercise does not require this.

## Practice & Apply

1. Choose an object in your home with a nonrectangular surface that will fit on a piece of grid paper. (Some ideas: a can of soup, your shoe, an iron.) Trace the surface onto the grid paper, and estimate its area. Answers will vary.

2. These parallelograms are drawn on a centimeter grid:

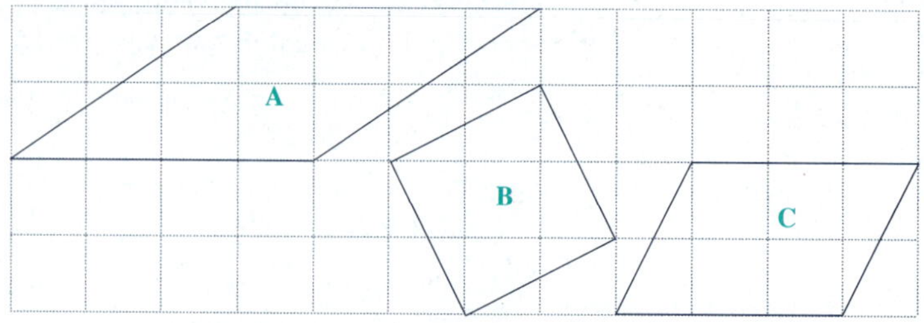


   a. Find the area of each parallelogram.

   b. Sketch a rectangle that has the same area as Parallelogram A.

   c. Sketch a rectangle that has the same area as Parallelogram C.

3. Find the area of this parallelogram: $60 \text{ in}^2$

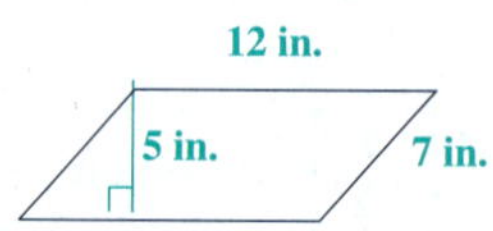


**2a.** Parallelogram A: 8 $cm^2$; Parallelogram B: 5 $cm^2$; Parallelogram C: 6 $cm^2$

**2b.** Possible rectangle:

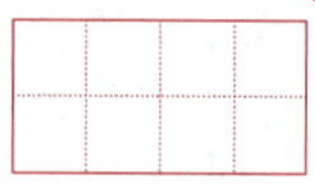

**2c.** See Additional Answers.

**4.** Yes; the length and width of the rectangle are the base and height.

4. Can you use the area formula for a parallelogram, $A = b \cdot h$, to find the area of a rectangle? If so, where are the base and height on the rectangle? If not, why not?

5. A parallelogram has an area of $42.6 \text{ cm}^2$. The height of the parallelogram is 8 cm. What is the length of the base? 5.325 cm

6. These triangles are drawn on a centimeter grid:

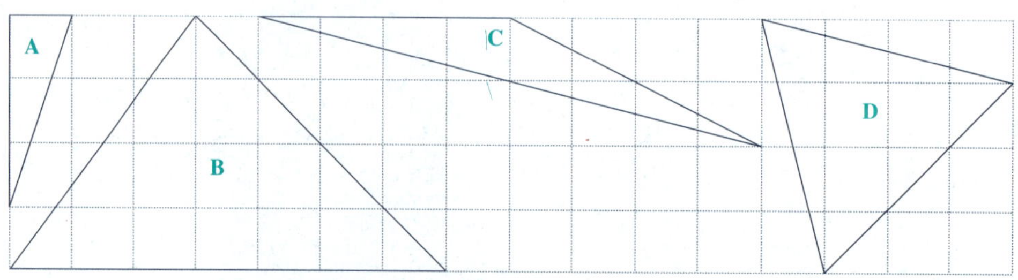


**6.** See Additional Answers.

   a. Find the area of each triangle.

   b. For each triangle, sketch a parallelogram with twice the area of the triangle.



impactmath.com/self_check_quiz

## Additional Answers

**2c.** Possible rectangle:

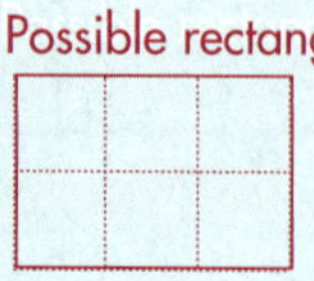

**6a.** Triangle A: 1.5 $cm^2$; Triangle B: 14 $cm^2$; Triangle C: 4 $cm^2$; Triangle D: 7.5 $cm^2$

*Quick Review*
Math Handbook
hot words hot topics
Hot Topics pp. 324–329, 340–345

• Additional Answers continued on page A669

Find the area of each triangle. Round your answers to the nearest hundredth.

**7.** $3.25 \text{ cm}^2$

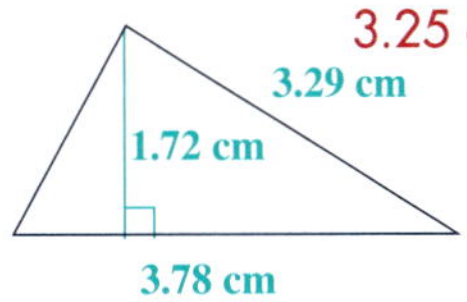


**8.** $2.20 \text{ cm}^2$

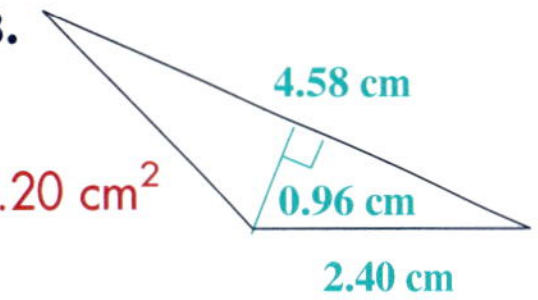


**9.** Consider this triangle.

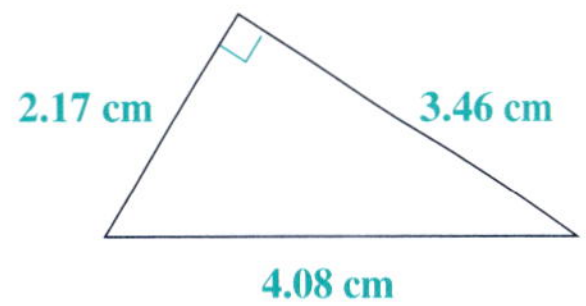


**a.** Which of the given measurements would you use to find the area of the triangle? Why?

**9a.** 2.17 cm and 3.46 cm; They are a base and height of the triangle.

**b.** What is the triangle's area? Round your answer to the nearest hundredth. $3.75 \text{ cm}^2$

**10.** The green parallelogram was created by shearing Parallelogram Z. *Shearing* a parallelogram means "sliding" one of its sides along a line parallel to the opposite side.

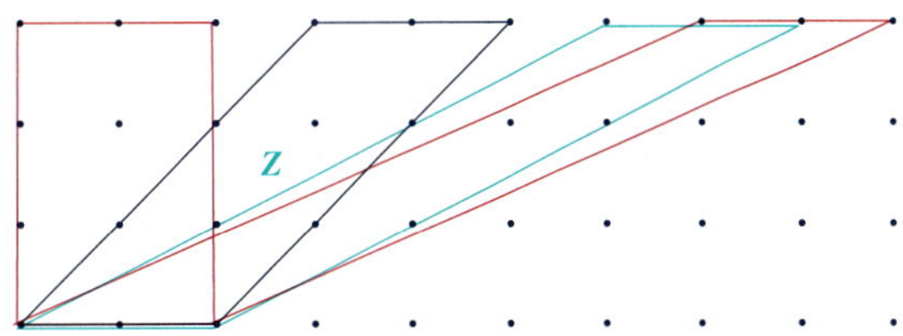


**a.** Create two more parallelograms by shearing Parallelogram Z. (In each case, "slide" the top of the parallelogram.)

**b.** Does shearing a parallelogram change its area? Explain.

**c.** Now shear Parallelogram Z to create a parallelogram with the smallest possible perimeter. What does this new parallelogram look like? It is a rectangle.

**d.** Meela sheared Parallelogram Z to create this figure. She says she has drawn the sheared parallelogram with the greatest possible perimeter. Do you agree with her? Explain.

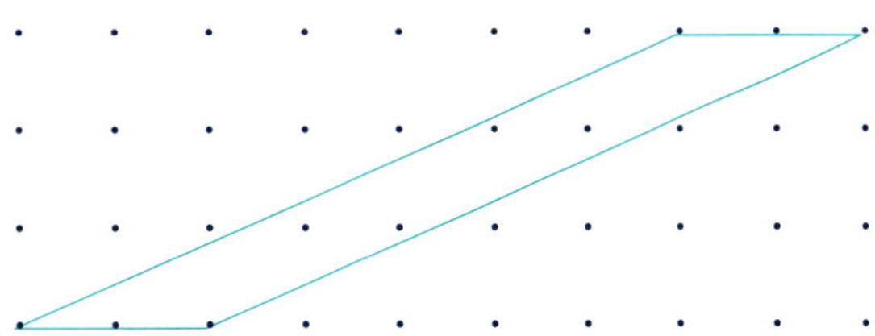

**10a.** See the figure for a possible answer.

**10b.** No; the base and height remain the same, so the area does not change.

**10d.** No; you can make a longer perimeter by adding more dots to the grid and stretching the top even farther to the right.



**Exercises 7–9:**
These exercises give practice in using the area formula for triangles. They will also help assess whether students have a clear understanding of the terms *base* and *height* for triangles.

**Exercise 10:**
This exercise should remind students of the problems in Problem Set G on page 521. Interested students may want to think about the possibility of shearing other types of polygons.

LESSON 8.4 Calculating Areas 529

# Assign and Assess

**Exercise 12:**
Students should be careful to use the *radius* of the circle to compute the area.

**Exercise 14c:**
Students may need to make a sketch to show the shape of the region in which the dog can play. The region will be three-quarters of a circle.

**Exercise 15:**
Students should carefully note which bases are associated with which heights.

In Exercises 11–13, use the $\pi$ key on your calculator to approximate $\pi$. (If your calculator does not have a $\pi$ key, use 3.14 as an approximation for $\pi$.) Round your answers to the nearest hundredth.

**11.** Calculate the area of a circle with radius 8.5 inches. 226.98 $in^2$

**12.** Calculate the area of a circle with diameter 15 feet. 176.71 $ft^2$

**13.** Calculate the area of a circle with circumference 90 feet. 644.58 $ft^2$

**14.** A dog is tied to a 15-foot leash in the center of a yard.

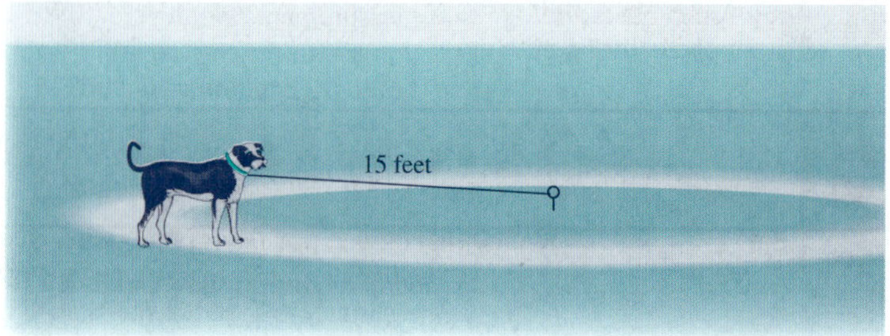

**a.** What is the shape of the area in which the dog can play? a circle

**b.** To the nearest square foot, what is the area of the space in which the dog can play? 707 $ft^2$

**c.** Suppose that, instead of being tied in the center of the yard, the dog is tied to the corner of the house. To the nearest square foot, what is the area of the space in which the dog can play? (The sides of the house are longer than the leash.) 530 $ft^2$

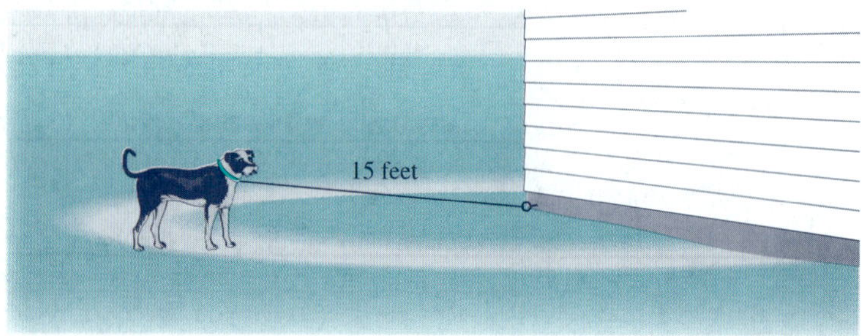


## Connect & Extend

**15.** This parallelogram has an area of 20.03 $cm^2$. Find the values of $a$, $b$, $c$, and $d$ to the nearest hundredth of a centimeter.

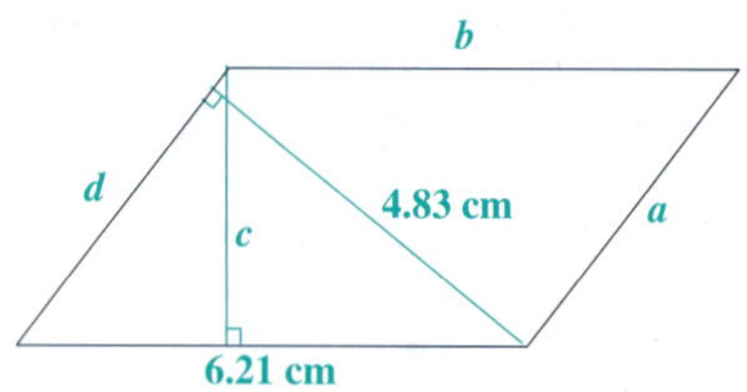


$a = d = 4.15$ cm, $b = 6.21$ cm, $c = 3.23$ cm



530 CHAPTER 8 Geometry and Measurement

**16c.** Answers will vary. The parallelogram with the smallest perimeter is a rectangle with length 15 cm and width 7 cm.

**16.** In this exercise, you will draw parallelograms.

**a.** Draw three different parallelograms with base length 15 cm and height 7 cm. Drawings will vary.

**b.** Which of your parallelograms has the least perimeter? Which has the greatest perimeter? Answers will vary.

**c.** Could you draw a parallelogram with the same base and height and an even smaller perimeter? If so, draw it. If not, explain why not.

**17.** A deck of cards has been pushed as shown. Notice that the sides of the deck are shaped like parallelograms.

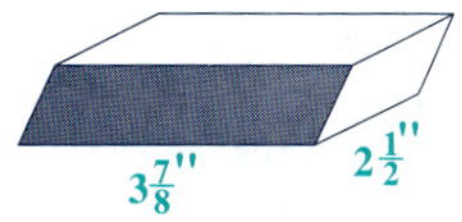


The deck contains 52 cards. Each card is $\frac{1}{48}$ of an inch thick, $3\frac{7}{8}$ inches long, and $2\frac{1}{2}$ inches wide. Find the area of the shaded parallelogram. $4\frac{19}{96}$ in.$^2$, or about 4.2 in.$^2$

**18.** Below is a floor plan for a museum, divided into four parallelograms and a rectangle. Find the area of the floor to the nearest hundredth of a square meter. 50.04 m$^2$

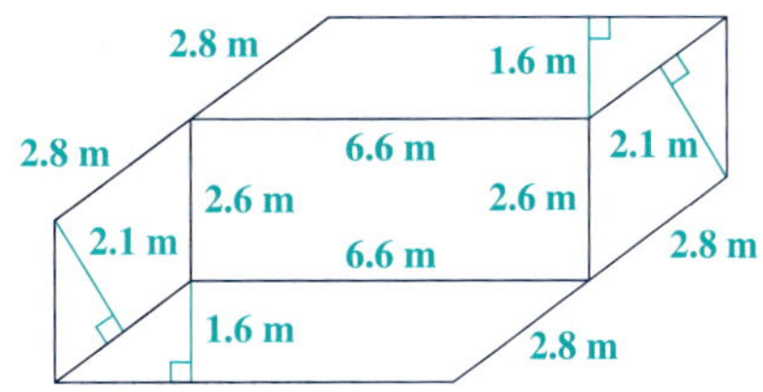


**19.** The area of this triangle is 782 square centimeters. Find $a$ and $b$ to the nearest tenth of a centimeter. $a$ = 53.0 cm, $b$ = 42.6 cm

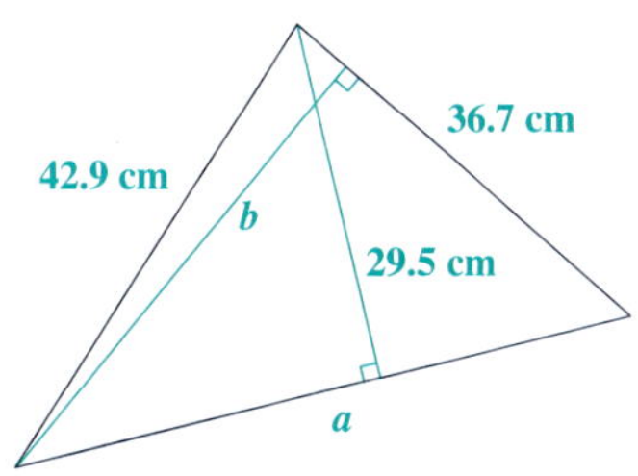




**Exercise 17:**
Some students may have difficulty visualizing this situation. Tell them not to worry about whether the corners of the cards are curved.

**Exercise 18:**
For each part of the figure, students should be careful in deciding which lengths correspond to heights and which to bases.

**Exercise 19:**
If students are uncertain about the segment to which 36.7 refers, tell them that it is the length of the shortest side of the largest rectangle.

LESSON 8.4 Calculating Areas 531

**Exercise 20:**
The student can use the information about the area and height to calculate the length of the corresponding base. The base is a side of the triangle, and all of the sides are of equal length.

**Exercise 21:**
In **Part a,** students can use the given figure to obtain more information about the lengths of parts of the figure. For example, they might redraw the figure as shown below.

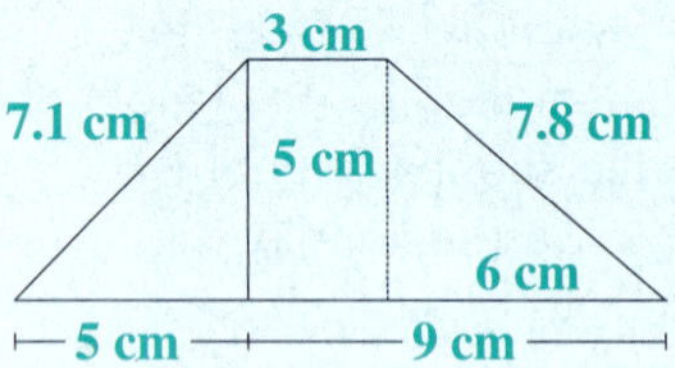


The area of the trapezoid is the area of the two triangles plus the area of the rectangle.

**Part c** will probably be a genuine challenge for most students. The two explanations given in the Additional Answers are not the only ones possible!

**20.** about 8.0 cm

### Remember

*Two lines or segments are parallel if they never meet no matter how far they are extended.*

**21b.** Possible answer: 5 cm, 5 cm, 3 cm, and 9 cm

**21c.** See Additional Answers.

**20.** In an *equilateral triangle,* all three sides are the same length. Suppose the area of an equilateral triangle is 27.7 $cm^2$ and the height is 6.9 cm. How long are each of the triangle's sides?

**21.** A *trapezoid* is a quadrilateral with exactly one pair of parallel sides, called *bases.* The *height* is the length of a perpendicular segment from one base to the other.

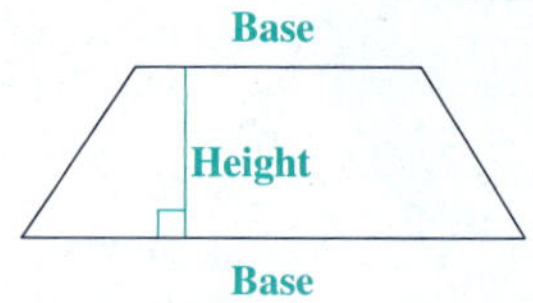


**a.** Find the area of this trapezoid: 42.5 $cm^2$

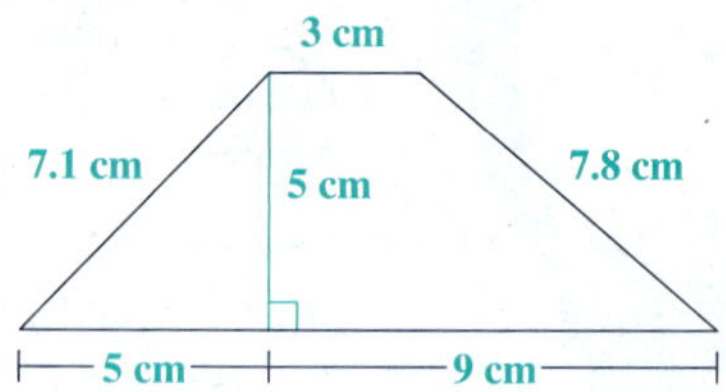


**b.** Which of the given measurements did you use to find the area?

**c.** **Challenge** Write a formula for the area of a trapezoid $A$ if the lengths of the bases are $B$ and $b$ and the height is $h$. Explain how you found your answer. (Hint: Divide the trapezoid into two triangles, or form a parallelogram from two copies of the trapezoid.)

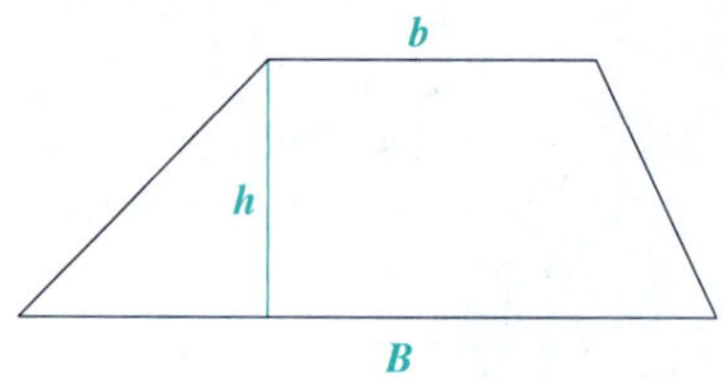




• Additional Answers on page A669

532 CHAPTER 8 Geometry and Measurement

## Remember

In a *regular polygon,* all sides are the same length and all angles have the same measure.

**22.** Any regular polygon can be divided into identical triangles. This hexagon is divided into six identical triangles.

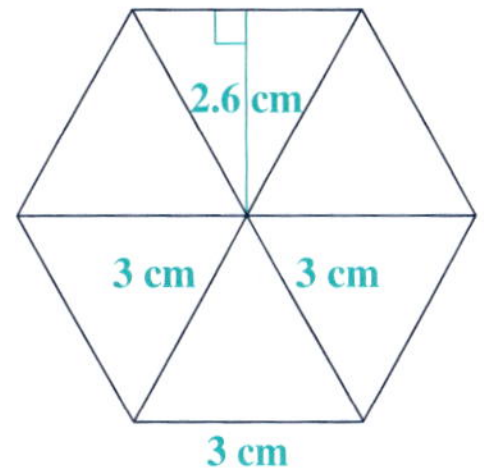


**a.** Find the area of each triangle and the area of the hexagon to the nearest tenth of a square centimeter. Explain how you found the areas. See Additional Answers.

**b.** This formula can be used to find the area of a regular polygon:

$$A = \frac{1}{2} \times \text{polygon perimeter} \times \text{height of one triangle}$$

Show that this formula gives you the correct area for the hexagon above.

**c.** Why do you think the formula works? See Additional Answers.

**d.** A stop sign is in the shape of a regular octagon. This sketch of an octagon has been divided into eight identical triangles.

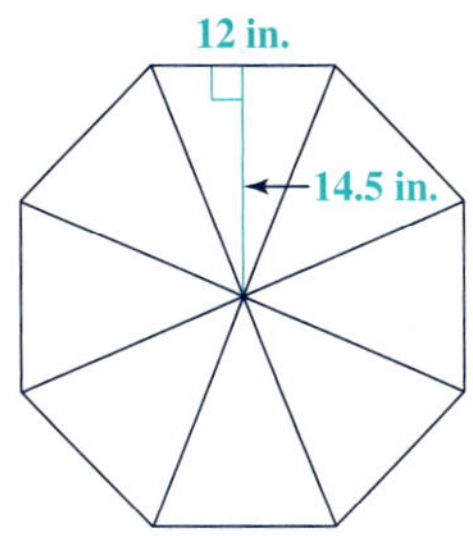


Use the formula from Part b to find the area of a stop sign.

**e.** Brett found the area of the stop sign by surrounding it with a square.

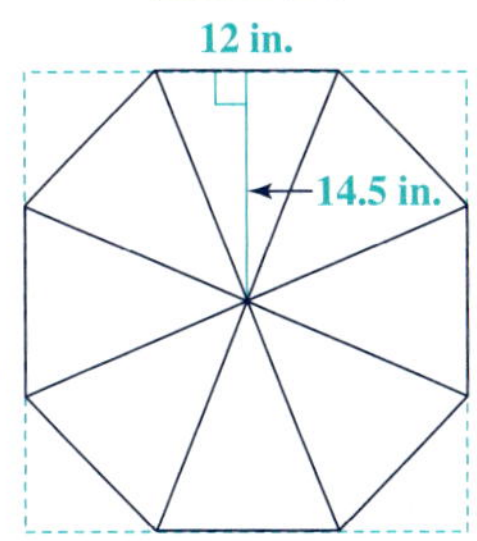


How long are the sides of the square? How long are the perpendicular sides of the small triangles in the corners of the square?

**f.** Explain how Brett might have calculated the area. Show that this method gives the same area you found in Part d.

He could have found the area of the square and then subtracted the area of the four triangles. The area of the square is $29^2$ in.$^2$, or 841 in.$^2$. The area of each triangle is $\frac{1}{2} \cdot 8.5$ in. $\cdot$ 8.5 in., or 36.1 in.$^2$, so the area of the octagon is $841 - 4 \cdot 36.1$, or 696.6 in.$^2$, which is very close to my answer from Part c. The difference might be due to rounding.

**22b.** $A = \frac{1}{2} \cdot (3 \cdot 6) \cdot 2.6 = 23.4$ cm$^2$, which is the same area I found in Part a.

**22d.** 696 in.$^2$.

**22e.** square's sides: 29 in.; triangles' sides: 8.5 in.



**Exercise 22c:**
Some students may prefer to write and simplify mathematical expressions to answer the question.

**Exercise 22e:**
Students should refer to the diagram as they read the questions posed. This will help them understand exactly what lengths they are being asked to find.

It should not be too difficult for students to see that the sides of the surrounding square are twice the height of one of the identical triangles.

### Additional Answers

**22a.** Area of triangle: 3.9 cm$^2$; Area of hexagon: 23.4 cm$^2$; Possible explanation: I found the area of a triangle using the formula $A = \frac{1}{2} \cdot b \cdot h$. I found the area of the hexagon by multiplying the area of the triangle by 6.

**22c.** Possible explanation: Because the perimeter is the sum of the six bases of the triangles. Instead of calculating the area of each triangle individually, the formula groups them all together and multiplies the sum by $\frac{1}{2}$ and the height.

LESSON 8.4 Calculating Areas 533

# Assign and Assess

**Exercise 23:**
Students can use 3.14 for $\pi$.

**Exercise 24:**
Students can easily perform the calculations for this exercise without a calculator. They can use 3.14 for $\pi$.

If students have difficulty visualizing three-dimensional objects, it may help to use a physical model. For **Part a,** a rectangular box for paper clips or a thick book may help. Students need to understand that the information in the diagram actually gives them the lengths of *all* edges of the box, including the edges that cannot be seen in the diagram. Realizing that opposite faces of the box are the same size and shape can save time with the calculations.

For **Part b,** students need to see that the length of the rectangle is equal to the circumference of each of the circular ends. Modeling the curved wall of the can with a piece of paper is easy and convincing. There should be no overlap where the edges meet.

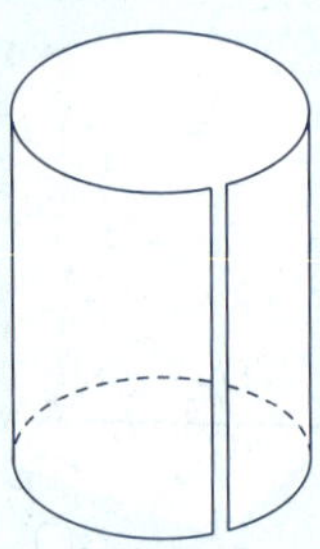

★ indicates multi-step problem

**Just the facts**

The world's highest fountain, located in Fountain Hills, Arizona, is capable of sending 8-ton streams of water 560 feet into the air, 10 feet higher than the Washington Monument.

★ **23.** The Smallville town council plans to build a circular fountain surrounded by a square concrete walkway. The fountain has a diameter of 4 yards. The walkway has an outer perimeter of 28 yards.

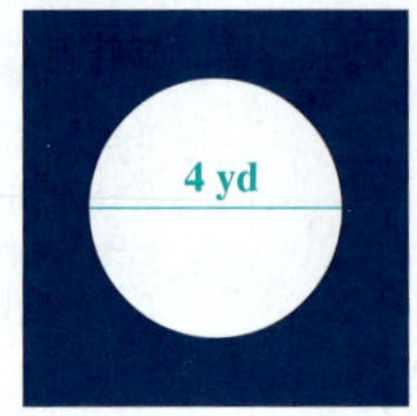


36.4 yd$^2$

Find the area of the walkway to the nearest tenth of a square yard.

**24.** The *surface area* of a three-dimensional figure is the sum of the areas of its faces. For example, this cube is made up of six faces, each with area 9 in.$^2$. So, its total surface area is $9 \cdot 6$, or 54, in.$^2$.

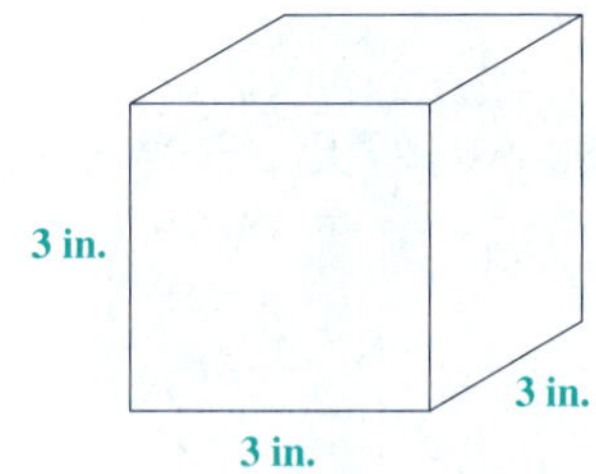


**a.** Find the surface area of this rectangular box: 27 ft$^2$

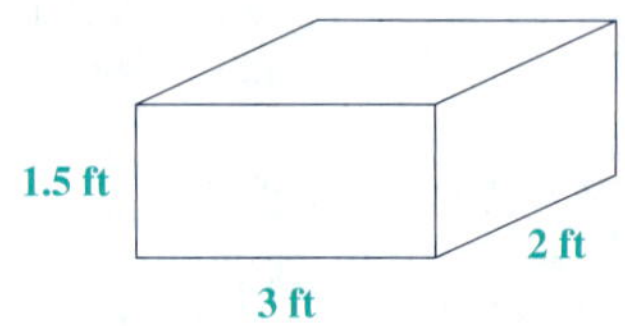


**b. Challenge** To find the surface area of a cylinder, you can imagine it as three separate pieces: the circular top and bottom, and the rectangle wrapped around them. Find the surface area of this cylinder. (Hint: You need to figure out what the length of the rectangle is. To do this, think about how this length is related to the circles.)

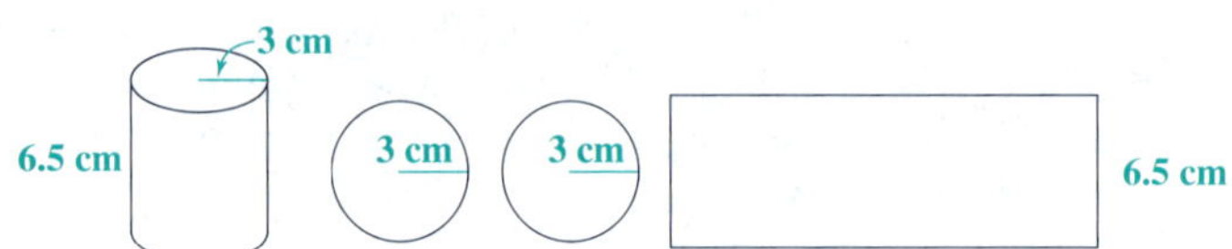


about 179 cm$^2$

## Mixed Review

**Geometry** Find the perimeter of each figure.

**25.** 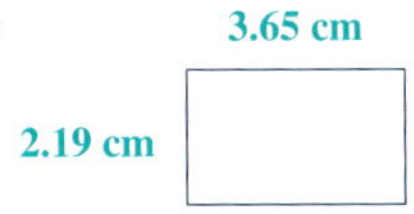

11.68 cm

**26.** 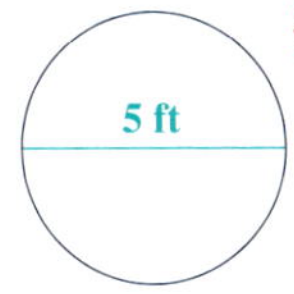

$5\pi$ ft, or about 15.7 ft

**27.** 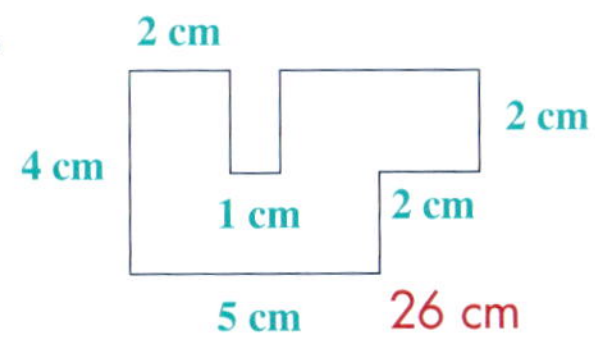

26 cm

**28.** 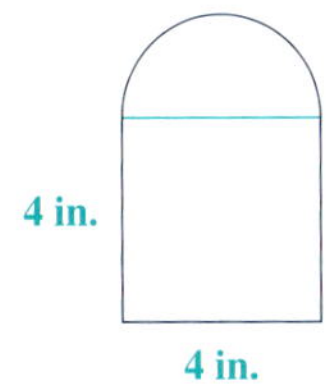

$12 + 2\pi$ in., or about 18.3 in.

Find each sum or difference.

**29.** $3\frac{1}{2} - 1\frac{5}{8}$ $1\frac{7}{8}$

**30.** $1\frac{7}{12} + 4\frac{2}{3}$ $6\frac{1}{4}$

**31.** $12\frac{6}{7} + 5\frac{5}{6}$ $18\frac{29}{42}$

**32.** 37.42 − 9.04

**33.** 553.89 + 332.7

**34.** 2,545 − 1,365.787

**35.** Hannah and Rosita each wrote rules for the number of toothpicks in each term of this sequence:

Both girls used $t$ to represent the number of toothpicks and $n$ to represent the term number. Use words or diagrams to explain why each rule is correct.

**a.** Hannah's rule: $t = 2 \cdot n + 2 \cdot n$

**b.** Rosita's rule: $t = 4 + 4 \cdot (n - 1)$

**36.** **Preview** Rachel set a thermometer in a beaker of liquid. She recorded the liquid's temperature every 5 minutes. From the data she collected, she wrote an equation to represent the temperature of the liquid $T$ in degrees Celsius after $m$ minutes.

$$T = 72 \times 0.97^m$$

Use Rachel's equation to approximate the liquid's temperature after 10 minutes. about 72.74°C

### In your own words

Explain the similarities and differences among the area formulas you learned in this lesson.

**32.** 28.38
**33.** 886.59
**34.** 1,179.213
**35.** See Additional Answers.



### Quick Check

**Informal Assessment**
Students should be able to:

- ✔ derive and apply a formula for the area of a parallelogram
- ✔ derive and apply a formula for the area of a triangle
- ✔ estimate areas of circles drawn on grids and derive a formula for the area of a circle

### Additional Answers

**35a.** Possible explanation: Each term can be split into two halves. The number of toothpicks in each half is equal to 2 times the term number. So, Term $n$ has $2 \cdot n + 2 \cdot n$ toothpicks.

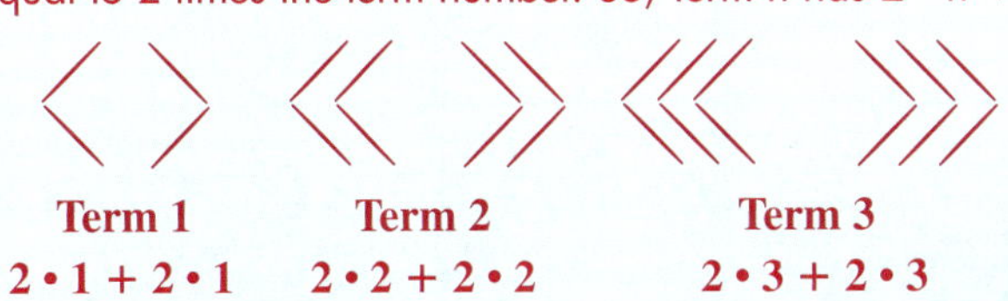


• *Additional Answers continued on page A669*
• *Quick Quiz on page A669*

# Teacher Notes

# 8.5 The Pythagorean Theorem

## Objectives

- To use right triangles and squares drawn on dot paper to arrive at the Pythagorean Theorem
- To prove the Pythagorean Theorem
- To use the Pythagorean Theorem to find missing lengths of sides of right triangles

## Overview (pacing: about 3 class periods)

In this lesson, students learn about the Pythagorean Theorem. They study several illustrative examples that use right triangles and squares drawn on dot paper. Students next examine a proof of the Pythagorean Theorem. Finally, they study some mathematical consequences and real-life applications of the theorem.

## Advance Preparation

For Investigation 1, students will need copies of Masters 74, 75, 76, and 84.

## Lesson Planner

| | Summary | Materials | On Your Own Exercises | Assessment Opportunities |
|---|---|---|---|---|
| Investigation 1 page T536 | Students learn the Pythagorean Theorem and examine a proof of the theorem. | • Master 74<br>• Master 75<br>• Master 84<br>• Master 76<br>• Scissors | Practice & Apply: 1–3, p. 544<br>Connect & Extend: 11, 12, pp. 546–548<br>Mixed Review: 15–28, p. 550 | Share & Summarize, pages T540, 540<br>Troubleshooting, page T540 |
| Investigation 2 page T540 | Students study applications of the Pythagorean Theorem, especially those that involve finding missing lengths of sides of a right triangle. | | Practice & Apply: 4–10, p. 545<br>Connect & Extend: 13, 14, pp. 548–549<br>Mixed Review: 15–28, p. 550 | Share & Summarize, pages T543, 543<br>Troubleshooting, page T543<br>Informal Assessment, page 550<br>Quick Quiz, page 550 |

# Introduce

1 Tell students that the Pythagorean Theorem, which they will study in this lesson, was discovered many centuries ago. It is one of the most famous facts in mathematics, and it has many uses in mathematics and real life. Tell them that it states a fact that is true for all right triangles. Discuss the meaning of the term *right triangle,* and sketch some right triangles on the board.

2 **Explore**
Students will be using right triangles and squares drawn on dot paper in the problems leading up to the Pythagorean Theorem. This Explore will serve as a warm-up. You may want to allow students to work in pairs. Each pair will need one copy of Master 74.

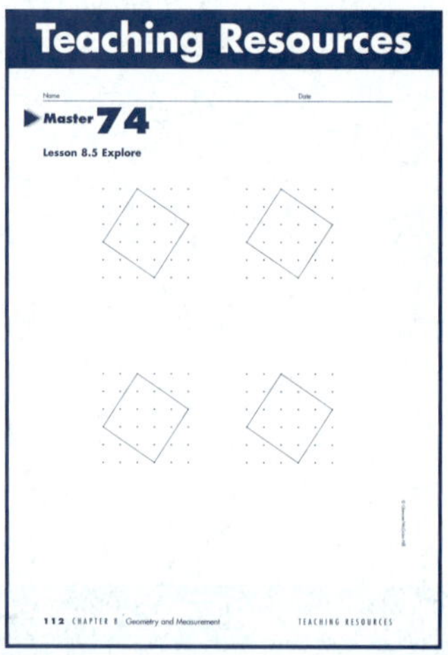


One way to find the area of the square is to draw segments that divide it into four right triangles and one unit square. The right triangles each have an area of $\frac{1}{2} \cdot 2 \cdot 3$, or 3 $cm^2$. The area of the unit square is 1 $cm^2$. So, the total area is $4 \cdot 3 + 1$, or 13 $cm^2$.

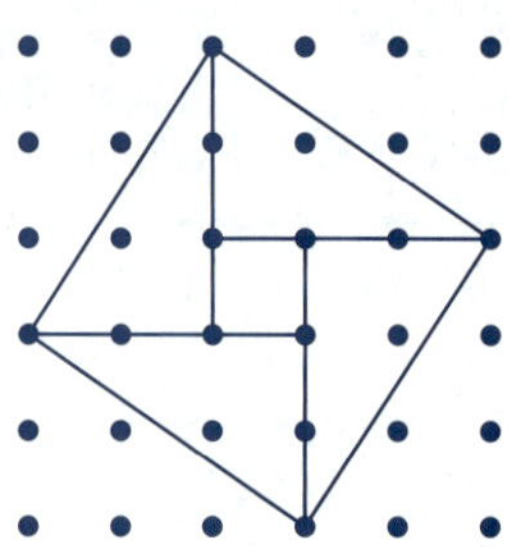

Another method is to enclose the given square in a larger square, as shown below. To find the area of the original square, find the area of the larger square and subtract the areas of the four right triangles that surround the smaller square.

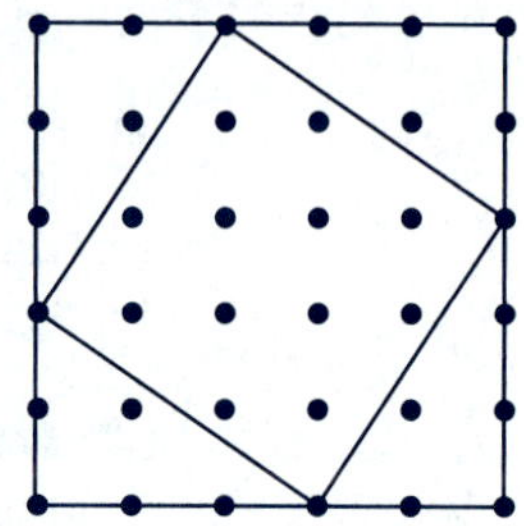

# Investigation 1

3 In this investigation, students learn the terms *hypotenuse* and *legs.* They examine several figures on dot paper that illustrate the Pythagorean Theorem. They learn what the theorem says and examine a proof.

Discuss the new vocabulary with the class. Emphasize that the term *hypotenuse* is used only in connection with right triangles. In a right triangle, the hypotenuse is always the side opposite the right angle. It is always the longest side of the right triangle.

# 8.5 The Pythagorean Theorem

VOCABULARY
**right triangle**

In this lesson, you will learn a famous mathematical fact known as the *Pythagorean Theorem.* The Pythagorean Theorem expresses a remarkable relationship between **right triangles**—triangles that have one 90° angle—and squares.

Before you investigate the theorem, you will practice finding the area of a square drawn on dot paper.

**1** *Introduce the term right triangle and how it applies to the Pythagorean Theorem.*

MATERIALS
copy of the square

## Explore

Find the exact area of this square. Describe the method you use.

13 sq units; See the teaching notes for possible strategies.

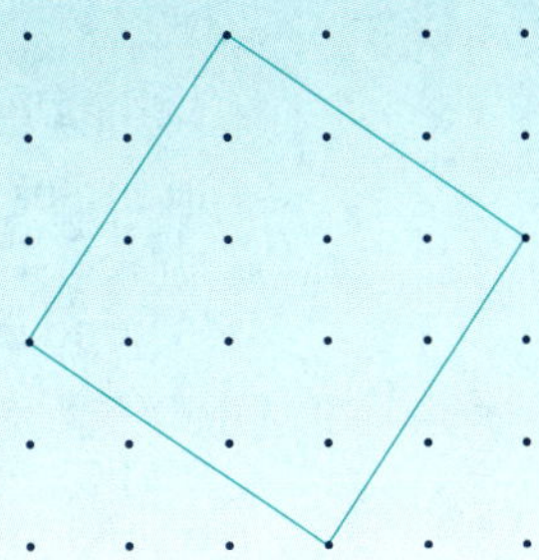

**2** *You may have students work in pairs on the Explore.*

## Investigation 1 Right Triangles and Squares

VOCABULARY
**hypotenuse**
**leg**

Every right triangle has one right angle. The side opposite the right angle is called the **hypotenuse.** The other two sides are called the **legs.**

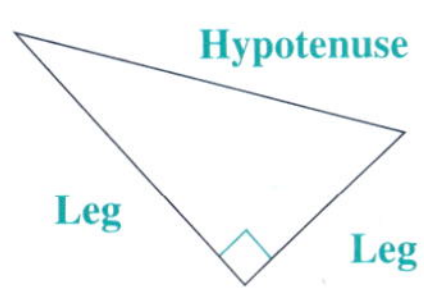


**3** *Introduce the terms hypotenuse and legs.*

The relationship expressed in the Pythagorean Theorem involves the areas of squares built on the sides of a right triangle. In the next problem set, you will try to discover that relationship.

# Develop

**1** **Problem Set A** **Suggested Grouping: Pairs**
In this problem set, students use Master 75 to examine right triangles and squares drawn on their sides. Studying the relationship between the areas of the squares leads students to the Pythagorean Theorem.

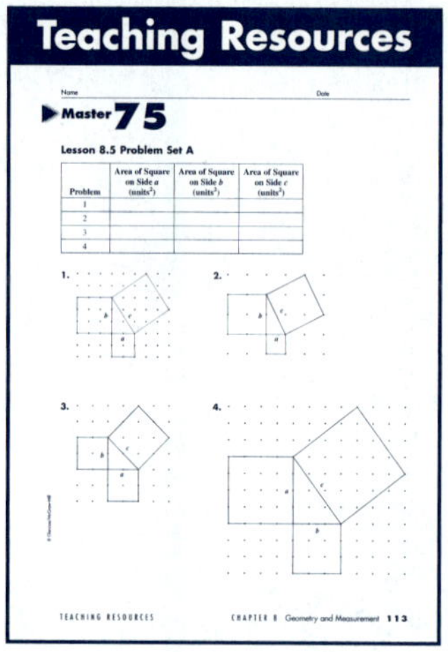


**2** In **Problems 1–4,** students can easily find the areas of the squares drawn on the legs of the right triangles. It is clear what the lengths of the sides are, and one need only apply the formula for the area of a square. For a square drawn on the hypotenuse, students can use techniques that worked for the square in the Explore on page 536.

Students have probably overlooked one subtle but important detail: how do we know that the quadrilateral drawn on each hypotenuse really is perfectly square? They certainly look square, but how can we be sure the interior angles are exactly 90°? A look at the figure for Problem 1 can resolve this question.

**3** The diagram below shows the figure for Problem 1. Notice that another right triangle has been included on the right side.

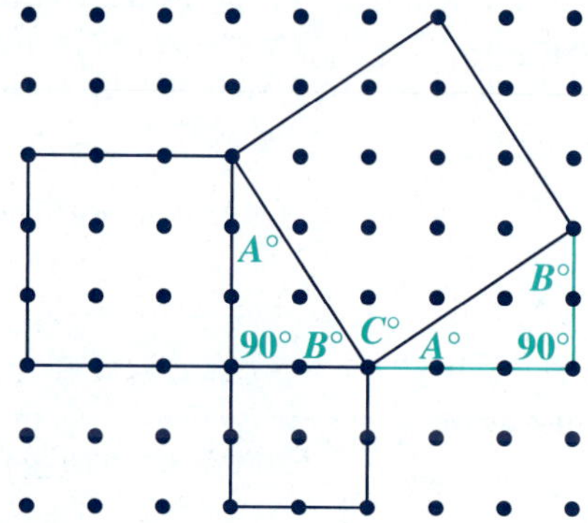


Notice the labels indicating angle measures in the right triangles. The two right triangles are the same size and shape, so the two acute angles have equal measures $A°$ and $B°$, as indicated. In the original right triangle, $A° + B° + 90° = 180°$, since the measures of the angles of a triangle have a sum of 180°. From this equation, it follows that $A° + B° = 90°$. At the point where the two right triangles touch, the sum of the three angles composing the 180° straight angle can be written as $A° + C° + B° = 180°$. Regrouping the addends gives $(A° + B°) + C° = 180°$. Since $A° + B° = 90°$, then $C° = 90°$. In a similar manner, all the other angles of the quadrilateral drawn on the hypotenuse are 90° angles. The side lengths of the quadrilateral are all the same, so the quadrilateral is indeed a square.

MATERIALS
- copies of the figures
- dot paper

**Just the facts**

The Pythagorean Theorem is named for Greek mathematician Pythagoras, who lived from approximately 580 to 500 B.C.

## Problem Set A

Problems 1–4 show right triangles with squares drawn on their sides. Find the exact area of each square, and record your results in a copy of the table.

| Problem | Area of Square on Side *a* (units²) | Area of Square on Side *b* (units²) | Area of Square on Side *c* (units²) |
|---|---|---|---|
| 1 | 4 | 9 | 13 |
| 2 | 1 | 4 | 5 |
| 3 | 4 | 4 | 8 |
| 4 | 16 | 9 | 25 |

1. 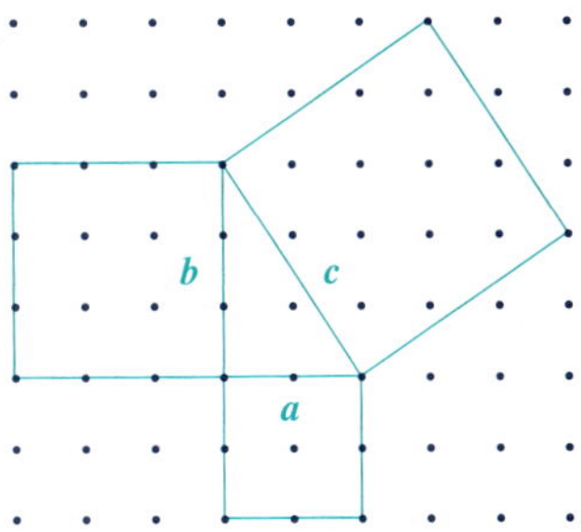


2. 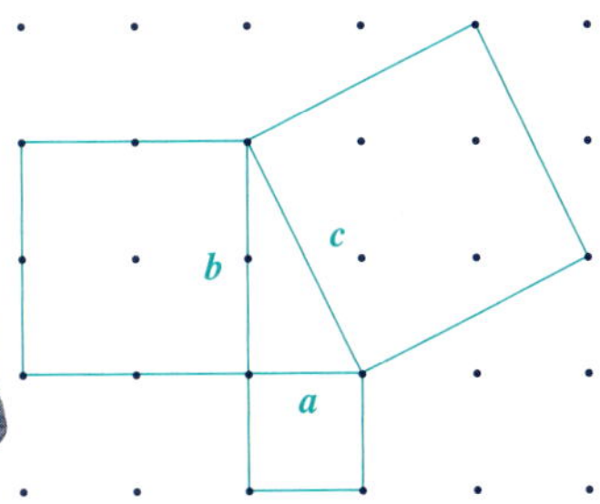


3. 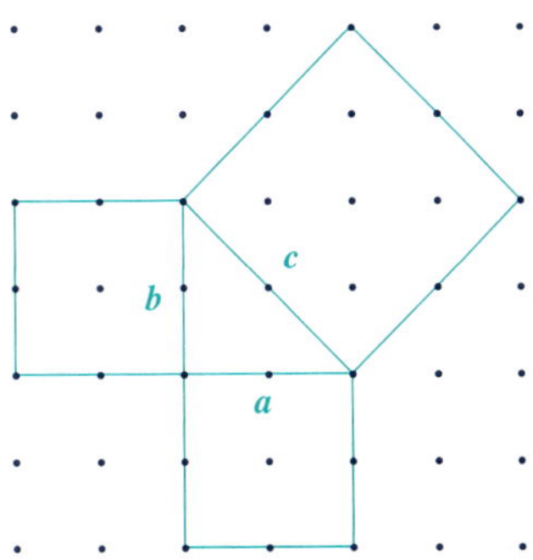


An Ionic column in Olympia, Greece

1 You may have students work in pairs.

2 Have students apply the area formula for the square.

3 Prove to students that the quadrilateral on the hypotenuse is a square.

LESSON 8.5 The Pythagorean Theorem 537

## Develop

1 For **Problem 6,** students should have a copy of Master 84 (1-centimeter dot paper). You will need to take a careful look at the diagrams students draw. They must be sure that the quadrilaterals they draw on the sides of the right triangle really are squares.

**Problem Set Wrap-Up** Be sure to discuss the answer for **Problem 5,** because the pattern students discover points the way to the Pythagorean Theorem. ■

2 When you talk about the Pythagorean Theorem, be sure to discuss both versions stated in the box on page 538. 3 Be sure students understand that although the words are different, the two versions state the same fact. Use the diagram to help students understand this.

### About the Mathematics

The relationship stated in the Pythagorean Theorem is remarkable for the fact that it is true for *all* right triangles. There is good historical evidence that its truth for *particular* right triangles (such as those drawn on dot paper in Problem Set A) was known in ancient times in Egypt, Babylonia, India, and China. Greek mathematicians were able to prove that it is true for all right triangles.

Many historians of mathematics think that Pythagoras himself was not aware of a general proof of the theorem that bears his name. Whether he was or not, it is clear that his successors were.

This theorem has deep connections with many parts of mathematics, both elementary and advanced.

4. See the table on p. 537.

4.

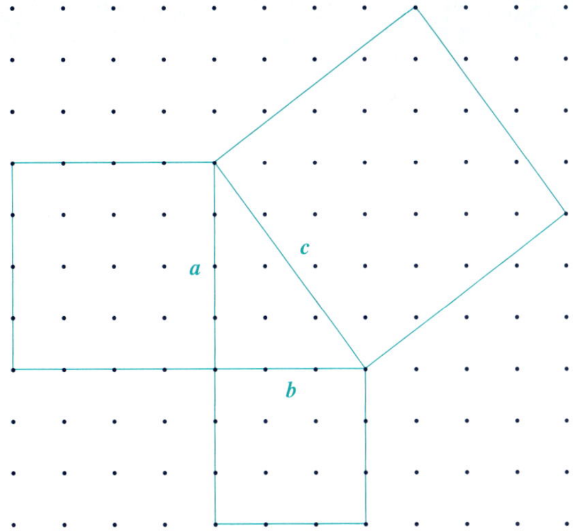


5. The area of the square on Side *a* plus the area of the square on Side *b* equals the area of the square on Side *c*.

5. Look for a pattern in your table. For the cases you considered, what is the relationship among the areas of the three squares?

6. Draw your own right triangle on dot paper. Does the relationship you described in Problem 5 hold for your triangle as well? yes

The problems in Problem Set A illustrate the Pythagorean Theorem.

**The Pythagorean Theorem**

In a right triangle, the area of the square built on the hypotenuse of the triangle is equal to the sum of the areas of the squares built on the legs.

The Pythagorean Theorem is often stated this way:

If $c$ is the length of the hypotenuse of a right triangle and $a$ and $b$ are the lengths of the legs, then $a^2 + b^2 = c^2$.

The diagram below illustrates this idea.

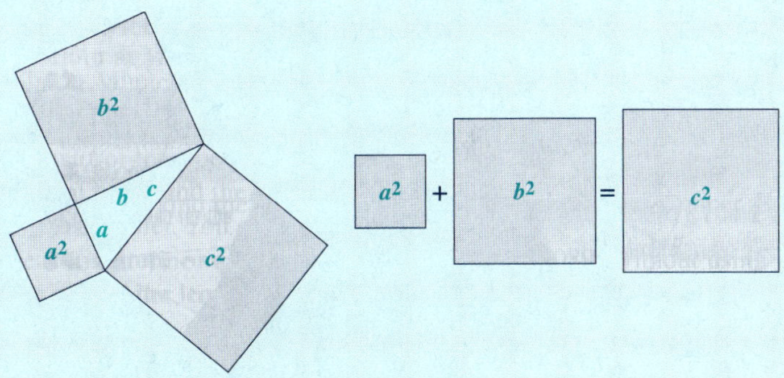


**Just the facts**

Pythagoras lived 2,500 years ago, but the famous theorem that bears his name was known even earlier. Records show that the Babylonians understood the theorem around 1500 B.C., more than 900 years before Pythagoras was born!



1 **Problem-Solving Strategies**

- Look for a pattern
- Draw a diagram

2 Discuss the Pythagorean Theorem.

3 Explain that the box contains two versions of the Pythagorean Theorem.

# Develop

1 **Problem Set B** **Suggested Grouping: Pairs**
In this problem set, students examine an actual proof of the Pythagorean Theorem.

The main idea used in the proof can be described as follows: Start with eight copies of the right triangle and the square drawn on its sides. (See Master 76.) Using these pieces, build two squares that you are sure have the same area. Remove identical pieces from each construction. The areas that remain from the two constructions will be equal.

**Problem Set Wrap-Up** Conduct a thorough discussion of the activity with the class. You may want to run through all the problems using a transparency on the overhead projector.

2 Some students may understand the activities in the problems and have a full appreciation of why proof is desirable. Other students may arrive at that stage later down the road. What is important is to give all students an opportunity to get there.

## Teaching Resources

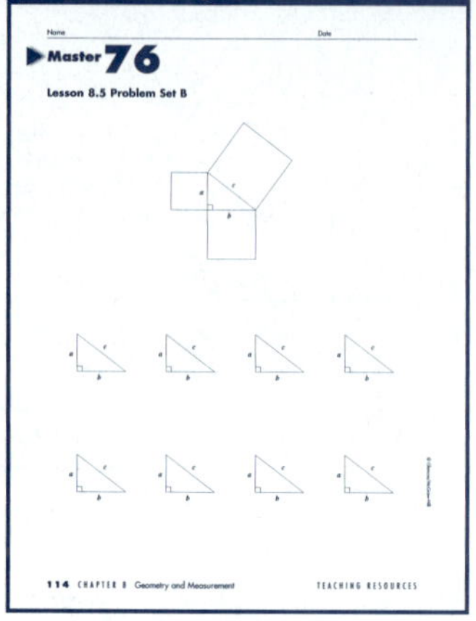
Master 76

Lesson 8.5 Problem Set B

114 CHAPTER 8 Geometry and Measurement

TEACHING RESOURCES

Throughout history, people have found new ways to *prove* the Pythagorean Theorem—that is, to show that it is always true. In Problem Set B, you will explore one such proof.

**1** You may have students work in pairs.

MATERIALS
- 8 copies of the triangle
- 1 copy of each square
- scissors

## Problem Set B

In this problem set, you will use paper triangles and squares to construct a proof of the Pythagorean Theorem.

- Start with a right triangle with squares drawn on its sides.
- Carefully cut out eight copies of the triangle and one copy of each square.

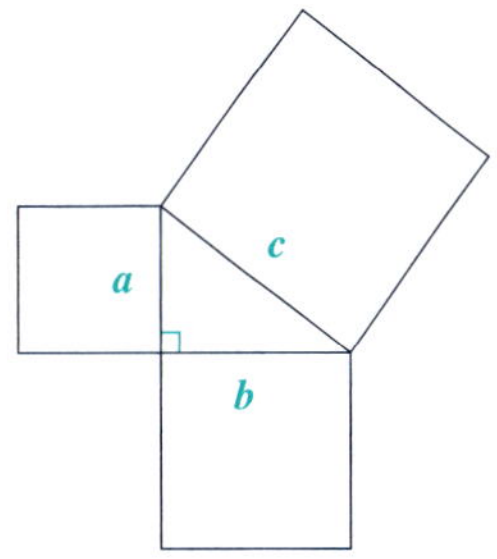


- Use four copies of the triangle, and the square from Side *c*, to make this square:

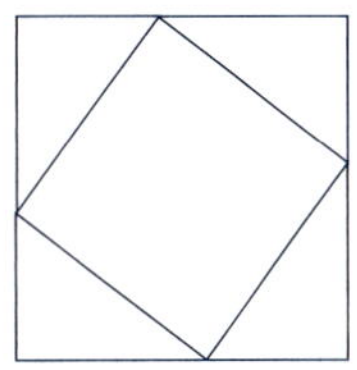

- Use four copies of the triangle, and the squares from Sides *a* and *b*, to make this square:

**1.** In both squares, the length of each side is $a + b$, the sum of the lengths of the legs of the right triangle. If two squares have the same side length, they must have the same area.

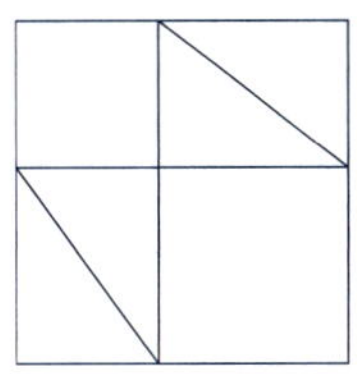

**2.** In the first square, the square from Side *c* remains. In the second, the squares from Sides *a* and *b* remain.

**1.** The two squares you made have the same area. Explain how you know this is true. See above.

Now take the four triangles away from each square you constructed.

**2.** Describe what is left of each square. See above.

**3.** Explain why the area of what is left must be the same for both squares.

**3.** I started with two squares with the same area and took away the same four pieces from each square. So, the area of what is left must be the same.

**4.** Explain how your work in this problem set shows that $a^2 + b^2 = c^2$.

The area that is left in the first square is $c^2$. The area that is left in the second square is $a^2 + b^2$. Since the areas are equal, $a^2 + b^2 = c^2$.

**2** Discuss the activity.



LESSON 8.5 The Pythagorean Theorem 539

1 **Share & Summarize**
You may want to call on volunteers to read their answers to the class. Discuss their answers to be sure the key points are clearly included.

2 **Troubleshooting** Be on the lookout for common misstatements of the Pythagorean Theorem. Students often come up with the statement, "The square of the length of the longest side of a triangle is equal to the sum of the squares of the other two sides." It would be nice to speak of the squares of the *lengths* of the other two sides, but not doing so is a minor slip. The defect is that the statement does not say that the triangle is a *right* triangle. The Pythagorean Theorem is true only for right triangles.

Other students may say, "In a right triangle, $a^2$ plus $b^2$ equals $c^2$." The trouble here is that if the statement were being made to someone who was hearing it for the first time, that person would have no way of knowing that $c$ is the length of the hypotenuse and that $a$ and $b$ are the lengths of the legs of the right triangle. This information should have been included.

### On Your Own Exercises

Practice & Apply: 1–3, p. 544
Connect & Extend: 11, 12, pp. 546–548
Mixed Review: 15–28, p. 550

# Investigation 2

In this investigation, students will apply the Pythagorean Theorem to find missing lengths and areas.

3 **Problem Set C** **Suggested Grouping: Individuals**
In this problem set, students use the Pythagorean Theorem to find missing areas of squares drawn on sides of right triangles. They then use the square root operation to find the lengths of the sides.

4 In **Problem 1,** students should note that the unknown area is the area of the square drawn on the hypotenuse. By the Pythagorean Theorem, this area is equal to 9 $\text{unit}^2$ + 16 $\text{unit}^2$, or 25 $\text{unit}^2$. The length of each side of the squares is found by taking the square root of the area values.

In **Problem 2,** the missing area, call it $A$, is the area of a square drawn on a leg of the right triangle. By the Pythagorean Theorem, $A + 2.25 = 6.25$. So $A = 4$ $\text{unit}^2$. To find the side lengths, take the square root of the area values.

**Problems 3 and 4** are similar to Problem 2. Note that the exact length of the hypotenuse is written as $\sqrt{13}$. If students give the length of this side as a decimal, the decimal should be preceded by the word *about*, since the decimal is an approximate value of $\sqrt{13}$.

The couch is too wide to fit if they carry it upright, but Marcus thinks it might fit if they tilt it like this:

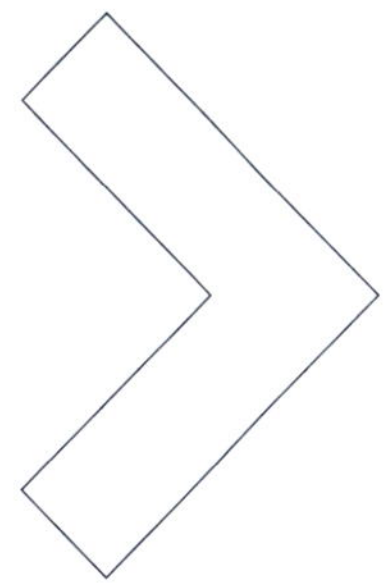

In this exercise, you will use the Pythagorean Theorem to figure out whether the couch will fit through the doorway.

**a.** Explain why the dashed segments shown here each have length 12 inches. Then find the length of Segment *c*, and explain how you found it. Round your answer up to the nearest tenth of an inch.

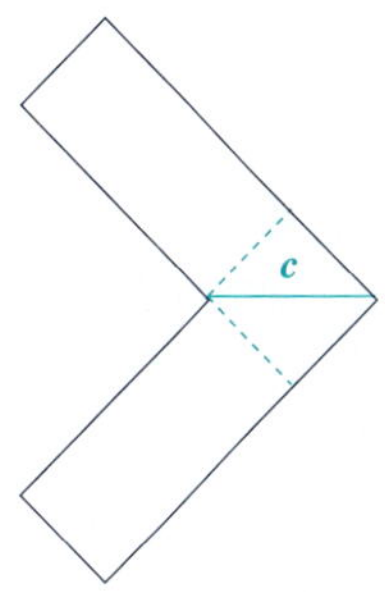


**14a.** The outside lengths are both 39 in. and the inside lengths are both 27 in. The difference between them is shown by the dashed segments, so they must be 12 in. long. Using the Pythagorean Theorem, $c^2 = 12^2 + 12^2 = 288$, so $c = \sqrt{288}$, or about 17.0 in.

**b.** Find the length of Segment *d* to the nearest tenth of an inch.
38.2 in.

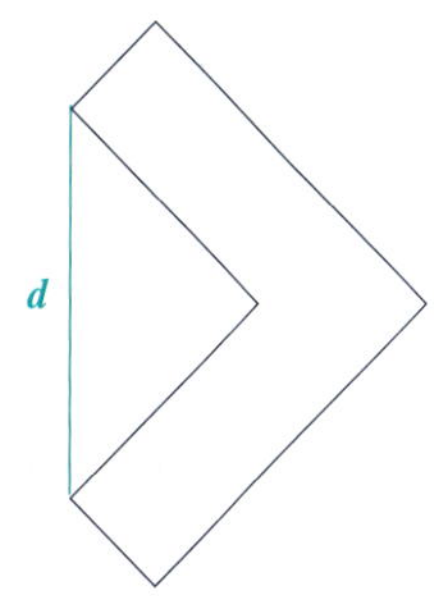


**c.** Segment *b* divides Segment *d* in half. Find the length of Segment *b* to the nearest tenth.
19.1 in.

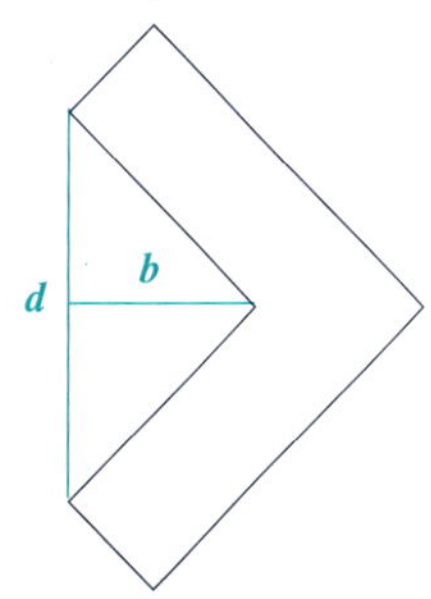


**d.** Will the couch fit through the doorway? Explain how you know.

**14d.** Yes; if the couch is tilted, its width will be $b + c$, which is about 36.1 in. This is less than the width of the doorway.



LESSON 8.5 The Pythagorean Theorem 549

## Quick Check

### Informal Assessment

Students should be able to:

- ✔ use right triangles and squares drawn on dot paper to arrive at the Pythagorean Theorem
- ✔ prove the Pythagorean Theorem
- ✔ use the Pythagorean Theorem to find missing lengths of sides of right triangles

### Quick Quiz

1. The triangle shown below is a right triangle, but the right angle has not been specially marked.

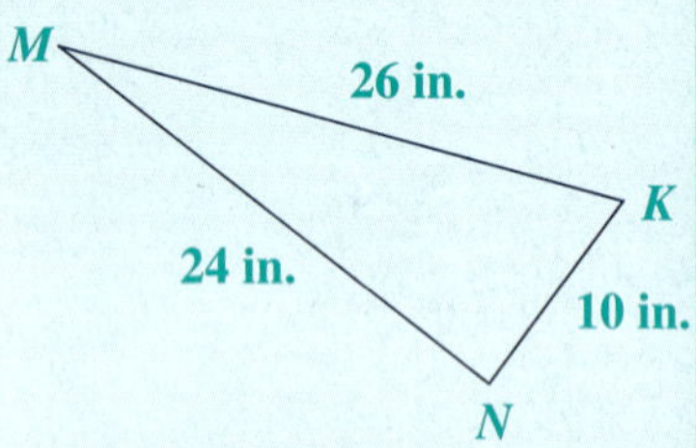


   a. What are the lengths of the legs of the triangle? **10 in. and 24 in.**

   b. Which angle is the right angle? **Angle N**

   c. What is the length of the hypotenuse? **26 in.**

## Mixed Review

15. 17
16. 19
18. 10
20. 81

Evaluate each expression.

**15.** $6 \cdot 7 - 5^2$ **16.** $\sqrt{121} - 2^2 + 4 \cdot 3$ **17.** $\frac{4^2 + 6^2}{13}$ 4

**18.** $\sqrt{6^2 + 8^2}$ **19.** $45 - 3 \cdot 5 + 3^2$ 39 **20.** $(\sqrt{16} + \sqrt{4} + \sqrt{9})^2$

**Geometry** Find the area of each figure.

**21.** $22.5 \text{ cm}^2$

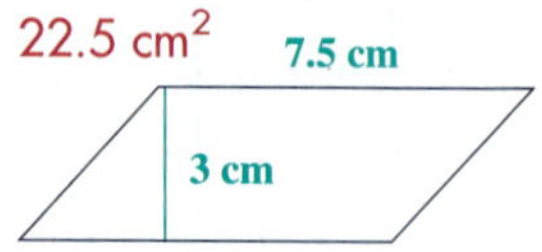


**22.**

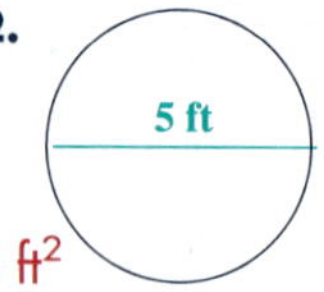


$6.25\pi \text{ ft}^2$, or about $19.6 \text{ ft}^2$

**23.** $22 \text{ cm}^2$

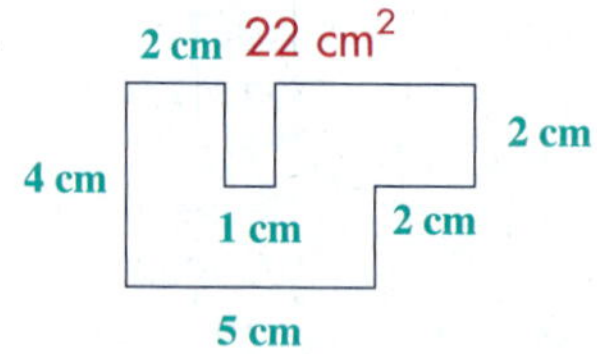


**24.**

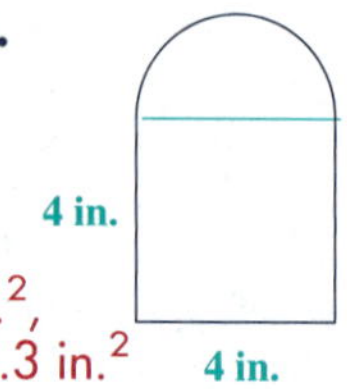


$16 + 2\pi \text{ in.}^2$, or about $22.3 \text{ in.}^2$

**25.** Luke ate 80% of the strawberries he picked. If he picked 30 strawberries, how many did he eat? 24

**26.** Ms. Friel's class has 25 students. Of these students, 23 went on the field trip to the museum. What percent of the class did not go to the museum? 8%

**27.** **Statistics** Jahmal surveyed students in the cafeteria about their favorite school lunch. He found that $66\frac{2}{3}\%$ like pizza best. If 80 students told Jahmal they like pizza best, how many students did he survey? 120

**28.** **Preview** Devon, Kyle, and Kristi are playing a game with this spinner. They take turns spinning. Each time the spinner lands on blue, Devon scores a point. Each time it lands on green, Kyle scores a point. And each time it lands on white, Kristi scores a point. Do you think the game is fair? Why or why not?

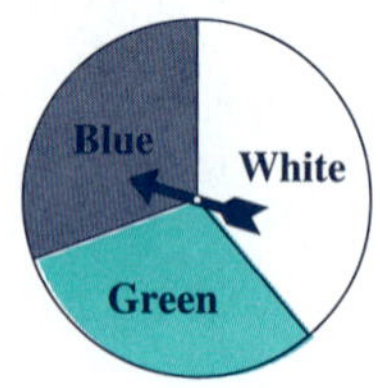


28. Possible answer: No; because the white section is larger than the other two, Kristi has a better chance of scoring points than do the other two players.



2. For the right triangle shown at the right, what is the missing length? Give the *exact* length. **$\sqrt{68}$ cm**

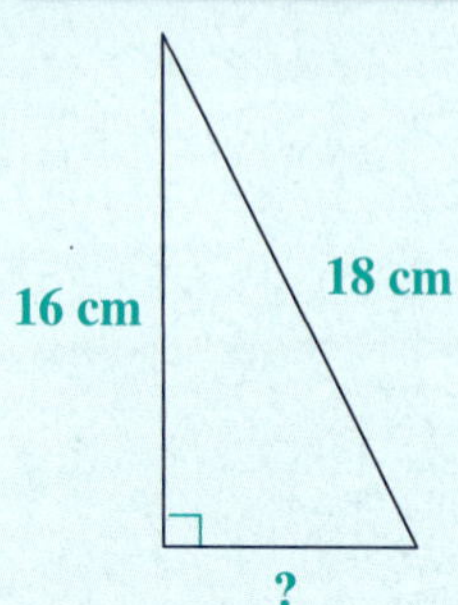


3. In the diagram at the right, the top end of the ladder is at a second-floor window that is 20 ft above ground level. The foot of the ladder is 3 ft from the wall. How long is the ladder? Round your answer to the nearest tenth of a foot. **20.2 ft**

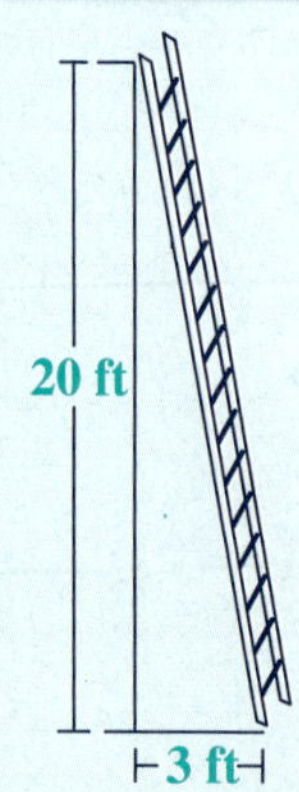

## Share & Summarize

State the Pythagorean Theorem in your own words. You might want to draw a picture to illustrate what you mean.

Possible answer: If you have a right triangle, you can make squares on each of its sides. The areas of the two smaller squares added together have the same area as the larger square.

1 Have volunteers read their answers to the class.

2 Clear up some misconceptions about the Pythagorean Theorem.

# Investigation 2 Using the Pythagorean Theorem

In this investigation, you will have a chance to practice using the Pythagorean Theorem.

3 You may have students work individually.

### Problem Set C

Find each missing area. Then use the areas of the squares to find the side lengths of the triangle.

4 Make sure students know which side of the triangle they are looking for.

**1.** 25 unit$^2$; 3 units, 4 units, 5 units
**2.** 4 unit$^2$; 1.5 units, 2 units, 2.5 units
**3.** 25 unit$^2$; 5 units, 12 units, 13 units
**4.** 4 unit$^2$; 2 units, 3 units, $\sqrt{13}$ units (about 3.6 units)

**1.**

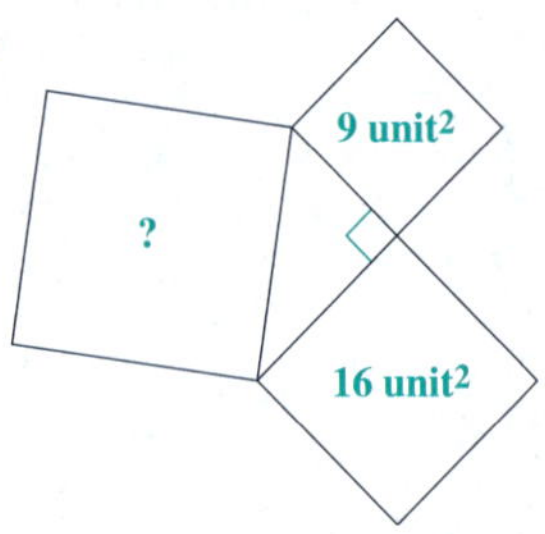


**2.**

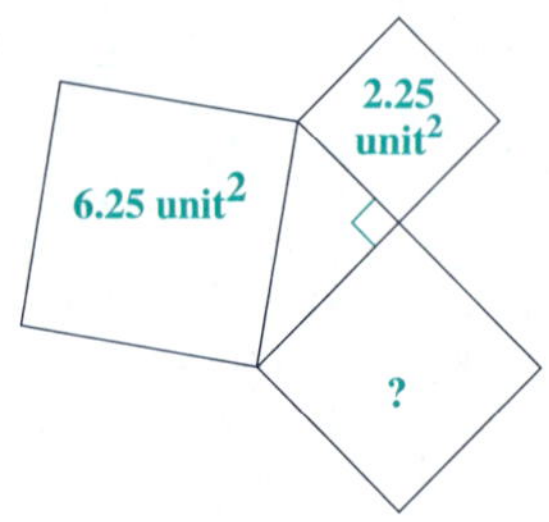


**3.**

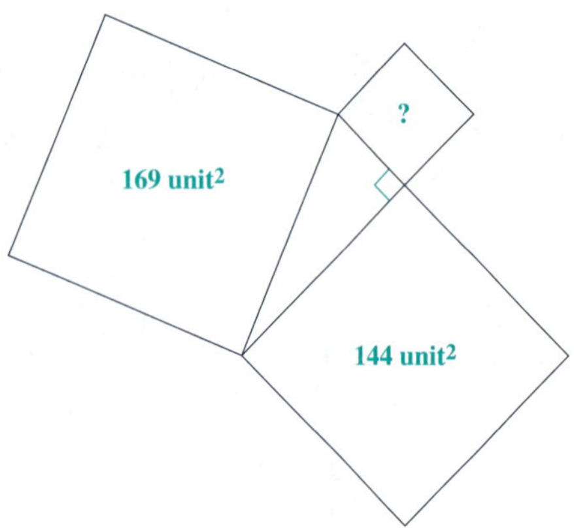


**4.**

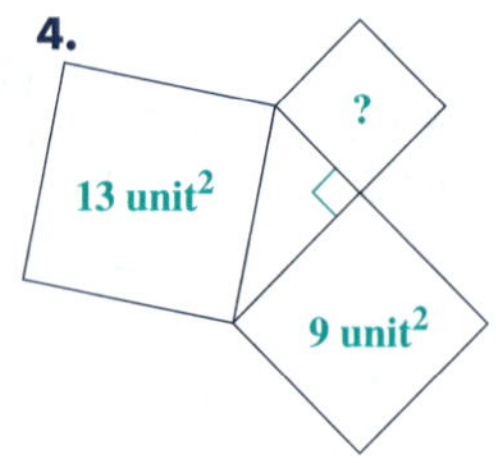

## Develop

1

**Example**
Discuss this Example with the class. Point out that here we are using the second way of stating the Pythagorean Theorem from the box on page 538.

If you know the lengths of any two sides of a right triangle, you can use the Pythagorean Theorem to find the length of the third side.

**EXAMPLE**

A right triangle has legs of length 1 inch and 2 inches. How long is the hypotenuse?

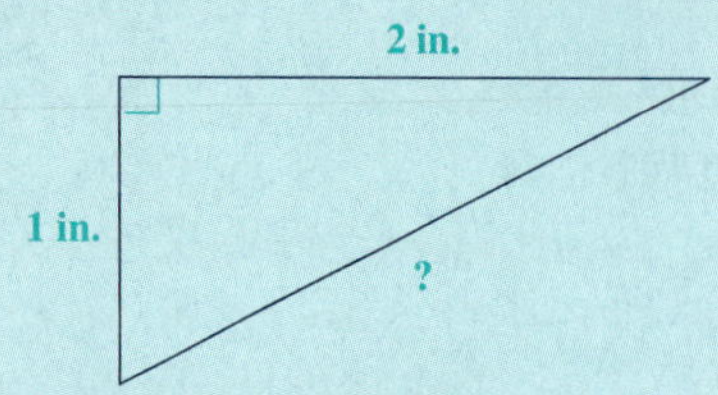


If $a$ and $b$ represent the lengths of the legs, and $c$ represents the length of the hypotenuse, the Pythagorean Theorem says that $a^2 + b^2 = c^2$.

In the triangle shown above, $a = 1$, $b = 2$, and $c$ is the length of the hypotenuse. So,

$$1^2 + 2^2 = c^2$$
$$1 + 4 = c^2$$
$$5 = c^2$$

Therefore, $c = \sqrt{5}$, or about 2.24 inches.

**1** Discuss the example with the class.

**Just the facts**

They may not have known the Pythagorean Theorem, but the Egyptians who built the pyramids knew that a triangle with side lengths of 3, 4, and 5 units must be a right triangle. They used a rope device like the one pictured here to check that they made perfect right angles at the corners of the pyramids.

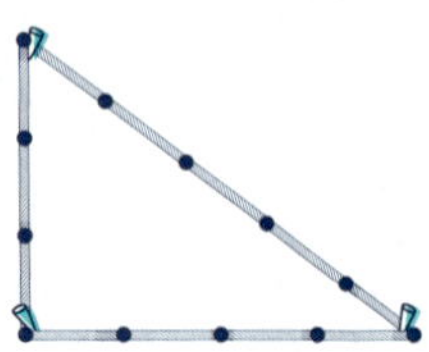

The Great Sphinx in Giza, Egypt



LESSON 8.5 The Pythagorean Theorem 541

# Develop

1 **Problem Set D** **Suggested Grouping: Individuals**
In this problem set, students look at more applications of the Pythagorean Theorem, including some real-life applications.

In **Problem 3,** the diagonal from Point *A* to Point *B* can be thought of as the hypotenuse of a right triangle whose legs have lengths 9 m and 12 m.

The method for **Problem 4** is similar to that for Problem 3.

## Problem Set D

Find each missing side length. Then find the area of the triangle.

**1.**

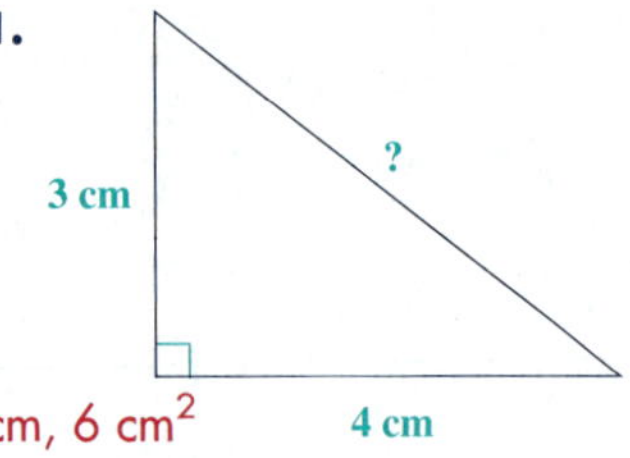


5 cm, 6 cm$^2$

**2.**

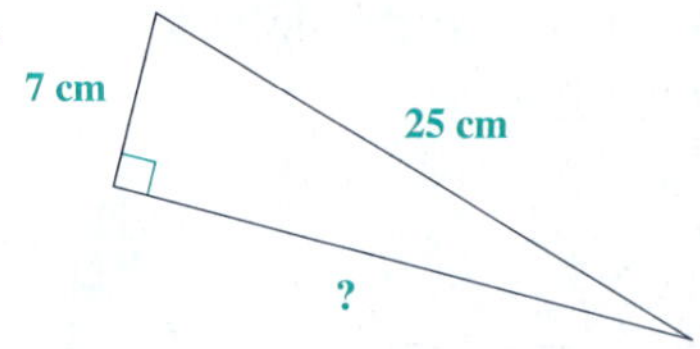


24 cm, 84 cm$^2$

**3.** A rectangular lawn measures 9 meters by 12 meters. Suppose you want to walk from Point $A$ to Point $B$.

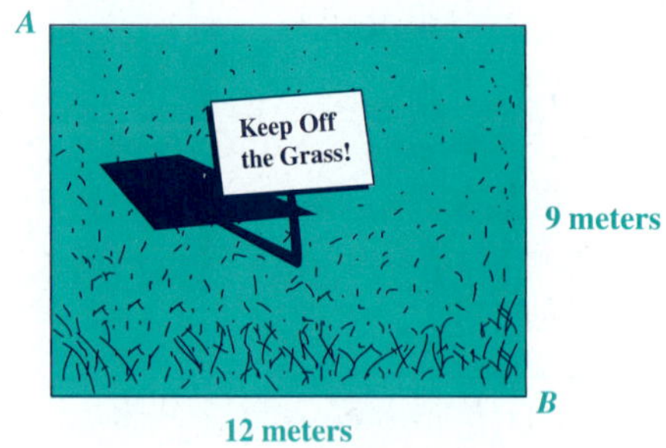


**a.** If you obey the sign and walk around the lawn, how far will you walk? 21 m

**b.** If you ignore the sign and walk directly across, how far will you walk? 15 m

**4.** A baseball diamond is a square measuring 90 feet on each side. What is the distance from home plate to second base?
about 127 ft

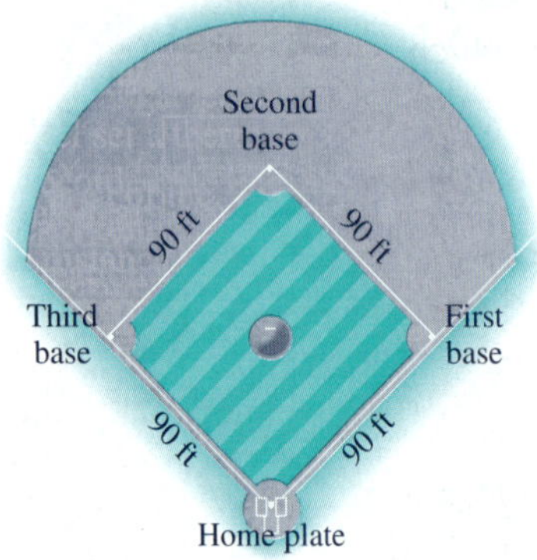


**1** You may have students work individually.

## Develop

**1** For **Problem 5,** students will probably find it helpful to sketch a diagram like the one shown below.

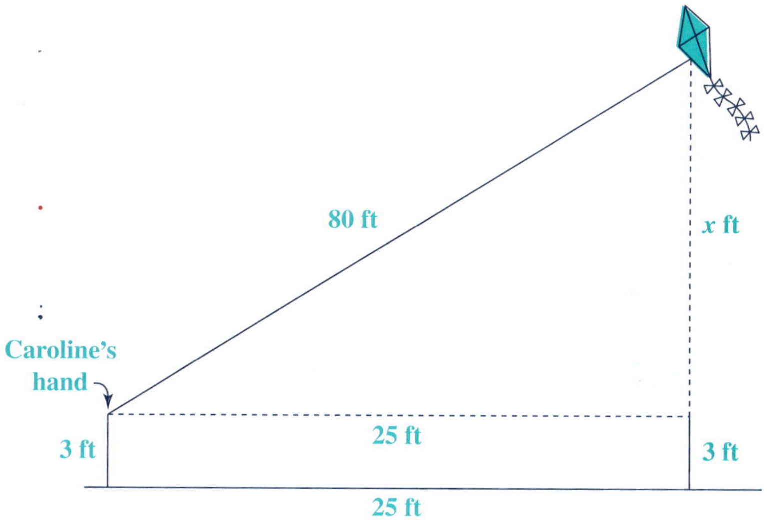


Students can use the Pythagorean Theorem to find the length *x* and then add 3 ft to find the height of the kite.

**2** In **Problem 6a,** students must multiply 11.5 by 2.5. In **Problem 6b,** ask students to sketch the right triangle they will use. Ask them to include the lengths of the legs in the diagram.

**Problem 7** takes a touch of insight. It uses the idea that the area of a triangle is the same no matter which side is chosen as a base.

**3**

### Share & Summarize

Have volunteers present their problems to the class. Check that each problem is clearly stated and solvable. Then ask the class for ideas about how to solve the problem.

**Troubleshooting** When students use the Pythagorean Theorem to find missing lengths, they may correctly find the square of the missing length but forget to take the square root. Students can usually catch the error if you ask whether the answer seems reasonable.

### On Your Own Exercises

Practice & Apply: 4–10, p. 545
Connect & Extend: 13, 14, pp. 548–549
Mixed Review: 15–28, p. 550

## Just the facts

In Thailand, *kite fighting* is a popular sport. A kite-fighting match involves a minimum of two heavy, star-shaped kites called *chulas* and at least four light, diamond-shaped kites called *pakpaos.* Although pakpaos are usually manipulated by a single person, it can take up to 20 people to handle a chula.

**5.** See Additional Answers.

**7.** $c = 13$ cm, $h \approx 4.6$ cm; To find $c$, I used the Pythagorean Theorem: $c^2 = 5^2 + 12^2 = 169$, so $c = 13$. To find $h$, I used the fact that the area of the triangle is $\frac{1}{2} \cdot 12 \cdot 5$, or 30. If the hypotenuse is the base, then $h$ is the height, so $\frac{1}{2} \cdot 13 \cdot h = 30$. So, $h = 30 \div 6.5$, or about 4.6 cm.

**5.** Caroline and Marcus are flying a kite. Caroline is holding the kite and has let out 80 feet of kite string. Marcus is standing 25 feet from Caroline and is directly under the kite.

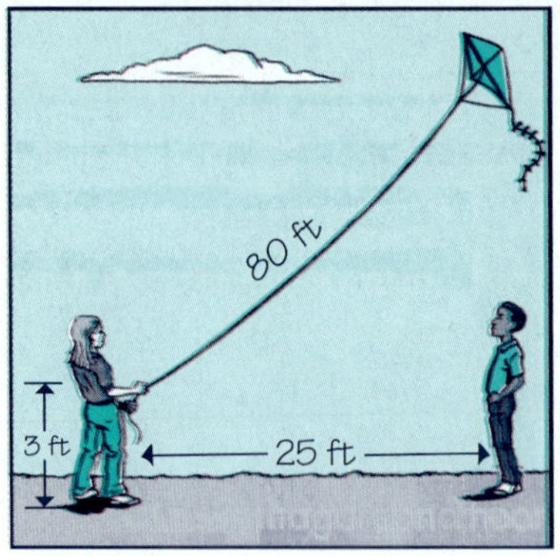


How far above the ground is the kite? Assume Caroline is holding the string 3 feet above the ground. Explain how you found your answer.

**6.** Safety regulations say that wheelchair ramps cannot be too steep. Suppose that, for every 1 foot a wheelchair ramp rises, it must cover a horizontal distance of at least 11.5 feet.

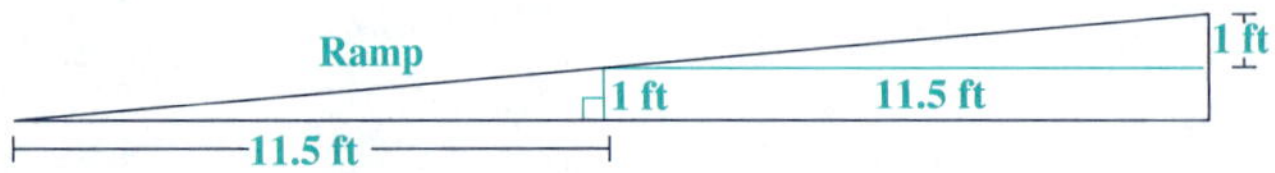


A ramp is being built to a restaurant entrance that is 2.5 feet above the ground.

**a.** How much horizontal distance does the ramp require? at least 28.75 ft

**b.** How long must the ramp be? at least 28.9 ft

**7. Challenge** Find the values of $c$ and $h$. Explain the method you used.

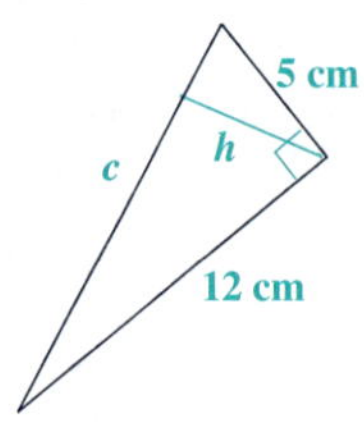


### Share & Summarize

Write your own problem that can be solved by using the Pythagorean Theorem. Then show how to solve your problem. Answers will vary.

**1** Problem-Solving Strategy
Draw a diagram

**2** Problem-Solving Strategy
Draw a diagram

**3** Have students share the application problems they created.



### Additional Answers

**5.** 79 ft; Possible explanation: The height of the kite is 3 feet plus the length of one leg of this right triangle:

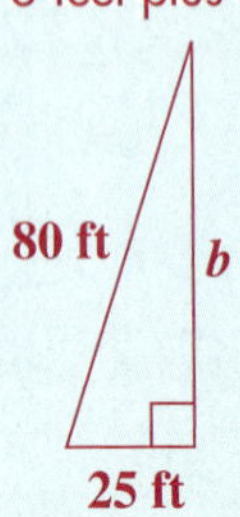


I can use the Pythagorean Theorem to find the missing length:

$25^2 + b^2 = 80^2$

$625 + b^2 = 6{,}400$

$b^2 = 6{,}400 - 625 = 5{,}775$

$b = \sqrt{5{,}775} \approx 76$

Adding 3 ft, the total height above the ground is about 79 ft.

# On Your Own Exercises

## On Your Own Exercises

Investigation 1, pp. 536–540
Practice & Apply: 1–3
Connect & Extend: 11, 12

Investigation 2, pp. 540–543
Practice & Apply: 4–10
Connect & Extend: 13, 14

Assign Anytime
Mixed Review: 15–28

**Exercises 1 and 2:** Watch for students who mistake the dimensions. In **Exercise 1** the rectangle is not a 2 × 1 rectangle. The upper-left side of the triangle in **Exercise 2** does not have a length of 2 units.

**Exercise 3:** When you discuss this exercise, you may want to check if any students can see how the Pythagorean Theorem might help explain the relationship in **Part d.**

## Practice & Apply

Find the exact area of each shape. They are drawn on 1-centimeter dot grids.

**1.**  4 cm$^2$

**2.** 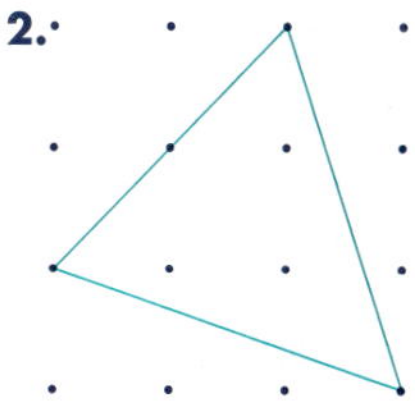 4 cm$^2$

**3.** In this exercise, you will look at the relationship among the areas of semicircles drawn on the sides of a right triangle.

**a.** Find the area, in square units, of each semicircle in this drawing:
area of semicircle on Side *a*: about 3.5 unit$^2$; area of semicircle on Side *b*: about 6.3 unit$^2$; area of semicircle on Side *c*: about 9.8 unit$^2$

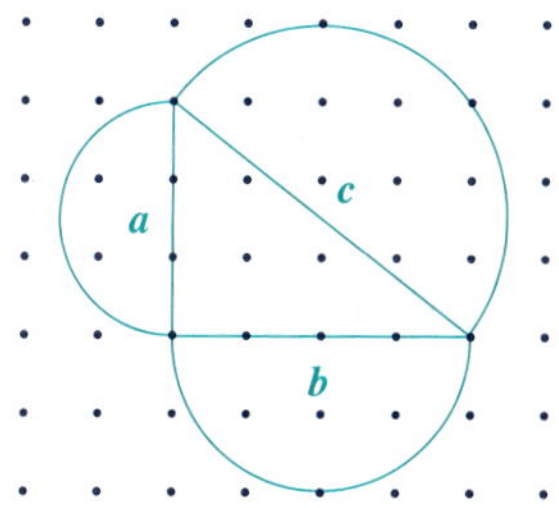


**b.** Find the area of each semicircle in this drawing:
area of semicircles on Sides *a* and *b*: about 1.6 unit$^2$; area of semicircle on Side *c*: about 3.1 unit$^2$

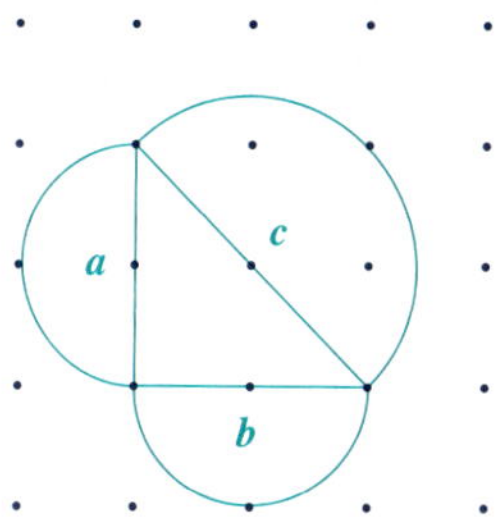


**c.** Find the area of each semicircle in this drawing:
area of semicircle on Side *a*: about 1.6 unit$^2$; area of semicircle on Side *b*: about 3.5 unit$^2$; area of semicircle on Side *c*: about 5.1 unit$^2$

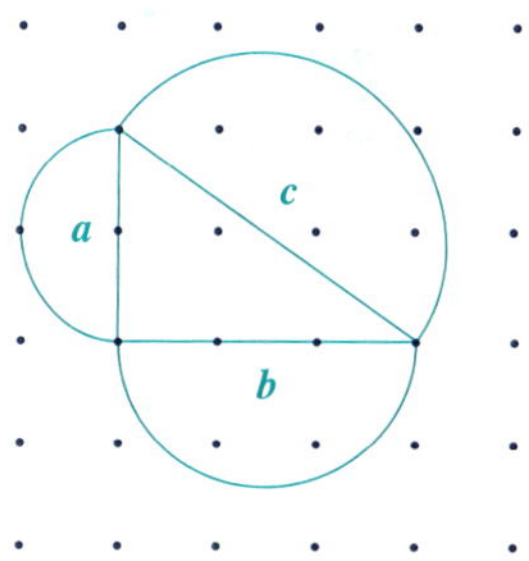


**d.** In the cases you looked at, is there a relationship among the areas of the three semicircles? If so, describe it.

The area of the semicircle on the hypotenuse is equal to the sum of the areas of the semicircles on the two legs.



impactmath.com/self_check_quiz

4. $\frac{5}{4}$ cm$^2$, or $1\frac{1}{4}$ cm$^2$

5. 10.25 cm$^2$

Find each missing area.

**4.**

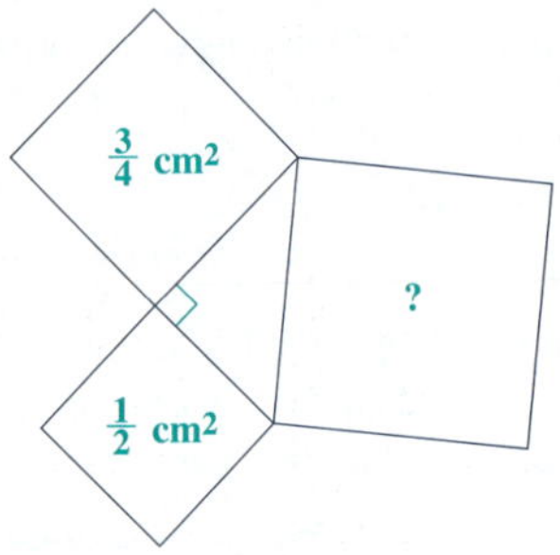


**5.**

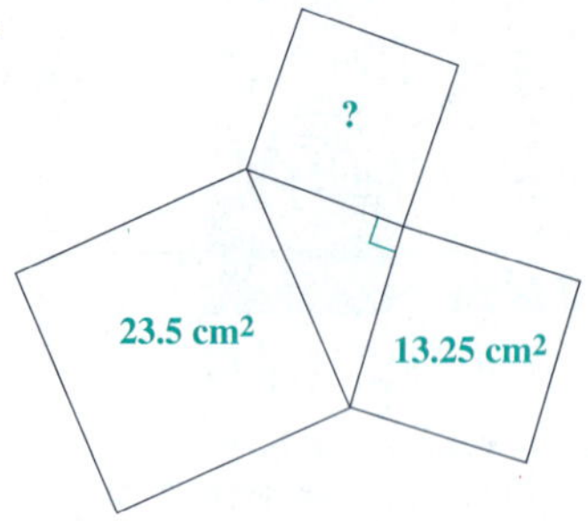


7. $\sqrt{500}$ cm, or about 22.4 cm

Find each missing side length.

**6.** 15 cm

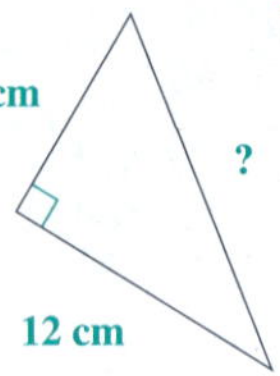


**7.**

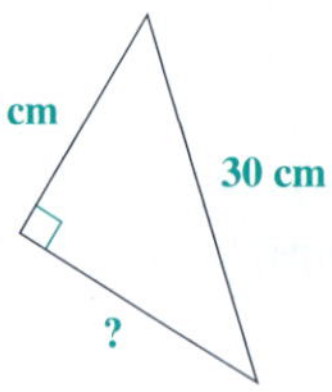


**8.** The legs of this right triangle are the same length. How long are they? $\sqrt{50}$ cm, or about 7.1 cm

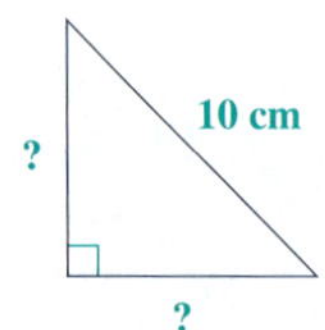


**9.** To ensure the safety of its workers, a painting company requires that the base of a ladder be 1 foot from a wall for every 4 feet it reaches up the wall.

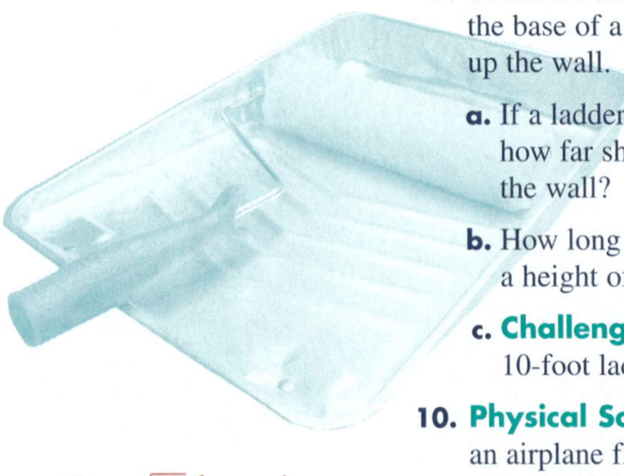

**a.** If a ladder reaches 8 feet up a wall, how far should its base be from the wall? 2 ft

**b.** How long must the ladder be to reach a height of 8 feet?

**c.** **Challenge** What is the highest a 10-foot ladder can reach? about 9.7 ft

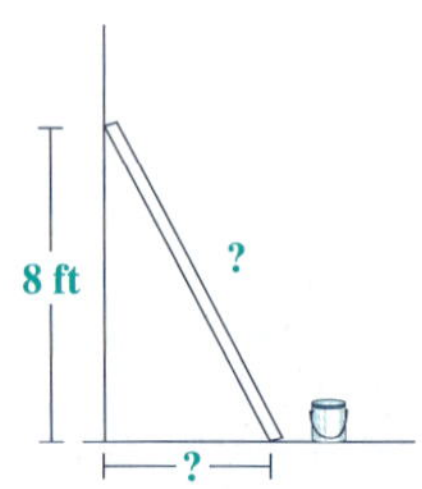


**10.** **Physical Science** During its initial climb, an airplane flew 14.4 miles, reaching an altitude of 5 miles. After the initial climb, how far was the airplane from its starting point, as measured along the ground?

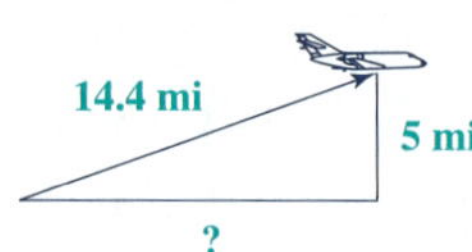


9b. $\sqrt{68}$ ft, or about 8.25 ft

10. about 13.5 mi



**Exercises 4–7:**
These exercises are straightforward applications of the Pythagorean Theorem.

**Exercise 8:**
If $x$ is the length (in centimeters) of each leg, then $x^2 + x^2 = 10^2$. So $2x^2 = 100$, which means that $x^2 = 50$. Taking the square root gives $x = \sqrt{50}$ cm.

**Exercise 9c:**
Students may need help getting started. If $x$ ft is the highest distance above ground the ladder can reach, then the base of the ladder will be $\frac{x}{4}$ ft from the wall. So $x^2 + \left(\frac{x}{4}\right)^2 = 10^2$.

LESSON 8.5 The Pythagorean Theorem 545

# Assign and Assess

**Exercise 11a:**
To find the areas of the squares drawn on the sides of the triangle that are not horizontal or vertical, students can use a combination of counting squares and using the Pythagorean Theorem. For example, they can count the number of squares for the area of the square on side *b* in Triangle I. Then, they would need to use the Pythagorean Theorem for the other two sides (*a* and *c*). To do this, students can drop a perpendicular line from the vertex to side *b*, forming two right triangles. For the left triangle, the base is 1 cm and the height is 3 cm. Thus, $1^2 + 3^2 = a^2$; $a^2 = 10$ cm$^2$. For the right triangle, the base is 3 cm and the height is 3 cm. Thus, $3^2 + 3^2 = c^2$; $c^2 = 18$ cm$^2$.

## Connect & Extend

### Just the facts

The word *acute* has many meanings. Here are just a few:

- having a sharp point
- shrewd
- sensitive
- severe or intense
- of great importance

**11.** You know that a *right triangle* has one angle that measures 90°. In an *acute triangle,* all three angles have measures less than 90°. In an *obtuse triangle,* one angle has a measure greater than 90°.

Acute Triangle    Obtuse Triangle

**a.** Three acute triangles are drawn on the centimeter dot grid below, and squares are drawn on their sides. Find the area of each square, and record your results in a table like this one.

| | Acute Triangles | | |
|---|---|---|---|
| **Triangle** | **Area of Square on Side *a*** | **Area of Square on Side *b*** | **Area of Square on Side *c*** |
| I | 10 cm$^2$ | 16 cm$^2$ | 18 cm$^2$ |
| II | 5 cm$^2$ | 8 cm$^2$ | 9 cm$^2$ |
| III | 4 cm$^2$ | 5 cm$^2$ | 5 cm$^2$ |

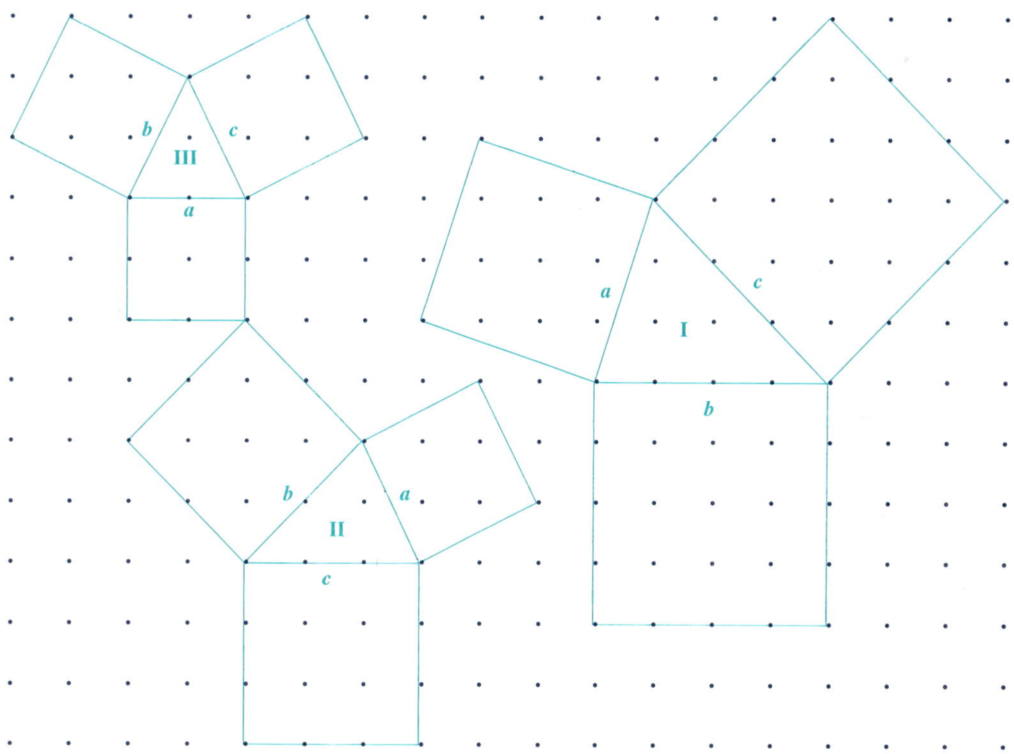

**b.** Follow the instructions in Part a for these three obtuse triangles.

| | Obtuse Triangles | | |
|---|---|---|---|
| **Triangle** | **Area of Square on Side *a*** | **Area of Square on Side *b*** | **Area of Square on Side *c*** |
| I | 2 $cm^2$ | 10 $cm^2$ | 16 $cm^2$ |
| II | 5 $cm^2$ | 5 $cm^2$ | 16 $cm^2$ |
| III | 2 $cm^2$ | 16 $cm^2$ | 26 $cm^2$ |

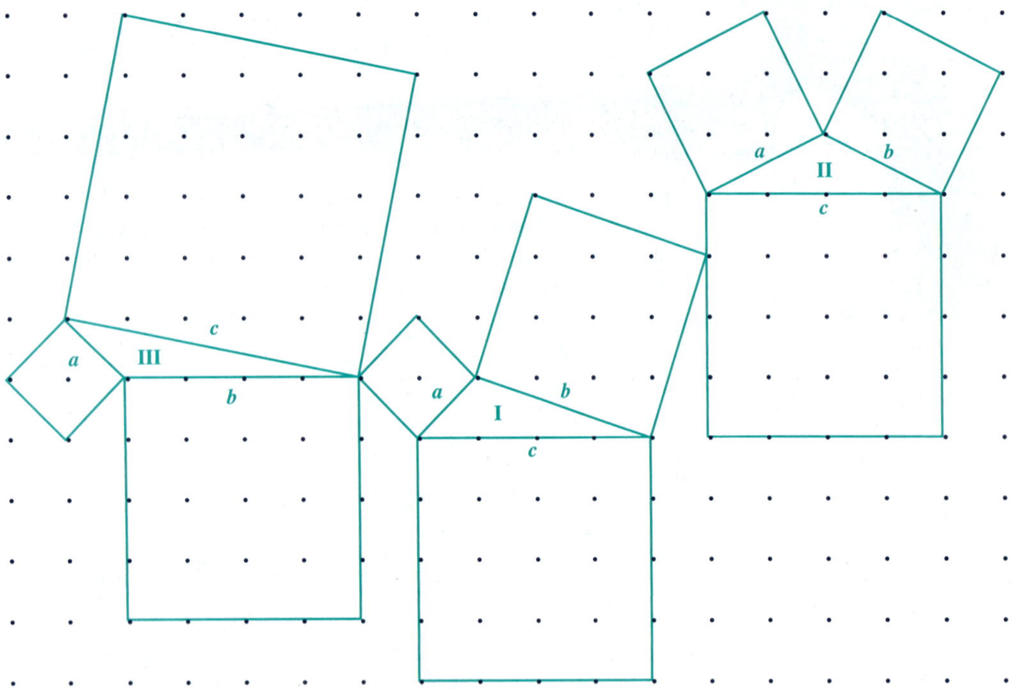


**Just the facts**

The word *obtuse* has several other meanings:

- blunt or dull
- slow to understand
- hard to comprehend

**c.** Based on your work in this exercise, which of the statements below do you think is true for acute triangles? Which do you think is true for obtuse triangles? See Additional Answers.

**i.** The sum of the areas of the two smaller squares is *less than* the area of the largest square. That is, $a^2 + b^2 < c^2$.

**ii.** The sum of the areas of the two smaller squares is *equal to* the area of the largest square. That is, $a^2 + b^2 = c^2$.

**iii.** The sum of the areas of the two smaller squares is *greater than* the area of the largest square. That is, $a^2 + b^2 > c^2$.



**Additional Answers**

**11c.** Statement iii, Statement i

LESSON 8.5 The Pythagorean Theorem 547

# Assign and Assess

**Exercise 12:**
Logic and the Pythagorean Theorem can often tell you when a triangle is *not* a right triangle. However, the Pythagorean Theorem cannot tell you that a triangle *is* a right triangle. For that, you need the *Converse of the Pythagorean Theorem:* If the sum of the squares of the lengths of two sides of a triangle is equal to the square of the length of the third side, then the triangle is a right triangle and the third side is the hypotenuse.

**Exercise 14:**
This exercise requires patience and careful reading.

**In your own words**

Explain what the Pythagorean Theorem is and how it is useful for finding lengths.

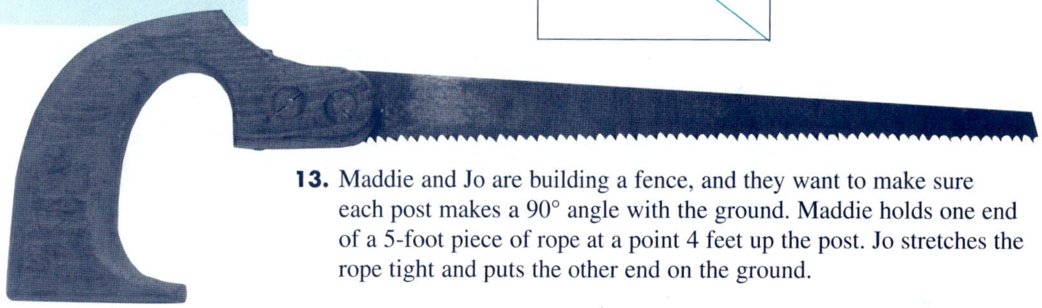

**12.** Mr. Mackenzie built a table. He had intended for the table to be rectangular, but he's not sure it turned out that way. He carefully measured the table and found that the side lengths are 60 inches and 45 inches, and the diagonal is 73.5 inches. Is the table a rectangle? Explain how you know.

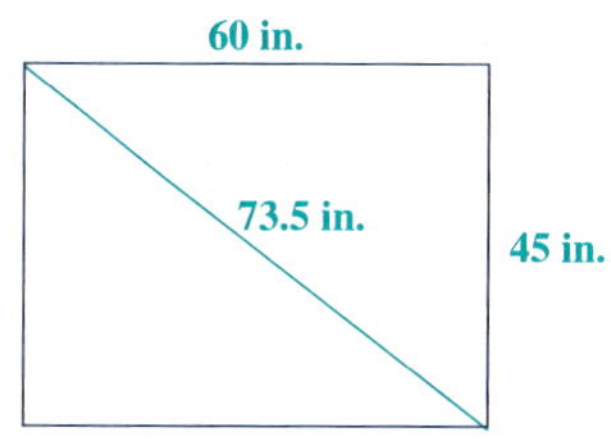


**12.** No; if it were a rectangle, the sides would make right angles, and the length of the diagonal squared would equal $45^2 + 60^2$, or 5,625 in. But $73.5^2 = 5{,}402.25$, so the table is not a rectangle.

**13.** Maddie and Jo are building a fence, and they want to make sure each post makes a 90° angle with the ground. Maddie holds one end of a 5-foot piece of rope at a point 4 feet up the post. Jo stretches the rope tight and puts the other end on the ground.

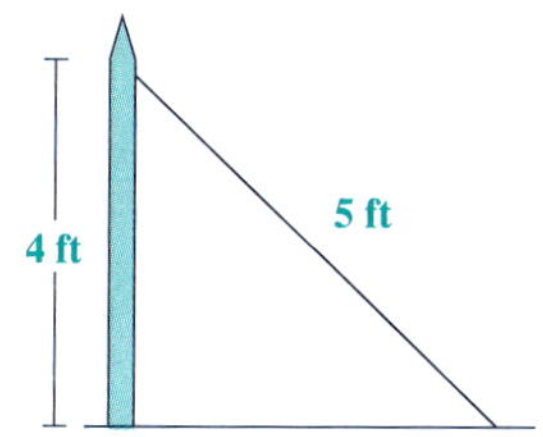


If the post makes a right angle with the ground, how far out will the string reach from the base of the post? 3 ft

**14.** Masako and Kai have found the perfect couch for their living room, but they are not sure whether it will fit through the doorway. The doorway measures 37 inches wide and 79 inches high.

They know they can take the legs off the couch. This is a side view of the couch with the legs removed.

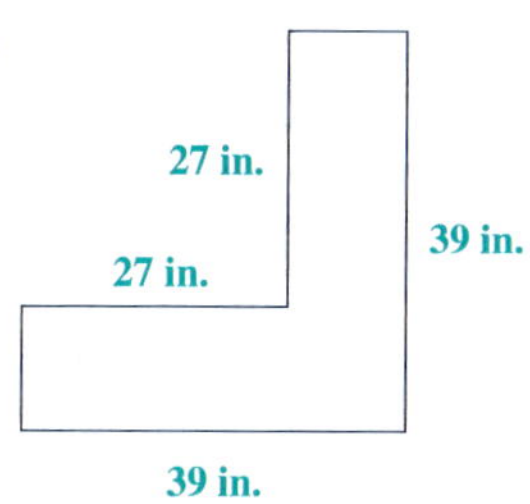




548 CHAPTER 8 Geometry and Measurement

# Review & Self-Assessment



VOCABULARY

**acute angle**
**area**
**base of a parallelogram**
**base of a triangle**
**chord**
**circumference**
**diameter**
**height of a parallelogram**
**height of a triangle**
**hypotenuse**
**inverse operations**
**leg**
**obtuse angle**
**parallelogram**
**perfect square**
**perimeter**
**perpendicular**
**radius**
**right angle**
**right triangle**
**square root**

MATERIALS

- protractor
- ruler
- string

**1a.** Possible answer: The angle on the left is clearly less than 90°.

**1b, 1c.** See Additional Answers.

## Chapter Summary

In this chapter, you explored ideas about geometry and measurement. You started by working with angles. You measured angles and drew angles with given measures. You looked at relationships among the angles formed by intersecting lines, and you found a rule for finding the angle sum for a polygon based on the number of angles (or sides) it has.

You then found the perimeters of polygons by adding side lengths, and you estimated the perimeters of curved objects by using string and by approximating with polygons. You also learned that the circumference of any circle divided by the diameter is equal to $\pi$, and you used this fact to find the formula for the perimeter of a circle.

You learned that the area of a shape is the number of square units that fit inside it. You estimated areas of shapes by counting squares, and you learned formulas for calculating areas of rectangles, parallelograms, triangles, and circles. You learned about the operation of *squaring* and how it is related to the areas of squares. You then learned about the inverse operation *taking the square root,* and you found or estimated the square roots of many numbers.

Finally, you investigated the Pythagorean Theorem, which expresses the relationship among the areas of squares drawn on the sides of a right triangle.

## Strategies and Applications

The questions in this section will help you review and apply the important ideas and strategies developed in this chapter.

### Measuring angles and drawing angles with given measures

**1.** Victor measured these angles with a protractor. He said both angles have measure of 130°.

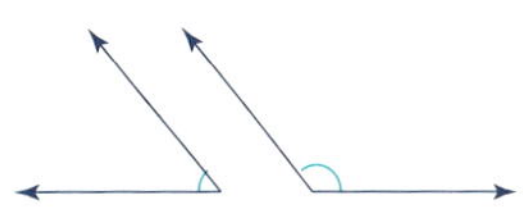

**a.** How do you know that Victor is incorrect?

**b.** What mistake do you think Victor made?

**c.** What advice would you give to help him measure angles correctly?

**2.** Draw an angle with measure 320°, and explain the steps you followed. See Additional Answers.





### Chapter Summary

This summary helps students recall the major topics of the chapter.

### Vocabulary

Students should be able to explain each of the terms listed in the vocabulary section.

### Problem-Solving Strategies and Applications

The questions in this section help students review and apply the important mathematical ideas and problem-solving strategies developed in the chapter. The questions are organized by mathematical highlights. The highlights correspond to those in "The Big Picture" chart on page 463a.

### Additional Answers

**1b.** Possible answer: When he measured the angle on the left, the ray passed through the mark labeled with both 50° and 130°. Victor must have read the wrong mark on the protractor.

**1c.** Possible answer: Before you measure the angle, think about whether it is greater than 90° or less than 90°. This will help you choose the correct angle measure.

**2.** Possible explanation: I drew a 40° angle. The angle "outside" the 40° has a measure of 320°, so I marked that angle.

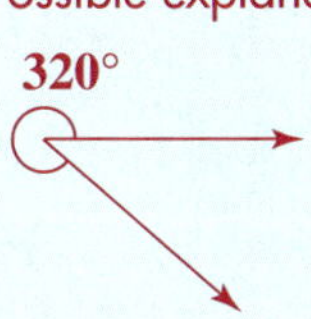

## Finding and estimating perimeters

**3.** Consider this shape:

**a.** Describe two methods for estimating the perimeter of the shape.

**b.** Use one of the methods you mentioned in Part a to estimate the shape's perimeter.

**3a.** Possible answer: Wrap a string around the figure, and then measure the length of the string. Draw a polygon to estimate the shape, and then measure and add the side lengths of the polygon.

**3b.** Answers will vary. The perimeter is approximately $3\frac{3}{4}$ in. or 9.5 cm.

**4.** Tell whether the diagram below provides enough information for you to find the perimeter of the shape. If it does, find the perimeter. If it does not, tell what additional information you would need to find the perimeter. yes, 14 cm

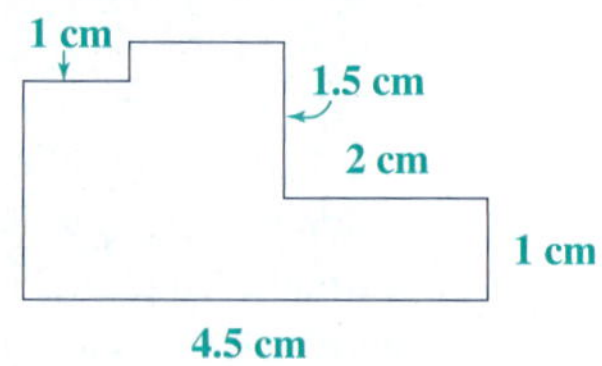


## Understanding π and the formula for the circumference of a circle

**5.** Explain what π is and how it is related to the circumference of a circle.

**5.** Possible answer: π is a number equal to the circumference of any circle divided by its diameter. π is a bit more than 3. Sometimes the number 3.14 is used to approximate π.

**6.** Describe how you can find the circumference of a circle if you know its radius. Double the radius, and multiply the result by π.

**Just the facts**

In 1999, with the aid of a computer, the value of π was found to 206 billion decimal places. If the computer were to print 1,000 digits per second, it would take about $6\frac{1}{2}$ years to print all of the decimal places!



552 CHAPTER 8 Geometry and Measurement

7. no; Possible explanation: Both shapes below have perimeter 8 units. The shape on the left has area 3 unit$^2$ and the shape on the right has area 4 unit$^2$.

8. about 14.3 cm$^2$; Possible answer: First I chose a side to be the base. I used my protractor to draw a segment that formed a right angle with the base and extended to the opposite side. I found the length of this segment and the length of the base, and multiplied them to get the area.

9a. The base is any side of the triangle. The height is the length of the segment from the vertex opposite the base to the base, perpendicular to the base.

9b. Multiply the base by the height and divide the result by 2.

9c. Possible answer: The area of a triangle is half the area of the parallelogram formed from two copies of the triangle. The base and height of the parallelogram correspond to a base and height of the triangle. Since the area of the parallelogram is the base times the height, the area of the triangle is half the base times the height.

### Finding and estimating areas

**7.** If two shapes have the same perimeter, must they also have the same area? Use words and drawings to help explain your answer.

**8.** Find the area of this parallelogram in centimeters, and explain the steps you followed:

**9.** In this chapter, you learned how to find the area of a triangle.

**a.** Describe what the base and height of a triangle are.

**b.** Explain how to find the area of a triangle if you know the lengths of the base and the height.

**c.** How is finding the area of a triangle related to finding the area of a parallelogram?

**10.** A CD has a diameter of about 12 cm. The hole in the center of a CD has a diameter of about 1.5 cm. Find the area of a CD—not including the hole—to the nearest tenth of a square centimeter. Explain how you found your answer.
See Additional Answers.

**Question 10:**
It may help students to make a sketch of a CD and show the given information in the sketch.

### Additional Answers

**10.** About 111.3 cm$^2$; I found the area of the outside circle and the area of the hole using the formula $A = \pi r^2$ (I had to remember to divide the diameters in half to get the radii.). The area of outside circle is $\pi \cdot 6^2 = 36\pi$, or about 113.1 cm$^2$. The area of the inside circle is $\pi \cdot 0.75^2 = 0.5625\pi$, or about 1.8 cm$^2$. The area of the CD is the area of the outer circle minus the area of the hole, which is about 113.1 − 1.8, or 111.3 cm$^2$.

Review and Self-Assessment **553**

**Question 12:**
It may help students first to think about squaring numbers. What happens when the number squared is 0? *The result is 0.* Between 0 and 1? *The result is between 0 and 1 but less than the original.* Equal to 1? *The result is 1.* Greater than 1? *The result is greater than 1 and greater than the original number.*

Students can recall that the square root operation is the inverse of the squaring operation and draw the appropriate conclusions.

### Understanding and applying the ideas of squaring and taking the square root

**11.** In this chapter, you learned about squaring a number and about taking the square root of a number.

**a.** Explain what it means to square a number. Give an example.

**b.** Explain what it means to take the square root of a number. Give an example. See below.

**c.** Explain how you know that squaring and taking the square root are inverse operations. Give an example if it helps you to explain your thinking.

**12.** How can you predict whether the square of a number will be *greater than, less than,* or *equal to* the original number?

**13.** Estimate $\sqrt{34}$ to the nearest tenth without using your calculator. Explain each step you take.

**11a.** Possible answer: Squaring a number means multiplying the number by itself. For example, $4^2 = 4 \cdot 4 = 16$.

### Understanding and applying the Pythagorean Theorem

**14.** The size of a television is given in terms of the length of the diagonal of its screen. For example, a 19-inch television has a screen with diagonal length 19 inches.

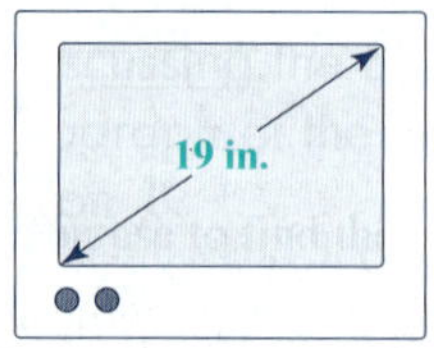


Ms. Perelli's television screen has a length of about 21.75 inches and a width of about 16.25 inches. What is the size of her television? Give your answer to the nearest inch, and explain how you found it.

**11c.** Possible answer: Squaring and taking the square root undo each other. For example, if I square 9, I get 81. If I take the square root of 81, I get 9 back. If I take the square root of 64, I get 8. If I square 8, I get 64 back.

**12–14.** See Additional Answers.

## Demonstrating Skills

Find the measure of each angle.

**15.** 23°

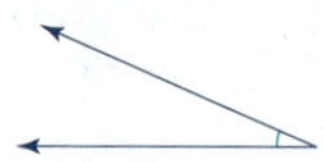

**16.** 245°

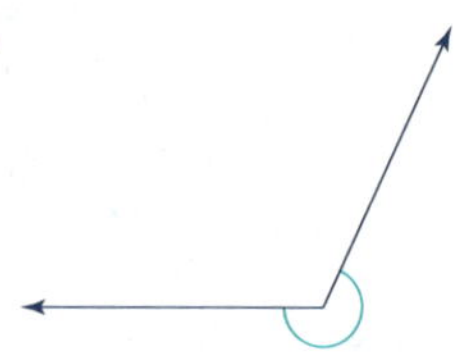

**17.** 100°

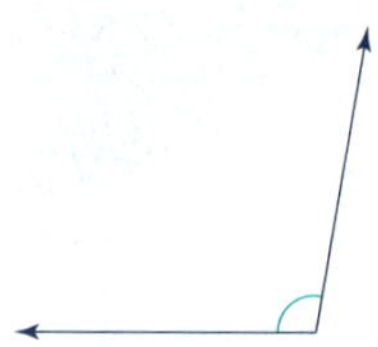

**18.** 340°

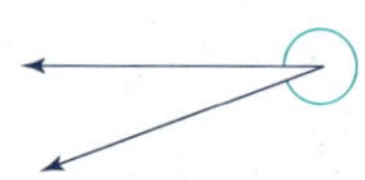

**11b.** Possible answer: Taking the square root of a number means finding the number that you can multiply by itself to get the original number. For example, $\sqrt{36} = 6$ because $6 \cdot 6 = 36$.



### Additional Answers

**12.** If the number is greater than 1, its square will be greater than the number. If it is between 0 and 1, its square will be less than the number. If it is 0 or 1, its square will be equal to the number.

**13.** $\sqrt{34}$ is between $\sqrt{25}$ and $\sqrt{36}$, so it must be between 5 and 6. Since 34 is much closer to 36 than to 25, its square root is closer to 6. I'll try 5.8: $5.8^2 = 33.64$, so 5.8 is too small. I'll try 5.9: $5.9^2 = 34.81$, so 5.9 is too large. So, $\sqrt{34}$ is between 5.8 and 5.9, and since 33.64 is closer to 34 than 34.81 is, 5.8 is a better estimate.

**14.** 27 in.; the length and width of the TV screen are the legs of a right triangle and the diagonal is the hypotenuse. Using the Pythagorean Theorem, $21.75^2 + 16.25^2 = \text{diagonal}^2$, and $21.75^2 + 16.25^2 = 737.125$. So, the length of the diagonal is $\sqrt{737.125}$, or about 27 in.

**19.** 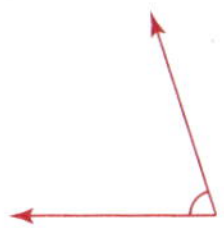

**20.** 

**21.** 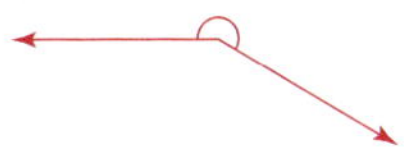

**22.** 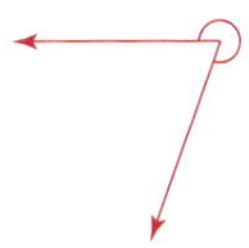

Draw an angle with the given measure.

**19.** 72° **20.** 160° **21.** 210° **22.** 295°

Find the perimeter and area of each figure.

**23.** 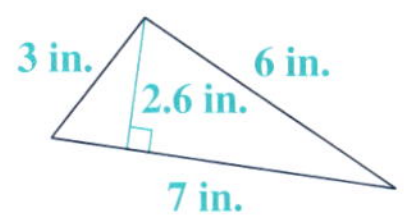


**24.** 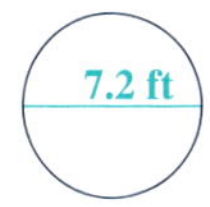


**25.** 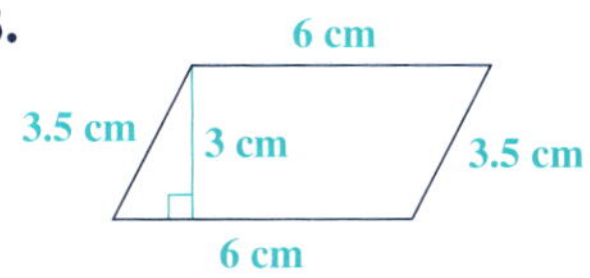


**26.** 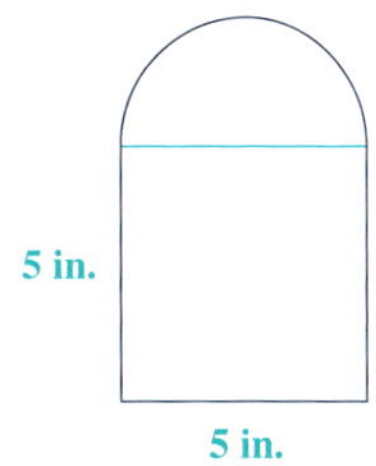


**23.** perimeter: 16 in.; area: 9.1 in.$^2$

**24.** perimeter: 7.2π, or about 22.6 ft; area: 12.96π, or about 40.7 ft$^2$

**25.** perimeter: 19 cm; area: 18 cm$^2$

**26.** perimeter: $15 + \frac{5\pi}{2}$, or about 22.9 in.; area: 25 + 3.125π, or about 34.8 in.$^2$

Find the value of each expression.

**27.** $5 \cdot 7 - 3^2 + 4$ 30 **28.** $(7^2 - 13 \cdot 3)^2$ 100 **29.** $5 + 4 + 6^2$ 45

Approximate each square root to the nearest hundredth.

**30.** $\sqrt{21}$ 4.58 **31.** $\sqrt{600}$ 24.49 **32.** $\sqrt{3}$ 1.73

In Questions 33 and 34, find the value of *b*.

**33.** 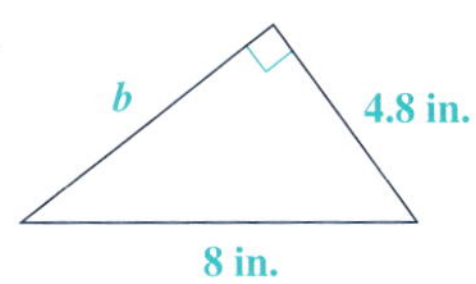

6.4 in.

**34.** 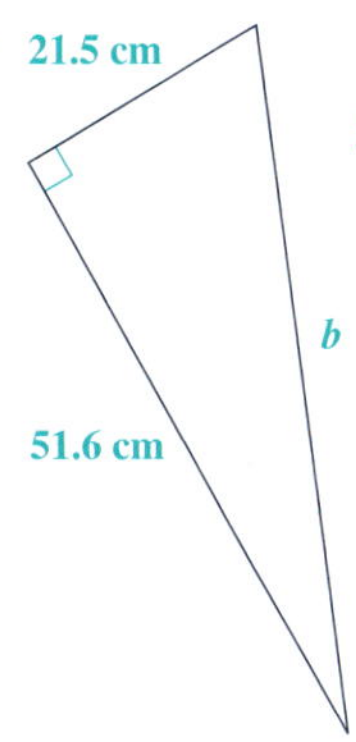

55.9 cm

**35.** Find the measures of Angles 1, 2, and 3.

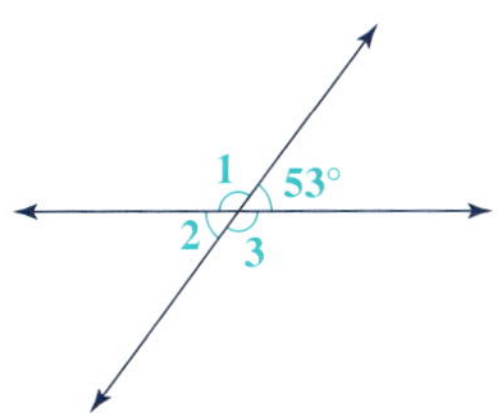


$m\angle 1 = m\angle 3 = 127°$, $m\angle 2 = 53°$



Review and Self-Assessment **555**

# CHAPTER 9

# Solving Equations

## Chapter Overview

This chapter begins with a look at true and false equations. Students learn that if an equation contains a variable, the values of the variable that make the equation true are called *solutions* of the equation.

Next, students learn about two methods for solving equations. The backtracking method is introduced by means of flowcharts. Students can use this method to find solutions for many equations in which the variable occurs only once. The second method, called the guess-check-and-improve method, allows students to find exact or approximate solutions for an even broader group of equations.

## the Big Picture

| Chapter 9 Highlights | Links to the Past | Links to the Future |
|---|---|---|
| Understanding equations and inequalities (9.1) | **Elementary Grades:** Reviewing number sentences, open sentences, and variables<br>**Chapter 7:** Using properties of fundamental operations to write expressions in equivalent forms | **Course 3, Chapter 4:** Understanding symbolic representations of inequalities |
| Solving equations by backtracking (9.2) | **Elementary Grades:** Solving addition/subtraction number stories by modeling with open sentences | **Course 2, Chapter 6:** Solving linear equations using the backtracking strategy<br>**Course 3, Chapter 4:** Solving equations by backtracking, and recognizing the limitations of this method |
| Solving equations using guess-check-and-improve (9.2) | **Elementary Grades:** Solving addition/subtraction number stories by modeling with open sentences | **Course 2, Chapter 6 and Course 3, Chapter 4:** Solving linear equations using the guess-check-and-improve strategy |
| Choosing a solution method for an equation (9.3) | **Elementary Grades:** Choosing the pan-balance method to solve simple equations | **Course 2, Chapter 6:** Solving equation using different methods<br>**Course 3, Chapter 4:** Choosing the most appropriate equation-solving method for a particular situation |

Time-Saving Tool
TeacherWorks™
All-In-One Planner and Resource Center

# Planning Guide

| Lesson Objectives | Pacing | Materials | NCTM Standards | Hot Topics |
|---|---|---|---|---|
| **9.1 Understanding Equations** page 557b<br>• To decide whether a number sentence is true or false<br>• To use the symbols $=$, $>$, and $<$ to write true number sentences<br>• To use trial-and-error to find solutions of equations | 3 class periods | * • Red and blue blocks<br>* • Yellow and green counters<br>• Master 77 | 1, 2, 6, 7, 8, 9, 10 | |
| **9.2 Backtracking** page 569b<br>• To construct flowcharts that represent input/output rules and equations<br>• To solve equations by using flowcharts for backtracking (working backward) | 4 class periods | | 1, 2, 6, 7, 8, 9, 10 | |
| **9.3 Guess-Check-and-Improve** page 585b<br>• To solve equations systematically by using guess-check-and-improve<br>• To find approximate solutions to equations, to the nearest tenth or hundredth<br>• To write and solve equations that describe mathematical puzzles or word problems<br>• To decide whether to use backtracking or guess-check-and-improve to solve equations | 4 class periods | | 1, 2, 6, 7, 8, 9, 10 | |

* Included in Impact Mathematics Manipulative Kit

**Key to NCTM Curriculum and Evaluation Standards:** 1=Number and Operations, 2=Algebra, 3=Geometry, 4=Measurement, 5=Data Analysis and Probability, 6=Problem Solving, 7=Reasoning and Proof, 8=Communication, 9=Connections, 10=Representation

# Assessment Opportunities

## Standard Assessment

*Impact Mathematics* offers three types of formal assessment. The Chapter 9 Review and Self-Assessment in the Student Edition serves as a self-assessment tool for students. In the Teacher's Guide, a Quick Quiz at the end of each lesson allows you to check students' understanding before moving to the next lesson. The Assessment Resources include blackline masters for chapter and quarterly tests.

- **Student Edition** Chapter 9 Review and Self-Assessment, pages 599–601
- **Teacher's Guide** Quick Quizzes, pages 569, 584, 598
- **Assessment Resources** Chapter 9 Test Form A, pages 212–213; Chapter 9 Test Form B, pages 214–215

## Ongoing Assessment

*Impact Mathematics* provides numerous opportunities for you to assess your students informally as they work through the investigations. Share & Summarize questions help you determine whether students understand the important ideas of an investigation. If students are struggling, Troubleshooting tips provide suggestions for helping them. On the Spot Assessment notes appear throughout the teaching notes. They give you suggestions for preventing or remedying common student errors. Assessment Forms in the Assessment Resources provide convenient ways to record student progress.

- **Student Edition** Share & Summarize, pages 560, 562, 573, 576, 578, 589, 591, 593
- **Teacher's Guide** On the Spot Assessment, pages T558, T559
  Troubleshooting, pages T560, T562, T573, T576, T578, T589, T591, T593
- **Assessment Resources** Chapter 9 Assessment Checklist, page 271

## Alternative Assessment, Portfolios, and Journal Ideas

The alternative assessment items in *Impact Mathematics* are perfect for inclusion in student portfolios and journals. The In Your Own Words feature in the Student Edition gives students a chance to write about mathematical ideas. The Performance Assessment items in the Assessment Resources provide rich, open-ended problems, ideal for take-home or group assessment.

- **Student Edition** In Your Own Words, pages 566, 583, 597
- **Assessment Resources** Chapter 9 Performance Assessment, pages 216–217

# Assessment Resources

The Assessment Resources provide a chapter test in two equivalent forms, along with additional performance items. The performance items can be used in a variety of ways. They are ideal for take-home assessment or in-class group assessment.

- Chapter 9 Test Form A, pages 212–213
- Chapter 9 Test Form B, pages 214–215
- Chapter 9 Performance Assessment, pages 216–217
- Chapter 9 Assessment Solutions, pages 218–221

Ch. 9 Test Form A

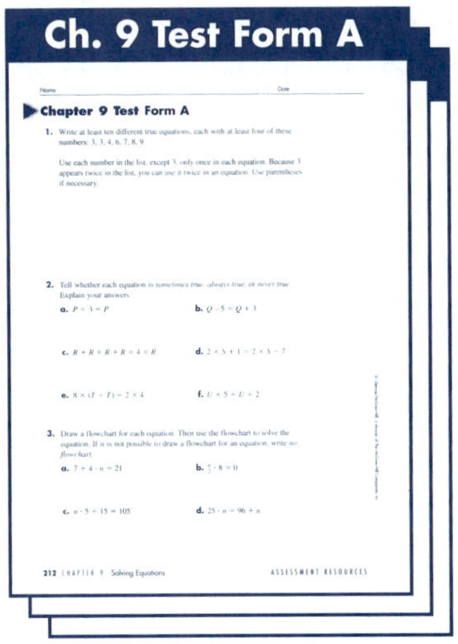

Ch. 9 Test Form B

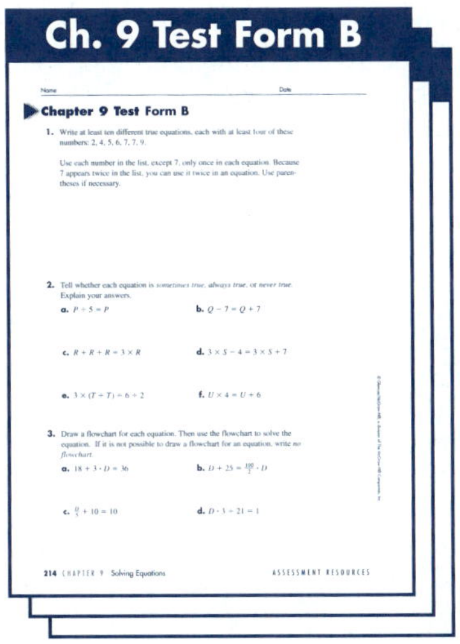

Ch. 9 Perf. Assess

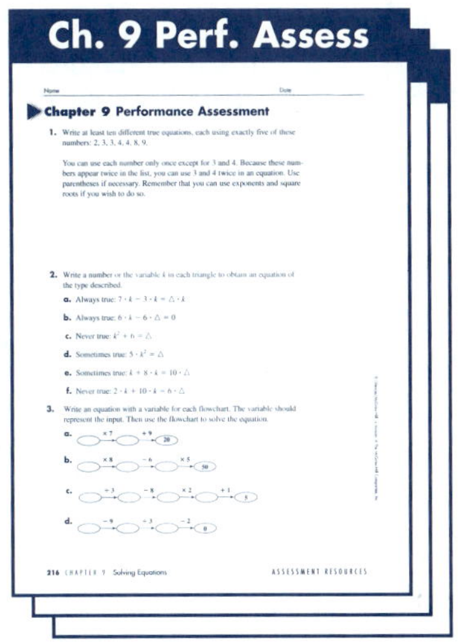

# Additional Resources

- **Math Skills Maintenance Workbook,** 10, 15, 16, 17, 18, 30, 31
- **Virtual Activities CD-ROM,** Solving Equations Using Addition and Subtraction
- **What's Math Got To Do With It? Videos,** Level 1, Video 3
- **StudentWorks™ CD-ROM**
- **Reading and Writing in the Mathematics Classroom**
- **Using the Internet in the Mathematics Classroom**

## ExamView® Pro

Use ExamView® Pro Testmaker CD-ROM to:

- Create Multiple versions of tests.
- Create Modified tests for Inclusion students with one mouse click.
- Edit existing questions and Add your own questions.
- Build tests aligned with state standards using built-in State Curriculum Correlations.
- Change English tests to Spanish with one mouse click and vice versa.

## Introduce

The problem on this page illustrates how solving equations can help people find answers to real-world problems.

Point out that Amy can use equations to determine the number of birdhouses she can afford to make, as well as the number she needs to sell in order to achieve her profit goal. Explain that finding the number that will make the expression on the left side of the equation equal in value to the right side is called *solving the equation*. Solving equations will give answers for Amy's questions. Tell students that in this chapter they will study two methods of solving equations.

## Think About It

To figure out how many birdhouses Amy needs to sell to earn $120, she can solve $120 = 8n$.

# CHAPTER 9

# Solving Equations

### Real-Life Math

**It's for the Birds** Amy makes and sells birdhouses at her town's craft fair. It costs $10 to rent a table at the fair, and Amy spends an average of $5 for supplies to make each birdhouse.

As she prepares for the event, she can use the equation $c = \$10 + \$5n$ to find the total cost of making the birdhouses. In the equation, $c$ represents the total cost and $n$ represents the number of birdhouses. Suppose Amy has a total of $75 to spend on rent and supplies. To determine how many birdhouses she can make, she can solve the equation $\$75 = \$10 + \$5n$.

**Think About It** Amy plans to sell her birdhouses for $8 each. So the equation $m = \$8n$ represents the amount of money she'll earn by selling the birdhouses. In the equation, $m$ represents the amount of money she'll earn, and $n$ represents the number of birdhouses. She hopes to earn $120. What equation should she solve?

# *Family Letter*

Dear Student and Family Members,

Our next chapter is about solving equations. Don't worry—you've been working with equations for years. An *equation* is a number sentence that includes an equals sign, which means that two expressions have the same value. Here are three examples:

$9 + 6 = 15$ $\qquad$ $9 + 6 = 5 \times 3$ $\qquad$ $7 + 8 = 18 - 3$

However, in this chapter, you will explore equations with variables (quantities that can change), such as $3 \times n = 18$.

You will learn a method called *backtracking* to solve equations. For example, consider the equation $4 \cdot n + 5 = t$. To find an output ($t$) with this equation, start with an input ($n$), multiply it by 4, and add 5. The following flowchart shows these steps.

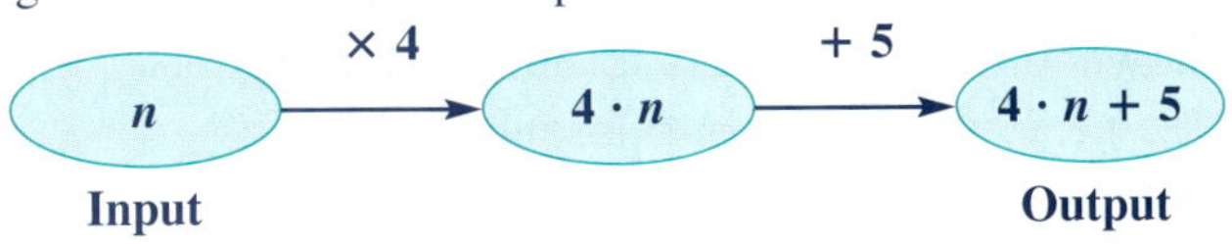


Here's the flowchart for an input of 3:

If you were given an output of 21, you could use the flowchart to work backward and determine that the input was 4.

You'll discover that some equations cannot be solved using backtracking. So, you'll explore another method, *guess-check-and-improve*.

**Vocabulary** Here's a list of the new vocabulary words associated with solving equations.

| | | |
|---|---|---|
| **backtracking** | **guess-check-and-improve** | **open sentence** |
| **equation** | **inequality** | **solution** |
| **flowchart** | | |

## What can you do at home?

Encourage your student to show you strategies for solving equations. You might even enjoy a game in which you each write a simple equation on a slip of paper, trade papers, and solve one another's equation. Once you have found the solution, talk about what you did to solve it.



Another version of the Family Letter, available in English and Spanish, is found in the Teaching Resources. You may want to send a copy of this letter home with your students.

**Teaching Resources**

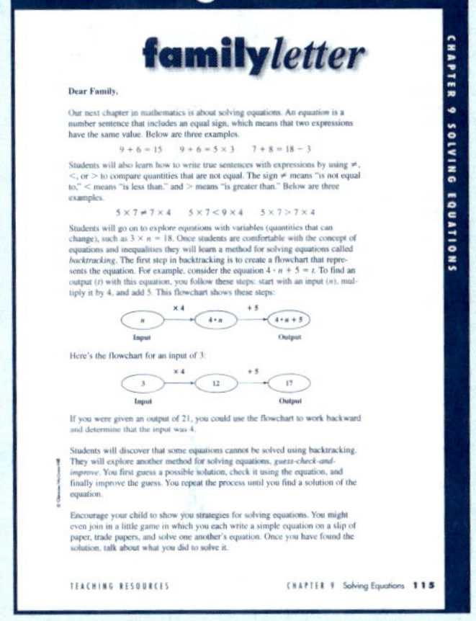

# Mathematical Background

Chapter 9 introduces students to solving equations by the backtracking method and the guess-check-and-improve method. Before they study these methods, students examine true and false equations and inequalities. They learn what is meant by a solution of an equation.

Students will solve various types of equations in this chapter. As they observe the characteristics of the equations they solve, they learn to look for clues that can help them select the most efficient solution method.

**Backtracking** The backtracking method is developed by the use of flowcharts. A flowchart starts with an input value and uses a sequence of operations that eventually results in an output value. Students call on their experience from *Think of a Number* games to gain insight into how to construct a flowchart. The numbers for the flowchart are written in ovals. Arrows from one oval to the next indicate what operations are being performed. For instance, think of these instructions for a *Think of a Number* game: Think of a number. Add 3. Multiply by 5. Subtract 15 to get the final result. The flowchart will look like this:

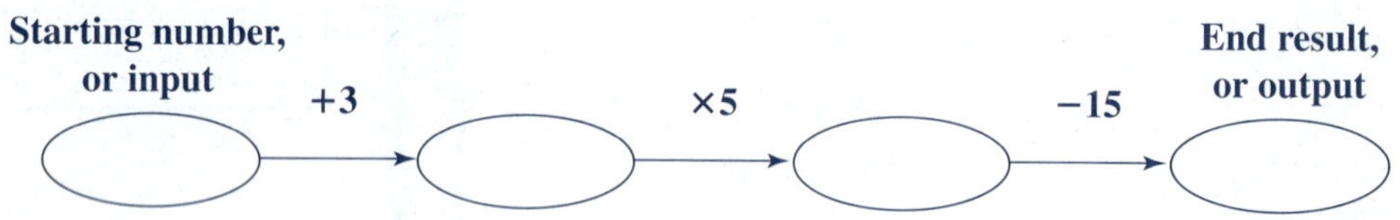


For any number put in the oval on the left, the flowchart tells what operations to perform to get a final result, or output. Students learn how to start with the final result and use inverse operations to work backward to the original input. They learn that many simple, one-variable equations can be represented by flowcharts similar to the one above. Knowing the output lets them solve the equation, that is, find the value of the variable that will make the equation true.

**Guess-Check-and-Improve** The guess-check-and-improve method can be applied to a much larger class of equations than the backtracking method. The idea is simple: guess the solution, and then successively refine the guesses to get closer and closer to the solution. In some cases, a guess will turn out to be the exact solution. In many instances, guessing and reasoning can be used together to arrive at the exact solution. Often, however, one is obliged to settle for an approximate solution. In many real-world applications, a sufficiently accurate approximation to the solution is all that is really needed.

The ideas in this chapter lay the groundwork for more extensive work on solving equations in Courses 2 and 3. In particular, this chapter will prepare students for solving equations by transforming them into equivalent equations, using the zero-product property, and using more sophisticated approximation techniques.

# 9.1 Understanding Equations

## Objectives

- To decide whether a number sentence is true or false
- To use the symbols =, >, and < to write true number sentences
- To use trial-and-error to find solutions of equations

## Overview (pacing: about 3 class periods)

In this lesson, students learn about the two main kinds of mathematical sentences that describe how two quantities are related: equations and inequalities. *Equations* state that two quantities are equal in value. *Inequalities* state that one quantity is greater than or less than another. Students see that, as with sentences in ordinary language, mathematical sentences may be true or false.

Students learn that equations containing variables are usually true for some values of the variables but false for others. If an equation contains a variable, the values of the variable that make the equation true are called *solutions* of the equation. Students explore trial-and-error methods for finding solutions of equations. The lab investigation prepares students for solving equations by backtracking, a method that is helpful for solving many types of simple equations.

## Advance Preparation

For the lab investigation, you will need to make copies of Master 77 so that each pair of students has a copy of the map used in the investigation.

## Lesson Planner

| | Summary | Materials | On Your Own Exercises | Assessment Opportunities |
|---|---|---|---|---|
| **Investigation 1** page T559 | Students learn to use inequality symbols to tell how two unequal numbers are related. | | Practice & Apply: 1–9, p. 565<br>Mixed Review: 34–44, p. 569 | On the Spot Assessment, page T558, T559<br>Share & Summarize, pages T560, 560<br>Troubleshooting, page T560 |
| **Investigation 2** page T560 | Students use trial-and-error to find solutions of equations that contain variables. | | Practice & Apply: 10–23, pp. 565–566<br>Connect & Extend: 24–33, pp. 566–569<br>Mixed Review: 34–44, p. 569 | Share & Summarize, pages T562, 562<br>Troubleshooting, page T562<br>Informal Assessment, page 569<br>Quick Quiz, page 569 |
| **Lab Investigation** page T563 | Students use the idea of undoing instructions in preparation for their study of the backtracking method for solving equations. | *• Red and blue blocks<br>*• Yellow and green counters<br>• Master 77 | | |

* Included in Impact Mathematics Manipulative Kit

# Introduce

1 Take time to discuss the meaning of the word *equation*. Students have probably used this word already. Make clear that an equation is a mathematical sentence that uses the equal sign (=) to state that two quantities have the same value.

## Explore

2 This game is best organized with students in groups of three or four.

3 In this Explore, students are challenged to use seven numbers that you call out to make as many true equations as they can. Read through the rules and examine the sample equations with the class. When you discuss the rules, you may wish to tell students that the square root operation is allowed. In order to include the squaring operation, the number 2 must be among the numbers that you call out for the game, since students will need to use 2 as an exponent.

4 Give students 5 minutes to work individually on creating equations. Then have them work in their groups to pool and compare equations. Ask each group to come up with a list of five equations that they think no one else wrote. These can be equations that students developed individually or new equations dreamed up collectively by the group.

Have a reporter from each group share the group's equations with the class. Look for duplication among groups. Discuss any equations that are particularly creative (using parentheses, complex fractions, and so on). Be sure that the equations do indeed work, that is, that the quantities joined by the equal sign do represent quantities that have the same value.

If you want to keep score as added motivation, you can give each group one point for an equation that is not on another group's list. The equation must, of course, be a true equation that meets all the rules of the game.

### On the Spot Assessment

5 Some groups may propose equations that seem to fit the rules of the game but turn out not to work because students have misapplied the order of operations rules. If that is the case, it would probably be a good idea to conduct a brief review of the order of operations rules.

# 9.1 Understanding Equations

VOCABULARY
**equation**

You have been working with equations for many years. An **equation** is a mathematical sentence stating that two quantities have the same value. An equals sign, =, is used to separate the two quantities. Here are three examples of equations:

$9 + 6 = 15 \qquad 9 + 6 = 5 \times 3 \qquad 7 + 5 = 15 - 3$

**Just the facts**

The equals sign was first used by Robert Recorde in his book *Whetstone of Witte,* published in 1557. Recorde states that he chose parallel line segments of equal length because "no two things can be more equal."

## Explore

In the game *Equation Challenge,* you will see how many equations you can make using a given set of numbers.

***Equation Challenge* Rules**

- Your teacher will call out seven single-digit numbers. Write the numbers on a sheet of paper. (Note: Your teacher may call the same number more than once. For example, the numbers might be 1, 2, 5, 3, 4, 9, and 3.)
- You have 5 minutes to write down as many correct equations as you can, using only the seven numbers, an equals sign, operation symbols, decimal points, and parentheses. Follow these guidelines when writing your equations:

  —Use each of the seven numbers only once in each equation.

  —You don't need to use all the numbers in each equation.

  —You can combine the numbers to make numbers with more than one digit. For example, you could combine 1 and 2 to make 12, 21, 1.2, or $\frac{1}{2}$.
- At the end of the 5 minutes, check your equations to make sure they are correct.

Here are some sample equations for the numbers 1, 2, 5, 3, 4, 9, and 3:

$9 + 5 = 14 \qquad \frac{35 + 1}{4} = 9 \qquad 9 \cdot 5 = 3 \cdot 3 \cdot (4 + 1)$

Play *Equation Challenge.* Be creative when writing your equations!

1 Introduce the term *equation.*

2 You may have students work in groups of 3 or 4 for the Explore.

3 Tell students that the square root operation is allowed.

4 Have students compare their equations.

5 Review the order of operations rules.

# Investigation 1

In this investigation, students explore the idea that equations and other number sentences can be true or false. This idea can be puzzling to students. Students may think that a false equation is not an equation at all. Their thinking often goes something like this: Everyone knows that $7 - 4$ does not give the same number as $8 - 3$. So it is wrong even to write $7 - 4 = 8 - 3$, and that means $7 - 4 = 8 - 3$ is not an equation.

1 Help students understand that the mathematical meaning of the word *equation* is based on the syntax, or form, of the mathematical statement. In the English language, "The sun is cold" is a perfectly good sentence grammatically, even though it happens to be false. Likewise, $7 - 4 = 8 - 3$ is an equation (two number expressions with the equal sign between them), even though it happens to be false.

2 Similar considerations apply to mathematical statements that use symbols of inequality ($\neq$, $<$, and $>$).

### Think & Discuss

The purpose of the Think & Discuss is to get students to think about ways to change false sentences into true sentences. Some students may see that a false equation can easily be made into a true statement by replacing the $=$ symbol with the $\neq$ symbol. They can also replace the $=$

3 symbol with either the $<$ symbol or the $>$ symbol, provided they know which expression in the original equation has the greater value.

Another way to change a false equation into a true equation is to change one or both of the number expressions. There are always many ways to do that.

4 **Problem Set A** **Suggested Grouping: Individuals**
In this problem set, students consider how false sentences can be changed into true sentences by changing the numbers or the relation symbol ($=$, $<$, or $>$).

5 In **Problems 1–4,** make sure students change only one number or symbol.

In **Problem 5,** watch for students who incorrectly evaluate $2 + 4^2 + 3$. Some students may compute $4^2$ as 8 rather than 16.

In **Problem 10,** students who say that the equation is true have probably misapplied the order of operations rules in evaluating $8 + 12 \div 4$. You may need to remind them that the division should be done before the addition.

### On the Spot Assessment

If students have difficulty remembering which inequality symbol to use ($>$ or $<$), you may want to remind them that the greater quantity is always at the wide end of the symbol. It may be helpful to put the symbols and some simple examples on chart paper and post the chart in a prominent place in the room.

**Problem Set Wrap-Up** For Problem 1, you may want to call on volunteers to read their answers to the class. You can examine their answers in a systematic way by writing the original equation on the board and placing a check mark above the number or symbol that was changed to get a true sentence. For example, 17 can be changed to any number less than 16. The $+$ symbol between 17 and 5 can be changed to $-$ or $\div$. The $<$ symbol can be changed to $\neq$ or $>$. For $3^2$, if you limit yourself to whole numbers, you can change 3 to any whole number greater than 3, or you can change the exponent 2 to any whole number greater than 2. The $+$ symbol between $3^2$ and 12 can be changed to $\times$. Finally, the number 12 can be changed to any number greater than 13.

You can discuss Problems 2–4 in a similar fashion.

# Investigation 1 Equations and Inequalities

VOCABULARY
**inequality**

As you know, an *equation* is a mathematical sentence stating that two quantities have the same value. A mathematical sentence stating that two quantities have *different* values is called an **inequality.** Inequalities use the symbols ≠, <, and >. The table below explains what these symbols mean.

| Symbol | What It Means | Example |
|---|---|---|
| $\neq$ | is not equal to | $7 + 2 \neq 5 + 1$ |
| $<$ | is less than | $4 + 5 < 20$ |
| $>$ | is greater than | $6 \cdot 9 > 6 + 9$ |

**Just the facts**

Texas became the 28th U.S. state in 1845. Alaska became the 49th U.S. state in 1959.

Just as sentences with words can be true or false, so can equations and inequalities. Consider these six sentences:

| | |
|---|---|
| Stop signs are yellow. | $4 = 32 \div 8$ |
| $5 \cdot 6 > 6 \cdot 5$ | The sun is hot. |
| Alaska is south of Texas. | $5 \times 4 = 27 - 3$ |

**Think & Discuss**

"4 = 32 ÷ 8" and "The sun is hot" are true. The rest are false.

Which of the sentences above are true, and which are false?

Find a way to make each of the false sentences true by changing or adding just one word, symbol, or number. See ①.

① Possible answers: Stop signs are red. Yield signs are yellow. Stop signs are not red. $5 \cdot 6 = 6 \cdot 5$. $5 \cdot 6 > 6 + 5$. Alaska is north of Texas. Mexico is south of Texas. Alaska is not south of Texas. $6 \times 4 = 27 - 3$. $5 \times 4 = 27 - 7$. $5 \times 4 < 27 - 3$.

## Problem Set A

See Additional Answers.

The sentences below are false. Make each sentence true by changing one symbol or number.

**1.** $17 + 5 < 3^2 + 12$

**2.** $14 + 5 = 12 + 11$

**3.** $23 - 11 = 22 \div 2$

**4.** $6 \cdot 5 > (4 \cdot 7) + 8$

Tell whether each sentence is true or false. If it is false, make it true by replacing the equal sign with < or >.

**5.** $5 + 13 = 2 + 4^2 + 3$

**6.** $7 + (2 \times 3) = (6 \times 2) + 1$

**7.** $24 \div 5 = 2 + 3$

**8.** $\frac{2}{5} = \frac{1}{2}$

**9.** $0.25 = \frac{1}{4}$

**10.** $8 + 12 \div 4 = (8 + 12) \div 4$



1 Help students understand the meaning of the word equation.

2 Introduce the term inequality.

3 Challenge students to use the < or > symbols.

4 You may have students work individually.

5 Be sure students change only one number or symbol.

### Additional Answers

**1.** Possible answers: $17 + 5 > 3^2 + 12$, $17 + 5 < 3^2 + 20$

**2.** Possible answers: $14 + 5 < 12 + 11$, $14 + 5 = 12 + 7$

**3.** Possible answers: $23 - 11 \neq 22 \div 2$, $23 - 11 > 22 \div 2$

**4.** Possible answers: $6 \cdot 5 < (4 \cdot 7) + 8$, $6 \cdot 10 > (4 \cdot 7) + 8$

**5.** false, $5 + 13 < 2 + 4^2 + 3$

**6.** true

**7.** false, $24 \div 5 < 2 + 3$

**8.** false, $\frac{2}{5} < \frac{1}{2}$

**9.** true

**10.** false, $8 + 12 \div 4 > (8 + 12) \div 4$

LESSON 9.1 Understanding Equations 559

1 **Share & Summarize**
You can use these questions to see whether students understand that it is the *form* of the number sentence that determines whether the sentence is an equation or an inequality. For **Question 1,** be on the lookout for students who say that an inequality is a false equation. The symbol between the two number expressions in an equation is =, regardless of whether the equation is true or false. For **Question 2,** the example and explanation are important. Students sometimes come to the hasty conclusion, for example, that a false equation can be made into a true sentence by replacing the equal sign with any one of the inequality symbols. The symbol $\neq$ will always do the job, but only one of the > and < symbols will work if the equal sign is the only item changed in the original sentence.

**Troubleshooting** If students have difficulty determining whether an equation is true, ask them to work out the value of each side of the equation separately to see if the two values are equal. If they have difficulty changing an untrue equation to make it true, have them work with each term and symbol in order from left to right in the equation, and ask if there is any way to make the equation true by changing that item. If they have difficulty deciding which inequality symbol to use, explain that they must first perform all the calculations on each side to determine which side has the greater value.

**Additional Examples** Decide whether each of the following is an equation. If it is, decide whether it is true. Make any false equations into two true sentences by changing only one number or symbol.

- $3 \times 15 \div 5 = 10 + 1$ an equation; false; possible true sentences: $3 \times 15 \div 5 < 10 + 1$, $3 \times 15 \div 5 = 10 - 1$
- $5 \times (3 + 4 - 2)$ not an equation
- $(16 \div 4) + 3 = \frac{48}{6}$ an equation; false; possible true sentences: $(16 \div 4) + 3 < \frac{48}{6}$, $(16 \div 4) + 4 = \frac{48}{6}$
- $(9 + 3) \times 5 = 120 \div 3$ an equation; false; possible true sentences: $(9 + 3) \times 5 > 120 \div 3$, $(9 + 3) \times 5 = 120 \div 2$
- $\frac{25 - 5}{6}$ not an equation

**On Your Own Exercises**

Practice & Apply: 1–9, p. 565
Mixed Review: 34–44, p. 569

# Investigation 2

In this investigation, students begin to work with equations that contain a variable. They learn that equations or inequalities that contain variables are called *open sentences*. Replacing the variable in an open equation with a specific value leads to an equation that contains only numbers. The truth or falsity of this new equation can easily be decided. The values of the variable that result in a true equation are the solutions of the equation. The number of solutions depends on the equation. Students will see instances of equations with more than one solution. They will also see that some open sentences may be true for all values of the variables. Other open sentences may be true for no values whatsoever.

2 Have a brief discussion of the term *open sentence*. Then go over the Think & Discuss with the class.

**About the Mathematics**

The term *open sentence* comes from symbolic logic, a discipline that is usually thought of as being closely aligned with mathematics. Open sentences in mathematics always contain one or more variables. It is sometimes helpful to think of variables as being analogous to indefinite pronouns in English. Compare the following:

He loves her.
$x = y + 4$

The English sentence is neither true nor false, but it will become so if you replace the pronouns with the names of specific individuals. Likewise, the equation is neither true nor false, but it will become so if you "close" the sentence by replacing $x$ and $y$ with specific numbers. Open sentences can also be closed by using *quantifiers*. Students will learn about quantifiers in later courses in mathematics.

3 **Think & Discuss**
The key idea for students to understand here is that an open sentence is not true or false. However you may be able to change it into a true sentence by replacing the variable with just the right number. Discuss the fact that for each of the three equations, only one number will make the open equation into a true equation. ■

4 Read the information at the bottom of the page with the class. Discuss what is meant by the term *solution*.

## Share & Summarize

1. Explain the difference between an equation and an inequality, and give an example of each. See Additional Answers.
2. Give an example of an equation or an inequality that is false. Then explain how you could change the sentence to make it true.
Possible answer: $4 \cdot 3 = 6 + 5$ is false because the left side has a value of 12 and the right side has a value of 11. You could make the sentence true by replacing = with >.

1 Make sure students know that inequalities are not false equations.

# Investigation 2 Equations with Variables

You can determine whether the equations in Investigation 1 are true by finding the value of each side. But what if an equation contains a variable? For example, consider this equation:

$$3 \times n = 18$$

VOCABULARY
**open sentence**

You can't tell whether this equation is true or false unless you know the value of $n$. An equation or inequality that can be true or false depending on the value of the variable is called an **open sentence.**

2 Introduce the term *open sentence.*

## Think & Discuss

For each open sentence, find a value of $n$ that makes it true and a value of $n$ that makes it false.

| $3 \times n = 18$ | $n \div 2 = 2.5$ | $n + 5 = 25$ |
|---|---|---|
| 6, any other value | 5, any other value | 20, any other value |

3 Discuss that only one number will make each open sentence true.

VOCABULARY
**solution**

Finding the values of the variable or variables that make an equation true is called *solving* the equation. A value that makes an equation true is a **solution** of the equation.

4 Introduce the term *solution.*

Consider this equation:

$$6 \cdot n - 1 = 29$$

Finding a solution of $6 \cdot n - 1 = 29$ is the same as answering this question:

For what value of $n$ is $6 \cdot n - 1$ equal to 29?

**Remember**
These are all ways to write "6 times $p$":
$6 \times p \quad 6 \cdot p \quad 6p$



**Additional Answers**
**Share & Summarize**

1. An equation states that two quantities have the same value. An equation uses an equal sign between the two quantities. For example, $8 + 8 = 4^2$ is an equation since both sides have a value of 16. An inequality is a mathematical sentence that uses $\neq$, <, or > to state that two quantities are not equal. For example, $7 + 8 < 7 \cdot 8$ is an inequality.

# Develop

1 Discuss the table at the top of page 561 to help students understand why Jing was convinced that $6 \cdot n - 1 = 29$ has one and only one solution.

2 **Problem Set B** **Suggested Grouping: Pairs**
In this problem set, students look for the number that will make an open sentence true.

3 For **Problems 1–6,** students are told that each equation has only one solution. Call on volunteers for each problem. Trial-and-error is a perfectly acceptable approach. Some students may have used reasoning. Ask students how they know their solutions work. They should understand that replacing the variable with the solution must give a true equation. In Problem 1, for example, a proof that 7 is the solution might proceed as follows. Replace $n$ with 7 in $6 \cdot n - 1 = 41$. The resulting equation is $6 \cdot 7 - 1 = 41$. Since $6 \cdot 7 - 1 = 42 - 1$ and $42 - 1 = 41$, the open equation $6 \cdot n - 1 = 41$ is true when $n = 7$.

In **Problem 7,** students replace $d$ with each of the trial numbers.

Try 1 in place of $d$.

$$\begin{aligned} 2 \cdot d + 3 &= 15 - 2 \cdot d \\ 2 \cdot 1 + 3 &= 15 - 2 \cdot 1 \\ 2 + 3 &= 15 - 2 \\ 5 &= 13 \quad \text{False} \end{aligned}$$

Try 2 in place of $d$.

$$\begin{aligned} 2 \cdot d + 3 &= 15 - 2 \cdot d \\ 2 \cdot 2 + 3 &= 15 - 2 \cdot 2 \\ 4 + 3 &= 15 - 4 \\ 7 &= 11 \quad \text{False} \end{aligned}$$

Try 3 in place of $d$.

$$\begin{aligned} 2 \cdot d + 3 &= 15 - 2 \cdot d \\ 2 \cdot 3 + 3 &= 15 - 2 \cdot 3 \\ 6 + 3 &= 15 - 6 \\ 9 &= 9 \quad \text{True} \end{aligned}$$

Try 4 in place of $d$.

$$\begin{aligned} 2 \cdot d + 3 &= 15 - 2 \cdot d \\ 2 \cdot 4 + 3 &= 15 - 2 \cdot 4 \\ 8 + 3 &= 15 - 8 \\ 11 &= 7 \quad \text{False} \end{aligned}$$

The trials show that only 3 is a solution of $2 \cdot d + 3 = 15 - 2 \cdot d$.

**Problem Set Wrap-Up** Ask students how they came up with their equations for **Problem 8.** A fairly systematic approach is necessary to get three equations that have 13 as a solution, since the likelihood of making up such equations at random is not that great. One approach that works is to write a numerical expression such as $13 + 1$ that involves 13, evaluate the expression, and write the related equation, $13 + 1 = 14$. Then replace 13 with a variable ($t$, $x$, or $k$, or any other variable).

4 If no one came up with an equation involving fractions, ask for such an equation. It should not take long for someone to find an equation that works.

## Think & Discuss

Ask students to think about these questions with a partner before discussing them as a class. In all of these examples, it is important that students realize that if a variable appears in an equation more than once, it must have the same value in each location. Thus, if students try the value 4 for $m$ in the first equation, they must replace $m$ with 4 in both places where $m$ occurs. The same is true for $n$ and $a$ in the second and third equations.

5 Students should be able to use their knowledge of arithmetic to explain why these three equations are always true. This is a good place to emphasize that all the rules of arithmetic apply to equations with variables.

### Access for all Learners

**Extra Help** If students are confused about why $m$, $n$, and $a$ must have the same value on both sides of the equation, explain that these equations could also be written using empty shapes instead of letter variables. This would give the following equations:

$\square^2 - 3\square = 0$
$\bigcirc + 3 = 3 + \bigcirc$
$\Delta \times 5 = \Delta + \Delta + \Delta + \Delta + \Delta$

Students are accustomed to putting the same number into each shape every time the shape appears in a single equation.

Jing tried several values for $n$. Here is what she found:

| $n$ | $6 \cdot n - 1$ | Test | Solution? |
|---|---|---|---|
| 3 | 17 | $6 \cdot n - 1 < 29$ | no |
| 4 | 23 | $6 \cdot n - 1 < 29$ | no |
| 5 | 29 | $6 \cdot n - 1 = 29$ | yes |
| 6 | 35 | $6 \cdot n - 1 > 29$ | no |
| 7 | 41 | $6 \cdot n - 1 > 29$ | no |
| 8 | 47 | $6 \cdot n - 1 > 29$ | no |

**Remember**
Equations with squared variables, like $m^2 - 3m = 0$ in the Think & Discuss, are called *quadratic equations.* They can be used to describe the path that a projectile—an object that is tossed, thrown, or shot into the air—travels along.

Jing found that 5 is a solution of $6 \cdot n - 1 = 29$, since $6 \cdot 5 - 1 = 29$. Jing noticed that the results kept increasing as $n$ increased, so she concluded that 5 must be the only solution.

1 Discuss why Jing thought that the equation has only one solution.

## Problem Set B

2 You may have students work in pairs.

Each of these equations has one solution. Solve each equation.

**1.** $6 \cdot n - 1 = 41$ 7

**2.** $6 \cdot n - 1 = 11$ 2

**3.** $2p + 7 = 19$ 6

**4.** $4 + 4 \cdot b = 20$ 4

**5.** $\frac{5}{4} = \frac{25}{d}$ 20

**6.** $m + 3 = 2m$ 3

**7.** Try the numbers 1, 2, 3, 4 in the equation $2 \cdot d + 3 = 15 - 2 \cdot d$ to test whether any of them is a solution. 3 is a solution.

**8.** Write three equations with a solution of 13. Check that your equations are correct by substituting 13 for the variable.
Possible answer: $t + 1 = 14$, $t \cdot 2 = 30 - 4$, $26 \div t = 2$

3 Make sure students understand that replacing the variable with the solution gives a true equation.

4 Ask students for an equation involving fractions for Problem 8.

All of the equations you have looked at so far have one solution. It is possible for an equation to have more than one solution or to have no solution at all.

For some equations, *every* number is a solution. Such equations are *always* true, no matter what the values of the variables are.

### Think & Discuss

Equations that include a squared variable often have two solutions. Find two solutions for the equation $m^2 - 3m = 0$. 0 and 3

Explain why each equation below is always true.

$n + 3 = 3 + n$ Changing the order in which two numbers are added doesn't change the sum.

$a \times 5 = a + a + a + a + a$ Multiplying a number by 5 is the same as adding the number 5 times.

5 Emphasize that all arithmetic rules apply to equations with variables.



LESSON 9.1 Understanding Equations 561

# Develop

1 **Problem Set C** **Suggested Grouping: Pairs**
In this problem set, students see that there are open equations that are true for only a few values of the variable or for no values of the variable whatsoever.

In **Problem 1,** students will probably try each of the values of $t$ given in the problem and leave it at that. You may want to ask students if they think there are values other than 2 and 4 that are solutions. Making a table of values for $t^2 + 8$ and for $6 \cdot t$ is a good way to think about this question. A table of values will probably convince students that if there are any other solutions, they are not greater than 4, because the value of $t^2 + 8$ appears to be increasing more rapidly than the value of $6 \cdot t$ when $t$ is greater than 4.

2 For **Problems 2–7,** encourage students to think about what each equation means. For example, in Problem 2, students might ask, "Is there some number I could subtract from itself to get 0?" A moment's thought should lead them to answer, "Sure, that works for every number. So the equation is always true."

For Problem 6, a table of values for $n \times 2$ and $n + 1$ can make it clear that when the value of $n$ is greater than 1, doubling will give a greater number than adding 1. You may want to ask students if they can use reasoning to see why this is so.

In Problem 7, students may recognize the use of the distributive property for multiplication (from their work in Chapter 7).

**Access** for all **Learners**

**Extra Challenge** Challenge students to make up three or four equations that have exactly two solutions. Ask them to explain the thinking they used in constructing their equations.

3 **Share & Summarize**
You can use these questions to assess how well students understand the concept of the solution of an open equation. You may want to call on volunteers to read their answers to each question. Use a class discussion of the answers to focus on the main ideas of the investigation.

**Troubleshooting** If students have difficulty determining whether an equation with variables is always true, suggest that they try the following additional examples. They can first check the equation using several values of the variable. If the equation is true for each of the values that they tested, they should find a way to mathematically explain why the equation will be true in all possible cases. They can use a similar approach for equations that are never true.

**Additional Examples**

Always true: $2 \times n - n = n$

$3 \times (n + 1) = 3 \times n + 3$

Never true: $p - 5 = p + (7 - 2)$

$a + 7 = a + 10$

## On Your Own Exercises

Practice & Apply: 10–23, pp. 565–566
Connect and Extend: 24–33, pp. 566–569
Mixed Review: 34–44, p. 569

## Problem Set C

1. Try the values 1, 2, 3, and 4 to test whether any are solutions of this equation: 2 and 4 are solutions.

$$t^2 + 8 = 6 \cdot t$$

Tell whether each equation below is *always true, sometimes true,* or *never true,* and explain how you know.

2. $m - m = 0$
3. $\frac{r}{3} = r$
4. $q + 7 = q - 7$
5. $p \div 7 = \frac{1}{7} \times p$
6. $n \times 2 = n + 1$
7. $(a + 3) \cdot 2 = 2a + 6$
8. **Challenge** Tell whether each equation has a whole-number solution. Explain how you know.
   a. $2 \cdot n - 1 = 37$
   b. $2n + 1 = 18$
   c. $3 \cdot n + 5 = n + 7$
   d. $n^2 + 2 = 1$

**2.** Always true; when you subtract a number from itself, the result is 0.

**3.** Sometimes true; the equation is true when $r$ is 0 but not when $r$ is 1 (or any other value).

**4.** Never true; if you add 7 to a number you always get a greater result than when you subtract 7 from the number.

**5–8.** See Additional Answers.

**1** Problem-Solving Strategy
Make a table

**2** Encourage students to think about what each equation means.

## Share & Summarize

1. Solve the equation $3p + 5 = 11$, and explain how you found the solution.
   2; Possible explanation: I tried $p = 1$ and saw that the left side was less than 11. So I tried $p = 2$ and it worked.
2. Give an example of an equation that is always true and an example of an equation that is never true. Possible answer: $\frac{1}{2} \cdot x = x \div 2$ is always true. $x + 10 < x$ is never true.

**3** Use these questions to assess whether students understand the concepts.

**Just the facts**

In *meteorology,* the science of weather, an equal sign represents fog.



### Additional Answers

**5.** Always true; finding $\frac{1}{7}$ of a number is the same as dividing the number by 7.

**6.** Sometimes true; the equation is true when $n$ is 1 but not when $n$ is 2 (or any other value).

**7.** Always true; $(a + 3) \cdot 2$ is the same as $(a + 3) + (a + 3)$, which is the same as $a + a + 3 + 3$, or $2a + 6$.

**8a.** Yes; this has a whole-number solution of 19.

**8b.** no; Possible explanation: If $n$ is a whole number, then $2n$ must be an even number and $2n + 1$ must be an odd number. So, $2n + 1$ can never equal 18.

**8c.** Yes; this has a whole-number solution of 1.

**8d.** no; Possible explanation: If $n$ is a whole number, then $n^2$ must be greater than or equal to 0, so $n^2 + 2$ must be greater than or equal to 2.

# Lab Investigation

1 **Grouping: Pairs**
This lab investigation will build readiness for the backtracking method for solving equations, which students will study in the next lesson.

2 **Materials and Preparation**
Each pair of students will need 1 red block, 1 blue block, 1 yellow counter, and 1 green counter. They will also need a copy of Master 77.

3 **Undoing Instructions**
Students should follow the instructions in this first part of the lab. **Question 1** is intended to make students think about how to undo the instructions before they actually attempt the steps. It is fine if they refer to the printed instructions they used to stack the blocks and counters. Most students will probably prefer to simply look at the stack and imagine disassembling it from the top down.

For **Question 3,** students should observe that there should be five steps in the "undoing" procedure, just as there were in the original instructions.

**Reversing Directions**
Each student should have a copy of the map on Master 77. Point out that streets running east-west have names preceded by *E* or *W* (for example *W Chestnut Ave* and *E Chestnut Ave*) while streets running north-south have names preceded by *N* or *S* (for example, *N K St* and *S K St*). If necessary, point out how students can tell from the map which direction to move to go east, west, north, or south.

4 Instructions that say *turn left* or *turn right* may be troublesome for some students. You may want to point out that the very next instruction will help them know that they turned the correct way, because it tells them whether they are then headed north, south, east, or west.

Have students follow the directions on page 563 to confirm that the directions will indeed get them from the starting point to the pool.

**Teaching Resources**

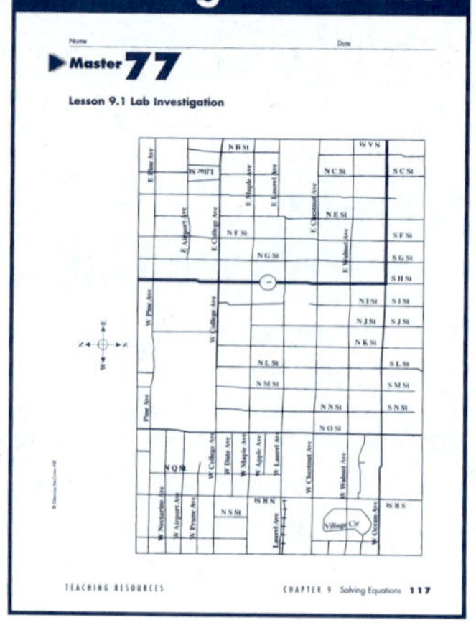

# Lab Investigation ▶ Just Undo It!

In this lab investigation, you will practice "undoing" sets of instructions. In Lesson 9.2, you will see how the strategies you use in this lab can help you solve equations.

MATERIALS
- 1 red block and 1 blue block
- 1 yellow counter and 1 green counter
- Lompoc, California, street map

## Undoing Instructions

Starting with a blank sheet of paper, follow these instructions:

- Draw a small X (in pencil) in the center of the paper.
- Put the red block on the X.
- Put the yellow counter on the red block.
- Put the blue block on the yellow counter.
- Put the green counter on the blue block.

**1.** Write a list of steps you think would undo these instructions, leaving you with a blank sheet of paper. Don't touch any of the items on the paper until you have finished writing your instructions.

**2.** Follow your steps from Question 1. Did your steps undo the instructions above? If not, rewrite them until they do. Answers will vary.

**3.** How do your steps compare with the original set of steps?

## Reversing Directions

Madeline lives in Lompoc, California, on the corner of Nectarine Avenue and R Street. Today she met her friend T.J. at the town pool. T.J. had given her these directions to get to the pool:

- Start at the corner of Nectarine Avenue and R Street.
- Walk 2 blocks east along Nectarine Avenue.
- Turn right from Nectarine Avenue onto O Street.
- Walk 4 blocks south on O Street.
- Turn left from O Street onto Maple Avenue.
- Walk 5 blocks east on Maple Avenue.
- Turn right from Maple Avenue onto J Street.
- Walk 4 blocks south on J Street.
- Turn left from J Street onto Ocean Avenue.
- Walk 7 blocks east on Ocean Avenue.
- The pool is at the corner of Ocean Avenue and C Street.

**1.**
- Remove the green counter.
- Remove the blue block.
- Remove the yellow counter.
- Remove the red block.
- Erase the X.

**3.** Possible answer: In my steps, the items are removed. In the original steps, the items are placed on the other items. The order in which items are removed is the reverse of the order in which they were added.



**1** You may have students work in pairs.

**2** Distribute materials to the students.

**3** Have students follow the instructions.

**4** Review the instructions *turn left* and *turn right* with students.

LESSON 9.1 Understanding Equations 563

You may want to mention to students that although the questions on page 564 may seem quite simple, the fact is that many adults also have difficulty reading maps and giving directions.

1 For **Question 4,** students will probably find it helpful to check the directions on page 563 and imagine how to reverse each step. They will need to keep in mind that an instruction such as "walk 7 blocks west" is the opposite of "walk 7 blocks east."

2 For **Question 6,** students need to understand that a "turn left" or "turn right" instruction is best followed by a statement telling which direction you are headed *after* the turn. In many real-life situations, it can be difficult operating without a map, but with a map of a city that contains east-west and north-south streets, knowing the compass direction makes it easier.

3 **What Did You Learn?**

**Question 9** gives students an opportunity to think about how to undo instructions. You may want to have a class discussion of what students have learned.

Now Madeline must reverse the directions to get home.

**4.** Without looking at the map, write a set of directions Madeline could follow to get home from the pool. See Additional Answers.

**5.** On the street map, carefully follow the directions you wrote in Question 4. Do you end up at the corner of Nectarine Avenue and R Street? If not, make changes to your directions until they work.

**6.** Write a set of directions to get from one place on the map to another. Word your steps like those on page 563.

- When you describe a turn, mention the street you start on, whether you turn left or right, and the street you end on.
- When you describe a walk along a street, mention the number of blocks you walk, the direction you walk, and the street name.

When you are finished, try your directions to make sure they are accurate. Directions will vary.

**7.** Exchange directions with your partner. Without looking at the map, write the steps that reverse your partner's directions. Then use the street map to test your directions. Answers will vary.

**8.** Describe some general strategies you find useful when reversing a set of directions.

## What Did You Learn?

**9.** In this lab investigation, you undid two types of instructions:

- Steps for stacking blocks and counters
- Directions for getting from one place to another

Describe how the methods you used to undo the instructions in each case were similar. Possible answer: In each case, you undo the actions in the reverse order.

**Just the facts**

It is believed that humans have been making maps since prehistoric times. Archeologists have discovered systems of lines drawn on cave walls and bone tablets that may be maps of hunting trails made by prehistoric peoples.

**5.** Answers will vary.

**8.** Possible answer: Start with the last step in the original directions and work toward the first step. Change east to west, west to east, north to south, and south to north. Change left turns to right turns and right turns to left turns, and change the order of the streets. For example, change a left turn from 1st Street onto 2nd Street into a right turn from 2nd Street onto 1st Street.

**1** Have students check the directions on page 563.

**2** Make sure students understand that the *turn left* and *turn right* instructions are best followed by a statement indicating the direction they are headed.

**3** Have students think about how to undo the instructions.



### Additional Answers

**4.**
- Start at the corner of Ocean Avenue and C Street.
- Walk 7 blocks west on Ocean Avenue.
- Turn right from Ocean Avenue onto J Street.
- Walk 4 blocks north on J Street.
- Turn left from J Street onto Maple Avenue.
- Walk 5 blocks west on Maple Avenue.
- Turn right from Maple Avenue onto O Street.
- Walk 4 blocks north on O Street.
- Turn left from O Street onto Nectarine Avenue.
- Walk 2 blocks west along Nectarine Avenue.
- Your house is at the corner of Nectarine Avenue and R Street.

# Teacher Notes

# On Your Own Exercises

## Practice & Apply

1. In a round of the game *Equation Challenge,* the following numbers were called out: 1, 2, 2, 4, 5, 5, and 9. Make at least four equations using these numbers.

In Exercises 2 and 3, the sentence is not true. Change or add one number or symbol to make it true.

2. $5 + 16 = 3 \times 8$
3. $8 \cdot 5 \neq 17 + 16 + 7$ Possible answers: $8 \cdot 5 = 17 + 16 + 7$, $8 \cdot 4 \neq 17 + 16 + 7$

In Exercises 4–9, tell whether the sentence is true or false. If it is false, make it true by replacing the equal sign with < or >.

4. $3 \times 11 = 42 - 9$ true
5. $(3 \times 5) + 4 = 4 + 5 + 1$ false, $(3 \times 5) + 4 > 4 + 5 + 1$
6. $\frac{1}{3} + \frac{4}{6} = \frac{3}{7} + \frac{16}{28}$ true
7. $0.95 = \frac{9}{10}$ false, $0.95 > \frac{9}{10}$
8. $3 \times 13 = 54 - 16$
9. $16 - 8 \div 4 = (16 - 8) \div 4$

**8.** false, $3 \times 13 > 54 - 16$
**9.** false, $16 - 8 \div 4 > (16 - 8) \div 4$

Solve each equation.

10. $x \cdot 12 = 48$ 4
11. $56 + m = 100$ 44
12. $6p + 10 = 28$ 3
13. $50 - 4 \cdot z = 30$ 5
14. Consider the equations $s + 13 = 20$ and $p + 13 = 20$.
    a. Solve each equation. 7, 7
    b. How do the solutions to the two equations compare? Explain why this makes sense. See Additional Answers.
15. Test the values 0, 1, 3, 4, and 6 to see whether any are solutions of $7m - m^2 + 10 = 16$. 1, 6

Tell whether each equation is *always true, sometimes true,* or *never true,* and explain how you know.

16. $5 \cdot m = \frac{25 \cdot m}{5}$
17. $5s = 25$
18. $t - 1 = t + 1$
19. $p^2 = p \cdot p$
20. $n + 6 = 7 \div n$
21. $7p = p \div 7$
22. Write three equations with a solution of 3.5.

**21, 22.** See Additional Answers.

**18.** Never true; the result of subtracting 1 from a number is always less than the result of adding 1 to the number.
**19.** Always true; a number squared is equal to the number times itself.
**20.** Sometimes true; the equation is true for $n = 1$ but not for $n = 7$ (or any other value of $n$).

**1.** Possible answer: $5 + 5 + 2 = 12$, $5 \div 2 = 2.5$, $\frac{1}{2} = \frac{2}{4}$, $5 + 4 - 2 = 9 - 2$, $(9 - 5) \div 4 = 2 \div 2$
**2.** Possible answers: $5 + 16 < 3 \times 8$, $5 + 16 = 3 \times 7$

**16.** Always true; both sides are equal to $5 \cdot m$ since $\frac{25 \cdot m}{5} = \frac{25}{5} \cdot m = 5 \cdot m$.
**17.** Sometimes true; the equation is true for $s = 5$ but not for $s = 25$ (or any other value of $s$).

### Just the facts

The solution to the equation in Exercise 10, $x \cdot 12 = 48$, is the number of 12-egg cartons needed to hold four dozen eggs.

impactmath.com/self_check_quiz



## Assign and Assess

### On Your Own Exercises

Investigation 1, pp. 559–560
Practice & Apply: 1–9

Investigation 2, pp. 560–562
Practice & Apply: 10–23
Connect & Extend: 24–33

Assign Anytime
Mixed Review: 34–44

**Exercise 11:**
Students can try several values of $m$ until they find the one that works, or they can use the related subtraction equation $m = 100 - 56$.

**Exercise 16:**
To explain why the equation is always true, students may find it helpful to rewrite the fraction $\frac{25 \cdot m}{5}$ as $\frac{25}{5} \cdot m$. Since $\frac{25}{5} = 5$, the expression $\frac{25 \cdot m}{5}$ becomes $5 \cdot m$.

### Additional Answers

**14b.** The solutions are the same because, except for the letter used to represent the variable, the equations are identical. The letter used to represent a variable doesn't affect the solution.

**21.** Sometimes true; the equation is true for $p = 0$ but not for $p = 7$ (or any other value of $p$).

**22.** Possible answer: $2t = 7$, $7 - t = 3.5$, $14 \div t = 4$

## Assign and Assess

**Exercise 23c:**
Students are often inclined to test only whole numbers when they look for solutions. You may need to remind them that the solution might be a fraction.

**Exercise 24:**
Students can connect the questions in this exercise with their work on solving equations by substituting each output for the variable *o* and then solving for *i*.

**23.** Of these three equations, one has no solution, one has one solution, and one has two solutions. Decide which is which, and find the solutions.

**a.** $p^2 + 6 = 5 \times p$ two solutions, 2 and 3

**b.** $3p + 5 = 3p - 5$ no solutions

**c.** $4 \cdot p + 5 = 7$ one solution, $\frac{1}{2}$

### Connect & Extend

**24.** Isabela and Jada were playing a game of *What's My Rule?* Here is Isabela's secret rule:

$$o = 37 - 4 \times i, \text{ where } o \text{ is the output and } i \text{ is the input}$$

**a.** What input value gives an output of 17? Check your answer by substituting it into the rule. 5

**b.** What input value gives an output of 5? Check your answer. 8

**c.** What input value gives an output of 0? Check your answer. $\frac{37}{4}$

**25.** Answers will vary.

**25.** Pretend you are playing a game of *What's My Rule?* Make up a secret rule for calculating an output value from an input value. Your rule should use one or two operations.

**a.** Write your rule in symbols.

**b.** Find the input value for which your rule gives an output of 25.

**c.** Find the input value for which your rule gives an output of 11.

**26.** Paul and Katarina were playing a game of *What's My Rule?* Here is Katarina's secret rule:

$$m = \frac{n}{10}, \text{ where } n \text{ is the input and } m \text{ is the output}$$

**a.** Write an equation you could solve to find the input value that gives an output of 1.5. $\frac{n}{10} = 1.5$

**b.** Solve your equation from Part a to find the input value *n*. 15

**c.** Write an equation you could solve to find the input value that gives an output of 20. $\frac{n}{10} = 20$

**d.** Solve your equation from Part c to find the input value *n*. 200

### In your own words

Explain the meaning of each of these words:

- equation
- inequality
- open sentence
- solution



566 CHAPTER 9 Solving Equations

In Exercises 27–31, use this idea:

You can think of an equation as a balanced scale. The scale at right represents the equation $3 \times 5 = 9 + 6$. The scale is balanced because both sides have the same value.

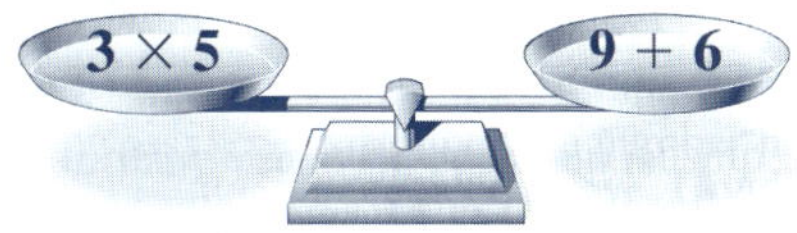


The second scale represents the equation $4 + n = 10$. To solve this equation, you need to find the value of $n$ that will make the scale balance.

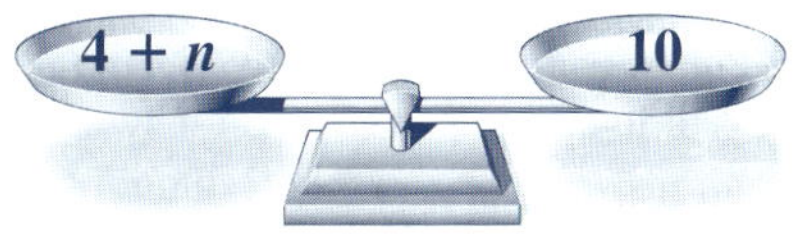


In Exercises 27–31, you will solve puzzles involving scales. Thinking about these puzzles may give you some ideas for solving equations.

**27.** This scale is balancing bags of peanuts and boxes of popcorn:

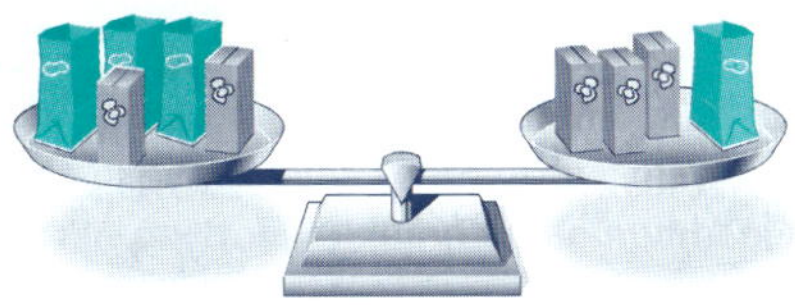

**a.** How many bags of peanuts will balance one box of popcorn? 2

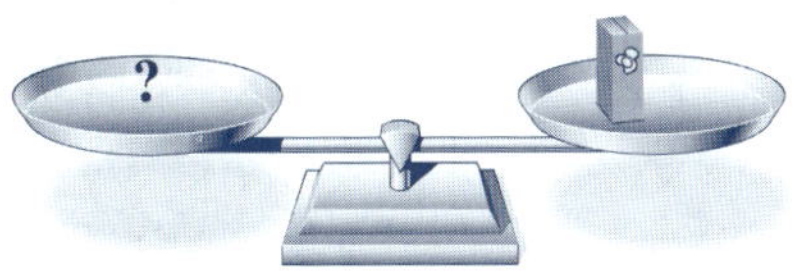


**b.** If a bag of peanuts weighs 5 ounces, how much does a box of popcorn weigh? 10 oz

**28.** Consider this scale:

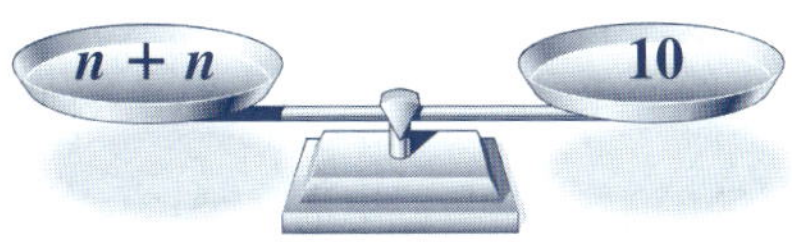


**28a.** $n + n = 10$, or $2n = 10$

**a.** Write the equation this scale represents.

**b.** What number will balance one $n$? 5

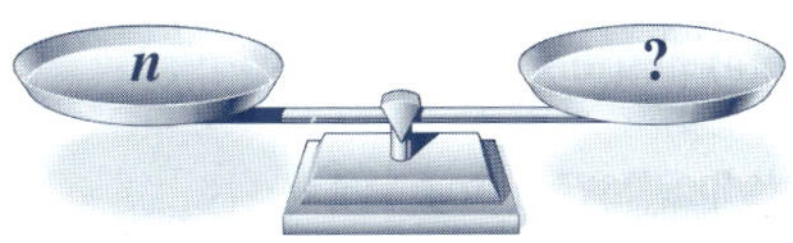




**Exercise 27:**
In **Part a,** students should see that removing one bag of peanuts and two boxes of popcorn from each side of the scale will leave the scale in balance.

Students can use their answer from Part a to find the answer to **Part b.**

LESSON 9.1 Understanding Equations 567

**Exercise 30:**
In **Part a,** students can imagine removing half the contents of each pan of the second scale to discover that 2 jacks balance 1 marble. Hence, the marble in the left pan of the first scale can be replaced with 2 jacks. This shows that 3 jacks balance 1 block. In the course of getting the answer for **Part a,** the students will also get the answer for **Part b.**

**Exercise 31:**
In **Part a,** one can combine the contents of the left pans of the scales on the left side of the first scale and then combine the contents of the right pans on the right side of the first scale. Remove 2 blocks and 1 spring from each side of the first scale to find that 4 marbles balance 1 spring.

For **Part b,** use the result from Part a. On the first scale, replace the 2 springs with 8 marbles. Then remove 3 marbles from each side to find that 1 block balances 5 marbles.

Some students may see how to write and manipulate equations to answer both parts.

In Exercises 29–31, refer to the information on page 567.

**29.** Consider this scale:

**29a.** $b + 7 = b + b$, or $b + 7 = 2b$

**a.** Write the equation this scale represents.

**b.** What number will balance one $b$? 7

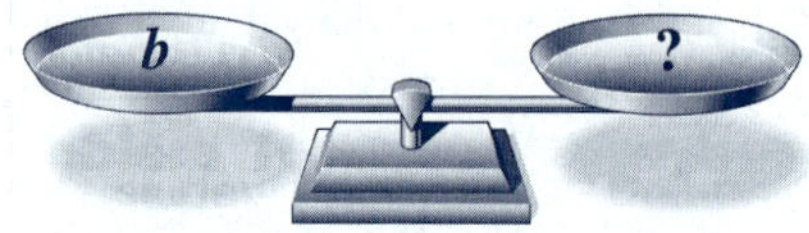


**30.** These two scales hold jacks, marbles, and a block:

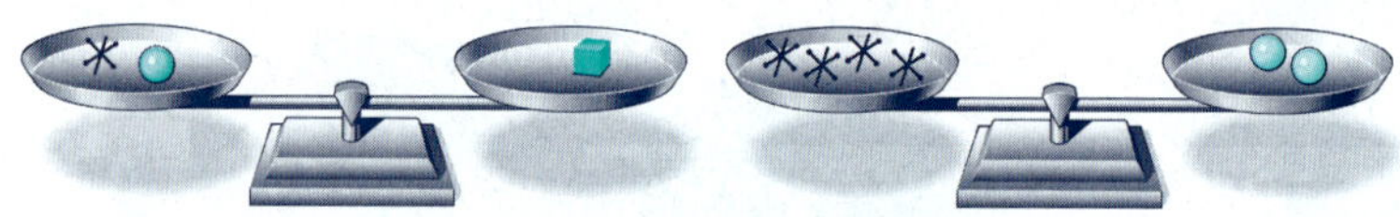

**a.** How many jacks will balance the block? 3

**b.** How many jacks will balance one marble? 2

**c.** If the block weighs 15 grams, how much does a jack weigh? How much does a marble weigh? 5 g, 10 g

**31. Challenge** These scales hold blocks, springs, and marbles:

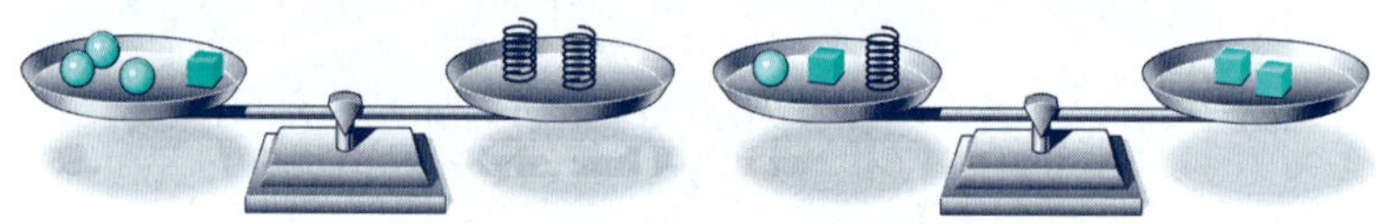

**a.** How many marbles will balance one spring? 4

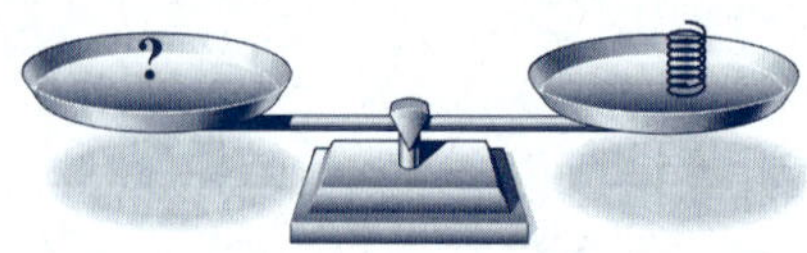


**b.** If a marble weighs 1 ounce, how much does a spring weigh? How much does a block weigh? 4 oz, 5 oz



568 CHAPTER 9 Solving Equations

## Just the facts

There are no negative temperatures on the Kelvin scale. 0 Kelvin is called *absolute zero*. It is thought to be the temperature at which all movement—including the motion of molecules and atoms—completely stops.

**32a.** 453 K

**32c.** Mercury; Possible explanation: The mean temperature on Mercury is much greater than the maximum temperature on Mars.

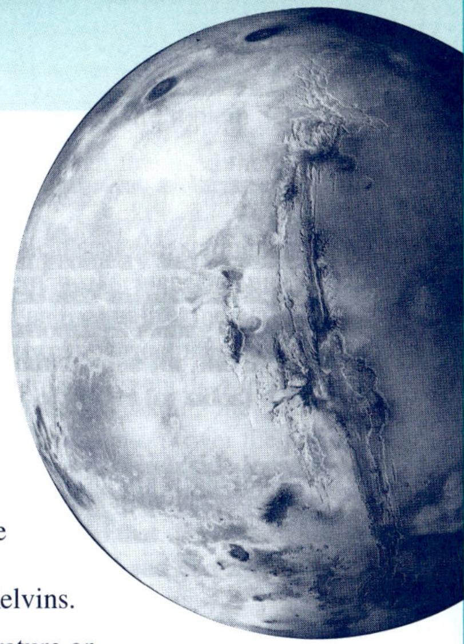

**32. Science** You are familiar with the Fahrenheit and Celsius temperature scales. The Kelvin scale is a temperature scale that is used frequently in science. This equation shows how Kelvin temperatures $K$ are related to Celsius temperatures $C$:

$$K = C + 273$$

**a.** The mean surface temperature on Mercury is about 180°C. Express this temperature in Kelvins.

**b.** The maximum surface temperature on Mars is about 290 Kelvin. Express this temperature in degrees Celsius. 17°C

**c.** Which planet is hotter, Mercury or Mars? Explain your answer.

**33. Economics** The unit of currency in South Africa is the rand. On September 4, 2000, one U.S. dollar was worth about 6.96 rand. This relationship can be expressed as an equation, where $R$ stands for the number of rand and $D$ stands for the number of dollars:

$$6.96 \times D = R$$

**a.** On that day, Jacob exchanged \$75 for rand. How many rand did he receive? 522

**b.** Jacob wanted to buy a small statue that cost 20 rand. He thought this was about \$2.85. His sister thought it was about \$139.20. Who was right? Explain your answer.

## Mixed Review

**33b.** Jacob; Possible explanation: $6.96 \times 2.85$ is about 19.84, so \$2.85 is about 20 rand.

**40, 41.** See Additional Answers.

**42.** $\sqrt{8}$ cm, or about 2.8 cm

**43.** 5 cm

**44.** 20 cm

Find each product or quotient without using a calculator.

**34.** $\frac{5}{6} \cdot \frac{7}{12}$ $\frac{35}{72}$ **35.** $\frac{5}{6} \div \frac{7}{12}$ $1\frac{3}{7}$ **36.** $23.7 \div 1{,}000$ 0.0237

**37.** $3.1 \cdot 50.7$ 157.17 **38.** $5\frac{3}{8} \cdot 2\frac{5}{9}$ $13\frac{53}{72}$ **39.** $4\frac{1}{4} \div \frac{3}{4}$ $5\frac{2}{3}$

**Number Sense** Order each set of numbers from least to greatest.

**40.** $\frac{2}{3}, 0.6, -\frac{5}{6}, \frac{3}{4}, -\frac{2}{3}, -0.1$ **41.** $\frac{5}{7}, \frac{31}{50}, \frac{3}{5}, \frac{50}{68}, \frac{7}{11}, \frac{17}{21}$

**Geometry** Find the missing length in each triangle.

**42.**

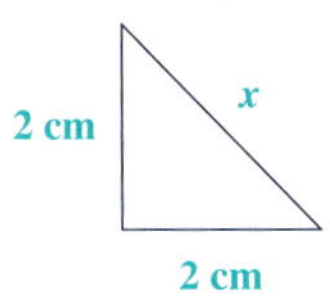


**43.**

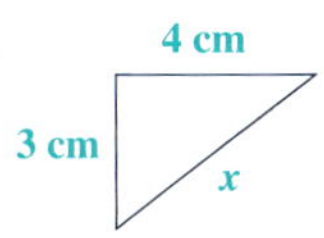


**44.**

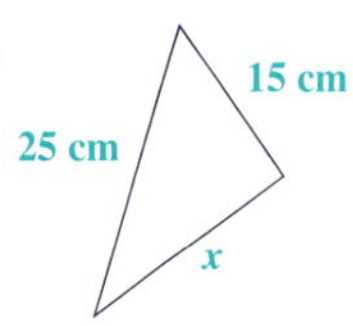




## Quick Check

### Informal Assessment

Students should be able to:

- ✔ decide whether a number sentence is true or false
- ✔ use the symbols =, >, and < to write true number sentences
- ✔ use trial-and-error to find solutions of equations

### Quick Quiz

1. Make at least two true equations using the numbers 2, 3, 3, 4, 6, 7, and 9. **Possible answers:** $6 - 2 = 9 - 4 + 3 + 3 - 7$, $(3 \times 3) \div 9 + 2 = 4 - (7 - 6)$, $3^2 + (4 - 3) = 9 + 7 - 6$

2. Tell whether each equation is true or false. If it is false, make it true by replacing the equal sign with < or >.

   a. $13 - 3 \cdot 3 = 30$ **false;** $13 - 3 \cdot 3 < 30$

   b. $8 + 3^2 = 27 - 10$ **true**

   c. $8 \div 4 - 2 = 8 - (5 - 3)$ **false;** $8 \div 4 - 2 < 8 - (5 - 3)$

   d. $5 \cdot 4 + 7 = 56 \div 8$ **false;** $5 \cdot 4 + 7 > 56 \div 8$

### Additional Answers

**40.** $-\frac{5}{6}, -\frac{2}{3}, -0.1, 0.6, \frac{2}{3}, \frac{3}{4}$

**41.** $\frac{3}{5}, \frac{31}{50}, \frac{7}{11}, \frac{5}{7}, \frac{50}{68}, \frac{17}{21}$

• *Quick Quiz continued on page A670*

# Teacher Notes

9.2

# Backtracking

## Objectives

- To construct flowcharts that represent input/output rules and equations
- To solve equations by using flowcharts for backtracking (working backward)

## Overview (pacing: about 4 class periods)

In this lesson, students learn to solve equations by working backward, a technique called *backtracking*. First, they learn to make flowcharts to represent rules and equations. Then, they use these flowcharts to help them see how to backtrack. They apply their new skills to solve many types of equations in which the variable appears only once. Ultimately, they write and solve equations for real-life situations.

## Advance Preparation

No advance preparation is needed for this lesson.

## Lesson Planner

| | Summary | Materials | On Your Own Exercises | Assessment Opportunities |
|---|---|---|---|---|
| Investigation 1 page T571 | Students make and use flowcharts to represent rules, to find the output for a given input, and to find the input for a given output. | | Practice & Apply: 1–6, p. 579<br>Connect & Extend: 18, p. 581<br>Mixed Review: 25–35, pp. 584–585 | Share & Summarize, pages T573, 573<br>Troubleshooting, page T573 |
| Investigation 2 page T574 | Students make and use flowcharts to solve equations by backtracking. | | Practice & Apply: 7–13, pp. 579–580<br>Connect & Extend: 19, 20, p. 582<br>Mixed Review: 25–35, pp. 584–585 | Share & Summarize, pages T576, 576<br>Troubleshooting, page T576 |
| Investigation 3 page T576 | Students write equations for real-life situations and use backtracking to solve the equations. | | Practice & Apply: 14–17, pp. 580–581<br>Connect & Extend: 21–24, pp. 583–584<br>Mixed Review: 25–35, pp. 584–585 | Share & Summarize, pages T578, 578<br>Troubleshooting, page T578<br>Informal Assessment, page 585<br>Quick Quiz, page 585 |

# Introduce

1 Have students read the introductory paragraphs and the cartoon. Then have them work on the Think & Discuss.

**Think & Discuss**

In this Think & Discuss, students use the method illustrated in the cartoon to solve the equation $3n + 7 = 40$. This equation is like the one for the cartoon, except the output number is now 40 instead of 43. Discuss the fact that the method used in the cartoon works perfectly well with this new output.

2 Solving equations by working backward is an approach that many students find easy and intuitively natural. You may want to work through some additional examples, each time changing the output in Jay's equation. Any output greater than or equal to 7 will do.

3 To wrap up your discussion of the Think & Discuss, point out that the technique used in the cartoon is called *back-tracking.*

# 9.2 Backtracking

Jay was playing *What's My Rule?* with his friend Marla. He figured out that this was the rule:

*To find the output, multiply the input by 3 and then add 7.*

**Remember**

$3 \times n$, $3 \cdot n$, and $3n$ are all ways to write "3 times $n$."

He wrote the rule in symbols, using $n$ to represent the input and $t$ to represent the output.

$$t = 3n + 7$$

Jay wanted to find the input that would give an output of 43. This is the same as solving the equation $3n + 7 = 43$.

1 Discuss the cartoon.

## Think & Discuss

Using Jay's rule, what input gives an output of 40? 11

Explain the reasoning you used to find the input. Check your answer by substituting it into the equation $3n + 7 = 40$. See ①.

① Possible explanation: 3 times the input, plus 7, equals 40. So, there must have been 33 before 7 was added. That means 3 times the input equals 33. The input must be 11.

2 Work through additional examples with the students.

Jay's method of solving $3n + 7 = 43$ involves working backward from the output value to find the input value. In this lesson, you will learn a technique for working backward called *backtracking.*

3 Introduce the term *backtracking.*

# Investigation 1

1 In this investigation, students use flowcharts to work forward from an input to the output. Then they use flowcharts to work backward from an output to the input.

Take some time to introduce the concept of a flowchart. Draw Hannah's flowchart on the board, starting with the input, *n*, and adding the operations, × 4 and + 5, one at a time.

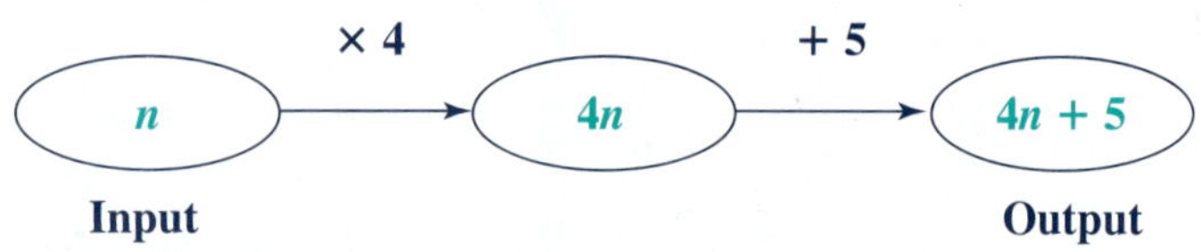


2 Then draw another copy of the flowchart, but leave the ovals blank. Have students choose a starting number. Write this number in the first oval. Then have students find the numbers that should go in the two remaining ovals.

3 **Problem Set A** **Suggested Grouping: Individuals**
This problem set gives students practice using flowcharts. Students use flowcharts to work forward from an input to the output.

4 If students have trouble constructing the flowchart for **Problem 3,** ask them what happens to the input first, and then what happens to each subsequent result.

# Investigation 1 Learning to Backtrack

Hannah's class was playing *What's My Rule?* Hannah found that the rule was $t = 4n + 5$, where $n$ is the input and $t$ is the output.

To find an output with this rule, you do the following:

*Start with an input.*
*Multiply it by 4.*
*Add 5.*

VOCABULARY
**flowchart**

Hannah drew a diagram, called a **flowchart,** to show these steps.

$$(n) \xrightarrow{\times 4} (4n) \xrightarrow{+5} (4n + 5)$$

**Input** **Output**

The oval at the left side of the flowchart represents the input. Each arrow represents a *mathematical operation.* The oval to the right of an arrow shows the result of a mathematical operation. The oval at the far right represents the output.

Here is Hannah's flowchart for the input value 3:

$$(3) \xrightarrow{\times 4} (12) \xrightarrow{+5} (17)$$

In the next problem set, you will practice working with flowcharts.

**Just the facts**

Computer programmers and engineers use complex flowcharts to represent the steps in computer programs, manufacturing operations, construction projects, and other procedures.

## Problem Set A

Copy each flowchart, and fill in the ovals.

**1.** $(8) \xrightarrow{\times 3} (24) \xrightarrow{-7} (17)$

**2.** $(5) \xrightarrow{\times 3} (15) \xrightarrow{-1} (14) \xrightarrow{\times 2} (28)$

**3.** In the rule $j = 7m - 2$, you can think of $m$ as the input and $j$ as the output.

**a.** Create a flowchart for the rule, but don't fill in the ovals.

**b.** Use your flowchart to find the value of $j$ when the value of $m$ is $\frac{5}{3}$. $\frac{29}{3}$ or $9\frac{2}{3}$

**3a.** $( \quad ) \xrightarrow{\times 7} ( \quad ) \xrightarrow{-2} ( \quad )$

1 Introduce the term flowcharts.

2 Have students practice filling in blank flowcharts.

3 You may have students work individually.

4 Have students discuss what happens to the inputs and each subsequent result in flowcharts.

LESSON 9.2 Backtracking 571

# Develop

For the equation in **Problem 4,** the division must be done before the addition. Since the order of the addends does not affect the final result, students will need to use a closely related, equivalent rule to create the flowchart: $d = a \div 10 + 3.2$. The flowchart for this equivalent rule can easily be seen to give the correct outputs:

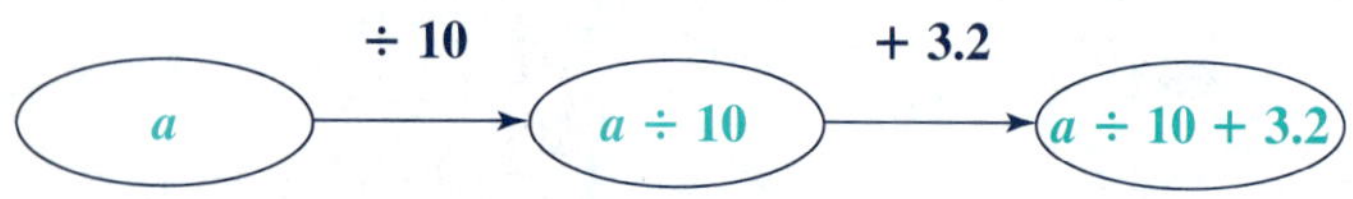


Without the ovals filled in, the flowchart is as follows:

1 **Problem Set Wrap-Up** Discuss **Problems 1 and 2** with the class. After you have done this, you may wish to ask students if they can write an equation for the rule exhibited by each flowchart.

For Problem 1, students can start with the variable $n$ in the first oval and then fill in the remaining ovals:

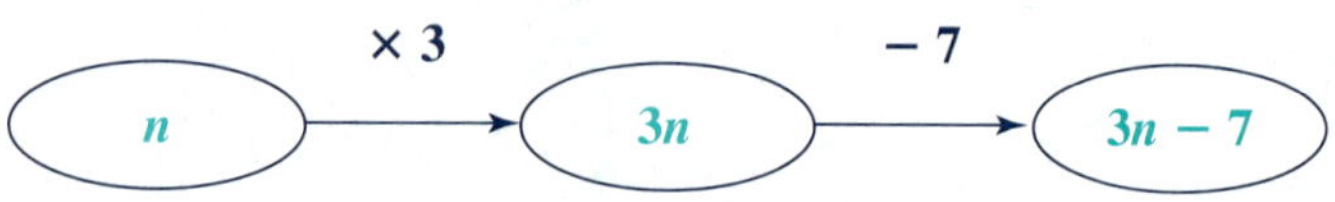


If $p$ is the output, then the rule is $p = 3n - 7$.

For Problem 2, the flowchart will look like this:

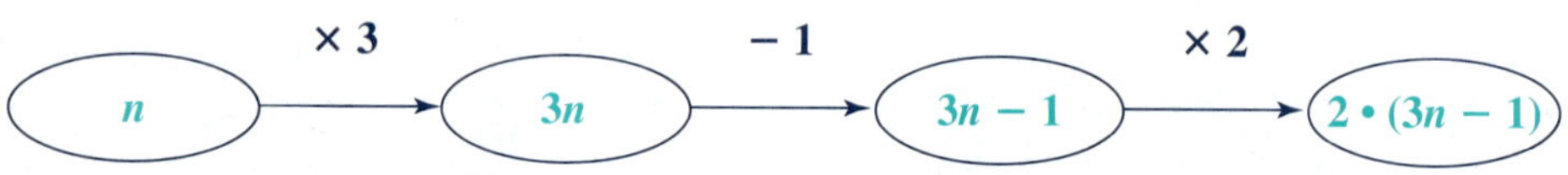


If $k$ is the output, then the rule is $k = 2 \cdot (3n - 1)$. ■

2 After discussing Problem Set A, read the paragraph above the Example with the class to be sure they understand what is meant by *backtracking.*

## Access for all Learners

**Language Diversity** For some students, you may need to point out that the word *backtrack* is made up of the two words *back* and *track*. Tell students that backtracking means following one's own tracks backward to return to the starting place. A phrase with a similar meaning is *retracing your steps*.

3 **Example**

Go through this Example with the class very carefully. By making a flow chart, students are using the draw a diagram problem-solving strategy. Be sure students understand that as they work their way back from output to input, they are using inverse operations to go from an oval on the right to the preceding oval on the left.

**4.** In the rule $d = 3.2 + a \div 10$, you can think of $a$ as the input and $d$ as the output.

**a.** Create a flowchart for this rule, but don't fill in the ovals. (Be sure to think about order of operations.)

**4a.** See Additional Answers.

**b.** Use your flowchart to find the value of $d$ when the value of $a$ is 111. 14.3

**1** Ask students to write a rule for each flowchart.

VOCABULARY
**backtracking**

In Problem Set A, you used flowcharts to *work forward*—starting with the input and applying each operation to find the output. You can also use flowcharts to *work backward*—starting with the output and undoing each operation to find the input. This process, called **backtracking,** is useful for solving equations.

**2** Review the meaning of *backtracking*.

**3** Problem-Solving Strategy
Draw a diagram

**EXAMPLE**

When playing a game of *What's My Rule?,* Hannah figured out that the secret rule was

$$t = \frac{n+1}{2}$$

Hannah wanted to find the input that gives an output of 33. That is, she wanted to solve this equation:

$$\frac{n+1}{2} = 33$$

To find the solution, she first made a flowchart:

Then she found the input by backtracking:

"Since 33 is the output, I'll put it in the last oval."

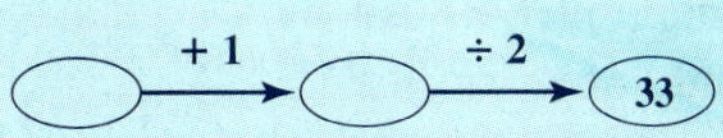


"Since the number in the second oval was divided by 2 to get 33, it must be 66."

"1 was added to the input to get 66, so the input must be 65."

Hannah checked her solution, 65, by substituting it into the original equation:

$$\frac{65+1}{2} = \frac{66}{2} = 33$$

**Remember**
Operations that undo each other are called *inverse operations.* To undo the division, Hannah used the inverse operation, multiplication. To undo the addition, she used the inverse operation, subtraction.



**Additional Answers**

**4a.**

1

## Problem Set B Suggested Grouping: Pairs

In this problem set, students use backtracking to solve equations. They use ideas from the Example on page 572.

2

For **Problem 2,** if students have difficulty writing the equation, have them copy the flowchart and leave all the ovals blank, as shown below.

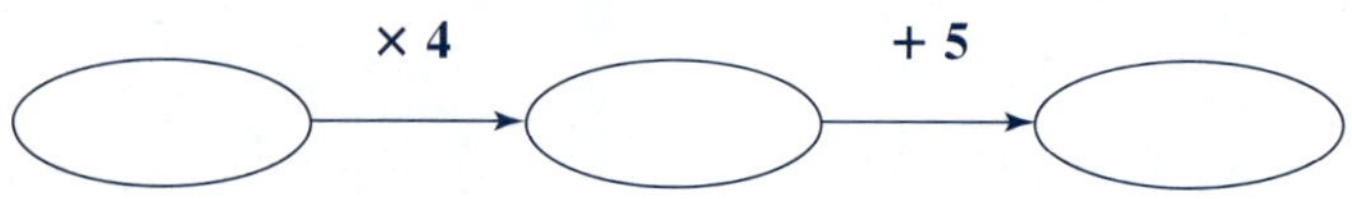


Next, have them select a variable, say *n,* to represent the input. Have them write it in the first oval.

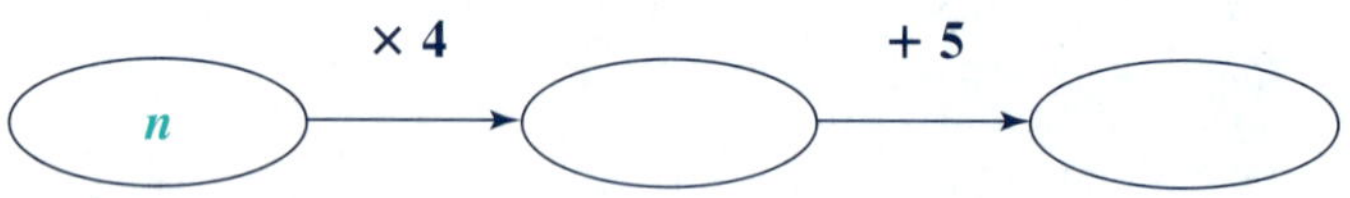


Then have students fill in the remaining ovals, using appropriate expressions involving *n*.

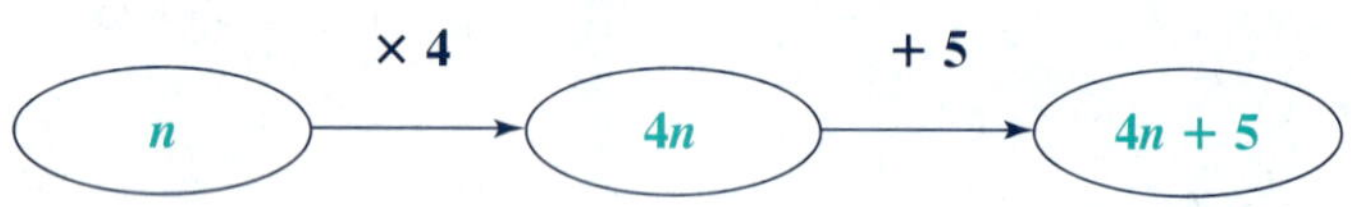


To write the equation Tyrone used, students can use the output expression $4n + 5$ and the value for that expression, 21, given in the problem. The equation is $4n + 5 = 21$.

When students use backtracking to solve the equation, they should start with the flowchart given in the problem and work to the left one step at a time.

### Access for all Learners

**Extra Help** If students are having difficulty working backward, they may not have caught on to the role of inverse operations. Use the flowchart in Problem 2 to provide some help. Draw the flowchart on the board. Below each arrow, draw an arrow that goes in the opposite direction, labeling it with the corresponding inverse operation:

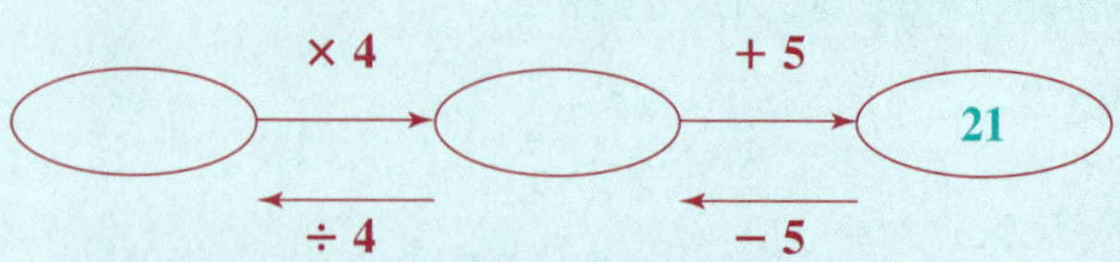


Students can use these arrows to help them work from right to left, from output to input.

3

For **Problem 5b,** students can copy the flowchart, leaving the ovals blank, and then put *k* in the oval for the input. They can then fill in the remaining ovals. When they do this, they need to realize that everything in the third oval must be multiplied by 2 to fill in the final oval. This is represented by putting parentheses around $5k - 1$. Help students see why they will get an incorrect result if they do not use parentheses. Ask students to substitute the answer they got by backtracking into each of the four equations to see which equations result in true equations. The two equations with parentheses will give true equations.

4

## Share & Summarize

Ask students to do these problems individually before you discuss them with the class. This will help you determine whether all students understand how to create flowcharts and use them for backtracking.

**Troubleshooting** If students could not complete the Share & Summarize correctly, work through some additional problems with them before they go on to the next investigation. It is important that they understand the basics of creating a flowchart and backtracking before going on to tackle more complex problems.

**Additional Examples**

1. Use a flowchart to help you solve each equation.

   a. $(n - 3) \times 7 = 21$ 6

   b. $4 \times n - 3 = 7$ $2\frac{1}{2}$

   c. $\frac{n}{2} + 5 = 16$ 22

• *Teaching notes continued on page A670*

## Problem Set B

1. Marcus used Hannah's rule, $t = \frac{n+1}{2}$, and got the output 53.

Use backtracking to find Marcus' input. Explain each step in your solution.

Tyrone solved three equations by backtracking. Below are the flowcharts he started with. For each flowchart, write the equation he was trying to solve. (Use any letter you want to represent the input variable.) Then backtrack to find the solution. Check your solutions.

2. 

$4n + 5 = 21$, 4

3. 

$3n - 8 = 26$, $11\frac{1}{3}$

4. 

$\frac{n}{6} + 10 = 97$, 522

5. Terry drew this flowchart:

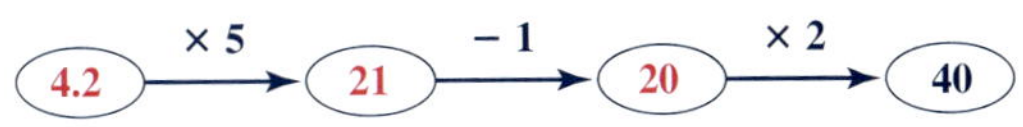


a. Copy Terry's flowchart, and fill in the ovals.

b. Which of these equations can be represented by Terry's flowchart? Explain how you know.

$5 \cdot k - 1 \cdot 2 = 40$ $\quad$ $(k \cdot 5 - 1) \cdot 2 = 40$

$2 \cdot (5k - 1) = 40$ $\quad$ $5k - 1 \cdot 2 = 40$

**1.** 105; The number in the second oval was divided by 2 to get 53, so it must be 106. To get 106, 1 was added to the input, so the input must be 105.

**5b.** $(k \cdot 5 - 1) \cdot 2 = 40$ and $2 \cdot (5k - 1) = 40$; You need to subtract 1 before you multiply by 2, so you need parentheses around the $5k - 1$. It doesn't matter whether the multiplication by 2 comes before or after the parentheses because order doesn't matter in multiplication. In $5 \cdot k - 1 \cdot 2 = 40$ and $5k - 1 \cdot 2 = 40$, the multiplication is done before the subtraction, which doesn't fit the flowchart.

## Share & Summarize

1. Explain what a flowchart is. Demonstrate by making a flowchart for the rule $t = 5n - 3$. See Additional Answers.

2. Use your flowchart from Question 1 to find the output for input $\frac{7}{10}$. $\frac{1}{2}$

3. Explain what backtracking is. Demonstrate how backtracking can be used to solve an equation by solving $5n - 3 = 45$. Check your solution. See Additional Answers.

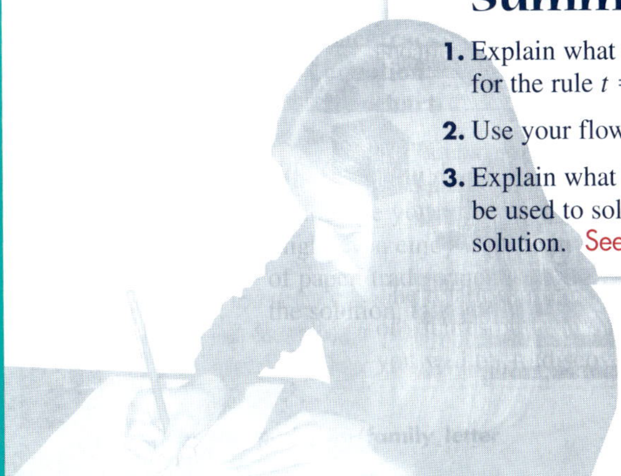



**1** You may have students work in pairs.

**2** Have students copy the flowchart and leave all the ovals blank.

**3** Explain why students will get an incorrect result if they do not use parenthesis.

**4** Have students work on the problems individually before discussing them as a class.

### Additional Answers
### Share & Summarize

1. Possible answer: A flowchart is a diagram that traces the steps you follow to evaluate a rule. In the rule $t = 5n - 3$, you find the output $t$ in two steps: multiply the input $n$ by 5, and then subtract 3. This is shown below. The arrows represent the steps, or the operations, and the results of the steps go in the ovals.

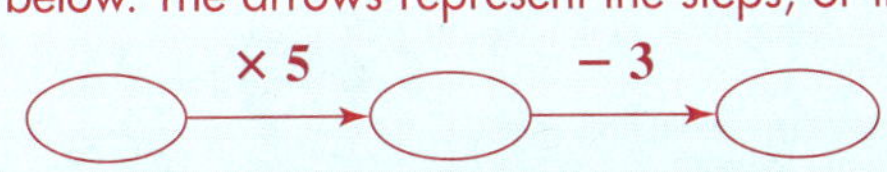


• Additional Answers continued on page A670

LESSON 9.2 Backtracking 573

# Investigation 2

In this investigation, students will use flowcharts and backtracking to solve more complex equations, including those that require parentheses.

1 **Problem Set C** **Suggested Grouping: Individuals**
This problem set will help students understand how the presence or absence of parentheses in an expression can change the meaning of the expression.

2 For **Problem 1a,** some students may wonder where the variable $P$ belongs in the flowcharts. Emphasize that $P$ is the output. It may help to show this by writing labels for the first and last ovals of the flowcharts, as shown below.

Miguel's rule:

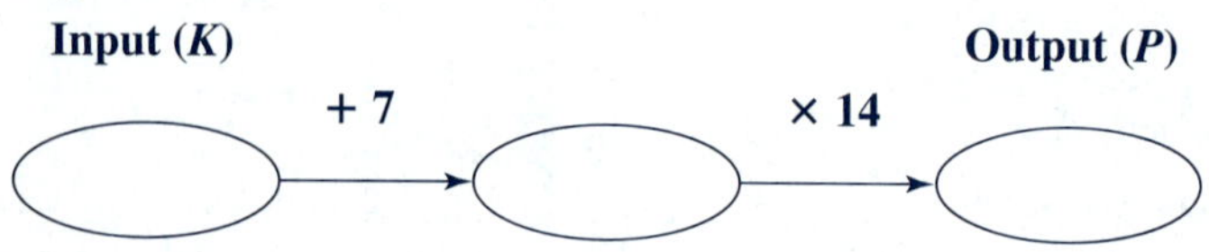


Althea's rule:

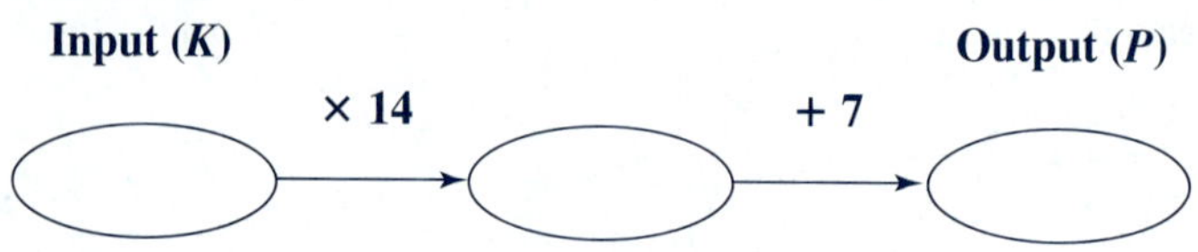


Each flowchart shows how to compute $P$ if you know $K$, and how to compute $K$ (by backtracking) if you know $P$.

3 For **Problem 2b,** students need to use parentheses around $3f - 6$. If students write the equation $3f - 6 \cdot 5 = 60$, remind them of the order of operations rules to help them understand that $3f - 6 \cdot 5$ does not mean the same as $(3f - 6) \cdot 5$.

# Investigation 2 Practicing Backtracking

**1a.** See Additional Answers.

**1c.** no; Possible explanation: In Miguel's rule, 7 is added to the input and the result is multiplied by 14. In Althea's rule, the input is multiplied by 14 and then 7 is added. The fact that the rules give different inputs for the same output also shows that the rules are not equivalent.

In this investigation, you will practice backtracking so you can use it to solve equations quickly and easily.

**1** You may have students work individually.

## Problem Set C

**1.** A group of students was playing *What's My Rule?* Miguel and Althea both thought they knew what the rule was. Both students used $K$ to represent the input and $P$ to represent the output.

Miguel's rule: $P = 14 \cdot (K + 7)$

Althea's rule: $P = 14 \cdot K + 7$

**a.** Make a flowchart for each rule, but don't fill in the ovals.

**b.** For each rule, use backtracking to find the input that gives the output 105. Miguel's rule: 0.5; Althea's rule: 7

**c.** Are these two rules equivalent? Explain why or why not.

**2** Emphasize that the variable $P$ is the output.

**2.** Gabriela and Erin were playing a game called *Think of a Number.*

Gabriela must figure out Erin's starting number.

**a.** Draw a flowchart to represent this game.

**b.** What equation does your flowchart represent? $(3f - 6) \cdot 5 = 60$

**c.** Use backtracking to solve your equation. Check your solution by following Gabriela's steps. 6

**3** Review the order of operations rules.

**2a.**

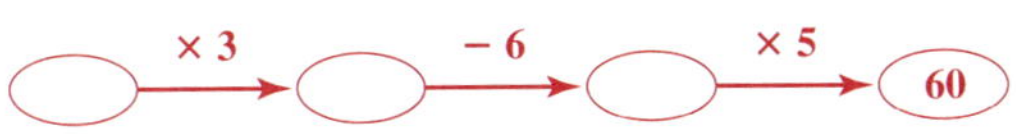




## Additional Answers

**1a.** Flowchart for Miguel's rule:

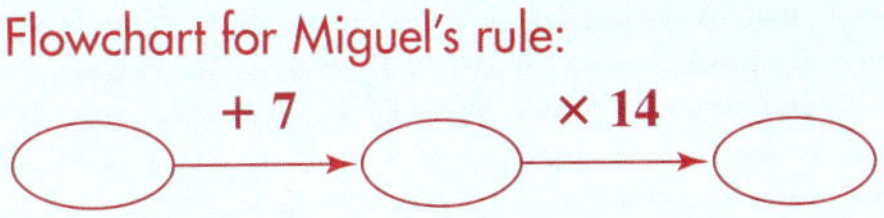


Flowchart for Althea's rule:

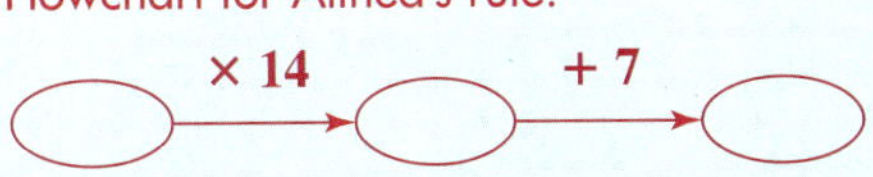

# Develop

## Think & Discuss

1 This Think & Discuss introduces an incorrect flowchart for a complex equation as a way of getting students to focus on the order of operations in an equation and a flowchart. If students do not volunteer the answer given, ask them which operation is used with $n$ first. They should see that $n + 1$ is the first calculation. Help them recognize that the arrow labeled "+ 1" should have come first in the flowchart.

## Problem Set D **Suggested Grouping: Pairs**

2 This problem set provides practice using flowcharts for relatively complex equations. Students can first work on each problem individually, and then compare results with their partners. If they get different results, they can work together until their answers agree.

3 Students may find **Problem 1** a bit tricky because the first and last flowcharts both start correctly, multiplying $n$ by 3. They differ, however, in the next two steps. If students have difficulty understanding which is the correct order of operations, you may wish to point out that the original equation can be rewritten as follows:

$$(1 + n \cdot 3) \div 4 - 11 = 10$$

## Think & Discuss

Luke wanted to solve this equation by backtracking:

$$\frac{2 \times (n + 1)}{3} - 1 = 5$$

He made this flowchart:

$$\bigcirc \xrightarrow{\times 2} \bigcirc \xrightarrow{+1} \bigcirc \xrightarrow{\div 3} \bigcirc \xrightarrow{-1} (5)$$

Does Luke's flowchart correctly represent the equation? Why or why not? No; the flowchart multiplies by 2 first and then adds 1. It should add 1 first and then multiply by 2.

Solve the equation, and explain how you found the solution. See ①.

① 8; Possible explanation: I fixed Luke's flowchart by reversing the × 2 and + 1 steps, and then I used backtracking.

**1** Make sure students understand that $n + 1$ is the first calculation.

The equation in the Think & Discuss involves several operations. You can often solve equations like this by backtracking, but you need to pay close attention to order of operations as you draw the flowchart.

## Problem Set D

**2** You may have students work in pairs.

**3** Show students that the original equation can be rewritten.

**1.** Conor, Althea, and Miguel each made a flowchart to represent this equation:

$$\frac{1 + n \cdot 3}{4} - 11 = 10$$

Tell whose flowchart is correct, and explain the mistakes the other students made.

**Conor's flowchart:**

$$\bigcirc \xrightarrow{\times 3} \bigcirc \xrightarrow{\div 4} \bigcirc \xrightarrow{+1} \bigcirc \xrightarrow{-11} (10)$$

**Althea's flowchart:**

$$\bigcirc \xrightarrow{+1} \bigcirc \xrightarrow{\times 3} \bigcirc \xrightarrow{\div 4} \bigcirc \xrightarrow{-11} (10)$$

**Miguel's flowchart:**

$$\bigcirc \xrightarrow{\times 3} \bigcirc \xrightarrow{+1} \bigcirc \xrightarrow{\div 4} \bigcirc \xrightarrow{-11} (10)$$

Miguel's flowchart is correct. In Conor's, the addition should come before the division. In Althea's, the multiplication should come before the addition.

LESSON 9.2 Backtracking 575

1 For **Problem 2,** if students have difficulty making the flowchart, you may want to see if they can think of the expression in relation to a *Think of a Number* game. The game would go as follows: Think of a number ($n$). Subtract 13 (this gives $n - 13$). Divide the result by 2 (this gives $\frac{n-13}{2}$). Finally, add 6 (this gives $\frac{n-13}{2} + 6$). The other player announces a final result of 15, so the equation is $\frac{n-13}{2} + 6 = 15$. The instructions correspond to the following flowchart:

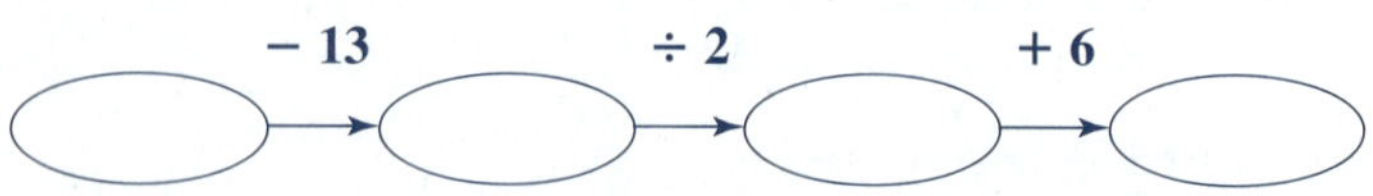


When the second player announces 15 as the number for the last oval, the person who called out the instructions uses backtracking to discover the value of $n$.

For **Problem 3,** students may find it helpful to use the approach suggested above for Problem 2.

**Problem Set Wrap-Up** A discussion of Problem 7 could be helpful. If most of your students are comfortable applying the rules of the order of operations, though, you may want to have them just continue to the Share & Summarize.

### 2 Share & Summarize

The purpose of this Share & Summarize is to make sure students understand the process of backtracking with flowcharts. They do not need to use a complicated example. Ask students to write down their best answer to the question, and then ask for volunteers to read their answers. Use a class discussion to see which explanations the class understands and agrees with.

**Troubleshooting** If students are uncertain how to set up a flowchart for a complex equation, ask them to consider each equation as a *Think of a Number* game. The suggestion for Problem 2 in Problem Set D illustrates the approach.

### On Your Own Exercises

Practice & Apply: 7–13, pp. 579–580
Connect & Extend: 19, 20, p. 582
Mixed Review: 25–35, pp. 584–585

# Investigation 3

In this investigation, students write equations to describe patterns and real-world situations. They use backtracking to solve problems involving the equations.

Have students examine the toothpick "ladders" in the diagrams. Then have them work on the Think & Discuss.

### 3 Think & Discuss

You will probably want to call on volunteers to explain why they think the rule $n = 3r + 2$ works. The answer given in the margin provides one explanation. Another approach is to note that there are 5 toothpicks in the first ladder. Thereafter, 3 toothpicks are added for each new rung. So when $r - 1$ new rungs have been added, there are a total of $5 + 3(r - 1)$ toothpicks. This expression can be simplified to obtain $3r + 2$.

**2, 3.** See Additional Answers.

In Problems 2 and 3, draw a flowchart to represent the equation. Then use backtracking to find the solution. Be sure to check your solution.

**2.** $\frac{n-13}{2} + 6 = 15$

**3.** $7\left(\frac{n+4}{7} + 1\right) = 84$

Solve each equation. Be sure to check your solutions.

**4.** $\frac{7z+2}{15} = 2$ 4

**5.** $(n \cdot 12 + 8) \cdot 100 = 2{,}100$ $1\frac{1}{12}$

**6.** $\frac{q-36}{6} + 16 = 83$ 438

**7.** $4 \cdot \left(\frac{b}{2} - 3\right) + 1 = 97$ 54

**Think & Discuss Answers**

① Possible explanation: For each rung, 3 toothpicks are added. A "ladder" with no rungs would be made from 2 toothpicks.

② A ladder with 36 rungs; the answer is the solution of the equation $3r + 2 = 110$. I can set up a flowchart and use it to solve this equation.

## Share & Summarize

See Additional Answers.

Give an example to demonstrate why it is important to pay close attention to order of operations when you make a flowchart.

**1** Encourage students to relate the expression to the *Think of a Number* game.

**2** Discuss students' answers.

# Investigation 3 Using Backtracking to Solve Problems

These "ladders" are made from toothpicks:

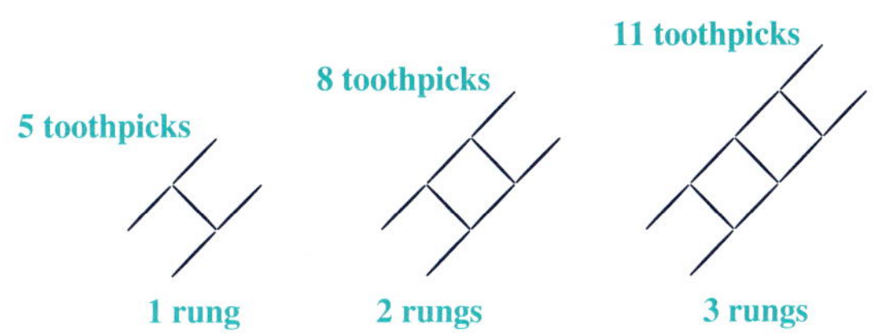


The rule for the number of toothpicks $n$ in a ladder with $r$ rungs is $n = 3r + 2$.

## Think & Discuss

Can you explain why the rule for the number of toothpicks $n$ in a ladder with $r$ rungs is $n = 3r + 2$? See ① in margin.

Suppose you have 110 toothpicks. What size ladder can you make? How can you use backtracking to help you find the answer? See ②.

**3** You may want to call on volunteers.



**Additional Answers**
**Problem Set D**

**2.**

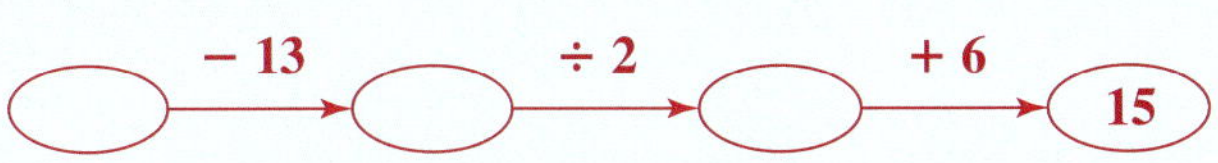


; 31

**3.**

; 73

• *Additional Answers continued on page A670*

# Develop

1 **Problem Set E** **Suggested Grouping: Individuals**
In this problem set, students write rules to describe patterns. They then use backtracking to solve equations and answer questions about the situations the patterns describe.

**Problem 1** uses the situation and rule from the Think & Discuss on page 576.

2 In **Problem 2a,** students can think of starting with 1 toothpick. To get 1 trapezoid, 4 more toothpicks must be added. To get the next trapezoid, another 4 toothpicks must be added, and so on.

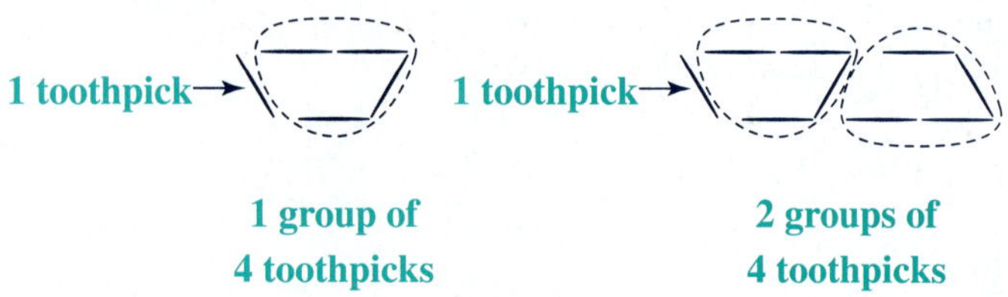


3 In **Problem 3a,** it may be necessary to give students a hint by pointing out that the numbers in the top row of the table are multiples of 3.

4 **Problem Set F** **Suggested Grouping: Pairs**
In this problem set, students get practice in setting up and solving equations for some simplified real-world situations.

5 In **Problem 1,** be sure that students note that the profit per apple is given in cents and that the amount Leong wants to earn is given in dollars. They should also understand that the amount he earns from an apple is the profit he makes from the sale of the apple, not what he charges for the apple.

## Problem Set E

1. Write and solve an equation to find the number of rungs on a toothpick ladder made with 53 toothpicks. $3r + 2 = 53, 17$
2. Look at this pattern of toothpick shapes:

1 trapezoid　　2 trapezoids　　3 trapezoids

   a. Write a rule for finding the number of toothpicks $n$ you would need to make a shape with $t$ trapezoids. $n = 4t + 1$

   b. Write and solve an equation to find the number of trapezoids in a shape with 125 toothpicks. $4t + 1 = 125, 31$

3. Look at the pattern in this table:

| $n$ | 0 | 3 | 6 | 9 | 12 |
|---|---|---|---|---|---|
| $y$ | 19 | 20 | 21 | 22 | 23 |

   a. Write a rule that relates $n$ and $y$. $y = n \div 3 + 19$

   b. Write and solve an equation to find the value of $n$ when $y$ is 55.

**3b.** $n \div 3 + 19 = 55$, 108

In the next problem set, you will see that backtracking can also be used to solve everyday problems.

## Problem Set F

1. Leong makes candy apples to sell at the farmer's market on Saturday. He makes a profit of 35¢ per apple.

   a. Write a rule Leong could use to calculate his profit if he knows how many candy apples he sold. Tell what each letter in your rule stands for.

   b. Leong wants to earn $8 so he can see a movie Saturday night. Write an equation Leong could solve to find the number of candy apples he must sell to earn $8.

   c. Use backtracking to solve your equation. How many candy apples does Leong need to sell?

   The solution is about 22.86, so Leong must sell at least 23 candy apples to earn $8.

**Problem Set F Answers**

**1a.** $P = 0.35A$, where $P$ stands for the profit and $A$ stands for the number of candy apples sold

**1b.** $8 = 0.35A$



**1** Problem-Solving Strategies
- Look for a pattern
- Write an equation
- Work backward

**2** Have students start with 1 toothpick.

**3** Tell students that the numbers for the $n$ row are multiples of 3.

**4** You may have students work in pairs.

**5** Be sure students know that the profit per apple is given in cents.

LESSON 9.2 Backtracking **577**

**1** In **Problem 3a,** students may prefer to write an equation that uses a variable for the distance traveled in miles.

**2**

### Share & Summarize

This is a multistep problem. First, students need to find a rule connecting the term to the number of toothpicks used. They will then need to backtrack to solve the equation.

**Troubleshooting** If students have trouble writing equations and solving them using backtracking, you may want to have them work with partners on additional examples.

**Additional Examples** For each toothpick pattern, write a rule for finding the number of toothpicks $n$ in Term $t$. Then write and solve an equation to find the numbers of the term that contains 46 toothpicks.

1.

Possible rule: $n = 4t + 2$; $4t + 2 = 46$, $t = 11$, so Term 11 contains 46 toothpicks.

2.

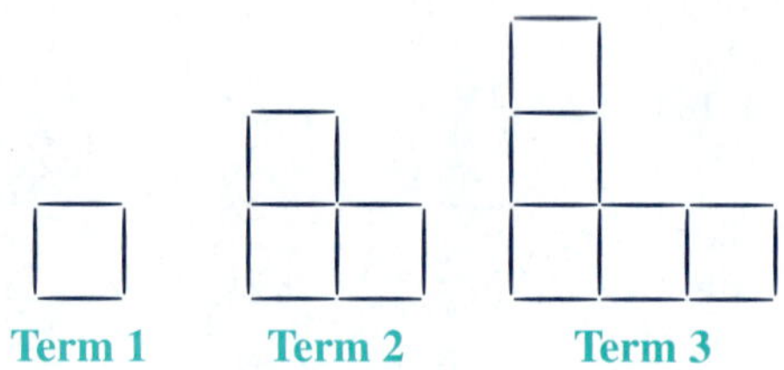


Possible rule: $n = 6t - 2$; $6t - 2 = 46$, $t = 8$, so Term 8 contains 46 toothpicks.

### On Your Own Exercises

Practice & Apply: 14–17, pp. 580–581
Connect & Extend: 21–24, pp. 583–584
Mixed Review: 25–35, pp. 584–585

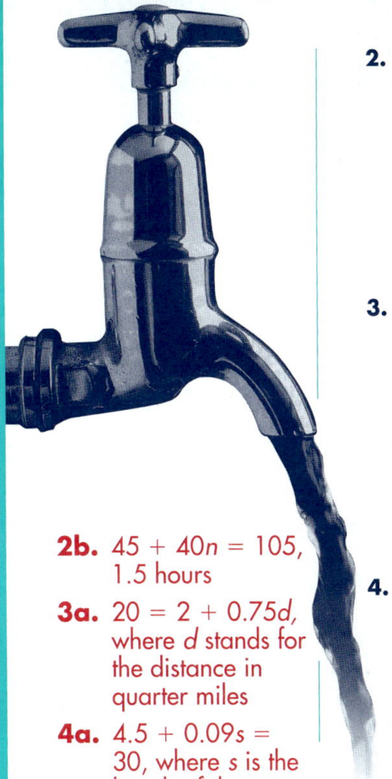

**2.** The plumbers at DripStoppers charge $45 for a house call, plus $40 for each hour of work.

**a.** Write a rule for the cost of having a DripStoppers plumber come to your home and do $n$ hours of work. $c = 45 + 40n$

**b.** DripStoppers sent Mr. Valdez a plumbing bill for $105. Write and solve an equation to find the number of hours the plumber worked at Mr. Valdez's home. Check your solution.

**3.** When you hire a taxi, you are usually charged a fixed amount of money when the ride starts, plus an amount that depends on how far you travel. Suppose a taxi charges $2 plus $0.75 for every quarter mile.

**a.** Write an equation you could solve to find out how far you can travel for $20.

**b.** Solve your equation. How far could you travel for $20? Check your answer. 24 quarter miles, or 6 miles

**4.** Caroline and Althea are making a kite. The materials for the main part of the kite cost $4.50, and the string costs 9¢ per yard.

**a.** Write an equation you could solve to find how long the string could be if the friends have $30 to spend on their project.

**b.** Solve your equation to find how long the string could be. about 283 yd

**2b.** $45 + 40n = 105$, 1.5 hours

**3a.** $20 = 2 + 0.75d$, where $d$ stands for the distance in quarter miles

**4a.** $4.5 + 0.09s = 30$, where $s$ is the length of the string in yards

## Share & Summarize

Jing used toothpicks to create this pattern:

For one of the terms, she needed 112 toothpicks. Which term was it? Explain how you found your answer. 55; Possible explanation: I wrote a rule for finding the number of toothpicks $t$ in Term $n$: $t = 2 + 2n$. To find the number of the term with 112 toothpicks, I needed to solve $2 + 2n = 112$. I used backtracking to find the solution.

**1** Have students write an equation that uses a variable for the distance.

**2** Problem-Solving Strategies

- Write an equation
- Work backward

# Teacher Notes

# On Your Own Exercises

## Practice & Apply

Copy each flowchart, and fill in the ovals.

**1.** (12) —× 2→ (24) —+ 12→ (36)

**2.** (2.7) —× 3→ (8.1) —− 1→ (7.1) —× 4→ (28.4)

Hannah solved three equations by backtracking. Below are the flowcharts she started with. For each flowchart, write the equation she was trying to solve. Then backtrack to find the solution. Check your solutions.

**3.** ( ) —× 16→ ( ) —− 4→ (28) $16n - 4 = 28$, 2

**4.** ( ) —÷ 3→ ( ) —+ 11→ (41) $\frac{n}{3} + 11 = 41$, 90

**5. Challenge** $(n + 1) \cdot 4 - 5 = 15$, 4

( ) —+ 1→ ( ) —× 4→ ( ) —− 5→ (15)

**6.** In a game of *What's My Rule?*, Rosita wrote the rule $b = 3a \div 4$, where $a$ is the input and $b$ is the output.

**a.** Make a flowchart for Rosita's rule. See below.

**b.** Use your flowchart to find the output when the input is 18. $13\frac{1}{2}$

**c.** Backtrack to solve the equation $3a \div 4 = 101$. $134\frac{2}{3}$

Make a flowchart to represent each equation, and then use backtracking to solve the equation. Be sure to check your solutions.

**7.** $4k + 11 = 91$ See below.

**8.** $4 \cdot (m - 2) = 38$ ( ) —− 2→ ( ) —× 4→ (38); 11.5

**6a.** ( ) —× 3→ ( ) —÷ 4→ ( )

**7.** ( ) —× 4→ ( ) —+ 11→ (91); 20

impactmath.com/self_check_quiz



## Assign and Assess

### On Your Own Exercises

Investigation 1, pp. 571–573
Practice & Apply: 1–6
Connect & Extend: 18

Investigation 2, pp. 574–576
Practice & Apply: 7–13
Connect & Extend: 19, 20

Investigation 3, pp. 576–578
Practice & Apply: 14–17
Connect & Extend: 21–24

Assign Anytime
Mixed Review: 25–35

**Exercise 5:**
Watch for students who say that the equation is $n + 1 \cdot 4 - 5 = 15$.

## Assign and Assess

**Exercise 11:** Some students may draw flowcharts that show multiplication by 2 followed by multiplication by 3. That order is also acceptable since multiplication is a commutative operation.

**Exercise 14a:** There are several ways to select 3 toothpicks in the first term and then add 2 toothpicks at a time to get the subsequent terms of the sequence.

**Exercise 15:** Students should note that each time the value of $x$ increases by 5, the value of $y$ increases by 2.5.

**9.** Neva and Jay were playing *Think of a Number.* Neva said:

*Think of a number. Subtract 1 from your number. Multiply the result by 2. Then add 6. What number do you get?*

Jay said he got 10. Neva must figure out Jay's starting number.

**a.** Draw a flowchart to represent this game.

**b.** What equation does your flowchart represent?

**c.** Use backtracking to find the number Jay started with. Check your solution by following Neva's steps. 3

**9a.** See below.
**9b.** $(n - 1) \cdot 2 + 6 = 10$

**10.** For a game of *What's My Rule?,* Mia and Desmond wrote the rule $y = 9(2x + 1) + 1$, where $x$ is the input and $y$ is the output.

**a.** Draw a flowchart for Mia and Desmond's rule.

**b.** Use your flowchart to solve the equation $9(2x + 1) + 1 = 46$. 2

**10a.** See Additional Answers.

Draw a flowchart to represent each equation. Then use backtracking to find the solution. Be sure to check your solutions.

**11–13.** See Additional Answers.

**11.** $\frac{3 \cdot m \cdot 2}{6} = 1$

**12.** $\frac{8p + 2}{5} - 5 = 19$

**13.** $3\left(\frac{5 + n}{3} - 4\right) = 15$

**14.** Look at the toothpick sequence below. The rule for the number of toothpicks $t$ needed to make Term $n$ is $t = 2n + 3$.

Term 1 Term 2 Term 3

**a.** Explain why this rule works for every term.

**b.** Write and solve an equation to find the number of the term that requires 99 toothpicks. $2n + 3 = 99$, 48

**14a.** See Additional Answers.

**15.** Look at the pattern in this table:

| $x$ | 0 | 5 | 10 | 15 | 20 |
|---|---|---|---|---|---|
| $y$ | 100 | 102.5 | 105 | 107.5 | 110 |

**a.** Write a rule that relates $x$ and $y$. $y = 100 + 0.5x$

**b.** Write and solve an equation to find the value of $x$ when $y$ is 197.5. $100 + 0.5x = 197.5$, 195

**9a.**

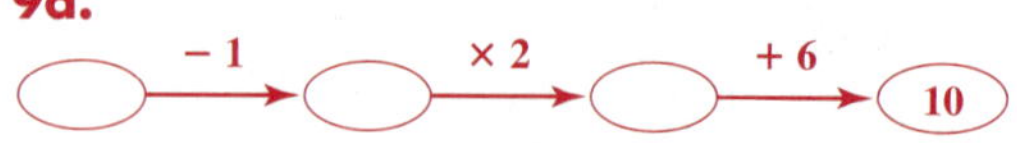




**Additional Answers**

**10a.**

**11.**

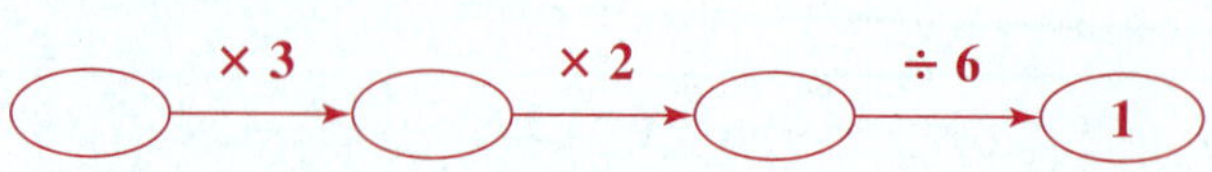

; 1

(Note: Labels on first and second arrows may be reversed.)

• Additional Answers continued on page A671

**Just the facts**

The earliest canoes had frames made from wood or whalebone and were covered with bark or animal skin.

**16.** **Economics** At Marshall Park, you can rent a canoe for \$5, plus \$6.50 per hour.

**a.** Write a rule for calculating the cost $C$ of renting a canoe for $h$ hours. Possible rule: $C = 5 + 6.5h$

**b.** Conor, Jing, and Miguel paid \$27.75 to rent a canoe. Write and solve an equation to find the number of hours the friends used the canoe. Check your solution. $5 + 6.5h = 27.75$, 3.5 hours

**17.** **Economics** Avocados cost \$1.89 each. Hannah plans to make a large batch of guacamole for a party, and she wants to buy as many avocados as possible. She has \$14.59 to spend.

**17a.** $1.89a = 14.59$, where $a$ is the number of avocados

**a.** Write an equation you could solve to find how many avocados Hannah can buy for \$14.59.

**17b.** See Additional Answers.

**b.** Solve your equation. How many avocados can Hannah buy?

## Connect & Extend

**18.** Evita wants to make a fence out of wooden poles. She drew a diagram to help her figure out how many poles she would need.

**a.** Write a rule connecting the number of poles $p$ to the number of sections $s$. $p = 1 + 3s$

**b.** The lumberyard has 100 poles in stock. Write an equation you could solve to find the number of fence sections Evita can build with 100 poles. $1 + 3s = 100$

**18c.** 33

**c.** Solve your equation. How many fence sections can Evita build?

**d.** If each pole is 2 yards long, how long will a 100-pole fence be? 66 yd

**Exercise 18:**
Watch for students who think that 4 poles are added for every new section of fence. The correct rule starts with 1 pole and adds 3 poles for each new section.



**Additional Answers**

**17b.** The solution is about 7.72, so Hannah can buy 7 avocados.

LESSON 9.2 Backtracking 581

# Assign and Assess

**Exercise 19c:**
If a student's estimate is not reasonably close to 3.5 h, ask the student to explain to you how he or she estimated the time to travel 220 mi.

**Exercise 20a:**
Some students may prefer to write the equation by using a fraction:
$\frac{n^2 + 3}{10} = 8.4$.

★ indicates multi-step problem

**19. Physics** A bus is traveling at an average speed of 65 miles per hour.

**a.** Copy and complete the table to show the distance the bus would travel in the given numbers of hours.

| Time (hours), $t$ | 1 | 2 | 3 | 4 | 5 |
|---|---|---|---|---|---|
| Distance (miles), $d$ | 65 | 130 | 195 | 260 | 325 |

**b.** On a grid like the one below, plot the points from your table. If it makes sense to do so, connect the points with line segments.

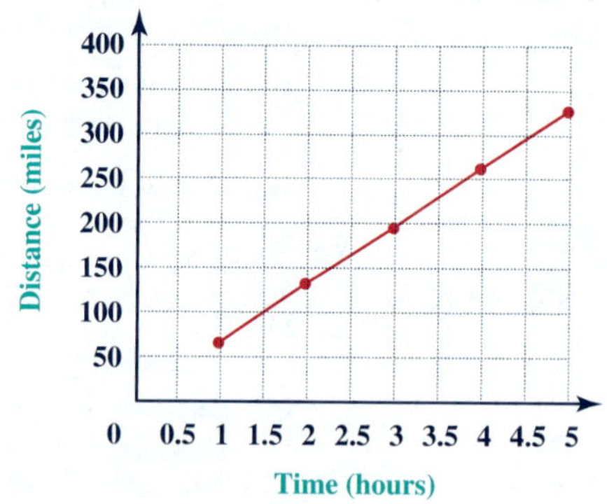


**c.** Use your graph to estimate how long it would take the bus to travel 220 miles. about 3.5 h

**d.** Write a rule that relates the time traveled $t$ (in hours) to the distance traveled $d$ (in miles). $d = 65t$

★ **e.** Write and solve an equation to find how long it would take the bus to travel 220 miles. How does the solution compare to your estimate from Part c?

**19e.** $65t = 220$; It would take about 3.38 h, or about 3 h 23 min. This is slightly less than my estimate.

**20.** Julie and Noah were playing *Think of a Number.* Noah said:

*Think of a number. Square it. Add 3. Divide your result by 10. What number do you get?*

Julie said she got 8.4. Noah must figure out Julie's starting number. He drew this flowchart to represent the game:

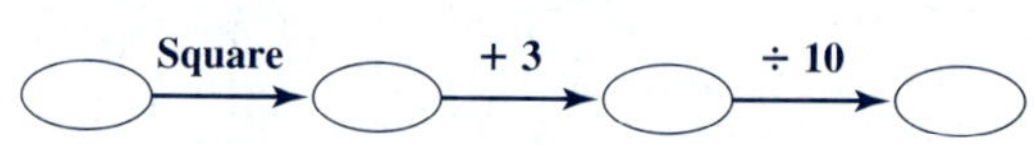


**a.** What equation does Noah need to solve to find Julie's starting number? $(n^2 + 3) \div 10 = 8.4$

**b.** Use backtracking to solve your equation. Check your solution by following Noah's steps. 9



582 CHAPTER 9 Solving Equations

**In** your **own words**

In this lesson, you used flowcharts to help solve equations. Explain how to make a flowchart for this equation:

$\left(\frac{d}{3} - 5\right) \times 2 + 4 = 30$

Also show how you would use your flowchart to find the value of *d*.

**21. Economics** Althea wants to buy a jacket at Donovan's department store. The store is having a "20% off" sale. In addition, Althea has a coupon for $5 off any item in the store.

**a.** Write a rule Althea could use to calculate the price $P$ she would pay for a jacket with an original price of $d$ dollars. (Note: The $5 is subtracted *after* the 20% discount is calculated.) $P = 0.8d - 5$

**b.** Althea pays $37.80 for a jacket. Write and solve an equation to find the original price of the jacket. $0.8d - 5 = 37.8$, $53.50

**22. Preview** This equation gives the height $h$ of a baseball, in feet, $t$ seconds after it is thrown straight up from ground level:

$$h = 40 \cdot t - 16 \cdot t^2$$

**a.** How high is the ball after 0.5 second? 16 ft

**b.** How high is the ball after 1 second? 24 ft

**c.** Based on what you learned in Parts a and b, estimate how long it would take the ball to reach a height of 20 feet. Explain your estimate.

**d.** Substitute your estimate for $t$ in the equation to find out how high the ball would be after that number of seconds. Were you close? Was your estimate too high or too low?

**23. Sports** Lana is making fishing lures. Each lure requires 2¢ worth of fishing line. Lana uses feathers and weights to make the lures. The feathers cost 17¢ apiece, and the weights cost 7¢ apiece.

**a.** Write a rule for the cost $C$ of a lure made with $f$ feathers and $w$ weights. $C = 0.17f + 0.07w + 0.02$

**b.** Lana doesn't want to spend more than 65¢ on each lure. Write an equation you could solve to find how many feathers she could use on a lure made with two weights.

**c.** Solve your equation to find how many feathers Lana can use on a lure with two weights.
$f \approx 2.88$, 2 feathers

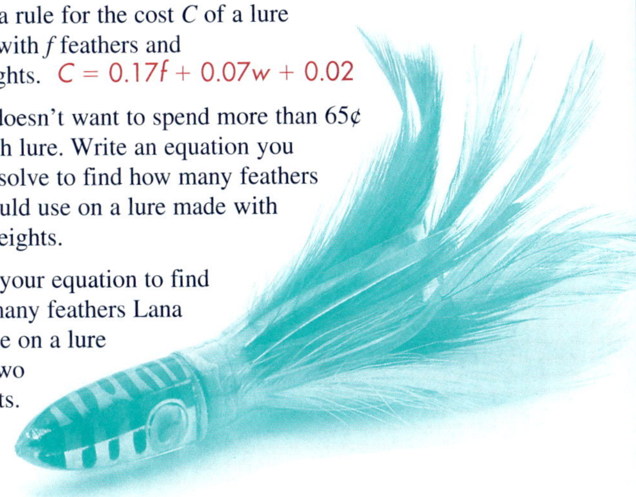

**22c.** Possible answer: About 0.75 second; 20 ft is halfway between 16 ft and 24 ft, so I think the ball would reach this height halfway between 0.5 s and 1 s.

**22d.** 21 ft; Possible answer: too high

**23b.** $0.65 = 0.17 \cdot f + 0.14 + 0.02$, or $0.65 = 0.17 \cdot f + 0.16$

**Exercise 23:**
In **Part a,** students might write an equation that expresses the cost of a lure in cents: $C = 17f + 7w + 2$. If so, the equation to solve in **Part b** is $65 = 17f + 16$.



LESSON 9.2 Backtracking 583

24. **Economics** Jordan has a part-time job as a telemarketer. He earns \$14.00 per hour plus 50¢ for every customer he calls.

   a. Write a rule for computing how much Jordan will earn on a three-hour shift if he calls $c$ customers. $D = 42 + 0.50c$

   b. Jordan would like to earn \$100 on his 3-hour shift. How many customers must he call? 116

## Mixed Review

Find the value of each expression without using a calculator.

25. $27 - 3^2 - 10$ 8

26. $4 + 4 \cdot 3 - 2$ 14

27. $\sqrt{49} + 3^2$ 16

28. $5^2 \div 5$ 5

29. $\frac{2}{3} \cdot \left(\frac{1}{3}\right)^2 - \frac{1}{27}$ $\frac{1}{27}$

30. $\sqrt{3^2 + 4^2}$ 5

31. **Sports** In a track-and-field competition, 10 women participated in the 100-meter run. Here are their times in seconds:

   12.2 11.3 13.5 11.5 11.7 12.6 15.5 11.8 13.4 11.5

   a. Find the mean and the median time. mean: 12.5 s; median: 12 s

   b. Make a stem-and-leaf plot of the times.

**Run Times**

| Stem | Leaf |
|---|---|
| 11 | 3 5 5 7 8 |
| 12 | 2 6 |
| 13 | 4 5 |
| 14 | |
| 15 | 5 |

**Key:** 12|6=12.6

**Geometry** Find each value of $x$.

32. Area = 4.5 ft²

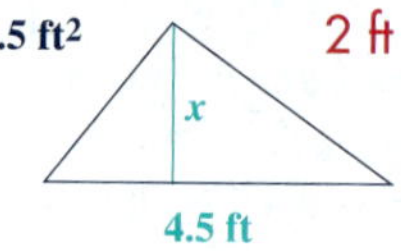


2 ft

33. Area ≈ $x$

$9\pi$ in.², or about 28.27 in.²

34. Area = 9.8 cm²

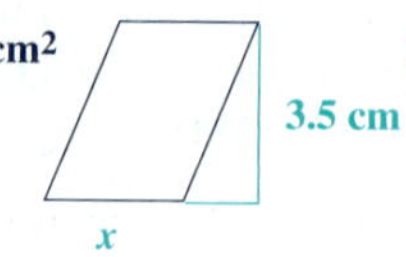


2.8 cm



584 CHAPTER 9 Solving Equations

35. Answers will vary.

**35. Statistics** In a landfill, bulldozers spread solid waste (trash and garbage) into layers, alternating with layers of dirt. The list below shows the number of solid-waste landfills that were in use in each region of the United States in a recent year.

| Northeast | South Atlantic | South | Midwest | West |
|---|---|---|---|---|
| Connecticut, 11 | Delaware, 3 | Alabama, 28 | Illinois, 61 | Alaska, 217 |
| Maine, 27 | Florida, 67 | Arkansas, 67 | Indiana, 32 | Arizona, 59 |
| Massachusetts, 106 | Georgia, 159 | Kentucky, 12 | Iowa, 77 | California, 278 |
| New Hampshire, 33 | Maryland, 25 | Louisiana, 29 | Kansas, 58 | Colorado, 72 |
| New Jersey, 14 | North Carolina, 114 | Mississippi, 14 | Michigan, 54 | Hawaii, 10 |
| New York, 42 | South Carolina, 37 | Oklahoma, 94 | Minnesota, 26 | Idaho, 37 |
| Pennsylvania, 47 | Virginia, 152 | Tennessee, 81 | Missouri, 30 | Montana, 82 |
| Rhode Island, 4 | West Virginia, 22 | Texas, 678 | Nebraska, 21 | Nevada, 56 |
| Vermont, 61 | | | North Dakota, 12 | New Mexico, 79 |
| | | | Ohio, 63 | Oregon, 88 |
| | | | South Dakota, 13 | Utah, 54 |
| | | | Wisconsin, 46 | Washington, 25 |
| | | | | Wyoming, 59 |

Source: United States Environmental Protection Agency Web site, www.epa.gov.

**a.** Find the region in which your state is located, and choose one of the other regions. Find the mean, median, and mode of each of those two regions.

**b.** Make a back-to-back stem plot of the two data sets.

**c.** Using the measures you found and the stem plot you made, write a paragraph comparing the numbers of municipal landfills in use in the two regions.



## Quick Check

**Informal Assessment**
Students should be able to:

- ✔ construct flowcharts that represent input/output rules and equations
- ✔ solve equations by using flowcharts for backtracking (working backward)

## Quick Quiz

1. Write an equation for each flowchart. Use backtracking to solve the equation.

a.

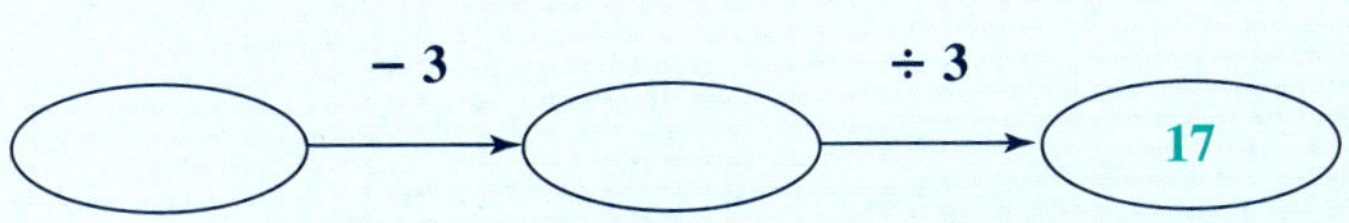


$(n - 3) \div 3 = 17$; 54

b.

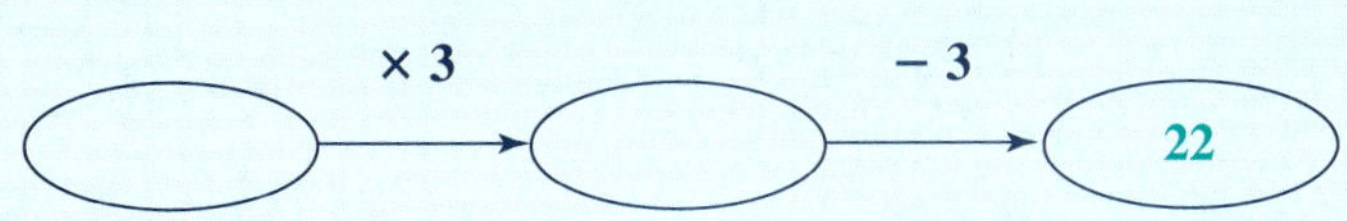


$n \times 3 - 3 = 22$; $\frac{25}{3}$

• ***Quick Quiz continued on page A671***

## Teacher Notes

# 9.3 Guess-Check-and-Improve

## Objectives

- To solve equations systematically by using guess-check-and-improve
- To find approximate solutions to equations, to the nearest tenth or hundredth
- To write and solve equations that describe mathematical puzzles or word problems
- To decide whether to use backtracking or guess-check-and-improve to solve equations

## Overview (pacing: about 4 class periods)

In this lesson, students examine several types of equations and learn to identify those that cannot be solved by backtracking. They learn to use the guess-check-and-improve method, which can be used to find exact and approximate solutions for a wide variety of equations.

## Advance Preparation

No advance preparation is required for this lesson.

## Lesson Planner

| | Summary | Materials | On Your Own Exercises | Assessment Opportunities |
|---|---|---|---|---|
| **Investigation 1** page T586 | Students learn to use the guess-check-and-improve method to solve equations that are not suitable for backtracking. | | Practice & Apply: 1–4, p. 594<br>Connect & Extend: 20–22, p. 596<br>Mixed Review: 30–39, p. 598 | Share & Summarize, pages T589, 589<br>Troubleshooting, page T589 |
| **Investigation 2** page T589 | Students write equations for number puzzles and real-life situations, and then solve the equations by using the guess-check-and-improve method. | | Practice & Apply: 5–10, pp. 594–595<br>Connect & Extend: 23, 24, p. 597<br>Mixed Review: 30–39, p. 598 | Share & Summarize, pages T591, 591<br>Troubleshooting, page T591 |
| **Investigation 3** page T592 | Students examine given equations and use the more appropriate method (backtracking or guess-check-and-improve) to solve each equation. | | Practice & Apply: 11–19, p. 595<br>Connect & Extend: 25–29, p. 597<br>Mixed Review: 30–39, p. 598 | Share & Summarize, pages T593, 593<br>Troubleshooting, page T593<br>Informal Assessment, page 598<br>Quick Quiz, page 598 |

# Introduce

This lesson deals with a method that can be used for finding exact or approximate solutions for most of the equations students are likely to encounter. This method, called the *guess-check-and-improve* method, has students guess at a solution and use observations about high and low guesses to close in on the solution. Students practice using the guess-check-and-improve method to solve a variety of equations. Finally, they learn to study an equation before attempting to solve it to help decide which of the methods they have studied is likely to be the most efficient solution method.

1 Read through the information at the top of page 586 with the class. Have students work with you to verify that for the given inputs, the equation $n = m \cdot (m + 1)$ gives the outputs shown in the table. Then have students work individually on the Think & Discuss.

### Think & Discuss

Discuss these questions with the class. Help students see how the following flowchart can be used to find the outputs shown in the table:

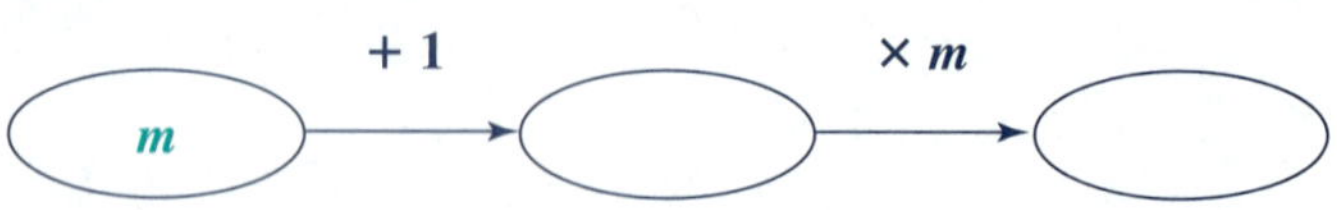


2 Then write 552 in the oval on the far right. Point out that the input variable $m$ occurs above one of the operation arrows. To backtrack from the output 552 to the input, students would have to divide by the value of $m$. But there is no way to find the value of $m$ until they have gotten all the way back to the first oval. Thus, backtracking will not work for this equation.

Discuss the fact that this difficulty will occur anytime the input variable is used above an operation arrow.

# Investigation 1

3 In this investigation, the equation from the Think & Discuss is used to introduce the guess-check-and-improve method of solving equations.

Tell students that they are going to take a close look at a method they have used on some other occasions to solve mathematical problems. Explain that this method can be used to find exact or approximate solutions for many kinds of equations, including equations for which backtracking does not work.

# 9.3 Guess-Check-and-Improve

Backtracking is useful for solving many types of equations. However, as you will see in this lesson, some equations are difficult or impossible to solve by backtracking.

Johanna and Rosita were playing *What's My Rule?* Johanna made this table to keep track of her guesses:

| Input | Output |
|---|---|
| 10 | 110 |
| 20 | 420 |
| 30 | 930 |
| 40 | 1,640 |

From her table, Johanna figured out that Rosita's secret rule was

$$n = m \cdot (m + 1)$$

where $m$ is the input and $n$ is the output.

Now the two friends want to figure out what input gives an output of 552. That is, they want to solve the equation $m \cdot (m + 1) = 552$.

1 Encourage students to verify the equation with the values in the table.

## Think & Discuss

Rosita suggests they solve the equation by backtracking. Try to solve the equation this way. Are you able to find the solution? Why or why not? See ①.

What advice would you give Rosita and Johanna to help them solve the equation? Possible answer: Guess numbers for $m$ until you find a solution. The solution is probably between 20 and 30, since 20 gives an output that is too low and 30 gives an output that is too high.

① no; Possible explanation: I made the flowchart below. When I tried to backtrack, I found that I had to divide 552 by $m$, but I didn't know what $m$ was, because that's what I was trying to find out. I couldn't go any further.

2 Explain why backtracking does not work here.

## Investigation 1 Using Guess-Check-and-Improve

VOCABULARY
**guess-check-and-improve**

As you have seen, backtracking does not work for every equation. In this lesson, you will learn another solution method. This method is called **guess-check-and-improve** because that is exactly what you do.

3 Introduce the term *guess-check-and-improve.*

The Example on page 587 shows how Rosita and Johanna used guess-check-and-improve to solve the equation $m \cdot (m + 1) = 552$.

# Develop

1 **Example**
This Example illustrates the kind of thinking typically used when applying the guess-check-and-improve method. Discuss the Example carefully with the class. Have students evaluate $m \cdot (m + 1)$ for each of the inputs in the table to confirm that the products given are correct. This can be thought of as a *check* step. Students can do the calculations using pencil and paper or a calculator. Pay special attention to the reasoning students use to decide what number to try next. This is an *improve* step. ■

2 Students will easily see that the guess-check-and-improve method may require several steps, since one must continue to guess, continue to check, and continue to improve the guess until one has the exact solution or an approximate solution that is close enough to the exact solution to meet the requirements of the problem. All this is summarized in the paragraph that follows the Example.

## About the Mathematics

Some of the problems in this investigation require solving *quadratic equations,* that is, equations in which the variable is squared in one term of the equation. Students will study this type of equation more fully in Course 3. We use quadratic equations here since they typically have more than one solution, and students need to develop the flexible and careful thinking that is needed to evaluate answers critically. (Could there be more than one solution? Are the solutions reasonable?)

Many quadratic equations have one or two negative numbers as solutions. Since we have not yet discussed operations with negative numbers, students will be looking only for positive solutions.

**EXAMPLE**

From the table Johanna had made during the game, she could see that the output for $m = 20$ was too low and the output for $m = 30$ was too high.

| Input | Output |
|---|---|
| 10 | 110 |
| 20 | 420 |
| 30 | 930 |
| 40 | 1,640 |

Using this information, the friends decided to try 25, the number halfway between 20 and 30. They checked their guess by substituting it for $m$ in the expression $m \cdot (m + 1)$:

$$\begin{aligned} m \cdot (m + 1) &= 25 \cdot (25 + 1) \\ &= 25 \cdot 26 \\ &= 650 \end{aligned}$$

The output 650 is too high. Johanna recorded the guesses and the results in the table.

| $m$ | $m \cdot (m + 1)$ | Comment |
|---|---|---|
| 20 | 420 | too low |
| 30 | 930 | too high |
| 25 | 650 | too high, but closer |

The friends now decided the solution must be between 20 and 25. The table below shows their next two guesses.

| $m$ | $m \cdot (m + 1)$ | Comment |
|---|---|---|
| 20 | 420 | too low |
| 30 | 930 | too high |
| 25 | 650 | too high, but closer |
| 22 | 506 | too low, but close |
| 23 | 552 | 23 is the solution! |

The solution to $m \cdot (m + 1) = 552$ is 23.

Review the process that Johanna and Rosita used:

- They *guessed* the solution.
- They *checked* their solution by substituting it into the equation.
- They used the result to *improve* their guess.

Now it's your turn to try guess-check-and-improve.



**1** Problem-Solving Strategy
Guess-check-and-improve

**2** Review the process with the students.

# Develop

1 **Problem Set A** **Suggested Grouping: Pairs or Individuals**

This problem set gives students practice in using the guess-check-and-improve method to solve equations. You may want to allow the use of calculators, since the reasoning processes are the real focus in this problem set.

2 For **Problem 1b,** ask students how they decided what number to use for their first guess. Students might reason that 100 is a good place to start. (The product is fairly large, so the value of *d* should be fairly large. Also, multiplying 100 by 103 is easy.) The product $100 \cdot (100 + 3)$ is 10,300, which is too large. A good second guess is 90, which turns out to be too small. But the results indicate a solution *between* 90 and 100. After that, students may pick a number halfway between 90 and 100. Some students may think about what the last digit of *d* should be for the product $d \cdot (d + 3)$ to be a whole number with 4 as the final digit. This idea is reasonable if one wants to check possible whole-number solutions. But the guesses and results so far do not rule out the possibility that the solution is not a whole number.

For **Problem 1d,** whole-number solutions are easily ruled out. Students may reason that since the product of $d$ and $d + 3$ must be a two-place decimal, the solution might be a one-place decimal.

3 **Problem 3** may be difficult for students. Some students are likely to use decimals for their guesses. Others may try simple fractions. Students who use the latter approach may arrive at the solution rather rapidly. There is no simple way to know whether one approach is better than the other.

For **Problem 5,** the focus is clearly on the reasoning process. If students' answers to **Parts a and b** display sound thinking, you may wish to ask for the solution of Miguel's equation. Observant students will see that the equation can be solved by guess-check-and-improve or backtracking.

4 **Problem Set Wrap-Up** Ask students to share their strategies for making their first guess for each of the problems in this set. After getting one student's approach, ask if anyone made a different first guess, and ask for the reasoning behind the guess. The important ideas for students to take away from this problem set are the reasoning strategies involved in choosing a good first guess and deciding how much larger or smaller to make the second guess.

**1a.** $d \cdot (d + 3) = 8{,}554$

**1b.** 91 (Note: $^-94$ is also a solution, but students are not expected to find it.)

**1c.** $d \cdot (d + 3) = 32.56$

**4.** 3.5 (Note: $^-5.5$ is also a solution, but students are not expected to find it.)

**Just the facts**

Imagine that a garage contains two tricycles and several bicycles, and that there are 20 bicycle wheels in all. You can solve the equation in Problem 2, $2n + 6 = 20$, to find the number of bicycles in the garage.

## Problem Set A

**1.** Conor, Marcus, and Jing are playing *What's My Rule?* Here is Jing's secret rule:

$$d \cdot (d + 3) = J, \text{ where } d \text{ is the input and } J \text{ is the output}$$

**a.** Conor gave Jing an input, and Jing calculated the output as 8,554. Write an equation you could solve to find Conor's input.

**b.** Find a solution of your equation using guess-check-and-improve.

**c.** Marcus gave Jing an input, and Jing calculated the output as 32.56. Write an equation you could solve to find Marcus' input.

**d.** Find a solution of your equation from Part c using guess-check-and-improve. 4.4 (Note: $^-7.4$ is also a solution, but students are not expected to find it.)

For each equation, use guess-check-and-improve to find a solution.

**2.** $2n + 6 = 20$ 7 **3.** $19 = \frac{4}{q} + 3$ 0.25 **4.** $s^2 + 2s = 19.25$

**5.** Miguel is trying to solve this equation using guess-check-and-improve: $25 - 3 \cdot d = 17.8$. The table shows his first two guesses. Miguel asked:

*"Why was the output for 8 lower than the output for 7? Shouldn't a greater input give a greater output?"*

| $d$ | $25 - 3 \cdot d$ | Comment |
|---|---|---|
| 7 | 4 | too low |
| 8 | 1 | still too low |

**a.** Answer Miguel's questions.

**b.** What input do you think Miguel should try next? Explain.

**5a, b.** See below.

**6.** Hannah and Luke were trying to solve $7.25t - t^2 = 12.75$. They made this table using guess-check-and-improve:

| $t$ | $7.25t - t^2$ | Comment |
|---|---|---|
| 5 | 11.25 | too low |
| 6 | 7.5 | too low |
| 4 | 13 | too high |
| 2 | 10.5 | too low |

**5a.** $3 \cdot d$ is being subtracted from 25. As the value of $d$ increases, the value of $3 \cdot d$ also increases, so the value of $25 - 3 \cdot d$ gets smaller.

Hannah thinks the solution must be between 2 and 4. Luke thinks it must be between 4 and 5.

**a.** Is there a solution between 2 and 4? If so, find it. If not, explain why not. yes, 3

**b.** Is there a solution between 4 and 5? If so, find it. If not, explain why not. yes, 4.25

**5b.** Possible answer: 2; He should try something smaller than 7. Since the output for 7 isn't even close to 17.8, he should try something much smaller than 7.

**1** Problem-Solving Strategy Guess-check-and-improve

**2** Discuss with students how they came up with their first guess.

**3** Challenge students to try simple fractions.

**4** Have students share strategies for making their first guess.

1 **Share & Summarize**
This Share & Summarize focuses on the reasoning involved in making guesses based on the information in the equation and the data from previous guesses. Encourage students to estimate two numbers that might have a solution between them.

**Troubleshooting** If students have had difficulty making reasonable initial guesses, review the problems in this set. Ask students to make an initial guess and to justify it with an estimate. Once they have made two guesses, ask them to justify their next guess based on their observations of the results from the first two guesses. You might offer some additional examples.

**Additional Examples** Use estimation to make a good first guess for the solution of the given equation. Then solve the equation.

1. $1.5s - 32 = 17.5$ **Possible answer: A good first guess is $s = 30$ because $1.5 \times 30 = 45$ and $45 - 32$ is close to 17.5; 33.**

2. $p \times (p + 0.75) = 26.64$ **Possible answer: A good first guess is 5 because $5 \times 5 = 25$, and therefore $5 \times 5.75$ will be a bit larger than 25; 4.8.**

### On Your Own Exercises

Practice & Apply: 1–4, p. 594
Connect & Extend: 20–22, p. 596
Mixed Review: 30–39, p. 598

# Investigation 2

In this investigation, students write equations to describe various situations. They solve problems about the situations by using guess-check-and-improve to find exact or approximate solutions of their equations.

2 **Problem Set B** **Suggested Grouping: Pairs**
In this problem set, students write equations to model problems. They then use guess-check-and-improve to find approximate solutions of the equations.

3 For **Problem 1c,** students will probably find it fairly easy to determine that the width $w$ must be between 8 and 9 meters. It will also be easy to see that the value of $w$ is closer to 8 than to 9. A table that shows values of $w$ and $w \cdot (w + 2)$ from $w = 8.0$ to $w = 8.5$ would make it clear that $w$ is about 8.3 meters, rounded to the nearest tenth.

## Share & Summarize

See Additional Answers.

Describe any strategies you have discovered for finding a solution efficiently by using guess-check-and-improve.

**1 Estimation** Have students estimate two numbers.

# Investigation 2 Solving Problems Using Guess-Check-and-Improve

In this investigation, you will solve problems by writing equations and then using guess-check-and-improve. As you work, you will find that you can't always give an exact decimal value for a solution. In such cases, you can approximate the solution.

**2** You may have students work in pairs.

### Problem Set B

1. The floor of Mr. Cruz's basement is shaped like a rectangle. The length of the floor is 2 meters greater than the width.

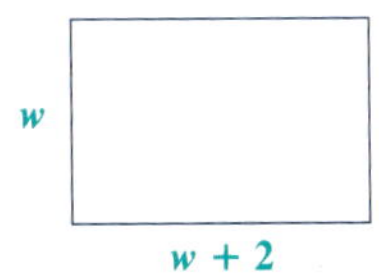


a. Write a rule to show the connection between the floor's area $A$ and the width $w$. $A = w \cdot (w + 2)$

b. The area of the basement floor is 85 square meters. Write an equation you could solve to find the floor's width.

**1b.** $w \cdot (w + 2) = 85$

c. Use guess-check-and-improve to find an approximate solution of your equation. Give the solution to the nearest tenth. 8.3

d. What are the dimensions of the basement floor? about 8.3 m by 10.3 m

**3 Problem-Solving Strategy** Make a table



**Additional Answers**
**Share & Summarize**

Possible answer: If I find an input that gives an output that is too high and an input that gives an output that is too low, I know there is a solution between those two inputs. I look to see which of the two inputs gives an output closer to the output I want, and I make my next guess closer to that input. For example, for $m \cdot (m + 1) = 40$, an input of 5 gives an output of 30, which is too low. An input of 6 gives an output of 42, which is too high. I know that the solution must be between 5 and 6, and since 42 is closer to 40 than to 30, I would make my next guess closer to 6 than to 5. I also look at the equation or at the pattern in the table to see if greater inputs give greater or lesser outputs. This helps me to decide whether my next input should be greater than or less than the previous input.

LESSON 9.3 Guess-Check-and-Improve 589

# Develop

1 For **Problem 2d,** students can use the guesses 4 and 5 to reason that since $6 \cdot 4^2 = 96$ and $6 \cdot 5^2 = 150$, the solution is between 4 and 5, and very close to 4. The number 4.1 is only one-tenth more than 4, and since $6 \cdot 4.1^2 = 100.86$, and 100.86 is much closer to 100 than is 96, the solution is 4.1, accurate to the nearest tenth. You may want to ask students how they would use a calculator with a square root key to check this result.

2 **Example**

This Example uses an equation that has the variable on both sides. Discuss the Example with the class. Be sure that students understand that the table is useful for showing how close the values of $m + 12$ and $3m$ are for various values of $m$.

3 Some students may point out that $m + 12 = 3m$ can be solved by first rewriting the equation as $12 = 3m - m$, or $2m = 12$. It is now easy to see that the solution is 6. This is a good observation, but use it to emphasize that even when there are other ways to solve an equation, guess-check-and-improve will also work.

**Just the facts**

Cubism is an art style that was developed in the early 1900s. Cubism today emphasizes flat, two-dimensional surfaces rather than three-dimensional perspective. Cubist paintings show many sides of an object at once.

**2c.** $6 \cdot L^2 = 100$

**2.** A *cube* is a three-dimensional shape with six identical square faces.

The *surface area* of a cube is the sum of the areas of its six faces.

This cube has edges of length *L*.

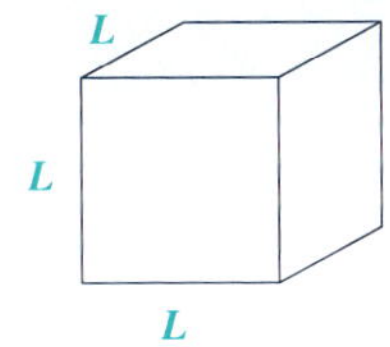


**a.** What is the area of one face of this cube? $L^2$

**b.** Write a rule for finding the surface area *S* of the cube. $S = 6 \cdot L^2$

**c.** Suppose the cube has a surface area of 100 square centimeters. Write an equation you could solve to find the cube's edge length.

**d.** Use guess-check-and-improve to find the edge length of the cube to the nearest 0.1 centimeter. 4.1 cm

In all the equations you have solved so far, the variable appears on only one side of the equation. You can also use guess-check-and-improve to solve equations in which the variable appears on *both* sides.

**EXAMPLE**

Solve this number puzzle: *12 more than a number is equal to 3 times the number. What is the number?*

If you let *m* stand for the number, you can write the puzzle as an equation:

$$m + 12 = 3m$$

To solve the equation, substitute values for *m* until the two sides of the equation are equal.

| $m$ | $m + 12$ | $3m$ | Comment |
|---|---|---|---|
| 1 | 13 | 3 | not the same |
| 2 | 14 | 6 | a bit closer |
| 5 | 17 | 15 | very close |
| 6 | 18 | 18 | Got it! |

So, 6 is the solution of the equation $m + 12 = 3m$.

Check that 6 is the solution of the original puzzle: 12 more than 6 is 18, which is equal to 3 times 6.

In the next problem set, you will practice solving equations in which the variable appears on both sides.

**1** Ask students how they would use a calculator to check the result.

**2** Discuss the Example with the class.

**3** Explain that the table shows how close the values of $m + 12$ and $3m$ are for different values of $m$.

# Develop

1 **Problem Set C** **Suggested Grouping: Pairs**
All of the equations for this problem set have the variable on both sides.

For all the problems in this problem set, using a table of values can be helpful. (See the Example on page 590.)

2 For **Problem 2,** if some students used guesses that were fractions and some used guesses that were decimals, you may want to discuss whether one approach was easier than the other.

3 For **Problem 3,** students will probably guess both solutions, since both are small whole numbers. You may want to observe that backtracking will not work for this equation, even if the equation is written in an equivalent form.

**Problem 5** requires students to write and solve an equation to solve a problem. **Parts a–c** walk students through setting up the equation.

### Share & Summarize

4 Have students work in pairs on this Share & Summarize activity. You may want to call on volunteers to present their number puzzles to the class. Use a class discussion to solve the puzzles, and discuss any features that made
5 certain puzzles especially challenging.

**Troubleshooting** If students have difficulty translating number puzzles presented in words into equations, you may want to consider some more of the puzzles that students wrote for the Share & Summarize. Discussing how to translate the puzzles into equations can be helpful for all students.

## On Your Own Exercises

Practice & Apply: 5–10, pp. 594–595
Connect & Extend: 23, 24, p. 597
Mixed Review: 30–39, p. 598

## Problem Set C

In Problems 1–3, write an equation to represent the number puzzle. Then find the solution using guess-check-and-improve.

**1.** 5 added to a number is the same as 1 subtracted from twice the number. What is the number? $5 + x = 2x - 1$, 6

**2.** 4 times a number, plus 1, is equal to 4 added to twice the number. What is the number? $4x + 1 = 4 + 2x$, 1.5

**3.** 2 added to the square of a number is equal to 5 times the number, minus 4. What is the number? (There are two solutions. Try to find them both.) $2 + x^2 = 5x - 4$, 2 or 3

**4.** Consider the equation $2m = 5m - 18$.

**a.** Make up a number puzzle that matches the equation.

**b.** Solve the equation, and check that the solution is also the answer to your puzzle. 6

**4a.** 2 times a number is equal to 18 less than 5 times the number. What is the number?

**5.** Peta and Ali went to the pet store to buy fish for their tanks. Peta bought three black mollies for $b$ dollars each and one peacock eel for \$12. Ali bought seven black mollies for $b$ dollars each and an Australian rainbow fish for \$2. Peta and Ali spent the same amount of money.

**a.** How much money did Peta spend? Your answer should be an expression containing the variable $b$. $3b + 12$

**b.** How much money did Ali spend? Your answer should be an expression containing the variable $b$. $7b + 2$

**c.** Using your two expressions, write an equation that states that Peta and Ali spent the same amount of money. $3b + 12 = 7b + 2$

**d.** Solve your equation to find the value of $b$. How much did each black molly cost? \$2.50

**e.** How much did each friend spend? \$19.50

Use guess-check-and-improve to find a solution of each equation.

**6.** $3n = 9 - 2n$ 1.8

**7.** $p + 1 = 5(p - 4)$ 5.25

**1** You may have students work in pairs.

**2** Discuss the differences between using fractions and decimals as guesses.

**3** Show students that backtracking will not work here in Problem 3.

## Share & Summarize

Make up a number puzzle like those you have worked with in this investigation. Exchange puzzles with your partner, and solve your partner's puzzle. Puzzles will vary.

**4** You may have students work in pairs.

**5** Discuss students' puzzles and the features that made them challenging.



LESSON 9.3 Guess-Check-and-Improve 591

# Investigation 3

In this investigation, students compare solution methods for a variety of equations. They look for characteristics that will help them decide in advance whether backtracking or guess-check-and-improve will be the more efficient approach for solving a given equation.

1 Read through the cartoon with the class. Then have students answer the Think & Discuss questions on the next page.

2 The equation shown in this last frame of the cartoon is a quadratic equation, which has two integer solutions, 6 and −16. Students are not expected to find negative solutions in this course.

# Investigation 3 Choosing a Method

Now you have two methods for solving equations: *backtracking* and *guess-check-and-improve.* The problems in this investigation will help you decide which solution method is more efficient for a particular type of equation.

In the cartoon, Marcus and Rosita are trying to solve some equations. Marcus uses backtracking for each equation, and Rosita uses guess-check-and-improve for each equation.

1 Problem-Solving Strategies

- Work backward
- Guess-check-and-improve

2 –16 is also a solution.

## Develop

1 **Think & Discuss**
You will probably want to call on volunteers to read their answers to the class. Discuss their answers, and use the discussion to elicit any further observations that might be helpful in comparing the two methods of solving equations.

Some students may point out that Marcus could have made flowcharts for the first two equations. For the second equation, Rosita had to make a table. Discuss what she had to do to get the numbers for the second column of her table. Students will quickly see that Rosita had to do quite a bit of multiplication.

2 **Problem Set D** **Suggested Grouping: Pairs**
In this problem set, students decide which method, backtracking or guess-check-and-improve, can be used to solve given equations. For each equation that can be solved by either method, students think about which method is the more efficient for that particular equation.

For **Problem 2,** students may notice that a good deal of multiplication might be necessary if guess-check-and-improve is used.

3 For **Problem 3,** some students may observe that backtracking will not work if the equation is left in the given form. However, backtracking will work if the equation is rewritten in the form $4 = 5k$ or $5k = 4$.

In **Problem 7,** Luke's statement is fine for the equations in Problems 1–6. However, backtracking will not work for an equation such as $\frac{1}{n-1} = 5$, even though the variable occurs only once.

4 **Share & Summarize**
You may want to call on volunteers to read their answers to the class. The discussion of the answers can serve as a good wrap-up for all the main ideas of this lesson.

**Troubleshooting** If students had trouble deciding which method to use, you may want to write a few more equations on the board and discuss with the class the advantages and disadvantages of each method.

### On Your Own Exercises

Practice & Apply: 11–19, p. 595
Connect & Extend: 25–29, p. 597
Mixed Review: 30–39, p. 598

**Think & Discuss Answers**

① $2.4 \cdot n + 4 = 10.36$; Marcus was able to solve the equation faster than Rosita. Rosita had to make a lot of guesses because the answer had two decimal places.

② $n \cdot (10 + n) = 96$; Marcus was not able to solve the equation using backtracking because the variable is in both parts of the product.

**Problem Set D Answers**

**2a.** 4.4

**2b.** Possible answer: Both methods work. Backtracking seems more efficient.

**3a.** 0.8

**3b.** Backtracking does not work.

**4a.** 8 (Note: $^-6$ is also a solution, but students are not expected to find it.)

**4b.** Backtracking does not work.

**5.** 3.5; I used guess-check-and-improve because backtracking doesn't work when the variable is on both sides.

**6.** 4; I used backtracking because there are lots of operations, and if I used guess-check-and-improve it would be tedious to check each guess.

## Think & Discuss

For which equation in the cartoon is backtracking a better method than guess-check-and-improve? Why? See ①.

For which equation is guess-check-and-improve a better method than backtracking? Why? See ②.

For which equation do both methods seem to work well? $3n + 7 = 19$

1. Discuss students' answers and other observations that may be helpful in comparing the two methods.

## Problem Set D

**1a.** 11

**1b.** Possible answer: Both methods work. Backtracking seems more efficient.

In Problems 1–4, do Parts a and b.

**a.** Find a solution of the equation using one method while your partner finds a solution using the other. Switch methods for each equation so you have a chance to practice both.

**b.** Discuss your work with your partner. Indicate whether both methods work, and tell which method seems more efficient.

2. You may have students work in pairs.

**1.** $3p - 8 = 25$ **2.** $1.6r + 3.96 = 11$

**3.** $k + 4 = 6k$ **4.** $(j - 2) \cdot j = 48$

Solve each equation. Tell which solution method you used, and explain why you chose that method.

3. Challenge students to rewrite the equation.

**5.** $4w + 1 = 2w + 8$ **6.** $5 \cdot \frac{2k + 4}{6} = 10$

**7.** Luke says, "When the variable appears only once in an equation, I use backtracking. If it occurs more than once, I use guess-check-and-improve." Discuss Luke's strategy with your partner. Would it be effective for the equations in Problem 1–6? Explain. See Additional Answers.

## Share & Summarize

Tell which solution method you would use to find a solution of each equation. Explain your choice.

4. Discuss students' answers.

**1.** $n^2 + n = 30$ **2.** $4 + (v - 3) = 8$

**3.** $3g = 6g - 7$ **4.** $2e - 7 = 4$

**1.** Possible answer: guess-check-and-improve because the variable appears twice

**2.** Possible answer: guess-check-and-improve because this is fairly easy to do in my head

**3.** Possible answer: guess-check-and-improve because the variable appears twice

**4.** Possible answer: backtracking because the variable appears only once



**Additional Answers**
**Problem Set D**

**7.** Possible answer: Yes; we found that we could not use backtracking to solve the equations in Problems 3, 4, and 5. In each of these, the variable occurs more than once.

# On Your Own Exercises

## On Your Own Exercises

Investigation 1, pp. 586–589
Practice & Apply: 1–4
Connect & Extend: 20–22

Investigation 2, pp. 589–591
Practice & Apply: 5–10
Connect & Extend: 23, 24

Investigation 3, pp. 592–593
Practice & Apply: 11–19
Connect & Extend: 25–29

Assign Anytime
Mixed Review: 30–39

**Exercise 5:**
You may want to allow the use of calculators for this exercise. If students are using pencil and paper, suggest that they use 3.14 as an approximate value for $\pi$.

## Practice & Apply

**1a.** $3 \cdot p + p^2 = 24.79$

**1b.** 3.7 (Note: $^-3.7$ is also a solution, but students are not expected to find it.)

**1c.** $3 \cdot p + p^2 = 154$

**1d.** 11 (Note: $^-14$ is also a solution, but students are not expected to find it.)

**1.** Hannah, Althea, and Jahmal are playing *What's My Rule?* Here is Althea's secret rule:

$3 \cdot p + p^2 = q$, where $p$ is the input and $q$ is the output

**a.** Hannah gives Althea an input, and Althea calculates the output as 24.79. Write an equation you could solve to find Hannah's input.

**b.** Find a solution of your equation using guess-check-and-improve.

**c.** Jahmal gives Althea an input, and Althea calculates the output as 154. Write an equation you could solve to find Jahmal's input.

**d.** Find a solution of your equation in Part c using guess-check-and-improve.

For each equation, use guess-check-and-improve to find a solution.

**2.** $16 - 5k = 2$ 2.8

**3.** $h \cdot (5 + h) = 26.24$

**4.** $y^2 + 72 = 17y$ (There are two solutions. Try to find both.)

**3.** 3.2 (Note: $^-8.2$ is also a solution, but students are not expected to find it.)

**4.** 8, 9

### Remember

The area of a circle is given by the formula $A = \pi \cdot r^2$, where $r$ is the radius of the circle.

**5. Geometry** Reina is planning her summer garden.

**a.** Reina wants to plant strawberries in a circular plot covering an area of 15 square meters. Write an equation you could solve to find the radius Reina should use to lay out the plot. $\pi \cdot r^2 = 15$

**b.** Use guess-check-and-improve to find the strawberry plot's radius to the nearest tenth of a meter. 2.2 m

**c.** Reina also wants to plant tomatoes in five identical circular plots with a total area of 25 square meters. Write an equation you could solve to find the radius of one of the tomato plots. $5 \cdot \pi \cdot r^2 = 25$

**d.** What should the radius of each tomato plot be, to the nearest tenth of a meter? 1.3 m

### Just the facts

Tomatoes are the world's most popular fruit—in terms of tons produced each year—followed by bananas, apples, oranges, and watermelons.

impactmath.com/self_check_quiz

**9a.** $2 \cdot m - 21.75 = \frac{1}{2} \cdot m$

**10a.** 3 times a number, minus 11, is equal to the number plus 3.

**14.** 2 ($^-1$ is also a solution, but students are not expected to find it.)

**Just the facts**

The macaw is a long-tailed parrot. In the rain forest, macaws have an average life span of 25 to 30 years. In captivity, they can reach ages of 70 years or more.

**16.** 12 ($^-12$ is also a solution, but students are not expected to find it.)

**17.** Possible answer: guess-check-and-improve because the variable appears twice

**18.** Possible answer: backtracking because the variable appears only once

**19.** Possible answer: guess-check-and-improve because the variable appears twice

In Exercises 6–8, write an equation to represent the number puzzle. Then find the solution using guess-check-and-improve.

**6.** 3 times a number, plus 5, is equal to 5 times the number. What is the number? $3s + 5 = 5s$, 2.5

**7.** 10 plus the square of a number is equal to 6 times the number, plus 2. What is the number? (This puzzle has two solutions. Find both of them.) $10 + M^2 = 6M + 2$, 4 or 2

**8.** 3 times a number, plus 1, is equal to 9 plus the number. What is the number? $3n + 1 = 9 + n$, 4

**9.** Marjorie said, "If you double my macaw's age and then subtract 21.75, your answer will be half my macaw's age."

**a.** Write an equation you could use to find how old Marjorie's macaw is.

**b.** Use guess-check-and-improve to find the macaw's age. Check your answer in Marjorie's original statement. $14\frac{1}{2}$

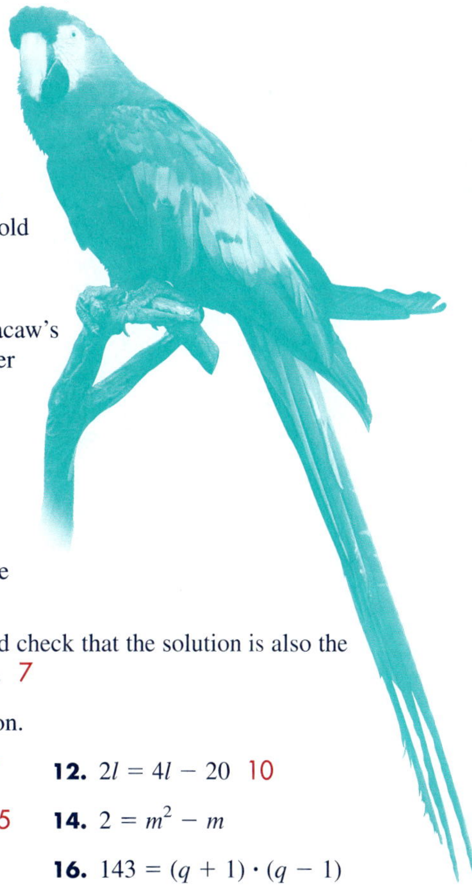

**10.** Consider the equation $3m - 11 = m + 3$.

**a.** Make up a number puzzle that matches the equation.

**b.** Solve the equation, and check that the solution is also the answer to your puzzle. 7

Find a solution of each equation.

**11.** $3.3h - 7 = 2.801$ 2.97

**12.** $2l = 4l - 20$ 10

**13.** $j \cdot (3 - 2j) = 1$ 1 or 0.5

**14.** $2 = m^2 - m$

**15.** $9 \times \frac{g \div 2 + 1}{5} = 12$ $11\frac{1}{3}$

**16.** $143 = (q + 1) \cdot (q - 1)$

In Exercises 17–19, tell whether you would use backtracking or guess-check-and-improve to find a solution, and explain your choice.

**17.** $n^2 + n = 30$

**18.** $47(2v - 3.3) = 85$

**19.** $s = 17.5s - 0.5$

**Exercises 11–16:**
You may want to ask students to draw a flowchart for any equation they solved by backtracking.

**Exercise 15:**
You may want to go over with the class how to draw a flowchart for this equation, as it requires care in following the order of operations.

**Exercise 18:**
You may want to ask students to find the solution of this equation by backtracking. The exact solution is $2\frac{521}{940}$.

Students will probably have little difficulty realizing that finding the exact solution by guess-check-and-improve would be extremely tedious.



LESSON 9.3 Guess-Check-and-Improve 595

**Exercise 22c:** Students will probably find it helpful to make a table showing their guesses and the value of $4.9 \cdot t^2$ for each value of $t$.

## Connect & Extend

**20b.** 17 ($^-16$ is also a solution, but students are not expected to find it.)

**20.** Aisha and Terrell were playing *Think of a Number.* Terrell said:

*Think of a number. Multiply the number by 1 less than itself. What number do you get?*

**20a.** $n \cdot (n - 1) = 272$, where $n$ is the number

Aisha said she got 272. Terrell must figure out Aisha's starting number.

**a.** What equation does Terrell need to solve to find Aisha's number?

**b.** Use guess-check-and-improve to find Aisha's number. Check your answer by following the steps to verify that you get 272.

**21. Geometry** The elevator in Rafael's apartment building has a square floor with an area of 6 square meters.

**a.** Write an equation you could solve to find the dimensions of the elevator floor. $s^2 = 6$, where $s$ is the length of a side

**b.** Use guess-check-and-improve to find the dimensions of the elevator floor to the nearest tenth of a meter. 2.4 m by 2.4 m

**22. Physical Science** When an object is dropped, the relationship between the distance it has fallen and the amount of time it has been falling is given by the rule

$$d = 4.9 \cdot t^2$$

where $d$ is the distance in meters and $t$ is the time in seconds.

**a.** Jing dropped a ball from a pier. It took the ball 1.1 seconds to hit the water. How many meters did the ball travel? about 5.9 m

**22b.** $4.9 \cdot t^2 = 300$

**b.** A bolt falls 300 meters down a mine shaft. Write an equation you could solve to find how long it took the bolt to fall.

**c.** Find how long it took the bolt to fall, to the nearest tenth of a second. 7.8 s



596 CHAPTER 9 Solving Equations

★ indicates multi-step problem

★ **23. Nutrition** Three blueberry muffins and a plain bagel have the same number of calories as two blueberry muffins and one bagel with cream cheese. A bagel has 150 calories, and cream cheese adds 170 calories. How many calories are in a blueberry muffin? 170

**24. Number Sense** This formula can be used to find the sum $S$ of the whole numbers from 1 to $n$:

$$S = \frac{n \cdot (n + 1)}{2}$$

For example, you can use the formula to find the sum of the whole numbers from 1 to 100:

$$\begin{aligned} S &= \frac{n \cdot (n + 1)}{2} \\ &= \frac{100 \cdot (100 + 1)}{2} \\ &= \frac{100 \cdot 101}{2} \\ &= \frac{10{,}100}{2} \\ &= 5{,}050 \end{aligned}$$

**a.** Use the formula to find the sum of the whole numbers from 1 to 9. Check your answer by calculating $1 + 2 + 3 + 4 + 5 + 6 + 7 + 8 + 9$. 45

**b.** If the sum of the numbers from 1 to $n$ is 6,670, what must $n$ be? 115

**c.** If the sum of the numbers from 1 to $n$ is 3,003, what must $n$ be? 77

For each equation, use guess-check-and-improve to find a solution.

**25.** $x^3 + x = 130$ 5

**26.** $4n^4 = 48$ about 1.86

**27.** $2a^4 - 4a^2 = 16$

**28.** $c^4 = \frac{1}{100}$

**29. Geometry** The *volume* of a three-dimensional shape is the amount of space inside of it. Volume is measured in cubic units, such as cubic centimeters and cubic inches. You can calculate the volume $V$ of a cylinder with radius $r$ and height $h$ using this formula: $V = \pi \cdot r^2 \cdot h$.

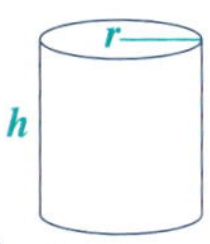


**a.** Calculate the volume of a cylinder with radius 2 centimeters and height 5 centimeters.

**b.** A soft drink can has a volume of 350 cubic centimeters and a height of 15 centimeters. Write an equation you could solve to find its radius. $\pi \cdot r^2 \cdot 15 = 350$

**c.** Solve your equation to find the can's radius to the nearest tenth of a centimeter. 2.7 cm

**27.** 2 ($^-2$ is also a solution, but students are not expected to find it.)

**28.** 0.32 ($^-0.32$ is also a solution, but students are not expected to find it.)

**29a.** $20\pi$ cubic centimeters, or about 62.8 cubic centimeters

**In your own words**

In this chapter, you have learned how to solve equations using two methods: backtracking and guess-check-and-improve. Compare these two methods, and explain why it is important to be able to solve equations in more than one way.



**Exercise 26:**
Students will find it helpful to make a table of values for $n$ and $4n^4$. You may want to allow the use of a calculator for this exercise and point out that students can enter a number and press $x^2$ twice to find the fourth power of the number. Discuss the fact that this works because $(n^2)^2 = n^2 \cdot n^2 = n \cdot n \cdot n \cdot n = n^4$.

**Exercise 28:**
You may want to suggest that students rewrite the equation using a decimal: $c^4 = 0.01$. You can also adapt the suggestions for Exercise 26 for this exercise.

LESSON 9.3 Guess-Check-and-Improve 597

## Quick Check

### Informal Assessment

Students should be able to:

- ✔ solve equations systematically by using guess-check-and improve
- ✔ find approximate solutions to equations, to the nearest tenth or hundredth
- ✔ write and solve equations that describe mathematical puzzles or word problems
- ✔ decide whether to use backtracking or guess-check-and-improve to solve equations

### Quick Quiz

Use guess-check-and-improve to solve each equation.

1. $11n - 8 = 135$ 13
2. $7x + 15 = 6x + 21$ 6
3. $\frac{p+9}{2} = 6.25$ 3.5
4. $3t^2 + 2t = 161$ 7
5. Find the solution of $x^3 = 12$ to the nearest tenth. 2.3

Examine each equation. Tell whether you think it would be better to use backtracking or guess-check-and-improve to find the solution. Explain your thinking.

6. $9 + \frac{3k-7}{12} = 200$ backtracking; Possible explanation: The variable occurs only once, and it is easy to make a flowchart for the equation.
7. $x \cdot (x^2 + 1) = 57$ guess-check-and-improve; Possible explanation: The variable is in two factors, one of which is a sum of a variable and the number 1. It is not possible to make a flowchart for the equation.

## Mixed Review

**36.** 35%

**39.** Possible story: Mary went into the bathroom to take a bath. She turned on the water and let the tub fill. She turned off the water, waited a couple of minutes, and then got in the tub. She was in the tub for several minutes and then got out. After a few minutes, she pulled the plug and let the water drain out.

Write each fraction as a decimal without using a calculator.

**30.** $\frac{99}{150}$ 0.66 **31.** $\frac{73}{90}$ $0.8\overline{1}$ **32.** $\frac{7}{18}$ $0.3\overline{8}$

**33.** $\frac{78}{110}$ $0.7\overline{09}$ **34.** $\frac{4}{11}$ $0.\overline{36}$ **35.** $\frac{63}{125}$ 0.504

**36.** Of the 80 acres on Ms. McDonald's farm, 28 acres are devoted to growing corn. What percent of the farm's area is devoted to corn?

**37.** Of the animals on Ms. McDonald's farm, 12.5% are goats. If Ms. McDonald has 7 goats, what is the total number of animals on her farm? 56

**38.** At the farmer's market, Ms. McDonald sold 60% of the 42 pounds of tomatoes she picked last week. How many pounds of tomatoes did she sell at the market? 25.2 lb

**39.** **Physical Science** The graph shows the level of water in a bathtub over a one-hour period. Write a story that explains all the changes in the water level.

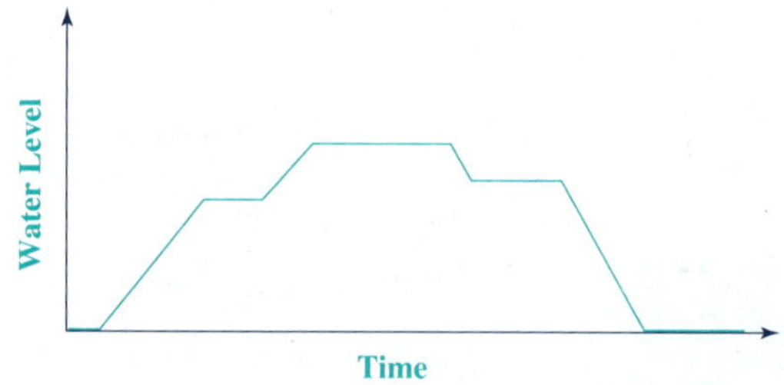

VOCABULARY
**backtracking**
**equation**
**flowchart**
**guess-check-and-improve**
**inequality**
**open sentence**
**solution**

## Chapter Summary

You started this chapter by looking at numeric equations and inequalities. Then you turned your attention to equations containing variables. You learned that a value of a variable that makes an equation true is called a *solution* of the equation. You saw that while many equations have only one or two solutions, some have every number as a solution, and others have no solution at all.

You then created flowcharts to represent rules, and you saw how *backtracking*—working backward from an output using a flowchart—can be used to solve some equations. You were then faced with equations that could not be solved by backtracking, and you learned how to use *guess-check-and-improve* to solve them. You also learned some strategies for determining which solution method to use for a given equation.

## Strategies and Applications

The questions in this section will help you review and apply the important ideas and strategies developed in this chapter.

### Understanding equations and inequalities

1. Explain what the symbols $=$, $>$, $<$, and $\neq$ mean. For each symbol, write a true mathematical sentence using that symbol.
2. Give an example of an open sentence, and give a value of the variable that makes your sentence true. Possible answer: $5 \cdot x = 45$, 9
3. Explain why the sentence $P + 5 = P$ is never true.
4. Explain why the sentence $2 \cdot (x + 3) = 2 \cdot x + 2 \cdot 3$ is always true.

### Solving equations by backtracking

5. Solve this equation by creating a flowchart and backtracking:

$$10 \cdot \frac{4 + 9x}{7} - 25 = 45$$

Explain each step in your solution.

**1.** $=$ means is equal to. Possible example: $7 + 3 = 2 \cdot 5$. $>$ means is greater than. Possible example: $7 \cdot 8 > 7 + 8$. $<$ means is less than. Possible example: $7 - 7 < 7 + 7$. $\neq$ means is not equal to. Possible example: $6 + 2 \neq 6 \div 2$.

**3.** When you add 5 to a number, the result is always greater than the number.

**4.** Possible explanation: $2 \cdot (x + 3) = (x + 3) + (x + 3) = x + x + 3 + 3 = 2 \cdot x + 2 \cdot 3$

**5.** Possible answer: First I made a flowchart. The output is 45.

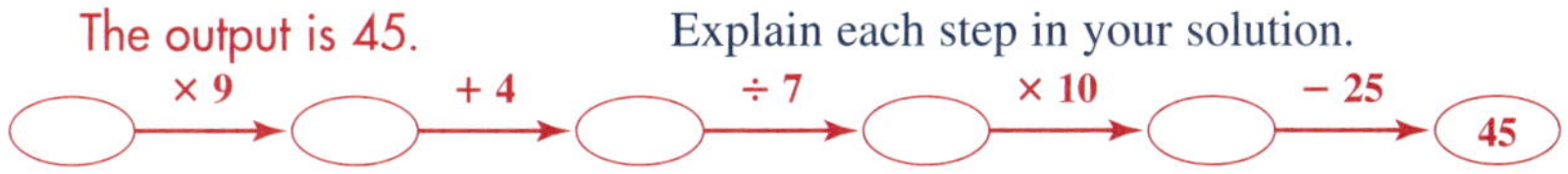


Then I worked backward from the output to find the input. 25 was subtracted from the number in the fifth oval to get 45, so that number must be 70. The number in the fourth oval was multiplied by 10 to get 70, so it must be 7. The number in the third oval was divided by 7 to get 7, so it must be 49. 4 was added to the number in the second oval to get 49, so it must be 45.

The number in the first oval was multiplied by 9 to get 45, so it must be 5. The solution is 5.

impactmath.com/chapter_test



**Chapter Summary**
This summary helps students recall the major topics of the chapter.

**Vocabulary**
Students should be able to explain each of the terms listed in the vocabulary section.

**Problem-Solving Strategies and Applications**
The questions in this section help students review and apply the important mathematical ideas and problem-solving strategies developed in the chapter. The questions are organized by mathematical highlights. The highlights correspond to those in "The Big Picture" chart on page 555a.

**Question 5:**
Students need to be careful with the order of operations when they make their flowchart for this equation.

**Question 6b:**
Students should realize that Russ can go on 7 rides because the answer of $7\frac{1}{3}$ rides is not realistic.

**Question 7a:**
Some students may think that addition must somehow figure in the equation since the word "total" was used in the description of the situation. Help them see why multiplication is the appropriate operation to use.

★ indicates multi-step problem

★ **6.** Admission to the town carnival is \$4.50 per person. Rides cost 75¢ each. Russ has \$10 to spend at the carnival, and he wants to go on as many rides as possible.

**a.** Write an equation you could solve to find the number of rides $r$ Russ can go on. $4.5 + 0.75r = 10$

**b.** Solve your equation by backtracking. How many rides can Russ go on? $7.\overline{3}$; Russ can go on 7 rides.

### Solving equations using guess-check-and-improve

★ **7.** The Smallville community garden is made up of 15 identical square plots with a total area of 264.6 $m^2$.

**a.** Write an equation you could solve to find the side length of each plot. $15 \cdot x^2 = 264.6$

**b.** Solve your equation using guess-check-and-improve. Make a table to record your guesses and the results. What is the side length of each plot? See Additional Answers.

**8a.** $x \cdot (x + 2) = 9 \cdot x + 8$

**8b.** 8 ($^-1$ is also a solution, but students are not expected to find it.)

**8.** A number multiplied by 2 more than the number is 9 times the number, plus 8.

**a.** Write an equation to represent the number puzzle.

**b.** Solve your equation using guess-check-and-improve.

### Choosing a solution method for an equation

**9.** Explain why you could not use backtracking to solve this equation: $(n + 3.5) \cdot n = 92$.

**10.** Tell which solution method you would use to solve the equation $6.34 + 10.97 \cdot y = 208.188$. Give reasons for your choice.

**9, 10.** See Additional Answers.

600 CHAPTER 9 Solving Equations

### Additional Answers

**7b.** 4.2 m; Possible table:

| $x$ | $15 \cdot x^2$ | Comment |
|---|---|---|
| 6 | 540 | too high |
| 4 | 240 | too low |
| 5 | 375 | too high |
| 4.4 | 290.4 | still too high |
| 4.2 | 264.6 | solution |

**9.** Possible answer: You could try to make a flowchart like this:

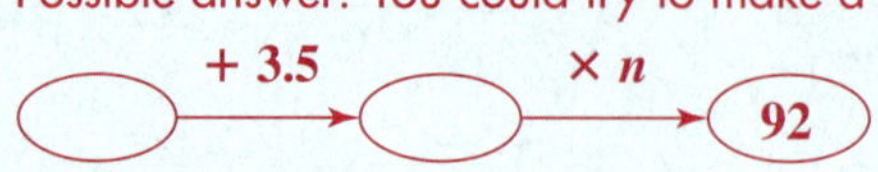


But when you tried to backtrack, the first step would be to find the number that was multiplied by $n$ to get 92. Since you don't know what $n$ is, this is impossible.

**10.** Possible answer: I would use backtracking. Since the numbers are all decimals, the solution is probably a decimal too, and it might take lots of guesses to find it. Since one side of the equation is a number, and the variable appears only once on the other side, this equation would be pretty easy to solve by backtracking.

12. false; Possible answer: $4^2 > 4 \cdot 2$

13. false; $30 \div (3 + 2) \neq 30 \div 3 + 30 \div 2$

14. false; $9 + 4 - 1 > 5 + 6$

15. Always true; the exponent of 3 means that $c$ is multiplied together 3 times.

16. never true; Possible explanation: If you subtract 5 from any number, the result is always smaller than when you add 1 to that number.

17. sometimes true: Possible explanation: The equation is true when $x = 9$ but not when $x = 10$ (or any other number).

18. Always true; when you subtract a number from itself, you always get 0.

19. sometimes true; Possible explanation: The equation is true when $p = 7$ but not when $p = 0$ (or any other number).

20. sometimes true; Possible explanation: The equation is true when $y = 0$ but not when $y = 1$ (or any other number).

22. 16 ($^-21$ is also a solution, but students are not expected to find it.)

## Demonstrating Skills

Tell whether each sentence is true or false. If it is false, make it true by changing one number or symbol.

**11.** $(4 + 5) \cdot 6 = 24 + 30$ true

**12.** $4^2 = 4 \cdot 2$

**13.** $30 \div (3 + 2) = 30 \div 3 + 30 \div 2$

**14.** $7 + 4 - 1 > 5 + 6$

Tell whether each equation is *always true*, *sometimes true*, or *never true*, and explain how you know.

**15.** $c^3 = c \cdot c \cdot c$

**16.** $n - 5 = n + 1$

**17.** $x - 9 = 0$

**18.** $2m - 2m = 0$

**19.** $\frac{13}{6 + p} = 1$

**20.** $5 \cdot (4 + y) = 20$

Find a solution of each equation.

**21.** $4.7x + 12.3 = 42.85$ 6.5

**22.** $m \cdot (m + 5) = 336$

**23.** $5n + 7 = 14$ 1.4

**24.** $\frac{5(x - 1)}{3} = 7.5$ 5.5

**25.** $6z - 29 = 7 + 3z$ 12

**26.** $7 \cdot \frac{3b + 4}{10} = 11.2$ 4



**Question 13:**
You may want to ask students to give the values of the expressions on the left and right sides of the equation.

# CHAPTER 10

# Understanding Probability

### Chapter Overview

In this chapter, students study basic concepts of probability. They use data in situations involving equally likely outcomes to compute experimental probabilities. They then compare experimental and theoretical probabilities. Students see how probability can help them decide whether a game is fair, and they use probability to devise strategies for scoring points in some simple games of chance. Finally, students compare and contrast situations involving independent and dependent events.

## the Big Picture

| Chapter 10 Highlights | Links to the Past | Links to the Future |
|---|---|---|
| Understanding probability (10.1) | **Elementary Grades:** Exploring probability | **Course 2, Chapter 10:** Using probability to make decisions and create strategies<br>**Course 3, Chapter 9:** Understanding the concept of a sample space and its application to probability |
| Identifying outcomes (10.1) | **Elementary Grades:** Collecting and recording data from an experiment | **Course 2, Chapter 10:** Finding the outcome of an event or experiment<br>**Course 3, Chapter 9:** Listing outcomes of a sample space |
| Finding probabilities of events (10.1) | **Elementary Grades:** Performing an experiment and estimating probabilities | **Course 2, Chapter 10:** Finding the probabilities of independent and dependent events<br>**Course 3, Chapter 9:** Determining the probability of an event |
| Using probabilities to analyze games (10.2) | **Elementary Grades:** Collecting and analyzing data samples to determine whether a game is fair | **Course 2, Chapter 10 and Course 3, Chapter 9:** Using probability to determine whether a game is fair |
| Working with situations in which the probabilities depend on previous results (10.3) | **Elementary Grades:** Finding probabilities in equally likely situations | **Course 2, Chapter 10 and Course 3, Chapter 9:** Recognizing and analyzing when previous outcomes influence later outcomes |

Time-Saving Tool TeacherWorks™ All-In-One Planner and Resource Center

# Planning Guide

| Lesson Objectives | Pacing | Materials | NCTM Standards | Hot Topics |
|---|---|---|---|---|
| **10.1 The Language of Chance** page 603b<br>• To associate a number from 0 to 1 with the likelihood that an event will occur<br>• To use data to find the experimental probability of an event<br>• To find the theoretical probability of an event for situations in which the individual outcomes are equally likely | 3 class periods | • Paper cups<br>• Coins<br>• Dice<br>• Master 78<br>• Scissors<br>• Tape<br>• Masters 81 or 82 (optional)<br>* • Counters | 1, 2, 4, 5, 6, 7, 8, 9, 10 | pp. 218–229 |
| **10.2 Analyzing Games** page 619b<br>• To conduct simple experiments and simulations to compare experimental and theoretical probabilities<br>• To make and use outcome tables to find theoretical probabilities<br>• To find probabilities by using the concept of geometric probability | 5 class periods | • Dice with faces labeled 5, 5, 5, 5, 0, 0<br>• Dice with faces labeled 1, 2, 3, 4, 4, 4<br>• Standard dice<br>* • Counters in two colors<br>• Master 79<br>• Master 80<br>• Grains of uncooked rice<br>• Inch rulers | 1, 2, 3, 4, 5, 6, 7, 8, 9, 10 | |
| **10.3 Making Matches** page 637b<br>• To make tree diagrams to show outcomes for situations involving independent choices<br>• To use tree diagrams to compute probabilities in situations that involve dependent choices | 3 class periods | * • Counters, blocks, or slips of paper<br>• Buckets or paper bags | 1, 2, 5, 6, 7, 8, 9, 10 | pp. 210–217 |

* Included in Impact Mathematics Manipulative Kit

**Key to NCTM Curriculum and Evaluation Standards:** 1=Number and Operations, 2=Algebra, 3=Geometry, 4=Measurement, 5=Data Analysis and Probability, 6=Problem Solving, 7=Reasoning and Proof, 8=Communication, 9=Connections, 10=Representation

# Assessment Opportunities

## Standard Assessment

*Impact Mathematics* offers three types of formal assessment. The Chapter 10 Review and Self-Assessment in the Student Edition serves as a self-assessment tool for students. In the Teacher's Guide, a Quick Quiz at the end of each lesson allows you to check students' understanding before moving to the next lesson. The Assessment Resources include blackline masters for chapter and quarterly tests.

- **Student Edition** Chapter 10 Review and Self-Assessment, pages 652–655
- **Teacher's Guide** Quick Quizzes, pages 619, 637, 651
- **Assessment Resources** Chapter 10 Test Form A, pages 226–229; Chapter 10 Test Form B, pages 230–233

## Ongoing Assessment

*Impact Mathematics* provides numerous opportunities for you to assess your students informally as they work through the investigations. Share & Summarize questions help you determine whether students understand the important ideas of an investigation. If students are struggling, Troubleshooting tips provide suggestions for helping them. On the Spot Assessment notes appear throughout the teaching notes. They give you suggestions for preventing or remedying common student errors. Assessment Forms in the Assessment Resources provide convenient ways to record student progress.

- **Student Edition** Share & Summarize, pages 607, 612, 622, 624, 625, 630, 642, 645
- **Teacher's Guide** On the Spot Assessment, pages T610, T623, T624, T629 Troubleshooting, pages T607, T612, T622, T624, T625, T630, T642, T645
- **Assessment Resources** Chapter 10 Assessment Checklist, page 272

## Alternative Assessment, Portfolios, and Journal Ideas

The alternative assessment items in *Impact Mathematics* are perfect for inclusion in student portfolios and journals. The In Your Own Words feature in the Student Edition gives students a chance to write about mathematical ideas. The Performance Assessment items in the Assessment Resources provide rich, open-ended problems, ideal for take-home or group assessment.

- **Student Edition** In Your Own Words, pages 617, 635, 650
- **Assessment Resources** Chapter 10 Performance Assessment, pages 234–238

# Assessment Resources

The Assessment Resources provide a chapter test in two equivalent forms, along with additional performance items. The performance items can be used in a variety of ways. They are ideal for take-home assessment or in-class group assessment.

- Chapter 10 Test Form A, pages 226–229
- Chapter 10 Test Form B, pages 230–233
- Chapter 10 Performance Assessment, pages 234–238
- Chapter 10 Assessment Solutions, pages 239–242

A semester test, with performance items is also provided in the Assessment Resources in two equivalent forms. This test covers material in Chapters 6–10.

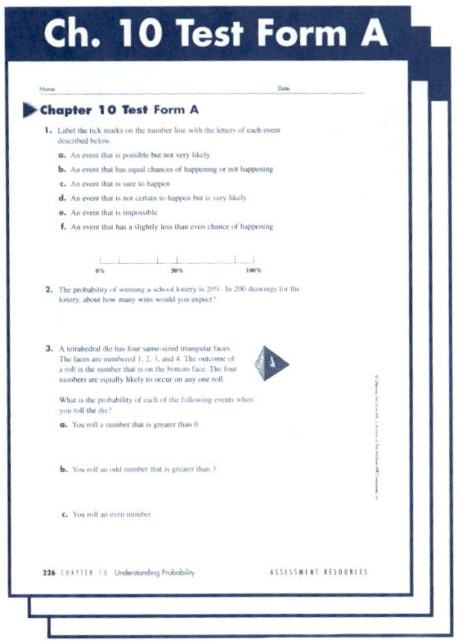


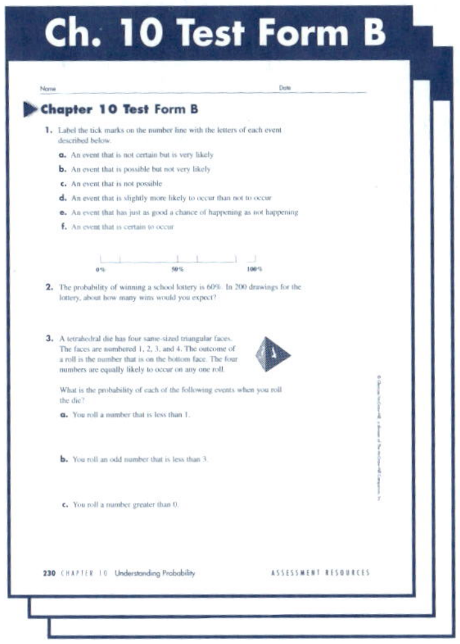


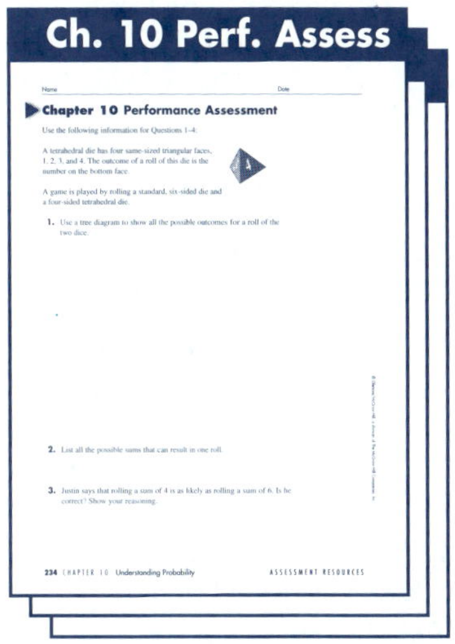


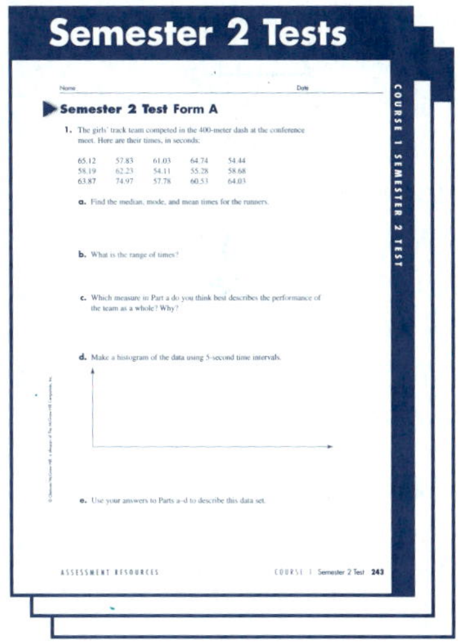


# Additional Resources

- **Math Skills Maintenance Workbook,** 4, 5
- **Investigations for the Special Education Student in the Mathematics Classroom,** 17
- **Virtual Activities CD-ROM,** Probability and Area
- **StudentWorks CD-ROM™**
- **Reading and Writing in the Mathematics Classroom**
- **Using the Internet in the Mathematics Classroom**

## ExamView® Pro

Use ExamView® Pro Testmaker CD-ROM to:

- Create Multiple versions of tests.
- Create Modified tests for Inclusion students with one mouse click.
- Edit existing questions and Add your own questions.
- Build tests aligned with state standards using built-in State Curriculum Correlations.
- Change English tests to Spanish with one mouse click and vice versa.

**Introduce**

The facts on this page give students some idea of the broad variety of situations in which chance and probability play a role.

Students have most likely heard about people winning fabulous lottery prizes, but they may not have realized how remote the chances are of picking the winning numbers. Tell students that in this chapter they will see how probability can help them analyze situations like those on this page. They will also see how probability can be used to study games and determine whether a game is fair.

**Think About It**

Students will need to locate the local weather report in the newspaper, on television, or on the Internet. Discuss the probability of rain tomorrow, which is usually given as a percent. Students may wish to write the percent as a fraction and as a decimal to more easily compare it to the other probabilities.

# CHAPTER 10

# Understanding Probability

## *Real-Life Math*

**Fat Chance!** A probability is a number between 0 and 1 that tells you the chance that an event will occur. The closer a probability is to 0, the less likely the event is to happen.

Just how likely is it that some everyday events happen?

- If you toss a coin to determine the answers of a 10-question true-false test, the probability you will get all the answers correct is $\frac{1}{1,024}$ or about 0.001.
- If you randomly dial a three-number combination on a dial lock with numbers from 0 to 29, the probability you will open the lock is $\frac{1}{27,000}$ or about 0.00004.

**Think About It** To better understand how unlikely these events are, compare these probabilities to the probability that it will rain in your community tomorrow.

# *Family Letter*

Dear Student and Family Members,

We close our exciting year of mathematics by exploring probability. Probability tells you that it is very unlikely for you to win the grand prize in a state lottery. Suppose you have to pick six different numbers from 1 to 54. To win the grand prize, all six numbers must match those selected in a random drawing. The probability that you will win is only 1 in 25,827,165 or 0.00000004.

The probability that some event will happen can be described by a number between 0 and 1.

- A probability of 0 means that the event has no chance of happening. (So the probability of winning a state lottery is very close to 0.)
- A probability of 1 means that the event is certain to happen.
- A probability of $\frac{1}{2}$ or 50% means that the event is just as likely to happen as not to happen.

For example, if a weather forecaster says that the probability of rain tomorrow is 90%, it's probably a good idea to take your umbrella, although it might not rain after all. If the probability of rain is 10%, it is unlikely that it will rain.

In this chapter, you'll use mathematical reasoning to calculate probabilities in simple situations like tossing coins or drawing names from a hat. You will also do some experiments in which you actually toss coins or draw names and then compare the results with the calculated probabilities.

**Vocabulary** Along the way, you'll learn these new vocabulary terms:

| | |
|---|---|
| **equally likely** | **simulation** |
| **experimental probability** | **theoretical probability** |
| **probability** | |

## What can you do at home?

Look for situations in everyday life that involve probability, such as the chance of rain or the odds in sports games. You might encourage your student's exploration by playing games of chance together. Have your student teach you the games we play in class and ask him or her to describe what part probability plays in each game.

impactmath.com/family_letter



Another version of the Family Letter, available in English and Spanish, is found in the Teaching Resources. You may want to send a copy of this letter home with your students.

**Teaching Resources**

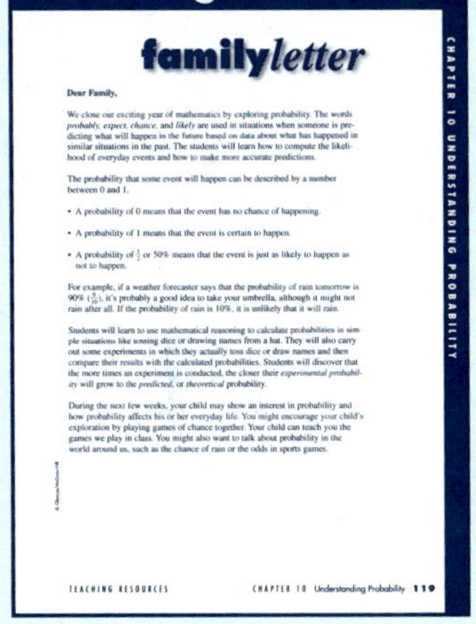

**familyletter**

Dear Family,

We close our exciting year of mathematics by exploring probability. The words *probably, expect, chance,* and *likely* are used in situations when someone is predicting what will happen in the future based on data about what has happened in similar situations in the past. The students will learn how to compute the likelihood of everyday events and how to make more accurate predictions.

The probability that some event will happen can be described by a number between 0 and 1.

- A probability of 0 means that the event has no chance of happening.
- A probability of 1 means that the event is certain to happen.
- A probability of $\frac{1}{2}$ or 50% means that the event is just as likely to happen as not to happen.

For example, if a weather forecaster says that the probability of rain tomorrow is 90% ($\frac{9}{10}$), it's probably a good idea to take your umbrella, although it might not rain after all. If the probability of rain is 10%, it is unlikely that it will rain.

Students will learn to use mathematical reasoning to calculate probabilities in simple situations like tossing dice or drawing names from a hat. They will also carry out some experiments in which they actually toss dice or draw names and then compare their results with the calculated probabilities. Students will discover that the more times an experiment is conducted, the closer their *experimental probability* will grow to the *predicted*, or *theoretical* probability.

During the next few weeks, your child may show an interest in probability and how probability affects his or her everyday life. You might encourage your child's exploration by playing games of chance together. Your child can teach you the games we play in class. You might also want to talk about probability in the world around us, such as the chance of rain or the odds in sports games.

TEACHING RESOURCES CHAPTER 10 Understanding Probability 119

CHAPTER 10 UNDERSTANDING PROBABILITY

# Mathematical Background

Chapter 10 gives students an opportunity to explore basic concepts of probability. They learn how data about a situation can be used to find an experimental probability for an event. They learn to calculate theoretical probabilities in situations that involve equally-likely outcomes. Next, they explore geometric probability. Finally, they use probability to analyze some situations in which one outcome can affect subsequent outcomes.

**Experimental vs. Theoretical Probability** The probability of an event is a number that tells how likely the event is to occur. In the first lesson of this chapter, students examine two ways of gauging probability. *Experimental probability* is based on data about a situation. The fraction that compares the number of times a particular event actually occurred to the total number of times it might have occurred is called the experimental probability of the event.

*Theoretical probability* is based on reasoning. When you roll a die, there are only six numbers that can be rolled. Each number on the die has the same chance of coming up as any other number. If you are interested in the chances of rolling a multiple of 3, you can simply write a fraction comparing the number of outcomes that represent successfully rolling a multiple of 3 to the total number of possible outcomes.

The way experimental and theoretical probabilities are defined ensures that they will be numbers from 0 to 1. It is important to note that experimental probabilities can change if you acquire more data about a situation. Theoretical probabilities, on the other hand, are fixed quantities that do not vary.

**Equally-Likely Outcomes** Situations in which the individual outcomes are equally likely to occur are especially simple and easy to study. Examples of such situations are rolling a die, flipping a coin, spinning a spinner that is divided into equal-sized sections, and drawing a card of a particular suit from a standard deck. Students learn to think about all the possible outcomes in such situations and to calculate the probabilities of various events that might occur.

**Games of Chance** Students then focus on how probability can be used to analyze games of chance. Some of the games use specially labeled number cubes, and others use standard dice. Tables can be used to examine all the possible outcomes and hence to calculate theoretical probabilities. One of the games that is examined theoretically involves an infinite number of possible outcomes. This final game uses a rectangular game board. Students drop a grain of rice on the board and collect data about the shapes in which the rice lands. They discover that the likelihood of the rice landing in a particular shape is the fraction of the total area of the game board occupied by that shape.

**Dependent Events** In the first two lessons of this chapter, students consider probability in situations involving independent events. In the third lesson, they turn their attention to simple situations in which the occurrence of one event is affected by previous events. Students learn to draw tree diagrams to show outcomes in such situations.

# 10.1 The Language of Chance

## Objectives

- To associate a number from 0 to 1 with the likelihood that an event will occur
- To use data to find the experimental probability of an event
- To find the theoretical probability of an event for situations in which individual outcomes are equally likely

## Overview (pacing: about 3 class periods)

In this lesson, students make intuitive judgments about the likelihood that various events will occur. They grade the likelihoods on a scale from 0 to 1. They consider how data about a situation can be used to find the experimental probability that a particular event will occur.

Next, students consider situations in which all the outcomes are equally likely. They find the theoretical probability of each outcome and the theoretical probability of events that can be described in terms of the outcomes. They consider simple experiments that allow them to compare experimental probabilities with theoretical probabilities. They see that with increasingly many trials, experimental probabilities tend to come closer and closer to theoretical probabilities.

In the lab investigation, students use experimental and theoretical probabilities to analyze a game of chance.

## Advance Preparation

For Investigation 1, students will need paper cups. For Investigation 2, each student will need a coin and a die. For the lab investigation, each group will need a top or a copy of Master 78 (copied onto heavy paper, if possible) for constructing one. They will also need 40 counters.

## Lesson Planner

| | Summary | Materials | On Your Own Exercises | Assessment Opportunities |
|---|---|---|---|---|
| Investigation 1 page T605 | Students use intuition to plot the likelihood of certain events on a number line from 0 to 1. They also use data about simple situations to find experimental probabilities for given events. | • Paper cups | Practice & Apply: 1–4, p. 615<br>Connect and Extend: 8, 9, pp. 616–617<br>Mixed Review: 16–31, p. 619 | Share & Summarize, pages T607, 607<br>Troubleshooting, page T607 |
| Investigation 2 page T608 | Students learn to recognize situations in which individual outcomes are equally likely. They find the probability of simple events in such situations. | • Coins<br>• Dice | Practice & Apply: 5–7, pp. 615–616<br>Connect & Extend: 10–15, pp. 617–618<br>Mixed Review: 16–31, p. 619 | On the Spot Assessment, page T610<br>Share & Summarize, pages T612, 612<br>Troubleshooting, page T612<br>Informal Assessment, page 619<br>Quick Quiz, page 619 |
| Lab Investigation page T613 | Students play a game of chance and use ideas about experimental and theoretical probability to analyze the game. | • Master 78<br>• Scissors<br>• Tape<br>• Masters 81 or 82 (optional)<br>*• Counters | | |

* Included in Impact Mathematics Manipulative Kit

# Introduce

1 Read the introduction to the lesson with the class. Ask students to give examples of predictions that they, or people they know, have made. Ask why they think people make predictions. They probably realize that many predictions are made in the hope of affecting future events. People also make predictions to help them set plans for the future.

2 Call attention to the italicized words (*probably, expect, chance,* and *likely*) at the end of the opening paragraph.

3 **Think & Discuss**
Use these questions for a class discussion. When you discuss other words and phrases people use to talk about chance, ask students to relate their examples to the five bulleted items in the second paragraph. For example, if someone says, "There's a 50-50 chance I'll stay at home on the weekend," this means that the person is just as likely to stay at home as not to stay at home. The expression "No way!" means that the event has no chance of happening. "You can bet on it" usually means that the person thinks the event is certain to happen.

When you discuss the likelihood of the six events listed in the Think & Discuss, ask students to explain why they ranked the events the way they did.

**Access for all Learners**

**Language Diversity** Words and phrases used to express the likelihood of events are often idiomatic and may not be clear to all students, especially students who are learning the English language. It may be helpful to list such words and phrases on the board and comment briefly on the meaning of each.

# 10.1 The Language of Chance

People often make comments like these:

- "It probably won't rain tomorrow."
- "I expect to have a lot of homework this week."
- "Our team has a good chance of winning the game."
- "It's not likely she will eat that entire cake!"
- "The chances are 50/50 that we'll go to the movies tonight."
- "There's a 40% chance of rain tomorrow."

The words *probably, expect, chance,* and *likely* are used when someone is making a prediction.

1 Ask students to give other examples of predictions.

2 Review the terms *probably, expect, chance,* and *likely.*

## Think & Discuss

What are some other words or phrases people use when predicting the chances of something happening? See ① in Additional Answers.

Listed below are six events. How likely do you think each event is? Talk about them with your class, and come to an agreement about whether each event

- has no chance of happening
- could happen but is unlikely
- is just as likely to happen as not to happen
- is likely to happen
- is certain to happen

3 Have students relate their examples to the bulleted list.

*Event 1:* Our class will have homework tonight. Answer depends on the class.

*Event 2:* It will snow tomorrow. Answer depends on climate and time of year.

*Event 3:* If I toss a penny in the air, it will land heads up. just as likely to happen as not to happen

*Event 4:* A gorilla will eat lunch with us today. no chance of happening

*Event 5:* You choose a name from a hat containing the names of all the students in your class, and you get a girl's name. Answer depends on the class.

*Event 6:* You draw a number from a hat containing the numbers from 1 to 5, and you get 8. no chance of happening

**Just the facts**

Some gorillas have been taught to communicate with humans by using sign language. One such gorilla, Koko, has a vocabulary of over 1,000 words!

### Additional Answers

① Possible answers: certain, maybe, unlikely, improbable, possible, impossible, odds, one in a million, frequent, rare

# Investigation 1

In this investigation, students move from informal descriptions of probability to more formal, mathematical descriptions. They learn that collecting data about a situation often makes it possible to calculate *experimental probabilities* for events. They conduct a simple experiment to gather data about how a paper cup lands when it is tossed into the air. They use their data to calculate the experimental probability of the cup landing in a particular way.

1 Discuss the word *probability* with the class. Point out that in mathematics, the probability of an event is a number that tells how likely it is that the event will happen. The probability that an event will happen can be stated using a number from 0 to 1. Remind students that numbers can
2 be expressed as percents. So, saying that a probability is a number from 0 to 1 means that probabilities can also be expressed as percents from 0% to 100%.

3 **Problem Set A** **Suggested Grouping: Pairs**
In this problem set, students think about how to rank events in terms of their likelihood.

4 For **Problem 1,** finding examples of impossible events and certain events will probably be easiest for students, though their examples will likely vary quite a bit. Allow time for pairs of students to share their responses for all parts of the problem with the class, but cut short any attempts at an extended debate over exactly how probable any one event is.

## Investigation 1 Probability in Everyday Life

**VOCABULARY**
**probability**

The **probability,** or chance, that an event will happen can be described by a number between 0 and 1:

- A probability of 0, or 0%, means the event has no chance of happening.
- A probability of $\frac{1}{2}$, or 50%, means the event is just as likely to happen as not to happen.
- A probability of 1, or 100%, means the event is certain to happen.

For example, the probability of a coin landing heads up is $\frac{1}{2}$, or 50%. This means you would expect a coin to land heads up $\frac{1}{2}$, or 50%, of the time.

**Remember**
*The probability P of an event E is the ratio that compares the number of favorable outcomes f to the number of possible outcomes n.*
$P(E) = \frac{f}{n}$

*1 Introduce the term probability.*

*2 Remind students that probabilities can be expressed as decimals, percents, and fractions.*

The more likely an event is to occur, the greater its probability. If a weather forecaster says the probability of rain is 90%, it's a good idea to take your umbrella when you go outside. Of course, it *might* not rain after all! On the other hand, if the forecaster says the probability of rain is 10%, you might want to leave your umbrella at home. Still, you *might* get wet!

You can represent the probability of an event by marking it on a number line like this one:

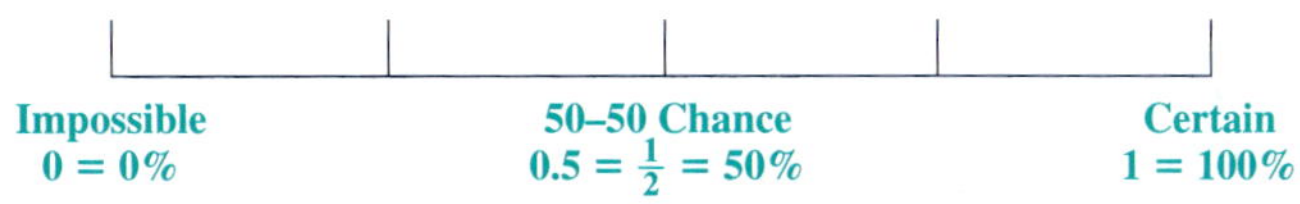

For example, the next number line shows the probabilities of tossing a coin and getting heads, of a goldfish walking across a room, and of Alaska getting snow this winter.

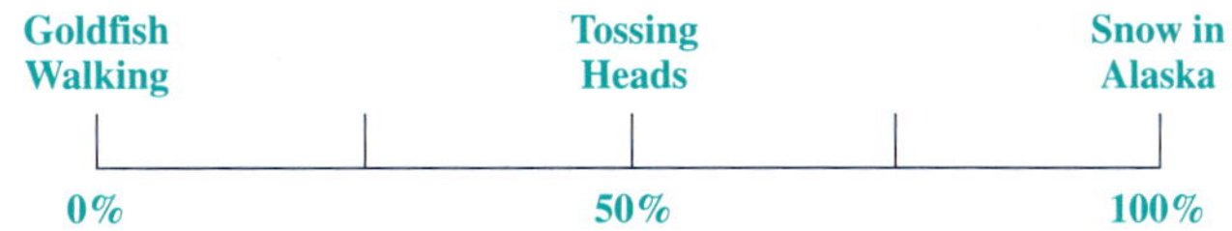


### Problem Set A

*3 You may have students work in pairs.*

*4 Have students share their answers with the class.*

1. Answers will vary.

1. Describe an event you think has the given chance of happening.
   a. The event has no chance of happening.
   b. The event could happen but is unlikely.
   c. The event is just as likely to happen as not to happen.
   d. The event is likely to happen.
   e. The event is certain to happen.



LESSON 10.1 The Language of Chance 605

## Develop

1 For **Problem 2,** students do not have to locate the events for Parts b and d of Problem 1 at the exact points marked b and d in the sample diagram. Unlikely events should fall somewhere in the interval from a to c, and likely events should fall somewhere in the interval from c to e.

2 **Problem Set B** **Suggested Grouping: Pairs**
In this problem set, students use win-loss data for two baseball teams to estimate the likelihood that one particular team will be the winner of the next game they play together.

3 Be sure students read **Problem 3** carefully. Students should compare the number of games won to the total number of games played. Some students may read too fast and compare the number of games won to the number of games lost. This gives the *odds* of the Rockets winning, which is not the same as the *probability* of the Rockets winning.

4 For **Problem 4,** students need to understand that the probability estimates are to be based on only the data about the first three games and then the last three games. Some students may want to argue that the probability estimate based on the first three games should be higher. They may argue that in fact we know the results for all six games and that the Rockets appear to be the better team. Help students understand that they are being asked to put themselves in the shoes of someone with a more limited knowledge of the scoring record.

5 **Problem 5** should help students understand that in real-life situations, probabilities based on data may change as additional data become available. ■

6 After reviewing the problems in Problem Set B, discuss the term *experimental probability* with the class. Experimental probabilities are based on data about a set of trials. The number of trials is limited, and the experimental probability can change if more trials are used. It is in this sense that experimental probabilities are estimates.

**2.** See the number line. Note: Placement of b and d may vary.

**2.** Copy this number line. Label the number line with your events from Problem 1. You can use the letters of the events for the labels.

1 Be sure students know that the events should be somewhere in those intervals.

In Problem Set A, you used your experience to estimate the chances that certain events would occur. For example, you know from experience that when you toss a coin, it lands tails up about half the time. In some situations, you can use data to help estimate probabilities.

## Problem Set B

2 You may have students work in pairs.

**Just the facts**

Little League baseball was started in 1939 in Williamsport, Pennsylvania, with three teams and 45 players. Today, more than 2.5 million children worldwide play Little League baseball.

On Saturday, Caroline's baseball team, the Rockets, is playing Jahmal's team, the Lions. Caroline decided to look at the scores from the last six times their teams played each other.

| Lions | 3 | 8 | 6 | 4 | 4 | 5 |
|---|---|---|---|---|---|---|
| Rockets | 5 | 2 | 4 | 5 | 7 | 6 |

**1.** How many times did the Rockets win? 4

**2.** Which team do you think is more likely to win the next game? the Rockets

**3.** Caroline can estimate the probability that her team will win by dividing the number of times the Rockets won by the number of games played. What probability estimate would she get based on the results of the six games? Give your answer as a fraction and as a percent. Possible answer: $\frac{2}{3}$, or $66\frac{2}{3}\%$

3 Make sure students read Problem 3 carefully.

**4.** Suppose Caroline knew the results of only the first three games. What would her probability estimate be? If she knew the results of only the last three games, what would her probability estimate be?

4 **Estimation** Make sure students base their estimates on data from the first and last three games.

**5.** These two teams have played six games against each other. Suppose they had played eight games and each team had won one more game. What probability estimate would you give for the Rockets winning the next game? $\frac{5}{8}$, or 62.5%

5 Show students that probabilities may change as additional data becomes available.

VOCABULARY
**experimental probability**

**4.** $\frac{1}{3}$, or $33\frac{1}{3}\%$; 1, or 100%

The probabilities you found in Problem Set B are examples of **experimental probabilities.** Experimental probabilities are always estimates, and they can vary depending on the particular set of data you use.

6 Introduce the term *experimental probability.*

Suppose you want to find an experimental probability when you have no data available. In such cases, you might perform *experiments* to create some data.

# Develop

1 **Problem Set C** **Suggested Grouping: Groups of 2 or 3**

In this problem set, students toss a paper cup in the air 30 times, record how the cup lands each time, and use the resulting data to find experimental probabilities.

2 For **Problem 1,** students toss their paper cups and tally the results. To keep track of the tosses, it will help not only to make a table and record how the cup lands each time, but also to keep a separate tally of the total number of tosses so that students will know when they have reached the goal of 30 tosses.

Leave it up to students whether to use fractions or percents for **Problem 4.**

For **Problem 5,** students should use fractions if they used a fraction for Problem 4. They should use percents if they used a percent for Problem 4.

3 For **Problem 6,** it can be instructive to take a little time to combine the results of all the groups to get a class experimental probability for each way the cup might land. Express the probabilities as percents, and discuss why doing so is helpful when you have a large number of trials. Students will have little trouble seeing that percents are much easier to comprehend and compare than complicated fractions.

Ask questions to help students compare the combined results with those of their individual groups:

> What similarities or differences do you notice? *answers will vary*
>
> If there are any large differences, what might have caused them? *size and weight of the cups, methods used to toss them*

**Problem Set Wrap-Up** Have groups compare the probabilities they found with other groups and discuss any similarities and differences.

Then ask groups to compare their results for Problems 4 and 5 to the class results found in Problem 6. Students can use calculators if necessary to express the group results as percents. Ask students whether their group results or the class experimental probabilities provide a better picture of how the cup is likely to land.

4 Finally ask students what total they would expect to get if they add all three of the experimental probabilities for Problems 4 and 5. Then have them check by doing the addition. The total of all the probabilities should be 1 or 100%. Discuss why this is so.

5 **Share & Summarize**

Use these questions to assess how well students grasp the meaning of probability, and how to calculate an experimental probability by using data about a situation.

**Troubleshooting** Watch for students who try to calculate experimental probabilities by dividing the number of times the event occurs by the number of times it does not occur. Emphasize that they should always divide by the total number of trials. You can use the situation in Problem Set A to help explain why this is important.

## On Your Own Exercises

Practice & Apply: 1–4, p. 615
Connect and Extend: 8, 9, pp. 616–617
Mixed Review: 16–31, p. 619

MATERIALS
paper cup

4. Answers will vary.
5. Answers will vary.
6. Possible answers: Combine the number of trials resulting in that position and divide by the total number of trials (30 · number of groups). *Or*, average the probabilities. (The two methods will yield the same probability.)

**Just the facts**

Many sport statistics can be thought of as experimental probabilities. For example, a basketball player's free-throw percentage is the probability he will score on his next free throw.

## Problem Set C

Much like a science experiment, a probability experiment involves trying something to see what happens. You may have some idea of what will occur, but the actual results can be surprising.

**1.** Toss a paper cup so that it spins in the air. Record how it lands: right side up, upside down, or on its side. This is one *trial* of the experiment.

Toss the cup 29 more times, for a total of 30 tosses. Record the landing position each time. You may want to use tally marks as shown below. Answers will vary.

| Right Side Up | Upside Down | On Its Side |
|---|---|---|
| II | I | IIII |

**2.** How many trials did you perform in your experiment? 30

**3.** How many times did the cup land right side up? Upside down? On its side? Answers will vary.

**4.** Find the portion of the trials for which the cup landed right side up, stating your answer as a fraction or a percent. Your answer is an experimental probability that the cup lands right side up when tossed.

**5.** Now find an experimental probability that the cup lands upside down and an experimental probability that it lands on its side.

**6.** Share your results with the class, and consider the results found by your classmates. Suggest at least one way you might use them to find a class experimental probability for the cup landing right side up.

### Share & Summarize

1. The event is certain to happen.
2. $\frac{16}{25}$, 64%

**1.** What does it mean to say that the probability of an event is 1?

**2.** Conor is in a basketball league. He was practicing free throws one afternoon, and he made 32 out of 50 shots. Estimate the probability that he makes a free throw, expressing it as a fraction and a percent.

**3.** Suppose you conducted $y$ trials of an experiment, and a particular event happened $x$ times. Find an experimental probability that the event will occur. $\frac{x}{y}$



**1** You may have students work in groups of 2 or 3.

**2** Problem-Solving Strategy
Make a table

**3** Combine the results of all the groups to get a class experimental probability.

**4** Have students add all three experimental probabilities together for Problem 6.

**5** Discuss students' answers.

LESSON 10.1 The Language of Chance 607

# Investigation 2

In this investigation, students are introduced to the concepts of equally likely outcomes and theoretical probability. In the context of a coin-tossing experiment, they compare theoretical and experimental probabilities.

1 Begin by discussing the new vocabulary with the class. Explain that the outcomes for an action are all the possible things that could happen for that particular action. For example, tossing a coin has two possible outcomes: heads or tails. Rolling a die or number cube has six possible outcomes: 1, 2, 3, 4, 5, or 6. Spinning a spinner with three equal-sized sections colored red, green, and blue will give you three outcomes: red, green, or blue. In all these cases, the outcomes for the situation are equally likely. None of the outcomes has a better chance of happening than any other outcome for that particular situation.

Explain that when a probability can be found without trying out the action, but by simply thinking about the different possibilities, it is called the *theoretical probability* of the event.

**About the Mathematics**

The notion of equally-likely outcomes makes sense for idealized situations in which there is no reason to suspect that any one outcome is more likely than another. It is important to keep in mind that this concept and the concept of theoretical probability are mathematical concepts. One must be careful not to assume that real-world situations mirror mathematical situations perfectly. For example, you might be tempted to assume that if you randomly pick a newborn child from a randomly selected location in the United States, then the chances that it is a boy are just the same as the chances that it is a girl. Birth statistics show that this is not quite so. They show that there is a slightly better chance that the baby will be a boy. In this situation, experimental probabilities make more sense than theoretical probabilities.

2 **Example**
Discuss this example carefully with the class. You might ask students for any ideas they have about how to make the drawing of a name as fair as possible. They may mention such things as these: make sure all the slips of paper are the same size and shape, mix the slips after putting them into the bowl, have the judge draw a slip without looking. The idea is to eliminate anything that would make it more likely for one name to be drawn than another.

3 In this situation, each person has the same chance of winning first prize as any other. Ask students for the probability that Althea will win. $\frac{1}{5}$ What is the probability that Luke will win? $\frac{1}{5}$ Ask for the probabilities that each of the other students will win, and use the answers to make the point that all of the outcomes are equally likely.

4 **Problem Set D** **Suggested Grouping: Pairs**
For this problem set, students should keep in mind that each name is as likely to be drawn as another name. So, to find the theoretical probability of drawing a name that fits a particular description, divide the number of names that fit that description by 5, which is the total number of names. Students can show their answers as a fraction or percent.

# Investigation 2 Theoretical Probability

**VOCABULARY**
**equally likely**
**theoretical probability**

In some situations, all of the possibilities for a situation—called the *outcomes*—have the same probability of occurring. For example, a coin toss has two possible outcomes: heads or tails. If the coin is fair, about half of the tosses will come up heads and half will come up tails. In situations like these, the outcomes are **equally likely.**

When outcomes are equally likely, you can calculate probabilities by reasoning about the situation. Since these **theoretical probabilities** do not depend on experiments, they are always the same for a particular event.

1 Introduce the terms *equally likely* and *theoretical probability.*

**Just the facts**

In 2001, the most popular names for girls born in the U.S. were Emily, Hannah, Madison, Samantha, Ashley, Sarah, Elizabeth, Kayla, Alexis, and Abigail.

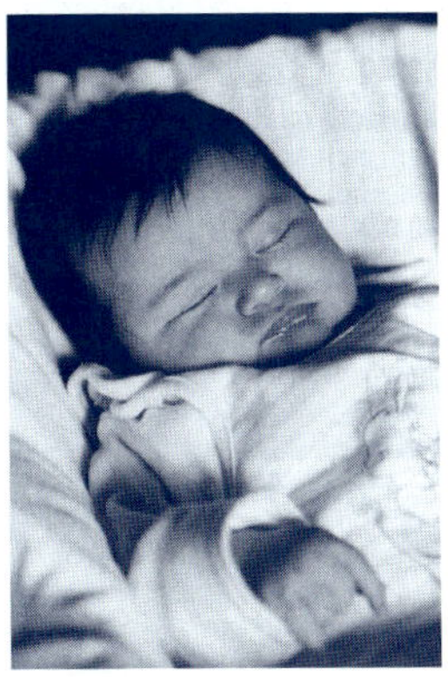

**EXAMPLE**

In a class competition, five students—Althea, Conor, Hannah, Luke, and Rosita—are tied for first place. To break the tie, they will write their names on slips of paper and place them in a bowl. A judge will choose one slip without looking, and the student whose name is on that slip will receive first prize. What is the probability that the name chosen has three syllables?

Althea | Conor | Hannah | Luke | Rosita

There are five names. The judge chooses a name *at random*—that is, in a way that all five names have the same chance of being selected. *Althea* and *Rosita* are the only names with three syllables.

Since there are five equally likely outcomes, and two of them are three-syllable names, you would expect a three-syllable name to be selected $\frac{2}{5}$ of the time. So, the probability of choosing a name with three syllables is $\frac{2}{5}$, or 40%.

2 Ask students how they can make the drawing of a name fair.

3 Tell students that all outcomes are equally likely.

4 You may have students work in pairs.

## Problem Set D

Think about the situation described in the Example. Decide how likely each of the following events is, and determine its theoretical probability. Explain your answers.

1. The name chosen does not begin with *R.*
2. The name chosen begins with *J.*
3. The name chosen has four letters or more.
4. The name chosen has exactly four letters.

**1.** $\frac{4}{5}$, or 80%; four of the five names do not begin with *R.*

**2.** 0, or 0%; none of the names begin with *J.*

**3.** 1, or 100%; all the names have at least four letters.

**4.** $\frac{1}{5}$, or 20%; of the five names, only one has exactly four letters.

1 For **Problem 7,** students should have no difficulty seeing that any name drawn either ends with *A* or does not end with *A*. When an event is certain to occur, its probability is 1, or 100%.

2 **Think & Discuss**

Discuss the cartoon with the class to be sure students understand why Rosita's statement is incorrect. Rosita is making the assumption that the behavior of the coin on one toss can affect the behavior on the next toss. Everyday experience clearly contradicts that assumption. Each toss is completely independent of the others.

### About the Mathematics

When a coin is tossed, the probability of getting heads is $\frac{1}{2}$, and the probability of getting tails is $\frac{1}{2}$. These are *theoretical* probabilities. Saying that each of these probabilities is $\frac{1}{2}$ is not the same as saying that for several tosses of a coin, half the outcomes must be heads and half must be tails. Tossing a coin an odd number of times could not possibly give a half-and-half result.

Rosita's statement suggests that she is confusing theoretical probability with experimental probability. This is a common confusion. There is a relationship between theoretical and experimental probability. The nature of that relationship is the subject of Problem Set E.

5. $\frac{2}{5}$, or 40%; two of the five names end with *A*.

6. $\frac{3}{5}$, or 60%; three of the five names do not end with *A*.

**5.** The name chosen ends with *A*.

**6.** The name chosen does *not* end with *A*.

**7.** Add the probabilities you found in Problems 5 and 6. Why does this sum make sense? See below.

When people talk about probabilities involved in games, rolling dice, or tossing coins, they usually mean *theoretical* rather than experimental probabilities. In the rest of this investigation, you will consider the relationship between these two types of probability.

1 Show students that any name drawn will either end with an A or will not end in an A.

## Think & Discuss

2 Discuss the cartoon with the class.

Discuss what the students are saying. Which students do you agree with? Answers will vary. Jing, Conor, and Miguel are correct.

**Just the facts**

In 2001, the most popular names for boys born in the U.S. were Jacob, Michael, Joshua, Matthew, Andrew, Joseph, Nicholas, Anthony, Tyler, and Daniel.

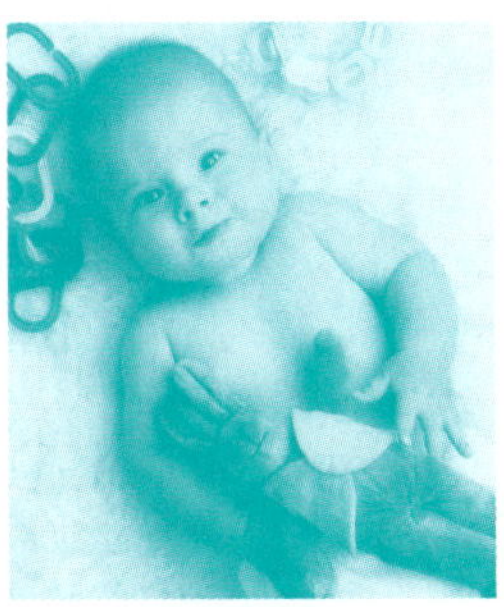

7. 1, or 100%; Possible explanation: The three names that weren't included for Problem 5 are the names for Problem 6. Since all five names are covered by the two problems, the probabilities must add to 1.

# Develop

1 **Problem Set E** **Suggested Grouping: Individuals**
The purpose of this problem set is to clarify the relationship between experimental and theoretical probability.

2 For **Problem 2b,** you may want to have students express the answer as a fraction and as a percent. The percent form will be useful when students answer Problem 8.

3 You will probably want to put the class results on the board for **Problem 4.**

## Access for all Learners

4 **Extra Challenge** You might have some students try this exercise, which will show the effect of increasing the number of times an experiment is performed.

Have students add a third column to their completed table from Problem 4. For each row, have them calculate the experimental probability for the data up to that point. When they are finished, the table will likely demonstrate an important idea: with an increasingly large number of tosses of a coin, the experimental probability of getting heads approaches the theoretical probability, which is $\frac{1}{2}$.

## On the Spot Assessment

If students make errors when expressing the probabilities as percents, you may need to review how to change a fraction to a percent. You may also need to review the concept of rounding to the nearest percent.

MATERIALS
coin

## Problem Set E

1. 6; you would expect to get heads $\frac{1}{2}$ of the 12 times, which is 6 out of 12.

**1.** If you toss a coin 12 times, how many times would you expect to get heads? Explain.

**2.** Conduct an experiment to find an experimental probability of getting heads. Toss a coin 12 times and record the results, writing H for each head and T for each tail.

**a.** How many heads did you get? Answers will vary.

**b.** Use your results to find an experimental probability of getting heads when you toss a coin. Answers will vary.

**3.** Compare the theoretical results with your experimental results.

**a.** Is your result in Part a of Problem 2 the same as the answer you computed in Problem 1? Answers will vary. Most students will find they are not exactly the same.

**b.** Is your experimental probability the same as the theoretical probability? Answers will vary.

**4.** Now combine your experimental results with those of other students. Make a table like this one, showing how many times out of 12 each student's coin came up heads. Tables will vary.

| Student | Number of Heads |
|---|---|
| James | 7 |
| Ali | 5 |
| | |
| | |
| | |
| | |

**5.** Calculate the total number of tosses your class made. How many heads would you expect for that number of tosses? See margin.

**6.** Now add the entries in the "Number of Heads" column. How closely does the result agree with your expectations? See below.

**7.** What percent of the total number of coin tosses came up heads? In other words, what is the experimental probability of getting heads based on the data for your entire class? Answers will vary.

**8.** Which experimental probability is closer to the theoretical probability: the one you calculated for Part b of Problem 2, or the one you calculated for Problem 7? See below.

It's normal for experimental probabilities to be different from theoretical probabilities. In fact, when you repeat an experiment a small number of times, the experimental and theoretical probabilities may not be close at all.

6. If the number of tosses was reasonably large, the result is probably close to (but not exactly) half of the total tosses.

8. Answers will vary, but Problem 7's result should be closer for most students.

5. Totals will vary. About half of the tosses would be expected to result in heads.

**Just the facts**

The United States produced half-cent coins from 1793 to 1857.

**Just the facts**

Lincoln is the only U.S. president on a coin who faces to the right.

1 You may have students work individually.

2 Encourage students express the answer as a fraction and as a percent.

3 Place the class results on the board.

4 Have students add a third column to the table for cumulative experimental probabilities.

# Develop

The paragraphs preceding Problem Set E expand on the observation made in Problem 8 of Problem Set E.

1 **Problem Set F** **Suggested Grouping: Individuals**
In this problem set, students roll a die 12 times and compute the experimental probability of rolling the number 3. They then combine their results with the results of other students. Finally, they compare the experimental probability based on the class results with the theoretical probability of rolling 3.

2 For **Problem 4,** make a table on the board to show the results for each student in the class. Add a third column that shows the experimental probability as a percent, rounded to the nearest percent. You might use a calculator to help you do this.

3 Ask for a volunteer to explain how to predict the number of 3s that will be obtained for a given number of rolls for **Problem 5c.**

**Problem Set Wrap-Up** To be sure students understand the implications of having equally likely outcomes when rolling a die, ask for the theoretical probability of rolling 4; of rolling 5; of rolling 1. $\frac{1}{6}, \frac{1}{6}, \frac{1}{6}$ Then ask how many times they would expect to get 5 if they rolled a die 120 times. 20 times

However, when you repeat an experiment a large number of times—for example, by combining all the coin tosses of everyone in your class or by performing more tosses—the experimental probability will usually grow closer to the theoretical probability.

The theoretical probability tells what is likely to happen *in the long run*—that is, if you try something a large number of times. It doesn't reveal exactly what will happen each time.

MATERIALS
die

## Problem Set F

1 You may have students work individually.

A standard die has six faces, each indicating a different number from 1 to 6. When you roll a die, the *outcome* is the number facing up.

**1.** Roll a die 12 times, and record each number you roll. How many 3s did you get? Results will vary.

**2.** Use your results to find an experimental probability of rolling 3.

**2.** Results will vary.

**3.** Collect results from others in your class, and make a table with these columns: Tables will vary.

| Student | Number of 3s |
|---|---|
| | |
| | |
| | |

2 Have students add a third column to the table that shows the probability as a percent.

**4.** Find the total number of rolls and the total number of 3s for all the students in your class. Then compute an experimental probability of rolling 3. Answers will vary but should be close to $\frac{1}{6}$.

**5.** Now think about the *theoretical* probability of rolling 3.

**a.** How many possible outcomes are there for a single roll? Are they all equally likely? 6, yes

**b.** On each roll of a die, what is the probability that you will roll 3?

**5b.** $\frac{1}{6}$

**c.** If you roll a die 12 times, how many 3s would you expect? 2

3 Have students explain how to predict the number of 3s on a given number of rolls.

**6.** Compare your theoretical results from Problem 5 with your own experimental results from Problem 2 and to your class experimental results from Problem 4. Which experimental result is closer to the theoretical result? Answers will vary. Most students will find that the class result is closer to the theoretical result.



LESSON 10.1 The Language of Chance 611

## Develop

### Share & Summarize

1 Have students work on these questions individually. Then call on volunteers to present their answers and explain their thinking to the class.

2 When you discuss **Question 2,** ask students whether the outcomes are equally likely and why. Students should be able to see that the outcomes are equally likely because the sections of the spinner are all the same size.

**Troubleshooting** Some students may have trouble calculating the number of times an event can be expected to occur when they know the probability of the event and the number of trials. They can find this number by multiplying the number of trials by the probability. Also check to make sure they understand that the product they find should be rounded to the nearest whole number.

### On Your Own Exercises

Practice & Apply: 5–7, pp. 615–616
Connect and Extend: 10–15, pp. 617–618
Mixed Review: 16–31, p. 619

## Share & Summarize

1 Discuss students' answers.

**1.** At Jenna's birthday party, her mother assigned each of the ten partygoers—including Jenna and her two sisters—a number from 1 to 10. To see who would play *Pin the Tail on the Donkey* first, Jenna's father pulled one of ten balls, numbered 1 to 10, from a box. The person whose number was selected would play first.

**a.** What is the probability that the number chosen was Jenna's or one of her sisters? $\frac{3}{10}$

**b.** What is the probability that the number chosen was *not* Jenna's or one of her sisters—that is, what is the probability that the number belonged to one of the seven guests? Explain how you found your answer.

**1b.** $\frac{7}{10}$; Possible explanation: Since the probability that the number was one of the sisters is $\frac{3}{10}$, the probability that it wasn't must be $1 - \frac{3}{10}$, or $\frac{7}{10}$.

2 Have students determine whether the outcomes are equally likely and why.

**2.** Chris has a spinner divided into five same-sized sections, numbered from 1 to 5. He spun the arrow 100 times, recording the result each time.

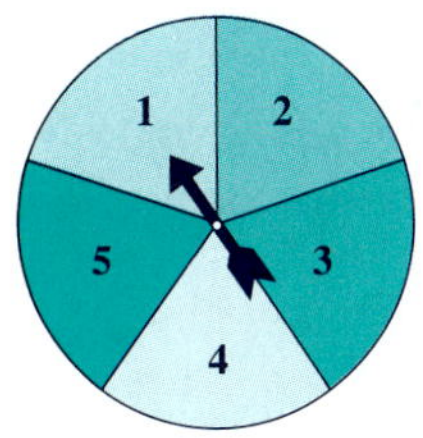


**a.** How many times would you expect the arrow to land on Section 4? 20

**b.** If the actual number of 4s Chris recorded was different from the number you answered in Part a, does that mean your calculation was wrong? Explain.

**2b.** No; the probability doesn't tell exactly what will happen each time.

**3.** Suppose you perform an experiment 20 times and a friend performs the same experiment 200 times. You both use your results to calculate an experimental probability. Whose experimental probability would you expect to be closer to the theoretical probability? Explain. My friend's; a theoretical probability predicts what will happen over many trials, and my friend performed many more trials than I did.

# Lab Investigation

**1** Suggested Grouping: Groups of 4

## 2 Materials and Preparation

Each group of four will need 40 counters. They will also need a four-cornered top, or a copy of Master 78 to construct a four-cornered top, and Masters 81 or 82 (optional). You may wish to assign construction of the top as homework. Alternatively, you might prefer to ask early finishers to construct enough tops for all the groups before students work on the lab investigation.

### Teaching Resources

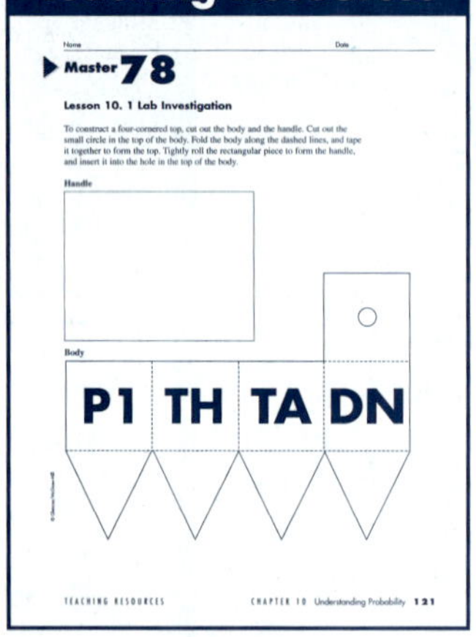
Name Date

Master 78

Lesson 10. 1 Lab Investigation

To construct a four-cornered top, cut out the body and the handle. Cut out the small circle in the top of the body. Fold the body along the dashed lines, and tape it together to form the top. Tightly roll the rectangular piece to form the handle, and insert it into the hole in the top of the body.

Handle

Body

P1 TH TA DN

TEACHING RESOURCES CHAPTER 10 Understanding Probability 121

## 3 Play the Game

Read through the rules with the class to ensure that students understand how the game is played. You should probably check the top that each group is using to ensure that it is properly constructed.

# Lab Investigation ▶ The Spinning Top Game

**1** You may have students work in groups of 4.

**2** Distribute materials to students.

**3** Go over the rules of the game with the class.

The Spinning Top game is played with a four-cornered top that contains four symbols. There are many variations of this game.

In this investigation, you will look at the probabilities for one of the most common versions of the Spinning Top game.

MATERIALS

- 4-cornered top with sides marked DN, TA, TH, and P1.
- counters

## Play the Game

You will play this game in a group of four. Here are the rules:

- Each player begins with ten counters.
- Each player puts a counter in the center of a table.
- Players take turns spinning the top. One of these four symbols will land face up:

  # DN TA TH P1

  The letter facing up tells the player what to do:

  | | |
  |---|---|
  | DN | Do nothing. |
  | TA | Take all the counters in the center. |
  | TH | Take half of the counters in the center. (Round down. For example, if the number of counters is five, take two.) |
  | P1 | Pay one by placing counter in the center. |

- Before each turn, each player puts another counter in the center. A player with no counters left is out of the game.
- The game continues until only one player has counters left or your teacher says time is up. The player with the most counters wins.



LESSON 10.1 The Language of Chance 613

# Develop

1 For **Question 1,** you may want to suggest that each group select a player to keep the tally of the symbols the players spin. Students will need to use the data they collect to answer subsequent questions.

For **Question 3,** students can use the lines on a sheet of notebook paper to help them draw the bars of their graph accurately. Alternatively, you may want to hand out copies of Masters 81 or 82.

**Teaching Resources**

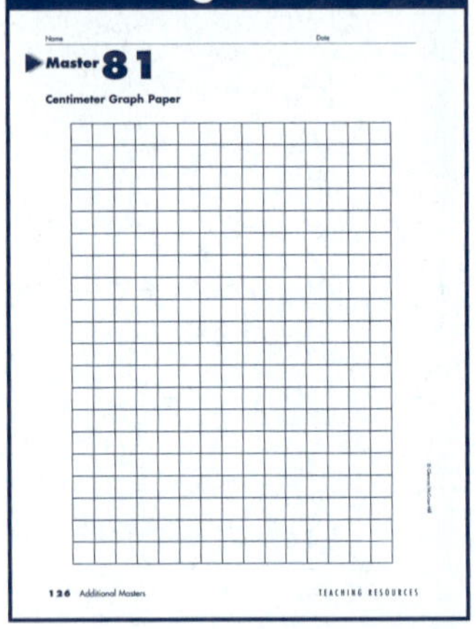


**Teaching Resources**

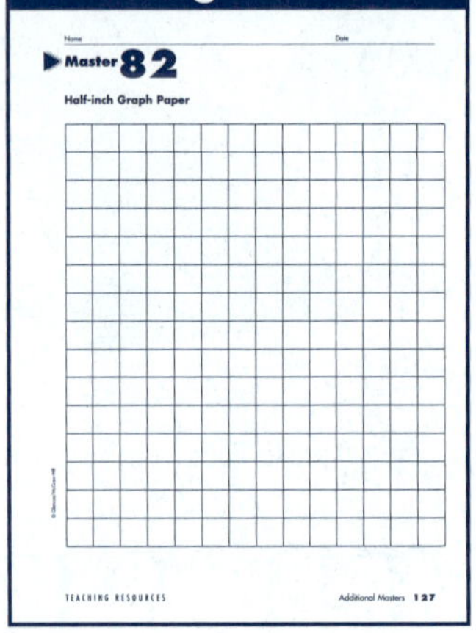


### 2 Calculate Probabilities

For **Question 4,** students will need to find the total number of spins of the top. The experimental probabilities can be expressed as fractions or percents. Percents can facilitate comparison of results. You may want to remind students that the sum of the probabilities should be 1 or 100%.

The probability in **Question 8** is the theoretical probability of losing a counter.

### 3 What Have You Learned?

You may want to discuss the answers for these questions with the class, calling on volunteers to read their answers.

If time permits, you might want to combine the results the groups obtained for Question 2 and compute the experimental probabilities for Question 4 for the entire class.

1. Play the game with your group, keeping a tally of the symbols the players spin.. Tables will vary.

| DN | TA | TH | P1 |
|---|---|---|---|
| II | | I | |

2. Answers will vary. If the number of turns is reasonably large, each symbol is likely to have appeared approximately the same number of times.

2. How many times did each symbol land face up in the whole game?

3. Draw a bar graph showing the number of times each symbol landed face up. Graphs will vary.

1 **Problem-Solving Strategies**

- Act it out
- Make a table

## Calculate Probabilities

Now that you have some experience playing the Spinning Top game, you can calculate the probabilities of certain outcomes.

4. Begin by calculating an experimental probability that each letter lands face up, based on the results of your game. Answers will vary.

2 Remind students that the sum of the probabilities should be 1, or 100%.

Assume each symbol is equally likely to land face up. **5.** $\frac{1}{4}$ for each

5. Calculate the theoretical probability of each symbol landing face up.

**6–10.** See Additional Answers.

6. How do the theoretical probabilities you calculated in Question 5 compare to your experimental probabilities from Question 4?

7. What is the (theoretical) probability of winning counters in a turn? Explain.

8. What is the probability of losing a counter in a turn? Explain.

9. Jahmal said the first player has a better chance to win all the counters in the center than the other three players do. Do you agree with Jahmal? Explain your answer.

10. Suppose the first player wins all the counters on his or her turn. What is the probability that the second player will also win all the counters? Explain your answer.

## What Have You Learned?

**11.** There are two ways to win counters but only one way to lose counters.

11. If you are equally likely to get each of the four sides of the top, why are you not equally likely to win counters as to lose counters?

12. Change the rules of the Spinning Top game so that the probability of winning counters on each turn is $\frac{1}{4}$ and the probability of losing counters is $\frac{1}{2}$. Answers will vary. Any change so that two of the symbols require the player to put in counters and one requires the player to take out counters will work.

3 Discuss students' answers.



### Additional Answers

6. Answers will vary.
7. $\frac{1}{2}$, or 50%; of the four possible letters, two (TA and TH) win counters for the player.
8. $\frac{1}{4}$, or 25%; of the four letters, one (P1) loses a counter for the player.
9. No; the top doesn't change, so the chances are the same with each spin.
10. $\frac{1}{4}$; each player has one chance in four to win all the counters. This is not changed by what the previous player did.

## Teacher Notes

# On Your Own Exercises



## Practice & Apply

**1.** Copy this number line. In Parts a–d, add a label to your number line indicating how likely you think the event is. Answers will vary.

**a.** I will listen to the radio tonight.

**b.** I will go to a movie sometime this week.

**c.** Everyone in my math class will get a perfect score on the next test.

**d.** I will wake up before 7:00 A.M. tomorrow.

**2.** Estimate the probability of each event, and explain your reasoning.

**a.** The school lunch will taste good tomorrow.

**b.** Everyone in our class will come to school next Monday.

**c.** A giraffe will come to school next Monday.

**3.** **Sports** Jahmal is practicing his archery skills. He hit the bull's-eye with 3 of the first 12 arrows he shot. Use these results to find an experimental probability that Jahmal will hit the bull's-eye on his next shot. $\frac{1}{4}$

**4.** Get a spoon (preferably plastic) and conduct this experiment:

For each trial, drop the spoon and record how it lands: right side up (so it would hold water) or upside down. Conduct 30 trials for your experiment. Use your results to find an experimental probability that the spoon will land right side up. Answers will vary.

**5.** A word is chosen at random from *book, paper, pencil,* and *eraser.*

**a.** What is the probability that the word has only one syllable? $\frac{1}{4}$

**b.** What is the probability that the word begins with *P*? $\frac{1}{2}$

**c.** What is the probability that the word ends with *P*? 0

**2a.** Possible answer: 0; the only school lunch that tastes good is pizza. Pizza day is Friday, but tomorrow is only Wednesday.

**2b.** Possible answer: $\frac{1}{4}$; we usually have a person absent each week, and Monday seems to be the most likely day for someone to be out.

**2c.** Possible answer: 0; they don't let giraffes in school!

impactmath.com/self_check_quiz



### On Your Own Exercises

Investigation 1, pp. 605–607
Practice & Apply: 1–4
Connect and Extend: 8, 9

Investigation 2, pp. 608–612
Practice & Apply: 5–7
Connect & Extend: 10–15

Assign Anytime
Mixed Review: 16–31

**Exercise 5:** The probabilities in this question are theoretical probabilities. Be sure students recall what is meant by the words *at random.* You may wish to refer students to the Example on page 608 if they need to review this idea.

### Quick Review Math Handbook

hot words hot topics

Hot Topics pp. 218–229

**Exercise 8:**
You may want to ask students what it means to say that there is a 20% chance of rain. Ask students how they could find an experimental probability of having rain at the beach.

**Just the facts**

The United States minted silver three-cent pieces from 1851 to 1873 and nickel three-cent pieces from 1865 to 1889.

**Remember**

A *prime number* is a whole number greater than 1 with only two factors, itself and 1.

**Connect & Extend**

**6.** Lupe tossed a coin 10 times and got 6 heads. Jing tossed a coin 1,000 times and got 530 heads.

**a.** Based on Lupe's results, what is an experimental probability of getting heads? Express your answer as a percent. 60%

**b.** Using theoretical probabilities, how many heads would you expect to get in 10 coin tosses? How far was Lupe's result from that number? 5, 1

**c.** Based on Jing's results, what is an experimental probability of getting heads? Express your answer as a percent. 53%

**d.** Using theoretical probabilities, how many heads would you expect to get in 1,000 coin tosses? How far was Jing's result from that number? 500, 30

**e.** **Challenge** The difference between the actual number of heads and the expected number of heads is much greater for Jing than for Lupe. How is it possible that Jing's experimental probability is closer to the theoretical probability? See Additional Answers.

**7.** Marika has a spinner divided into 10 equal sections, numbered from 1 to 10. Think about a single spin of the arrow.

**a.** What is the probability that the arrow will point to Section 1? $\frac{1}{10}$

**b.** What is the probability that the arrow will point to an odd number? $\frac{5}{10}$, or $\frac{1}{2}$

**c.** What is the probability that the arrow will point to an even number? $\frac{5}{10}$, or $\frac{1}{2}$

**d.** What is the probability that the arrow will point to a prime number? $\frac{4}{10}$, or $\frac{2}{5}$

**8.** **Earth Science** Imelda and a group of her friends planned a beach outing for a certain day. The local weather service said there was a 20% chance of rain on that day. When the day came, it rained and the trip was canceled.

Imelda said the weather service had been wrong when they gave the 20% rain prediction. They said it wasn't going to rain, but it did.

**a.** Do you agree with Imelda? Why or why not? See below.

**b.** If the weather service prediction didn't mean that it wouldn't rain, what do you think it meant? Possible answer: It meant there was a one-in-five chance it would rain.

**8a.** Possible answer: No; the weather service did not say it would not rain.



**Additional Answers**

**6e.** Possible answer: The experimental probability is the percent of tosses that were heads. Jing made many more tosses than Lupe, so each toss accounts for a much smaller percent of the total tosses. Lupe's results were only 1 head different from what you would expect, but that 1 head is 10% of his total tosses. Jing's results were 30 heads different from what you would expect, but those 30 heads are only 3% of her total tosses.

616 CHAPTER 10 Understanding Probability

★ indicates multi-step problem

**In your own words**

*Give an example to illustrate the difference between an experimental probability and a theoretical probability.*

**9.** See Additional Answers.

★**9.** Miguel's radio alarm clock goes off at 6:37 every morning. He complained that, almost every morning, he wakes up to commercials rather than music. Describe an experiment he could conduct to estimate the probability that he will wake up to a commercial. Explain how you would use the result to find an experimental probability.

**10.** Describe a situation for which the probability that something will occur is $\frac{1}{6}$. Possible answer: rolling a die and getting 1

**11.** A whole number is chosen at random from the numbers 1 to 10.

**a.** What is the probability that the number is odd? $\frac{5}{10}$, or $\frac{1}{2}$

**b.** What is the probability that the number is prime? $\frac{4}{10}$, or $\frac{2}{5}$

**c.** What is the probability that the number is a perfect square? $\frac{3}{10}$

**d.** What is the probability that the number is a factor of 36? $\frac{6}{10}$, or $\frac{3}{5}$

**12.** A whole number is chosen at random from the numbers 10 to 20.

**a.** What is the probability that the number is odd? $\frac{5}{11}$

**b.** What is the probability that the number is prime? $\frac{4}{11}$

**c.** What is the probability that the number is a perfect square? $\frac{1}{11}$

**d.** What is the probability that the number is a factor of 36? $\frac{2}{11}$

**13. Sports** Two sixth graders and two seventh graders are having a checkers tournament. They decide to choose randomly who the first two players will be and who will use which color. Consider the possible arrangements. For example, the first game might be a seventh grader playing black and a sixth grader playing red.

**a.** What other possible arrangements are there? See below.

Assume each of the arrangements is equally likely to occur.

**b.** What is the probability that the seventh graders will play each other in the first game? $\frac{1}{4}$

**c.** What is the probability that at least one seventh grader will play in the first game? $\frac{3}{4}$

**d.** What is the probability that exactly one sixth grader will play in the first game? $\frac{1}{2}$

**13a.** two sixth graders playing each other, two seventh graders playing each other, and a sixth grader playing black and a seventh grader playing red



**Exercise 10:**
Discuss all the situations that students propose. Some of the responses may require careful analysis to be sure the probability is indeed $\frac{1}{6}$.

**Exercise 13:**
Suggest that students make an organized list that shows all the possible arrangements. The example in the opening paragraph indicates that no attempt is being made to distinguish between the two sixth graders or between the two seventh graders.

**Additional Answers**

**9.** Possible answer: For a certain number of days, Miguel could keep count of how many days he wakes up to a commercial. To find an experimental probability, he would divide the number of days he woke up to a commercial by the total number of days.

**Exercise 15a:** The theoretical probability of drawing the same number on two consecutive draws is $\frac{1}{4}$. In Lesson 10.2, students will learn to calculate theoretical probabilities in such situations.

**14.** No; the fact that the other people lost does not affect Ruben's chances. He still has one chance out of four of winning a prize.

**14.** Ruben's class went to a carnival. There was a game of chance he wanted to play, but his teacher told him that only one player out of four wins a prize at that game. Ruben got in line anyway, behind three other people. As Ruben waited, he noticed that none of the three people ahead of him won a prize. He was very excited, because he was sure that meant he would win. Was he right? Explain.

**15. Preview** Cut or tear a sheet of paper to create four slips of paper, as identical to each other as possible. Number the slips of paper from 1 to 4, fold them once, and put them in a hat or bag.

Then conduct this experiment at least 40 times:

- Without looking, draw a slip of paper and look at the number you drew.
- Put the slip back in the hat or bag, mix them up, and draw another.

Keep a tally of how many times the numbers were the same and how many times they were different.

**a.** Based on your results, find the probability that two chosen numbers will be the same.

**15a.** Answers will vary but should be close to $\frac{1}{4}$.

**b.** Is the probability you found an experimental probability or a theoretical probability? experimental



**618** CHAPTER 10 Understanding Probability

## Mixed Review

Find a solution of each equation. **16.** 5 **17.** 12 **18.** 7.5

**16.** $6 + 7m = 41$ **17.** $1.9z + 14.3 = 37.1$ **18.** $3 + 3p = 5p - 12$

**19.** $x^2 - 5x = 14$ **20.** $4(v - 2) = 20$ 7 **21.** $\frac{2(3c + 12)}{6} = 8$ 4

Find the prime factorization of each number. **24.** $2^4 \cdot 3^2 \cdot 13$

**22.** 3,740 **23.** 19,551 **24.** 1,872

Use the fact that $783 \cdot 25 = 19{,}575$ to find each product without using a calculator.

**25.** $7.83 \cdot 25$ **26.** $78.3 \cdot 2.5$ 195.75 **27.** $7{,}830 \cdot 250$

Use the fact that $7{,}848 \div 12 = 654$ to find each quotient without using a calculator.

**28.** $7{,}848 \div 0.12$ **29.** $7.848 \div 12$ 0.654 **30.** $78.48 \div 1.2$ 65.4

**19.** 7 ($^-2$ is also a solution but students are not expected to find it.)
**22.** $2^2 \cdot 5 \cdot 11 \cdot 17$
**23.** $3 \cdot 7^3 \cdot 19$
**25.** 195.75
**27.** 1,957,500
**28.** 65,400

**31. Life Science** The table shows the number of bald eagle pairs in the contiguous United States in even-numbered years between 1982 and 1998.

**a.** Graph the data on a grid like the one below. See the graph.

**b.** Describe the overall change in the population of bald eagle pairs.

**c.** Use your graph to predict the number of bald eagle pairs in 1995 and in 2006.

| Year | Bald Eagle Pairs |
|---|---|
| 1982 | 1,480 |
| 1984 | 1,757 |
| 1986 | 1,875 |
| 1988 | 2,475 |
| 1990 | 3,035 |
| 1992 | 3,749 |
| 1994 | 4,449 |
| 1996 | 5,094 |
| 1998 | 5,748 |

Source: U.S. Fish and Wildlife Service

**Just the facts**

The *contiguous* United States are all those states that border on another, which means all U.S. states are contiguous except for Alaska and Hawaii.

**U.S. Bald Eagle Pairs**

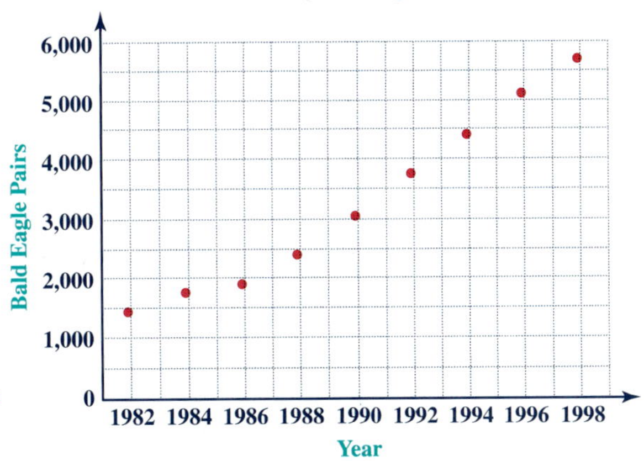


**31b.** Possible answer: The population has grown over the years, slowly at first and then more quickly.

**31c.** Possible answer: about 4,800; about 7,500



## Quick Check

### Informal Assessment

Students should be able to:

✔ associate a number from 0 to 1 with the likelihood that an event will occur

✔ use data to find the experimental probability of an event

✔ find the theoretical probability of an event for situations in which the individual outcomes are equally likely

### Quick Quiz

1. The spinner below has 7 equal-sized sections numbered 1 to 7.

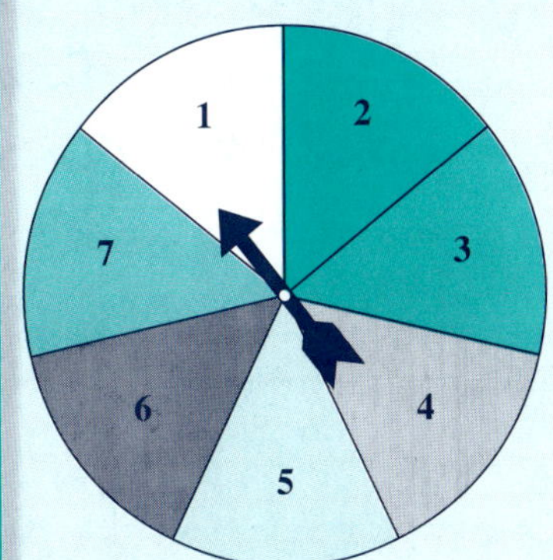


a. How many outcomes are possible if you spin the spinner once? 7

b. Are all of the outcomes equally likely? *yes*

2. Mario spun the spinner shown in Question 1 twenty times and got the following numbers:

| | | | | | | | | | |
|---|---|---|---|---|---|---|---|---|---|
| 1 | 3 | 7 | 3 | 2 | 2 | 2 | 5 | 6 | 1 |
| 6 | 4 | 6 | 5 | 1 | 4 | 5 | 5 | 3 | 1 |

Based on these results, what is the experimental probability for spinning the number 5? $\frac{1}{5}$, *or 20%*

3. What is the theoretical probability of spinning 5 with the spinner shown in Question 1? $\frac{1}{7}$

## Teacher Notes

# 10.2 Analyzing Games

## Objectives

- To conduct simple experiments and simulations to compare experimental and theoretical probabilities
- To make and use outcome tables to find theoretical probabilities
- To find probabilities by using the concept of geometric probability

## Overview (pacing: about 5 class periods)

In this lesson, students explore experimental and theoretical probabilities in the context of games that use special number cubes and dice. They also explore the concept of geometric probability.

## Advance Preparation

For Investigation 1, each pair of students will need specially labeled cubes (one with faces labeled 5, 5, 5, 5, 0, 0 and one with faces labeled 1, 2, 3, 4, 4, 4). For Investigation 3, each pair will need a copy of Master 79 (the *Rolling Differences* game board) and 40 counters, 20 in each of two colors. For Investigation 4, each group of students will need grains of uncooked rice and a copy of Master 80 (the *Rice Drop* game board).

## Lesson Planner

| | Summary | Materials | On Your Own Exercises | Assessment Opportunities |
|---|---|---|---|---|
| **Investigation 1** page T621 | Students play a game using specially labeled number cubes. They make a table to find which player has the greater probability of winning. | • Dice with faces labeled 5, 5, 5, 5, 0, 0<br>• Dice with faces labeled 1, 2, 3, 4, 4, 4 | Practice & Apply: 1–3, p. 631<br>Connect & Extend: 11–14, pp. 633–634<br>Mixed Review: 21–28, p. 637 | Share & Summarize, pages T622, 622<br>Troubleshooting, page T622 |
| **Investigation 2** page T623 | Students make a table to show the sums that can be rolled using a pair of standard dice. They use the table to find which player has the greater probability of winning a game called *Three and Seven.* | • Standard dice | Practice & Apply: 4, 5, p. 631<br>Connect & Extend: 15, 16, p. 634<br>Mixed Review: 21–28, p. 637 | On the Spot Assessment, pages T623, T624<br>Share & Summarize, pages T624, 624<br>Troubleshooting, page T624 |
| **Investigation 3** page T624 | Students make a table of the differences that can be obtained in a dice game called *Rolling Differences.* They use the table and probabilities to devise a possible winning strategy. | • Standard dice<br>• Master 79<br>*• Counters in 2 colors | Practice & Apply: 6–8, p. 632<br>Connect & Extend: 17, 18, p. 635<br>Mixed Review: 21–28, p. 637 | Share & Summarize, pages T625, 625<br>Troubleshooting, page T625 |
| **Investigation 4** page T626 | Students explore the concept of geometric probability. | • Grains of uncooked rice<br>• Master 80<br>• Inch rulers | Practice & Apply: 9, 10, pp. 632–633<br>Connect & Extend: 19, 20, p. 636<br>Mixed Review: 21–28, p. 637 | On the Spot Assessment, page T629<br>Share & Summarize, pages T630, 630<br>Troubleshooting, page 630<br>Informal Assessment, page 637<br>Quick Quiz, page 637 |

* Included in Impact Mathematics Manipulative Kit

## Introduce

1 You can use the two introductory paragraphs to help students understand what the word *event* means in discussions about probability. Suppose you are rolling a standard die. There are six possible outcomes. A set of outcomes of a specific type is called an *event*. The event *rolling an even number* consists of the even outcomes 2, 4, and 6. The number of outcomes for which the event occurs (3) divided by the number of possible outcomes (6) is the probability of the event ($\frac{1}{6}$).

If a die is rolled twice, the number of possible outcomes is much greater than 6. Tell students that in this lesson, they will learn to make tables that can help them analyze situations in which there are a large number of possible outcomes.

### Explore

The purpose of this Explore is to help students find outcomes when two coins are flipped or two dice are rolled. Have students work in groups of three or four.

2 You may want to handle the first question in a class discussion to clear up one possible point of confusion about listing and counting outcomes. Some students may think that when two coins are flipped, HT and TH should be considered the same, since in each case there is one head and one tail. This may seem all the more correct when the coins look exactly alike. If students imagine marking both sides of one coin to make it possible to distinguish between the coins, then it is easy to see that HT and TH are different outcomes. So, when two coins are flipped at the same time, there are four possible outcomes: HH, HT, TH, and TT.

3 Students will readily see that the same kind of thing happens with rolling two dice. To keep all of the outcomes straight, think of one die as being white and the other as being red. Then rolling the numbers 2 and 5 can occur not in just one way, but in two ways: 2 on white and 5 on red, or 2 on red and 5 on white.

4 With these ideas in mind, students are ready to work on the second and third questions in the Explore. Listing the possible outcomes for rolling two dice takes patience and a systematic approach. When most groups feel they have managed to list all the possible outcomes, have groups share their results with the class. Pay special attention to the procedures students used for listing the outcomes in a systematic way that accounts for all possibilities.

# 10.2 Analyzing Games

① 1-1, 1-2, 1-3, 1-4, 1-5, 1-6, 2-1, 2-2, 2-3, 2-4, 2-5, 2-6, 3-1, 3-2, 3-3, 3-4, 3-5, 3-6, 4-1, 4-2, 4-3, 4-4, 4-5, 4-6, 5-1, 5-2, 5-3, 5-4, 5-5, 5-6, 6-1, 6-2, 6-3, 6-4, 6-5, 6-6; 36

**Just the facts**

The Booker T. Washington commemorative half dollar was the first U.S. coin to feature an African American.

In Lesson 10.1, you considered *events* such as rolling a particular number on a die or getting heads when you flip a coin. Sometimes more than one outcome can cause an event to occur.

1 Introduce the term *event*.

For example, when you roll a die, there are three outcomes for which the event *rolling a prime number* occurs: 2, 3, and 5. To calculate the probability of an event, you have to know the number of possible outcomes and which outcomes cause the event to happen.

For a die roll, you can easily see that there are six possible outcomes. For some situations, it is more difficult to identify the number of possibilities.

## Explore

When you flip two coins, one possible outcome is HH. List all the possible outcomes. How many are there? HH, HT, TH, TT; 4

When you roll two dice, one possible outcome is 1-1. List all the possible outcomes. How many are there? See ①.

In the case of the dice, how can you be sure you have found all the possible outcomes? See ② in Additional Answers.

2 Discuss the first question with the class.

3 Help students see that when rolling two dice, a 2, 5 outcome is different than a 5, 2 outcome.

4 Have groups share their results from the Explore with the class.

For some experiments, figuring out all the possible outcomes can be a real problem in organization! You might forget one possibility or count the same one twice. Making a table can help you systematically count all the outcomes.

This table shows the possible outcomes when you flip two coins:

| | | Coin A | |
|---|---|---|---|
| | | H | T |
| Coin B | H | HH | TH |
| | T | HT | TT |



**Additional Answers**

② Possible answer: I thought about what might happen if I rolled 1 on the first die. There are six possibilities for the second die, giving 1-1, 1-2, 1-3, 1-4, 1-5, and 1-6. Then I did the same for rolling 2 on the first die, then 3, and so on until I'd considered all six possibilities for the first die.

# Investigation 1

In this investigation, students play a game using specially labeled dice. They use experimental and theoretical probability to analyze the game and decide which player has the greater chance of winning.

1 **Problem Set A** **Grouping: Pairs**

In this problem set students play *Who's Greater?*, a game that uses a die with faces labeled 5, 5, 5, 5, 0, 0, and a die with faces labeled 1, 2, 3, 4, 4, 4. They record data for the numbers they roll. After the winner is declared, they study the data to see what they reveal about the experimental probability of scoring with the die labeled 5, 5, 5, 5, 0, 0.

2 You may want to read through the rules for *Who's Greater?* with the entire class. Once students understand how to play the game, have them work on the problems.

3 For **Problem 1,** tell students to keep a complete record of the numbers they roll. To do this, they can make a table that has the following format:

| Turn | Die A | Die B |
|---|---|---|
| 1 | 5 | 3 |
| 2 | 0 | 1 |
| 3 | 5 | 4 |

**Tally**

**Die A**

||

**Die B**

|

The tally to the right of the table will let students keep track of the points they have scored. Together, the table and tally will show all the data needed for the remaining problems in the problem set.

For **Problem 3,** students can divide the number of points scored by the player using Die A by the combined scores of the two players.

4 **Problem Set Wrap-Up** Have pairs of students share their results in a class discussion. Call for a show of hands for students who won using Die A and then for students who won using Die B. Call on volunteers to read their answers for **Problem 4.** In Problem Set B, students will see how to find the theoretical probability that Die A will score a point.

## Investigation 1 *Who's Greater?*

You have probably played games of chance with standard dice. In this investigation, you will play a game with different kinds of dice.

MATERIALS

- die with faces labeled 5, 5, 5, 5, 0, 0
- die with faces labeled 1, 2, 3, 4, 4, 4

### Problem Set A

Answers will vary.

*Who's Greater?* is a game for two players. To play, you need two dice. Die A should be marked 5 on four of its faces and 0 on the other two. Die B should have 1, 2, and 3 on three of its faces and 4 on the remaining faces.

Each player uses one of the dice. A player uses the same die for the entire game. On a turn, each player rolls his or her die, and the player who rolls the greater number scores 1 point. The first person to score 10 points wins.

1. Play *Who's Greater?* with your partner. Record the die each player used and the final score of the game.
2. Which die did the winner use?
3. Use your results to calculate an experimental probability that, on a single turn, the player using Die A will score a point.
4. Which die would you prefer to play this game with? Give reasons for your choice.

**Just the facts**

Cube-shaped dice, with markings similar to those on modern dice, have been found in Egyptian tombs dating from 2000 B.C.

1 You may have students work in pairs.

2 Go over the rules of the game with the class.

3 Problem-Solving Strategy
Make a table

4 Discuss students' answers.



LESSON 10.2 Analyzing Games 621

1 **Problem Set B** **Suggested Grouping: Pairs**
In this problem set, students make a table to show which player will score in turns of rolling the dice for the *Who's Greater?* game. They use the table to calculate the theoretical probability that Die A will score. They then play more rounds of the game and compare their experimental results to the theoretical probabilities.

2 You may want to work through **Problem 1** with the whole class to ensure that students understand how the table works.

When you discuss **Problem 3,** ask students for the theoretical probability that Die B will score a point. Have students explain how to find the answer. Some students may use the same method they used in Problem 3. Others may reason that they can simply use the fact that $1 - \frac{2}{3} = \frac{1}{3}$ to conclude that the probability is $\frac{1}{3}$.

3 For **Problem 6,** you will probably want to discuss the results with the entire class. Ask whether they think *Who's Greater?* is a fair game. Ask them to explain their thinking.

4 **Share & Summarize**
Students can work on this Share & Summarize with a partner. Call on volunteers to present their answers to the class. Engage the class in checking that the dice and tables do indicate a probability of $\frac{5}{6}$ that Die A will score a point. Ask students to explain how they decided which numbers to use on the dice.

**Troubleshooting** If students had difficulty working with the table in Problem Set B, give some additional practice by stating a number that could have been rolled with Die A and a number that could have been rolled with Die B. Ask students which die scored on that roll, and have them explain how this is shown in the table. This may help you determine whether students are confused by the repeated numbers on the dice. You might also have them create a similar table for two new dice: Die C, labeled 3, 3, 5, 5, 6, 6; and Die D, labeled 2, 3, 4, 6, 6, 6.

## On Your Own Exercises

Practice & Apply: 1–3, p. 631
Connect & Extend: 11–14, pp. 633–634
Mixed Review: 21–28, p. 637

M A T E R I A L S

- die with faces labeled 5, 5, 5, 5, 0, 0
- die with faces labeled 1, 2, 3, 4, 4, 4

## Problem Set B

You can use this table to analyze the results of *Who's Greater?* The numbers on the faces of Die A are listed across the top; those of Die B are listed down the side. The entry "A" means that when Die A shows 5 and Die B shows 1, the roller of Die A scores a point.

**Die A**

| Die B | 5 | 5 | 5 | 5 | 0 | 0 |
|---|---|---|---|---|---|---|
| **1** | A | A | A | A | B | B |
| **2** | A | A | A | A | B | B |
| **3** | A | A | A | A | B | B |
| **4** | A | A | A | A | B | B |
| **4** | A | A | A | A | B | B |
| **4** | A | A | A | A | B | B |

1. What combination does the cell with "B" in it represent?
2. Copy and complete the table to show the winning die for each roll.
3. Calculate the theoretical probability that Die A will score a point on a turn. Do this by counting the number of outcomes in which Die A scores and dividing by the total number of outcomes. $\frac{2}{3}$
4. Based on the theoretical probabilities, which die would you prefer to play *Who's Greater?* with? Explain.
5. Will the player who uses the die in your answer to Problem 4 always win the game? Explain. See Additional Answers.
6. Play the game again a few times with your partner. Trade dice so that each of you has a chance to play with both Die A and Die B.
   a. For each game, record which die the winner used. Combine your results with the other pairs in your class. Answers will vary.
   b. How do your results compare with what you expected would happen? Results are likely to show that Die A wins more often than Die B, as expected.

1. Die A shows 0 and Die B shows 2.
2. See the table.
4. Die A; the chances of scoring a point with it are greater.

**Just the facts**

The ancient Egyptian game *senet* is one of the earliest known dice games. The dice first used to play senet were made of sticks or the knucklebones of animals.

### Share & Summarize

Design a pair of dice so the probability that the player using Die A scores a point on a turn is $\frac{5}{6}$. Use a table to show that your dice give the desired probability. See Additional Answers.

1 You may have students work in pairs.

2 Discuss Problem 1 with the class.

3 Ask students if they think the game is fair.

4 Discuss how students decided which numbers to use on the dice.



**Additional Answers**

5. No; even though the player with Die A has a better chance, it's possible that 0 will actually be rolled more often than 5—and then the player with Die B will win.

• *Additional Answers continued on page A671*

# Investigation 2

In this investigation, students play a game with two standard dice. They explore the possible sums of the two dice, and they make and test a conjecture about which player is more likely to score a point. They play the game, record results, and calculate experimental probabilities for rolling a sum of 3 and a sum of 7.

1 **Problem Set C** **Suggested Grouping: Individuals**
In this problem set, students think about the possible sums they could obtain with a roll of two dice. They make a conjecture as to whether there is a better chance of getting a sum of 3 or a sum of 7.

2 For **Problem 4,** some students may decide to make a table showing all possible sums. Others may simply list the ways to get a sum of 3 and a sum of 7. You may want to call on volunteers to read their answers.

3 **Problem Set D** **Grouping: Pairs**
In this problem set, pairs of students play the *Three and Seven* game and record their results. They use their data to calculate experimental probabilities for rolling each of the possible sums. They compare results with those of other students to see whether they can decide which sum, 3 or 7, makes a player more likely to win.

4 In **Problem 1,** students play a game of *Three and Seven.* Remind students that each player will roll his or her die 40 times. The table will be long but easy to fill in.

In **Problem 3,** students should calculate experimental probabilities for *all* the possible sums, not just for the sums 3 and 7.

5 For **Problem 4,** you will probably want to list the results for all the pairs of students on the board. There is no need to record actual scores, only which player won (the one who used a sum of 3 or the one who used a sum of 7).

# Investigation 2 Dice Sums

*Three and Seven* is a game two people play with a pair of standard dice, with faces numbered from 1 to 6. Each player rolls one die, and the two numbers are added.

- If the sum is 3, Player 1 scores a point.
- If the sum is 7, Player 2 scores a point.
- If the sum is any other number, neither player scores.

The player with the most points after 40 turns wins.

**Problem Set C Answers**

**1.** 2; both would show 1.

**2.** 12; both would show 6.

**Just the facts**

There are left-handed dice and right-handed dice. The arrangements of dots on the two types are mirror images of each other. In both, opposite faces add to 7.

## Problem Set C

**1.** What is the least sum you can get in this game? What number would be on each die to get this sum?

**2.** What is the greatest sum you can get? What number would be on each die to get this sum?

**3.** What other sums are possible? each whole number from 3 to 11

**4.** Do you think one player has a better chance of winning the game, or are the players' chances the same? Explain. Answers will vary.

If you play the game, you can collect evidence to test your answer to Problem 4.

MATERIALS
2 standard dice

## Problem Set D

**1.** Play *Three and Seven* with a partner. If you can't agree who will be Player 1 and who will be Player 2, toss a coin to decide. Keep track of all the sums you roll. Record your results in a table like this one:

| Turn | Die A | Die B | Sum |
|---|---|---|---|
| 1 | | | |
| 2 | | | |

Answers will vary.

**2.** Is there a sum that appears more often than any other? Answers will vary.

**3.** Use your results to calculate an experimental probability for each of the possible sums.

**4.** Compare your results with other students. How many games did Player 1 win? How many games did Player 2 win?

**5.** Which player seems to have the better chance of winning? Player 2

**Problem Set D Answers**

**3.** Answers will vary. (The theoretical probabilities are 2 and 12: $\frac{1}{36}$; 3 and 11: $\frac{2}{36}$; 4 and 10: $\frac{3}{36}$; 5 and 9: $\frac{4}{36}$; 6 and 8: $\frac{5}{36}$; 7: $\frac{6}{36}$.)

**4.** Answers will vary, but Player 2 will probably have won almost all games.

**1** Problem-Solving Strategy
Make and test a conjecture

**2** Problem-Solving Strategies
- Make a table
- Make an organized list

**3** You may have students work in pairs.

**4** Remind students that each person has to roll the die 40 times.

**5** List the results of all the pairs.



LESSON 10.2 Analyzing Games 623

## Develop

### Think & Discuss

1 Students can work on this Think & Discuss with a partner. Students will probably find it easy to create this table. It is, after all, simply an addition table for the whole numbers from 1 to 6. You may, nevertheless, want to check that students are setting up the table correctly.

> **On the Spot Assessment**
>
> Check that students realize how many possible outcomes there are for a roll of two dice. Some students may have failed to consider all the outcomes. You may need to explain that a 1 on Die A and a 5 on Die B is thought of as a different outcome from a 5 on Die A and a 1 on Die B. The distinction is similar to that for a flip of two coins. (See the notes for the Explore on page T620).

After students have completed their work, you can call on a volunteer to put his or her table on the board. The class can check that all the entries are correct.

2 **Problem Set E** **Suggested Grouping: Pairs**
In this problem set, students find the theoretical probabilities of rolling a sum of 3 and a sum of 7. They use their results to decide which player has the better chance of winning the *Three and Seven* game.

3 For **Problems 1 and 2,** check to be sure students have considered all the outcomes that give a sum of 3 and all that give a sum of 7.

> **On the Spot Assessment**
>
> If students overlooked any of the relevant outcomes in Problems 1 or 2, have them refer to their table for the Think & Discuss. Ask how many outcomes could give a sum of 4; a sum of 5; a sum of 8. 3; 4; 5 Remind students that in counting outcomes, they should treat a 3 on Die A and a 5 on Die B as a different outcome from a 5 on Die A and a 3 on Die B.

4 **Share & Summarize**
Ask one or two volunteers to read their answers to the class. Discuss the answers to clear up any points that are confusing or incorrect. Stress that in considering the outcomes, one must be concerned both about what numbers were rolled and on which die each number occurred.

**Troubleshooting** If students are having difficulty understanding how to find theoretical probabilities, reinforce the fact that they need to create tables showing all the possible outcomes to help them find the probabilities of different outcomes. You might work through a simpler game as a class, such as a game called *Four and Five* played with two spinners, each with four equal-sized sections labeled 1 to 4. Help the class create a table showing each of the possible sums, and then use the table to calculate the theoretical probabilities of rolling a sum of 4 and of rolling a sum of 5.

**On Your Own Exercises**

Practice & Apply: 4, 5, p. 631
Connect & Extend: 15, 16, p. 634
Mixed Review: 21–28, p. 637

# Investigation 3

In this investigation, students play a game called *Rolling Differences.* They use a table and probability to analyze the game, and devise a strategy that will give a player a good opportunity to win.

## Think & Discuss See Additional Answers.

To find all the possible rolls of two dice that will produce a sum of 3 or a sum of 7, it is helpful to use a table. Describe or create a table that might be useful to you.

## Problem Set E

Answer these questions about the game *Three and Seven.*

1. On a single turn, what is the probability that the sum will be 3?
2. On a single turn, what is the probability that the sum will be 7?
3. Which player has a better chance of winning? Explain.

## Share & Summarize

Hannah thinks there are 11 equally likely outcomes in the *Three and Seven* game because there are 11 possible sums. How would you help her understand the mistake in her thinking?

**Problem Set E Answers**

1. $\frac{2}{36}$, or $\frac{1}{18}$
2. $\frac{6}{36}$, or $\frac{1}{6}$
3. Player 2; the probability of getting a sum of 7 is much greater than the probability of getting a sum of 3.

**Share & Summarize Answer**

Possible answer: There are really 36 equally likely outcomes, because there are 36 ways a pair of dice can land. Some of these have the same sum. For example, 4-3, 3-4, and 5-2 all have a sum of 7. So even though there are 11 possible sums, they are not all equally likely.

1 Problem-Solving Strategy
Make a table

2 You may have students work in pairs.

3 Make sure students consider all of the outcomes.

4 Discuss any answers that are confusing or incorrect.

# Investigation 3 Rolling Differences

*Rolling Differences* is a dice game played by two people. To play, you need a pair of standard dice and a game board like the one below. Each player also needs 20 counters of a single color. Your counters should be a different color from your opponent's.

| 0 | 1 | 2 | 3 | 4 | 5 |
|---|---|---|---|---|---|



• Additional Answers on page A671

## Develop

1 Give a copy of Master 79 to each pair of students who will work together on Problem Set F. Read the rules at the top of page 625 with the class. Have students refer to their copy of Master 79 as you go over the rules. If a student mentions the idea of calculating a difference by subtracting the greater number rolled from the lesser, explain that in this game they will always find the *positive* difference.

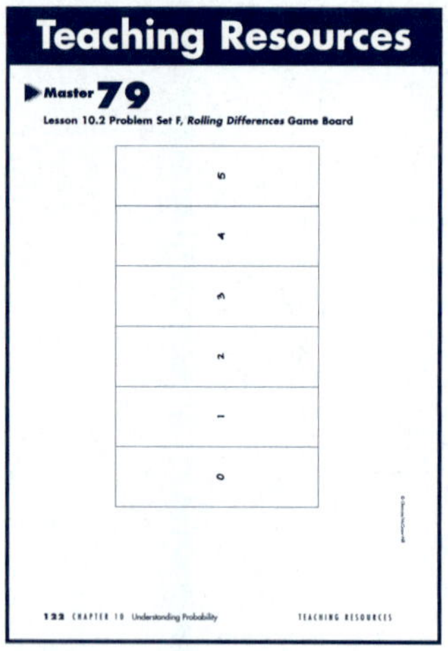


2 **Problem Set F** **Suggested Grouping: Pairs**
In this problem set, students play one or two games of *Rolling Differences* and record data about the differences rolled. They think about strategies for achieving a good score.

3 For **Problem 2,** students should realize that the differences that are rolled most frequently indicate where it would be best to put the most counters.

4 **Problem Set G** **Suggested Grouping: Pairs**
In this problem set, students make a table of differences for a roll of two dice. They use the table to find the theoretical probability of rolling a given difference. They use the probabilities to decide on the best ways to distribute their counters for the *Rolling Differences* game.

For **Problem 1,** check that students understand that there are 36 possible outcomes and that these can result in differences from 0 to 5.

5 For **Problem 3,** you may want to call on some volunteers to discuss what distribution of counters they think would be best and explain their thinking.

6 **Share & Summarize**
This question is closely related to Problem 3 in Problem Set G. You may want to ask students exactly how they would distribute 20 counters according to the probabilities for the difference.

**Troubleshooting** If students had trouble analyzing *Rolling Differences,* you might work through the analysis of a simpler version of the game as a class. For example, the game could be played using two spinners with four equal-sized sections labeled 1 to 4, a gameboard with four equal sections labeled 0 to 3, and 10 counters for each player.

### On Your Own Exercises

Practice & Apply: 6–8, p. 632
Connect & Extend: 17, 18, p. 635
Mixed Review: 21–28, p. 637

Here are the rules for *Rolling Differences:*

- Each player distributes 20 counters in the six sections of the game board in any way he or she chooses. (It's fine to leave some sections blank or even to put all counters in the same section.)
- On a player's turn, he or she rolls the dice and finds the difference of the numbers rolled. If the player has one or more counters in the section labeled with that difference, he or she removes one.
- The first player to remove all his or her counters from the board wins.

**MATERIALS**

- *Rolling Differences* game board
- 2 dice
- 40 counters (20 in each of 2 colors)

## Problem Set F

1. Play *Rolling Differences* one or two times, keeping a record of the differences rolled. Which difference was rolled least often? **1.** Answers will vary.
2. What do you think is a good strategy for distributing the counters on the game board? Why? See Additional Answers.

## Problem Set G

In this problem set, you will analyze the probability of rolling each difference with two dice.

1. Make a table that will help you find the number of ways to get each possible difference on a roll of two dice. See Additional Answers.
2. Copy and complete the table to show the probability of rolling each difference.
3. Using your table, determine what you think is the best way to arrange the counters.

| Difference | Ways to Get the Difference | Probability |
|---|---|---|
| 0 | 6 | $\frac{6}{36}$, or $\frac{1}{6}$ |
| 1 | 10 | $\frac{10}{36}$, or $\frac{5}{18}$ |
| 2 | 8 | $\frac{8}{36}$, or $\frac{2}{9}$ |
| 3 | 6 | $\frac{6}{36}$, or $\frac{1}{6}$ |
| 4 | 4 | $\frac{4}{36}$, or $\frac{1}{9}$ |
| 5 | 2 | $\frac{2}{36}$, or $\frac{1}{18}$ |

**Just the facts**

Dominoes were invented in China in the 12th century. Each domino was created to represent one of the possible results of throwing two dice.

**3.** See Additional Answers.

### Share & Summarize

In the game *Rolling Differences,* do you think putting all the counters on the most likely difference is a good strategy? Explain.

Possible answer: The probability of rolling the most likely difference is actually only about 28%. This means that other results have a good chance to occur as well. So you are not likely to win if you put all your counters in that cell.



1. Go over the rules of the game with the class.
2. You may have students work in pairs.
3. Tell students that the differences rolled most frequently show where it is best to place the counters.
4. You may have students work in pairs.
5. Ask students how to best arrange the counters.
6. Discuss students' answers.

**Additional Answers**
**Problem Set F**

**2.** Possible answer: A difference of 1 seems to occur most often, so put lots of counters in Section 1. (Some students may want to put all of the counters in Section 1, while others may say that even though this difference is most likely, other differences also occur, so some counters should be in other sections.)

• *Additional Answers continued on page A672*

# Investigation 4

In this investigation, students drop a grain of rice onto a game board on which are drawn closed, two-dimensional figures of various shapes and sizes. Students find experimental and theoretical probabilities that the grain of rice will land on a particular shape.

1

## Problem Set H Suggested Grouping: Groups of 3 or 4

In this problem set, students play the *Rice Drop* game and record their results. Individual students keep a record of their results and calculate experimental probabilities that the grain of rice will land on different parts of the board. The students in each group then combine their results and compute a set of group experimental probabilities.

2

Read through the game rules on page 626 with the class, and hand out a copy of Master 80 to each group. When students are confident that they understand the rules, they can work on the problems on page 627.

# Investigation 4 Geometric Probability

To calculate theoretical probabilities for the situations you have worked with so far, you divided the number of outcomes that mean an event occurred by the total number of outcomes. In this investigation, you will look at a situation for which you need to use a different strategy.

1 You may have students work in groups of 3 or 4.

MATERIALS
- *Rice Drop game board*
- *grain of uncooked rice*

## Problem Set H

*Rice Drop* is a game of chance. To play, a larger version of the game board at right should be placed on a hard, flat surface such as a table or desk. The game board is the area inside the outer border, not the entire page.

2 Go over the rules of the game with the class.

Here are the game rules:

- On a player's turn, he or she holds a grain of rice about 1 foot above the game board, near the center.
- The player drops the grain of rice and watches where it lands. If it bounces outside the outer border, the drop does not count—the player must try again.
- If the grain of rice lands in one of the four figures (square, circle, triangle, or parallelogram), the player scores 1 point. A grain that lands on the edge of a figure should be counted as inside the figure if half or more of the grain is inside.
- Each player gets 10 chances to score. (Remember, if the grain of rice bounces off the board, the drop does not count.) The player with the greatest score wins.

**Just the facts**

Rice is the staple food (providing more than one-third of a person's caloric intake) for about 60% of the world's population.

# Develop

1 Before students begin the game, discuss the chart in **Problem 1.** Emphasize that each student should keep his or her own chart. The completed chart should have a total of 10 tally marks. Take this opportunity to remind students that if the grain of rice bounces outside the outer border, that trial does not count, and no tally mark is made in the table.

For **Problem 2f,** students need to understand that "any figure" means any of the shaded figures. In other words, the number of outcomes for the event *landed in a shaded area* is the total number of tally marks in the first four columns of the chart.

2 For **Problem 3,** you may want to ask students to express the experimental probabilities both as fractions and as percents. Using percents will make it easier to compare and discuss results with other groups.

3 **Think & Discuss**

You may want to call on volunteers to share their answers to these questions with the class.

You might extend the second question in the following manner:

> How can you calculate the fraction of the game board covered by the figures? **Divide the sum of the areas of all the figures by the total area of the game board.**
>
> How can you find the areas of the figures? **Use the area formulas for the square ($A = s^2$), rectangle ($A = lw$), circle ($A = \pi r^2$), and parallelogram ($A = bh$) after taking the appropriate measurements of the figures.**

This is a preview of what students will investigate in Problem Set J on page 629.

1. Answers will vary.

1. Play the game with your group. On your turn, record the results of your drops by making tally marks in a chart like this one:

| Circle | Square | Triangle | Nonsquare Parallelogram | No Figure |
|---|---|---|---|---|
| | | | | |

2. Use the results of your 10 drops to estimate the chances that the grain of rice will land in  Answers will vary.
   a. the circle  b. the square
   c. the triangle  d. the nonsquare parallelogram
   e. no figure  f. any figure
3. Now combine the results of your group, and calculate new estimates for the probabilities in Problem 2.  Answers will vary.
4. Which set of probabilities do you think are more reliable: those from Problem 2 or Problem 3? Explain.

4. Problem 3; since there were more drops to work with, the experimental probabilities are probably closer to the theoretical probabilities.

## Think & Discuss

Do you think a grain of rice is as likely to land on the circle as it is to land outside any of the figures?  Answers will vary.

Can you think of a way you might calculate the theoretical probability of scoring a point on a single drop?  Possible answer: Calculate the fraction of the game board covered by the figures.

**Just the facts**

Hundreds of thousands of varieties of rice are thought to exist.

1 Make sure students have a total of 10 tally marks in their charts.

2 Have students express the experimental probabilities as fractions and percents.

3 Have students share their answers with the class.



LESSON 10.2 Analyzing Games 627

# Develop

1 Before students work on Problem Set I, read the paragraph at the top of the page with the class. You may want to ask students whether the grain of rice in the *Rice Drop* game is just as likely to land in one spot as another is reasonable. Some students may point out that dropping the rice from a point above the middle of the board may create a bias in favor of points near the middle of the board. The height of the rice before it is dropped can also affect the outcome. Explain that to find a *theoretical* probability for the rice landing in a shape, we have to assume that these biases have been eliminated and that the rice lands randomly within the outer border.

2 **Problem Set I** **Suggested Grouping: Individuals or Pairs**

In this problem set, students consider how the area of a shaded region on a game board is related to the likelihood that a grain of rice landing in a random spot will lie in the shaded region.

In **Problems 1 and 2,** students will see that the area of the shaded section of the game board is a key consideration in calculating the theoretical probability that the rice will land in that section.

3 In **Problem 3,** students will see that reducing the area of the shaded section of the game board will reduce the theoretical probability that the rice will land in that section.

4 For **Problem 4,** students should reason that separating the shaded portion of the game board into two parts does not affect the theoretical probability that the rice will land in a shaded section. The key idea is that the total shaded area of the game board is unchanged.

For the rest of this investigation, assume that the rice lands in a completely random spot on the game board. That is, assume the rice is just as likely to land in one spot on the game board as in another.

## Problem Set I

**1.** Suppose you use a game board divided into four equal rectangles, like this one. What is the probability that the rice lands on the shaded rectangle? $\frac{1}{4}$

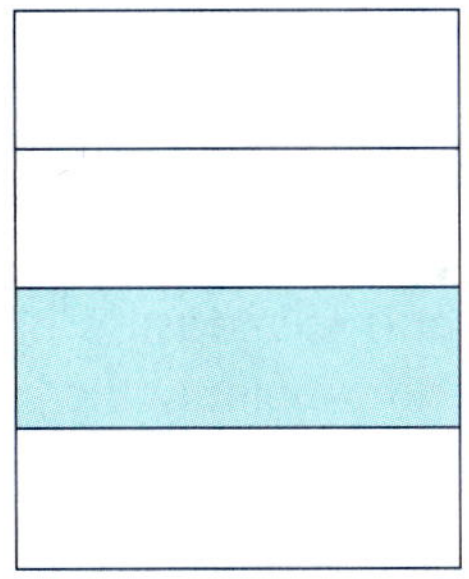

**2.** The game board at right is also divided into four equal sections. What is the probability that the rice lands on the shaded rectangle? $\frac{1}{4}$

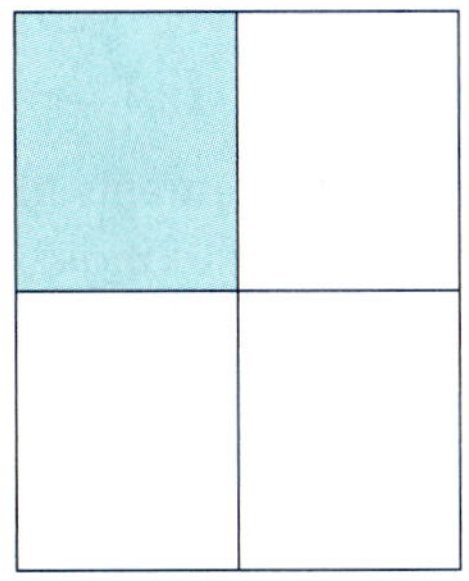

**3.** To create this game board, a square was removed from the shaded rectangle of the game board in Problem 2. Is the probability that the rice lands in the shaded figure *less than, greater than,* or *the same as* your answer to Problem 2? Explain your reasoning.

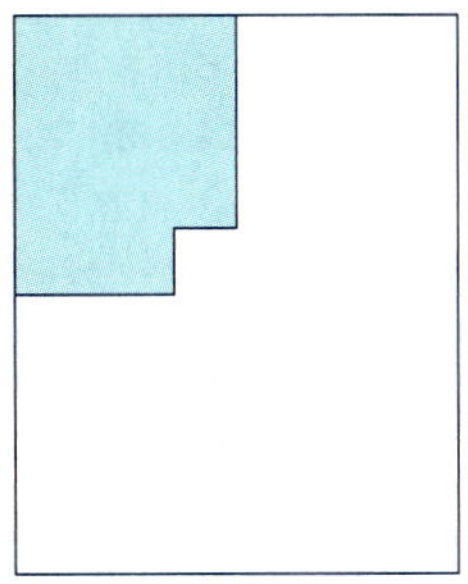

**4.** Now the square that was removed from the rectangle in Problem 2 has been returned, but in a different place. Is the probability that the rice lands in a shaded figure on this board *less than, greater than,* or *the same as* your answer to Problem 2? Explain your reasoning.

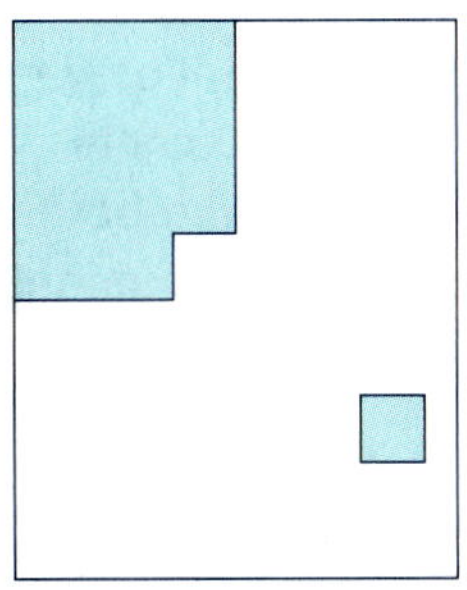

**3.** less than; Possible explanation: In Problem 2, if the rice had landed on the removed square, it would have been in the figure, but here it wouldn't be. There are fewer places for the rice to land and be in the figure now.

**4.** the same; Possible explanation: If the rice has the same chance of landing on the spot where the square used to be as it does of landing where the square is now, the probability is the same.

**Just the facts**

The average Cambodian eats more than 350 pounds of rice each year. The average American eats about 20 pounds of rice each year.

**1** Discuss whether the grain of rice has an equal chance of landing anywhere on the board.

**2** You may have students work individually or in pairs.

**3** Show students that reducing the shaded section on the game board will reduce the chance that the grain of rice will fall in a shaded section.

**4** Show students that separating the shaded section on the game board will not reduce the chance that the grain of rice will fall in a shaded section.

# Develop

1 **Problem Set J** **Suggested Grouping: Pairs**
In this problem set, students calculate the theoretical probability that a grain of rice will land in a given shape on the *Rice Drop* game board.

2 For **Problem 1,** students need to measure the length and width of the outer border of the game board and the square in inches. Encourage students to measure as accurately as they can. They will need to multiply the dimensions of each figure to find the area.

For **Problem 2,** students should divide the area of the square by the area of the game board to find what part of the game board is taken up by the square.

3 For **Problem 3,** students need to take measurements to find the areas of the circle, parallelogram, and triangle. You might refer them to Chapter 8 if they have difficulty recalling how to find the area of any of these figures. You may need to remind students that they can use 3.14 as an approximate value of $\pi$. After making the necessary measurements and calculating the area of each figure, they will need to divide the area of each figure by the area of the game board and express the quotient as a percent.

In **Problem 4,** students can divide the total area of the shapes by the area of the game board. Alternatively, they can add the probabilities for the individual shapes. These procedures are valid because the shapes do not overlap.

4 In **Problem 5a,** some students may say that if the grain of rice is dropped from a point above the center of the board, then spots closer to the center of the board are more likely. This is a plausible conjecture. You might want to ask students if they have any ideas for testing the conjecture.

### On the Spot Assessment

Students who have difficulty with Problem 3 may have forgotten how to calculate the areas of some of the figures. Ask them what measurements they will need to calculate area. Then ask them what formula they will use for each figure.

MATERIALS

- *Rice Drop* game board
- inch ruler

**1.** board: 59.5 $in.^2$; square: 9 $in.^2$

**3.** about 15.1% for the triangle and parallelogram; about 16.2% for the circle

**Just the facts**

It takes approximately 300 to 600 gallons of water to produce 1 pound of rice!

## Problem Set J

Rosita thought she could find the theoretical probabilities for the original *Rice Drop* game board using the areas of the figures and the area of the board.

**1.** Find the area of the game board and the area of the square.

**2.** Use your answer to Problem 1 to find the probability that the rice lands in the square. Express your answer as a percent. about 15.1%

**3.** Find the probability that the rice lands in each of the remaining figures. Express your answers as percents.

**4.** What is the probability that a player will score on a single drop? Explain how you found your answer. See Additional Answers.

**5.** **Challenge** The theoretical probabilities you found in this problem set assume that the rice lands randomly. Compare the theoretical probabilities you found in this problem set to the experimental probabilities you found in Problem Set H.

**5a.** Answers will vary.

**a.** Do you think the rice lands in a completely random spot, or are some spots more likely than others? Explain why you think so.

**b.** What could you do to test whether your answer to Part a is correct? Possible answer: Conduct another experiment with many more trials.

**1** You may have students work in pairs.

**2** Encourage students to measure carefully.

**3** Remind students how to find the area of any of these figures.

**4** **Problem-Solving Strategy**
Make and test a conjecture

**Additional Answers**
**Problem Set J**

**4.** about 61.5%; Possible explanations: I added the areas of all the figures and divided by the area of the board. *Or,* I added the probabilities for the four figures.

LESSON 10.2 Analyzing Games 629

1 **Share & Summarize**

This Share & Summarize is an excellent opportunity to bring together the key ideas of the investigation. After students have had an appropriate amount of time to work on these questions, call on volunteers to present their answers and explain their thinking.

**Troubleshooting** If students need more practice calculating geometrical probabilities, you might have them work in pairs to create new *Rice Drop* game boards. They can calculate the probability of scoring a point on a single drop with their own board and then with their partner's board, and then check answers and compare calculations with their partner.

## On Your Own Exercises

Practice & Apply: 9, 10, pp. 632–633
Connect & Extend: 19, 20, p. 636
Mixed Review: 21–28, p. 637

★ indicates multi-step problem

# Share & Summarize

1 Have students share their answers and explain their thinking.

★**1.** Suppose this was the game board for a game of *Rice Drop*. The measurements are given in inches.

Assuming that the rice lands in a completely random place, what is the probability of scoring a point on a single drop?

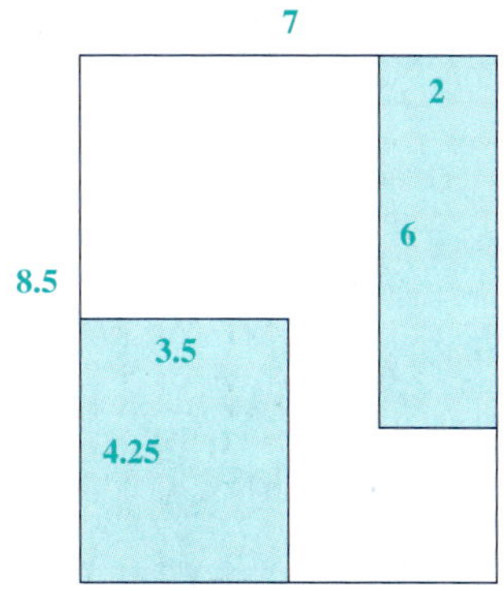


**1.** about 0.452, or 45.2%

**2.** Miguel and Althea were playing *Rice Drop* using a checkerboard. They decided that a drop scored a point if the rice landed on a green square. A drop that doesn't land on the board doesn't count.

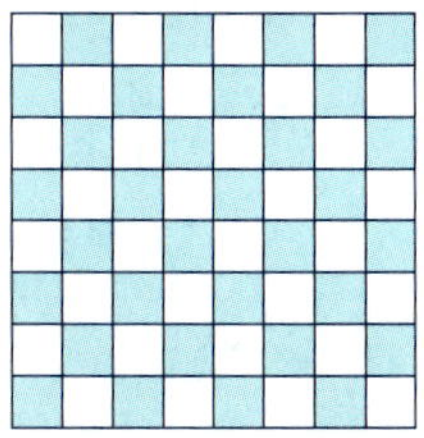

Assuming that the rice lands in a completely random place, what is the probability that a drop scores a point? Explain how you found your answer.

**2.** $\frac{1}{2}$; half the squares are green, half the squares are white, and the squares are all the same size, so the rice should land on green half the time.

**3.** Luke likes playing darts. He throws a dart at the dartboard shown here. The points earned for each ring are shown.

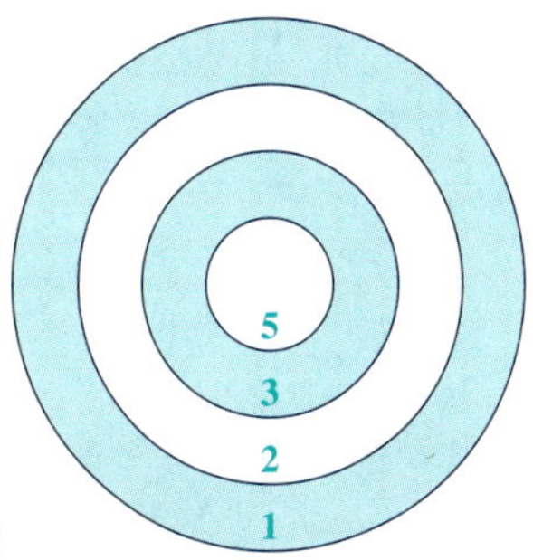


★**a.** Assuming Luke's dart hits the dartboard in a random place, what is the probability that Luke scores at least 3 points? (The radius of the inner circle is equal to the width of each ring.) $\frac{1}{4}$

**b.** Do you think the assumption in Part a is reasonable? Explain. No; darts is a game of skill, not chance. If Luke plays a lot, he'd be able to get close to the center more often.

## Teacher Notes

# On Your Own Exercises

★ indicates multi-step problem

## Practice & Apply

★ **1.** Suppose two standard dice are rolled. What is the probability that one die shows 5 and the other shows 1, 2, 3, or 6?

**1.** $\frac{8}{36}$, or $\frac{2}{9}$

**2.** Althea and Rosita are playing a game. Each girl rolls one standard die, and the girl with the greater number scores a point. If the numbers are the same, neither girl scores.

**a.** How many possible combinations of dice rolls are there? 36

**b.** On a single turn, what is the probability that the girls will roll the same number? $\frac{6}{36}$, or $\frac{1}{6}$

**c.** What is the probability that the girls will roll different numbers?

**2c.** $\frac{30}{36}$, or $\frac{5}{6}$

**d.** What is the probability that Althea will score a point? $\frac{15}{36}$, or $\frac{5}{12}$

★ **3. Technology** Some calculators have a *random number generator.* María uses her calculator to choose a whole number from 3 to 7. Rashid uses his to choose a whole number from 1 to 9. In both cases, each possible outcome is equally likely. Which of the two friends has a better chance of getting the greater number? Explain. You may want to create a chart to help you.
See Additional Answers.

**4.** Ramón says the probability of getting a sum of 12 when rolling two dice is equal to the probability of getting a sum of 6. To explain his reasoning, he pointed out that each number has the same probability to be rolled on each die.

**a.** Explain why Ramón is incorrect.

**4a.** Possible answer: The probability of each number is the same for each die, but the number of possible combinations for each sum aren't the same. There is only one way to get a sum of 12 (6 on each die) but several ways to get a sum of 6 (for example, 1 and 5, 2 and 4).

**b.** What is the probability of rolling a sum of 6? A sum of 12? A sum of 1? $\frac{5}{36}$, $\frac{1}{36}$, 0

**5.** Suppose you have two sets of eight cards. The cards in each set are numbered from 1 to 8. You pick one card at random from each set and find the sum of the numbers.

**a.** What is the probability that the sum is 1? 0

**b.** What is the probability that the sum is 3? $\frac{2}{64}$, or $\frac{1}{32}$

★ **c.** Suppose you play a game in which Player A scores a point when the sum of the two cards is 3 and Player B scores a point when the sum is 9. Which player would you rather be? Explain.

Player B; there are only two ways to get 3 (1 + 2 and 2 + 1) but many ways to get 9 (1 + 8, 8 + 1, 2 + 7, 7 + 2, and so on).

impactmath.com/self_check_quiz



• ***Additional Answers on page A672***

## Assign and Assess

### On Your Own Exercises

Investigation 1, pp. 621–622
Practice & Apply: 1–3
Connect & Extend: 11–14

Investigation 2, pp. 623–624
Practice & Apply: 4, 5
Connect & Extend: 15, 16

Investigation 3, pp. 624–625
Practice & Apply: 6–8
Connect & Extend: 17, 18

Investigation 4, pp. 626–630
Practice & Apply: 9, 10
Connect & Extend: 19, 20

Assign Anytime
Mixed Review: 21–28

**Exercise 1:**
Students can list and count the desired outcomes and then divide the result by the total number of possible outcomes. Students must recall that outcomes such as 1, 5 and 5, 1 are considered to be two different outcomes.

**Exercise 3:**
The situation in this exercise is similar to the *Who's Greater?* game in Investigation 1. If students have difficulty getting started on this exercise, they may find it helpful to refer to the table they made in Problem Set B.

**Exercise 5:**
Students may find it helpful to make a table of all the sums when working on this exercise.

## Assign and Assess

**Exercise 6:**
Students may find it helpful to refer to the table they created for Problem 1 in Problem Set G.

**Exercise 9:**
All the probabilities in this exercise are theoretical. The probability of the rice landing in any one of the shaded figures is the area of that figure divided by the area of the 8-by-10 rectangle.

In **Part a,** students can add the areas of the triangles and divide the result by 80 square units. Another approach is to add the probabilities for the individual triangles. Similar methods apply for **Parts b and c.**

For **Part d,** the answer can be found by subtracting the answer for Part c from 100%.

**6a.** $\frac{1}{2}$; Possible explanation: There are 6 possibilities that give 0, 8 that give 2, and 4 that give 4, for a total of 18 out of 36 possibilities.

**6b.** $\frac{1}{2}$; Possible explanation: The difference is either even or odd. Since the probability of an even difference is $\frac{1}{2}$, the probability of an odd difference is $1 - \frac{1}{2}$.

**6c.** $\frac{2}{3}$; Possible explanation: There are 6 ways to get 0, 10 ways to get 1, and 8 ways to get 2, for a total of 24 out of 36 possibilities.

**Just the facts**

Wild rice is the only cereal grain native to the United States.

**6.** Suppose someone rolls two standard dice.

**a.** Find the probability that the difference rolled is even. Explain.

**b.** Find the probability that the difference is odd. Explain.

**c.** Find the probability that the difference is less than 3. Explain.

**d.** Find the probability that the difference is 3 or more. Explain.

**e.** Find the probability that the difference is 6 or more. Explain.

**6d, e.** See Additional Answers.

**7.** Suppose someone rolls two dice, each with faces numbered 2, 4, 6, 8, 10, and 12. Consider the differences between two rolled numbers.

**a.** How many possible differences are there? What are they? **7a.** 6; 0, 2, 4, 6, 8, 10

**b.** Are the probabilities for all the possible differences the same? Explain. **7b, c.** See Additional Answers.

**c.** Suppose you were playing a game of *Rolling Differences* with these dice instead of standard dice. The spaces on the game board would be labeled with the possible differences, rather than 0 to 5. How would you arrange your counters in the spaces? Explain.

**8.** Imagine a game called *Rolling Sums* that is just like *Rolling Differences,* but in which you *add* the numbers to determine which counter to remove.

**a.** How many sections should there be on the game board? 11

**b.** What is the greatest number that should appear on the board? 12

**c.** What is the least number that should appear on the board? 2

**d.** Describe a strategy you would use to arrange your counters on the board. See Additional Answers.

**9.** Darnell and Camila were playing *Rice Drop* using this game board. The dimensions are in inches. Assume the rice will land on a completely random spot.

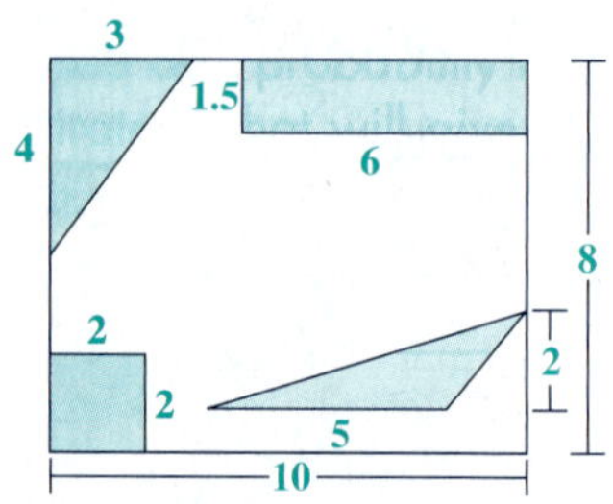


**a.** Find the probability that the rice lands in a triangle. 13.75%

**b.** Find the probability that the rice lands in a quadrilateral. 16.25%

**c.** Find the probability that the rice lands in a shaded figure. 30%

**d.** Find the probability that the rice does not land in a shaded figure. 70%



### Additional Answers

**6d.** $\frac{1}{3}$; Possible explanation: Since the probability of getting less than 3 is $\frac{2}{3}$, the probability of getting 3 or more is $1 - \frac{2}{3}$.

**6e.** 0; Possible explanation: The greatest possible difference is 5.

**7b.** no; Possible explanation: As shown in the table at the right, some differences can be made in more ways than others. For example, there are ten ways to roll a difference of 2, but only two ways to roll a difference of 10.

| Die B \ Die A | 2 | 4 | 6 | 8 | 10 | 12 |
|---|---|---|---|---|---|---|
| 2 | 0 | 2 | 4 | 6 | 8 | 10 |
| 4 | 2 | 0 | 2 | 4 | 6 | 8 |
| 6 | 4 | 2 | 0 | 2 | 4 | 6 |
| 8 | 6 | 4 | 2 | 0 | 2 | 4 |
| 10 | 8 | 6 | 4 | 2 | 0 | 2 |
| 12 | 10 | 8 | 6 | 4 | 2 | 0 |

• *Additional Answers continued on page A672*

★ indicates multi-step problem

**10.** about $0.001\overline{3}$ or 0.13%

★**10.** A small area created by buildings built close together is often called a *courtyard.* Suppose a group of six friends are standing in a courtyard like the one shown here when it starts to rain.

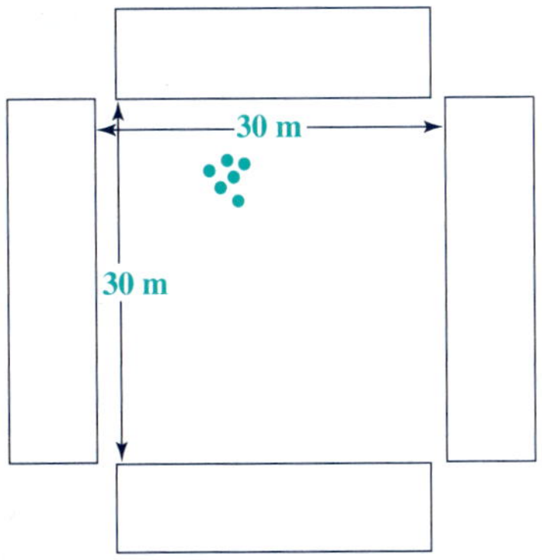


What is the probability that the first raindrop hits one of the friends? (Assume each person occupies a circle about 50 cm in diameter.)

## Connect & Extend

★**11.** Marcus and Miguel played *Who's Greater?* with two different dice. Marcus' die had 100 on one face and 0 on each of the other five faces. Miguel's was numbered from 1 to 6.

Before playing, Marcus said: "I will probably win because the greatest number on my die is 100, which is much greater than any of the numbers on your die."

Miguel said: "I have the better chance of winning because my die has more faces with numbers that are greater than the numbers on your die."

Who is right? Explain your answer.

**11, 12.** See Additional Answers.

**Just the facts**

*A 4-sided die is called a tetrahedron. An 8-sided die is an octahedron. A 12-sided die is a dodecahedron. And a 20-sided die is an icosahedron.*

★**12.** Charo and Irene have dice with 4 sides (numbered 1 to 4), 8 sides (numbered 1 to 8), 12 sides (numbered 1 to 12), and 20 sides (numbered 1 to 20). With each of the dice, all the possible outcomes are equally likely.

The friends want to play *Who's Greater?* with these dice. To play, each chooses a die and rolls it. Whoever rolls the greater number scores a point. If the numbers are the same, no one scores.

Irene chose a 12-sided die. Which die should Charo choose? Explain.

**13.** Possible answer: Label the faces of one die 0, 0, 0, 7, 8, and 9, and the faces of the other die 1, 2, 3, 4, 5, and 6.

**13.** Design two *different* six-sided dice that have an equal probability of beating each other at *Who's Greater?*



**Exercise 10:**
The areas that students use should be expressed in the same units, either square centimeters or square meters. Students should also note that the exercise gives the *diameter* occupied by each person. To find the area of the circle, they need to divide by 2 to find the radius.

**Exercise 12:**
Although it is possible to make a table to help with this exercise, the reasoning approach illustrated in the answer is by far the most efficient approach.

### Additional Answers

**11.** See page A672.

**12.** The 20-sided die; if Charo used a 12-sided die, the players would have an equal chance of winning. The 20-sided die adds 8 numbers that are greater than any of the numbers on Irene's die, so it gives Charo the best chance of scoring.

• ***Additional Answers continued on page A672***

**Exercise 14:**
Students may not think of making a table for this exercise. If necessary, give students who have trouble getting started a hint by suggesting that they try making a table. A table will be helpful for assuring that no possibilities have been overlooked.

**Exercise 15:**
Students should not have too much trouble with this exercise if they are careful to note that they are told the total number of possible outcomes.

**Exercise 16:**
Students who have difficulty solving this exercise by using reasoning alone will probably need to make a table to help them consider the possible combinations.

★ indicates multi-step problem

**14. Economics** Omry and Shaked are twin brothers. For their birthday, they each received gift certificates from the local candy store, worth 25¢, 50¢, $1, $5, and $10.

★**a.** Suppose both boys have only one gift certificate left, but neither can remember which one. They want to buy and share a chocolate that costs $2.25. Assuming the gift certificates are equally likely to be remaining, what is the probability that they have enough to buy the chocolate? Explain your answer. See Additional Answers.

**b.** What is the probability that each boy has a $5 gift certificate left? $\frac{1}{25}$

**15.** Suppose you roll three standard dice and add the results. There are 216 equally likely outcomes.

**a.** What is the least possible sum? What is the probability of rolling this sum? 3, $\frac{1}{216}$

**b.** What is the greatest possible sum? What is the probability of rolling this sum? 18, $\frac{1}{216}$

**c.** Find all the ways to get a sum of 4, and find the probability of rolling that sum. 1-1-2, 1-2-1, 2-1-1; $\frac{3}{216}$, or $\frac{1}{72}$

**16.** Mr. Shu's suitcase has a lock consisting of two number wheels with the numbers 0, 2, 4, 5, 6, and 8 on each. To open the lock, the wheels must be set so the product of the two numbers is 32.

Locked

Unlocked

**16b.** See Additional Answers.

**16c.** 0; it has the highest probability of being chosen.

**16d.** They will be equally secure. Possible explanation: Each number pair has a $\frac{1}{36}$ chance of being guessed. This is the same probability as getting a product of 4, 25, 36, or 64.

**a.** Suppose a stranger finds the suitcases and attempts to open it by trying different combinations. What is the probability that he will open the suitcase on the first try? $\frac{2}{36}$, or $\frac{1}{18}$

**b.** Mr. Shu can reset the lock to open with a different product. Which product would give a more secure lock than 32? Explain.

**c.** Which product would you suggest Mr. Shu *not* use? Why?

**d.** Mrs. Shu's suitcase has two number wheels identical to those on Mr. Shu's suitcase. Her suitcase opens when the numbers form a particular pair, such as 0-5 or 6-2, rather than a particular product. Assuming Mr. Shu sets his lock as you suggested in Part b, whose suitcase will be more secure, Mr. Shu's or Mrs. Shu's? Explain.



**Additional Answers**

**14a.** $\frac{16}{25}$; Possible explanation: The table at the right shows that there are 16 out of 25 possibilities for which the total value is $2.25 or more.

**16b.** Possible answers: 4, 25, 36, or 64; for each of these products, the proba bility of guessing the combination is only $\frac{1}{36}$.

**Omry's Gift Certificates**

| Shaked's Gift Certificates | $0.25 | $0.50 | $1.00 | $5.00 | $10.00 |
|---|---|---|---|---|---|
| **$0.25** | $0.50 | $0.75 | $1.25 | $5.25 | $10.25 |
| **$0.50** | $0.75 | $1.00 | $1.50 | $5.50 | $10.50 |
| **$1.00** | $1.25 | $1.50 | $2.00 | $6.00 | $11.00 |
| **$5.00** | $5.25 | $5.50 | $6.00 | $10.00 | $15.00 |
| **$10.00** | $10.25 | $10.50 | $11.00 | $15.00 | $20.00 |

**In your own words**

*Give an example to illustrate how analyzing probabilities can help you plan strategies for playing games. You may choose a game from this lesson or make up one of your own.*

**17.** Avril has a die numbered from 1 to 6. Chelsea has a spinner divided into ten identical sections numbered from 1 to 10. Avril rolls her die, and Chelsea spins her spinner.

**a.** One possible outcome is 1-10, meaning Avril rolled a 1 on the die and Chelsea spun a 10 on the spinner. How many possible outcomes are there? 60

**b.** What is the probability that the difference of the two numbers is 10? Explain your answer. 0; the greatest possible difference is 9.

**c.** What is the probability that the difference is 7? Explain your answer. See margin.

**d.** What is the probability that the difference is greater than 7? Explain your answer. See below.

**18.** Avril has a die numbered from 1 to 6. Chelsea has a spinner divided into ten identical sections, numbered from 1 to 10. They decide to play the following game: On each turn, they spin the spinner and roll the die. Avril scores a point if the difference is 5, and Chelsea scores a point if the difference is 7.

**a.** Who has the better chance of scoring a point? Explain. See below.

**b.** How could the girls change the scoring rules so that they have the same chance of winning? See below.

**c.** Suppose you were playing the original game and could choose any number as the difference you earn a point on. Which number would you choose? Why? 1; it is the most likely difference.

**17c.** $\frac{1}{20}$; Possible explanation: There are three ways to get a difference of 7 (3-10, 2-9, and 1-8), so the probability is $\frac{3}{60}$, or $\frac{1}{20}$.

**17d.** $\frac{1}{20}$; Possible explanation: There are three ways for the difference to be 8 or 9 (1–10, 2–10, and 1–9), so the probability is $\frac{3}{60}$, or $\frac{1}{20}$.

**18a.** Avril; the probability of getting 5 is $\frac{6}{60}$ and the probability of getting 7 is $\frac{3}{60}$.

**18b.** Possible answer: Avril scores 1 point when the difference is 5 and Chelsea scores 2 points when the difference is 7.



**Exercises 17 and 18:** If students can use reasoning instead of a table to solve these exercises, they will save a good deal of time.

LESSON 10.2 Analyzing Games 635

# Assign and Assess

**Exercise 19:**
This exercise involves geometric probability. Here, however, one uses length rather than area.

**Exercise 20:**
Most students are likely to find this exercise a bit challenging.

In **Part b,** students should divide the area covered by Carla's car by the area of the region within which the acorns might fall. Note that this latter area does *not* include the ground space taken up by the base of the tree trunk.

In **Part d,** students need to divide the area of the moon roof by the ground space covered by the car.

★ indicates multi-step problem

★ **19. Technology** Marcus videotaped all six episodes of his favorite hour-long television show. His friend missed an episode, and Marcus said he would loan her the tape. The number line illustrates where on the tape the show had been recorded.

Marcus often rewinds the tape and watches different parts of previous episodes. He doesn't remember where on the tape he last stopped watching the show. Before giving the tape to his friend, Marcus put it in his VCR and pushed the play button. What is the probability that the tape started somewhere within the show his friend wanted to see? $\frac{1}{6}$

**Just the facts**

**The white oak tree is the state tree of Illinois. White oaks can live up to 400 years.**

**20.** While visiting a friend, Carla parked her car under a large oak tree. She left her moon roof open. (A *moon roof* is a small window in the roof.)

The tree drops acorns in a circular area around its trunk, as shown in the diagram. Assume the acorns fall randomly within the circle, which has a radius of 20 feet. No acorns fall where the trunk is. The trunk has a radius of 5 feet.

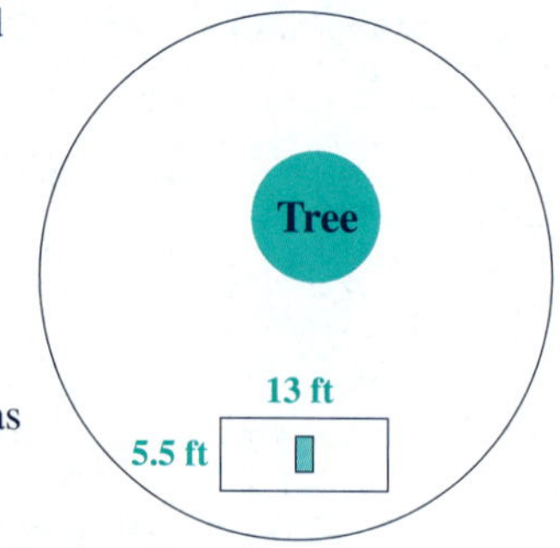


**a.** Find the area of the region in which an acorn might fall.

**b.** While Carla was at her friend's house, an acorn fell from the tree. Find the probability that the acorn hit Carla's car (including the moon roof). about 6.1%

**20a.** about 1,178 ft$^2$
**20c.** about 0.3%

**c.** The dimensions of the moon roof are 32 inches by 16 inches. Find the probability that the acorn fell through the moon roof.

**d. Challenge** Suppose the acorn hit Carla's car. Find the probability that it fell through the moon roof. (Hint: Should this probability be *more than* or *the same as* the probability you answered in Part c?) about 5%



636 CHAPTER 10 Understanding Probability

## Mixed Review

**21.** mean: 79; median: 73; mode 99; mean or median

**22.** mean: 65; median: 85; mode: none; median

**23.** mean: 75.5; median: 77; mode: none; mean or median

**24.** $m\angle a = 156°$, $m\angle b = 24°$, $m\angle c = 156°$

**Statistics** Find the mean, median, and mode of each set of test scores. Then tell which measure you think best represents the data.

**21.** 85, 99, 73, 64, 99, 80, 69, 72, 70

**22.** 0, 90, 93, 6, 85, 97, 84

**23.** 52, 94, 73, 81, 65, 88

**Geometry** Find the value of each lettered angle without measuring.

**24.**

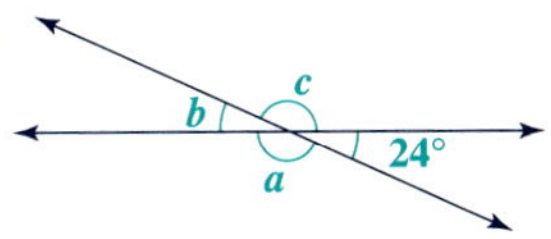


**25.**

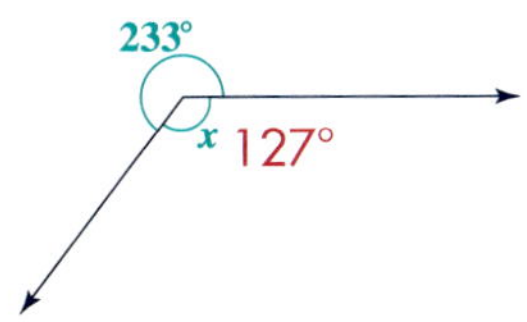


**26.** 65°

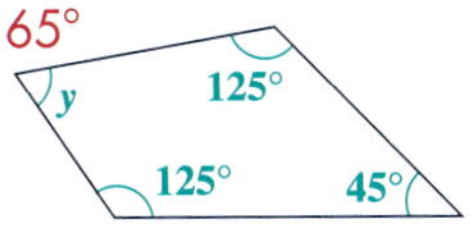


**27. Economics** Mr. Morales hired Althea to distribute takeout menus for his new restaurant to homes in the area. He told Althea he would pay her $14.50, plus 4¢ for every menu she delivered.

**27a.** $A = 14.50 + 0.04 \cdot m$

**a.** Write a rule you can use to calculate the amount Althea will earn, *A*, if you know the number of menus she delivers, *m*.

**27b.** $14.50 + 0.04 \cdot m = 23$

**b.** Althea wants to earn $23 to buy a pair of earrings for her mother. Write an equation you could solve to find the number of menus should would need to deliver to earn enough money.

**c.** Solve your equation. How many menus does Althea need to deliver? 212.5; 213 menus

**28. Geometry** Make sketches to show how you can arrange seven identical square tiles to form three shapes with different perimeters. You must use all the tiles in each of your shapes. Give the perimeter of each shape.

Possible answer:

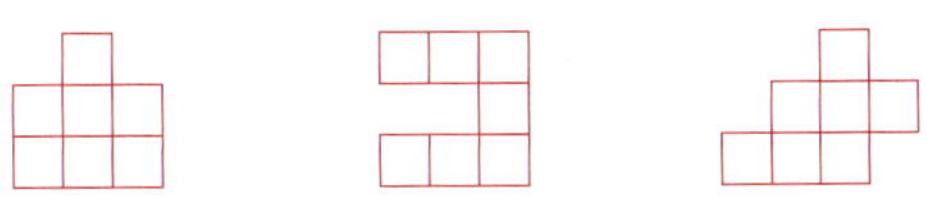

perimeter 12 units | perimeter 16 units | perimeter 14 units



## Quick Check

### Informal Assessment

Students should be able to:

- ✔ conduct simple experiments and simulations to compare experimental and theoretical probabilities
- ✔ make and use outcome tables to find theoretical probabilities
- ✔ find probabilities by using the concept of geometric probability

### Quick Quiz

1. Ricardo has a die with faces labeled 1, 3, 5, 7, 9, 11. Jimmy has a die with faces labeled 2, 4, 6, 8, 10, 12. Each friend rolls his die. The person who rolls the lesser number scores 1 point.
   a. What is the probability that Ricardo scores a point in a turn? $\frac{7}{12}$
   b. What is the probability that Jimmy scores a point in a turn? $\frac{5}{12}$

2. Ricardo and Jimmy use the dice from Question 1. If each person rolls his die and the sum is a multiple of 7, Ricardo scores a point. If the sum is a multiple of 3, Jimmy scores a point. If the sum is a multiple of both 3 and 7, they both score a point. If the sum is not a multiple of 3 or 7, neither person scores.
   a. What is the probability that Ricardo scores a point in a turn? $\frac{5}{36}$
   b. What is the probability that Jimmy scores a point in a turn? $\frac{1}{3}$
   c. What is the probability that they both score a point in a turn? $\frac{1}{18}$

• *Quick Quiz continued on page A673*

# Teacher Notes

# 10.3 Making Matches

## Objectives

- To make tree diagrams to show outcomes for situations involving independent choices
- To use tree diagrams to compute probabilities in situations that involve dependent choices

## Overview (pacing: about 3 class periods)

In the first two lessons of this chapter, students have considered probability for situations in which the events that occur are independent. This means that the occurrence of one event does not affect the occurrence of another. In this lesson, students explore situations in which one event can affect subsequent events. The situations studied here are comparatively simple. Students will explore dependent events in more detail in Courses 2 and 3.

## Advance Preparation

For Investigation 1, each pair of students will need counters, blocks, or slips of paper (two in each of two colors) and a bucket or paper bag.

## Lesson Planner

| | Summary | Materials | On Your Own Exercises | Assessment Opportunities |
|---|---|---|---|---|
| **Investigation 1** page T639 | Students use simple simulations and tree diagrams to investigate probability situations in which the occurrence of one event can affect subsequent events. | *• Counters, blocks, or slips of paper<br>• Buckets or paper bags | Practice & Apply: 1–3, p. 646<br>Connect & Extend: 6–8, pp. 648–649<br>Mixed Review: 12–23, p. 651 | Share & Summarize, pages T642, 642<br>Troubleshooting, page T642 |
| **Investigation 2** page T643 | Students examine how the principles studied in Investigation 1 can be extended to situations that involve matching cards from a standard deck. | | Practice & Apply: 4, 5, p. 647<br>Connect & Extend: 9–11, p. 650<br>Mixed Review: 12–23, p. 651 | Share & Summarize, pages T645, 645<br>Troubleshooting, page T645<br>Informal Assessment, page 651<br>Quick Quiz, page A674 |

* Included in Impact Mathematics Manipulative Kit

# Introduce

1 Remind students that when they flip a coin, what happens on one flip does not affect what happens on the next flip. The same is true for rolling a die or spinning a spinner. Tell students that in this lesson, they will consider situations in which the occurrence of one event can affect what happens in the next event.

2 **Think & Discuss**
Discuss the Think & Discuss questions with the class. The important thing for students to understand is that in the case of rolling a die, the probability of rolling a particular number does not change from one roll to the next, since the number of possible outcomes is unchanged from one roll to the next. In the case of drawing markers without looking at them, the probabilities do not stay the same from one stage to the next, since choosing a marker at one stage reduces the number of possible outcomes for the next stage.

# 10.3 Making Matches

In the probability games you have considered so far, the result of one round or trial does not affect the result of another. For example, if you toss a coin and get heads, your chances of getting heads when you flip the coin again are still 50%. These are called *independent events.*

In this lesson, you will work with situations in which what happens in one case *does* affect what can happen in the next, or *dependent events.*

1 Tell students that they will study events that will influence probabilities of other events.

2 Discuss the questions with the class.

## Think & Discuss

Suppose you roll a die several times.

- What is the probability of getting 6 on the first roll? What is the probability of getting 4? $\frac{1}{6}$, $\frac{1}{6}$
- Suppose the first time you roll the die, you get 6. You roll a second time. What is the probability of getting 6 on the second roll? What is the probability of getting 4? $\frac{1}{6}$, $\frac{1}{6}$

Now imagine that you and some friends are making a poster for a school party. Your teacher gives you six markers, in six different colors, and each of you chooses one without looking. Two of the colors are red and green.

- If you choose first, what is the probability that you will get the red marker? What is the probability that you will get the green marker? $\frac{1}{6}$, $\frac{1}{6}$
- Suppose you chose first and got the red marker. What is the probability that the second person will get the red marker? What is the probability that the second person will get the green marker? 0, $\frac{1}{5}$

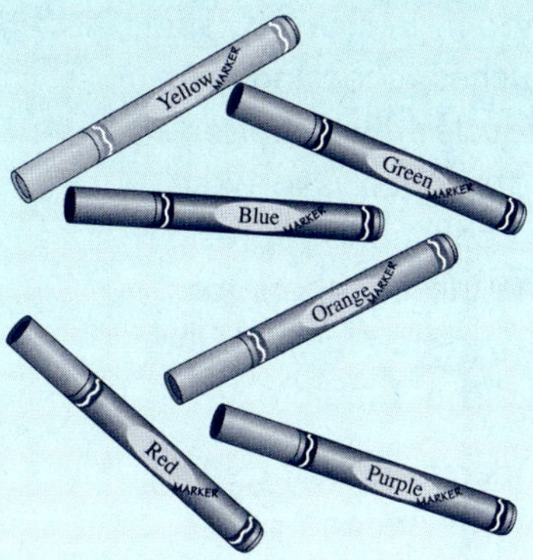


- Why are the probabilities for the dice situation different from those for the marker situation? See ①.

① Possible answer: When the die is rolled the second time, it still has the same six sides and they're still equally likely. When the second person takes a marker, the first marker chosen is gone, leaving only five equally likely possibilities.

# Investigation 1

1 In this investigation, students use a straightforward simulation to examine experimental probabilities for a situation in which choices at one stage can affect choices at the next stage. They learn to make and use tree diagrams to find theoretical probabilities in such situations.

Read the first paragraph on page 639 with the class, and discuss what is meant by *simulation*. Students should have little difficulty seeing that selecting colored markers and selecting colored slips of paper are quite similar. The total number of markers and their colors matches the total number of slips of paper. In both cases, objects are selected at random.

2 **Problem Set A** **Suggested Grouping: Pairs**
In this problem set, students use the make a model and act it out problem-solving strategies. Students use manipulatives to simulate selecting socks randomly from a drawer. They find an experimental probability for selecting two socks of the same color.

3 In **Problem 1,** students should not look as they draw items from the bucket or paper bag. If they look, their selection will not be completely random. In **Part a,** students can tally matches and nonmatches, checking from time to time to see if they have conducted the required number of trials. A slightly simpler approach is to write the numbers from 1 through 16 in a column to represent the trial numbers. They can circle the trial numbers for which the "socks" matched and draw a line through the trial numbers for which the "socks" did not match.

4 In **Problem 4,** students need to understand that getting a matching pair depends entirely on the *second* sock drawn. At the stage when the second sock is drawn, there are three socks left. Two of these have the same color, and the third has the same color as the sock that was drawn first. So, there is a $\frac{2}{3}$ chance that the next sock drawn will *not* match the first and only a $\frac{1}{3}$ chance that it will match the first.

**Problem Set Wrap-Up** Have students share their answers to Problem 4. If students have difficulty understanding the reasoning for Problem 4, it may help to make a table of all the possibilities. In the following table, the numbers 1 and 2 are used to distinguish between the two socks of the same color.

| First Sock | Second Sock | Match? |
|---|---|---|
| Brown 1 | Black 1 | no |
| Brown 1 | Black 2 | no |
| Brown 1 | Brown 2 | yes |
| Brown 2 | Black 1 | no |
| Brown 2 | Black 2 | no |
| Brown 2 | Brown 1 | yes |
| Black 1 | Black 2 | yes |
| Black 1 | Brown 1 | no |
| Black 1 | Brown 2 | no |
| Black 2 | Black 1 | yes |
| Black 2 | Brown 1 | no |
| Black 2 | Brown 2 | no |

There are 12 possibilities, and 4 of them result in a *yes* in the last column. So the probability of a match is $\frac{4}{12}$, or $\frac{1}{3}$.

## Investigation 1 Matching Colors

VOCABULARY
**simulation**

2. $\frac{1}{3}$; of the three socks left in the drawer, two are black and one is brown.

A **simulation** is an experiment in which you use different items to represent the items in a real situation. For example, to simulate choosing markers and looking at their colors, you can write the color of each marker on a slip of paper and put all the slips into a bag. You can simulate choosing markers by drawing slips from the bag. Mathematically, the situations are identical.

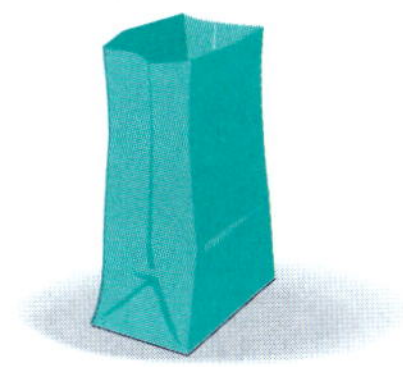

Using a simulation can help with some of the problems in this investigation.

1 Introduce the term simulation.

MATERIALS

- counters or blocks (2 in each of 2 colors) or slips of paper
- bucket or bag

### Problem Set A

2 Problem-Solving Strategies

- Make a model
- Act it out

Ken woke up early and found that a storm had knocked out the power in his neighborhood. He has to dress in the dark. Ken has four socks in his drawer, two black and two brown. Color is the only difference between them. As long as both socks are the same color, Ken doesn't care which he wears. He takes two socks out of the drawer.

**1.** Answers will vary.

**1.** Simulate this situation, using counters, blocks, or slips of paper with colors written on them. If you use counters or blocks, you may have to let other colors stand for the sock colors. For example, a red block might represent a brown sock, and a blue block might represent a black sock. Use a bucket or bag to represent Ken's sock drawer.

   **a.** Without looking, pick two "socks," one at a time, from the "drawer." Record whether the socks match. Then put the socks back, mix them up, and try again. Repeat this process 16 times, and record the results.

   **b.** Use your results to find an experimental probability that Ken will choose matching socks.

3 Make sure students do not look into the bag as they are drawing items.

**2.** If the first sock Ken picks is brown, what is the (theoretical) probability that the second sock will also be brown? Explain. See margin.

**3.** If the first sock is black, what is the (theoretical) probability that the second sock will also be black? $\frac{1}{3}$

**4.** Ken says that since he has two colors of socks, he has a 50% chance of getting a matching pair. Do you think he is correct? If not, what do you think the actual probability is? Explain.
no; Possible explanation: No matter which sock Ken chooses first, he will have a one-in-three chance of getting a matching sock. So, the probability is $\frac{1}{3}$, not $\frac{1}{2}$.

4 Make sure students understand that getting a matching pair depends on the second sock drawn.

**Just the facts**

On November 9, 1965, a massive power failure caused the largest blackout in U.S. history. The blackout affected over 30 million people in nine states along the East Coast.

LESSON 10.3 Making Matches 639

## Develop

1 When you discuss this page, you might observe that it was a bit difficult in Problem Set A to keep track of all the possibilities when the socks were being drawn. Tell students that a *tree diagram* can make the job a little easier.

Discuss the diagrams on page 640 to help students understand the basic idea of constructing a tree diagram. Then go straight to page 641 to show how the tree diagram allows one to read off the outcomes for tossing two coins.

In Lesson 10.2, you saw how you could use a table to keep track of the possible outcomes for two coin tosses. You can also draw a *tree diagram* to show all the possibilities. The possible results for the first coin can be shown like this:

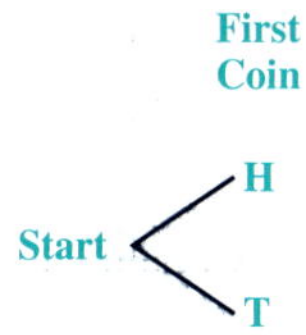


The possibilities for the second coin can be shown as branches from each of the first two branches:

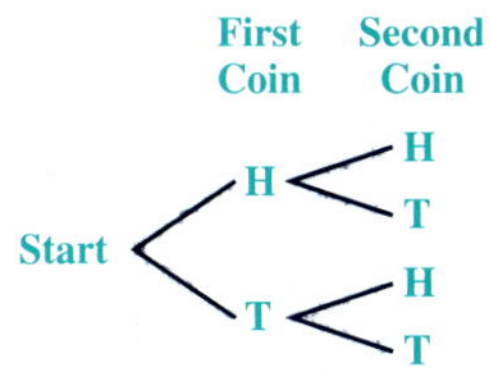


**1** Introduce the term *tree diagram.*

After finishing the discussion of how to create the tree diagram for the coin tosses, have students work on Problem Set B.

1 **Problem Set B** **Suggested Grouping: Pairs**
In this problem set, students make tree diagrams for situations that involve two successive choices. They use the tree diagrams to help them find theoretical probabilities related to each situation.

2 For **Problem 1,** students will probably have no difficulty understanding that there should be six branches for the first choice. You may want to suggest that students use the first letter of each color name (*R* for red, *O* for orange, and so on) instead of writing the full name of each color.

For **Problem 2d,** students should see from their tree diagrams that when the first branch is for red, there is only one second branch that corresponds to green. Likewise, when the first branch is green, only one second branch corresponds to red. They also need to see that the second branches are in six groups of five branches each, for a total of $6 \times 5$, or 30 second branches. Therefore, there are 30 possible two-color combinations, of which only 2 are red-green. This means that the probability of choosing red and green (in either order) is $\frac{2}{30}$, or $\frac{1}{15}$.

3 For **Problem 3,** students need to understand that using color names and numbers makes it possible to distinguish all the possible outcomes when the socks are selected. If they wish, they may abbreviate the labels. For example, they might use B1 and B2 for the black socks, and b1 and b2 for the brown socks.

For **Problem 4,** be sure students can show you exactly how the answer is related to the tree diagram.

For **Problem 5,** students should see that the answer can be found by using the answer for Problem 4.

4 **Problem Set Wrap-Up** Students may find it helpful to hear each other's explanations for solving the problems in this set. Bring the class together, and encourage a few students to briefly explain their ideas.

You can read off the possible outcomes by following the branches, beginning with Start. For example, following the top set of branches gives the outcome HH.

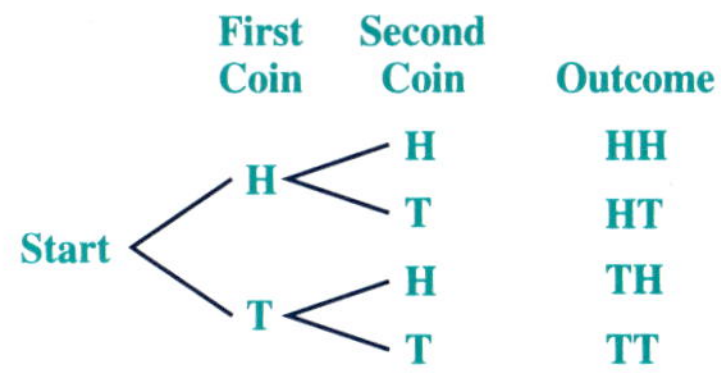


## Problem Set B

1. Suppose you choose one of six markers from a bag: red, orange, yellow, green, blue, or purple. Draw a tree diagram showing the possible colors for the marker. See the answer to Problem 2c.

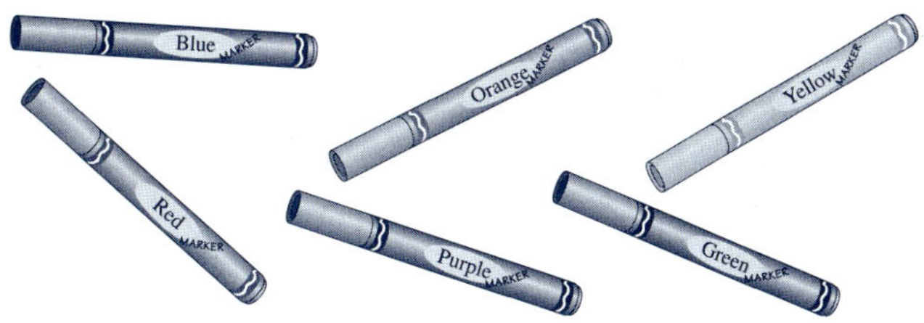


2. Now consider what happens when you choose a second marker.
   a. Suppose the first marker you chose was red. What are the possible choices for the second marker?
   b. Add a new set of branches to the "red" branch of your tree diagram for Problem 1, showing the possibilities for your second choice. See the answer to Problem 2c.
   c. Complete your tree diagram by adding branches that show the possibilities when each of the other colors are picked first.
   d. What is the probability that, if you choose the two colors at random, you will get red and green (chosen in either order)? $\frac{1}{15}$
3. Draw a tree diagram to show the possible choices of sock colors for Ken if he chooses two socks from a drawer containing two brown and two black socks. Since there are two socks of each color, label the socks as brown 1, brown 2, black 1, and black 2.
4. How many possible sock pairs are there? How many of them have matching colors? 12, 4
5. What is the probability that Ken will choose a matching pair? $\frac{1}{3}$

**2a.** orange, yellow, green, blue, and purple

**2c.** See Additional Answers.

**3.** See Additional Answers.

**Just the facts**

The knitting machine was invented in 1589 by William Lee of England. Queen Elizabeth I refused to give Lee a patent for his machine because she felt the stockings it produced were too coarse. (It may also be that she didn't want to put people who knit by hand out of business.)

**1** Problem-Solving Strategy
Draw a diagram

**2** Show students that their should be 6 branches for the first choice.

**3** Encourage students to abbreviate the labels.

**4** Discuss students' answers.



• *Additional Answers on page A673*

LESSON 10.3 Making Matches 641

## 1 Problem Set C Suggested Grouping: Pairs

In this problem set, students use logical reasoning or draw tree diagrams to solve problems about situations involving dependent choices.

Students can use the following tree diagram to help solve **Problem 4.**

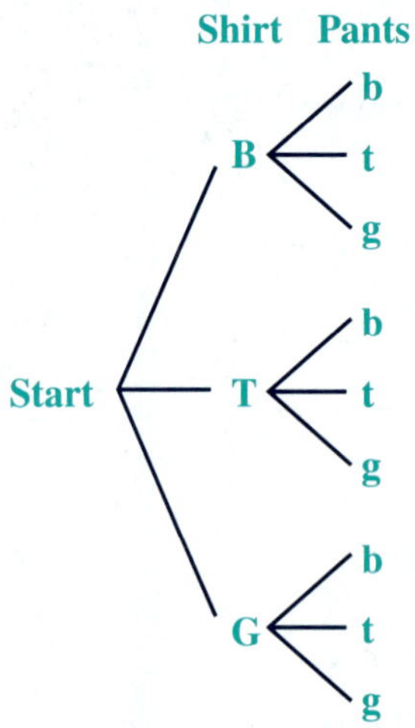


For **Problem 5,** students can use the following tree diagram to help solve the problem. Note that B1 and B2 are used to distinguish the two blue shirts, and b1 and b2 are used to distinguish the two pairs of blue pants.

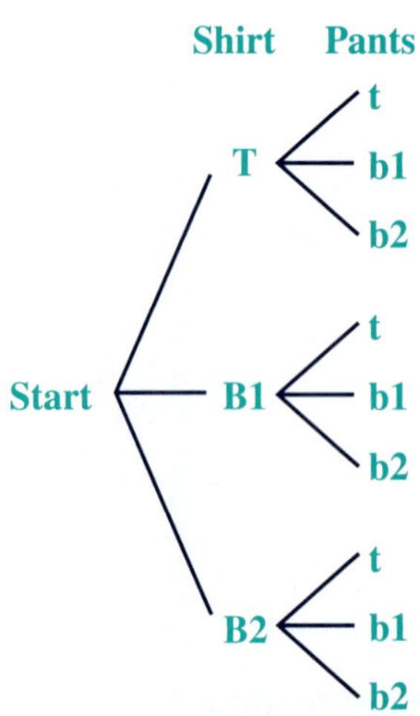


## 2 Share & Summarize

You may wish to ask students to volunteer their answers to these questions to the class. Discuss the answers, and correct any mistaken thinking.

**Troubleshooting** If students have difficulty understanding when one choice affects another, you may wish to use simple simulations to show the effect (or lack of effect) that a first choice can have on a second choice. You can use simulation procedures similar to those in Problem Set A and vary the situations by referring to Problem Sets B and C.

### On Your Own Exercises

Practice & Apply: 1–3, p. 646
Connect & Extend: 6–8, pp. 648–649
Mixed Review: 12–23, p. 651

★ indicates multi-step problem

1. $\frac{1}{2}$; there are two pairs of pants, one tan and one blue, so the probability of getting blue pants is $\frac{1}{2}$.

3. yes; Possible explanation: There are four possible combinations, and in two of them the shirt and pants match.

4. $\frac{1}{3}$; Possible explanation: Whatever shirt he chooses, one of the three pairs of pants is the same color.

5. $\frac{5}{9}$; Possible explanation: There are nine possible combinations of shirts and pants. Of these, four consist of a blue shirt and blue pants, and one consists of a tan shirt and tan pants. So, in five of the nine combinations, the shirt and pants match.

2.

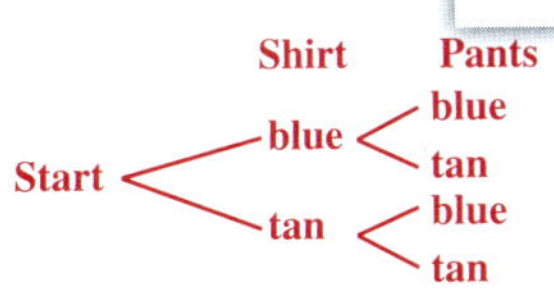


## Problem Set C

After choosing socks, Ken has to choose pants and a shirt. His school requires uniforms of blue, tan, or green. He has two pairs of pants, one blue and one tan. He has two shirts, also one blue and one tan. Now he takes one shirt and one pair of pants.

1. Suppose the shirt is blue. What is the probability that the pants will also be blue? Explain how you found your answer.
2. Draw a tree diagram showing the possible choices of shirts and pants. See below.
3. Ken says the probability that he will choose matching shirt and pants is 50%. Is he right? How do you know?
4. ★ Suppose Ken has a third shirt and a third pair of pants, both green. Now what is the probability that he will chose a shirt and pants of the same color? Explain.
5. ★ Suppose Ken has one tan shirt and two blue shirts, and one tan pair of pants and two blue pairs of pants. Find the probability that the shirt and pants will match, and show how you found your answer.

## Share & Summarize

1. When Ken is choosing socks in the dark, does the color of the first sock affect the chance of choosing a particular color next? Why or why not?
2. When Ken is choosing shirts and pants in the dark, does the color of the shirt chosen affect the chance of choosing pants of a particular color? Why or why not? No; when he chooses a shirt, the collection of pants is unchanged.

1. Yes; whatever color the first sock is, there are fewer socks of that color left in the drawer, so that color has a smaller chance of being chosen second.

**1 Problem-Solving Strategies**

- Use logical reasoning
- Draw a diagram

**2 Problem-Solving Strategy**

Solve a simpler problem

# Investigation 2

1 In this investigation, students extend the ideas of Investigation 1 to situations that involve more numerous possibilities for the choices. They analyze card games that involve choosing cards from a standard deck (without actually playing the games). There can often be many possibilities in such situations, and students will see that using logical reasoning is often advantageous in solving probability problems about such situations.

2 **Problem Set D** **Suggested Grouping: Pairs**
In this problem set, students analyze a game in which players score a point depending on whether they draw cards from the same or different suits of a standard deck of cards.

3 You may want to read the rules with the class to be sure all students understand how the game is played.

4 For **Problem 2,** students need to understand the importance of replacing the card after the first draw. You may want to ask students what the probability would have been if Player 1 had not replaced his or her card after the first draw. $\frac{12}{51}$, *or* $\frac{4}{17}$

5 **Problem Set Wrap-Up** Ask students how they got their answers for **Problem 4.** The main point for them to understand is that Player 2 has three times the chance of scoring as Player 1 has.

You may also want to have students explain the thinking that led them to their answers for **Problem 5.**

# Investigation 2 Matching Cards

**Just the facts**

The cards described here are English playing cards. In other countries, different suits are used. Traditional German playing cards use the suits hearts, leaves, bells, and acorns. Spanish playing cards use the suits coins, cups, swords, and clubs.

In an ordinary deck of playing cards, there are four suits:

Clubs Diamonds Spades Hearts

There are 13 cards in each suit, one for each of the numbers from 1 to 10, and three *face cards:* jack, queen, and king. Clubs and spades are black while diamonds and hearts are red.

Many kinds of games—involving various combinations of chance and skill—are played with decks of cards. In this investigation, you will work with some simple games of chance that involve choosing cards from a deck.

1 **Problem-Solving Strategy** Use logical reasoning

## Problem Set D

2 You may have students work in pairs.

In the first game, a deck of cards is shuffled and placed on a table face down. For one round of the game, players do the following:

3 Go over the rules of the game with the class.

- Player 1 chooses a card from the deck without looking and writes down its suit (spades, hearts, diamonds, or clubs).
- Player 1 puts the card back and shuffles the deck.
- Player 2 chooses a card without looking. If it has the same suit as the first card, Player 1 scores a point. Otherwise, Player 2 scores a point.
- Player 2 returns the card and shuffles the deck.

The winner is the player with more points at the end of 20 rounds.

1. What is the probability that Player 1 will choose a heart? $\frac{1}{4}$
2. If Player 1 chooses a heart, what is the probability that Player 2 will also choose a heart? Explain.
3. What is the probability that Player 2 chooses a card of the same suit as Player 1's card, no matter what that suit was? How do you know?
4. What would you expect the score to be after 20 rounds? See below.
5. Think of a way to change the scoring rules to give both players the same chance of winning. Possible answer: Player 1 earns 3 points each time the suits match, and Player 2 earns 1 point each time they don't match.

4 Explain the importance of replacing the card after the first draw.

5 Tell students that Player 2 has a three times better chance of scoring as Player 1.

**2.** $\frac{1}{4}$; since Player 1 put the card back, the deck has the same cards it had before, and Player 2 has the same chance as Player 1 to get a heart.

**3.** $\frac{1}{4}$; whatever suit Player 1 chose, there are 13 cards of that suit out of a total of 52 for Player 2 to choose from.

**4.** Player 1: 5 points; Player 2: 15 points



LESSON 10.3 Making Matches 643

# Develop

1 **Problem Set E** **Suggested Grouping: Pairs**
In this problem set, students study a second game in which the players draw a card and name the suit of the card. In this game, however, Player 1 does not replace his or her card before Player 2 draws.

2

For **Problem 5,** ask for volunteers to read their answers to the class. Have the class examine the reasoning carefully. The answer given in the margin is valid, but it may not occur to students to use numbers other than whole numbers when awarding points.

3

**Problem Set Wrap-Up** You may want to briefly review the answers before having students continue to Problem Set F.

4 **Problem Set F** **Suggested Grouping: Pairs**
In this problem set, students think about the total number of outcomes for the game in Problem Set D. They also consider what is involved in drawing tree diagrams for the first and second card games.

In **Problems 1–3,** the tree diagram is assumed to have enough branches to account for the identities of the individual cards, not just the suits of the cards.

## Problem Set E

The second card game is similar to the first. The only difference is that Player 1 does not put the card back before Player 2 chooses. After both players have chosen, the cards are returned to the deck. Player 1 scores a point if the two cards have the same suit, and Player 2 scores a point if they have different suits.

**1.** What is the probability that Player 1 will choose a heart? $\frac{1}{4}$

**2.** Suppose Player 1 chooses a heart.

  **a.** What is the probability that Player 2 also chooses a heart? $\frac{12}{51}$

  **b.** Is your answer to Part a different from your answer to Problem 2 of Problem Set D? Why or why not?

**3.** What is the probability that Player 2 chooses a card of the same suit as Player 1, no matter what suit Player 1 chooses? $\frac{12}{51}$

**4.** Is this game *more fair, less fair,* or *just as fair* as the game in Problem Set D? Explain. See below.

**5.** **Challenge** Find a way to assign points so that the game will be fair. Explain how you devised your point system. See below.

## Problem Set F

Suppose you want to draw a tree diagram to show the possible choices for the first card game, in which Player 1 replaces the card before Player 2 chooses.

**1.** How many branches would you need to show the possibilities for the first card? 52

**2.** How many branches would you have to add to show the possibilities for the second card? 52

**3.** How many total branches would your tree diagram have? $52 \cdot 52$, or 2,704

As you have probably realized, this tree diagram would be very large! Since the game concerns only the suits of the cards, and since the four suits are equally likely for each draw, you can draw a simplified tree diagram showing the four possible suits for each draw.

**Just the facts**

In Italy, Germany, Spain, Switzerland, and other countries, traditional decks of playing cards do not contain queens.

**2b.** Yes; since Player 1 does not return the card, the deck is not the same as it was. There are 51 cards left, and 12 of these are hearts.

**Problem Set E Answers**

**4.** Less fair; in this game, Player 1 has only a $\frac{12}{51}$ chance of scoring a point, which is less than the $\frac{1}{4}$ chance in the first game.

**5.** Possible answer: Player 1 has a probability of $\frac{12}{51}$ of scoring in a round and Player 2 has a probability of $\frac{39}{51}$. 39 is $12 \times 3.25$. If Player 1 earns 3.25 points each time the suits match and Player 2 earns 1 point each time they don't match, the players have the same chance of winning.

**1** You may have students work in pairs.

**2** Tell students that they will not replace their card after the first draw.

**3** Challenge students to use decimals or fractions when awarding points.

**4** You may have students work in pairs.

1

### Share & Summarize

You may want to call on volunteers to read their answers to the class.

**Troubleshooting** If students are having difficulty understanding situations in which one choice can affect a second choice, it may help to consider some additional situations. For example, write the numbers 1, 2, 3, 4, 6, and 9 on separate slips of paper. Have students consider drawing two slips, one after another, and ask such questions as the following:

> For the first draw, what is the probability of drawing an even number? $\frac{1}{2}$
>
> If you draw an even number on the first draw and do not replace the slip, what is the probability of drawing an even number on the second draw? $\frac{2}{5}$
>
> Why are the probabilities different? *There are fewer even numbers and fewer total slips for the second draw.*

### On Your Own Exercises

Practice & Apply: 4, 5, p. 647
Connect & Extend: 9–11, p. 650
Mixed Review: 12–23, p. 651

For example, suppose the first card chosen is a heart. Here is the part of the tree diagram showing the possible suits for the second card:

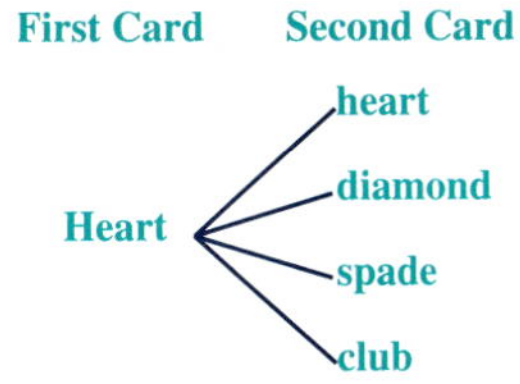


**4.** Draw a tree diagram showing all the possible suit combinations for the first game. See Additional Answers.

**5.** Hearts and diamonds are red while clubs and spades are black. What is the probability that the two cards have the same color? $\frac{1}{2}$

**6.** Can you use a simplified tree diagram for the second game, in which Player 1 keeps the card instead of returning it to the deck before Player 2 chooses? Explain.

**6.** No; for the second card, the four suits are not equally likely.

**Share & Summarize Answers**

**1a.** no

**1b.** Yes; when one king is removed from a deck, there are fewer kings left.

**1c.** no

**2.** Possible answer: pulling a numbered slip out of a hat and getting an even number, and then pulling out another slip and getting an even number

## Share & Summarize

In some probability situations, one event can affect the probability of another.

**1.** For each pair of events, decide whether the first event affects the probability of the second. If your answer is "yes," explain why.

| | First Event | Second Event |
|---|---|---|
| **a.** | getting heads on the flip of a coin | getting heads on the second flip of the coin |
| **b.** | getting a king when choosing a card from a deck | getting a king when choosing a second card without returning the first |
| **c.** | drawing a certain name from names written on slips of paper and chosen at random from a hat | drawing a second name if the first slip is returned to the hat before the second choice |

**2.** Make up your own sequence of two events for which the first event affects the probability of the second.

**3.** Now make up your own sequence of two events for which the first event *does not* affect the probability of the second.

Possible answer: rolling a die and getting an even number, and then rolling the die again and getting an even number

**1** Discuss students' answers.



• Additional Answers on page A673

LESSON 10.3 Making Matches 645

# On Your Own Exercises

## On Your Own Exercises

Investigation 1, pp. 639–642
Practice & Apply: 1–3
Connect & Extend: 6–8

Investigation 2, pp. 643–645
Practice & Apply: 4, 5
Connect & Extend: 9–11

Assign Anytime
Mixed Review: 12–23

**Exercise 3b:**
The tree diagram shows not only which marbles can be selected but also the order in which they can be selected. You may wish to see whether students can explain how reasoning can be used to predict the total number of possible outcomes, independently of the tree diagram. Some students may see that there are six possibilities for the first choice and that, for each of these, there are five possibilities for the second choice. The total number of possibilities is therefore $6 \times 5$, or 30.

## Practice & Apply

**1b.** $\frac{1}{2}$; from Part a, the probability is $\frac{1}{2}$ that the first card is a 2. If it is, the second card must be a 1.

**2a.** 123, 132, 213, 231, 312, 321

**2d.** $\frac{2}{6}$, or $\frac{1}{3}$

**3b.**

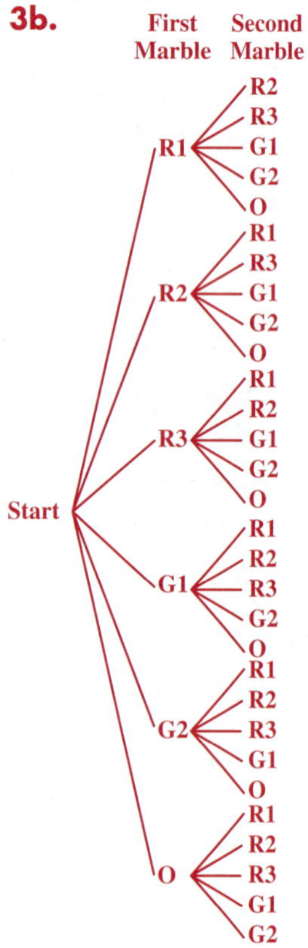


**1.** Suppose there are two cards, numbered 1 and 2. The cards are mixed, and placed face down.

**a.** You arrange the cards in a row and then turn them over. What is the probability that the first card will be a 2? $\frac{1}{2}$

**b.** What is the probability that the cards will form the number 21? Explain.

**2.** Suppose you have three cards, numbered 1, 2, and 3. The cards are shuffled and placed face down in a row.

**a.** List all the three-digit numbers that can be created from these three cards.

**b.** What is the probability that the cards will form the number 213 when they are turned over? $\frac{1}{6}$

**c.** What is the probability that the cards will form a number between 200 and 300? $\frac{2}{6}$, or $\frac{1}{3}$

**d.** What is the probability that the cards will form an even number?

**e.** What is the probability that the cards will form a number less than 300? $\frac{4}{6}$, or $\frac{2}{3}$

**3.** Manuel and Leila are splitting a box of marbles. The box contains three red, two green, and one orange marble. Each friend chooses a marble at random.

**a.** If Leila chooses first and gets a red marble, what is the probability that Manuel's marble will also be red? $\frac{2}{5}$

**b.** Draw a tree diagram showing all the possible combinations when each friend chooses one marble. Label the red marbles R1, R2, and R3; the green marbles G1 and G2; and the orange marble O.

**c.** What is the probability that the two marbles will be the same color? $\frac{8}{30}$, or $\frac{4}{15}$



impactmath.com/self_check_quiz

**Quick Review**
**Math Handbook**
hot words hot topics
Hot Topics pp. 210–217

**4.** Jahmal and Hannah are playing a game with an ordinary deck of playing cards. For each turn, Hannah chooses a card and returns it to the deck. She shuffles the deck, and then Jahmal chooses a card.

**a.** If Hannah picks the 5 of clubs, what is the probability that Jahmal will pick the 5 of clubs? $\frac{1}{52}$

**b.** If Hannah picks a black card (either a spade or a club), what is the probability that Jahmal will pick a red card (either a heart or a diamond)? How do you know? $\frac{1}{2}$; half the cards are red.

**c.** If Hannah picks the 6 of spades, what is the probability that Jahmal will pick a king? $\frac{1}{13}$

**d.** If Hannah picks a red queen, what is the probability that Jahmal will pick a red queen? $\frac{1}{26}$

**5.** Jahmal and Hannah are playing a game with an ordinary deck of playing cards. For each turn, the deck is shuffled and the cards are spread out face down. At the same time, Hannah and Jahmal each choose a card.

**a.** If Hannah picks the 5 of clubs, what is the probability that Jahmal also picks the 5 of clubs? 0

**b.** If Hannah picks a black card (either a spade or a club), what is the probability that Jahmal picks a red card (either a heart or a diamond)? How do you know?

**5b.** $\frac{26}{51}$; if one black card is removed, there are 26 red cards out of 51.

**c.** If Hannah picks the 6 of spades, what is the probability that Jahmal picks a king? $\frac{4}{51}$

**d.** If Hannah picks a red queen, what is the probability that Jahmal picks a red queen? $\frac{1}{51}$



LESSON 10.3 Making Matches 647

**Exercise 6c:**
Ask students how they arrived at the answer for this part of the exercise.

**Exercise 7:**
If students need a hint to get started, you might ask them how this exercise is similar to Exercise 6.

## Connect & Extend

**6.** Ramesh has created a game using these six cards:

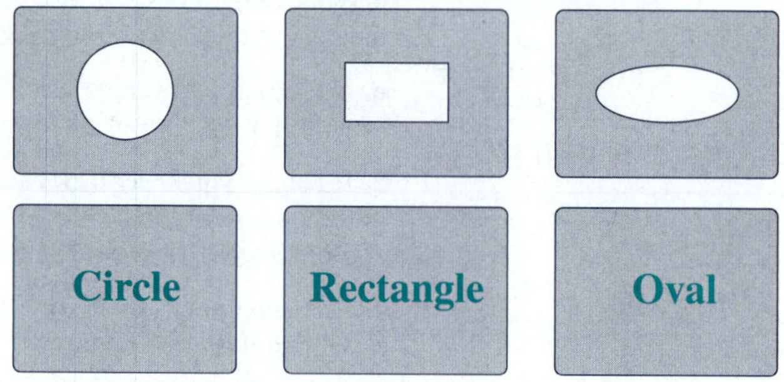


The three shape cards are placed face up. The three word cards are shuffled, and one word card is placed face down next to each shape card. A player scores 1 point for each word card that matches a shape card.

**a.** Write all the possible arrangements of the three word cards. Use C to stand for the circle card, R to stand for the rectangle card, and O to stand for the oval card. How many possibilities are there?

**6a.** CRO, COR, OCR, ORC, RCO, ROC; 6

**b.** What is the probability that a player will match all three word cards correctly? $\frac{1}{6}$

**c.** What is the probability that a player will match at least one word card correctly? $\frac{4}{6}$, or $\frac{2}{3}$

**7.** Shaunda has written letters to four friends: Caroline, Raul, Jing, and Ernest. She has four envelopes, each with the name and address of one of the friends. Shaunda's little brother wants to help, so he puts one letter in each envelope. Since he can't read yet, he puts the letters in the envelopes at random.

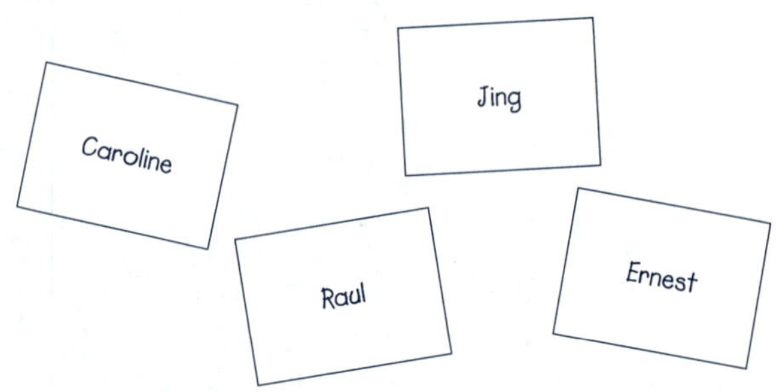


**a.** How many ways can the letters and envelopes be paired? 24

**b.** What is the probability that everyone receives the right letter? $\frac{1}{24}$

8. **Life Science** A person's *genes* determine many things about that person, including how he or she looks. For example, a person has two genes that determine eye color. The gene for blue eyes is *recessive* and the gene for brown eyes is *dominant.* This means that if a person has one blue-eye gene and one brown-eye gene, he or she has brown eyes. A person with two brown-eye genes also has brown eyes. To have blue eyes, both genes must be blue.

   A child gets one eye-color gene from each parent. Assume the chances of passing either gene to a child are equal. For example, a father with one blue-eye gene and one brown-eye gene has a 50% chance of passing the blue-eye gene to his child.

   **a.** Suppose two people are having a child. One has a blue-eye gene and a brown-eye gene, and the other has two brown-eye genes. What is the probability that the child will have blue eyes? (Hint: You can find the possible gene combinations for the child by making a table or a tree diagram.) 0

   **b.** Suppose the two parents both have one blue-eye gene and one brown-eye gene. What is the probability that the child will have blue eyes? $\frac{1}{4}$

   **c.** Now suppose one of the parents has two blue-eye genes and the other has one blue-eye gene and one brown-eye gene. What is the probability that the child will have blue eyes? $\frac{1}{2}$

**Exercise 8:**
When you discuss **Part a** as a class, ask for a volunteer to go to the board to show how a table or tree diagram for this situation would look.

You might also ask students how a table or tree diagram can help in answering **Parts b and c.**



LESSON 10.3 Making Matches **649**

**Assign** and **Assess**

**Exercise 9b:**
If students are having difficulty with this exercise, ask them to describe the cards that have a value greater than or equal to the value of the 6 of diamonds. Also ask whether Maria's choice affects David's.

**Exercise 11e:**
Ask students to explain why the probability can be found by calculating $\frac{4}{52} \times \frac{1}{51}$.

**In your own words**

*Describe a situation in which one event affects the probability of another.*

**9.** Maria and David are playing a game with an ordinary deck of playing cards. For each turn, the deck is shuffled and placed face down. Maria chooses a card and records it. She returns the card to the deck and shuffles the cards, and then David chooses a card. The player with the higher card scores a point. (Aces are the lowest cards, and kings are the highest.) If the cards have the same value, neither player scores.

**a.** Suppose Maria chooses a king. What is the probability that she will score a point? $\frac{48}{52}$, or $\frac{12}{13}$

**b.** Suppose Maria chooses the 6 of diamonds. What is the probability that she will *not* score a point? $\frac{32}{52}$, or $\frac{8}{13}$

**10.** Maria and David are playing a game with an ordinary deck of playing cards. For each turn, the deck is shuffled and placed face down. David chooses a card and keeps it. Then Maria chooses a card. The player with the higher card earns a point. (Aces are the lowest cards, and kings are the highest.) If the cards have the same value, neither player scores.

**a.** Suppose David picks a king. What is the probability that he will score a point? $\frac{48}{51}$, or $\frac{16}{17}$

**b.** Suppose David picks the 6 of diamonds. What is the probability that he will *not* score a point? $\frac{31}{51}$

**11.** Althea shuffled a standard deck of playing cards and turned over the first two cards.

**a.** What is the probability that the first card was an ace? $\frac{1}{13}$

**b.** Suppose the first card was an ace. What is the probability that the second card was a 2 with the same suit as the ace? $\frac{1}{51}$

**c.** How many possible combinations of two cards are there in a standard deck?

**11c.** $52 \cdot 51$, or 2,652

**d.** How many of those combinations include an ace and then a 2 (in that order) of the same suit? 4

**e.** What is the probability of getting an ace and then a 2 of the same suit? $\frac{4}{2{,}652}$, or $\frac{1}{663}$

**f.** What is the probability of getting an ace and a 2 in *any* order, but of the same suit? $\frac{8}{2{,}652}$, or $\frac{2}{663}$



650 CHAPTER 10 Understanding Probability

★ indicates multi-step problem

## Mixed Review

Find the value of each expression in simplest form.

**12.** $\frac{5}{6} + \frac{4}{9}$ $1\frac{5}{18}$ **13.** $\frac{19}{26} - \frac{17}{39}$ $\frac{23}{78}$ **14.** $\frac{45}{56} \cdot \frac{32}{35}$ $\frac{36}{49}$

**15.** $\frac{14}{15} \div \frac{2}{5}$ $2\frac{1}{3}$ **16.** $11\frac{19}{21} + 6\frac{1}{7}$ $18\frac{1}{21}$ **17.** $5\frac{1}{4} - 2\frac{5}{12}$ $2\frac{5}{6}$

**18.** $3\frac{5}{8} \cdot 1\frac{3}{4}$ $6\frac{11}{32}$ **19.** $9\frac{1}{3} \div \frac{5}{9}$ $16\frac{4}{5}$ **20.** $\frac{11}{14} \div 1\frac{3}{14}$ $\frac{11}{17}$

**21.** 12 ft by 4 ft

★**21.** **Geometry** A rectangle has an area of 48 square feet and a perimeter of 32 feet. What are the dimensions of the rectangle?

**22.** Which has the greater area: a circle with diameter 11 meters, or a square with side length 10 meters? the square

**23.** **Economics** The Book Bin is having a clearance sale.

**a.** All dictionaries are marked $33\frac{1}{3}\%$ off. Ramesh bought a French dictionary with a sale price of $18. What was the dictionary's original price? $27

**b.** Novels are on sale for 20% off. Althea bought a novel with an original price of $11.95. What was the sale price? $9.56

**c.** Travel books are all marked down by a certain percent. Miguel bought a book about African safaris. The book was originally priced at $27.50, but Miguel paid only $16.50. What percent did Miguel save? 40%

### Quick Check

**Informal Assessment**
Students should be able to:

- ✔ make tree diagrams to show outcomes for situations involving independent choices
- ✔ use tree diagrams to compute probabilities in situations that involve dependent choices

• *Quick Quiz on page A674*

# Chapter 10 Review

## Chapter Summary

This summary helps students recall the major topics of the chapter.

## Vocabulary

Students should be able to explain each of the terms listed in the vocabulary section.

## Problem-Solving Strategies and Applications

The questions in this section help students review and apply the important mathematical ideas and problem-solving strategies developed in the chapter. The questions are organized by mathematical highlights. The highlights correspond to those in "The Big Picture" chart on page 601a.

**Question 2c:** Leah calculated the experimental probability correctly. However, Althea's theoretical probability is more accurate in the sense that the theoretical probability is better than the experimental probability for making predictions.

# Chapter 10 Review & Self-Assessment

VOCABULARY
**equally likely**
**experimental probability**
**probability**
**simulation**
**theoretical probability**

## Chapter Summary

Probability is useful in many areas of life, from playing games to making plans based on weather predictions. In this chapter, you learned how to find *experimental* and *theoretical probabilities* for events in which the possible outcomes are *equally likely.*

You examined probabilities in several types of situations, including some in which the possible outcomes were not easy to determine. For certain games of chance, you came up with strategies for play based on your knowledge of probabilities. You also used *simulation* and tree diagrams to examine situations in which the number of possible outcomes were affected by what had happened before.

## Strategies and Applications

The questions in this section will help you review and apply the important ideas and strategies developed in this chapter.

### Understanding probability

Althea took the king of hearts and the king of clubs from a standard deck of cards, leaving only 50 cards. She told Leah that the probability of selecting a queen was now 8%, but she didn't tell her how many or what cards she had removed.

1. What does it mean that the probability was 8%?

**1.** Possible answers: 8% of the possible draws are queens; if you select a card several times (replacing the chosen card each time), you would expect to get a queen about 8% of the time.

2. Leah selected a card from Althea's deck, looked at it, and then put it back. Althea shuffled the cards. They repeated this process until Leah had chosen a card 100 times.
   a. How many times would you expect Leah to have picked a queen?
   b. Leah chose a queen 7 times. She said that this means Althea was wrong and that the actual probability is 7%. Althea and Leah both calculated the probabilities they gave. Is either incorrect in her calculation? Explain.
   c. Whose is the more accurate probability, Leah's or Althea's? Althea's

**2a.** 8

**2b.** No; Althea gave a theoretical probability based on what she knew were the possible outcomes. Leah gave an experimental probability based on the actual outcome of several trials.



impactmath.com/chapter_test

**2d.** no; Possible explanation: The experimental probability for the 1,000 draws is 8.8%, which is closer to 8% than 7% is. Although 8 is greater than 1, 8 is a smaller portion of 1,000 than 1 is of 100.

**4.** Possible answer: An experimental probability is based on what happened in several trials, while a theoretical probability is based on what you expect to happen in the long run. For example, if you toss a coin 50 times and divide the number of heads you get by 50, that's an experimental probability. Because there are two equally likely possibilities when flipping a coin, you'd expect 25 of the 50 flips to be heads, so $\frac{25}{50}$, or $\frac{1}{2}$, is the theoretical probability.

**5a.** Because there are six equally likely possible outcomes: 1, 2, 3, 4, 5, and 6.

**5b.** Because three of the outcomes are prime: 2, 3, and 5.

**5c.** There are six squares. Assuming the rice is equally likely to land in each, there are six equally likely outcomes, just as in Josh's die roll. Since three of the squares are shaded, there are three possibilities that make the event "the rice lands in a shaded square" happen. Josh had three possibilities that make the event "getting a prime number" happen.

**d.** Leah kept selecting cards until she had 1,000 trials. She chose a queen 88 times. Althea said, "The difference between the 88 queens you selected and the 80 you should have expected was 8, but the difference was only 1 when you drew 100 cards. Your experimental probability will be less accurate for the 1,000 draws than it was for the 100 draws."

Is Althea correct? Explain.

## Identifying outcomes

**3.** Name two strategies for identifying the outcomes of a probability situation. Illustrate each strategy by using them to find the number of outcomes for spinning this spinner twice: See Additional Answers.

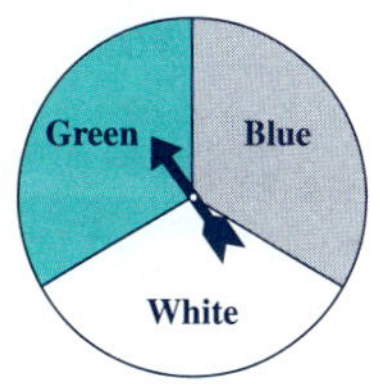


## Finding probabilities of events

**4.** Explain the difference between a theoretical probability and an experimental probability. You may want to illustrate with an example.

**5.** Josh said, "Suppose you roll a standard die. To calculate the probability of getting a prime number, you have to divide 3 by 6, giving 0.5."

**a.** Why did Josh choose 6 for the divisor?

**b.** Why did Josh choose 3 for the dividend?

**c.** Consider this *Rice Drop* game board. Explain why the procedure for calculating the probability that the rice lands in a shaded square is the same as the one Josh used for getting a prime number on a die roll.

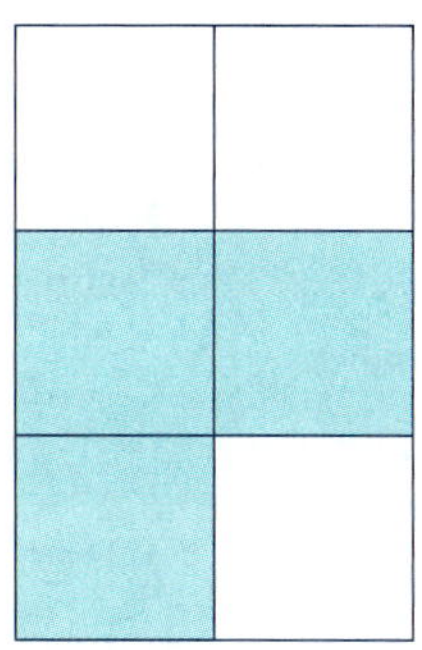



## Additional Answers

**3.** Possible answer: creating a table and making a tree diagram

| | | First Spin | | |
|---|---|---|---|---|
| | | **B** | **W** | **G** |
| **Second Spin** | **B** | BB | BW | BG |
| | **W** | WB | WW | WG |
| | **G** | GB | GW | GG |

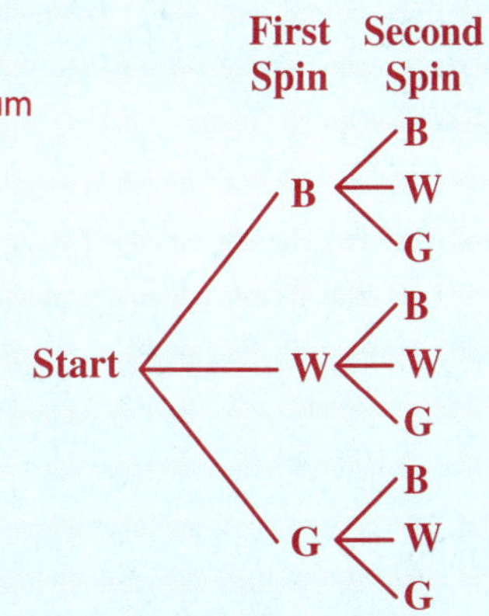


Review and Self-Assessment 653

**Question 9b:**
You may want to ask students whether the arrangement of the spaces on the wheel has any effect on the probability of winning a particular prize. Ask them how they know their answer is correct.

★ indicates multi-step problem

### Using probabilities to analyze games

★ **6.** A bag contains five slips of paper numbered 1 to 5. In the game *Find the Difference,* each player chooses one of the cards below. Players take turns drawing two numbers from the bag. If the difference of the numbers is on the player's card, the player covers that difference. The numbers are returned to the bag after each turn. The first player to cover all his numbers wins.

**Card A**

| 1 | 2 |
|---|---|
| 3 | 4 |

**Card B**

| 1 | 2 |
|---|---|
| 2 | 1 |

**Card C**

| 3 | 4 |
|---|---|
| 4 | 3 |

Which card gives a player the best chance of winning? Explain.
See Additional Answers.

### Working with situations in which the probabilities depend on previous results

**7.** Craig and Kenna were playing a board game in which they rolled two dice. Rolling doubles (that is, rolling the same number on both dice) lets you take an extra turn. Kenna rolled two 3s and then two 5s. As she was getting ready to take another extra turn, Craig said, "The chances of you getting doubles again are next to nothing!" Is Craig correct? Explain your answer.

**8.** Describe a probability experiment in which the result of one trial changes the probabilities for the next trial's result. You might want to use dice, cards, spinners, or slips of paper drawn from a bag in your experiment. Answers will vary.

**7.** no; Possible explanation: The dice don't "remember" what happened on previous rolls. Kenna's chances are the same for the next roll as they were for the previous rolls: $\frac{1}{6}$.

**9a.** key chain: $\frac{14}{200}$, or 0.07; troll doll: $\frac{16}{200}$, or 0.08; baseball cap: $\frac{13}{200}$, or 0.065; stuffed animal: $\frac{6}{200}$, or 0.03; beach ball: $\frac{3}{200}$, or 0.015.

**9b.** See Additional Answers.

### Demonstrating Skills

**9.** At a fund-raising carnival, Marcus operated a game in which each player spun a wheel. The section on which the wheel stopped would indicate what prize, if any, the player won.

The table shows how many equal-sized spaces listed each type of prize as well as how many people won each prize by the end of the day.

| | Key Chain | Troll Doll | Baseball Cap | Stuffed Animal | Beach Ball | No Prize |
|---|---|---|---|---|---|---|
| Number of Spaces | 5 | 4 | 3 | 2 | 1 | 45 |
| Number of Winners | 14 | 16 | 13 | 6 | 3 | 148 |

**a.** Find an experimental probability of winning each prize.

**b.** Find the theoretical probability of winning each prize.



### Additional Answers

**6.** Possible answer: This list shows the possible pairs and the difference for each pair:

| Difference | Pair |
|---|---|
| 1 | (1, 2), (2, 3), (3, 4), (4, 5), (2, 1), (3, 2), (4, 3), (5, 4) |
| 2 | (1, 3), (2, 4), (3, 5), (3, 1), (4, 2), (5, 3) |
| 3 | (1, 4), (2, 5), (4, 1), (5, 2) |
| 4 | (1, 5), (5, 1) |

Here is the probability of each difference: probability of 1: $\frac{8}{20}$; probability of 2: $\frac{6}{20}$; probability of 3: $\frac{4}{20}$; probability of 4: $\frac{2}{20}$. Since 1 and 2 are the most likely differences, Card B is the best choice.

**9b.** key chain: $\frac{5}{60}$, or about 0.083; troll doll: $\frac{4}{60}$, or about 0.067; baseball cap: $\frac{3}{60}$, or 0.05; stuffed animal: $\frac{2}{60}$, or about 0.033; beach ball: $\frac{1}{60}$, or about 0.017

Use this information for Questions 10 and 11:

At the beginning of a computer game called *Geometry Bug,* players take turns choosing circles, squares, and triangles on the screen. After all the shapes have been chosen, a small "bug" appears and flies over the shapes. The bug lands on a random place on the screen. If it lands on one of the players' shapes, that player scores a point. The winner is the player with the most points after 50 landings.

Rosa and Cari were playing with this screen. Rosa's shapes are green and Cari's are white. The screen is 8 inches wide and 5 inches high. The circles have a radius of $\frac{1}{2}$ inch. The squares have a side length of 1 inch. The triangles are right triangles with legs 1 inch long.

**Questions 10 and 11:** You may need to remind students that the legs of a right triangle are the sides that form the right angle. You might also remind them that they can use 3.14 as an approximate value for $\pi$.

**10.** Consider probabilities for a single bug landing. Write your answers as decimals to the nearest thousandth.

**a.** What is the probability that the bug lands in Cari's square? 0.025

**10b.** about 0.039

**b.** What is the probability that the bug lands in one of Rosa's circles?

**10c.** Rosa: about 0.114; Cari: about 0.102

**c.** Find each player's probability of scoring on a single bug landing.

**11.** The girls decide to play with an optional rule. When the bug lands on a shape, the shape is removed from the board. For example, if the bug lands on Cari's square, Cari scores a point but the square disappears.

**11a.** 0.0375

**a.** Find the probability that the bug lands on one of Cari's triangles.

**b.** Suppose the first shape the bug landed on was Cari's triangle in the bottom right. Now what is the probability that the bug lands on any of Cari's shapes? about 0.089

**12.** A group of six friends wanted to play a game that requires three teams. To decide which two people would play on each team, Luke put six cubes in a bag. Two of the cubes were white, two were black, and two were red. Each person took a cube without looking, and the two with the same color formed a team.

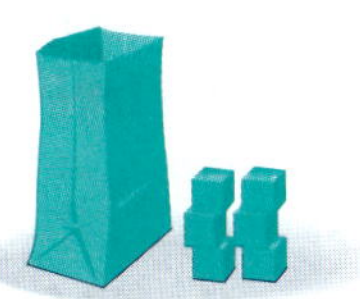

**12a.** See Additional Answers.

**a.** Draw a tree diagram to show the first *two* drawings of the cubes. To make the labeling easier, use W1 and W2 for the white cubes, B1 and B2 for the black cubes, and R1 and R2 for the red cubes.

**12b.** $\frac{6}{30}$, or $\frac{1}{5}$

**b.** Jing drew a cube first and then Jahmal. Use your tree diagram to find the probability that Jing and Jahmal will be on the same team.



**Additional Answers**

**12a.**

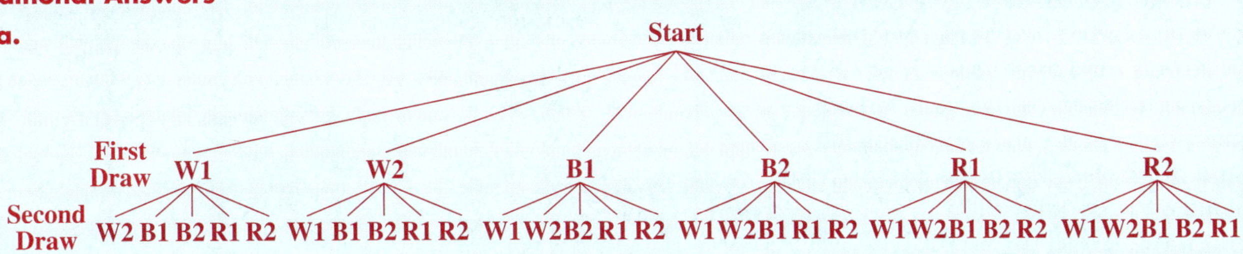


Review and Self-Assessment **655**

# Chapter 6 Mathematical Background

## Page 341a Notes, continued

**Measures of distribution** Mean, median, and mode can describe what is typical, but they do not reveal the distribution of a set of data. The measures maximum, minimum, and range all provide this information. The *maximum* is the greatest number in a data set, the *minimum* is the least number, and the *range* is calculated by subtracting the minimum from the maximum.

**Graphical representations** Measures of central tendency and distribution describe only some aspects of a data set. None of these measures reflects the shape of the data, which gives information about where most of the values in the data set are, where there are no values, and how many items have similar values. A graph represents data visually and can show something about the nature of the data between the maximum and minimum.

**Bar graphs** There are two basic types of bar graphs. One uses a bar to show the value of each item in the data set. This kind of display is used to compare data items. A second type of bar graph, a frequency bar graph, organizes data based on the frequency of each of the values. The values in the data set are put on the horizontal axis; and the frequency of each is indicated on the vertical axis.

More detailed bar graphs are the double bar graph and the stacked bar graph. These two displays graph data from more than one data set. For example, one data set may give the number of members in girls' favorite bands while the second data set gives the same information for boys. The data are graphed side by side in a double bar graph. The data are placed in one bar for a stacked bar graph, with a different color or pattern used in the bar to show what part of the data is related to each data set.

**Histograms** When values are broadly distributed or there are a large number of data values, the data can often be shown on a histogram. In a histogram, the data are grouped into intervals, and the frequency of values within each interval is displayed as adjacent vertical bars. The histogram is a type of frequency graph that shows how data are distributed, but at the cost of losing information about the original data set.

**Line plots** A simplified version of a frequency bar graph is a line plot. This graph has a horizontal axis, or number line, containing all the data values. Each piece of data is represented by an X placed above the corresponding value on the horizontal axis.

**Stem-and-leaf plot** A stem-and-leaf plot maintains information about the individual data items and provides a picture of the data. This graph is useful when there are many items in a data set.

To make a stem-and-leaf plot, look at the place values in each number. The least place value becomes the "leaf," and the remaining place value or values become the "stems."

A stem-and-leaf plot of this data set is shown below: 55, 43, 30, 48, 79, 38, 32, 79, 45, 35, 62, 55, 39, 47, 42, 80, 79.

| Stem | Leaf |
|---|---|
| 3 | 0 2 5 8 9 |
| 4 | 2 3 5 7 8 |
| 5 | 5 5 |
| 6 | 2 |
| 7 | 9 9 9 |
| 8 | 0 |

Key: 4|3 = 43

A stem-and-leaf plot makes it easy to see the range as well as bunches and gaps in the data. A stem-and-leaf plot is like a histogram in that it uses intervals to group the data but, unlike a histogram, it also keeps and displays the exact values of the individual data points.

# Lesson 6.1

## Page 345 Additional Answers

**2a.** Possible answer: The list could be the three cities whose temperatures the graphs show. Graph 1 might show temperatures for Miami because the temperatures are fairly high for all the months. Graph 3 might show temperatures for Fairbanks because the temperatures are very low in some months and never get very high. Graph 2 might show temperatures for Chicago because the temperatures are more moderate.

• continued on next page

**Miami Temperatures**

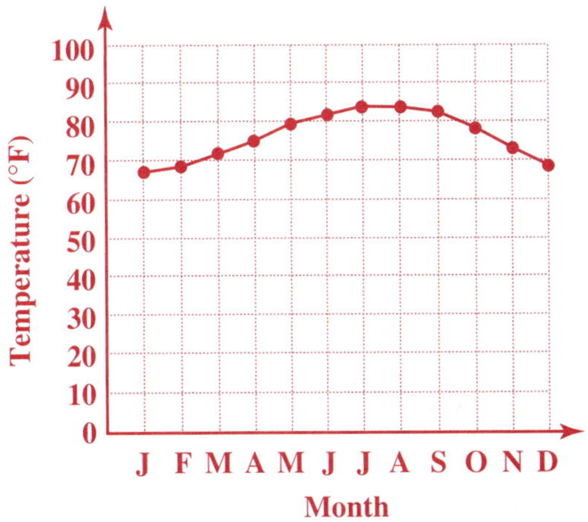


**Chicago Temperatures**

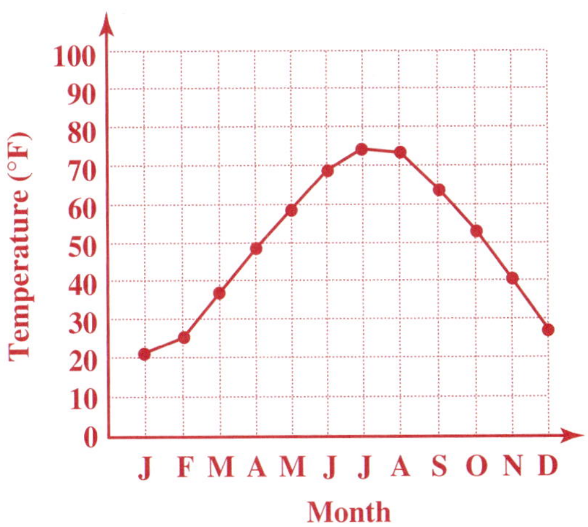


**Fairbanks Temperatures**

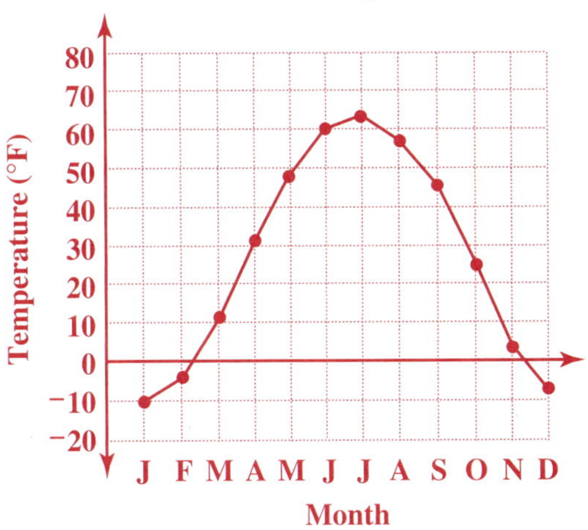


## *Page 349 Additional Answers*

### Problem Set D

**2.**

**Total Emissions**

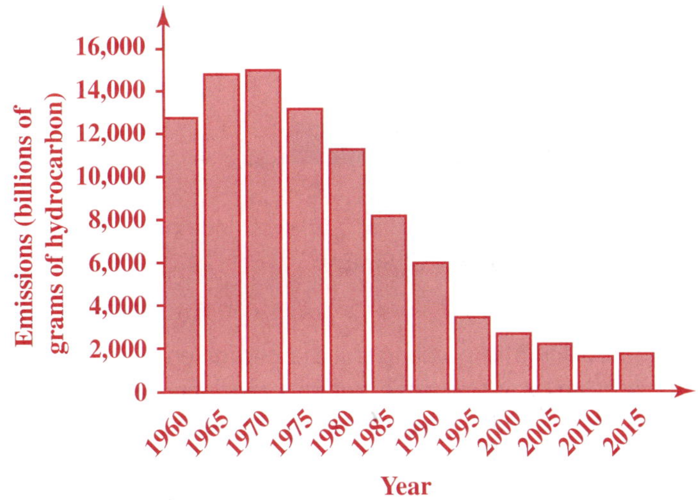


### Share & Summarize

Possible letter:

Dear Citizens for Safe Air:

We have analyzed information about hydrocarbon emissions in the United States from cars for the past several decades. We found that even though cars have been improved and now produce fewer hydrocarbons than in the 1960s and 1970s, cars are driven more. When you look at total hydrocarbon emissions, you can see an increase in the estimate from 2010 to 2015. We think the problem with hydrocarbons from cars may not be over. The United States needs to keep finding ways to reduce hydrocarbons from each car or to reduce the number of miles that people drive.

Sincerely,

Data Inc.

## *Page T350 Notes, continued*

**Problem Set Wrap-Up** Have students share their answers to **Problems 1–4**. You might also have them discuss how they solved each problem. Use students' responses to lead into the discussion of the histogram. ■

4 Introduce the term *histogram.* You might want to provide students with an example of a histogram, such as the one that follows, so they can see that the data are organized in intervals and that no space is used between bars when there are data in consecutive intervals.

• *continued on next page*

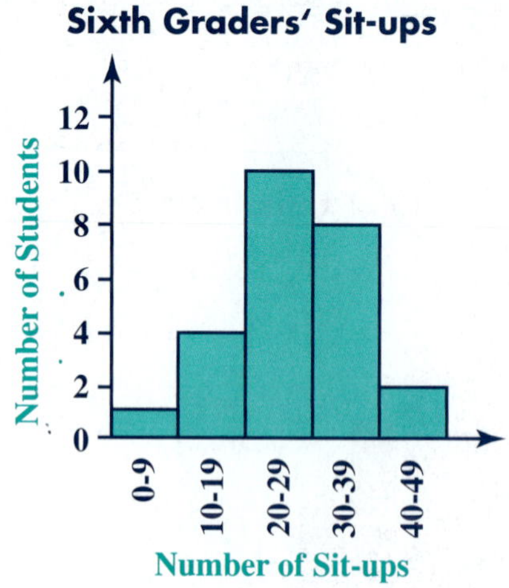


In addition to pointing out the characteristics of a histogram, you can have students read the histogram. Check students' understanding by asking such questions as,

- How many students did fewer than 10 sit-ups? *one*
- How many students did 30–39 situps? *eight*

## Page 352 Additional Answers

### Share & Summarize

1. Possible answer: A histogram shows the frequency or percent of data values in a particular interval. You could use a histogram to show the number of CDs at a music store that fall in various price ranges or to show the number of phone calls a help line receives at different times of the day.
2. Possible answer: The bar graph shows the country each skier was from and the time for each individual skier. The bar graph allows you to easily compare times of individual skiers. The histogram does not show individual times or the countries the skiers were from, but it does show the number of times in each interval and gives a picture of how the data values are distributed.

## Page 354 Additional Answers, continued

**3c.**

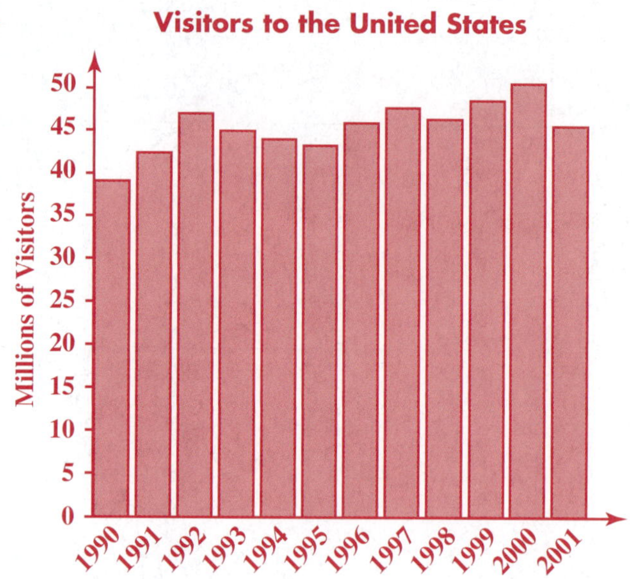


## Page 356 Additional Answers, continued

**5f.**

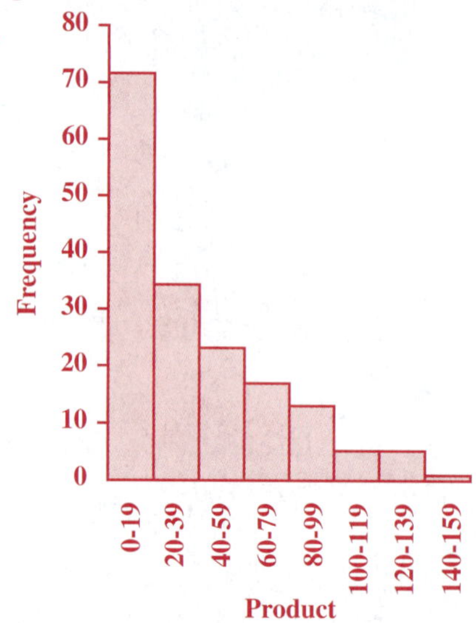


## Page 357 Additional Answers, continued

**7a.**

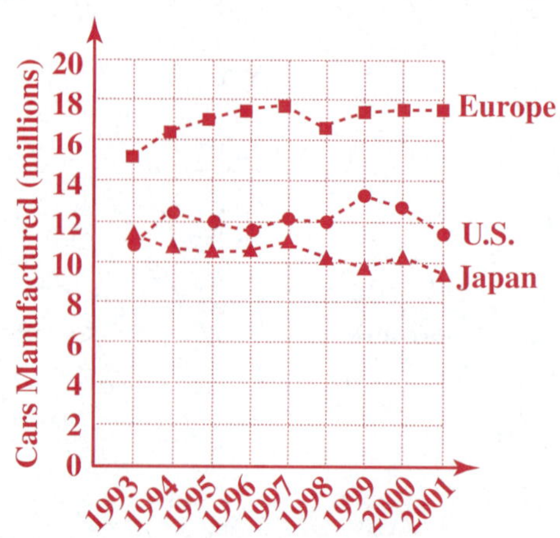

## Page 361 Quick Quiz, continued

2. Explain the difference between a histogram and a bar graph. Possible answer: A bar graph often shows categorical data for individual items in a data set. A histogram shows the frequencies for categorical data given in ranges of values.

3. The bar graph shown does not have a title. Which of the following might be a title for this bar graph? Explain your answer.

   a. Points Scored by Players

   b. Temperature of Boiling Water as It Cools

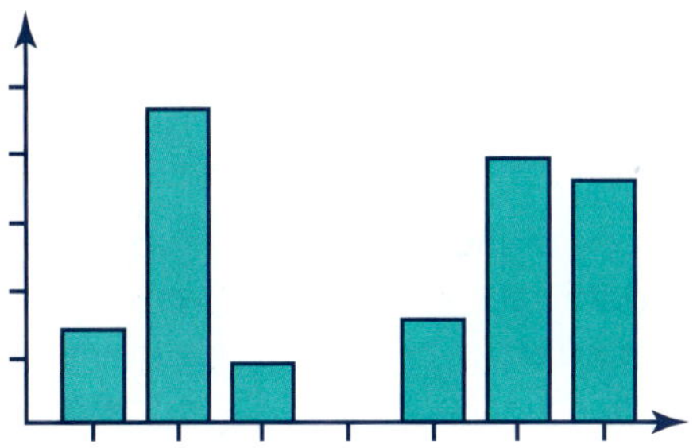

   a. Points Scored by Players. Temperature would probably be shown as a line graph. If it were shown as a bar graph, the bars would have a decreasing pattern.

4. Make a histogram of the data below.

| Player | Number of Points |
|---|---|
| John | 24 |
| Rae | 33 |
| Marta | 18 |
| Juan | 26 |
| Yoko | 55 |
| Greg | 32 |
| Ann | 48 |
| Sam | 26 |
| Randy | 28 |

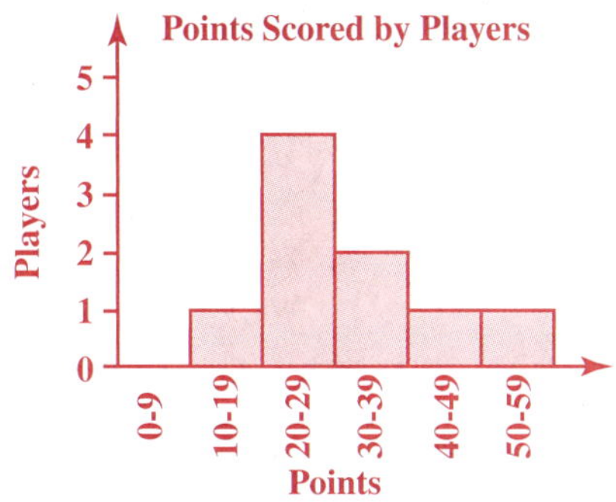


# Lesson 6.2

## Page 373 Additional Answers, continued

**1c.** could be true or false; Possible explanation: Consider these two cases:

- Case 1: Arturo's class: 50, 55, 60; Lee's class: 51, 52, 62
- Case 2: Arturo's class: 50, 55, 60; Lee's class: 48, 52, 59

In Case 1, the median of Arturo's class is greater, and the shortest and tallest students are shorter than the shortest and tallest students in Lee's class. In Case 2, Arturo's class has the greater median, and each person in Lee's class is shorter than the corresponding person in Arturo's class.

**2a.** could be true or false; Possible explanation: Suppose the heights for Marta's class are 58, 58, 60, 60 and the heights for Grace's class are 56, 56, 56, 62. Marta's class has a greater mean (59 vs. 57.5), but Grace's class has the tallest person. If the heights for Marta's and Grace's classes are 60, 61, 62, 63 and 58, 59, 60, 61, respectively, then Marta's class has both the greater mean and the tallest person.

**2c.** could be true or false; Possible explanation: Suppose Marta's class has heights 50, 55, 60 and Grace's class has heights 45, 52, 62. Then the mean of Marta's class is greater, but the tallest student in his class is shorter than the tallest student in Grace's class. On the other hand, if Grace's class has heights 45, 52, 59, then each person in Grace's class is shorter than the corresponding person in Marta's class.

**2d.** could be true or false; Possible explanation: Marta's class could have heights 56, 56, 56, 70, 70 and Grace's could have heights 56, 56, 56, 56, 56. The middle student in Marta's class is not taller than the middle student in Grace's class. If the heights for Marta's and Grace's classes are 60, 61, 62 and 58, 59, 60, respectively, then Marta's class has both the greater mean and the tallest middle person.

## Page 380 Additional Answers, continued

**1b.** Possible answer: Going from left to right, the stacks of X's get taller until reaching their highest point at $7\frac{1}{4}$ and then get shorter again. There is a gap at $6\frac{3}{4}$. The shape indicates that a couple of small sizes and a couple of large sizes were sold, but most of the sizes were somewhere in the middle, with $7\frac{1}{4}$ being the most popular.

## Page 387 Additional Answers, continued

**18a.** *Star Wars*

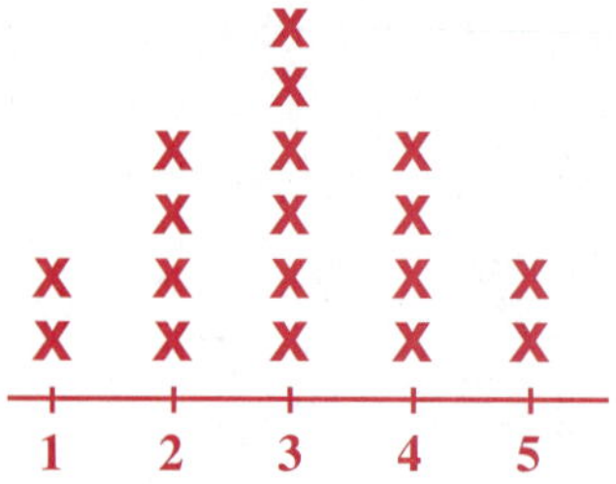


*The Sound of Music*

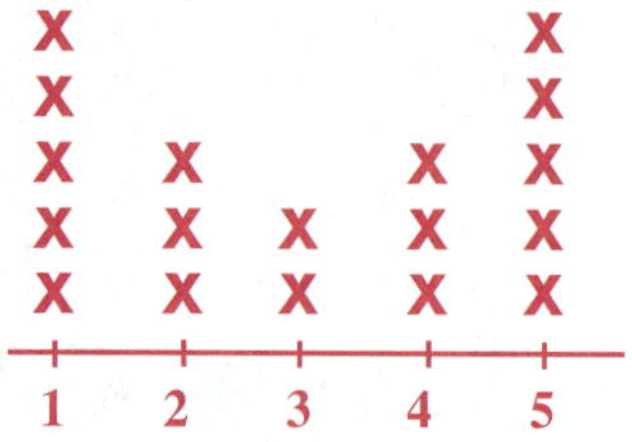


*The Wizard of Oz*

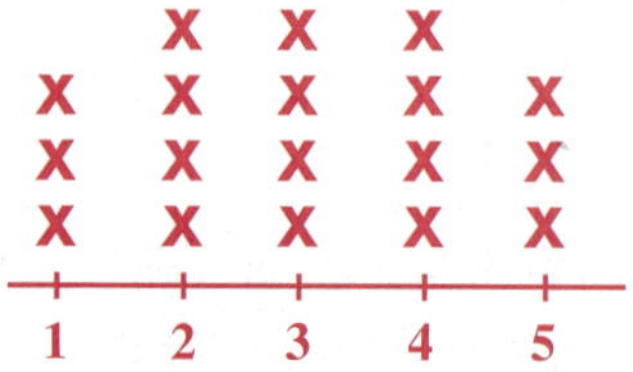


**18c.** No; although the distributions of ratings for the three movies are quite different, the means and medians are the same. Reporting those averages does not indicate the differences.

**18d.** Possible answer: I would give the mode. *The Sound of Music* has two modes, 1 and 5, which indicates that some students really loved it while others didn't like it at all. *The Wizard of Oz* has three modes, 2, 3, and 4, indicating that the ratings were fairly spread out. *Star Wars* has one mode, 3, which indicates that most students thought it was a pretty good movie.

## Page 388 Additional Answers, continued

**19c.** Displays will depend on the values students considered to be outliers. For the outliers given in Part a, the plots are as follows:

**Morning Glories with Peat**

| Stem | Leaf |
|---|---|
| 12 | 9 |
| 13 | 9 |
| 14 | 2 3 5 5 7 8 |
| 15 | 1 4 6 |
| 16 | 8 |

Key: 4|3=43

**Morning Glories with Topsoil**

| Stem | Leaf |
|---|---|
| 8 | 7 |
| 9 | |
| 10 | 5 |
| 11 | 6 |
| 12 | 7 |
| 13 | 2 |
| 14 | 3 5 9 |
| 15 | 5 7 |
| 16 | 7 |

Key: 11|6=116

**Zinnias with Peat**

| Stem | Leaf |
|---|---|
| 4 | 2 3 5 7 7 7 9 |
| 5 | 0 2 2 6 |

Key: 4|3=43

**Zinnias with Topsoil**

| Stem | Leaf |
|---|---|
| 2 | 3 7 |
| 3 | 2 3 5 6 8 9 |
| 4 | 2 |
| 5 | 0 1 4 |

Key: 3|2=32

## Page 389 Quick Quiz

### Quick Quiz

1. This data set shows the number of times a student went to the art museum over the last six years.

   2 4 2 5 2 6

   a. Find the mode, median, mean, and range of the data set. mode: 2; median: 3; mean: 3.5; range: 4

   b. Explain what each of the measures in Part a tells you about the data set. Possible answer: The mode tells the most typical number of visits per year. The median is the middle value of the data set, so there are the same number of visits greater than 3 as there are less than 3. The mean tells how many visits there would be if all the visits were spread out over the six years. The range tells the difference between the greatest and least numbers of visits.

   c. Create a line plot of the data set.

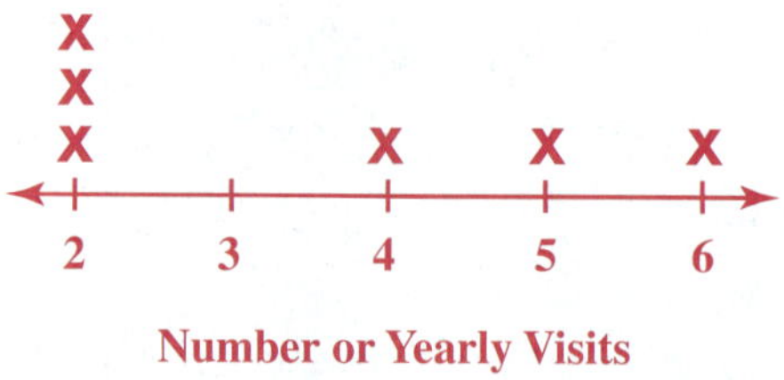


2. Create a stem-and-leaf plot for the data set shown below. Then find the mode, median, and range of the data.

   16 22 26 17 31 21 15 22

| Stem | Leaf |
|---|---|
| 1 | 5 6 7 |
| 2 | 1 2 2 6 |
| 3 | 1 |

Key: 2|6=26

mode: 22; median: 21.5; range: 16

• continued on next page

3. How would replacing the greatest value in a data set with a much greater value affect the median and the mean of the data set? *The median would not change (assuming the data set had more than two values), but the mean would increase.*

4. Two members of a large cross-country team had injuries that affected their times. A newspaper wanted to know the average time of the team members in the race. In reporting the typical time of the team, someone suggested they should use the median time rather than the mean. Would this be a good idea? Explain your answer. *Possible answer: Yes, it would be a good idea. Since there are many members in the race, the slow times would influence the mean but not the median. The median would give a truer picture of the team's capabilities.*

# Lesson 6.3

## Page T391 Notes, continued

**Problem Set Wrap-Up** Discuss each problem in the problem set. You might have one group present their answers to a problem and encourage others to add their ideas. Students should reach a consensus about which data are needed to answer the questions, but since they are not actually performing calculations, they may not be able to agree on which measure will best describe the data.

Encourage students to explain why they chose a specific measure. This may lead to a discussion about the strengths and weaknesses of the different measures and displays. As long as each group's reasoning is sound, there is no need to have all groups in class agree on the same measures. In addition, groups are not bound by their decisions. They can decide to use different measures when they actually perform the analysis.

Have students save their answers so they can refer to them in the next investigation.

## Page 391 Additional Answers, continued

**5.** Possible answer:

**a.** I will need the list of activities boys said were their favorites and the list girls said were their favorites.

**b.** I will find the top two or three activities for boys and the top two or three activities for girls.

**6.** Possible answer:

**a.** For each student, I will need the favorite activity and the list of times that each student spent on each activity.

**b.** I will see whether each student spends the most time on the activity he or she likes best. I will calculate the percent of students who spend the most time on their favorite activity.

## Page 399 Additional Answers, continued

**1b.** Possible answers:

Question 1: A histogram showing time intervals on the horizontal axis and the number of students who use the Internet for that amount of time on the vertical axis

Question 2: A bar graph with a bar for each activity and the height of each bar representing the mean, median, or mode time spent on that activity

Question 3: A circle graph with each section representing the percent of students who chose a particular activity as their favorite

## Page 400 Additional Answers, continued

**Question 2 continued:**

**2b.** The typical times students spend in each activity are as follows: chatting, about 90 min; playing games, about 90 min; doing homework, about 45 min; surfing the Web, about 90 min; using e-mail, about 10 min.

**2c.**

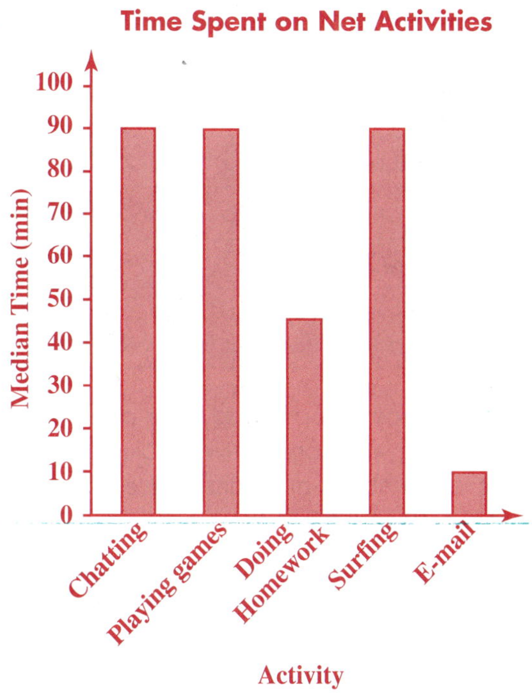


• continued on next page

Question 3:

**2a.** The favorite activity among the 15 students is playing games, with 7 of the 15 students choosing it as their favorite. None of the students listed doing homework as their favorite activity.

**2b.** The mode favorite activity is playing games, followed by chatting, surfing the web, and using e-mail.

**2c.**

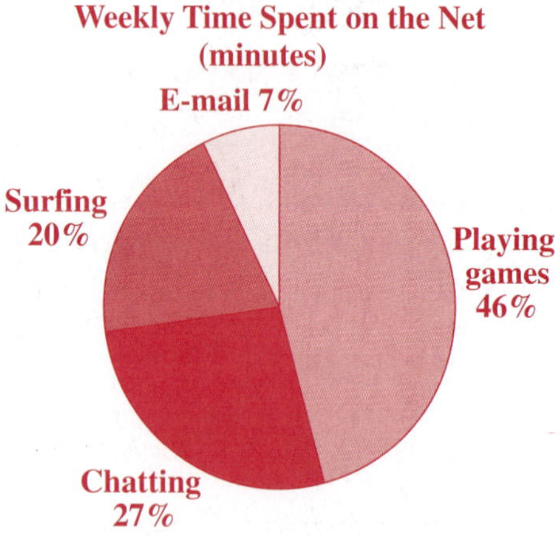


# Chapter 7 Mathematical Background

## Page 409a Notes, continued

For example, the following series of steps (including one specific example and the steps translated into symbols) will always give you an answer of 10:

| | | |
|---|---|---|
| Think of a number. | 6 | $n$ |
| Add 5. | 11 | $n + 5$ |
| Double the result. | 22 | $2(n + 5)$ |
| Subtract twice the number you started with. | 10 | $2(n + 5) - 2n = 2n + 10 - 2n = 10$ |

Finally, students consider how to write convincing arguments to explain why consecutive whole numbers, consecutive even numbers, and consecutive odd numbers have certain properties.

The use of letters to represent variable quantities and to help study the relationships among those quantities is one of the key differences between ordinary arithmetic and algebra. The algebraic style of thinking is introduced and systematically developed in Courses 1 through 3. This kind of thinking is important for the mathematical development of each student. Like most new ways of thinking, it may at first seem somewhat strange. But once students grasp the basic ideas, these ideas will shed light on all they have learned before as well as all they have yet to learn.

# Lesson 7.1

## Page 412 Additional Answers, continued

**4.** Possible answer:

**a.** Term 1 has one square, and one square is added with each term, so the number of squares always equals the term number.

**b.** You can imagine that the pattern starts with one vertical toothpick. For Term 1, one vertical toothpick is added to this starting toothpick. For Term 2, two vertical toothpicks are added to the starting toothpick, and so on. So, the number of vertical toothpicks equals 1 plus the term number.

**c.** The horizontal toothpicks are the tops and bottoms of the squares. Each square has one top toothpick and one bottom toothpick, for a total of two. The number of squares in each term is the term number. So, you multiply the term number by 2 to get the number of toothpicks on the top and bottom.

## Page 413 Additional Answers, continued

**2b.** Possible table:

| Term Number, $n$ | 1 | 2 | 3 | 4 |
|---|---|---|---|---|
| Number of Toothpicks | 4 | 10 | 16 | 22 |

**2c.** Possible rules:

$t = 6n - 2$, where $t$ is the number of toothpicks

$s = 2n - 1$, where $s$ is the number of squares

$o = 4n$, where $o$ is the number of outer toothpicks

$j = 2n - 2$, where $j$ is the number of inner toothpicks

**3a.** Possible variables: number of toothpicks, number of triangle vertices, number of inner toothpicks

**3b.** Possible table:

| Term Number, $n$ | 1 | 2 | 3 | 4 |
|---|---|---|---|---|
| Number of Inner Toothpicks | 0 | 1 | 2 | 3 |

**3c.** Possible rules:

$t = 2n + 1$, where $t$ is the number of toothpicks

$v = n + 2$, where $v$ is the number of triangle vertices

$j = n - 1$, where $j$ is the number of inner toothpicks

• continued on next page

**4a.** Possible variables: number of dots, number of dots in the top row

**4b.** Possible table:

| **Term Number, $n$** | 1 | 2 | 3 | 4 |
|---|---|---|---|---|
| **Number of Dots** | 1 | 3 | 5 | 7 |

**4c.** Possible rules:

$d = 2n - 1$, where $d$ is the number of dots

$t = n - 1$, where $t$ is the number of dots in the top row

**5b.** Possible sequence:

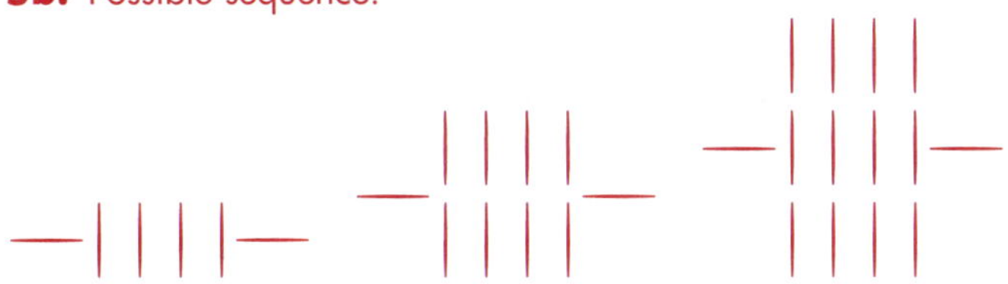

Term 1 Term 2 Term 3

## *Page 417 Additional Answers, continued*

**2c.** Possible answer: Each term is made up of two halves, each containing a number of toothpicks equal to the term number plus 1. So, the number of toothpicks in Term $n$ is $2 \cdot (n + 1)$.

Term 1 — 2 groups of 2

Term 2 — 2 groups of 3

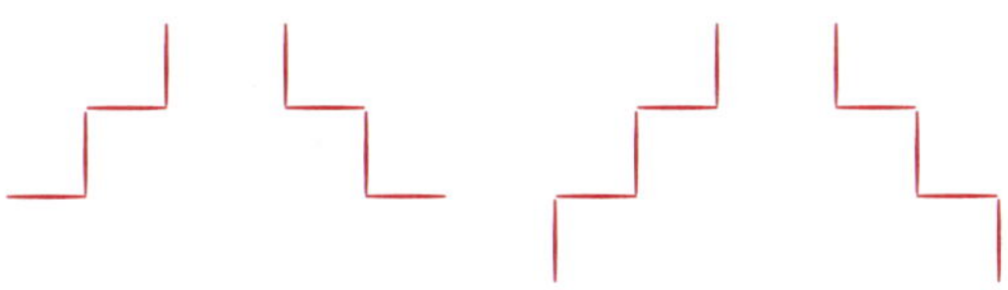

Term 3 — 2 groups of 4

Term 4 — 2 groups of 5

**3b.** Possible explanation for $t = 2 \cdot (n + 2)$: Each term has two halves. The number of toothpicks in each half is equal to the term number plus 2. So, the number of toothpicks in Term $n$ is $2 \cdot (n + 2)$.

Term 1 — 2 groups of 3

Term 2 — 2 groups of 4

Term 3 — 2 groups of 5

Possible explanation for $t = 2n + 4$: Each term has two toothpicks on the left and two on the right, for a total of four. The top and bottom each have a number of toothpicks equal to the term number. So, the number of toothpicks in Term $n$ is $2n + 4$.

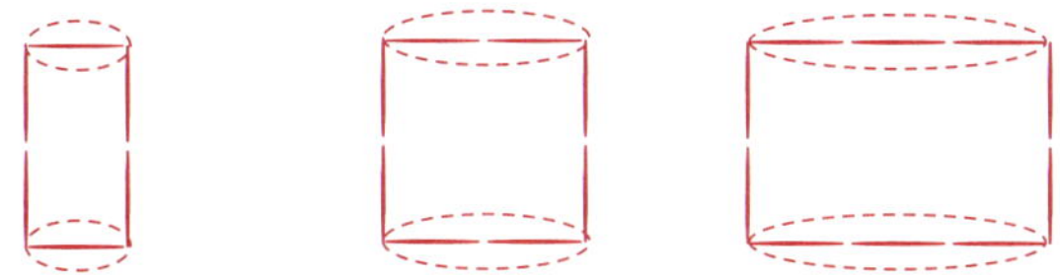

Term 1 — 4 plus 2 groups of 1

Term 2 — 4 plus 2 groups of 2

Term 3 — 4 plus 2 groups of 3

**4b.** Possible explanation for $d = 2 + 4n$: Each term has two dots in the center and four "arms," each with a number of dots equal to the term number. So, Term $n$ has $2 + 4n$ dots.

Term 1 — 2 plus 4 groups of 1

Term 2 — 2 plus 4 groups of 2

Term 3 — 2 plus 4 groups of 3

Possible explanation for $d = 2 \cdot (2n + 1)$: Each term has two halves. Each half has one dot at the "point" and two "arms," each with a number of dots equal to the term number. So, Term $n$ has $2 \cdot (2n + 1)$ dots.

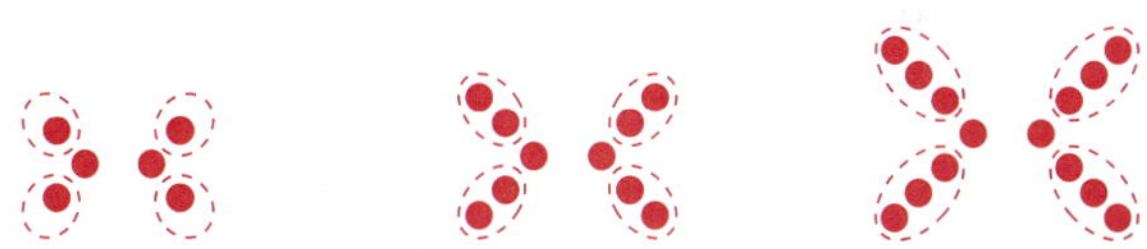

Term 1 — 2 groups of (2 groups of 1 plus 1)

Term 2 — 2 groups of (2 groups of 2 plus 1)

Term 3 — 2 groups of (2 groups of 3 plus 1)

## *Page 418 Additional Answers, continued*

**5.** Conor's rule is correct. Possible explanation: Each term has one toothpick, plus (term number) groups of five toothpicks. So, Term $n$ has $5n + 1$ toothpicks.

Term 1 — 1 group of 5, plus 1

Term 2 — 2 groups of 5, plus 1

Term 3 — 3 groups of 5, plus 1

• *continued on next page*

Jahmal's rule is incorrect. For Term 2, his rule gives $t = 3 \cdot 2 + 3 = 9$, but Term 2 actually has 11 toothpicks.

Rosita's rule is correct. Possible explanation: The one-hexagon shape has 6 toothpicks. The two-hexagon shape has $2 \cdot 6$ toothpicks, minus 1 because the hexagons share a side. The three-hexagon shape has $3 \cdot 6$ hexagons, minus 2 for the two shared sides. The $h$-hexagon shape will have $h - 1$ shared sides, so the total number of toothpicks is $6h - (h - 1)$.

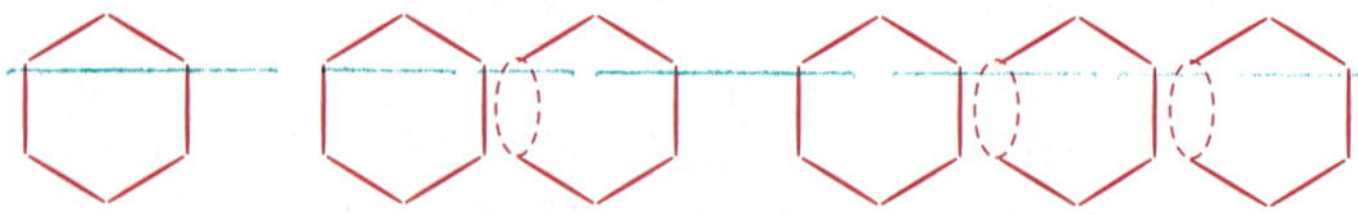

| Term 1 | Term 2 | Term 3 |
|---|---|---|
| 6 toothpicks | 2 groups of 6, minus 1 shared toothpick | 3 groups of 6, minus 2 shared toothpicks |

## Page 423 Additional Answers, continued

**5b.** Possible explanation for $t = 2 \cdot n + 1$: Each term has one vertical toothpick plus two slanted sides. The number of toothpicks in each slanted side is equal to the term number. So, the number of toothpicks in Term $n$ is $2 \cdot n + 1$.

| Term 1 | Term 2 |
|---|---|
| 2 groups of 1, plus 1 | 2 groups of 2, plus 1 |

| Term 3 | Term 4 |
|---|---|
| 2 groups of 3, plus 1 | 2 groups of 4, plus 1 |

Possible explanation for $t = 3 + 2 \cdot (n - 1)$: You can separate the three rightmost toothpicks from the rest of the term. The rest of the term has two slanted segments, each with a number of toothpicks equal to 1 less than the term number. So, the number of toothpicks in Term $n$ is $2 \cdot (n - 1) + 3$.

| Term 1 | Term 2 |
|---|---|
| 3 | 2 groups of 1, plus 3 |

| Term 3 | Term 4 |
|---|---|
| 2 groups of 2, plus 3 | 2 groups of 3, plus 3 |

**6c.** correct; Possible explanation: Each term has five center dots and four "arms." The number of dots in each arm is equal to the term number minus 1. So, the number of dots in Term $j$ is $4 \cdot (j - 1) + 5$.

| Term 1 | Term 2 | Term 3 | Term 4 |
|---|---|---|---|
| 5 | 4 groups of 1, plus 5 | 4 groups of 2, plus 5 | 4 groups of 3, plus 5 |

## Page 429 Quick Quiz

### Quick Quiz

1. Consider the toothpick sequence of U-shapes whose first three terms are shown below.

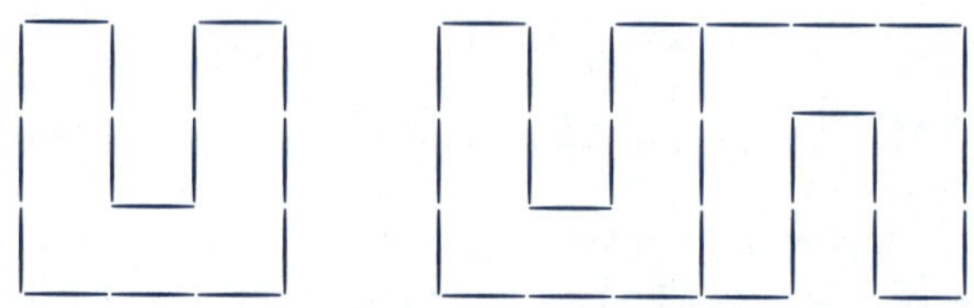

Term 1 Term 2

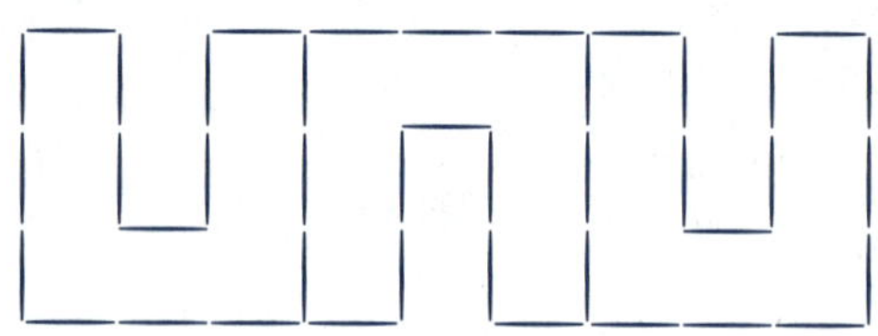

Term 3

a. Write two rules that tell how the number of toothpicks $t$ in a term is related to the term number $n$. *Possible rules: $t = 3 + 13n$, $t = 16 + 13(n - 1)$*

b. Use each of your rules to predict the number of toothpicks in Term 4. Do the two predictions agree? *55, 55; yes*

2. Select one of the rules you wrote for Question 1a and explain why it always works. *Possible answer: You can make Term 1 by putting 3 vertical toothpicks in a column on the left and then adding 13 more toothpicks to complete the U-shape. To get Term 2, add another 13 toothpicks on the right side of*

• *continued on next page*

Term 1 to form an upside-down U-shape. Continue adding groups of 13 toothpicks on the right to form the next terms. You started with 3 vertical toothpicks and, for Term $n$, you have also used $n$ groups of 13 toothpicks. So $t = 3 + 13n$.

3. Miguel wrote the rules $d = 5 + 4n$ and $d = 9n$ to describe how the number of dots $d$ in each term of a dot sequence is related to the term number $n$. Tell whether his rules are equivalent. Explain how you know. no; Possible explanation: The rule $d = 5 + 4n$ gives 13 dots for Term 2, but the rule $d = 9n$ gives 18 dots for Term 2.

4. Write a rule for the following table. Use the given letters to represent the variables.

| **m** | 1 | 2 | 3 | 10 | 15 |
|---|---|---|---|---|---|
| **k** | 2 | 7 | 12 | 47 | 72 |

Possible rule: $k = 5m - 3$

# Lesson 7.2

## Page 437 Additional Answers, continued

**2c.** Possible answers: To find the number of green mints, multiply the number of pink mints by 4. *Or,* to find the number of pink mints, divide the number of green mints by 4.

**2d.** $g = 4p$, or $p = \frac{1}{4} \cdot g$, or $p = g \div 4$, where $g$ is the number of green mints and $p$ is the number of pink mints

**3a.** number of fish, number of hamsters

**3b.** Possible table:

| **Hamsters** | 1 | 2 | 3 | 4 | 5 |
|---|---|---|---|---|---|
| **Fish** | 6 | 12 | 18 | 24 | 30 |

**3c.** Possible answers: To find the number of fish, multiply the number of hamsters by 6. *Or,* to find the number of hamsters, divide the number of fish by 6.

**3d.** $f = 6h$, $h = \frac{1}{6} \cdot f$, $h = f \div 6$, where $h$ is the number of hamsters and $f$ is the number of fish

**4a.** manager's earnings, worker's earnings

**4b.** Possible table:

| **Manager's Earnings** | \$50.00 | \$100.00 | \$150.00 | \$200.00 | \$250.00 |
|---|---|---|---|---|---|
| **Worker's Earnings** | \$7.14 | \$14.29 | \$21.43 | \$28.57 | \$35.71 |

**4c.** Possible answers: To find the worker's earnings, divide the manager's earnings by 7 or multiply the manager's earnings by $\frac{1}{7}$. *Or,* to find the manager's earnings, multiply the worker's earnings by 7.

**4d.** $w = \frac{m}{7}$, or $w = \frac{1}{7} \cdot m$, or $m = 7w$, where $m$ represents the amount the manager earns and $w$ represents the amount the assembly worker earns

**5a.** cost of a child's ticket, cost of an adult's ticket

**5b.** Possible table:

| **Adult's Ticket Cost** | \$5 | \$6 | \$7 | \$8 | \$9 |
|---|---|---|---|---|---|
| **Child's Ticket Cost** | \$2 | \$3 | \$4 | \$5 | \$6 |

**5c.** Possible answers: To find the cost of a child's ticket, subtract \$3 from the cost of an adult's ticket. *Or,* to find the cost of an adult's ticket, add \$3 to the cost of a child's ticket.

**5d.** $c = a - 3$, or $a = c + 3$, where $a$ represents the price of an adult's ticket and $c$ represents the price of a child's ticket

**6a.** number of vertical toothpicks, total number of toothpicks

**6b.** Possible table:

| **Vertical Toothpicks** | 1 | 2 | 3 | 4 | 5 |
|---|---|---|---|---|---|
| **Total Toothpicks** | 6 | 8 | 10 | 12 | 14 |

**6c.** Possible answer: To find the total number of toothpicks, multiply the number of vertical toothpicks by 2 and add 4.

**6d.** $t = 2 \cdot v + 4$, where $t$ represents the number of toothpicks in the term and $v$ represents the number of vertical toothpicks

## Page 449 Quick Quiz, continued

2. In a certain school, there are four times as many students who buy lunch as bring their lunch from home.
   a. Using letters for the variables, write a rule that represents this situation. State what each variable in your rule represents. $4h = b$, where $h$ is the number who bring their lunch from home and $b$ is the number who buy their lunch
   b. Write another rule that represents this situation. Possible rule: $h = \frac{b}{4}$

• continued on next page

3. The width of a tablecloth is 14 inches less than the length. If $W$ is the width of the tablecloth and $L$ is the length, which of the following rules are correct?

A. $W + 14 = L$

B. $L + 14 = W$

C. $L - 14 = W$

D. $L \div W = 14$

E. $L - W = 14$

A, C, E

# Lesson 7.3

## Page 456 Additional Answers, continued

**3a.** Possible answer: $2 + 4 + 6 = 12$, $10 + 12 + 14 = 36$, $24 + 26 + 28 = 78$

**3b.** Possible rule: To find the sum, multiply the middle number by 3.

**3c.** Possible answer: If $m$ is the middle even number, then the three numbers are $m - 2$, $m$, and $m + 2$. The sum is $m - 2 + m + m + 2 = 3 \cdot m$, which is 3 times the middle number.

### Share & Summarize

Possible answer: If you experiment with a lot of numbers, you might *think* it is always true, but you can't be sure without trying every set of numbers, which is impossible. With symbols, the letter (like $n$) stands for any number at all, so if it works for $n$, it will work for any number.

## Page 457 Additional Answers, continued

**2a.** See the table for a possible answer.

**2b.** The result is always 1 more than the starting number.

**2c.**

| Step in Words | Step in Symbols |
|---|---|
| Think of a number. | $n$ |
| Divide by 2. | $\frac{n}{2}$ |
| Add 1. | $\frac{n}{2} + 1$ |
| Multiply by 4. | $4 \times \left(\frac{n}{2} + 1\right) = 2 \cdot n + 4$ |
| Subtract your original number. | $2 \cdot n + 4 - n = n + 4$ |
| Subtract 3. | $n + 4 - 3 = n + 1$ |

**3a.** See the table for a possible answer.

**3b.** You always get half of your starting number.

**3c.**

| Step in Words | Step in Symbols |
|---|---|
| Think of a number. | $n$ |
| Add 2. | $n + 2$ |
| Multiply by 5. | $5 \cdot (n + 2) = 5 \cdot n + 10$ |
| Subtract 10. | $5 \cdot n + 10 - 10 = 5 \cdot n$ |
| Subtract 2 times your original number. | $5 \cdot n - 2 \cdot n = 3 \cdot n$ |
| Divide by 6. | $0.5 \cdot n$ |

## Page 458 Additional Answers, continued

**6.** Possible answer:

| Step in Words | Step in Symbols |
|---|---|
| Think of a number. | $n$ |
| Add 1. | $n + 1$ |
| Multiply by 2. | $2 \cdot (n + 1) = 2 \cdot n + 2$ |
| Subtract 2. | $2 \cdot n + 2 - 2 = 2 \cdot n$ |
| Divide by 2. | $2 \cdot n \div 2 = n$ |
| Subtract your original number. | $n - n = 0$ |

**7c.** Possible answer: U = 6, H = 1, E = 2; U = 7, H = 1, E = 4; U = 9, H = 1, E = 8

**7d.** U can't be 1, 2, 3, or 4 because these numbers would not give a two-digit answer. U can't be 0 because 0 + 0 is 0, so the sum would have to be U as well.

**7e.** H can't be anything but 1. When you add two one-digit numbers, you can never get an answer greater than 18, so the first digit must be 1.

**8e.** Paul is 3 times as old as Ann will be next year. Since $24 = 3 \cdot 8$, Ann will be 8 next year, so she must be 7 this year.

## Page 461 Additional Answers, continued

**4b.** Possible explanation for $t = 2 + 4n$: Each term has two horizontal toothpicks, plus "diamond" shapes made from four toothpicks each. The number of diamond shapes is equal to the term number. So, Term $n$ has two toothpicks plus $n$ groups of four, or $2 + 4n$ toothpicks.

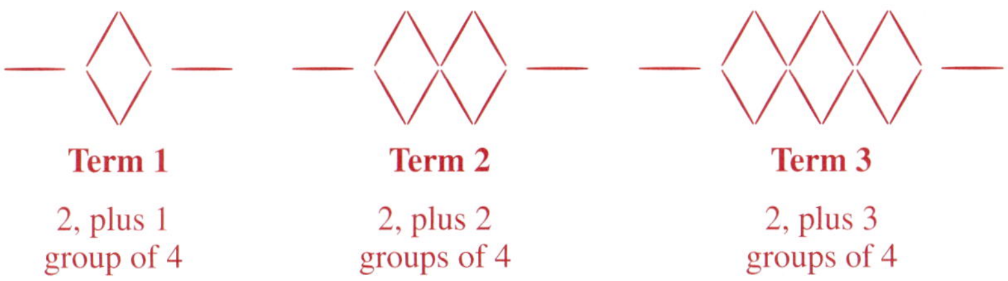


Possible explanation for $t = 2 \cdot (1 + 2n)$: Each term has two halves. Each half is made of one horizontal toothpick, plus two groups of slanted toothpicks (one group on the top and one group on the bottom). The number of toothpicks in each of the two groups is equal to the term number. So, Term $n$ has two groups of (1 plus two groups of $n$) toothpicks, or $2 \cdot (1 + 2n)$ toothpicks.

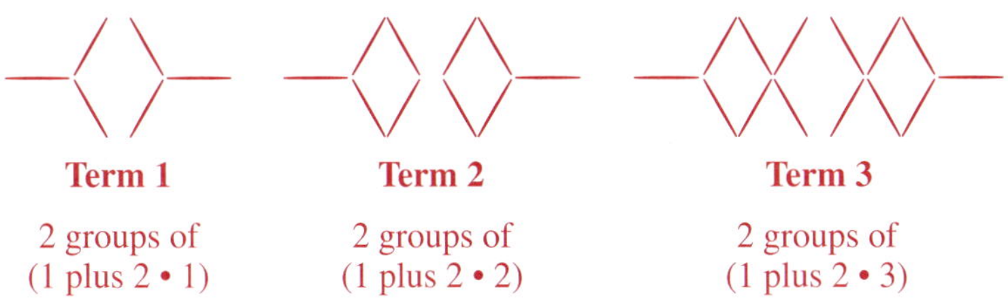


## Page 462 Additional Answers

**8c.** Possible answer: When I work through the steps using $n$ as my starting number, I always get $n$ back.

| Step in Words | Step in Symbols |
|---|---|
| Think of a number. | $n$ |
| Add 3. | $n + 3$ |
| Multiply by 4. | $4 \cdot (n + 3) = 4n + 12$ |
| Subtract 12. | $4n + 12 - 12 = 4n$ |
| Subtract your original number. | $4n - n = 3n$ |
| Divide by 3. | $3n \div 3 = n$ |

# Chapter 8 Mathematical Background

## Page 465a Notes, continued

**Squaring and Square Roots** The operations of squaring a number and finding the square root of a number have natural geometric interpretations. The area of a square is equal to the square of a length of a side. The square root of the area of a square is equal to the length of a side of the square. The squaring and square root operations are used in the final lesson of the chapter, where students study the Pythagorean Theorem. This famous theorem explains how the side lengths of any right triangle are related. Students examine a proof of the Pythagorean Theorem and go on to explore some of its many applications.

# Lesson 8.2

## Page 491 Additional Answers, continued

**10a.** Possible answer: inner perimeter: 6.7 cm; outer perimeter: 17.6 cm

**10b.** Possible answer: inner perimeter: 14.3 cm; outer perimeter: 14.2 cm

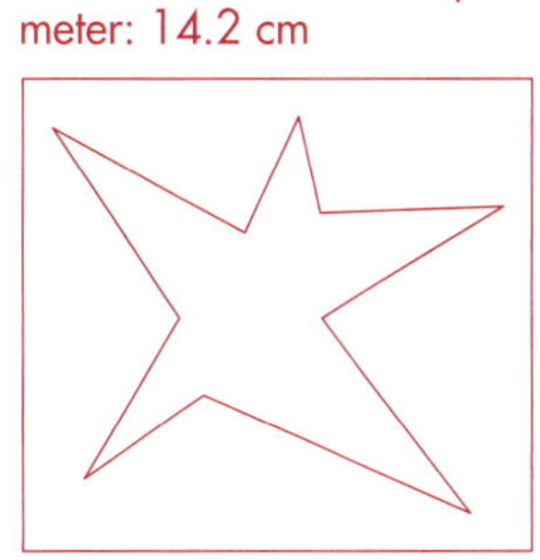

• continued on next page

**10c.** Possible answer: inner perimeter: 17.7 cm; outer perimeter: 17.7 cm

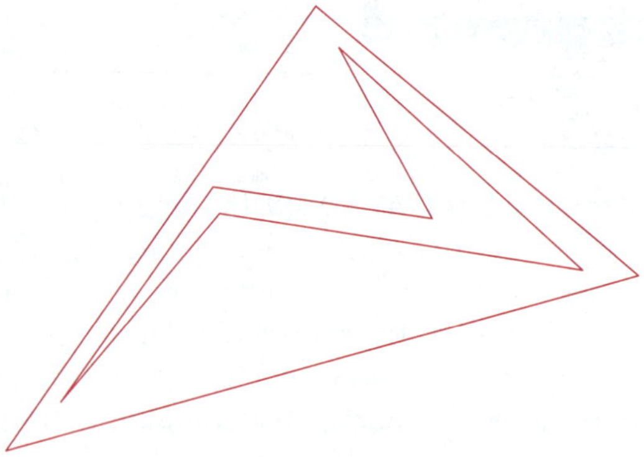

**10d.** The outer shape; it has all of the space in the inner shape plus more space outside the inner shape.

## Page 493 Additional Answers

**14.** Yes; it increases because there are two extra segments on the "outside" of the shape. In the circle, these segments were inside. Since each of these segments is a 6.5-inch radius, the perimeter of the new shape is 2 · 6.5 in., or 13 in., greater than the perimeter of the circle.

**24a.**

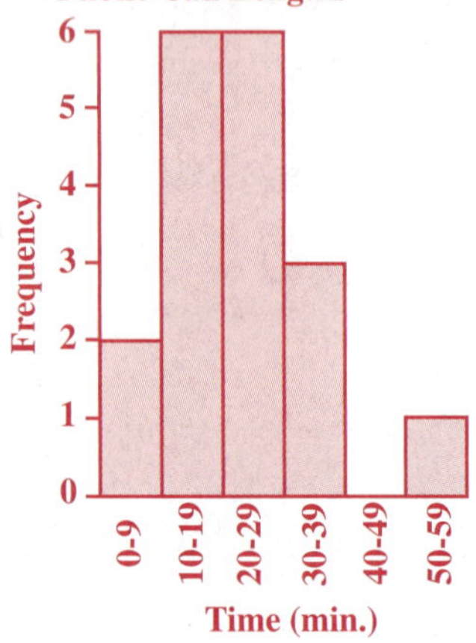


**24b.** Possible answer: There was one call between 50 and 60 min, but the rest were less than 40 min long, with a majority (12 of the 18) between 10 and 30 min long.

## Page 493 Quick Quiz, continued

2. Draw a figure that has two straight sides and two curved sides. Use a 6-sided polygon to approximate the perimeter of your figure. *Check students' work.*

3. What is the diameter of a circle that has a radius of $13\frac{1}{8}$ in.? $26\frac{1}{4}$ in.

4. Find the circumference of the circle shown below. Round your answer to the nearest tenth of a centimeter. 47.1 cm

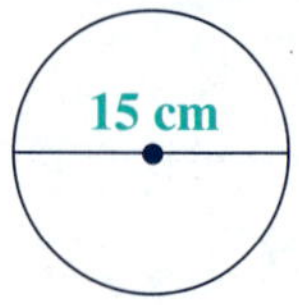


# Lesson 8.3

## Page 513 Quick Quiz

### Quick Quiz

1. Find the area of the following shape:

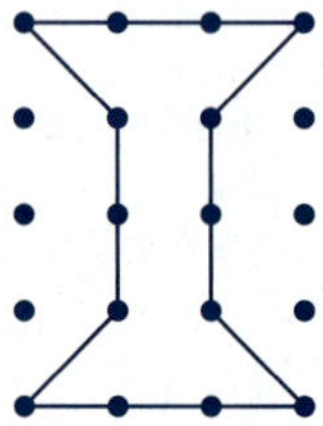

*6 sq units*

2. Find the area of a rectangle that is 24.3 cm long and 12 cm wide. 291.6 $cm^2$

3. Find the length of a rectangle that has a width of 18 in. and an area of 558 $in.^2$. 31 in.

4. Write the area of the square shown below by using exponential notation. Then compute the area.

*$A = (8.4\text{ cm})^2$; 70.56 $cm^2$*

5. Write the first three perfect squares that are greater than 900. *961; 1,024; 1,089*

6. Evaluate $5 \times (8 + 7)^2 - 4^2$. *1,109*

7. What is the value of $\sqrt{196}$? 14

8. Between what two consecutive whole numbers is $\sqrt{95}$? 9 and 10

# Lesson 8.4

## Page 528 Additional Answers, continued

**6b.** Possible parallelograms (they will be drawn on a 1-cm grid):

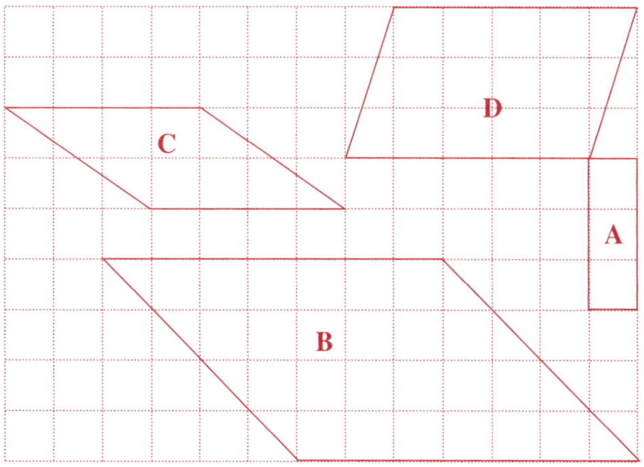


## Page 532 Additional Answers

**21c.** Possible formulas: $A = \frac{1}{2} \cdot B \cdot h + \frac{1}{2} \cdot b \cdot h$, or $A = \frac{1}{2} \cdot h \cdot (B + b)$

Possible explanations:

- I divided the trapezoid into two triangles, each with height $h$. The area of one triangle is $\frac{1}{2} \cdot B \cdot h$, and the area of the other is $\frac{1}{2} \cdot b \cdot h$. The total area is the sum of these areas.

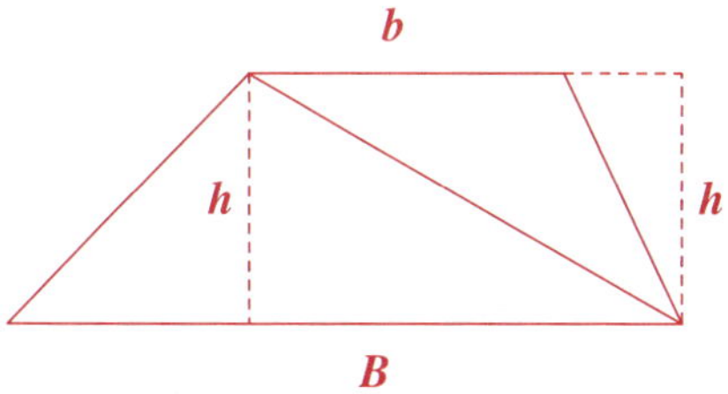


- I put two copies of the trapezoid together to form a parallelogram with base $B + b$ and height $h$. The area of the parallelogram is $h \cdot (B + b)$. The area of the trapezoid is half this area.

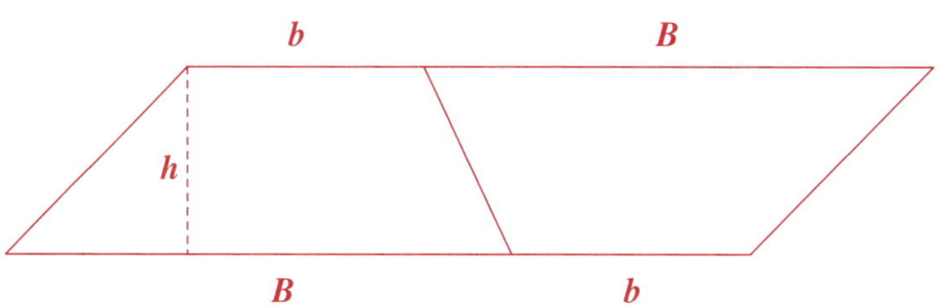


## Page 535 Additional Answers, continued

**35b.** Possible explanation: Term 1 has 4 toothpicks, Term 2 has 4 toothpicks plus 1 group of 4, Term 3 has 4 toothpicks plus 2 groups of 4, Term 4 has 4 toothpicks plus 3 groups of 4, and so on. So, Term $n$ has 4 plus $n - 1$ groups of 4, or $4 + 4 \cdot (n - 1)$ toothpicks.

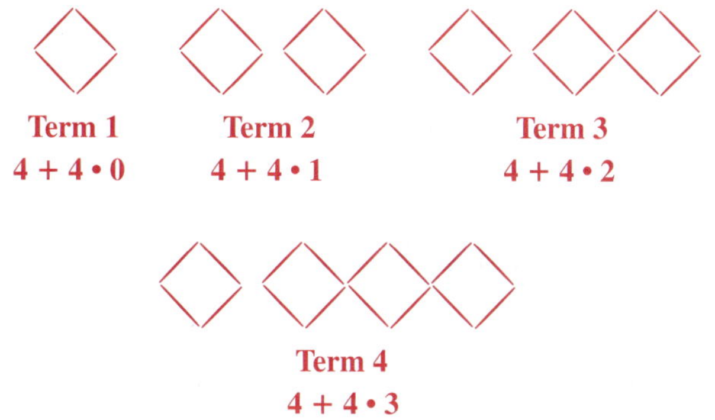


## Page 535 Quick Quiz

### Quick Quiz

1. Find the area of the parallelogram.

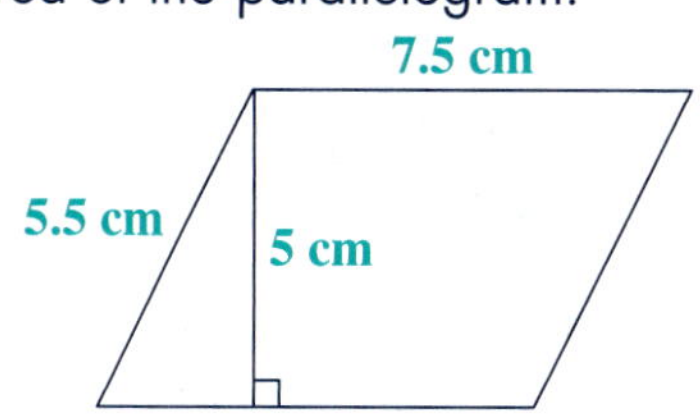


$37.5 \text{ cm}^2$

2. Find the area of triangle $ABC$. Round your answer to the nearest tenth of a square centimeter.

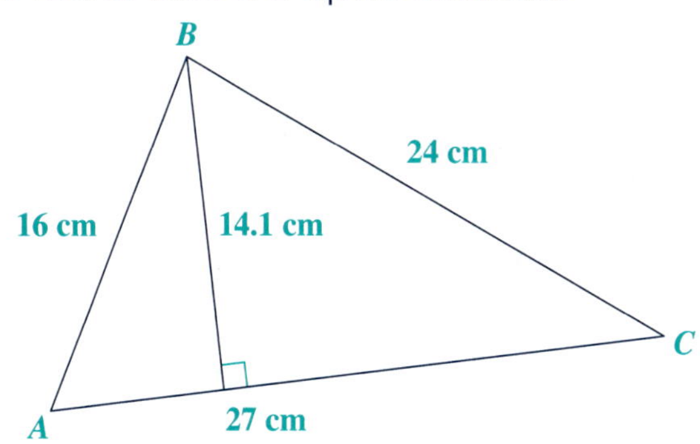


$190.4 \text{ cm}^2$

• continued on next page

3. Find the approximate area of the circle. Use 3.14 for $\pi$.

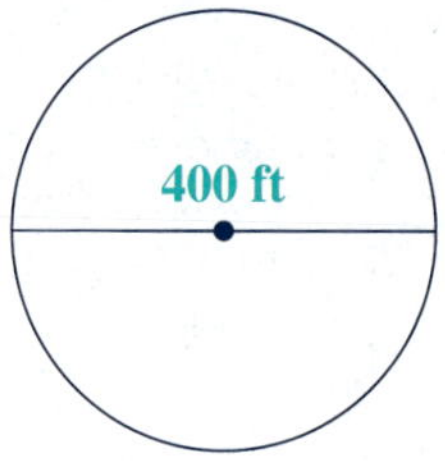


125,600 $ft^2$

# Lesson 9.1

## Page 569 Quick Quiz, continued

3. Solve each equation.

   a. $4 \cdot x = 28$ 7

   b. $n + 15 = 20$ 5

   c. $3 \cdot k + 7 = 13$ 2

   d. $8x = x^2 + 15$ (There are two solutions.) 3, 5

4. Tell whether each equation is *always true, sometimes true,* or *never true,* and explain how you know.

   a. $p + 4 = 7 + p$ Never true; 7 is greater than 4, so the sum of 7 and $p$ is always greater than the sum of $p$ and 4.

   b. $m^2 + 9 = 13$ Sometimes true; the equation is true for $m = 2$, but it is not true for $m = 1$.

   c. $7x = 5x + 2x$ Always true; $5x + 2x = (x + x + x + x + x) + (x + x)$, and the sum of seven $x$'s can be written as $7 \cdot x$, or $7x$.

# Lesson 9.2

## Page T573 Notes, continued

2. Write an equation for each flowchart, using $n$ as the input. Then use backtracking to find the value of the input.

   a.

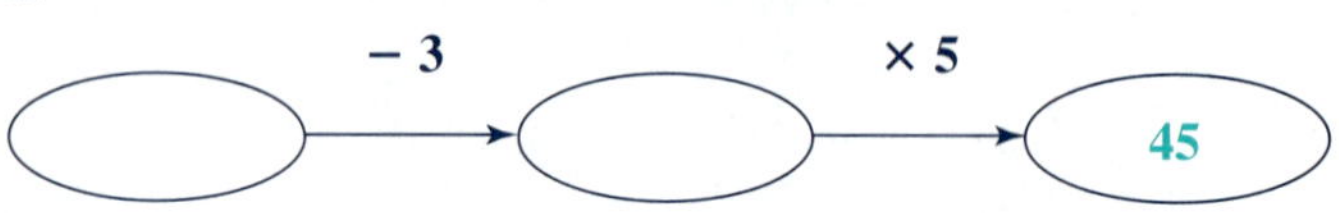


   $(n - 3) \times 5 = 45$; 12

   b.

   $n \times 5 - 3 = 57$; 12

   c.

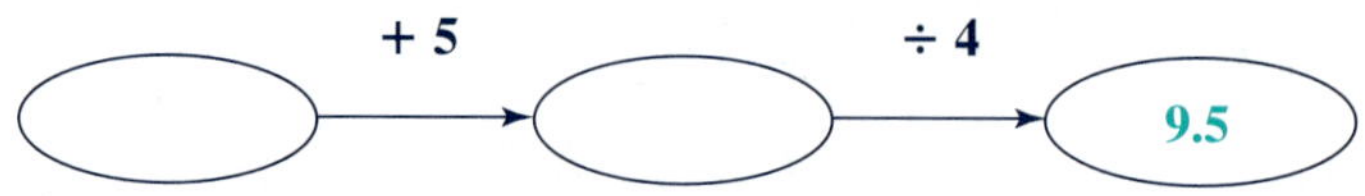


   $(n + 5) \div 4 = 9.5$; 33

## On Your Own Exercises

Practice & Apply: 1–6, p. 579
Connect & Extend: 18, p. 581
Mixed Review: 25–35, pp. 584–585

## Page 573 Additional Answers, continued

**3.** Possible answer: Backtracking is a method for finding the input that gives a certain output by working backward using a flowchart. In the equation $5n - 3 = 45$, you can think of 45 as the output. To solve the equation, find the value of the input $n$. Start by writing the output, 45, in the last oval:

Since 3 was subtracted from the number in the second oval to get 45, that number must be 48:

Since the input was multiplied by 5 to get 48, it must be equal to $\frac{48}{5}$, or 9.6.

## Page 576 Additional Answers, continued

### Share & Summarize

Possible answer: Suppose you want to use backtracking to solve the equation $6 + n \cdot 3 = 21$. If you don't pay attention to order of operations, you might make a flowchart that does the addition before the multiplication:

If you backtrack using this flowchart, you get 1, which is not the solution since $6 + 1 \cdot 3$ is equal to 9, not to 21.

## Page 580 Additional Answers, continued

**12.**

; 14.75

**13.**

; 22

**14a.** There are 3 toothpicks in the beginning of each term:

For each term, an additional 2 toothpicks are added in the shape of the letter V:

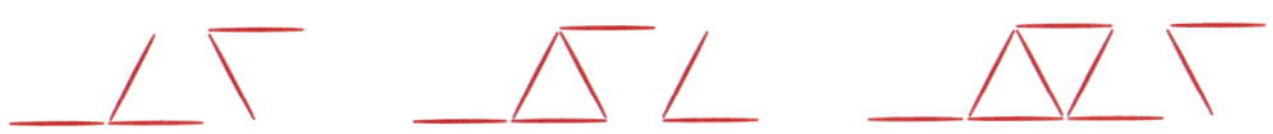

So, the number of toothpicks in Term $n$ is $2n + 3$.

## Page 585 Quick Quiz, continued

2. Make a flowchart for the following equation. Then solve the equation by using backtracking.

$$\frac{2 \cdot (n - 2)}{4} - 7 = 5$$

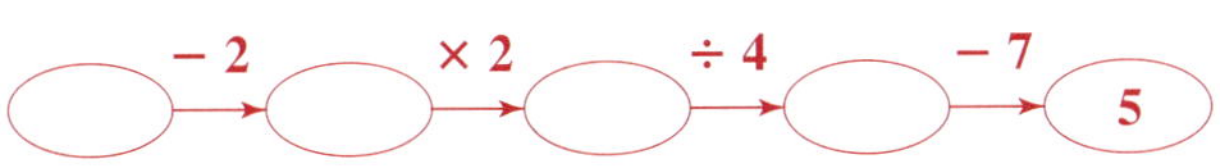


; 26

3. A store has CDs on sale for $9 each. Jana has a coupon that will give her $4 off the total cost of her purchase. She has $32 to spend. How many CDs can she buy? 4 CDs

# Lesson 10.2

## Page 622 Additional Answers, continued

### Share & Summarize

Possible answer: Put 3 on each face of Die A. Put 5 on one of the faces of Die B and 2 on the other five faces.

**Die A** (columns), **Die B** (rows)

| Die B \ Die A | 3 | 3 | 3 | 3 | 3 | 3 |
|---|---|---|---|---|---|---|
| **2** | A | A | A | A | A | A |
| **2** | A | A | A | A | A | A |
| **2** | A | A | A | A | A | A |
| **2** | A | A | A | A | A | A |
| **2** | A | A | A | A | A | A |
| **5** | B | B | B | B | B | B |

## Page 624 Additional Answers

### Think & Discuss

Possible answer: List the possible rolls for Die A along the top of the table and the possible rolls for Die B along the side. In each cell, enter the sum for the corresponding combination.

**Die A** (columns), **Die B** (rows)

| Die B \ Die A | 1 | 2 | 3 | 4 | 5 | 6 |
|---|---|---|---|---|---|---|
| **1** | 2 | 3 | 4 | 5 | 6 | 7 |
| **2** | 3 | 4 | 5 | 6 | 7 | 8 |
| **3** | 4 | 5 | 6 | 7 | 8 | 9 |
| **4** | 5 | 6 | 7 | 8 | 9 | 10 |
| **5** | 6 | 7 | 8 | 9 | 10 | 11 |
| **6** | 7 | 8 | 9 | 10 | 11 | 12 |

## Page 625 Additional Answers, continued

**Problem Set G**

**1.** Possible table:

| | | Die A | | | | | |
|---|---|---|---|---|---|---|---|
| | | **1** | **2** | **3** | **4** | **5** | **6** |
| **Die B** | **1** | 0 | 1 | 2 | 3 | 4 | 5 |
| | **2** | 1 | 0 | 1 | 2 | 3 | 4 |
| | **3** | 2 | 1 | 0 | 1 | 2 | 3 |
| | **4** | 3 | 2 | 1 | 0 | 1 | 2 |
| | **5** | 4 | 3 | 2 | 1 | 0 | 1 |
| | **6** | 5 | 4 | 3 | 2 | 1 | 0 |

**3.** Possible answer: The best strategy is to distribute the counters according to the probabilities: the most on 1, next most on 2, and so on.

## Page 631 Additional Answers, continued

**3.** Their chances of getting the greater number are the same, as shown in the table. Each has a $\frac{20}{45}$, or $\frac{4}{9}$, chance of getting the greater number.

| | | Rashid | | | | | | | | |
|---|---|---|---|---|---|---|---|---|---|---|
| | | **1** | **2** | **3** | **4** | **5** | **6** | **7** | **8** | **9** |
| **María** | **3** | M | M | | R | R | R | R | R | R |
| | **4** | M | M | M | | R | R | R | R | R |
| | **5** | M | M | M | M | | R | R | R | R |
| | **6** | M | M | M | M | M | | R | R | R |
| | **7** | M | M | M | M | M | M | | R | R |

## Page 632 Additional Answers, continued

**7c.** Possible answer: Since these dice are really just standard dice multiplied by 2, I would use the same strategy as for the standard game, only placing the most counters on the space with the most likely difference (2) and so on.

**8d.** Possible answer: Since 7 is the most likely sum, I would put the most counters on that section. I would also put several counters on the other likely sums: 5, 6, 8, and 9. I might put one or two counters on each of the other sections as well.

## Page 633 Additional Answers, continued

**11.** Miguel; Possible explanation: The table shows that for a single roll, the chance Miguel will score is $\frac{30}{36}$, which is much higher than Marcus' chance to score.

| | | Marcus' Die | | | | | |
|---|---|---|---|---|---|---|---|
| | | **0** | **0** | **0** | **0** | **0** | **100** |
| **Miguel's Die** | **1** | M | M | M | M | M | R |
| | **2** | M | M | M | M | M | R |
| | **3** | M | M | M | M | M | R |
| | **4** | M | M | M | M | M | R |
| | **5** | M | M | M | M | M | R |
| | **6** | M | M | M | M | M | R |

## *Page 637 Quick Quiz, continued*

3. A grain of rice is dropped and hits the square shown below at a random point inside the square. What is the probability that the rice lands in the triangle? $\frac{1}{16}$

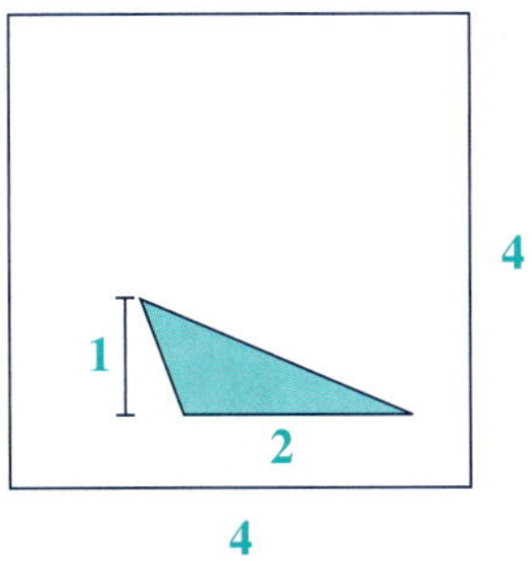


# Lesson 10.3

## *Page 641 Additional Answers*

**2c.**

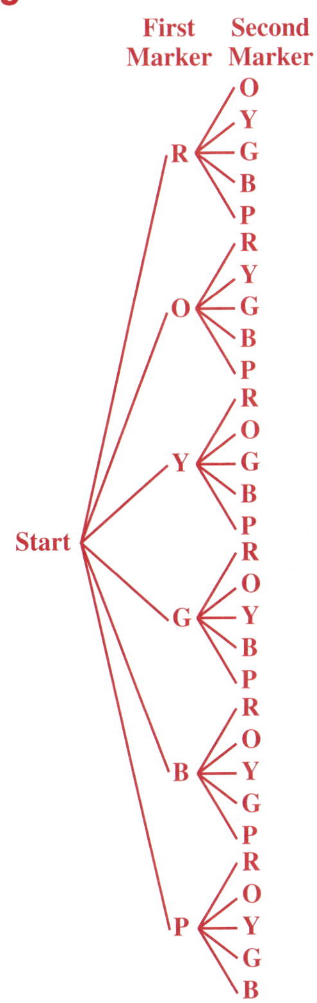


**3.**

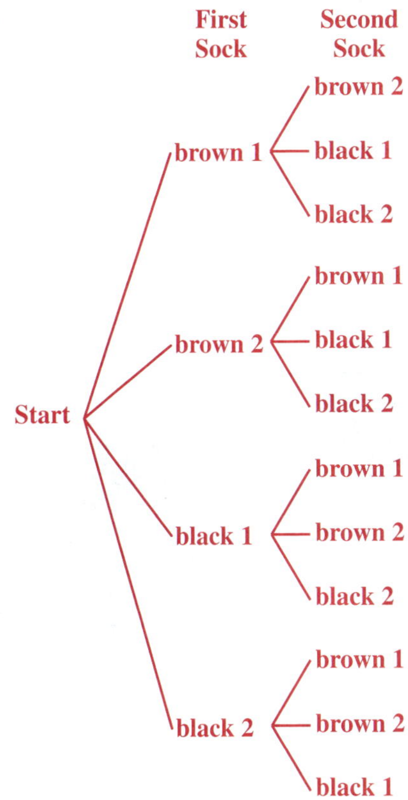


## *Page 645 Additional Answers*

**Problem Set F**

**4.**

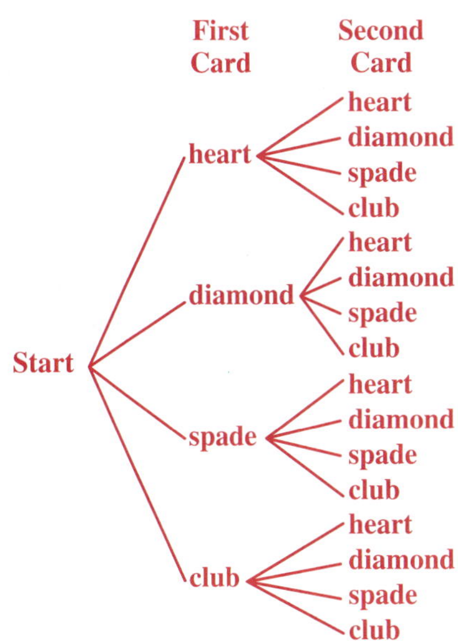

## Page 651 Quick Quiz

### Quick Quiz

1. Diana spins Spinner 1 and then Spinner 2 to get random letter-number combinations.

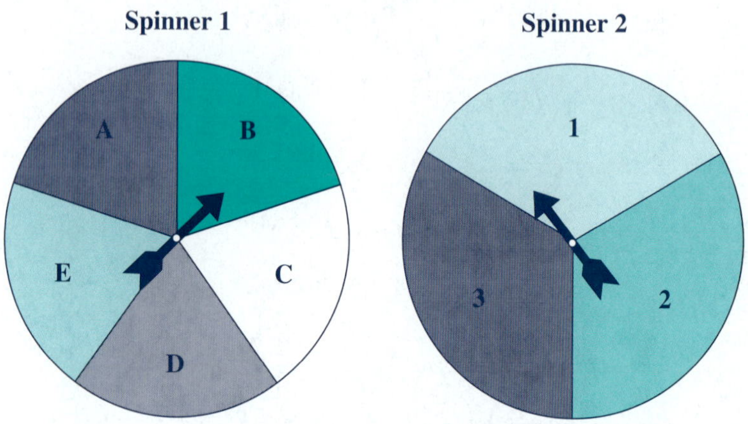


a. Draw a tree diagram that shows all the possible outcomes.

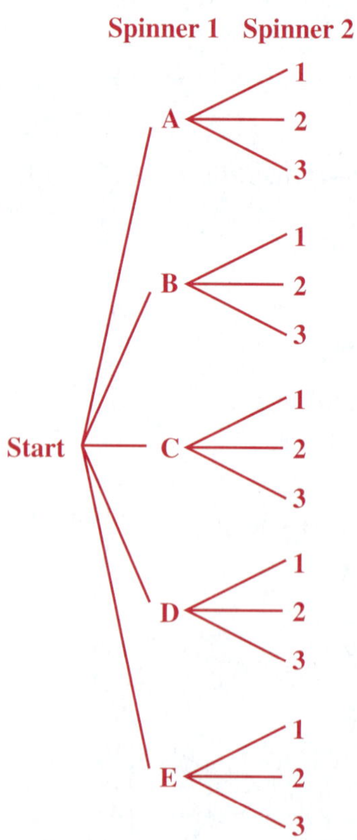


b. What is the probability of spinning a letter-number combination that begins with the letter C? $\frac{1}{5}$

2. The 4 of spades is drawn randomly from a standard deck of 52 playing cards and is not replaced. What is the probability that the next card drawn is the 4 of hearts? $\frac{1}{51}$

3. A bag contains three red blocks and two blue blocks. All the blocks are the same size. Luci reaches into the bag and removes two blocks. What is the probability that both blocks are red? $\frac{3}{10}$

# GLOSSARY/ GLOSARIO

## English

**absolute value** The *absolute value* of a number is its distance from 0 on the number line, and is indicated by drawing a bar on each side of the number. For example, $|^{-}20|$ means "the absolute value of $^{-}20$." Since $^{-}20$ and 20 are each 20 units from 0 on the number line, $|20| = 20$ and $|^{-}20| = 20$. [page 143]

**acute angle** An angle that measures less than 90°. Each of the angles shown below is an *acute angle.* [page 469]

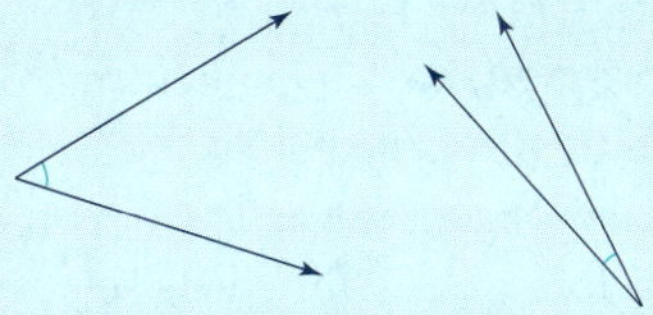

**angle** Two rays with the same endpoint. For example, the figure below is an *angle.* [page 47]

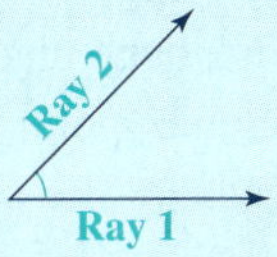


**area** The amount of space inside a two-dimensional shape. [page 494]

**axes** The horizontal line and vertical line that are used to represent the variable quantities on a graph. For example, in the graph below, the horizontal axis represents width and the vertical axis represents length. [page 279]

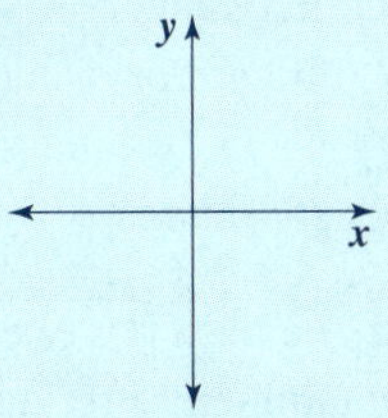


**backtracking** The process of using a flowchart to work backward, starting with the output and undoing each operation to find the input. [page 572]

**base of a parallelogram** Any of the sides of a parallelogram. The *base of a parallelogram* is used in computing its area. (See the figures in the glossary entry for height of a parallelogram.) [page 515]

**base of a triangle** Any of the sides of a triangle. The *base of a triangle* is used in computing its area. (See the figures in the glossary entry for *height of a triangle.*) [page 518]

## Español

**valor absoluto** El *valor absoluto* de un número es su distancia desde 0 en la recta numérica, lo cual se indica trazando una barra en cada lado del número. Por ejemplo, $|^{-}20|$ significa "el valor absoluto de $^{-}20$". Como $^{-}20$ y 20 se encuentran a 20 unidades de 0 en la recta numérica, $|20| = 20$ y $|^{-}20| = 20$.

**ángulo agudo** Ángulo que mide menos de 90°. Cada uno de los siguientes es un *ángulo agudo.*

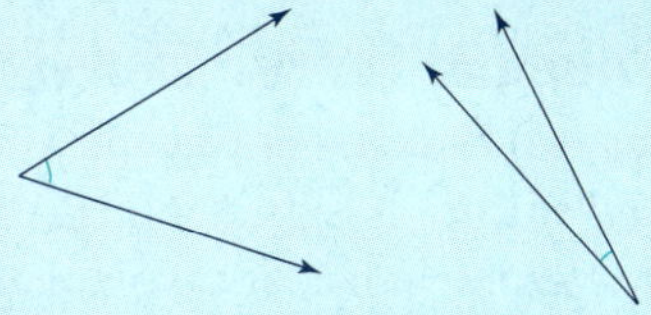

**ángulo** Dos rayos que parten del mismo punto. Por ejemplo, la siguiente figura muestra un *ángulo.*

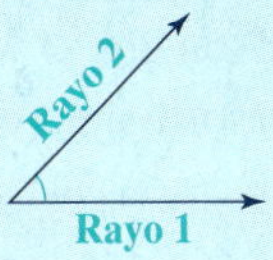


**área** La cantidad de espacio dentro de una figura bidimensional.

**ejes** La recta horizontal y la recta vertical que se usan para representar las cantidades variables en una gráfica. Por ejemplo, en la siguiente gráfica el eje horizontal representa longitud.

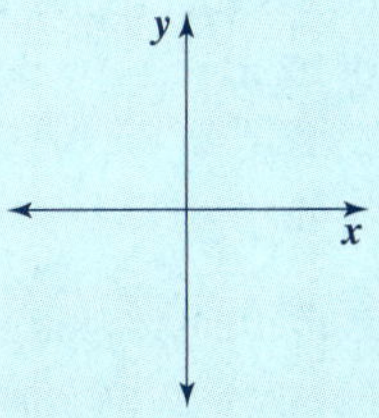


**vuelta atrás** El proceso de usar un flujograma para trabajar en sentido inverso, comenzando con la salida y anulando cada operación hasta llegar a la entrada.

**base de un paralelogramo** Cualquiera de los lados de un triángulo. La *base de un paralelogramo* se usa para calcular su área. (Ver las figuras en el inciso del glosario para *altura de un paralelogramo.*)

**base de un triángulo** Cualquiera de los lados de un triángulo. La *base de un triángulo* se usa para calcular su área. (Ver las figuras en el inciso del glosario para *altura de un triángulo.*)

| English | Español |
|---|---|

## English

**chord** A segment connecting two points on a circle. (See the figure in the glossary entry for radius. This figure shows a chord of a circle.) [page 486]

**circumference** The perimeter of a circle (distance around a circle). [page 486]

**common factor** A common factor of two or more numbers is a number that is a factor of all the numbers. For example, 5 is a *common factor* of 15, 25, and 40. [page 82]

**common multiple** A *common multiple* of a set of numbers is a multiple of all the numbers. For example, 24 is a common multiple of 3, 8, and 12. [page 85]

**composite number** A whole number greater than 1 with more than two factors. For example, 12 is a *composite number* since it is greater than 1 and has more than two factors. In fact, it has six factors: 1, 2, 3, 4, 6, and 12. [page 80]

**concave polygon** A polygon that looks like it is "collapsed" or has a "dent" on one or more sides. Any polygon with an angle measuring more than 180° is concave. The figures below are *concave polygons*. [page 50]

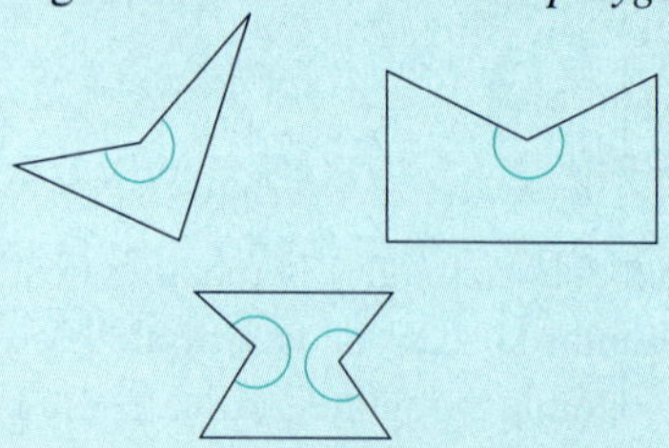

**coordinates** Numbers that represent the location of a point on a graph. For example, if a point is 3 units to the right and 7 units up from the origin, its *coordinates* are 3 and 7. [page 302]

**diameter** A chord that passes through the center of a circle. *Diameter* also refers to the distance across a circle through its center. (See the figure in the glossary entry for radius. This figure shows a diameter of a circle.) [page 486]

**distribution** The *distribution* of a data set shows how the data are spread out, where there are gaps, where there are lots of values, and where there are only a few values. [page 351]

**equally likely** Outcomes of a situation or experiment that have the same probability of occurring. For example, if one coin is tossed, coming up heads and coming up tails are *equally likely* outcomes. [page 608]

## Español

**cuerda** Segmento que conecta dos puntos en un círculo. (Ver la figura en el inciso del glosario para *radio*. Dicha figura muestra una cuerda de un círculo.)

**circunferencia** El perímetro de un círculo (distancia alrededor del círculo).

**factor común** Un factor común de dos o más números es un número que es un factor de todos los números. Por ejemplo, 5 es un *factor común* de 15, 25 y 40.

**múltiplo común** Un *múltiplo común* de un conjunto de números es un múltiplo de todos los números. Por ejemplo, 24 es un múltiplo común de 3, 8 y 12.

**número compuesto** Un número entero mayor que 1 con más de dos factores. Por ejemplo, 12 es un *número compuesto* dado que es mayor que 1 y tiene más de dos factores. De hecho, tiene seis factores: 1, 2, 3, 4, 6 y 12.

**polígono cóncavo** Polígono que parece que se hubiera "hundido" o que tiene una hendidura en uno más de sus lados. Cualquier polígono con un ángulo mayor que 180° es cóncavo. Las siguientes figuras son *polígonos cóncavos.*

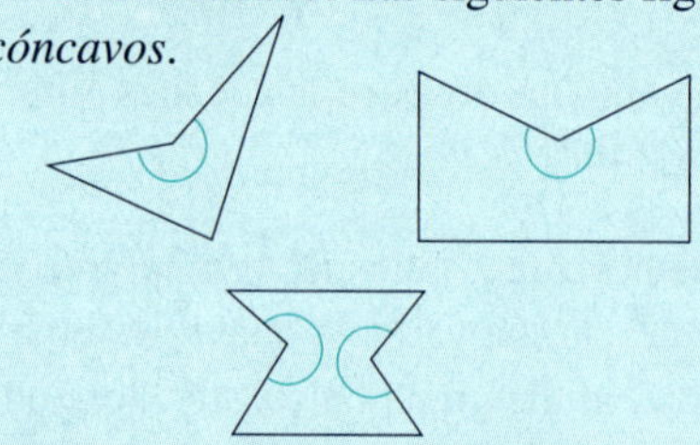

**coordenadas** Números que representan la posición de un punto en una gráfica. Por ejemplo, si un punto se encuentra a 3 unidades a la derecha y 7 unidades hacia arriba del origen, sus *coordenadas* son 3 y 7.

**diámetro** Cuerda que pasa por el centro de un círculo. El *diámetro* también se refiere a la distancia a través de un círculo, pasando por su centro. (Ver las figuras en el inciso del glosario para radio. Dicha figura muestra un diámetro de un círculo.)

**distribución** La *distribución* de un conjunto de datos muestra la extensión de los datos, las brechas entre los datos, los lugares donde hay muchos valores y donde hay pocos valores.

**equiprobables** Resultados de una situación o experimento que tienen la misma posibilidad de ocurrir. Por ejemplo, si se lanza una moneda, es *equiprobable* que la moneda caiga mostrando cara o escudo.

# English

**equation** A mathematical sentence stating that two quantities have the same value. An equal sign, =, is used to separate the two quantities. For example, $5 + 8 = 3 + 10$ is an *equation.* [page 558]

**equivalent fractions** Fractions that describe the same portion of a whole, or name the same number. For example, $\frac{3}{4}$, $\frac{9}{12}$, and $\frac{30}{40}$ are *equivalent fractions.* [page 100]

**experimental probability** A probability based on experimental data. An *experimental probability* is always an estimate and can vary depending on the particular set of data that is used. [page 606]

**exponent** A small, raised number that tells how many times a factor is multiplied. For example, in $10^3$, the exponent 3 tells you to multiply 3 factors of 10: $10 \cdot 10 \cdot 10 = 1{,}000$. [page 81]

**factor** A *factor* of a whole number is another whole number that divides into it without a remainder. For example, 1, 2, 3, 4, 6, 8, 12, and 24 are factors of 24. [page 78]

**factor pair** A *factor pair* for a number is two factors whose product equals that number. For example, the factor pairs for 24 are 1 and 24, 2 and 12, 3 and 8, and 4 and 6. [page 78]

**flowchart** A diagram, using ovals and arrows, that shows the steps for going from an input to an output. For example, the diagram below is a *flowchart.* [page 571]

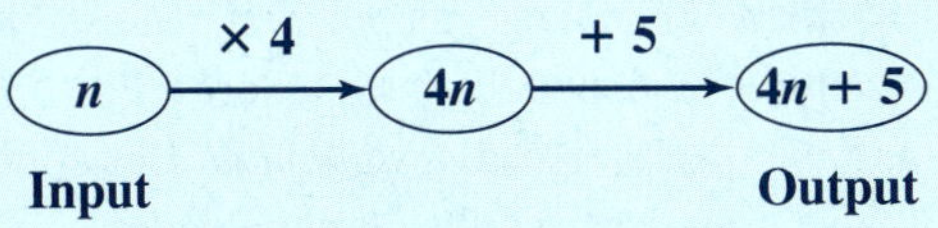


**greatest common factor, GCF** The *greatest common factor* (often abbreviated *GCF*) of two or more numbers is the greatest of their common factors. For example, the greatest common factor (or GCF) of 24 and 36 is 12. [page 83]

**guess-check-and-improve** A method for solving an equation that involves first guessing the solution, then checking the guess by substituting into the original equation, and then using the result to improve the guess until the correct solution is found. [page 586]

# Español

**ecuación** Enunciado matemático que establece que dos cantidades tienen el mismo valor. Se usa un signo de igualdad, =, para comparar las dos cantidades. Por ejemplo, $5 + 8 = 3 + 10$ es una *ecuación.*

**fracciones equivalentes** Fracciones que describen la misma parte de un todo o que representan el mismo número. Por ejemplo, $\frac{3}{4}$, $\frac{9}{12}$ y $\frac{30}{40}$ son *fracciones equivalentes.*

**probabilidad experimental** Probabilidad que se basa en datos experimentales. Una *probabilidad experimental* es siempre una estimación y puede variar según el conjunto de datos en particular que se usen.

**exponente** Número pequeño y elevado que indica cuántas veces se multiplica un factor. Por ejemplo, en $10^3$, el exponente 3 te indica que multipliques 3 factores de 10: $10 \cdot 10 \cdot 10 = 1{,}000$

**factor** Un factor de un número entero es otro número entero que lo divide sin que quede un residuo. Por ejemplo, 1, 2, 3, 4, 6, 8, 12 y 24 son factores de 24.

**par de factores** Un *par de factores* para un número consta de dos factores cuyo producto es igual al número. Por ejemplo, los pares de factores para 24 son 1 y 24; 2 y 12; 3 y 8; y 4 y 6.

**flujograma** Diagrama que usa óvalos y flechas para mostrar los pasos a seguir desde un dato de entrada hasta uno de salida. Por ejemplo, el siguiente diagrama es un *flujograma.*

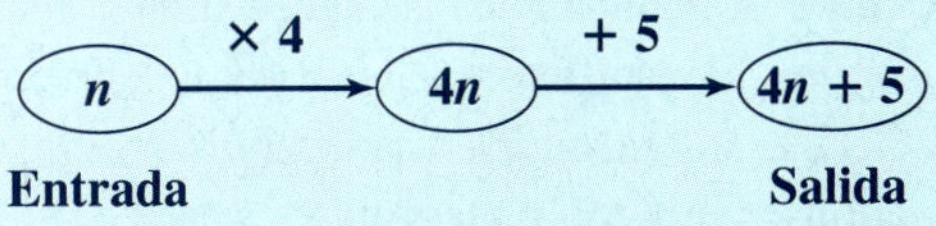


**máximo común divisor, MCD** El *máximo común divisor* (a menudo abreviado, MCD) de dos números es el mayor de sus factores comunes. Por ejemplo, el máximo común divisor (o MCD) de 24 y 36 es 12.

**conjetura, verifica y mejora** Método para resolver una ecuación que implica primero hacer una conjetura, verificar la conjetura y sustituirla en la ecuación original y luego usar el resultado para mejorar la conjetura hasta hallar la solución correcta.

| English | Español |
|---|---|

**height of a parallelogram** The distance from the side opposite the base of a parallelogram to the base. The *height of a parallelogram* is always measured along a segment perpendicular to the base (or to the line containing the base). The figures below show a base and the corresponding height for two parallelograms. [page 515]

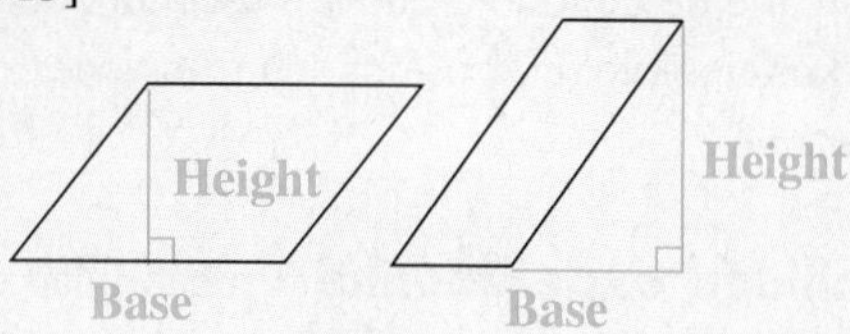


**altura de un paralelogramo** La distancia desde el lado opuesto a la base de un paralelogramo, hasta la base. La *altura de un paralelogramo* se mide siempre a lo largo de un segmento perpendicular a la base (o a la recta que contiene la base). Las siguientes figuras muestran una base y la altura correspondiente de dos paralelogramos.

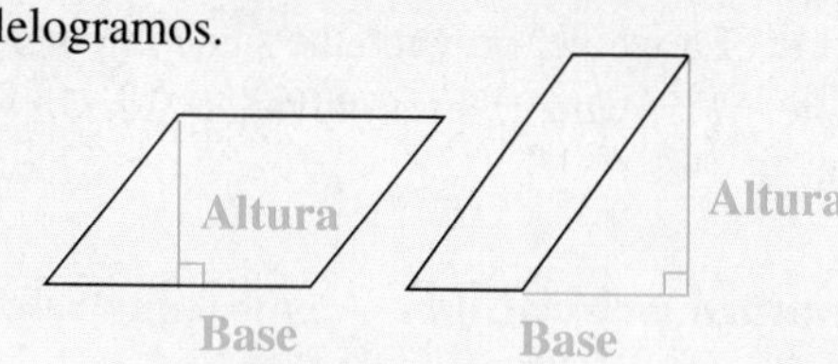


**height of a triangle** The distance from the base of a triangle to the vertex opposite the base. The *height of a triangle* is always measured along a segment perpendicular to the base (or the line containing the base). The figures below show a base and the corresponding height for two triangles. [page 518]

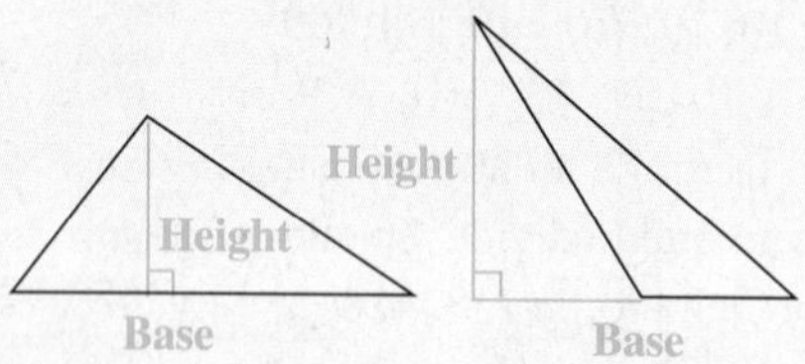


**altura de un triángulo** La distancia desde la base de un triángulo hasta el vértice opuesto a la base. La *altura de un triángulo* se mide siempre a lo largo de un segmento perpendicular a la base (o de la recta que contiene la base). Las siguientes figuras muestran una base y la altura correspondiente de dos triángulos.

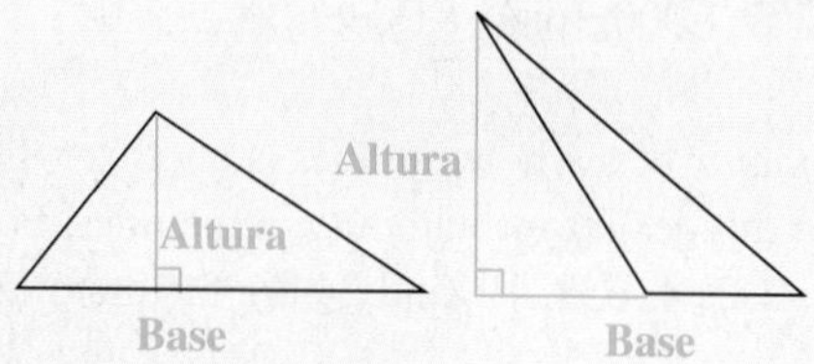


**histogram** A bar graph in which data are divided into equal intervals, with a bar for each interval. The height of each bar shows the number of data values in that interval. [page 350]

**histograma** Gráfica de barras en la cual los datos se dividen en intervalos iguales, con una barra para cada intervalo. La altura de cada barra muestra el número de valores de los datos en ese intervalo.

**hypotenuse** The side opposite the right angle in a right triangle. The *hypotenuse* is the longest side of a right triangle. (See the figure in the entry for *leg* below.) [page 536]

**hipotenusa** El lado opuesto al ángulo recto en un triángulo rectángulo. La *hipotenusa* es el lado más largo de un triángulo rectángulo. (Ver a continuación la figura en el inciso del glosario para cateto.)

**inequality** A mathematical sentence stating that two quantities have different values. For example, $5 + 9 > 12$ is an *inequality.* The symbols $\neq$, $<$, and $>$ are used in writing inequalities. [page 559]

**desigualdad** Enunciado matemático que establece que dos cantidades tienen distintos valores. Por ejemplo, $5 + 9 > 12$ es una *desigualdad.* Los símbolos $\neq$, $<$ y $>$ se usan para escribir desigualdades.

**inverse operations** Two operations that "undo" each other. For example, addition and subtraction are inverse operations, and multiplication and division are inverse operations. [page 504]

**operaciones inversas** Dos operaciones que se "anulan" entre sí. Por ejemplo, la adición y la sustracción son operaciones inversas y la multiplicación y la división son operaciones inversas.

**least common multiple, LCM** The *least common multiple* (often abbreviated *LCM*) of two or more numbers is the smallest of their common multiples. For example, the least common multiple (or LCM) of 6 and 15 is 30. [page 85]

**mínimo común múltiplo, mcm** El *mínimo común múltiplo* (a menudo abreviado mcm) de dos o más números es el menor de sus múltiplos comunes. Por ejemplo, el mínimo común múltiplo de 6 y 15 es 30.

## English

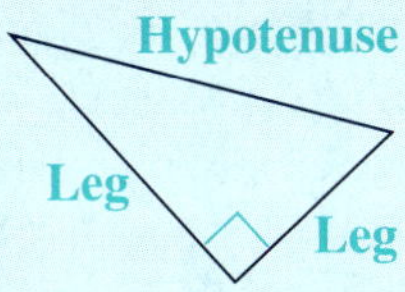


**leg** One of the sides of a right triangle that is not the hypotenuse, or one of the shorter two sides of a right triangle. The figure at the right shows the two legs and hypotenuse of a right triangle. [page 536]

**line graph** A graph in which points are connected with line segments. [page 289]

**line plot** A number line with X's indicating the number of times each data value occurs. [page 363]

**line symmetry** A polygon has *line symmetry* (or reflection symmetry) if you can fold it in half along a line so that the two halves match exactly. The polygons below have *line symmetry.* The lines of symmetry are shown as dashed lines. [page 51]

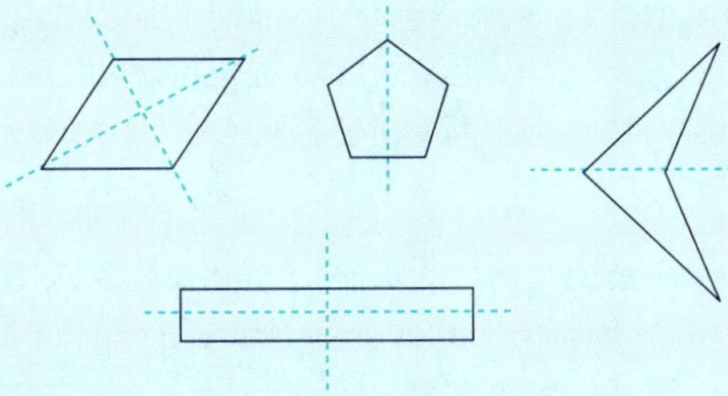

**lowest terms** A fraction is in *lowest terms* if its numerator and denominator are relatively prime. For example, $\frac{5}{6}$ is in lowest terms because the only common factor of 5 and 6 is 1. [page 100]

**mean** The number you get by distributing the total of the values in a data set among the members of the data set. You can compute the *mean* by adding the values and dividing the total by the number of values. For example, for the data 5, 6, 6, 8, 8, 8, 9, 10, 12, the total of the values is 72 and there are 9 values, so the mean is $72 \div 9 = 8$. [page 371]

**median** The middle value when all the values in a data set are ordered from least to greatest. For example, for the data set 4.5, 6, 7, 7, 8.5, 10.5, 12, 12, 14.5, the *median* is 8.5. [page 363]

**mixed number** A whole number and a fraction. For example $12\frac{3}{4}$ is a *mixed number.* [page 98]

**mode** The value in a data set that occurs most often. For example, for the data set 4.5, 6, 7, 7, 7, 8.5, 10.5, 12, 12, the *mode* is 7. [page 363]

## Español

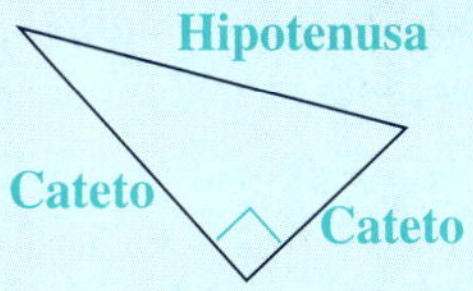


**cateto** Uno de los lados de un triángulo rectángulo que no es la hipotenusa o uno de los dos lados más cortos de un triángulo rectángulo. La figura muestra los dos catetos y la hipotenusa de un triángulo rectángulo.

**gráfica lineal** Gráfica en la cual los puntos se conectan con segmentos de recta.

**esquema lineal** Recta numérica que contiene equis que indican el número de veces que ocurre cada valor de los datos.

**simetría lineal** Un polígono tiene *simetría lineal* (o simetría de reflexión) si se puede doblar por la mitad a lo largo de una línea de modo que las dos mitades coincidan exactamente. Los siguientes polígonos tienen *simetría lineal.* Los ejes de simetría se muestran como líneas punteadas.

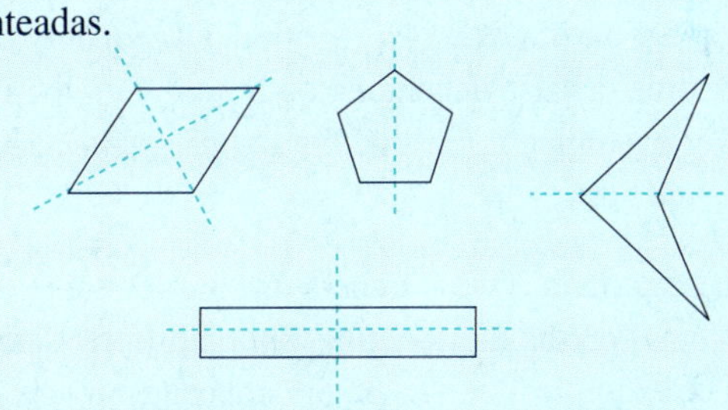

**en términos reducidos o reducida** Una fracción está *en términos reducidos* o *reducida* si su numerador y denominador son primos relativos. Por ejemplo, $\frac{5}{6}$ está reducida dado que el único factor común de 5 y 6 es 1.

**media** El número que se obtiene al distribuir el total de los valores en un conjunto de datos entre los miembros del conjunto de datos. Se puede calcular la *media* sumando los valores y luego dividiendo el total entre el número de valores. Por ejemplo, para los datos 5, 6, 6, 8, 8, 8, 9, 10, 12, el total de los valores es 72 y hay 9 valores, de modo que la media es $72 \div 9 = 8$.

**mediana** El valor central cuando todos los valores en un conjunto de datos se ordenan de menor a mayor. Por ejemplo, para el conjunto de datos 4.5, 6, 7, 7, 8.5, 10.5, 12, 12, 14.5, la *mediana* es 8.5.

**número mixto** Un número entero y una fracción. Por ejemplo, $12\frac{3}{4}$ es un *número mixto*.

**moda** El valor en un conjunto de datos que ocurre con más frecuencia. Por ejemplo, para el conjunto de datos 4.5, 6, 7, 7, 7, 8.5, 10.5, 12, 12, la *moda* es 7.

# English

**multiple** A *multiple* of a whole number is the product of that number and another whole number. For example, 35 is a multiple of 7 since $35 = 7 \times 5$. [page 85]

**negative number** A number that is less than 0. For example, $^{-}18$ (read "negative eighteen") is a *negative number.* [page 142]

**obtuse angle** An angle that measures more than 90° and less than 180°. Each of the angles shown below is an *obtuse angle.* [page 469]

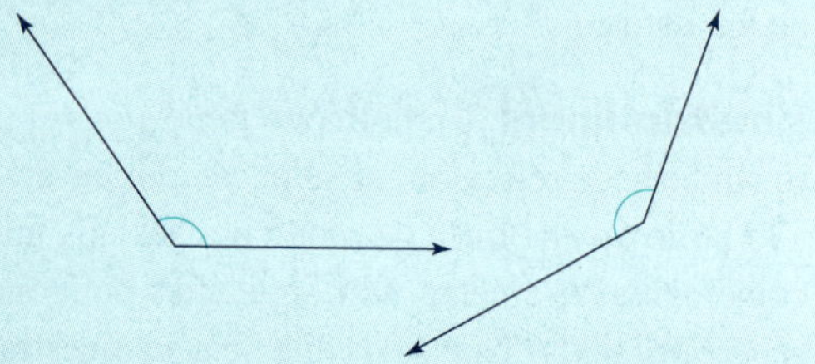

**open sentence** An equation of inequality that can be true or false depending on the value of the variable. For example, $5 + n = 20$ is an *open sentence.* [page 560]

**opposites** Two numbers that are the same distance from 0 on the number line, but on different sides of 0. For example 35 and $^{-}35$ are opposites. [page 143]

**ordered pair** A pair of numbers that represent the coordinates of a point, with the horizontal coordinate of the point written first. For example, the point with horizontal coordinate 3 and vertical coordinate 7 is represented by the ordered pair (3, 7). [page 302]

**order of operations** A convention for reading and evaluating expressions. The order of operations says that expressions should be evaluated in this order:

- Evaluate any expressions inside parentheses and above and below fraction bars.
- Evaluate all exponents, including squares.
- Do multiplications and divisions from left to right.
- Do additions and subtractions from left to right.

For example, to evaluate $5 + 3 \times 7$, you multiply first and then add: $5 + 3 \cdot 7 = 5 + 21 = 26$.

To evaluate $10^2 - 6 \div 3$, you evaluate the exponent first, then divide, then subtract:

$10^2 - 6 \div 3 = 100 - 6 \div 3 = 100 - 2 = 98$.
[pages 19 and 501]

# Español

**múltiplo** Un *múltiplo* de un número entero es el producto de ese número y otro número entero. Por ejemplo, 35 es un múltiplo de 7 dado que $35 = 7 \times 5$.

**número negativo** Un número menor que 0. Por ejemplo, $^{-}18$ (que se lee "dieciocho negativo") es un *número negativo*.

**ángulo obtuso** Ángulo que mide más de 90° y menos de 180°. Cada uno de los siguientes ángulos es un *ángulo obtuso.*

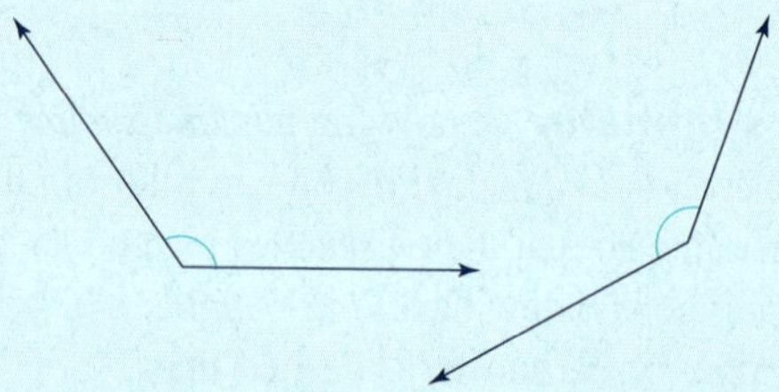

**enunciado abierto** Ecuación de desigualdad que puede ser verdadera o falsa dependiendo del valor de la variable. Por ejemplo, $5 + n = 20$ es un *enunciado abierto*.

**opuestos** Dos números equidistantes de 0 en la recta numérica, pero en lados opuestos de 0. Por ejemplo, 35 y $^{-}35$ son opuestos.

**par ordenado** Un par de números que representa las coordenadas de un punto, en el cual la coordenada horizontal del punto se escribe primero. Por ejemplo, el punto con la coordenada horizontal 3 y coordenada vertical 7 se representa con el par ordenado (3, 7).

**orden de las operaciones** Una convención para leer y evaluar expresiones. El orden de las operaciones indica que las expresiones se deben evaluar en el siguiente orden:

- Evalúa cualquier expresión entre paréntesis y sobre y debajo de barras de fracciones.
- Evalúa todos los exponentes, incluyendo los cuadrados.
- Efectúa las multiplicaciones y las divisiones de izquierda a derecha.
- Efectúa las sumas y las restas de izquierda a derecha.

Por ejemplo, para evaluar $5 + 3 \times 7$, multiplica primero y luego suma. $5 + 3 \cdot 7 = 5 + 21 = 26$.

Para evaluar $10^2 - 6 \div 3$, evalúa primero el exponente, luego divide y por último resta.

$10^2 - 6 \div 3 = 100 - 6 \div 3 = 100 - 2 = 98$.

## English

**origin** The point where the axes of a graph meet. The *origin* of a graph is usually the 0 point for each axis. [page 279]

**outlier** A value that is much greater than or much less than most of the other values in a data set. For example, for the data set 6, 8.2, 9.5, 11.6, 14, 30, the value 30 is an *outlier.* [page 375]

**parallelogram** A quadrilateral with opposite sides that are the same length. The opposite sides of a *parallelogram* are parallel. Each of the figures shown below is a parallelogram. [page 515]

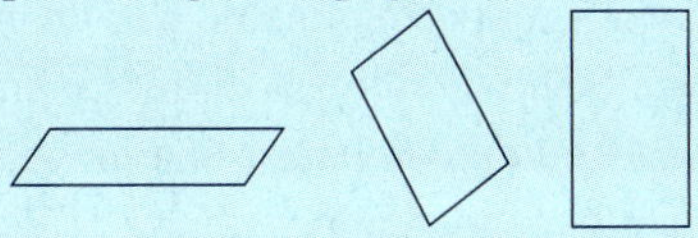

**percent** *Percent* means "out of 100." A percent represents a number as a part out of 100 and is written with a percent sign. For example, 39% means 39 out of 100, or $\frac{39}{100}$, or 0.39. [page 227]

**perfect square** A number that is equal to a whole number multiplied by itself. In other words, a *perfect square* is the result of squaring a whole number. For example, 1, 4, 9, 16, and 25 are perfect squares since these are the results of squaring 1, 2, 3, 4, and 5, in that order. [page 500]

**perimeter** The distance around a two-dimensional shape. [page 482]

**perpendicular** Two lines or segments that form a right angle area are said to be *perpendicular.* For example, see the figures below. [page 469]

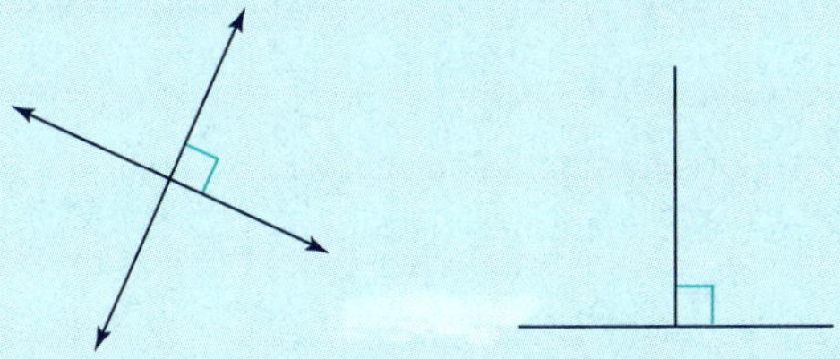

Perpendicular Lines Perpendicular Segments

**polygon** A flat (two-dimensional) geometric figure that has these characteristics:

- It is made of straight line segments.
- Each segment touches exactly two other segments, one at each of its endpoints.

The shapes below are *polygons.* [page 42]

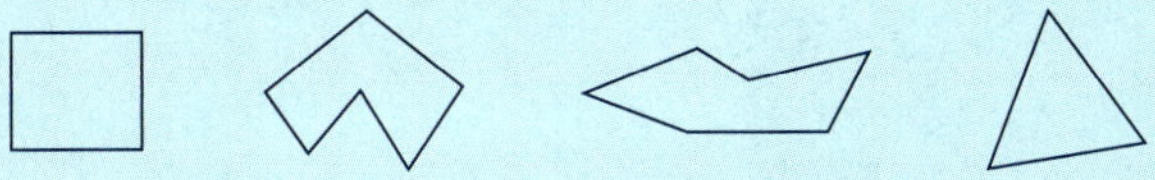

## Español

**origen** El lugar donde se encuentran los ejes de una gráfica. El *origen* de una gráfica es por lo general el punto 0 de cada eje.

**valor atípico** Un valor que es mucho mayor o mucho menor que la mayoría de los otros valores en un conjunto de datos. Por ejemplo, para el conjunto de datos 6, 8.2, 9.5, 11.6, 14, 30, el valor 30 es un *valor atípico*.

**paralelogramo** Cuadrilátero cuyos lados opuestos tienen la misma longitud. Los lados opuestos de un *paralelogramo* son paralelos. Cada una de las siguientes figuras es un paralelogramo.

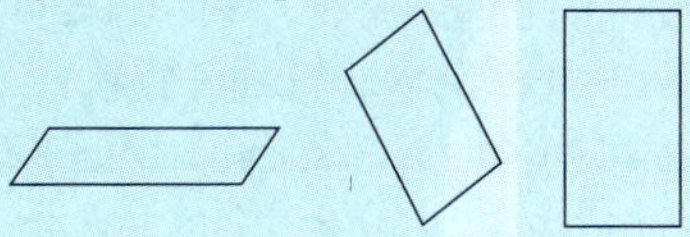

**por ciento, porcentaje** *Por ciento* significa "de cada 100". Un por ciento representa un número como una parte de 100 y se escribe con un signo de porcentaje. Por ejemplo, 39% significa 39 de cada 100 ó $\frac{39}{100}$ ó 0.39.

**cuadrado perfecto** Número que es igual a un número entero multiplicado por sí mismo. Es decir, un *cuadrado perfecto* es el resultado de elevar al cuadrado un número entero. Por ejemplo, 1, 4, 9, 16 y 25 son cuadrados perfectos, dado que son el resultado de elevar al cuadrado 1, 2, 3, 4 y 5, en ese orden.

**perímetro** La distancia alrededor de una figura bidimensional.

**perpendicular** Se dice que dos rectas o segmentos que forman un ángulo recto son *perpendiculares*. Por ejemplo, observa las siguientes figuras.

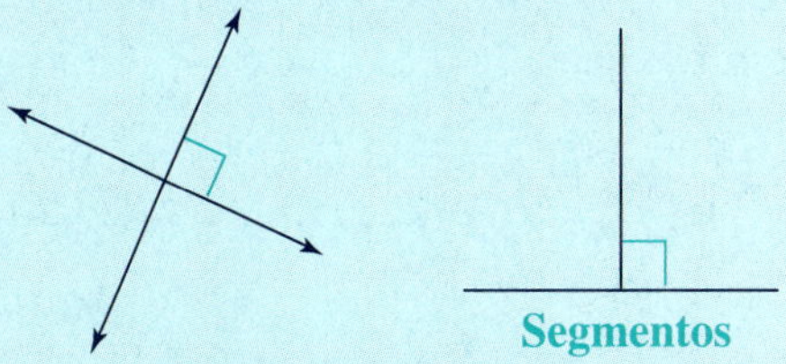

Rectas perpendiculares Segmentos perpendiculares

**polígono** Figura geométrica plana (bidimensional) que posee las siguientes tres características:

- Está compuesta de segmentos de recta
- Cada segmento interseca exactamente otros dos segmentos, uno en cada uno de sus extremos.

Las siguientes figuras son *polígonos*.

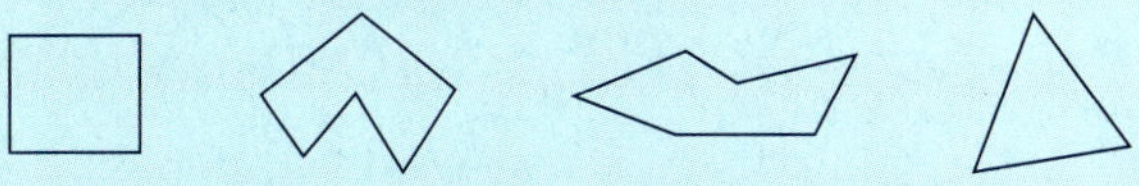

# English

**positive number** A number that is greater than 0. For example, 28 is a *positive number.* [page 142]

**prime factorization** The *prime factorization* of a composite number shows that number written as a product of prime numbers. For example, the prime factorization of 98 is $2 \cdot 7 \cdot 7$. [page 80]

**prime number** A whole number greater than 1 with only two factors: itself and 1. For example, 13 is a *prime number* since it only has two factors, 13 and 1. [page 80]

**probability** The chance that an event will happen, described as a number between 0 and 1. For example, the probability of tossing a coin and getting heads is $\frac{1}{2}$ or 50%. A *probability* of 0 or 0% means the event has no chance of happening, and a probability of 1 or 100% means the event is certain to happen. [page 605]

**radius (plural: radii)** A segment from the center of a circle to a point on a circle. *Radius* also refers to the distance from the center to a point on a circle. The figure below shows a *chord,* a *diameter,* and a *radius* of a circle. [page 486]

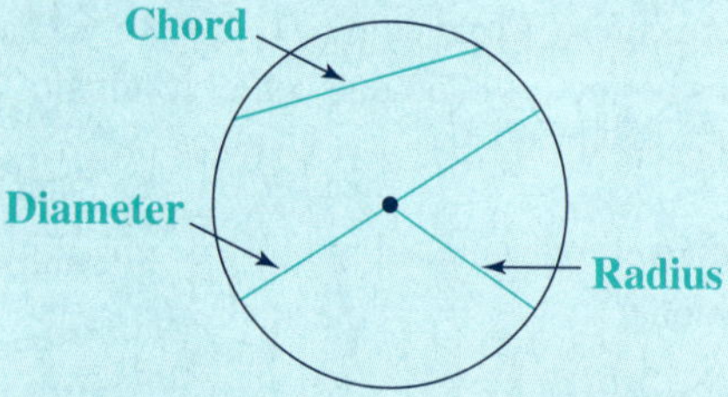


**range** The difference between the minimum and maximum values of a data set. For example, for the data set 4.5, 6, 7, 7, 7, 8.5, 10.5, 12, 12, the *range* is $12 - 4.5 = 7.5$. [page 363]

**reciprocal** Two numbers are *reciprocals* if their product is 1. For example, the reciprocal of $\frac{5}{7}$ is $\frac{7}{5}$. [page 188]

**regular polygon** A polygon with sides that are all the same length and angles that are all the same size. The shapes below are *regular polygons.* [page 50]

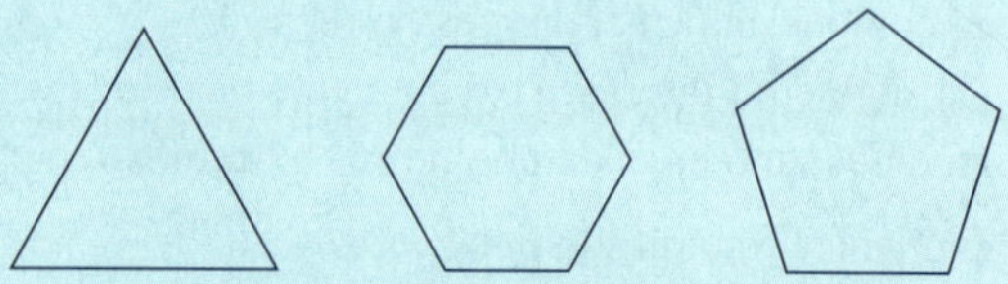

# Español

**número positivo** Número mayor que 0. Por ejemplo, 28 es un *número positivo.*

**factorización prima** La *factorización prima* de un número compuesto muestra ese número escrito como un producto de números primos. Por ejemplo, la factorización prima de 98 es $2 \cdot 7 \cdot 7$.

**número primo** Número entero mayor que 1 cuyos únicos factores son 1 y el número mismo. Por ejemplo, 13 es un *número primo* dado que sólo tiene dos factores: 13 y 1.

**probabilidad** La oportunidad de que un evento ocurra, descrita como un número entre 0 y 1. Por ejemplo, la probabilidad de lanzar una moneda y que ésta caiga mostrando cara es de $\frac{1}{2}$ ó 50%. Una probabilidad de 0 ó 0% significa que el evento no tiene oportunidad de ocurrir y una *probabilidad* de 1 ó 100% significa que el evento ocurrirá con seguridad.

**radio** Un segmento desde el centro del círculo hasta un punto del mismo. *Radio* también se refiere a la distancia desde el centro hasta un punto del círculo. La siguiente figura muestra una *cuerda*, un *diámetro* y un *radio* de un círculo.

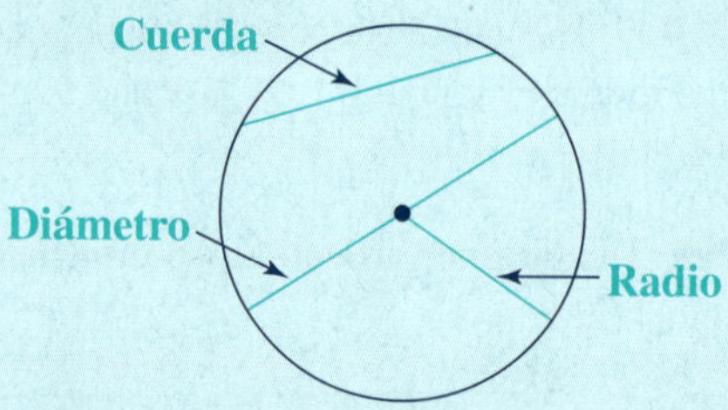


**rango** La diferencia entre los valores mínimo y máximo en un conjunto de datos. Por ejemplo, para el conjunto de datos 4.5, 6, 7, 7, 7, 8.5, 10.5, 12, 12, el *rango* es $12 - 4.5 = 7.5$.

**recíproco** Dos números son *recíprocos* si su producto es 1. Por ejemplo, el recíproco de $\frac{5}{7}$ es $\frac{7}{5}$.

**polígono regular** Polígono cuyos lados son todos de la misma longitud y cuyos ángulos tienen la misma medida. Las siguientes figuras son *polígonos regulares.*

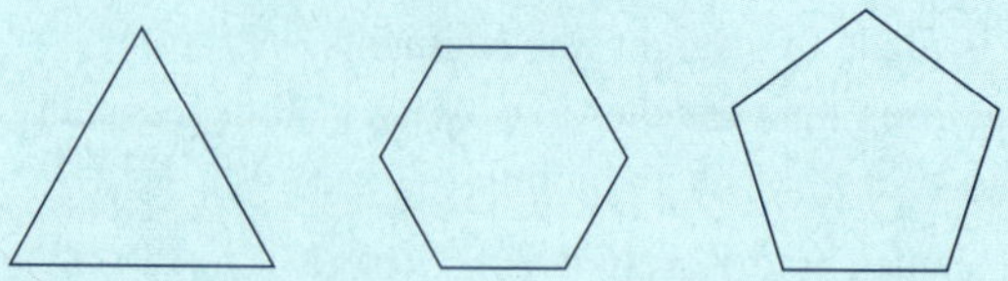

# English

**relatively prime** Two or more numbers are *relatively prime* if their only common factor is 1. For example, 7 and 9 are relatively prime. [page 83]

**repeating decimal** A decimal with a pattern of digits that repeat without stopping. For example, 0.232323 ... is a *repeating decimal.* Repeating decimals are usually written with a bar over the repeating digits, so 0.232323 ... can be written as $0.\overline{23}$. [page 132]

**right angle** An angle that measures exactly 90°. Right angles are often marked with a small square at the vertex. Each angle shown below is a *right angle.* [page 469]

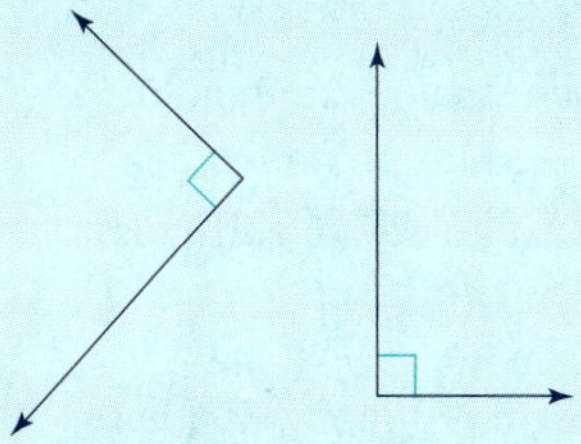

**right triangle** A triangle that has one 90° angle. [page 536]

**sequence** An ordered list. For example, 2, 5, 8, 11, ... is a *sequence.* [page 6]

**simulation** An experiment in which you use different items to represent the items in a real situation. For example, to simulate choosing markers and looking at their colors, you can write the color of each marker on a slip of paper and put all the slips into a bag. You can simulate choosing markers by drawing slips from the bag. Mathematically, the situations are identical. [page 639]

**solution** A value of a variable that makes an equation true. For example, 4 is the *solution* of the equation $3n + 7 = 19$. [page 560]

**square root** The number you need to square to get a given number. For example, the *square root* of 36 is 6. [page 504]

**stem-and-leaf plot** A visual display of data that groups data, but also allows you to read individual values. To make a *stem-and-leaf plot* of data that are two-digit numbers, use the tens digits as the stems and the ones digits as the leaves. [page 366]

# Español

**relativamente primo** Dos o más números son *relativamente primos* si su único factor común es 1. Por ejemplo, 7 y 9 son relativamente primos.

**decimal periódico** Decimal con un patrón de dígitos que se repiten indefinidamente. Por ejemplo, 0.232323 ... es un *decimal periódico.* Los decimales periódicos por lo general se escriben con una barra sobre los dígitos que se repiten, de este modo, 0.232323 ... se puede escribir como $0.\overline{23}$.

**ángulo recto** Ángulo que mide exactamente 90°. Los ángulos rectos por lo regular se marcan con un cuadrado en el vértice. Cada ángulo siguiente es un *ángulo recto.*

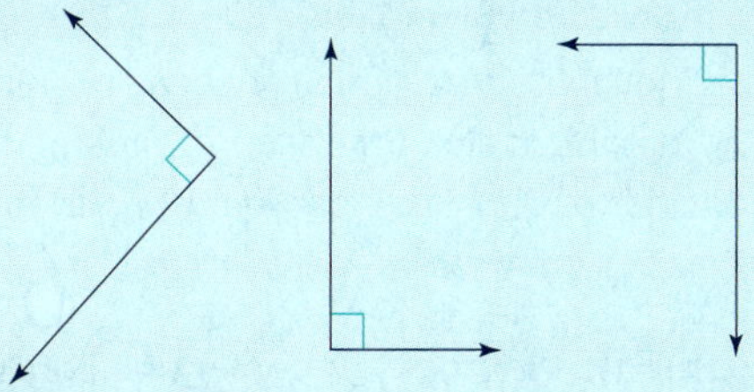

**triángulo rectángulo** Triángulo que posee un ángulo de 90°.

**sucesión** Lista ordenada. Por ejemplo, 2, 5, 8, 11, ... es una *sucesión.*

**simulacro** Un experimento que usa diferentes artículos para representar una situación real. Por ejemplo, en un simulacro para elegir marcadores y verificar sus colores, puedes escribir el color de cada marcador en tiras de papel y colocar las tiras en una bolsa. Saca tiras de la bolsa para llevar a cabo el experimento de escoger marcadores. Matemáticamente, las situaciones son idénticas.

**solución** Un valor de una variable que hace verdadera una ecuación. Por ejemplo, 4 es la *solución* de la ecuación $3n + 7 = 19$.

**raíz cuadrada** El número que debes elevar al cuadrado para obtener un número dado. Por ejemplo, la *raíz cuadrada* de 36 es 6.

**diagrama de tallo y hojas** Representación visual de datos, la cual agrupa los datos, pero también permite leer valores individuales. Para trazar un *diagrama de tallo y hojas* de datos de dos dígitos, usa los dígitos de las decenas como el tallo y los dígitos de las unidades como las hojas.

| English | Español |
|---|---|
| **term** An item in a sequence. For example, 8 is a *term* in the sequence 2, 5, 8, 11, .... [page 6] | **término** Un artículo en una sucesión. Por ejemplo, 8 es un *término* de la sucesión 2, 5, 8, 11, .... |
| **theoretical probability** Probability calculated by reasoning about the situation. Since *theoretical probabilities* do not depend on experiments, they are always the same for a particular event. [page 608] | **probabilidad teórica** Probabilidad que se calcula mediante el razonamiento de la situación. Dado que las *probabilidades teóricas* no dependen de experimentos, son siempre idénticas para un evento en particular. |
| **triangle inequality** The *triangle inequality* states that the sum of the lengths of any two sides of a triangle is greater than the length of the third side. For example, the triangle inequality tells you that it is impossible to have a triangle with sides of lengths 4, 5, and 10, since 4 + 5 is not greater than 10. [page 56] | **desigualdad del triángulo** La *desigualdad del triángulo* establece que la suma de las longitudes de cualquier par de lados de un triángulo es mayor que la longitud del tercer lado. Por ejemplo, la desigualdad del triángulo indica que es imposible tener un triángulo cuyos lados miden 4, 5 y 10 de longitud, dado que 4 + 5 no es mayor que 10. |
| **variable** A quantity that varies, or changes. For example, in a problem about the sizes of buildings, the height and width of the buildings would be variables. [page 278] | **variable** Cantidad que varía o cambia. Por ejemplo, en un problema sobre el tamaño de edificios, la altura y el ancho de los edificios serían variables. |
| **vertex (plural: vertices)** A corner of a polygon, where two sides meet. *Vertices* are usually labeled with capital letters, such as *A, B,* and *C* for the vertices of a triangle. [page 44] | **vértice** La esquina de un polígono, donde se encuentran dos de sus lados. Por lo general, los *vértices* se designan con letras mayúsculas, como por ejemplo, *A, B* y *C* para los vértices de un triángulo. |

# INDEX

## A



## B



## C

## H



## I



## K



## L

## M



## N



## O

## W



## Z

# PHOTO CREDITS

**Chapter 1** **2–3 (bkgd),** PhotoDisc; **2 (tr),** PhotoDisc; **2 (bl),** Jeff J. Daly/Visuals Unlimited; **7,** PhotoDisc; **11,** PhotoDisc; **15,** Mark Steinmetz; **21,** Bischel Studios; **24,** Laura Sifferlin; **32,** Mali Apple; **35,** Digital Vision; **36,** Mali Apple; **39,** PhotoDisc; **46 (tl),** PhotoDisc; **46 (tr),** PhotoDisc; **46 (cr),** CORBIS; **62,** Aaron Haupt; **70,** Aaron Haupt

**Chapter 2** **74–75 (bkgd),** CORBIS; **74 (tr),** PhotoDisc; **76,** PhotoDisc; **82,** Matt Meadows; **87,** CORBIS; **97,** PhotoDisc; **107,** Apple.com; **110,** CORBIS; **114,** CORBIS; **125,** Courtesy D. Carr & H. Craighead/Cornell University; **130,** Bogart Photography; **138,** CORBIS; **141,** Aaron Haupt; **143,** PhotoDisc; **146,** PhotoDisc

**Chapter 3** **152–153 (bkgd),** Ken Frick; **152 (b),** Courtesy Habitat for Humanity/Columbus Ohio Chapter; **158,** Aaron Haupt; **161,** PhotoDisc; **164,** Doug Martin; **167,** PhotoDisc; **169,** CORBIS; **177,** PhotoDisc; **179,** Matt Meadows; **186,** CORBIS; **189,** Aaron Haupt; **195,** Tim Courlas; **199,** PhotoDisc; **202,** Doug Martin; **211,** CORBIS; **212,** PhotoDisc; **215,** Aaron Haupt; **216,** Tim Courlas; **219,** CORBIS; **222,** PhotoDisc

**Chapter 4** **224–225 (bkgd),** PhotoDisc; **224 (b),** CORBIS; **229,** CORBIS; **230,** PhotoDisc; **242,** PhotoDisc; **245,** PhotoDisc; **246,** CORBIS; **251,** Elaine Shay; **261,** Doug Martin; **265,** CORBIS; **269,** PhotoDisc; **271,** Christopher Ena/AP/Wide World Photos; **273,** CORBIS

**Chapter 5** **276–277 (bkgd),** AP/Wide World Photos; **276 (l),** AP/Wide World Photos; **285,** CORBIS; **289,** PhotoDisc; **291,** CORBIS; **292,** PhotoDisc; **298,** Ann Summa; **306–307,** Mali Apple; **314,** PhotoDisc; **316,** PhotoDisc; **318,** Courtesy ABDO Publishing Co., **329,** PhotoDisc; **332,** CORBIS; **337,** Mali Apple

**Chapter 6** **340–341 (bkgd),** PhotoDisc; **340 (b),** Aaron Haupt; **344,** PhotoDisc; **345 (b),** CORBIS; **352,** Aaron Haupt; **360,** Aaron Haupt; **374,** PhotoDisc; **375,** Aaron Haupt; **378,** CORBIS; **384,** CORBIS; **388,** Mark Burnett; **390,** CORBIS; **391,** Doug Martin; **395,** PhotoDisc; **402,** Inga Spence/Visuals Unlimited; **404,** Chuck Savage/CORBIS

**Chapter 7** **408–409 (bkgd),** CORBIS; **408 (tr),** PhotoDisc; **416,** Leslye Borden/PhotoEdit, Inc.; **417,** PhotoDisc; **419,** Aaron Haupt; **422,** Tim Courlas; **425,** Mali Apple; **440,** Mali Apple; **440,** Mali Apple; **444,** PhotoDisc; **459,** CORBIS

**Chapter 8** **464–465 (bkgd),** CORBIS; **464 (l),** DUOMO/CORBIS; **471,** Doug Martin; **475,** Doug Martin; **483,** Doug Bryant; **492 (tr),** 2002 Cordon Art B.V. -Baarn-Holland. All rights reserved.; **492 (bl),** Doug Martin; **493,** CORBIS; **500,** PhotoDisc; **510,** Mali Apple; **514,** Jack Demuth; **519,** PhotoDisc; **527,** Mark Ransom; **528,** PhotoDisc; **530,** PhotoDisc; **531,** Fran Brown; **534,** Wesley Treat; **552,** Mark Ransom; **553,** PhotoDisc

**Chapter 9** **557–558 (bkgd),** CORBIS; **556 (b),** CORBIS; **556 (tr),** PhotoDisc; **563,** Creative Publications; **564,** Creative Publications; **569,** NASA; **571,** Aaron Haupt; **573,** Mali Apple; **575,** PhotoDisc; **579,** Aaron Haupt; **584,** PhotoDisc; **591,** CORBIS; **598,** PhotoDisc; **600,** PhotoDisc

**Chapter 10** **602–603 (bkgd),** Brand X Pictures; **602 (bl),** Geoff Butler; **602 (tr),** PhotoDisc; **604,** PhotoDisc; **608,** PhotoDisc; **609,** PhotoDisc; **610 (tl),** Courtesy American Numismatic Society; **610 (bl),** PhotoDisc; **615,** Aaron Haupt; **626–627,** Matt Meadows; **629,** CORBIS

Unlisted photographs are property of Glencoe/McGraw-Hill